Results and Problems in Cell Differentiation

Volume 63

More information about this series at http://www.springer.com/series/400

Malgorzata Kloc
Editor

Oocytes

Maternal Information and Functions

Springer

Editor
Malgorzata Kloc
Department of Surgery
The Houston Methodist Hospital
Houston, TX
USA

The Houston Methodist Research Institute
Houston, TX
USA

Department of Genetics
MD Anderson Cancer Center
The University of Texas
Houston, TX
USA

ISSN 0080-1844 ISSN 1861-0412 (electronic)
Results and Problems in Cell Differentiation
ISBN 978-3-319-60854-9 ISBN 978-3-319-60855-6 (eBook)
DOI 10.1007/978-3-319-60855-6

Library of Congress Control Number: 2017948675

Printed on acid-free paper

This Springer imprint is published by Springer Nature
The registered company is Springer International Publishing AG
The registered company address is: Gewerbestrasse 11, 6330 Cham, Switzerland

Preface

The oocytes are female germline cells that after maturation become fertilizable female gametes/eggs. Animal oocytes are produced during a lengthy process of oogenesis that manufactures and assembles highly specialized oocyte infrastructure and culminates with entry into meiosis. Oocytes are very unique in combining generic cellular features with specialized structures, maternal information and functions, necessary to support the development of future embryo and ensuring hereditary continuity of maternal nuclear and cytoplasmic information.

The first part of this volume describes how invertebrate and vertebrate oocytes communicate and exchange information with somatic cells, extracellular environment and symbiotic organisms. The second part describes how structural, cellular, organellar and molecular polarity of oocytes is established and what the developmental consequences of polar distribution of maternal information are. The third part concentrates on epigenetic, transcriptional and translational mechanisms regulating heredity and expression of maternal information. The fourth part focuses on oocyte/egg-specific features and functions of generic cellular organelles and components, such as centrioles, mitochondria, lipids and vitellogenin, and the role of cohesin and condensin, which convey proper topology of chromosomes through the ubiquitous cell cycle, in the meiotic chromosomal events and age-related chromosome segregation errors. Finally, the last part of the volume describes the origin and evolution of the subpopulation of maternal genes and analyses how non-inheritable maternal information present in the oocyte can be used for genetic manipulation and engineering.

Houston, TX Malgorzata Kloc

Abstract

This book combines the most recent knowledge on the maternal, i.e. oocyte/egg-specific, molecules and processes. The volume covers the most recent advances in a plethora of subjects such as maternal localized RNAs, functions and mechanisms of RNA localization, transcriptional repression of maternal messages, maternal inheritance and maternal role of CRISPR/Cas9-based genome editing (the CRISPR/Cas system is a prokaryotic immune system that confers resistance to foreign genetic elements such as plasmids and phages and provides a form of acquired immunity), chromatin remodelling and epigenetic modifications, maternal function of nucleosomes, maternal mitochondria and energy supply, role of bacterial symbionts and their maternal transmission, acquisition of oocyte polarity and evolution of maternal effect genes, germ plasm and oosome origin and functions, mechanisms of oocyte activation and soma-germ cell communication. There is no other book in existence, which would combine such a comprehensive list of subjects in one volume. This book is also exceptional and unique in providing the information that is a cross-section through oocytes from various invertebrate and vertebrate species. This will give the readers a completely new and invaluable perspective on all covered subjects.

Contents

Part I
Oocyte Interactions with Environment

Chapter 1
Exogenous Molecule and Organelle Delivery in Oogenesis

Malgorzata Kloc and Jacek Z. Kubiak

Abstract Recent discoveries on the delivery of small- and large-size molecules and organelles to the oocytes/eggs from external sources, such as surrounding somatic cells, body fluids, and sperm, change our understanding of female germ cells' (oocytes and eggs) self-containment and individuality. In this chapter, we will summarize present-day knowledge on sources and presumptive functions of different types of exogenous molecules and organelles delivered to the animal oocytes and eggs.

1.1 Internal Versus External Source of Oocyte-/Egg-Specific Molecules

One of the oldest and most firmly established paradigms of developmental biology is a strict division between the germ line and soma. Such tenet assumes that the germ cells, which form during lengthy process of gametogenesis, are transcriptionally and translationally self-sufficient and independent from the surrounding somatic cells. This concept of self-containment and self-reliance is especially pronounced in female germ cells that have been believed to be the sole source of maternally produced and inherited molecules, organelles, and maternal determinants (Kloc et al. 1998, 2004, 2008, 2016a, b; Tworzydlo et al. 2010a, b). Although in certain types of oogenesis (e.g., in mouse), the growing oocytes receive

M. Kloc (✉)
The Houston Methodist Research Institute, Houston, TX, USA

Department of Surgery, The Houston Methodist Hospital, 6550 Fannin St., Houston, TX 77030, USA
e-mail: mkloc@houstonmethodist.org

J.Z. Kubiak
CNRS UMR 6290, Cell Cycle Group, Institute of Genetics and Development of Rennes, Rennes, France

University of Rennes 1, Faculty of Medicine, Rennes, France

Department of Regenerative Medicine, Military Institute of Hygiene and Epidemiology (WIHE), Warsaw, Poland

M. Kloc (ed.), *Oocytes*, Results and Problems in Cell Differentiation 63,
DOI 10.1007/978-3-319-60855-6_1

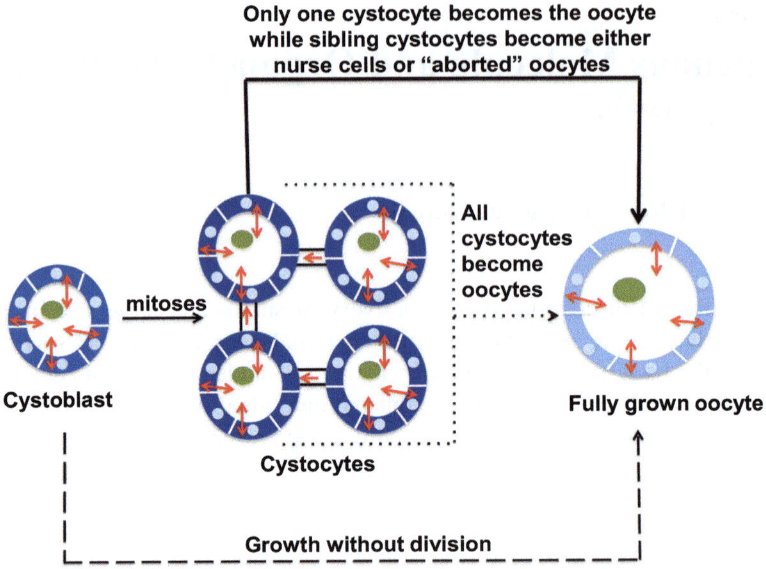

Fig. 1.1 Cross talk between somatic cells and germ cells during oogenesis. Oogenesis starts with the cystoblasts that originate from the primordial germ cells. There are three types of oogenesis depending on the fate of the cystoblast. (1) The cystoblast does not divide but grows and develops into a single oocyte/egg (some insect species with panoistic ovaries). (2) The cystoblast divides several times with incomplete cytokinesis giving rise to the cyst of sister cystocytes connected by intercellular bridges (*ring canals*). The cystocyte, which has the highest number of bridges and inherited the oldest bridge, becomes the oocyte/egg, and the rest of the cystocytes become the nurse cells (insects with meroistic ovaries) or "aborted" oocytes (mouse), which play a "nursing" role by transferring their molecular and organellar components to the oocyte, and eventually disintegrate. (3) All cystocytes become the oocytes/eggs (*Xenopus*). The cystoblast is located in the somatic cell niche. The cystocytes are surrounded by prefollicular cells (*dark blue*) and the oocyte is surrounded by follicular cells (*light blue*). The somatic cells exchange reciprocally with the cystocytes and oocytes/eggs various molecules (see the main text). In addition nurse cells and/or "aborted" oocytes deliver molecules and organelles to the cystocytes and oocyte via cytoplasmic bridges

additional molecular and organellar supplies from the "aborted" oocytes (Lei and Spradling 2016) or the accompanying nurse cells (e.g., in many insect species; Jaglarz et al. 2008; Tworzydlo et al. 2010a, b), both the "aborted" oocytes and the nurse cells are the oocyte's siblings (cystocytes) derived from the division of a common precursor cell (cystoblast; Brubacher and Huebner 2011; Kloc et al. 2016a, b; Lei and Spradling 2016; Fig. 1.1). Thus, all these cells belong to the germ line, and the materials they produce and deliver to the oocyte are also germ line-derived (Fig. 1.1). However, over the years, studies have shown various instances of delivery of exogenous, somatic-derived molecules to the female and male germ cells (El-Hayek and Clarke 2016; Kloc et al. 2016a, b; Chap. 2). In addition, there are numerous examples of bidirectional communication between female germ cells and surrounding somatic cells (Buccione et al. 1990; Gilchrist et al. 2004; Guo et al. 2016; Monniaux 2016; Moussaddykine et al. 2012; Russell et al. 2016; Wigglesworth et al. 2013). These discoveries blur the germ cell/soma demarcation line forcing us to rethink the independence and individuality of germ cells and to envisage the germ cells and surrounding somatic cells not as independent entities but as a functional "syncytium."

1.2 Exogenous Molecules, Their Sources and Routs of Delivery to the Oocyte/Egg

During the whole process of oogenesis (oogonial divisions, formation of the oocyte, subsequent oocyte growth, and egg maturation), the cells of female germ line remain in intimate contact with somatic (prefollicular and follicular cells) component (Fig. 1.2; Ackert et al. 2001; Buccione et al. 1990), and the ovary, as a whole, is bathed with body fluids. These two milieus are constantly exchanging a multitude of components with the germ cells. The routs employed in these exchanges are determined by the size and identity of the cargo: small molecules are transferred through the gap junctions, while large ones by endocytosis and/or gametic synapses and tunneling nanotubes.

1.2.1 Exchange Through the Gap Junctions

The somatic cells are connected to the oocyte and between themselves by the gap junctions. The gap junctions are the intercellular channels formed by the apposition of two cylindrical hemichannels (composed of various connexin, innexin, or pannexin proteins) situated in the neighboring cells (Fig. 1.2; Evans 2015). The gap junctions transfer to the oocyte small molecules such as less than 1000 Da proteins, small metabolites, ions, nucleotides, amino acids, and meiotic maturation regulatory molecules and secondary messengers (e.g., cAMP and cGMP; Ackert et al. 2001; Wigglesworth et al. 2013). Majority of data on gap junction transfer between the somatic cells and oocyte comes from the studies of mammalian oocytes. It was shown in mammals that the gap junctions between oocyte and cumulus cells participate in transfer of ribonucleosides (Heller and Shultz 1980). The meiotic arrest of mammalian oocyte is imposed by cGMP produced and delivered by somatic cells, while reciprocally the oocyte transfers various paracrine factors to the somatic cells regulating their cGMP levels (Wigglesworth et al. 2013). This interchange between germ line and somatic cells plays a key role in regulation of mammalian reproduction.

It seems that the gap junctions play a similar role in invertebrate gametogenesis. In *Drosophila* ovary, the gap junctions between somatic cells contain innexins 2 and 3, while those between follicular cells and germ line cells contain innexins 2 and 4. The experimental interference with innexin 2 disrupts follicular

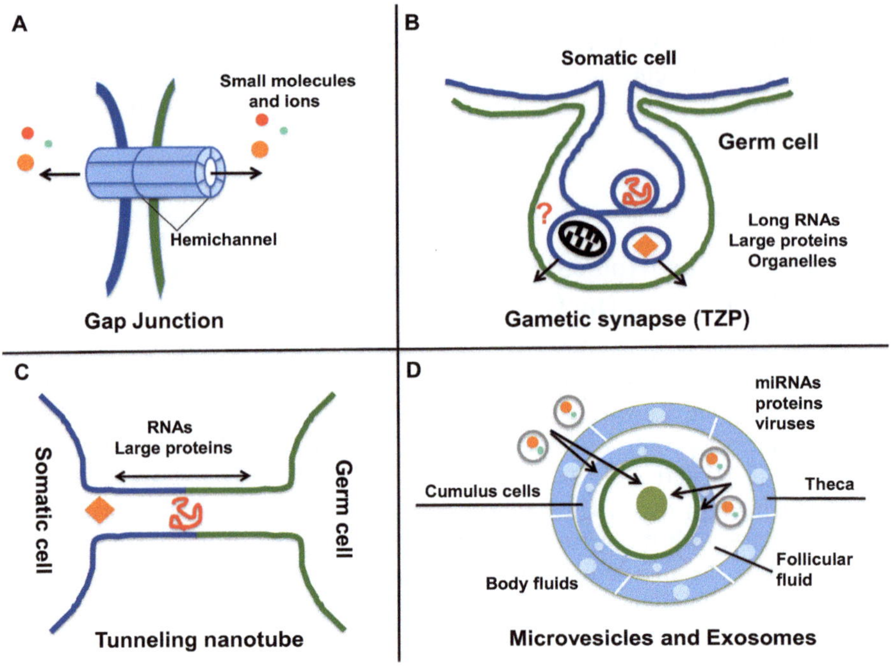

Fig. 1.2 Different routes of molecule and organelle delivery in oogenesis. (**a**) The gap junctions are located between the germ cells and somatic cells and also between neighboring somatic cells. Gap junctions are intercellular channels formed by the apposition of two cylindrical hemichannels (composed of various connexin, innexin, or pannexin proteins). They transfer small molecules such as ions, secondary messengers, and peptides between cells. (**b**) The gametic synapses (transzonal projections, TZP) have been observed between cumulus cells in bovine oocytes (Macaulay et al. 2014, 2016). They resemble synapse of neural cells. Gamete synapse has a membrane; thus, there is no direct connection between the cytoplasm of oocyte and somatic cells. The transfer of large-size molecules such as long noncoding (Lnc) RNAs and mRNAs and potentially (this was not proven, marked in figure by "?") organelles such as mitochondria occurs through the vesicles budding from the end of the synapse. (**c**) Tunneling nanotubes (TNTs) have been described between many different types of somatic cells. So far the only example of tunneling nanotubes between somatic cells and oocytes are TNT-like projections in insect oogenesis (Tworzydlo et al. 2010a, b). TNTs transfer molecules not larger than 200 nm (which is their largest diameter). (**d**) Vesicular transport via microvesicles and exosomes from body fluids (surrounding the whole ovary) or follicular fluid surrounding follicle (in mammals follicle is composed of egg enveloped by cumulus/granulosa cells and multilayered theca outside) delivers yolk proteins, enzymes, hormones, retrotransposons, and miRNAs

cell differentiation, oocyte growth, and choriogenesis and prevents dissipation of nurse cells (Bohrmann and Zimmermann 2008). In addition, the centripetal follicular cells (cFC) that reside at the anterior region of the oocyte contain distinct subpopulation of innexin-3-containing gap junctions, which regulate ion exchange and tension-dependent signaling that in turn regulates tissue integrity (Kruger and Bohrmann 2015). Recent studies showed that similar to *Drosophila*, the gonad of *Caenorhabditis elegans* has two different types of gap junctions with different composition and functions. Gap junctions, which regulate proliferation and differentiation of germ cells, are composed of innexins 8, 9, 14, and 21, and those which regulate oocyte maturation contain innexins 8, 9, 14, and 22 (Starich et al. 2014). Although genetic studies in *C. elegans* and *Drosophila* showed that these differently composed gap junctions are necessary for their specific functions, we still do not know why they require different molecular composition. Interestingly, in *Drosophila* the gap junctions are also involved in the transmission of retrotransposons from somatic cells to the oocytes. Certain lines of *Drosophila* flies contain retrovirus ZAM (belonging to the *gypsy* family of retrotransposons) that is transferred from follicular cells to the oocyte using exocytosis/endocytosis pathway involved in vitellogenin uptake (see below) and also requires signaling through the gap junctions (Brasset et al. 2006).

1.2.2 Transfer from Body Fluids

In many invertebrates and vertebrates, the oocyte growth phase includes accumulation of large amounts of yolk precursor proteins, such as vitellogenin, in the oocyte/egg cytoplasm. After fertilization the yolk accumulated in the egg will be used to support embryonic development. In vertebrates the vitellogenin is produced by the liver, and after being released into the blood, it is endocytotically taken up by the growing oocyte. In frog *Xenopus laevis*, the oocyte growth has been divided into six stages: Stages I–II are previtellogenic oocytes and stages III–VI are vitellogenic oocytes that accumulate vitellogenin and other yolk proteins (e.g., EP45, Marteil et al. 2010). Experimental studies in frogs showed that follicular cells surrounding the oocytes regulate their vitellogenin uptake and that defolliculated (denuded) oocytes will uptake vitellogenin supplied in the culture medium indefinitely (Wallace et al. 1981). In frogs, insects, and crabs, the endocytic uptake of yolk protein during vitellogenic phase of oocyte growth is regulated by the gap junctions and cAMP signaling from the somatic cells (Anderson and Woodruff 2001; Luque et al. 2013; Medeiros et al. 2004; Monaco et al. 2007). Interestingly, recent studies indicate that yolk proteins such as vitellogenin, phosvitin (Pv), lipovitellin (Lv), and yolk-derived small peptides that are stored in egg as the nutrients for embryo development have also potent antimicrobial, antiviral, and immune defense activities in developing fish embryos (Liu et al. 2011; Sun et al. 2013; Tong et al. 2010; Zhang et al. 2015), *Caenorhabditis elegans* (Fischer et al. 2013), and metazoan coral *Euphyllia ancora* (Du et al. 2016). They also regulate senescence and aging in honeybees (Aurori et al. 2014). In addition, in *Drosophila*, the vesicular trafficking of yolk granules during vitellogenesis participates in transfer of retrotransposons ZAM (infectious entities similar to vertebrate retrovirus) from the follicular cells to the oocyte (see above; Leblanc et al. 2000). Minervini et al. (2010) showed experimentally that the sequence located in the 5′UTR of the *Drosophila* ZAM plays a role of the "insulator" (chromatin boundary DNA element) that organizes chromatin into regulatory domains and modulates range of enhancer activity through a "positional-blocking mechanism." The most recent studies showed that in *Caenorhabditis elegans* double-stranded RNA (dsRNA) present in extracellular space is transported within vesicles into the oocytes along with yolk. Subsequently this extracellular dsRNA is transferred between generations to cause gene silencing (Marré et al. 2016).

In mammals, ovarian follicles built of single oocyte surrounded by granulosa cells are bathed in ovarian follicular fluid (Fig. 1.2). Follicular fluid contains component of blood plasma and products secreted by surrounding the oocyte granulosa and thecal cells. Beside various hormones (FSH, GH, LH, estrogens, and androgens), proteins, peptides, amino acids, nucleotides, mucopolysaccharides, and enzymes, the follicular fluid contains subpopulation of microvesicles and exosomes rich in various microRNAs (Hung et al. 2015; Kloc et al. 2016a, b; da Silveira et al. 2012; Sohel et al. 2013), which upon delivery to the oocytes and eggs can affect their transcription, translation, and control the stability of maternal transcripts (Dallaire and Simard 2016; Grossman and Shalgi 2016). Exosomes are small (30–150 nm) endocytic membranous vesicles. They form inside the cell through endosome/multivesicular body (MVB) pathway, in which they acquire various cytoplasmic, membrane-bound, and/or nuclear components (Kloc et al. 2016a, b). Recently, the exosomal miRNAs, which can regulate ubiquitin, neurotrophin, and insulin as well as MAPK signaling, have been found in bovine follicular fluid. They had been shown to regulate levels of mRNAs in follicular cells in vitro (Sohel et al. 2013). It is reasonable to assume that some of these miRNA will end up in the oocytes and eggs where they can regulate maternal, and following fertilization also zygotic, transcripts (Dallaire and Simard 2016; Grossman and Shalgi 2016). Recently it has also been shown that mouse oocytes reciprocally regulate, via paracrine signaling, expression of miRNAs in ovarian granulosa cells (Sumitomo et al. 2016).

1.2.3 Transfer of Large-Size Molecules Through the Gametic Synapses and Tunneling Nanotubes

Until recently the common belief was that only small-size molecules are delivered from the somatic sources to developing oocytes and eggs. However, recent studies indicate that also large-size molecules, such as messenger RNAs and long noncoding (Lnc) RNAs, are transported from surrounding somatic cells to the oocyte (Fig. 1.2; Kloc et al. 2016a, b; Macaulay et al. 2014, 2016). The long noncoding (Lnc) RNAs have length greater than 200 nucleotides and presumably do not have protein-coding potential and are non-translatable (Bassett et al. 2014; Hombach and Kretz 2016). Because the LncRNAs do not produce proteins and are probably multifunctional, their specific functions are very hard to authenticate experimentally (Bassett et al. 2014). Some of the presumed functions of LncRNAs include epigenetic regulation (Qiu et al. 2016); imprinting and X chromosome dosage compensation (*H19* and *Xsist* LncRNAs; Chen et al. 2016a, b; Gendrel and Heard

2014; Li et al. 2016); regulation of neighboring or distant gene expression in *cis* or *trans*, respectively (Peschansky et al. 2016; Yu et al. 2016); being precursors of certain classes of small RNAs (Bassett et al. 2014); sponging the miRNA and competing with endogenous mRNA in transcriptional silencing (Zhang et al. 2016); influencing chromatin conformation (Sun et al. 2016); or simply being a nonfunctional by-product of transcription of LncRNA locus (Bassett et al. 2014).

Macaulay et al. (2014, 2016) have used combination of transmission electron microscopy, confocal microscopy, and colored transcript detection to follow the transfer of LncRNAs, mRNAs, and transfected synthetic transcripts (plasmids) from the cumulus cells to the bovine oocytes. They showed that these transcripts were transferred to the oocytes from surrounding cumulus cells via specialized structures extending from cumulus cells toward the oocyte membrane, which they named the transzonal projections (TZPs) or gametic synapses (Fig. 1.2; Chap. 2). TZPs differ from the 50–200 nm diameter tunneling nanotubes (TNTs; Fig. 1.2), which are known to connect and transfer large-size molecules and organelles between various somatic cells (Kloc et al. 2016a, b). TZPs have larger diameter (200 μm), and in contrast to the TNTs, their membrane is not fused with the oocyte membrane. Instead, at the area of contact with the oocyte, the TZPs form the vesicles reminiscent of synaptic vesicles between neural projections (Macaulay et al. 2014, 2016). Using RNA-seq and bioinformatics analysis, these authors showed that among hundreds of identified RNAs, there were *H19* and *Xsist* LncRNAs (which function in imprinting and X chromosome silencing) and 63 different mRNAs related to cytoskeletal and oocyte maturation functions. They also showed that removal of cumulus cells, which are the sources of these transcripts, reduces oocyte maturation rates (Macaulay et al. 2014, 2016). Although we still do not know if such a transfer of large exogenous molecules occurs only during preparation for oocyte maturation or also during earlier stages of oocyte development, the discovery of TNT-like structures between follicular cells and oocyte in developing insect ovaries (Tworzydlo et al. 2010a, b) suggests that the large-size molecule exchange may occur in all stages of oogenesis to support consecutive stages of oocyte development and growth.

1.3 Transfer of Sperm-Derived Components and Epigenetic Heredity

Another unexpected source of delivery of exogenous molecules and organelles to the egg is the sperm. The commonly accepted phenomenon of Mendelian heredity is based on the transfer of genetic information contained within egg and sperm DNA to the next generation. In Mendelian heredity scenario, the contribution from the sperm is limited to the delivery of paternal DNA. However, there are many examples of non-Mendelian heredity when various environmental factors induce heritable phenotypic (epigenetic) variations that are not caused by mutations in

DNA sequence but depend on DNA methylation pattern and/or posttranslational modification of histones (Rassoulzadegan and Cuzin 2015). There is also evidence that the paternal contribution from sperm is not limited to DNA and that RNAs and organelles present in the sperm may cause heritable epigenetic variations (Barroso et al. 2009; Chen et al. 2016a, b; Rassoulzadegan and Cuzin 2015).

1.3.1 Sperm-Derived Mitochondria and Centrioles

It is known that in some animals, the sperm, during fertilization, delivers not only genomic DNA but also mitochondria and centrioles (Fig. 1.3; Barroso et al. 2009). Although in many animals sperm delivers many (up to 100 in mammals) mitochondria, they usually contain "self-destruction switch" and are eliminated during, or immediately after, fertilization through the process of mitophagy. However, if some of the mitochondria or their DNA escape elimination, they can impose paternally derived, and possibly epigenetic, effects on the egg and developing embryo (Al Rawi et al. 2011; Baumann 2016; van der Bliek 2016).

Other organelles often delivered by sperm are the centrioles, which motile sperm uses to produce axoneme. In the majority of animal species, female centrioles are lost during oogenesis, and, at fertilization, sperm delivers one or two paternal centrioles, which will be used to organize mitotic apparatus in the zygote. However, in some insects, snails, and rodents, both sperm and egg do not have centrioles or

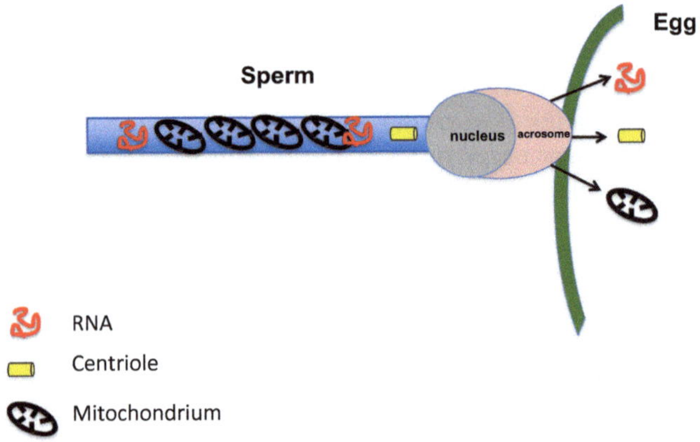

Fig. 1.3 Delivery of sperm-derived molecules and organelles during fertilization. During fertilization sperm delivers not only genetic information (DNA) but also mitochondria, centrioles, and various coding and noncoding RNAs. Although majority of mitochondria reaching the egg undergo mitophagy, the ones that survive and also mitochondrial DNA release from dying mitochondria can cause epigenetic modification of the zygotic genome. It is possible that paternally derived centrioles may also impose epigenetic changes on the embryonic genome. Some of the sperm-derived mRNAs are transcribed in the zygote into protein, and those together with sperm-derived noncoding RNAs may also influence developmental program of the embryo

the paternal centriole is rapidly degraded following fertilization, and zygote produces centriole de novo, from the maternal proteins (Pimenta-Marques et al. 2016; Gueth-Hallonet et al. 1993; Manandhar et al. 2005; Ross and Normark 2015). For decades scientists questioned whether centrioles contain their own heredity traits. If they did then the paternal centrioles could theoretically impose their "heredity program" on the zygote and developing embryos. However, after decades of controversy, recent studies firmly established that centrioles do not contain their own genetic material and that the old (providing template) "mother" centriole is not necessary for the formation of a daughter centriole (Rodrigues-Martins et al. 2007). In spite of these findings, some researchers believe that the paternal centrioles, when delivered to the egg cytoplasm, may still somehow influence establishment of paternally derived epigenetic traits in the zygote and developing embryo (Ross and Normark 2015).

1.3.2 Sperm-Derived RNAs

It has been known for some time that sperm contains a multitude of noncoding RNAs, coding mRNAs, and proteins that are delivered to the egg during fertilization (Chen et al. 2016a, b; Jodar et al. 2016; Johnson et al. 2015; Stoeckius et al. 2014; Yan 2014). RNA analysis of the one-cell hybrid embryos originating from the crosses of males and females from different *C. elegans* strains showed that they expressed 160 paternal mRNAs and multiple paternal siRNAs and piRNAs (Stoeckius et al. 2014). Fang et al. (2014) found that each mouse sperm contained 4885 mRNA transcripts. Some of these transcripts such as Wnt4 and Foxg1 were delivered to the egg during fertilization, and Wnt4, but not Foxg1, was translated into protein in zygotes. This suggests that the transmission of paternal (sperm-derived) mRNAs may influence embryonic development.

Recently, Cropley et al. (2016) showed that in mice, the F1 generation sons of obese males had defects in glucose and lipid metabolism that were transmitted to their own, F2, sons. These authors prepared small RNA library from the sperm and after its sequencing showed that sperm of F1 males had a unique profile of several small RNA species, including the tRNA-derived (so-called diet-responsive) fragments (Sharma et al. 2016). The authors postulate that this non-Mendelian inheritance of metabolic defects occurs via transmission of sperm-derived noncoding RNA (Cropley et al. 2016; Sharma et al. 2016; Watson 2016).

In response to the accumulating data on the importance of sperm-derived RNAs for non-Mendelian and epigenetic heredity, Schuster et al. (2016) developed a simple method for isolation of sperm RNA from several mammalian species and established SpermBase (www.spermbase.org) database, which contains information on sperm-derived RNAs.

Studies presented above indicate that paternally derived coding and noncoding RNAs can affect zygote and embryonic development (Liu et al. 2012) and also participate in the establishment of epigenetic (non-Mendelian) transgenerational inheritance (Fig. 1.3; Chen et al. 2016a, b; Kumar et al. 2013).

1.4 Transfer from Somatic Cells and Testicular Fluid to Sperm

Another phenomenon that abolishes the concept of a strict delineation between germ line and soma is the transfer of information between somatic cells, testicular fluid, and sperm. In the testes, developing male germ cells are in close contact with somatic Sertoli cells, which are functional equivalent of follicular and nurse cells in the ovary, and together they are bathed in testicular fluid. Both Sertoli cells and male germ cells secrete various proteins into the testicular fluid. Analysis of this "secrotome" using the "integrative omics" (combined proteomics, transcriptomics, genomics, and interactomics analyses) methodology showed that, similar to the follicular fluid in the ovaries, there is an extensive cross talk between the germ and Sertoli cells mediated by the components of testicular fluid (Chalmel et al. 2014). Also, similar to the extensive physical contact between follicular cells and oocyte/egg, the Sertoli cells form the tight junctions, adherent junctions, tubulolobular complexes (TBC), and cytoplasmic bridges with the male germ cells, and the assortment of these connections changes during different phases of reproductive cycle (Ahmed et al. 2016; Segretain and Decrossas 1991).

1.5 Conclusions

Taken together, presented above studies indicate that there is bidirectional exchange between germ line and soma and that paternally derived coding and noncoding RNAs can affect both zygote and embryonic development and also participate in the establishment of epigenetic (non-Mendelian) transgenerational inheritance. This clearly shows striking mutual dependence of germ line and soma and argues for the existence of functional germ/soma syncytium.

References

Ackert CL, Gittens JE, O'Brien MJ, Eppig JJ, Kidder GM (2001) Intercellular communication via connexin43 gap junctions is required for ovarian folliculogenesis in the mouse. Dev Biol 233:258–270

Ahmed N, Yufei H, Yang P, Muhammad Yasir W, Zhang Q, Liu T, Hong C, Lisi H, Xiaoya C, Chen Q (2016) Cytological study on Sertoli cells and their interactions with germ cells during annual reproductive cycle in turtle. Ecol Evol 6:4050–4064

Al Rawi S, Louvet-Vallée S, Djeddi A, Sachse M, Culetto E, Hajjar C, Boyd L, Legouis R, Galy V (2011) Postfertilization autophagy of sperm organelles prevents paternal mitochondrial DNA transmission. Science 334:1144–1147

Anderson KL, Woodruff RI (2001) A gap junctionally transmitted epithelial cell signal regulates endocytic yolk uptake in Oncopeltus fasciatus. Dev Biol 239:68–78

Aurori CM, Buttstedt A, Dezmirean DS, Mărghitaş LA, Moritz RF, Erler S (2014) What is the main driver of ageing in long-lived winter honeybees: antioxidant enzymes, innate immunity, or vitellogenin? J Gerontol A Biol Sci Med Sci 69:633–639

Barroso G, Valdespin C, Vega E, Kershenovich R, Avila R, Avendaño C, Oehninger S (2009) Developmental sperm contributions: fertilization and beyond. Fertil Steril 92:835–848

Bassett AR, Akhtar A, Barlow DP, Bird AP, Brockdorff N, Duboule D, Ephrussi A, Ferguson-Smith AC, Gingeras TR, Haerty W, Higgs DR, Miska EA, Ponting CP (2014) Considerations when investigating lncRNA function in vivo. eLife 3:e03058

Baumann K (2016) Development: eliminating paternal mitochondria. Nat Rev Mol Cell Biol 17:464

Bohrmann J, Zimmermann J (2008) Gap junctions in the ovary of Drosophila melanogaster: localization of innexins 1, 2, 3 and 4 and evidence for intercellular communication via innexin-2 containing channels. BMC Dev Biol 8:111–122

Brasset E, Taddei AR, Arnaud F, Faye B, Fausto AM, Mazzini M, Giorgi F, Vaury C (2006) Viral particles of the endogenous retrovirus ZAM from Drosophila melanogaster use a pre-existing endosome/exosome pathway for transfer to the oocyte. Retrovirology 3:25–33

Brubacher JL, Huebner E (2011) Evolution and development of polarized germ cell cysts: new insights from a polychaete worm, Ophryotrocha labronica. Dev Biol 357:96–107

Buccione R, Schroeder AC, Eppig JJ (1990) Interactions between somatic cells and germ cells throughout mammalian oogenesis. Biol Reprod 43:543–547

Chalmel F, Com E, Lavigne R, Hernio N, Teixeira-Gomes AP, Dacheux JL, Pineau C (2014) An integrative omics strategy to assess the germ cell secretome and to decipher sertoli-germ cell crosstalk in the mammalian testis. PLoS One 9(8):e104418

Chen CK, Blanco M, Jackson C, Aznauryan E, Ollikainen N, Surka C, Chow A, Cerase A, McDonel P, Guttman M (2016a) Xist recruits the X chromosome to the nuclear lamina to enable chromosome-wide silencing. Science 354:468–472. pii:aae0047 [Epub ahead of print]

Chen Q, Yan W, Duan E (2016b) Epigenetic inheritance of acquired traits through sperm RNAs and sperm RNA modifications. Nat Rev Genet. doi:10.1038/nrg.2016.106

Cropley JE, Eaton SA, Aiken A, Young PE, Giannoulatou E, Ho JW, Buckland ME, Keam SP, Hutvagner G, Humphreys DT, Langley KG, Henstridge DC, Martin DI, Febbraio MA, Suter CM (2016) Male-lineage transmission of an acquired metabolic phenotype induced by grand-paternal obesity. Mol Metab 5:699–708

da Silveira JC, Veeramachaneni DN, Winger QA, Carnevale EM, Bouma GJ (2012) Cell-secreted vesicles in equine ovarian follicular fluid contain miRNAs and proteins: a possible new form of cell communication within the ovarian follicle. Biol Reprod 86:71–81

Dallaire A, Simard MJ (2016) The implication of microRNAs and endo-siRNAs in animal germline and early development. Dev Biol 416:18–25

Du X, Wang X, Wang S, Zhou Y, Zhang Y, Zhang S (2016) Functional characterization of Vitellogenin_N domain, domain of unknown function 1943, and von Willebrand factor type D domain in vitellogenin of the non-bilaterian coral Euphyllia ancora: implications for emergence of immune activity of vitellogenin in basal metazoan. Dev Comp Immunol 67:485–494. pii:S0145-305X(16)30348–2

El-Hayek S, Clarke HJ (2016) Control of oocyte growth and development by intercellular communication within the follicular niche. Results Probl Cell Differ 58:191–224

Evans WH (2015) Cell communication across gap junctions: a historical perspective and current developments. Biochem Soc Trans 43:450–459

Fang P, Zeng P, Wang Z, Liu M, Xu W, Dai J, Zhao X, Zhang D, Liang D, Chen X, Shi S, Zhang M, Wang L, Qiao Z, Shi H (2014) Estimated diversity of messenger RNAs in each murine spermatozoa and their potential function during early zygotic development. Biol Reprod 90:94

Fischer M, Regitz C, Kull R, Boll M, Wenzel U (2013) Vitellogenins increase stress resistance of Caenorhabditis elegans after Photorhabdus luminescens infection depending on the steroid-signaling pathway. Microbes Infect 15:569–578

Gendrel AV, Heard E (2014) Noncoding RNAs and epigenetic mechanisms during X-chromosome inactivation. Annu Rev Cell Dev Biol 30:561–580

Gilchrist RB, Ritter LJ, Armstrong DT (2004) Oocyte-somatic cell interactions during follicle development in mammals. Anim Reprod Sci 82–83:431–446

Grossman H, Shalgi R (2016) A role of microRNAs in cell differentiation during gonad development. Results Probl Cell Differ 58:309–336

Gueth-Hallonet C, Antony C, Aghion J, Santa-Maria A, Lajoie-Mazenc I, Wright M, Maro B (1993) Gamma-tubulin is present in acentriolar MTOCs during early mouse development. J Cell Sci 105:157–166

Guo J, Shi L, Gong X, Jiang M, Yin Y, Zhang X, Yin H, Li H, Emori C, Sugiura K, Eppig JJ, Su YQ (2016) Oocyte-dependent activation of MTOR in cumulus cells controls the development and survival of cumulus-oocyte complexes. J Cell Sci 129:3091–3103

Heller DT, Schultz RT (1980) Ribonucleotide metabolism by mouse oocytes: metabolic cooperativity between the fully grown oocyte and cumulus cells. J Exp Zool 214:355–364

Hombach S, Kretz M (2016) Non-coding RNAs: classification, biology and functioning. Adv Exp Med Biol 937:3–17

Hung WT, Hong X, Christenson LK, McGinnis LK (2015) Extracellular vesicles from bovine follicular fluid support cumulus expansion. Biol Reprod 93:117

Jaglarz MK, Krzeminski W, Bilinski SM (2008) Structure of the ovaries and follicular epithelium morphogenesis in Drosophila and its kin. Dev Genes Evol 218:399–411

Jodar M, Sendler E, Krawetz SA (2016) The protein and transcript profiles of human semen. Cell Tissue Res 363:85–96

Johnson GD, Mackie P, Jodar M, Moskovtsev S, Krawetz SA (2015) Chromatin and extracellular vesicle associated sperm RNAs. Nucleic Acids Res 43:6847–6859

Kloc M, Larabell C, Chan AP, Etkin LD (1998) Contribution of METRO pathway localized molecules to the organization of the germ cell lineage. Mech Dev 75:81–93

Kloc M, Bilinski S, Dougherty MT, Brey EM, Etkin LD (2004) Formation, architecture and polarity of female germline cyst in Xenopus. Dev Biol 266:43–61

Kloc M, Jaglarz M, Dougherty M, Stewart MD, Nel-Themaat L, Bilinski S (2008) Mouse early oocytes are transiently polar: three-dimensional and ultrastructural analysis. Exp Cell Res 314:3245–3254

Kloc M, Kubiak JZ, Bilinski SM (2016a) Gametic synapses, nanotubes and sperm RNAs – redefining the origin of maternal determinants. Mech Dev 141:1–3

Kloc M, Kubiak JZ, Li XC, Ghobrial RM (2016b) Noncanonical intercellular communication in immune response. World J Immunol 6:67–74

Krüger J, Bohrmann J (2015) Bioelectric patterning during oogenesis: stage-specific distribution of membrane potentials, intracellular pH and ion-transport mechanisms in Drosophila ovarian follicles. BMC Dev Biol 15:1

Kumar M, Kumar K, Jain S, Hassan T, Dada R (2013) Novel insights into the genetic and epigenetic paternal contribution to the human embryo. Clinics (Sao Paulo) 68(Suppl 1):5–14

Leblanc P, Desset S, Giorgi F, Taddei AR, Fausto AM, Mazzini M, Dastugue B, Vaury C (2000) Life cycle of an endogenous retrovirus, ZAM, in Drosophila melanogaster. J Virol 74:10658–10669

Lei L, Spradling AC (2016) Mouse oocytes differentiate through organelle enrichment from sister cyst germ cells. Science 352:95–99

Li C, Hong T, Webb CH, Karner H, Sun S, Nie Q (2016) A self-enhanced transport mechanism through long noncoding RNAs for X chromosome inactivation. Sci Rep 6:31517

Liu M, Pan J, Ji H, Zhao B, Zhang S (2011) Vitellogenin mediates phagocytosis through interaction with FcγR. Mol Immunol 49:211–218

Liu WM, Pang RT, Chiu PC, Wong BP, Lao K, Lee KF, Yeung WS (2012) Sperm-borne microRNA-34c is required for the first cleavage division in mouse. Proc Natl Acad Sci USA 109:490–494

Luque ME, Serrano Mde L, Mónaco ME, Villecco EI, Sánchez SS (2013) Involvement of cAMP and calmodulin in endocytic yolk uptake during Xenopus laevis oogenesis. Zygote 21:1–9

Macaulay AD, Gilbert I, Caballero J, Barreto R, Fournier E, Tossou P, Sirard MA, Clarke HJ, Khandjian ÉW, Richard FJ, Hyttel P, Robert C (2014) The gametic synapse: RNA transfer to the bovine oocyte. Biol Reprod 91:90

Macaulay AD, Gilbert I, Scantland S, Fournier E, Ashkar F, Bastien A, Saadi HA, Gagné D, Sirard MA, Khandjian ÉW, Richard FJ, Hyttel P, Robert C (2016) Cumulus cell transcripts transit to the bovine oocyte in preparation for maturation. Biol Reprod 94:16

Manandhar G, Schatten H, Sutovsky P (2005) Centrosome reduction during gametogenesis and its significance. Biol Reprod 72:2–13

Marré J, Traver EC, Jose AM (2016) Extracellular RNA is transported from one generation to the next in Caenorhabditis elegans. Proc Natl Acad Sci USA 113:12496–12501

Marteil G, D'Inca R, Pascal A, Guitton N, Midtun T, Goksøyr A, Richard-Parpaillon L, Kubiak JZ (2010) EP45 accumulates in growing Xenopus laevis oocytes and has oocyte-maturation-enhancing activity involved in oocyte quality. J Cell Sci 123:1805–1813

Medeiros MN, Mendonça LH, Hunter AL, Paiva-Silva GO, Mello FG, Henze IP, Masuda H, Maya-Monteiro CM, Machado EA (2004) The role of lipoxygenase products on the endocytosis of yolk proteins in insects: participation of cAMP. Arch Insect Biochem Physiol 55:178–187

Minervini CF, Ruggieri S, Traversa M, D'Aiuto L, Marsano RM, Leronni D, Centomani I, De Giovanni C, Viggiano L (2010) Evidences for insulator activity of the 5'UTR of the Drosophila melanogaster LTR-retrotransposon ZAM. Mol Gen Genomics 283:503–509

Mónaco ME, Villecco EI, Sánchez SS (2007) Implication of gap junction coupling in amphibian vitellogenin uptake. Zygote 15:149–157

Monniaux D (2016) Driving folliculogenesis by the oocyte-somatic cell dialog: lessons from genetic models. Theriogenology 86:41–53

Moussaddykine S, Assou S, Déchaud H, Hamamah S (2012) Other actors in the oocyte and follicular growth: the role of microRNAs in the cumulus-oocyte dialog. Gynecol Obstet Fertil 40:170–173

Peschansky VJ, Pastori C, Zeier Z, Wentzel K, Velmeshev D, Magistri M, Silva JP, Wahlestedt C (2016) The long non-coding RNA FMR4 promotes proliferation of human neural precursor cells and epigenetic regulation of gene expression in trans. Mol Cell Neurosci 74:49–57

Pimenta-Marques A, Bento I, Lopes CA, Duarte P, Jana SC, Bettencourt-Dias M (2016) A mechanism for the elimination of the female gamete centrosome in Drosophila melanogaster. Science 353(6294):aaf4866

Qiu JJ, Ren ZR, Yan JB (2016) Identification and functional analysis of long non-coding RNAs in human and mouse early embryos based on single-cell transcriptome data. Oncotarget, Aug 16. doi:10.18632/oncotarget.11304 [Epub ahead of print]

Rassoulzadegan M, Cuzin F (2015) Epigenetic heredity: RNA-mediated modes of phenotypic variation. Ann N Y Acad Sci 1341:172–175

Rodrigues-Martins A, Riparbelli M, Callaini G, Glover DM, Bettencourt-Dias M (2007) Revisiting the role of the mother centriole in centriole biogenesis. Science 316:1046–1050

Ross L, Normark BB (2015) Evolutionary problems in centrosome and centriole biology. J Evol Biol 28:995–1004

Russell DL, Gilchrist RB, Brown HM, Thompson JG (2016) Bidirectional communication between cumulus cells and the oocyte: old hands and new players? Theriogenology 86:62–68

Schuster A, Tang C, Xie Y, Ortogero N, Yuan S, Yan W (2016) SpermBase: a database for sperm-borne RNA contents. Biol Reprod 95:99. 116.142190

Segretain D, Decrossas B (1991) The Sertoli-spermatid processes in the mouse: new cellular structure. Biol Cell 71:321–324

Sharma U, Conine CC, Shea JM, Boskovic A, Derr AG, Bing XY, Belleannee C, Kucukural A, Serra RW, Sun F, Song L, Carone BR, Ricci EP, Li XZ, Fauquier L, Moore MJ, Sullivan R, Mello CC, Garber M, Rando OJ (2016) Biogenesis and function of tRNA fragments during sperm maturation and fertilization in mammals. Science 351:391–396

Sohel MH, Hoelker M, Noferesti SS, Salilew-Wondim D, Tholen E, Looft C, Rings F, Uddin MJ, Spencer TE, Schellander K, Tesfaye D (2013) Exosomal and non-exosomal transport of extra-cellular microRNAs in follicular fluid: implications for bovine oocyte developmental competence. PLoS One 8(11):e78505

Starich TA, Hall DH, Greenstein D (2014) Two classes of gap junction channels mediate soma-germline interactions essential for germline proliferation and gametogenesis in Caenorhabditis elegans. Genetics 198:1127–1153

Stoeckius M, Grün D, Rajewsky N (2014) Paternal RNA contributions in the Caenorhabditis elegans zygote. EMBO J 33:1740–1750

Sumitomo J, Emori C, Matsuno Y, Ueno M, Kawasaki K, Endo TA, Shiroguchi K, Fujii W, Naito K, Sugiura K (2016) Mouse oocytes suppress miR-322-5p expression in ovarian granulosa cells. J Reprod Dev 62:393–399

Sun C, Hu L, Liu S, Hu G, Zhang S (2013) Antiviral activity of phosvitin from zebrafish Danio rerio. Dev Comp Immunol 40:28–34

Sun M, Nie F, Wang Y, Zhang Z, Hou J, He D, Xie M, De W, Wang Z, Wang J (2016) LncRNA HOXA11-AS promotes proliferation and invasion of gastric cancer by scaffolding the chromatin modification factors PRC2, LSD1 and DNMT1. Cancer Res 76:6299–6310. pii: canres.0356.2016 [Epub ahead of print]

Tong Z, Li L, Pawar R, Zhang S (2010) Vitellogenin is an acute phase protein with bacterial-binding and inhibiting activities. Immunobiology 215:898–902

Tworzydlo W, Biliński SM, Kocárek P, Haas F (2010a) Ovaries and germline cysts and their evolution in Dermaptera (Insecta). Arthropod Struct Dev 39:360–368

Tworzydlo W, Kloc M, Bilinski SM (2010b) Female germline stem cell niches of earwigs are structurally simple and different from those of Drosophila melanogaster. J Morphol 271:634–640

van der Bliek AM (2016) GENETICS. Demystifying the demise of paternal mitochondrial DNA. Science 353:351–352

Wallace RA, Misulovin Z, Etkin LD (1981) Full-grown oocytes from Xenopus laevis resume growth when placed in culture. Proc Natl Acad Sci USA 78:3078–3082

Watson ED (2016) Transferring fragments of paternal metabolism to the offspring. Cell Metab 23:401–402

Wigglesworth K, Lee KB, O'Brien MJ, Peng J, Matzuk MM, Eppig JJ (2013) Bidirectional communication between oocytes and ovarian follicular somatic cells is required for meiotic arrest of mammalian oocytes. Proc Natl Acad Sci USA 110:E3723–E3729

Yan W (2014) Potential roles of noncoding RNAs in environmental epigenetic transgenerational inheritance. Mol Cell Endocrinol 398:24–30

Yu Y, Yarrington RM, Chuong EB, Elde NC, Stillman DJ (2016) Disruption of promoter memory by synthesis of a long noncoding RNA. Proc Natl Acad Sci USA 113:9575–9580

Zhang S, Dong Y, Cui P (2015) Vitellogenin is an immunocompetent molecule for mother and offspring in fish. Fish Shellfish Immunol 46:710–715

Zhang Z, Cheng J, Wu Y, Qiu J, Sun Y, Tong X (2016) LncRNA HOTAIR controls the expression of Rab22a by sponging miR-373 in ovarian cancer. Mol Med Rep 14:2465–2472

Chapter 2
Control of Mammalian Oocyte Development by Interactions with the Maternal Follicular Environment

Hugh Clarke

Abstract Development of animal germ cells depends critically on continuous contact and communication with the somatic compartment of the gonad. In females, each oocyte is enclosed within a follicle, whose somatic cells supply nutrients that sustain basal metabolic activity of the oocyte and send signals that regulate its differentiation. This maternal microenvironment thus plays an indispensable role in ensuring the production of fully differentiated oocytes that can give rise to healthy embryos. The granulosa cells send signals, likely membrane-associated Kit ligand, which trigger oocytes within resting-stage primordial follicles to initiate growth. During growth, the granulosa cells feed amino acids, nucleotides, and glycolytic substrates to the oocyte. These factors are necessary for the oocyte to complete its growth and are delivered via gap junctions that couple the granulosa cells to the oocyte. In a complementary manner, growing oocytes also release growth factors, notably growth-differentiation factor 9 and bone morphogenetic protein 15, which are necessary for proper differentiation of the granulosa cells and for these cells to support oocyte growth. During the late stages of oocyte growth, cyclic GMP that is synthesized by the granulosa cells and diffuses into the oocyte is required to prevent its precocious entry into meiotic maturation. Finally, at the early stages of maturation, granulosa cell signals promote the synthesis of a subset of proteins within the oocyte that enhance their ability to develop as embryos. Thus, the maternal legacy of the follicular microenvironment is witnessed by the fertilization of the ovulated oocyte and subsequent birth of healthy offspring.

H. Clarke (✉)
Department of Obstetrics and Gynecology, Research Institute of the McGill University Health Centre, McGill University, Room E.M0.2218, Glen Research Building, 100 Boul Decarie, Montreal, QC, Canada, H4A 3J1
e-mail: Hugh.clarke@mcgill.ca

M. Kloc (ed.), *Oocytes*, Results and Problems in Cell Differentiation 63,
DOI 10.1007/978-3-319-60855-6_2

2.1 Introduction

Maternal control of embryonic development typically describes the concept that factors which accumulate in the oocyte during its growth and development before fertilization influence the development of the embryo after fertilization. Many, though by no means all, of these factors are messenger mRNAs, reflecting that embryos are transcriptionally inactive during the early stages of postfertilization development, so new protein synthesis relies entirely on mRNAs that were synthesized by the oocyte. Thus, the maternal rather than embryonic genome determines the mRNA population of the early embryo. Here we describe maternal control of a different nature, focusing primarily on studies using the mouse as a model organism. Germ cells develop within a microenvironment that is created by the somatic cells of the gonad. In females, the cells of the ovarian follicle—principally the granulosa cells which surround and enclose the oocyte throughout its growth and development—create this environment. Because individual oocytes and their follicles undergo growth throughout reproductive life, and many aspects of growth can be faithfully recapitulated in vitro, much has been learned about the interactions between the female germ cell and its environment during this period. Such studies have revealed a continuous and multidimensional exchange of signals that not only drives oocyte development but also permits the oocyte to remodel the somatic microenvironment to meet its evolving needs.

2.2 Generating Primordial Follicles

The early life histories of oocytes and the granulosa cells are separate stories, reflecting their different embryological origin. Oocytes are descended from the primordial germ cells (PGCs), which arise from posterior epiblast under the influence of signals sent by adjacent tissues, notably bone morphogenetic protein (BMP) 4 from the extra-embryonic ectoderm (Gunesdogan and Surani 2016). Following migration to the genital ridge, the PGCs of female embryos differentiate into oogonia, which undergo multiple rounds of mitotic proliferation before entering meiosis (Pepling 2012; Jorgensen 2013; Grive and Freiman 2015). The granulosa cells are not born so far from home but instead thought to be derived from cells at the surface of the developing gonad (Mork et al. 2012; Hummitzsch et al. 2015). By near the time of birth, each oocyte has become enclosed by a small number of granulosa cells in a structure termed a primordial follicle. The signals that trigger formation of the primordial follicles are not fully understood. However, when oocytes lack the bHLH-type transcription factor, FIGLA, the follicles do not form (Soyal et al. 2000). Similarly, when *Sohlh2*, which also encodes a bHLH transcription factor, is deleted in oocytes, primordial follicles are able to form but disappear shortly after birth (Choi et al. 2008). These and other reports (Reddy et al. 2009; Saatcioglu et al. 2016) in which other genes have been deleted in the oocyte highlight that, even at the primordial follicle stage, normal follicular structure and function is not sustained in the absence of a healthy oocyte. Conversely, oocytes that do not lie within a primordial follicle will undergo apoptosis (Pepling and Spradling 2001). These observations underscore the indispensable role of signaling between the germ line and soma in females, a mutual dependence that differs strikingly different from the male, where seminiferous tubules can remain organized when no germ cells are present.

Within the primordial follicle, the oocyte and granulosa cells are closely apposed as revealed by electron micrographs, and cell adhesion complexes appear to physically couple the oocyte to the granulosa cells and the granulosa cells to each other (Jorgensen 2013). Oocytes and granulosa cells express a variety of junctional proteins, including cadherins and nectins that characterize adherens junctions (Mora et al. 2012). Even at this early stage, oocytes express mainly E-cadherin, whereas granulosa cells express mainly N-cadherin (Mora et al. 2012), meaning that the intercellular complexes are likely heterotypic. In addition, gap junctional structures have been detected using electron microscopy both between granulosa cells and between oocytes and granulosa cells (Mitchell and Burghardt 1986). In contrast, protein components of desmosomes have not been detected in the primordial follicles (Mora et al. 2012). Desmosomes generate very strong intercellular adhesion, and it was suggested that their absence facilitates the changes in the association between the oocyte and granulosa cells that will occur when the primordial follicle enters the growth phase, as described below (Mora et al. 2012).

2.3 Growth and Maturation of the Oocyte and Its Follicle

Prior to ovulation and fertilization, the oocyte undergoes a prolonged period of growth, followed by a briefer stage termed meiotic maturation (Fig. 2.1). Because available evidence indicates that no new functional oocytes are created after birth under physiological conditions (Zhang et al. 2012, 2015; Lei and Spradling 2013; Zarate-Garcia et al. 2016), the entry of primordial follicles into the growth phase must be regulated. Thus, some primordial follicles will remain in this arrested condition for up to decades before initiating growth. Oocytes from aged females, however, develop poorly as embryos (Nelson et al. 2013; Ben-Meir et al. 2015; Haverfield et al. 2016; Meldrum et al. 2016). Whether this reduced oocyte quality reflects damage that accumulates during its prolonged arrest within the primordial follicle or arises during the growth and maturation process is unclear; however, in both mice and humans, primordial follicles of aged females show increased DNA damage as reflected by immunological detection of γH2A.X (Titus et al. 2013). This suggests that there may be an age-associated loss of oocyte quality within the primordial follicles.

The growth phase of oogenesis lasts about 3 weeks in the mouse and 3–4 months in human. The most obvious feature of the growth phase is the enormous increase—greater than 100-fold—in the volume of the oocyte. This increase

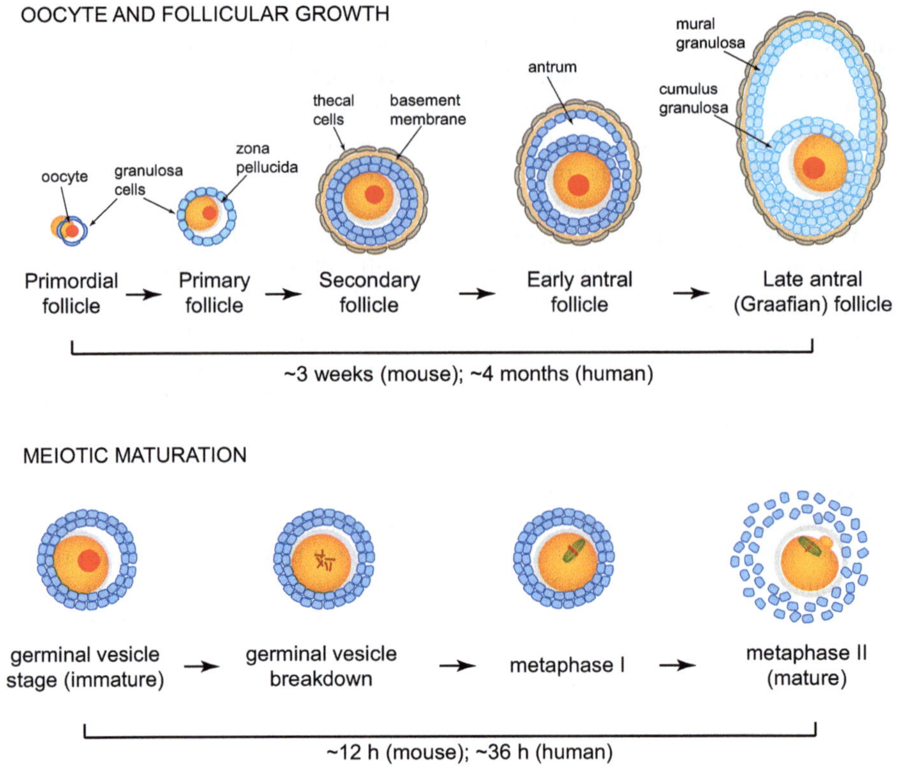

Fig. 2.1 Oocyte and follicular growth and meiotic maturation. *Upper* Near the time of birth, each oocyte becomes surrounded by a small number of squamous granulosa cells in a primordial follicle. During reproductive life, individual follicles enter the growth phase and are termed primary follicles. The granulosa cells assume a cuboidal morphology and begin to proliferate mitotically, continuing to fully cover the surface of the growing oocyte. At this stage, the oocyte elaborates the zona pellucida, which physically separates it from the granulosa cell bodies. A basement membrane appears outside the granulosa cell layer and thecal cells are recruited external to this membrane. As the granulosa cells continue to proliferate, they generate multiple layers. A fluid-filled cavity appears, which separates the granulosa into the cumulus granulosa, adjacent to the oocyte, and the mural granulosa, adjacent to the follicle wall. Preovulatory follicles contain a large antrum and may be termed Graafian. *Lower* Luteinizing hormone, which binds to receptors on the mural granulosa cells, triggers meiotic maturation of the oocyte. The nuclear (germinal vesicle) membrane breaks down and the chromosomes condense and become assembled on the first meiotic spindle. After translocation of the spindle to the oocyte cortex, the first meiotic division occurs and half of the chromosomes are discarded in the first polar body. The oocyte chromosomes become aligned on the second meiotic spindle (metaphase II). Fertilization will trigger the completion of the second meiotic division. Not shown are cytoplasmic events that occur during maturation. Adapted from Res Probl Cell Diff 58:191–224

reflects the accumulation of mitochondria and other organelles, mRNA, and proteins that will support early embryonic development after fertilization (Sánchez and Smitz 2012). The growing oocyte also undergoes major ultrastructural rearrangements, including the accumulation of cortical granules that will play an essential role in preventing polyspermic fertilization (Ducibella et al. 1994), the assembly of an internal lattice structure (Yurttas et al. 2008; Kim et al. 2014), establishment of oocyte-specific DNA methylation patterns (Lucifero et al. 2004; Smallwood et al. 2011; Tomizawa et al. 2012; Stewart et al. 2015), and the accumulation of proteins implicated in mRNA translational control into a subcortical complex (Li et al. 2008; Flemr et al. 2010).

Growth of the oocyte is accompanied by growth of the follicle. Approximately coincident with the initiation of oocyte growth, the squamous granulosa cells that enclose it in the non-growing primordial follicle become cuboidal in shape and begin to proliferate mitotically (Hirshfield 1991; Da Silva-Buttkus et al. 2008). At this stage, the follicle is termed primary. A layer of theca cells is then assembled external to the granulosa cells, from which they are separated by a basement membrane (Hirshfield 1991; Liu et al. 2015). As the oocyte continues to grow, the granulosa cells proliferate so that they fully cover the expanding surface of the oocyte and also generate multiple layers around it. Studies using Ki67 as a marker of cell proliferation indicate that, in multilayered follicles, granulosa cells in the layer closer to the oocyte proliferate more rapidly than those in the outer layer adjacent to the basement membrane (Da Silva-Buttkus et al. 2008). This suggests that oocyte factors may promote granulosa cell proliferation (Gilchrist et al. 2006), as discussed further below. As the follicle continues to grow, a fluid-filled cavity termed the antrum develops. This separates the granulosa cells into two populations—the cumulus granulosa that surround the oocyte and the mural granulosa that line the follicular wall.

The final stage of oocyte development is termed meiotic maturation (Conti et al. 2012; Holt et al. 2013; Adhikari and Liu 2014; Coticchio et al. 2015). It is physiologically triggered by a release of luteinizing hormone, which also induces ovulation and thereby coordinates the completion of oocyte development with potential fertilization. During maturation, cyclin-dependent kinase (CDK) 1 becomes active, advancing the cell cycle from late G2 to M-phase. Thus, the nuclear membrane is disassembled, a process termed germinal vesicle breakdown, the chromosomes become fully condensed, and the first meiotic spindle is assembled. The spindle migrates from the center of the oocyte to its periphery and the first meiotic division occurs accompanied by cytokinesis to generate the first polar body (Bennabi et al. 2016). The chromosomes remaining in the oocyte then become organized on the second meiotic spindle at metaphase II. In addition to these nuclear events, cytoplasmic changes occur also as a subset of mRNAs become translationally activated, whereas others become silenced (Kang and Han 2011; Susor et al. 2015), the cortical granules complete their migration to the oocyte periphery (Cheeseman et al. 2016), and other ultrastructural events occur such as aggregation of mitochondria around the spindle (Van Blerkom 2010). During maturation of the oocyte, the surrounding cumulus granulosa cells secrete a matrix that separates individual cells from each other and also from the oocyte (Russell and Salustri 2006). This process, termed cumulus expansion, breaks physical contact between the oocyte and somatic cells of the follicle.

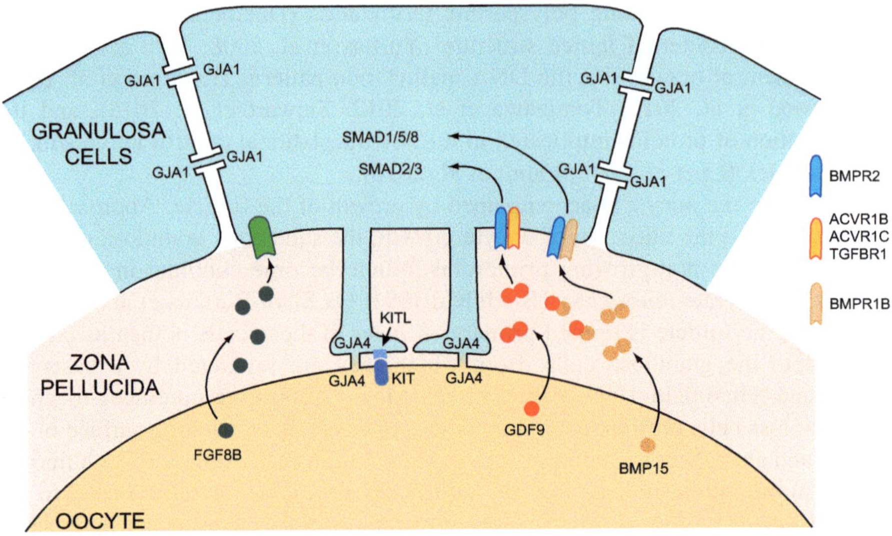

Fig. 2.2 Pathways of oocyte-granulosa cell communication. Gap junctions composed of GJA1 (connexin43) subunits link granulosa cells to each other, whereas those composed of GJA4 (connexin37) subunits likely link them to the oocyte, whose major gap junctional component is GJA4. KITL produced by the granulosa cells is also able to activate KT on the oocyte membrane. GDF9 and BMP15 produced by the bind to receptors, either as homodimers or as heterodimers, on the granulosa cells. FGF8B is also secreted by the oocyte and acts on the granulosa cells. Adapted from Res Probl Cell Diff 58:191–224

2.4 Mechanisms of Granulosa Cell-Oocyte Communication

Oocytes and granulosa cells communicate using both gap junctions, which allow direct transfer of molecules between coupled cells, and secreted factors, which by binding to membrane-associated receptors activate specific signaling pathways within the target cell (Fig. 2.2).

2.4.1 Gap Junctions

Gap junctions are intercellular channels that permit the exchange of molecules up to approximately 1 kDa between coupled cells (Evans 2015; Winterhager and Kidder 2015). The basic unit of the gap junction is the connexon, which is composed of six connexin proteins that become arranged in a ring in the plasma membrane surrounding a "hollow" core. Connexons of adjacent cells become associated to generate the gap junction. A gap junctional plaque is a group of connexons that are localized together in the plasma membrane. Gap junctions have long been known to couple the oocyte to the granulosa cells and the granulosa cells to each other, and gene-knockout studies confirmed their essential role in the development of both compartments during follicular growth. These studies focused on two connexins: connexin-37 (*Gja4*), the main connexin expressed in mouse oocytes, and connexin-43 (*Gja1*), the main connexin expressed in granulosa cells. In mice lacking *Gja4*, there is no detectable gap junctional communication between the oocyte and granulosa cells (Simon et al. 1997; Veitch et al. 2004; Gittens and Kidder 2005). The oocytes are able to begin growth, but fail to reach full size, and do not acquire the ability (termed meiotic competence) to undergo meiotic maturation (Carabatsos et al. 2000). Although the absence of connexin-37 does not apparently disturb gap junctional communication between granulosa cells, the follicle does not progress beyond the late pre-antral stage (Gittens and Kidder 2005). Thus, impaired development of the oocyte also impairs follicular development.

Absence of connexin-43 leads to a complementary phenotype—oocyte-granulosa cell communication is retained, but communication between granulosa cells is severely impaired (Ackert et al. 2001; Veitch et al. 2004; Gittens and Kidder 2005). The granulosa cells of mice lacking *Gja1* are unable to generate a second layer of cells around the growing oocytes; hence the follicles are defined as remaining at the primary stage. As when connexin-37 is absent, the oocytes are unable to reach full-size, manifest multiple ultrastructural abnormalities, and do not acquire meiotic competence. Thus, reduced communication between granulosa cells impairs oocyte development.

It is not fully established whether connexin-37 or connexin-43 forms the granulosa cell component of the gap junctions with the oocyte. Although antibodies specific for connexin-43 stain punctae in the zona pellucida, it is not known whether these correspond to gap junctions between the oocyte and granulosa cells or between cytoplasmic processes, termed transzonal projections (discussed below) of granulosa cells that penetrate into the zona pellucida. Experiments using primary granulosa cells in culture indicated that granulosa cells lacking *Gja4* (connexin-37) could not form gap junctions with the oocyte (Veitch et al. 2004). When $Gja4^{-/-}$ granulosa cells were aggregated with $Gja4^{+/+}$ oocytes, however, gap junctional coupling was observed, and the oocytes grew apparently normally (Gittens and Kidder 2005). It may be that granulosa cells normally employ connexin-37 to couple with the oocyte and connexin-43 to couple with other granulosa cells (Gittens and Kidder 2005). This would require selective trafficking of each connexin to its appropriate location, a process that has been described for certain connexins (Evans 2015). Alternatively, heterotypic gap junctions have been identified in other cell types (Koval et al. 2014).

2.4.2 Secreted Factors

These experiments highlight the importance of signaling from the granulosa cells to the oocyte. It is not yet known whether the oocyte provides support to the granulosa cells via the gap junctions. The oocyte does, however, regulate granulosa cell differentiation and activity via secreted growth factors (Gilchrist et al. 2008; Su et al. 2009). This role of the oocyte was hinted at many years ago by the observation that when the oocyte was removed from the follicle, the granulosa cells rapidly underwent a process termed luteinization, which normally occurs only after ovulation has expelled the oocyte from the follicle (el-Fouly et al. 1970). Its role was clearly revealed, however, in studies focused on a process termed cumulus expansion. At the time of ovulation, the cumulus cells secrete hyaluronic acid, a glycosylaminoglycan that expands to separate the cumulus cells from each other and embed them in a gelatinous matrix (Eppig 1991). When the oocyte was microsurgically removed from cumulus-oocyte complexes obtained from preovulatory follicles, however, the remaining shell of cumulus cells was unable to undergo expansion (Buccione et al. 1990; Vanderhyden et al. 1990). Co-culture of the shells with oocytes or with oocyte-conditioned medium, however, restored the ability to expand. Thus, oocytes secrete an expansion-enabling factor (Dragovic et al. 2007). Moreover, oocytes also maintain the cumulus granulosa cell lineage, as assessed by expression of marker mRNAs, preventing expression of mural granulosa cell markers (Diaz et al. 2007; Emori et al. 2013; Wigglesworth et al. 2015).

Two factors secreted by growing oocytes are growth-differentiation factor (GDF) 9 and BMP15, both members of the transforming growth factor β superfamily of growth factors. GDF9 binds to a receptor dimer composed of BMPR2-TGFBR1 and signals through SMAD2/3, whereas BMP15 binds to a BMPR2-BMPR1B dimer and signals through SMAD1/5/8. GDF9:GDF9 homodimers are thought to be the major form that signals in rodents, whereas GDF9:BMP15 heterodimers are the major active form in other species including humans (Dragovic et al. 2007; Peng et al. 2013; Mottershead et al. 2015). The granulosa cells are believed to be the principal or sole target of oocyte-derived GDF9 and BMP15; notably, oocyte-specific deletion of *Smad4*, which is required for all SMAD-dependent signaling, has no significant phenotypic effect (Li et al. 2011).

Mice lacking *Gdf9* are unable to ovulate and are sterile (Dong et al. 1996). Strikingly, during folliculogenesis, no more than a single layer of granulosa cells is generated around the growing oocyte. This defect appears to be due to an increase in the intrafollicular level of inhibin A, because multiple layers of granulosa cells are restored in *Gdf9*$^{-/-}$; *Inha*$^{-/-}$ mice (Wu et al. 2004). Although mice lacking *Bmp15* are fertile, *Gdf9*$^{+/-}$:*Bmp15*$^{-/-}$ individuals manifest severe follicular defects (Su et al. 2004), highlighting the shared role of these oocyte-derived growth factors in folliculogenesis.

In addition to these defects in the granulosa cells, however, oocyte development is also compromised in the absence of GDF9 (Dong et al. 1996; Carabatsos et al. 1998). Although the oocytes grow slightly larger than wild-type oocytes, they show ultrastructural abnormalities as well as an impaired ability to undergo meiotic maturation. As GDF9 likely does not act directly on the oocyte, these results suggest that the GDF9-deprived granulosa cells are unable to interact normally with the oocyte. Consistent with this, oocyte-derived factors promote the expression in granulosa cells of mRNAs encoding glycolytic enzymes (BMP15 and fibroblastic growth factor 15) (Sugiura et al. 2007) and those required for synthesis of cholesterol (GDF9, BMP15) (Su et al. 2008), and for amino acid transport (Eppig et al. 2005), which enable the synthesis of products transferred from the granulosa cells to the oocyte. Recent studies have revealed that the oocyte promotes these activities in part by suppressing expression of *Ddit4l*, a negative regulator of MTOR. Thus, oocyte-derived factors likely increase protein synthesis in the neighboring granulosa cells (Guo et al. 2016).

2.5 Transzonal Projections: Bridges for Granulosa Cell-Oocyte Communication

Metazoan oocytes are surrounded by a protective extracellular matrix composed of glycoproteins. In mammals, this coat is termed the zona pellucida and is made of three (rodents) or four (primates) glycosylated proteins that are secreted by the oocyte (Wassarman and Litscher 2012, 2013). The zona pellucida is not present at the primordial follicle stage, so the oocyte and adjacent granulosa cells are in direct contact. Upon initiation of oocyte growth, the encoding genes become transcriptionally activated within the oocyte. The zona pellucida is not elaborated as a continuous structure enveloping the oocyte. Rather, as revealed through electron microscopy, it is assembled as aggregates or clumps that are subsequently knit together to form the continuous structure (Wassarman and Litscher 2012). As the oocyte continues to grow, the zona pellucida thickens to reach a final diameter of about 7 μm in mouse and 20 μm in humans (Litscher and Wassarman 2014). As a result, the granulosa cell bodies become physically separated from the oocyte, and it may be asked how the cell types are able to maintain the physical contact that is necessary for oocyte growth and development.

In all metazoan species that have been studied, including a wide range of mammals, thin cytoplasmic projections extend from the granulosa cells through the zona pellucida and contact the oocyte plasma membrane (Fig. 2.3) (Anderson and Albertini 1976; Albertini and Rider 1994; Motta et al. 1994; Makabe et al. 2006; Li and Albertini 2013). These projections mainly originate from the granulosa cells in the layer immediately adjacent to the oocyte but have also been observed to arise from granulosa cells in more distal layers (Jaffe and Egbert 2016). Both these structures, termed transzonal projections (TZPs), and their granulosa cell origin were identified over a century ago, and it was proposed at the time that they might allow the granulosa cells to feed nutrients to the oocyte (Hadek 1965). TZPs frequently form a bulbous foot-like structure at their site of attachment to the oocyte surface, but have also been described to penetrate deeply into invaginations of the oocyte plasma membrane (Motta et al. 1994; Makabe et al. 2006).

The gap junctions connecting the granulosa cells to the oocyte are located at the tips of the TZPs (Fig. 2.2), and gap junctions likely also couple TZPs to each other (Hadek 1965). Although gap junctions may be able to serve an adhesive function

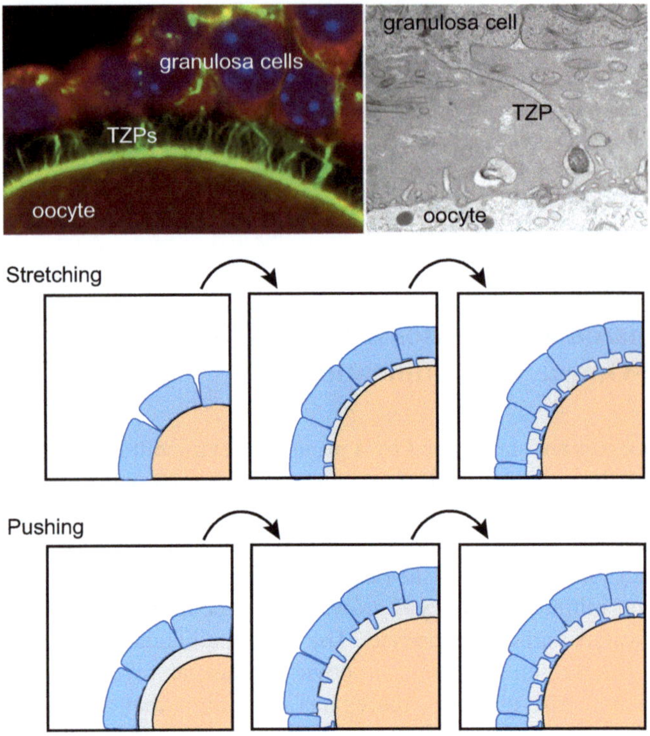

Fig. 2.3 Transzonal projections (TZPs) physically link the granulosa cells to the oocyte. *Upper* A confocal image of a portion of a granulosa cell-oocyte complex stained using fluorescent phalloidin, which binds to filamentous actin, reveals the TZPs and oocyte cortex. Granulosa cell nuclei are stained *blue*. The electron micrograph on the *right* shows a TZP extending from the granulosa cell through the zona pellucida to the oocyte plasma membrane. *Lower* Two models of TZP formation. The stretching model proposes that sites of cell adhesion that are established between the oocyte and granulosa cells before the deposition of the zona pellucida remain during and after its deposition. This generates a stretched cytoplasmic filament that becomes the TZP. The pushing model proposes that TZPs are elaborated from the granulosa cells after deposition of the zona pellucida and grow toward the oocyte where they establish contact with its plasma membrane. The models are not mutually exclusive

(Evans 2015), it seems probable that other proteins maintain the intercellular contact at the TZP tips; notably, TZPs are present in the ovarian follicles of mice lacking *Gja4* (Simon et al. 1997). Intriguingly, the number of TZPs is reduced by about 30% when *Ptk2* (protein tyrosine kinase) is deleted from oocytes (McGinnis and Kinsey 2015). This is accompanied by reduced gap junctional coupling as assayed by connexin-37 immunoreactivity in the zona pellucida and dye-transfer from the oocyte to the adjacent granulosa cells. As PTK2 promotes the formation of intercellular junctions by stabilizing complexes containing nectins or cadherins, it may help to stabilize the connection of the TZPs to the oocyte.

How TZPs arise remains unknown, and (at least) two models may be envisioned through which they could be generated (Fig. 2.3). The first (Chiquoine 1960;

Hadek 1965) is rooted in the physical contact between the oocyte and granulosa cells that exists in the primordial follicle. It is proposed that the sites of intercellular adhesion remain as the zona pellucida is assembled. As the granulosa cell bodies are pushed away from the oocyte, cytoplasmic filaments are generated where the granulosa cells remains tethered to the oocyte surface. As these filaments elongate, they become the TZPs. The second model posits that the TZPs are elaborated from the granulosa cells in a manner similar to the growth of filopodia. Although TZPs are within the size range observed for filopodia (diameter approximately 100 nm), little direct evidence yet supports either model. However, several observations are worth noting. First, as the oocyte grows, the number of TZPs that project to it increases substantially (Makabe et al. 2006). This is consistent with the consideration that the number of granulosa cells adjacent to the oocyte must increase in order for them to continue to fully cover the expanding oocyte surface. Second, because the zona pellucida becomes fully formed at an early stage of oocyte growth, a substantial fraction of the granulosa cell population in the oocyte-adjacent layer will be born after the zona has been elaborated. These granulosa cells likely project TZPs, which appear to be uniformly distributed around the surface of oocyte. It is not immediately obvious how the first model can accommodate these observations. Intriguingly, at the time of primordial follicle formation, the pre-granulosa cells extend filopodia that enclose the oocyte (Lechowska et al. 2011; Pepling 2012). Moreover, oocytes can occasionally be observed in the adrenal gland where, although they produce an extracellular coat resembling a zona pellucida, no TZPs are observed to project from the adjacent adrenal cells (Zamboni and Upadhyay 1983). These results suggest that the granulosa cells may have an inherent and distinct predisposition to generate filopodia.

2.6 Granulosa Cell-Oocyte Communication at the Initiation of Oocyte and Follicular Growth

An early morphological indicator that a primordial follicle and its enclosed oocyte have entered the growth phase is the transition of the granulosa cells from a squamous to cuboidal morphology. This suggests that oocyte growth might be initiated by signals sent by the granulosa cells, and several lines of evidence point to Kit ligand (KITL) as a strong candidate (Fig. 2.4). KITL is constitutively expressed by the granulosa cells (Hutt et al. 2006; Thomas and Vanderhyden 2006) and interacts with the Kit receptor (KIT) expressed by oocytes (Manova et al. 1990; Kidder and Vanderhyden 2010). Oocyte-specific expression of a constitutively active form of KIT (Kit^{D818V}) causes most oocytes within primordial follicles to begin growing shortly after birth (Saatcioglu et al. 2016). Conversely, deletion of its kinase prevents oocytes from beginning to grow, even when the adjacent granulosa cells become cuboidal and began to proliferate. Within the granulosa cells, deletion of *Tsc1*, which encodes a

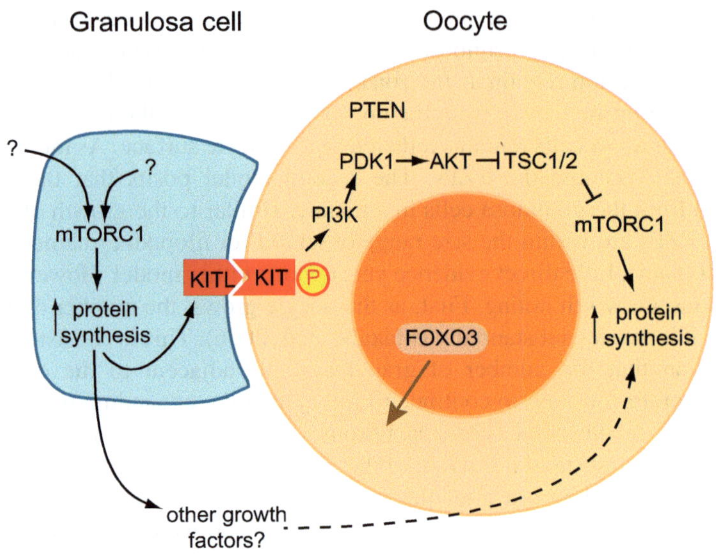

Fig. 2.4 Granulosa cell-oocyte communication at the initiation of oocyte growth. In response to an intra- or extracellular signal of unknown origin, the granulosa cells of a primordial follicle may increase their rate of protein synthesis. This leads to an increase in the production of KITL, possibly in particular the membrane-associated variant. The granulosa cell-derived KITL activates its receptor, KIT, on the oocyte plasma membrane. This in turn increases protein synthesis on the oocyte via PI3-kinase signaling. Reduced nuclear activity of the transcription factor, FOXO3, is a key downstream event driving oocyte growth. A role for other granulosa cell-derived factors is not excluded. Adapted from Res Probl Cell Diff 58:191–224

repressor of MTORC1 and whose deletion would be expected to increase protein synthesis, leads to an increase in KITL and widespread initiation of oocyte growth (Zhang et al. 2014). Conversely, deletion of *Rptor* in the granulosa cells, which would decrease protein synthesis, prevents the squamous-cuboidal transition in these cells as well as the initiation of oocyte growth (Zhang et al. 2014). Taken together, these results suggest that an increase in KIT activity in the oocyte triggers its growth.

The mechanism by which KIT activation triggers oocyte growth remains to be fully established. KIT typically signals through phosphatidylinositol kinase-3, leading to activation of AKT and an increase in global protein synthesis. Consistent with this, deletion in oocytes of *Pten*, which antagonizes this pathway, promotes oocyte growth (John et al. 2008; Reddy et al. 2008). Growth also depends on other molecular signals, however. For example, deletion of *Nobox* in the mouse prevents normal oocyte growth, strongly implicating this transcription factor (Rajkovic et al. 2004). Additionally, the transcription factor, FOXO3, translocates from the nucleus to the cytoplasm at an early stage of growth (albeit clearly after growth has begun), and genetic deletion of *Foxo3* causes most oocytes in primordial follicles to begin to grow (Castrillon et al. 2003; John et al. 2008). Taking these results together, a plausible proposal for the transition of primordial follicles into the growth phase may be that an increase in protein synthesis within the granulosa cells, which is associated with cuboidalization in some manner, leads to increase in the production and availability of KITL. This activates, or increases the activation of, KIT on the oocyte membrane, leading to an increase in MTOR-regulated protein synthesis and consequent oocyte growth.

Although the initiation of oocyte growth normally is accompanied by cuboidalization of the granulosa cells, it is worth noting that this morphological change is apparently not indispensable. Expression of a constitutively active *Kit* mutant (Saatcioglu et al. 2016) or deletion of *Tsc1* or *Tsc2* in the oocyte (Adhikari et al. 2009, 2010) triggers its growth, even though the granulosa cells retain their squamous morphology. Similarly, deletion of the oocyte-specific transcription factor *Sohlh2* or oocyte-specific deletion of transcription factor *Lhx8* also triggers initiation of oocyte growth, while the adjacent granulosa cells remain squamous (Pangas et al. 2006; Choi et al. 2008; Ren et al. 2015). Nevertheless, the squamous-cuboidal transition may be required for oocytes to grow efficiently or beyond a certain size and in any case remains a reliable marker of the physiological activation of oocyte growth.

Although an increase in the rate of protein synthesis in the adjacent granulosa cells likely plays a central role in triggering oocyte growth, the source and the nature of the signal that increases protein synthesis in the granulosa cells of one primordial follicle but not those of its neighbor remains unknown. Indeed, it could be hypothesized that the oocyte is the source of the signal. A clue may lie in observations that, when ovary-like structures have been generated by aggregating oocytes and granulosa cells, many of the reconstituted follicles rapidly initiate growth (Eppig et al. 2000; Hayashi et al. 2012). This strongly suggests that the regulated entry of primordial follicles into the growth pool depends on cellular interactions in the intact ovary that are not recapitulated in the de novo-generated structures.

2.7 Granulosa Cell-Oocyte Communication During Oocyte and Follicular Growth

The granulosa cells play an essential role not only in initiating oocyte growth but also in sustaining it. This was dramatically illustrated many years ago when granulosa-oocyte complexes (GOCs) were isolated from ovarian follicles and cultured intact or following physical separation of the two cell types (Eppig 1979). Oocytes within intact GOCs continued to grow, whereas as oocytes not in direct contact with the granulosa cells grew little or not at all, even when the granulosa cells were provided in co-culture. Subsequent studies showed that the rate of oocyte growth was proportional to the number of granulosa cells that were associated with it (Brower and Schultz 1982). These results highlight the indispensable role played by the granulosa cells as well as the crucial importance of physical contact between the two cell types. As discussed above, gap junctions couple the granulosa cells to the oocyte and to each other. Gap junctions permit the transfer of molecules up to about 1 kDa between cells, and these include factors essential for normal oocyte growth and development (Eppig 1991). For example, the oocyte expresses only low levels of mRNAs encoding factors required to convert glucose to pyruvate and thus relies on pyruvate transferred from the granulosa cells. Via the gap junctions, the granulosa cells also provide the oocyte with ribonucleotides and some amino acids that it cannot efficiently take up from extracellular sources. More broadly, it has been suggested that, because the surface/volume ratio of the oocyte steadily decreases as it grows, the coupling with the granulosa cells provides oocyte access to the cell surface of the granulosa cells, thereby mitigating the effect of its growth (Eppig 1991).

Although oocytes cannot reach full size when gap junctional coupling with the granulosa cells is lost, they nonetheless undergo substantial growth, reaching about half the volume of wild-type oocytes (Carabatsos et al. 2000; Gittens and Kidder 2005). Since oocytes grow only very little or not at all when the granulosa are physically removed, this suggests that contact-dependent but gap junction-independent signaling from the granulosa cells also drives oocyte growth. The nature of these signals remains to be established. However, KITL promotes oocyte growth in vitro (Packer et al. 1994; Thomas et al. 2008). Importantly, the membrane-associated form of the ligand, which is generated by alternative splicing of the mRNA, is considerably more effective than the soluble form (Thomas et al. 2008). In vivo, both isoforms of the mRNA are present, but decline at the end of follicular growth, when oocyte growth has ceased, via an oocyte-regulated process (Joyce et al. 1999, 2000). These results suggest that the granulosa cells promote oocyte growth by extracellular, albeit membrane-associated, signaling as well as through gap junctional coupling.

In addition to providing factors that sustain the metabolic activity of the growing oocyte and enable its continual increase in volume, the granulosa cells also transmit signals that regulate its developmental progression. The best-known of these regulate the concentration of cyclic AMP (cAMP) in the oocyte. Activation of cAMP-dependent protein kinase A inhibits entry of the oocyte into M-phase, by promoting phosphorylation of sites on CDK1 that inhibit its activity; thus, a high level of cAMP prevents the initiation of meiotic maturation, whereas a sufficient decrease in cAMP concentration within the oocyte permits maturation (Holt et al. 2013; Adhikari and Liu 2014). Based on evidence including that blocking gap junctional communication between the oocyte and cumulus granulosa within the follicle can trigger maturation in the absence of LH (the physiological trigger) (Sela-Abramovich et al. 2006; Richard and Baltz 2014), it was long thought that these permitted the granulosa cells to deliver cAMP to the oocyte. Later studies demonstrated, however, that meiotic arrest depends on cAMP generated by the oocyte itself (Bornslaeger and Schultz 1985; Mehlmann et al. 2004).

Studies over the last decade have established that cGMP, rather than cAMP, is the maturation-inhibiting molecule transferred by the granulosa via the gap

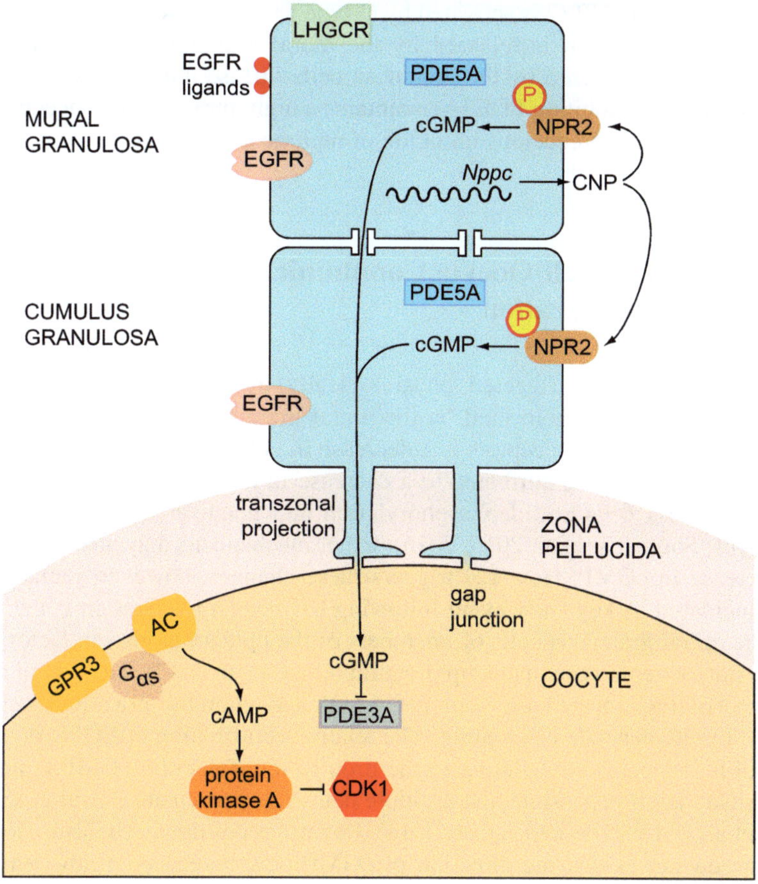

Fig. 2.5 Granulosa cell-oocyte communication regulates meiotic maturation. C-type natriuretic peptide (CNP) is manufactured and secreted by mural granulosa cells where it associates with the guanylyl cyclase natriuretic peptide receptor (NPR) 2 on the mural and cumulus granulosa cell membranes, triggering the synthesis of cyclic GMP (cGMP). Via the gap junctions that couple the granulosa cells to the oocyte, cGMP diffuses to the oocyte, where it inhibits the activity of phosphodiesterase (PDE) 3A. Owing to PDE inhibition, the concentration of cAMP, which is synthesized by the oocyte, remains high enough to activate cAMP-dependent protein kinase A. This inhibits cyclin-dependent kinase (CDK) 1. When luteinizing hormone (LH) binds to its receptor on the mural granulosa cells, production of cGMP decreases, leading to a drop in its concentration within the oocyte. PDE3A becomes active, degrades cAMP, thereby leading to an increase in CDK1 activity. This triggers meiotic maturation. Adapted from Res Probl Cell Diff 58:191–224

junctions (Fig. 2.5) (Norris et al. 2009; Conti et al. 2012; Jaffe and Egbert 2016). cGMP acts by inhibiting the activity of the major cAMP-directed phosphodiesterase, PDE3A, that is present in the oocyte. cGMP is manufactured by the granulosa cells through the activity of the guanylyl cyclase natriuretic peptide receptor (NPR) 2, located on the plasma membrane of both mural and cumulus granulosa cells

(Zhang et al. 2010). NPR2 is activated by C-type natriuretic peptide (CNP, encoded by the *Nppc* gene) that is released by the mural granulosa cells. Thus, cGMP constitutively synthesized by the granulosa cells diffuse into the oocytes via gap junctions and, by inhibiting PDE3A, maintains a high intra-oocyte concentration of cAMP and prevents precocious initiation of meiotic maturation.

2.8 Granulosa Cell-Oocyte Communication During Meiotic Maturation

Meiotic maturation is triggered physiologically by luteinizing hormone (LH), which binds to receptors located on the mural granulosa cells. The consequence of activation of the LH receptors is a decrease in cGMP concentration throughout the follicle. This can be attributed to a decrease in the production of CNP by the mural granulosa cells and dephosphorylation and inactivation of NPR2 (Egbert et al. 2014; Shuhaibar et al. 2015, 2016). Other mechanisms may also contribute to the decrease in cGMP. How LH triggers these changes, however, remains to be fully elucidated. A key early event following LH receptor activation is the release from the mural granulosa cells of, members of the epidermal growth factor (EGF) family, notably epiregulin and amphiregulin (Conti et al. 2012). These bind to EGF receptor (EGFR) located on both the mural and cumulus granulosa cells and activate EGFR-dependent signaling. The relative contribution of EGFR-dependent and EGFR-independent signaling to the drop in intrafollicular cGMP is not fully established, and it is possible that multiple pathways participate (Conti et al. 2012; Wang et al. 2013; Jaffe and Egbert 2016). Regardless of the mechanism, however, the consequence is that the direction of cGMP flow reverses—it now leaves the oocyte and returns to the granulosa cells to maintain a uniform concentration across the coupled cells. The resulting decrease in oocyte cGMP derepresses PDE3A, leading to a decrease in intra-oocyte cAMP and activation of CDK1. Thus, the gap junctions linking the oocyte to the granulosa cells play a crucial role, not only in preventing precocious maturation but also in enabling oocyte maturation in response to LH.

Recent studies have uncovered an additional role for the cumulus granulosa cells during maturation. Oocytes of mice lacking amphiregulin ($Areg^{-/-}$) show an increased incidence of spindle abnormalities at metaphase II and a reduced ability to develop as embryos (Chen et al. 2013). Moreover, providing amphiregulin to cumulus cell-enclosed oocytes during maturation in vitro increases the fraction that are able to develop as embryos. This important observation confirms that amphiregulin exerts this beneficial effect during oocyte maturation. Further studies revealed that amphiregulin increases AKT phosphorylation in the oocyte and that this effect depends on the presence of the cumulus cells. Finally, amphiregulin increases the synthesis of proteins encoded by mRNAs that are translationally activated during maturation. Taken together these results indicate that amphiregulin-stimulated cumulus cells send signals to the oocyte that increase the translation of a subset of mRNAs during maturation and increase the developmental competence of the oocytes. The nature of the signals and the mechanism by which they are transmitted to the oocyte are important questions that remain to be answered.

2.9 Newly Identified Pathways of Granulosa Cell-Oocyte Communication

Recent studies described below have identified new and unanticipated mechanisms of intercellular communication within the follicle (Fig. 2.6). Membrane-bound structures termed extracellular vesicles (ECV) have been identified in a variety of normal and pathological contexts (Machtinger et al. 2016; Navakanitworakul et al. 2016). Two classes are currently recognized, distinguished by their origin. Exosomes range from 10 to 100 nm in diameter and arise from the fusion of multivesicular bodies with the plasma membrane. Microvesicles range from 100 to 1000 nm and are budded off from the plasma membrane. Because the origin of vesicles cannot be reliably determined after they have been released from the cell, they have been collectively termed ECV. ECV can contain a diverse array of macromolecules including mRNAs, proteins, and microRNAs and thus hold the potential to alter the physiology of the cells with which they fuse. Several groups have independently identified and characterized ECV in follicular fluid of mammals including humans (da Silveira et al. 2012; Sohel et al. 2013; Santonocito et al. 2014; Hung et al. 2015; Navakanitworakul et al. 2016). Similar to ECVs in other contexts, the follicular ECVs contain abundant small RNAs whose size is consistent with miRNAs. The composition of the ECVs changes during follicular growth, and, importantly, the follicular ECV can be taken up by cumulus cells and are able to promote expansion of the cumulus cell matrix, identifying a potential biological function for them (Hung et al. 2015).

These observations raised the intriguing possibility that ECVs might also be a vehicle for information transfer between the oocyte and granulosa cells. Indeed, recent work has identified small vesicles at the tips of TZPs (Macaulay et al. 2014, 2016). Moreover, live-cell imaging has demonstrated that cargo including RNAs can be transported along the TZPs toward the tips from where the vesicles presumably bud. These observations suggest that such vesicles could provide a mechanism to deliver macromolecules from the cumulus granulosa cells to the oocyte. In support of this possibility, when oocytes were incubated on a monolayer of granulosa cells expressing EGFP, the encoding *Egfp* mRNA was subsequently detected in the oocytes (Macaulay et al. 2014). Future studies should uncover the extent of the role played by ECV in delivering cargo between and within the somatic and germ-line compartments of the follicle.

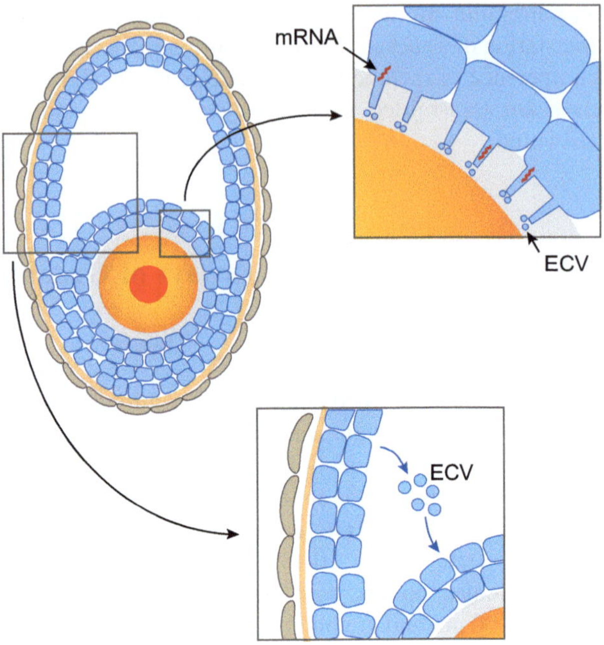

Fig. 2.6 New pathways of granulosa cell-oocyte communication. Extracellular vesicles derived from the granulosa cells are present in follicular fluid and can fuse with cumulus granulosa cells, providing a means to transfer macromolecules between granulosa cells, including between the mural and cumulus subpopulations. Vesicles that may be related to ECV have also been observed at the base of TZPs where they meet the oocyte surface and may similarly transfer macromolecules from the granulosa cells to the oocyte

2.10 Conclusion

The developing oocyte relies on its somatic environment—specifically, the granulosa cells that enclose it—throughout its development from a resting state in the primordial follicle until it is ovulated in preparation for fertilization. The granulosa cells provide nutrients that sustain the metabolic activity of the oocyte and send signals that regulate its developmental progression. In turn, the oocyte sends signals to the granulosa cells that regulate their differentiation and function. This bidirectional communication between the germ cell and its somatic support cells relies on both intra- and extracellular pathways and dynamically remodels the microenvironment of the follicle corresponding to the needs of the oocyte, thus ensuring that ovulation releases a fully formed oocyte that is competent to develop as a healthy embryo.

Acknowledgments Work in my laboratory is supported by the Eunice Kennedy Shriver National Institute of Child Health and Human Development of the National Institutes of Health (R21HD086407), Canadian Institutes of Health Research, the Natural Sciences and Engineering Research Council of Canada, and the Research Institute of the McGill University Health Centre. Research reported in this publication is solely the responsibility of the authors and does not necessarily represent the official views of the National Institutes of Health.

References

Ackert CL, Gittens JE, O'Brien MJ, Eppig JJ, Kidder GM (2001) Intercellular communication via connexin43 gap junctions is required for ovarian folliculogenesis in the mouse. Dev Biol 233:258–270

Adhikari D, Liu K (2014) The regulation of maturation promoting factor during prophase I arrest and meiotic entry in mammalian oocytes. Mol Cell Endocrinol 382:480–487

Adhikari D, Flohr G, Gorre N, Shen Y, Yang H, Lundin E, Lan Z, Gambello MJ, Liu K (2009) Disruption of Tsc2 in oocytes leads to overactivation of the entire pool of primordial follicles. Mol Hum Reprod 15:765–770

Adhikari D, Zheng W, Shen Y, Gorre N, Hamalainen T, Cooney AJ, Huhtaniemi I, Lan Z-J, Liu K (2010) Tsc/mTORC1 signaling in oocytes governs the quiescence and activation of primordial follicles. Hum Mol Genet 19:397–410

Albertini DF, Rider V (1994) Patterns of intercellular connectivity in the mammalian cumulus-oocyte complex. Microsc Res Tech 27:125–133

Anderson E, Albertini DF (1976) Gap junctions between the oocyte and companion follicle cells in the mammalian ovary. J Cell Biol 71:680–686

Ben-Meir A, Burstein E, Borrego-Alvarez A, Chong J, Wong E, Yavorska T, Naranian T, Chi M, Wang Y, Bentov Y, Alexis J, Meriano J, Sung HK, Gasser DL, Moley KH, Hekimi S, Casper RF, Jurisicova A (2015) Coenzyme Q10 restores oocyte mitochondrial function and fertility during reproductive aging. Aging Cell 14:887–895

Bennabi I, Terret ME, Verlhac MH (2016) Meiotic spindle assembly and chromosome segregation in oocytes. J Cell Biol 215:611–619

Bornslaeger EA, Schultz RM (1985) Adenylate cyclase activity in zona-free mouse oocytes. Exp Cell Res 156:277–281

Brower PT, Schultz RM (1982) Intercellular communication between granulosa cells and mouse oocytes: existence and possible nutritional role during oocyte growth. Dev Biol 90:144–153

Buccione R, Vanderhyden BC, Caron PJ, Eppig JJ (1990) FSH-induced expansion of the mouse cumulus oophorus in vitro is dependent upon a specific factor(s) secreted by the oocyte. Dev Biol 138:16–25

Carabatsos MJ, Elvin J, Matzuk MM, Albertini DF (1998) Characterization of oocyte and follicle development in growth differentiation factor-9-deficient mice. Dev Biol 204:373–384

Carabatsos MJ, Sellitto C, Goodenough DA, Albertini DF (2000) Oocyte-granulosa cell heterologous gap junctions are required for the coordination of nuclear and cytoplasmic meiotic competence. Dev Biol 226:167–179

Castrillon DH, Miao L, Kollipara R, Horner JW, DePinho RA (2003) Suppression of ovarian follicle activation in mice by the transcription factor Foxo3a. Science 301:215–218

Cheeseman LP, Boulanger J, Bond LM, Schuh M (2016) Two pathways regulate cortical granule translocation to prevent polyspermy in mouse oocytes. Nat Commun 7:13726

Chen J, Torcia S, Xie F, Lin CJ, Cakmak H, Franciosi F, Horner K, Onodera C, Song JS, Cedars MI, Ramalho-Santos M, Conti M (2013) Somatic cells regulate maternal mRNA translation and developmental competence of mouse oocytes. Nat Cell Biol 15:1415–1423

Chiquoine AD (1960) The development of the zona pellucida of the mammalian ovum. Am J Anat 106:149–169

Choi Y, Yuan D, Rajkovic A (2008) Germ cell-specific transcriptional regulator sohlh2 is essential for early mouse folliculogenesis and oocyte-specific gene expression. Biol Reprod 79:1176–1182

Conti M, Hsieh M, Musa Zamah A, Oh JS (2012) Novel signaling mechanisms in the ovary during oocyte maturation and ovulation. Mol Cell Endocrinol 356:65–73

Coticchio G, Dal Canto M, Mignini Renzini M, Guglielmo MC, Brambillasca F, Turchi D, Novara PV, Fadini R (2015) Oocyte maturation: gamete-somatic cells interactions, meiotic resumption, cytoskeletal dynamics and cytoplasmic reorganization. Hum Reprod Update 21:427–454

Da Silva-Buttkus P, Jayasooriya GS, Mora JM, Mobberley M, Ryder TA, Baithun M, Stark J, Franks S, Hardy K (2008) Effect of cell shape and packing density on granulosa cell proliferation and formation of multiple layers during early follicle development in the ovary. J Cell Sci 121:3890–3900

da Silveira JC, Veeramachaneni DN, Winger QA, Carnevale EM, Bouma GJ (2012) Cell-secreted vesicles in equine ovarian follicular fluid contain miRNAs and proteins: a possible new form of cell communication within the ovarian follicle. Biol Reprod 86:71

Diaz FJ, Wigglesworth K, Eppig JJ (2007) Oocytes determine cumulus cell lineage in mouse ovarian follicles. J Cell Sci 120:1330–1340

Dong J, Albertini DF, Nishimori K, Kumar RM, Lu N, Matzuk M (1996) Growth differentiation factor-9 is required during early ovarian folliculogenesis. Nature 383:531–535

Dragovic RA, Ritter LJ, Schulz SJ, Amato F, Thompson JG, Armstrong DT, Gilchrist RB (2007) Oocyte-secreted factor activation of SMAD 2/3 signaling enables initiation of mouse cumulus cell expansion. Biol Reprod 76:848–857

Ducibella T, Duffy P, Buetow J (1994) Quantification and localization of cortical granules during oogenesis in the mouse. Biol Reprod 50:467–473

Egbert JR, Shuhaibar LC, Edmund AB, Van Helden DA, Robinson JW, Uliasz TF, Baena V, Geertsi A, Wunder F, Potter LR, Jaffe LA (2014) Dephosphorylation and inactivation of NPR2 guanylyl cyclase in granulosa cells contributes to the LH-induced decrease in cGMP that causes resumption of meiosis in rat oocytes. Development 141:3594–3604

el-Fouly MA, Cook B, Nekola M, Nalbandov AV (1970) Role of the ovum in follicular luteinization. Endocrinology 87:286–293

Emori C, Wigglesworth K, Fujii W, Naito K, Eppig JJ, Sugiura K (2013) Cooperative effects of 17beta-estradiol and oocyte-derived paracrine factors on the transcriptome of mouse cumulus cells. Endocrinology 154:4859–4872

Eppig JJ (1979) A comparison between oocyte growth in coculture with granulosa cells and oocytes with granulosa cell-oocyte junctional contact maintained in vitro. J Exp Zool 209:345–353

Eppig JJ (1991) Intercommunication between mammalian oocytes and companion somatic cells. Bioessays 13:569–574

Eppig JJ, Wigglesworth K, Hirao Y (2000) Metaphase I arrest and spontaneous parthenogenetic activation of strain LTXBO oocytes: chimeric reaggregated ovaries establish primary lesion in oocytes. Dev Biol 224:60–68

Eppig JJ, Pendola FL, Wigglesworth K, Pendola JK (2005) Mouse oocytes regulate metabolic cooperativity between granulosa cells and oocytes: amino acid transport. Biol Reprod 73:351–357

Evans WH (2015) Cell communication across gap junctions: a historical perspective and current developments. Biochem Soc Trans 43:450–459

Flemr M, Ma J, Schultz RM, Svoboda P (2010) P-body loss is concomitant with formation of a messenger RNA storage domain in mouse oocytes. Biol Reprod 82:1008–1017

Gilchrist RB, Ritter LJ, Myllymaa S, Kaivo-Oja N, Dragovic RA, Hickey TE, Ritvos O, Mottershead DG (2006) Molecular basis of oocyte-paracrine signalling that promotes granulosa cell proliferation. J Cell Sci 119:3811–3821

Gilchrist RB, Lane M, Thompson JG (2008) Oocyte-secreted factors: regulators of cumulus cell function and oocyte quality. Hum Reprod Update 14:159–177

Gittens JE, Kidder GM (2005) Differential contributions of connexin37 and connexin43 to oogenesis revealed in chimeric reaggregated mouse ovaries. J Cell Sci 118:5071–5078

Grive KJ, Freiman RN (2015) The developmental origins of the mammalian ovarian reserve. Development 142:2554–2563

Gunesdogan U, Surani MA (2016) Developmental competence for primordial germ cell fate. Curr Top Dev Biol 117:471–496

Guo J, Shi L, Gong X, Jiang M, Yin Y, Zhang X, Yin H, Li H, Emori C, Sugiura K, Eppig JJ, Su YQ (2016) Oocyte-dependent activation of MTOR in cumulus cells controls the development and survival of cumulus-oocyte complexes. J Cell Sci 129:3091–3103
Hadek R (1965) The structure of the mammalian egg. Int Rev Cytol 18:29–71
Haverfield J, Nakagawa S, Love D, Tsichlaki E, Nomikos M, Lai FA, Swann K, FitzHarris G (2016) Ca(2+) dynamics in oocytes from naturally-aged mice. Sci Rep 6:19357
Hayashi K, Ogushi S, Kurimoto K, Shimamoto S, Ohta H, Saitou M (2012) Offspring from oocytes derived from in vitro primordial germ cell-like cells in mice. Science 338:971–975
Hirshfield AN (1991) Development of follicles in the mammalian ovary. Int Rev Cytol 124:43–101
Holt JE, Lane SI, Jones KT (2013) The control of meiotic maturation in mammalian oocytes. Curr Top Dev Biol 102:207–226
Hummitzsch K, Anderson RA, Wilhelm D, Wu J, Telfer EE, Russell DL, Robertson SA, Rodgers RJ (2015) Stem cells, progenitor cells, and lineage decisions in the ovary. Endocr Rev 36:65–91
Hung WT, Hong X, Christenson LK, McGinnis LK (2015) Extracellular vesicles from bovine follicular fluid support cumulus expansion. Biol Reprod 93:117
Hutt KJ, McLaughlin EA, Holland MK (2006) Kit ligand and c-Kit have diverse roles during mammalian oogenesis and folliculogenesis. Mol Hum Reprod 12:61–69
Jaffe LA, Egbert JR (2016) Regulation of mammalian oocyte meiosis by intercellular communication within the ovarian follicle. Annu Rev Physiol. doi:10.1146/annurev-physiol-022516-034102
John GB, Gallardo TD, Shirley LJ, Castrillon DH (2008) Foxo3 is a PI3K-dependent molecular switch controlling the initiation of oocyte growth. Dev Biol 321:197–204
Jorgensen JS (2013) Defining the neighborhoods that escort the oocyte through its early life events and into a functional follicle. Mol Reprod Dev 80:960–976
Joyce IM, Pendola FL, Wigglesworth K, Eppig JJ (1999) Oocyte regulation of kit ligand expression in mouse ovarian follicles. Dev Biol 214:342–353
Joyce IM, Clark AT, Pendola FL, Eppig JJ (2000) Comparison of recombinant growth differentiation factor-9 and oocyte regulation of KIT ligand messenger ribonucleic acid expression in mouse ovarian follicles. Biol Reprod 63:1669–1675
Kang MK, Han SJ (2011) Post-transcriptional and post-translational regulation during mouse oocyte maturation. BMB Rep 44:147–157
Kidder GM, Vanderhyden BC (2010) Bidirectional communication between oocytes and follicle cells: ensuring oocyte developmental competence. Can J Physiol Pharmacol 88:399–413
Kim B, Zhang X, Kan R, Cohen R, Mukai C, Travis AJ, Coonrod SA (2014) The role of MATER in endoplasmic reticulum distribution and calcium homeostasis in mouse oocytes. Dev Biol 386:331–339
Koval M, Molina SA, Burt JM (2014) Mix and match: investigating heteromeric and heterotypic gap junction channels in model systems and native tissues. FEBS Lett 588:1193–1204
Lechowska A, Bilinski S, Choi Y, Shin Y, Kloc M, Rajkovic A (2011) Premature ovarian failure in nobox-deficient mice is caused by defects in somatic cell invasion and germ cell cyst breakdown. J Assist Reprod Genet 28:583–589
Lei L, Spradling AC (2013) Female mice lack adult germ-line stem cells but sustain oogenesis using stable primordial follicles. Proc Natl Acad Sci USA 110:8585–8590
Li R, Albertini DF (2013) The road to maturation: somatic cell interaction and self-organization of the mammalian oocyte. Nat Rev Mol Cell Biol 14:141–152
Li L, Baibakov B, Dean J (2008) A subcortical maternal complex essential for preimplantation mouse embryogenesis. Dev Cell 15:416–425
Li X, Tripurani SK, James R, Pangas SA (2011) Minimal fertility defects in mice deficient in oocyte-expressed Smad4. Biol Reprod 86:1–6
Litscher ES, Wassarman PM (2014) Evolution, structure, and synthesis of vertebrate egg-coat proteins. Trends Dev Biol 8:65–76
Liu C, Peng J, Matzuk MM, Yao HH (2015) Lineage specification of ovarian theca cells requires multicellular interactions via oocyte and granulosa cells. Nat Commun 6:6934

Lucifero D, Mann MR, Bartolomei MS, Trasler JM (2004) Gene-specific timing and epigenetic memory in oocyte imprinting. Hum Mol Genet 13:839–849

Macaulay AD, Gilbert I, Caballero J, Barreto R, Fournier E, Tossou P, Sirard MA, Clarke HJ, Khandjian EW, Richard FJ, Hyttel P, Robert C (2014) The gametic synapse: RNA transfer to the bovine oocyte. Biol Reprod 91:90

Macaulay AD, Gilbert I, Scantland S, Fournier E, Ashkar F, Bastien A, Saadi HA, Gagne D, Sirard MA, Khandjian EW, Richard FJ, Hyttel P, Robert C (2016) Cumulus cell transcripts transit to the bovine oocyte in preparation for maturation. Biol Reprod 94:16

Machtinger R, Laurent LC, Baccarelli AA (2016) Extracellular vesicles: roles in gamete maturation, fertilization and embryo implantation. Hum Reprod Update 22:182–193

Makabe S, Naguro T, Stallone T (2006) Oocyte-follicle cell interactions during ovarian follicle development, as seen by high resolution scanning and transmission electron microscopy in humans. Microsc Res Tech 69:436–449

Manova K, Nocka K, Besmer P, Bachvarova R (1990) Gonadal expression of c-kit encoded at the W locus of the mouse. Development 110:1057–1069

McGinnis LK, Kinsey WH (2015) Role of focal adhesion kinase in oocyte-follicle communication. Mol Reprod Dev 82:90–102

Mehlmann LM, Saeki Y, Tanaka S, Brennan TJ, Evsikov AV, Pendola FL, Knowles BB, Eppig JJ, Jaffe LA (2004) The Gs-linked receptor GPR3 maintains meiotic arrest in mammalian oocytes. Science 306:1947–1950

Meldrum DR, Casper RF, Diez-Juan A, Simon C, Domar AD, Frydman R (2016) Aging and the environment affect gamete and embryo potential: can we intervene? Fertil Steril 105:548–559

Mitchell PA, Burghardt RC (1986) The ontogeny of nexuses (gap junctions) in the ovary of the fetal mouse. Anat Rec 214:283–288

Mora JM, Fenwick MA, Castle L, Baithun M, Ryder TA, Mobberley M, Carzaniga R, Franks S, Hardy K (2012) Characterization and significance of adhesion and junction-related proteins in mouse ovarian follicles. Biol Reprod 86:153

Mork L, Maatouk DM, McMahon JA, Guo JJ, Zhang P, McMahon AP, Capel B (2012) Temporal differences in granulosa cell specification in the ovary reflect distinct follicle fates in mice. Biol Reprod 86:37

Motta P, Makabe S, Naguro T, Correr S (1994) Oocyte follicle cells association during development of human ovarian follicle. A study by high resolution scanning and transmission electron microscopy. Arch Histol Cytol 57:369–394

Mottershead DG, Sugimura S, Al-Musawi SL, Li JJ, Richani D, White MA, Martin GA, Trotta AP, Ritter LJ, Shi J, Mueller TD, Harrison CA, Gilchrist RB (2015) Cumulin, an oocyte-secreted heterodimer of the transforming growth factor-beta family, is a potent activator of granulosa cells and improves oocyte quality. J Biol Chem 290:24007–24020

Navakanitworakul R, Hung WT, Gunewardena S, Davis JS, Chotigeat W, Christenson LK (2016) Characterization and small RNA content of extracellular vesicles in follicular fluid of developing bovine antral follicles. Sci Rep 6:25486

Nelson SM, Telfer EE, Anderson RA (2013) The ageing ovary and uterus: new biological insights. Hum Reprod Update 19:67–83

Norris RP, Ratzan WJ, Freudzon M, Mehlmann LM, Krall J, Movsesian MA, Wang H, Ke H, Nikolaev VO, Jaffe LA (2009) Cyclic GMP from the surrounding somatic cells regulates cyclic AMP and meiosis in the mouse oocyte. Development 136:1869–1878

Packer AI, Hsu YC, Besmer P, Bachvarova RF (1994) The ligand of the c-kit receptor promotes oocyte growth. Dev Biol 161:194–205

Pangas SA, Choi Y, Ballow DJ, Zhao Y, Westphal H, Matzuk MM, Rajkovic A (2006) Oogenesis requires germ cell-specific transcriptional regulators Sohlh1 and Lhx8. Proc Natl Acad Sci USA 103:8090–8095

Peng J, Li Q, Wigglesworth K, Rangarajan A, Kattamuri C, Peterson RT, Eppig JJ, Thompson TB, Matzuk MM (2013) Growth differentiation factor 9:bone morphogenetic protein 15 heterodimers are potent regulators of ovarian functions. Proc Natl Acad Sci USA 110: E776–E785

Pepling ME (2012) Follicular assembly: mechanisms of action. Reproduction 143:139–149

Pepling ME, Spradling AC (2001) Mouse ovarian germ cell cysts undergo programmed breakdown to form primordial follicles. Dev Biol 234:339–351

Rajkovic A, Pangas SA, Ballow D, Suzumori N, Matzuk MM (2004) NOBOX deficiency disrupts early folliculogenesis and oocyte-specific gene expression. Science 305:1157–1159

Reddy P, Liu L, Adhikari D, Jagarlamudi K, Rajareddy S, Shen Y, Du C, Tang W, Hamalainen T, Peng SL, Lan Z-J, Cooney AJ, Huhtaniemi I, Liu K (2008) Oocyte-specific deletion of Pten causes premature activation of the primordial follicle pool. Science 319:611–613

Reddy P, Adhikari D, Zheng W, Liang S, Hamalainen T, Tohonen V, Ogawa W, Noda T, Volarevic S, Huhtaniemi I, Liu K (2009) PDK1 signaling in oocytes controls reproductive aging and lifespan by manipulating the survival of primordial follicles. Hum Mol Genet 18:2813–2824

Ren Y, Suzuki H, Jagarlamudi K, Golnoski K, McGuire M, Lopes R, Pachnis V, Rajkovic A (2015) Lhx8 regulates primordial follicle activation and postnatal folliculogenesis. BMC Biol 13:39

Richard S, Baltz JM (2014) Prophase I arrest of mouse oocytes mediated by natriuretic peptide precursor C requires GJA1 (connexin-43) and GJA4 (connexin-37) gap junctions in the antral follicle and cumulus-oocyte complex. Biol Reprod 90:137

Russell DL, Salustri A (2006) Extracellular matrix of the cumulus-oocyte complex. Semin Reprod Med 24:217–227

Saatcioglu HD, Cuevas I, Castrillon DH (2016) Control of oocyte reawakening by Kit. PLoS Genet 12:e1006215

Sánchez F, Smitz J (2012) Molecular control of oogenesis. Biochim Biophys Acta 1822:1896–1912

Santonocito M, Vento M, Guglielmino MR, Battaglia R, Wahlgren J, Ragusa M, Barbagallo D, Borzi P, Rizzari S, Maugeri M, Scollo P, Tatone C, Valadi H, Purrello M, Di Pietro C (2014) Molecular characterization of exosomes and their microRNA cargo in human follicular fluid: bioinformatic analysis reveals that exosomal microRNAs control pathways involved in follicular maturation. Fertil Steril 102:1751–1761. (e1751)

Sela-Abramovich S, Edry I, Galiani D, Nevo N, Dekel N (2006) Disruption of gap junctional communication within the ovarian follicle induces oocyte maturation. Endocrinology 147:2280–2286

Shuhaibar LC, Egbert JR, Norris RP, Lampe PD, Nikolaev VO, Thunemann M, Wen L, Feil R, Jaffe LA (2015) Intercellular signaling via cyclic GMP diffusion through gap junctions restarts meiosis in mouse ovarian follicles. Proc Natl Acad Sci USA 112:5527–5532

Shuhaibar LC, Egbert JR, Edmund AB, Uliasz TF, Dickey DM, Yee SP, Potter LR, Jaffe LA (2016) Dephosphorylation of juxtamembrane serines and threonines of the NPR2 guanylyl cyclase is required for rapid resumption of oocyte meiosis in response to luteinizing hormone. Dev Biol 409:194–201

Simon A, Goodenough D, Li E, Paul D (1997) Female infertility in mice lacking connexin 37. Nature 385:525–529

Smallwood SA, Tomizawa S, Krueger F, Ruf N, Carli N, Segonds-Pichon A, Sato S, Hata K, Andrews SR, Kelsey G (2011) Dynamic CpG island methylation landscape in oocytes and preimplantation embryos. Nat Genet 43:811–814

Sohel MM, Hoelker M, Noferesti SS, Salilew-Wondim D, Tholen E, Looft C, Rings F, Uddin MJ, Spencer TE, Schellander K, Tesfaye D (2013) Exosomal and non-exosomal transport of extracellular microRNAs in follicular fluid: implications for bovine oocyte developmental competence. PLoS One 8:e78505

Soyal SM, Amleh A, Dean J (2000) FIG alpha, a germ cell-specific transcription factor required for ovarian follicle formation. Development 127:4645–4654

Stewart KR, Veselovska L, Kim J, Huang J, Saadeh H, S-i T, Smallwood SA, Chen T, Kelsey G (2015) Dynamic changes in histone modifications precede de novo DNA methylation in oocytes. Genes Dev 29:2449–2462

Su Y-Q, Wu X, O'Brien MJ, Pendola FL, Denegre JN, Matzuk MM, Eppig JJ (2004) Synergistic roles of BMP15 and GDF9 in the development and function of the oocyte-cumulus cell complex in mice: genetic evidence for an oocyte-granulosa cell regulatory loop. Dev Biol 276:64–73

Su Y-Q, Sugiura K, Wigglesworth K, O'Brien MJ, Affourtit JP, Pangas SA, Matzuk MM, Eppig JJ (2008) Oocyte regulation of metabolic cooperativity between mouse cumulus cells and oocytes: BMP15 and GDF9 control cholesterol biosynthesis in cumulus cells. Development 135:111–121

Su YQ, Sugiura K, Eppig JJ (2009) Mouse oocyte control of granulosa cell development and function: paracrine regulation of cumulus cell metabolism. Semin Reprod Med 27:32–42

Sugiura K, Su Y-Q, Diaz FJ, Pangas SA, Sharma S, Wigglesworth K, O'Brien MJ, Matzuk MM, Shimasaki S, Eppig JJ (2007) Oocyte-derived BMP15 and FGFs cooperate to promote glycolysis in cumulus cells. Development 134:2593–2603

Susor A, Jansova D, Cerna R, Danylevska A, Anger M, Toralova T, Malik R, Supolikova J, Cook MS, Oh JS, Kubelka M (2015) Temporal and spatial regulation of translation in the mammalian oocyte via the mTOR-eIF4F pathway. Nat Commun 6:6078

Thomas F, Vanderhyden B (2006) Oocyte-granulosa cell interactions during mouse follicular development: regulation of Kit ligand expression and its role in oocyte growth. Reprod Biol Endocrinol 4:19

Thomas FH, Ismail RS, Jiang J-Y, Vanderhyden BC (2008) Kit ligand 2 promotes murine oocyte growth in vitro. Biol Reprod 78:167–175

Titus S, Li F, Stobezki R, Akula K, Unsal E, Jeong K, Dickler M, Robson M, Moy F, Goswami S, Oktay K (2013) Impairment of BRCA1-related DNA double-strand break repair leads to ovarian aging in mice and humans. Sci Transl Med 5:172ra121

Tomizawa S, Nowacka-Woszuk J, Kelsey G (2012) DNA methylation establishment during oocyte growth: mechanisms and significance. Int J Dev Biol 56:867–875

Van Blerkom J (2010) Mitochondrial function in the human oocyte and embryo and their role in developmental competence. Mitochondrion 11:797–813

Vanderhyden B, Caron P, Buccione R, Eppig J (1990) Developmental pattern of the secretion of cumulus expansion-enabling factor by mouse oocytes and the role of oocytes in promoting granulosa cell differentiation. Dev Biol 140:307–317

Veitch GI, Gittens JE, Shao Q, Laird DW, Kidder GM (2004) Selective assembly of connexin37 into heterocellular gap junctions at the oocyte/granulosa cell interface. J Cell Sci 117:2699–2707

Wang Y, Kong N, Li N, Hao X, Wei K, Xiang X, Xia G, Zhang M (2013) Epidermal growth factor receptor signaling-dependent calcium elevation in cumulus cells is required for NPR2 inhibition and meiotic resumption in mouse oocytes. Endocrinology 154:3401–3409

Wassarman PM, Litscher ES (2012) Influence of the zona pellucida of the mouse egg on folliculogenesis and fertility. Int J Dev Biol 56:833–839

Wassarman PM, Litscher ES (2013) Biogenesis of the mouse egg's extracellular coat, the zona pellucida. Curr Top Dev Biol 102:243–266

Wigglesworth K, Lee KB, Emori C, Sugiura K, Eppig JJ (2015) Transcriptomic diversification of developing cumulus and mural granulosa cells in mouse ovarian follicles. Biol Reprod 92:23

Winterhager E, Kidder GM (2015) Gap junction connexins in female reproductive organs: implications for women's reproductive health. Hum Reprod Update 21:340–352

Wu X, Chen L, Brown CA, Yan C, Matzuk MM (2004) Interrelationship of growth differentiation factor 9 and inhibin in early folliculogenesis and ovarian tumorigenesis in mice. Mol Endocrinol 18:1509–1519

Yurttas P, Vitale AM, Fitzhenry RJ, Cohen-Gould L, Wu W, Gossen JA, Coonrod SA (2008) Role for PADI6 and the cytoplasmic lattices in ribosomal storage in oocytes and translational control in the early mouse embryo. Development 135:2627–2636

Zamboni L, Upadhyay S (1983) Germ cell differentiation in mouse adrenal glands. J Exp Zool 228:173–193

Zarate-Garcia L, Lane SI, Merriman JA, Jones KT (2016) FACS-sorted putative oogonial stem cells from the ovary are neither DDX4-positive nor germ cells. Sci Rep 6:27991

Zhang M, Su YQ, Sugiura K, Xia G, Eppig JJ (2010) Granulosa cell ligand NPPC and its receptor NPR2 maintain meiotic arrest in mouse oocytes. Science 330:366–369

Zhang H, Zheng W, Shen Y, Adhikari D, Ueno H, Liu K (2012) Experimental evidence showing that no mitotically active female germline progenitors exist in postnatal mouse ovaries. Proc Natl Acad Sci USA 109:12580–12585

Zhang H, Risal S, Gorre N, Busayavalasa K, Li X, Shen Y, Bosbach B, Brännström M, Liu K (2014) Somatic cells initiate primordial follicle activation and govern the development of dormant oocytes in mice. Curr Biol 24:2501–2508

Zhang H, Panula S, Petropoulos S, Edsgard D, Busayavalasa K, Liu L, Li X, Risal S, Shen Y, Shao J, Liu M, Li S, Zhang D, Zhang X, Gerner RR, Sheikhi M, Damdimopoulou P, Sandberg R, Douagi I, Gustafsson JA, Liu L, Lanner F, Hovatta O, Liu K (2015) Adult human and mouse ovaries lack DDX4-expressing functional oogonial stem cells. Nat Med 21:1116–1118

Chapter 3
Transovarial Transmission of Symbionts in Insects

Teresa Szklarzewicz and Anna Michalik

Abstract Many insects, on account of their unbalanced diet, live in obligate symbiotic associations with microorganisms (bacteria or yeast-like symbionts), which provide them with substances missing in the food they consume. In the body of host insect, symbiotic microorganisms may occur intracellularly (e.g., in specialized cells of mesodermal origin termed bacteriocytes, in fat body cells, in midgut epithelium) or extracellularly (e.g., in hemolymph, in midgut lumen). As a rule, symbionts are vertically transmitted to the next generation. In most insects, symbiotic microorganisms are transferred from mother to offspring transovarially within female germ cells. The results of numerous ultrastructural and molecular studies on symbiotic systems in different groups of insects have shown that they have a large diversity of symbiotic microorganisms and different strategies of their transmission from one generation to the next. This chapter reviews the modes of transovarial transmission of symbionts between generations in insects.

3.1 Introduction

Many animals are host to obligate symbiotic microorganisms (bacteria, fungi, protozoans), which may play various roles, e.g., supplement their unbalanced diet (in clams, tube worms living in deep-sea hydrothermal and cold vents, hematophagous insects and leeches, plant sap-sucking insects), digest complex carbohydrates (in termites), emit bioluminescence (in squids, fish), increase resistance against parasites and pathogens (in aphids, carpenter ants), and stimulate immune system (gut bacteria in mammals) (reviewed in Abt and Artis 2009; Baumann 1998; Buchner 1965; Dale and Moran 2006; Dasch et al. 1984; Kikuchi 2009). These microorganisms may be inherited in three ways: (1) vertically (maternally), i.e., from mother to offspring; (2) horizontally, i.e., between specimens of the same population; and (3) environmentally, i.e., through the acquisition of free-living microorganisms from the environment (Bright and Bulgheresi 2010; Kikuchi

T. Szklarzewicz (✉) • A. Michalik
Department of Developmental Biology and Morphology of Invertebrates, Institute of Zoology and Biomedical Research, Jagiellonian University, Gronostajowa 9, 30-387 Kraków, Poland
e-mail: teresa.szklarzewicz@uj.edu.pl

M. Kloc (ed.), *Oocytes*, Results and Problems in Cell Differentiation 63,
DOI 10.1007/978-3-319-60855-6_3

2009; Kikuchi and Fukatsu 2003). In most animals, the symbionts are transmitted to the next generation vertically; however, the mechanism of the vertical transmission in particular groups of animals may vary. In some heteropteran bugs, gut symbionts are transmitted to the progeny through the contamination of the egg surface with microorganisms, the deposition of capsules containing bacteria onto the egg surface, and feeding on mother's excrements by the newly hatched larva (Fukatsu and Hosokawa 2002; Hosokawa et al. 2005; Kikuchi et al. 2007). In most animals, which have obligate symbiotic microorganisms, they are transmitted to the next generation transovarially, i.e., through the infestation of female germ cells.

It is estimated that at least 15–20% (according to Ishikawa 2003, more than 70%) of species of insects harbor obligate bacteria and/or yeast-like symbionts (reviewed, e.g., in Baumann 2005; Buchner 1965; Douglas 1989, 1998; Ishikawa 1989; Kikuchi 2009; Moran and Baumann 2000). Results of recently conducted molecular studies have confirmed an earlier hypothesis based on morphological observations and experiments with sterile insects and eggs (Buchner 1965) that in most insects their symbionts are responsible for the synthesis of substances missing in the diet, required for the proper growth and reproduction of the host insect (Douglas 2009; McCutcheon and Moran 2007; McCutcheon et al. 2009; Wu et al. 2006). Microorganisms living in the body of hematophagous insects such as tsetse flies, triatomid bugs, lice, and bedbugs supply them with vitamin B (Dasch et al. 1984; Ishikawa 2003). Most plant sap-sucking hemipterans (aphids, scale insects, whiteflies, psyllids, cicadas, leafhoppers, treehoppers, spittlebugs, planthoppers) harbor symbiotic microorganisms, which provide them with the amino acids missing in the plant sap (Douglas 2009; Ishikawa 2003; Wilkinson and Ishikawa 2001). As a rule, symbiotic microorganisms are localized intracellularly in the insect body: in fat body cells, in the cells of the midgut epithelium, in the specialized giant cells of mesodermal origin termed bacteriocytes (the older term "mycetocytes"). Microorganisms living inside bacteriocytes are termed "mycetomic symbionts."

Paul Buchner, who is regarded as the father of studies on the symbiotic associations between insects and microorganisms, distinguished two types of symbionts: primary symbionts (later termed P-symbionts) and accessory symbionts (later termed secondary symbionts, S-symbionts, facultative symbionts) (Buchner 1965). Because the association between primary symbionts and insects is the result of a single, ancient infection of ancestor of this group of insects, these microorganisms are present in all individuals of the insect taxon (bacterium *Buchnera aphidicola* in aphids, bacterium *Carsonella ruddii* in psyllids, bacterium *Portiera aleyrodidarum* in whiteflies). The primary symbionts are, as a rule, harbored in bacteriocytes and are always transovarially transmitted between generations (Buchner 1965). The function of P-symbionts is to supplement the diet of the host insect with essential nutrients. In contrast to primary symbionts, secondary symbionts may occur in only some insect populations. They may be localized intracellularly in bacteriocytes (or in other cells, e.g., fat body cells) or extracellularly (e.g., in the hemolymph). These symbionts may be inherited both transovarially and horizontally (Fukatsu et al. 2000; Moran and Telang 1998;

Oliver et al. 2006, 2010). The results of studies on aphids have indicated that their S-symbionts belong to different taxa and may play various roles, such as increase of the resistance to heat stress or parasites. It is likely that they may also take over the function of the lost primary symbiont, bacterium *Buchnera aphidicola*, and provide nutrients (Burke et al. 2009; Lukasik et al. 2013: Koga et al. 2003; Montllor et al. 2002; Oliver et al. 2003; Scarborough et al. 2005; Vorburger et al. 2010). The above facts indicate that the presence of secondary symbionts in insects results from more recent, multiple, independent infections. A more complex situation has been found in hemipterans belonging to Auchenorrhyncha, which encompasses groups such cicadas, leafhoppers, treehoppers, planthoppers, and spittlebugs. As a rule, auchenorrhynchans are characterized by the occurrence of two types of symbiotic microorganisms (termed by Takiya et al. (2006) "coprimary symbionts"), both of which are engaged in the synthesis of amino acids. For example, in the glassy-winged sharpshooter, *Homalodisca vitripennis* (Cicadomorpha, Cicadellidae: Cicadellinae) one of the two symbionts synthesizes eight out of ten essential amino acids, whereas the second one is responsible for the production of the remaining two amino acids (McCutcheon and Moran 2007). The symbiotic systems of Auchenorrhyncha are unusual because, during the evolution of some lineages of this hemipteran group, ancient symbiotic microorganisms, such as Bacteroidetes bacterium *Sulcia muelleri* and betaproteobacterial symbionts, have been replaced by other bacteria or yeast-like symbionts (McCutcheon et al. 2009; Koga et al. 2013; Takiya et al. 2006).

Results of earlier histological observations (Buchner 1965; Ries 1931; Schneider 1940) and more recent ultrastructural studies (Cheng and Hou 2001; Cheung and Purcell 1999; Eberle and McLean 1982; Kobiałka et al. 2016; Koga et al. 2012; Michalik et al. 2014a, b, 2016; Pyka-Fosciak and Szklarzewicz 2008; Sacchi et al. 1988, 2008; Swiatoniowska et al. 2013; Szklarzewicz and Moskal 2001; Szklarzewicz et al. 2006, 2010, 2013, 2016; Żelazowska and Biliński 1999) have revealed that insects use a variety of mechanisms of transovarial transmission of mycetomic symbionts from mother to offspring; however, regardless of the method of the transmission, the beginning of the migration of symbionts from bacteriocytes to ovaries correlates with the stage of ovary development. This observation strongly supports the hypothesis presented by Eberle and McLean (1982), which postulates that the infestation of ovaries by symbiotic microorganisms is stimulated by a factor produced by the developing ovary.

3.2 Localization of Transovarially Transmitted Symbionts in the Insect Body

In most insects living in symbiotic association with microorganisms, the transovarially transmitted symbionts are harbored in the cytoplasm of giant, polyploid cells termed bacteriocytes, which are grouped in large organs termed bacteriomes (Fig. 3.1a). Previously bacteriomes have been termed "mycetomes"; however, currently, the term "mycetome" designates only the organ containing yeast-like symbionts. In most insects, bacteriomes are localized between the body wall and the ovaries. In insects, which possess more than one type of symbiont (in aphids and scale insects with primary and secondary symbionts and in auchenorrhynchans with coprimary symbionts), these microorganisms are usually harbored in separate bacteriocytes (Fig. 3.1a). A few exceptions to this rule are (1) aphid *Aphis viburni* (Aphidomorpha, Aphidoidea) in which secondary symbionts occupy their own bacteriocytes and also coreside in bacteriocytes with the primary symbiont, bacterium *Buchnera aphidicola* (Fig. 3.1b) (Michalik et al. 2014a), (2) several species of adelgids (Aphidomorpha, Adelgoidea) in which betaproteobacterial and gammaproteobacterial symbionts occur in the same bacteriocytes (Toenshoff et al. 2012), and (3) whiteflies (Aleyrodomorpha) in which primary and secondary symbionts inhabit the same bacteriocytes (Costa et al. 1993; Gottlieb et al. 2008; Szklarzewicz and Moskal 2001).

In the scale insects *Icerya purchasi* and *Palaeococcus fuscipennis* (Monophlebidae) bacteriomes are composed of extremely large bacteriocytes surrounded by a single layer of small epithelial cells (Fig. 3.1c). They contain two types of symbiotic bacteria: large pleomorphic bacteria (primary symbionts) and small rod-shaped bacteria (secondary symbionts) (Niżnik and Szklarzewicz 2007; Szklarzewicz et al. 2006). The pleomorphic bacteria are harbored in the cytoplasm of bacteriocytes, whereas the rod-shaped bacteria are present in epithelial cells (Fig. 3.1c). Interestingly, in another genus of the family Monophlebidae, *Drosicha*, bacteriomes have different organization. They are composed of several lobes. The lobes' periphery contains smaller bacteriocytes with pleomorphic bacteria, whereas central syncytium contains enterobacterial symbionts (Matsuura et al. 2009). The syncytial organization of bacteriomes/mycetomes is also typical for psyllids (Kot et al. 2012, 2014) and planthoppers belonging to the families Flatidae and Delphacidae (Michalik et al. 2009). In psyllids, the bacteriomes consist of externally located bacteriocytes containing the primary symbiont (gammaproteobacterium *Carsonella ruddii*) and internal syncytium containing secondary symbionts (Fig. 3.1d), which belong to distinct clusters of Gammaproteobacteria (Fukatsu and Nikoh 1998; Kot et al. 2012, 2014; Spaulding and von Dohlen 1998; Thao et al. 2000a, b). In planthoppers from the Flatidae and Delphacidae families, the yeast-like symbionts are harbored in the syncytial mycetome, which is surrounded by a single layer of small epithelial cells (Cheng and Hou 2001; Michalik et al. 2009).

In 2001, von Dohlen and coworkers described in mealybugs (Pseudococcidae: Pseudococcinae) an unusual phenomenon (termed "nested symbiosis") of symbiotic bacteria within other bacteria. The ultrastructural, histochemical, and molecular methods showed that Pseudococcinae mealybugs are host to betaproteobacterial symbionts (named *Tremblaya princeps* by Thao et al. 2002), which in turn harbor gammaproteobacterial symbionts (Fig. 3.1e, f) (von Dohlen et al. 2001; Gatehouse et al. 2011; Husnik et al. 2013; Kono et al. 2008; McCutcheon and von Dohlen 2011; Thao et al. 2002). The molecular analyses have shown that both these

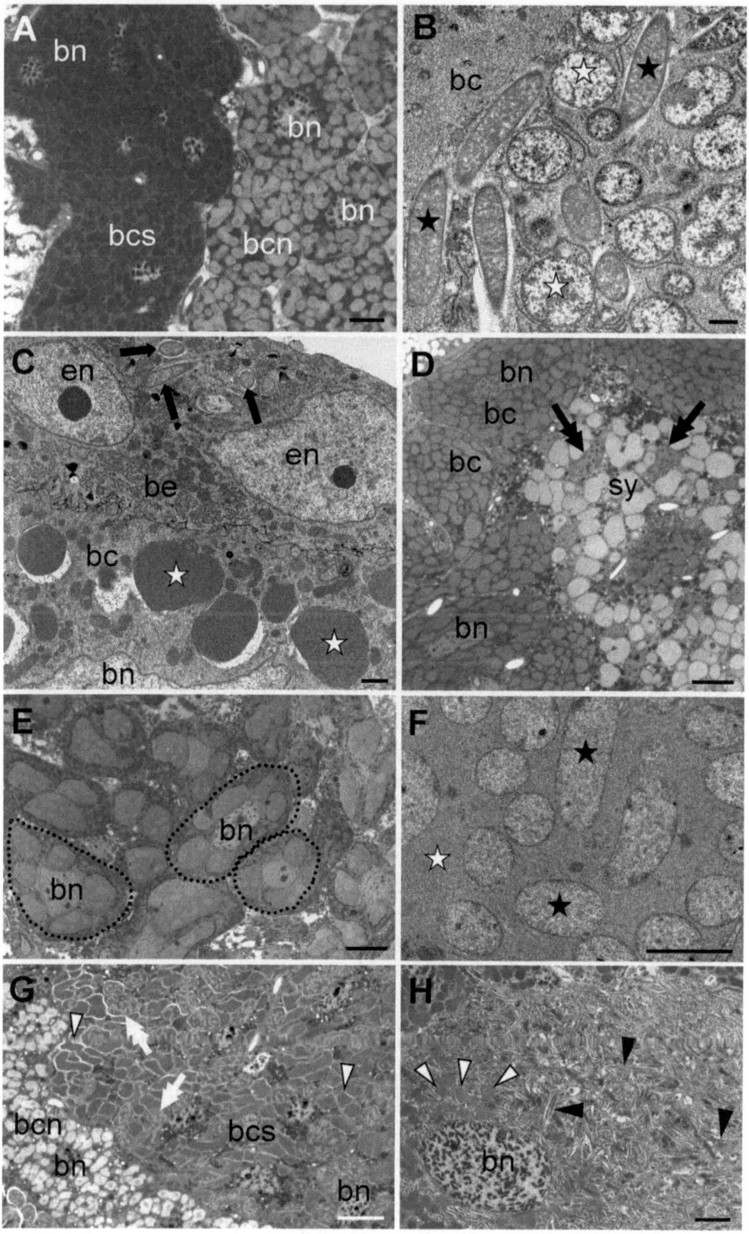

Fig. 3.1 Distribution of symbiotic microorganisms in the insect body (**a**) *Deltocephalus pulicaris* (Hemiptera, Cicadomorpha, Cicadellidae: Deltocephalinae). Fragment of the bacteriome composed of external bacteriocytes with *Sulcia muelleri* bacterium and internal bacteriocytes with the betaproteobacterial symbiont *Nasuia deltocephalinicola*. Methylene blue, scale bar = 20 μm. (**b**) *Aphis viburni* (Hemiptera, Aphidomorpha, Aphididae: Aphidinae). Primary symbiont *Buchnera aphidicola* (*white asterisks*) and secondary symbionts (*black asterisks*) in the same bacteriocyte. TEM, scale bar = 2 μm. (**c**) *Palaeococcus fuscipennis* (Hemiptera, Coccomorpha, Monophlebidae). Fragment of the bacteriocyte with pleomorphic bacteria (primary symbionts) (*white asterisks*). *Black arrows* indicate rod-shaped bacteria (secondary symbionts) in the

symbionts, like the coprimary symbionts of auchenorrhynchans (see above), contribute to the synthesis of amino acids for their host insect (Husnik et al. 2013; Husnik and McCutcheon 2016; McCutcheon and von Dohlen 2011). The similar phenomenon of "nested symbiosis" has recently been found in two species of Auchenorrhyncha: in *Macrosteles laevis* (Cicadomorpha, Cicadellidae: Deltocephalinae) and in *Cicadella viridis* (Cicadomorpha, Cicadellidae: Cicadellinae) (Kobiałka et al. 2016; Michalik et al. 2014b). In the bacteriocytes of these species, the bacteria contain other bacteria: gammaproteobacterium *Sodalis* inside of the bacterium *Sulcia* in *Cicadella viridis* and gammaproteobacterium *Arsenophonus* inside of the bacterium *Sulcia* in *Macrosteles laevis* (Fig. 3.1g) (Kobiałka et al. 2016; Michalik et al. 2014b). Moreover, in *Cicadella viridis,* before migration into the ovaries, the *Sodalis* bacteria enter the cells of *Sulcia* bacteria (Michalik et al. 2014b). This unusual phenomenon takes place in bacteriocytes containing both *Sulcia* and *Sodalis* bacteria (Fig. 3.1h). According to Michalik et al. (2014b), the symbiosis in *Cicadella viridis* represents the initial stage of the acquisition of bacterium *Sodalis*. These authors suggest that the occurrence of bacteria *Sodalis* inside of bacteria *Sulcia* may be associated with the high virulence of bacterium *Sodalis*. Thus, the "hiding" of *Sodalis* within *Sulcia* bacteria may protect them from the immune system of host insect.

In the heteropteran bug *Chilacis typhae* (Artheneidae) (Kuechler et al. 2011) and scale insect *Marchalina hellenica* (Buchner 1965), symbiotic bacteria do not occur in bacteriomes but in enlarged cells of the midgut epithelium. In both *Chilacis typhae* (Kuechler et al. 2011) and *Marchalina hellenica* (Buchner 1965; Szklarzewicz et al. 2013), symbionts leave the cells of the midgut epithelium and

---

Fig. 3.1 (continued) epithelial cells. TEM, scale bar = 1 μm. (**d**) *Cacopsylla mali* (Hemiptera, Psyllomorpha, Psyllidae). Fragment of the bacteriome composed of external bacteriocytes with primary symbiont *Carsonella ruddii* and internal syncytium with secondary symbionts. *Double black arrows* indicate nuclei of the syncytium. Methylene blue, scale bar = 10 μm (courtesy of Marta Kot, Jagiellonian University, Kraków, Poland). (**e**) *Trionymus aberrans* (Hemiptera, Coccomorpha, Pseudococcidae: Pseudococcinae). Fragment of the bacteriome. Bacteriocytes (encircled by a *black dotted line*) are filled with large pleomorphic bacteria *Tremblaya princeps* which contain small, rod-shaped bacteria in the cytoplasm. Methylene blue, scale bar = 10 μm. (**f**) *Trionymus thulensis* (Hemiptera, Coccomorpha, Pseudococcidae: Pseudococcinae). Fragment of the pleomorphic bacterium *Tremblaya princeps* (*white asterisk*), which contains numerous rod-shaped bacteria (*black asterisks*). TEM, scale bar = 1 μm. (**g**) *Macrosteles laevis* (Hemiptera, Cicadomorpha, Cicadellidae: Deltocephalinae). Fragment of the bacteriome containing bacteriocytes with the bacterium *Nasuia deltocephalinicola* and bacteriocytes with the bacterium *Sulcia muelleri* (*white arrowheads*), as well as the bacterium *Sulcia muelleri* with the bacterium *Arsenophonus* inside its cells (*double white arrows*). Methylene blue, scale bar = 20 μm. (**h**) *Cicadella viridis* (Hemiptera, Cicadomorpha, Cicadellidae: Cicadellinae). Fragment of the bacteriocyte with the bacterium *Sulcia muelleri* (*white arrowheads*) and bacterium *Sodalis* (*black arrowheads*). Methylene blue, scale bar = 20 μm. *bc* bacteriocyte, *be* epithelium surrounding the bacteriome, *bcs* bacteriocyte with the bacterium *Sulcia muelleri*, *bcn* bacteriocyte with the bacterium *Nasuia deltocephalinicola*, *bn* bacteriocyte nucleus, *en* epithelial cell nucleus, *sy* syncytial region of the bacteriome containing the secondary symbionts infest the ovaries. Because these symbionts are transovarially inherited, Kuechler et al. (2011) suggested that such a symbiosis might represent the intermediate state between the gut symbiosis and bacteriocyte symbiosis. On the other hand, in lice, the bacteriomes tightly adhere to the midgut epithelium (Buchner 1965; Eberle and McLean 1983; Ries 1931). In males, the bacteriocytes do not leave the bacteriomes, whereas in females, they migrate from the midgut bacteriome to the apical parts of lateral oviducts, where they form the large accumulations termed "ovarial ampullae" (named also "intraovarian bacteriomes") (Buchner 1965; Eberle and McLean 1983; Ries 1931; Żelazowska and Biliński 1999).

In several species of heteropteran bugs (belonging to the families Blissidae (*Ischnodemus sabuleti*), Lygaeidae (*Arocatus longiceps*, *Kleidocerys resedae*, *Nysius ericae*, and *Nithecus jacobaeae*), and Artheneidae (*Chilacis typhae*), bacteriocytes occur not only within bacteriomes/midgut epithelium but also within the ovarioles. In all these insects, the basal part of the tropharium containing early previtellogenic oocytes has a ring-shaped infection zone composed of several bacteriocytes (Fig. 3.2a, b, c) (Kuechler et al. 2010, 2011, 2012; Matsuura et al. 2012; Schneider 1940; Swiatoniowska et al. 2013).

In some insects, symbiotic microorganisms, e.g., yeast-like symbionts in leafhoppers (Hemiptera, Cicadomorpha, Cicadellidae) (Fig. 3.2d) and bacteria *Burkholderia* in scale insects *Acanthococcus aceris* and *Gossyparia spuria* (both Eriococcidae) (Fig. 3.2e), do not colonize specialized cells but are dispersed in the fat body (Kobiałka and Szklarzewicz, in preparation; Michalik et al. 2016). According to Michalik et al. (2016), the lack of bacteriocytes and the presence of symbiotic microorganisms in the fat body of scale insects indicate an initial (evolutionary young) stage of symbiosis.

3.3 Modes of Transovarial Transmission of Mycetomic Symbionts

3.3.1 Infestation of Embryos in Viviparous Aphids

Aphids represent widely distributed plant pests which are classified into the following three superfamilies: Adelgoidea, Phylloxeroidea, and Aphidoidea (Wojciechowski et al. 2015). All generations of Adelgoidea and Phylloxeroidea are oviparous, whereas Aphidoidea (termed "modern aphids") have a complex life cycle with parthenogenetic viviparous females and sexual oviparous females (Dixon 1985; Moran 1992).

The symbiotic systems of aphids belong to the most explored systems within insects. These studies showed that the representatives of Aphidoidea, as a rule, harbor a primary symbiont—gammaproteobacterium *Buchnera aphidicola*—as well as secondary symbionts belonging to different systematic groups (with the exception of some representatives of the subfamily Hormaphidinae which, instead

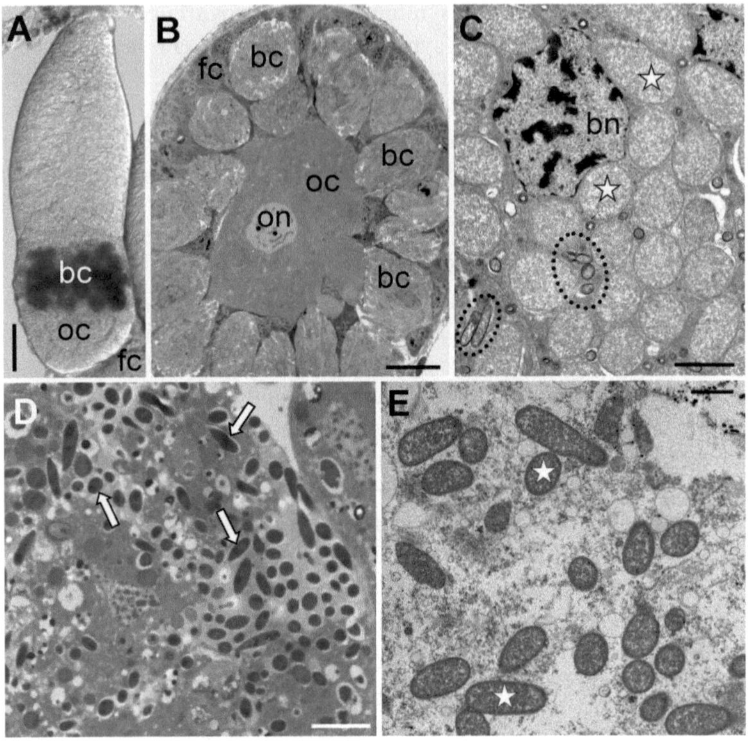

Fig. 3.2 Distribution of symbiotic microorganisms in the insect body (**a**) *Nysius ericae* (Hemiptera, Heteroptera, Lygaeidae: Orsilinae). Fragment of the ovariole with the bacteriocytes in the ring-shaped infection zone. Differential interference contrast, scale bar = 100 μm. (**b**) *Nithecus jacobaeae* (Hemiptera, Heteroptera, Lygaeidae: Orsilinae). Fragment of the ovariole, which contains the infection zone (cross section). Methylene blue, scale bar = 20 μm. (**c**) *Nysius ericae* (Hemiptera, Heteroptera, Lygaeidae: Orsilinae). Fragment of the bacteriocyte residing in the infection zone filled with large, elongated bacteria (*white asterisks*) and small, rod-shaped bacteria (encircled by a *black dotted line*). TEM, scale bar = 2 μm. (**d**) *Graphocraerus ventralis* (Hemiptera, Cicadomorpha, Cicadellidae: Deltocephalinae). Fragment of the fat body with yeast-like symbionts (*white arrows*). Methylene blue, scale bar = 20 μm (courtesy of Michał Kobiałka, Jagiellonian University, Kraków, Poland). (**e**). *Gossyparia spuria* (Hemiptera, Coccomorpha, Eriococcidae). Bacteria *Burkholderia* (*white asterisks*) dispersed in the fat body cells. TEM, scale bar = 1 μm. *bc* bacteriocyte, *bn* bacteriocyte nucleus, *fc* follicular epithelium, *oc* oocyte, *on* oocyte nucleus

of bacteria, have yeast-like symbionts) (for further details see, e.g., Baumann et al. 1995; Buchner 1965; Douglas 1998; Fukatsu and Ishikawa 1992, 1996; Hongoh and Ishikawa 2000; Ishikawa 2003; Michalik et al. 2014a; Nováková et al. 2013; Oliver et al. 2010; Wegierek et al. 2017).

Viviparous females of Aphidoidea are characterized by an unusual developmental strategy termed "the telescoping of generations" (Kindlmann and Dixon 1989; Moran 1992). In the females of these aphids, embryos develop, inside of which the next generation of embryos forms (Fig. 3.3a). Bacteriocytes are present in both the

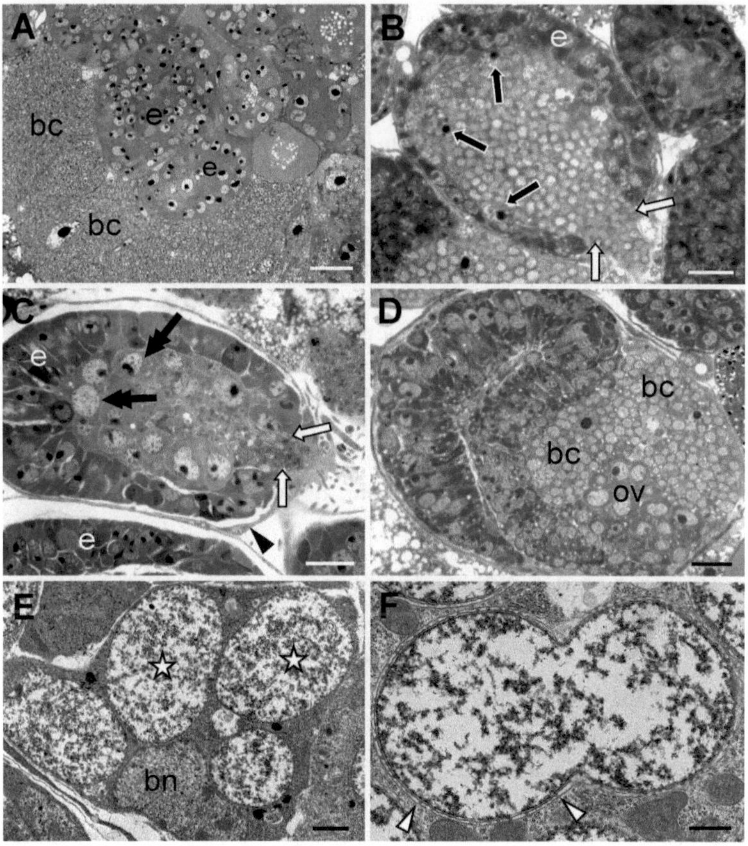

Fig. 3.3 The successive stages of transovarial transmission of symbionts in viviparous aphids (Hemiptera, Aphidomorpha, Aphidoidea, Aphididae) (**a**) *Aphis viburni* (Aphidinae). Fragment of the embryo. Bacteriocytes with bacterium *Buchnera aphidicola* (primary symbiont) are present in the vicinity of next-generation embryos. Methylene blue, scale bar = 20 μm. (**b**) *Cavariella theobaldi* (Aphidinae). Symbiotic bacteria infest the embryo at the blastula stage (longitudinal section). *White arrows* indicate the broad stream of bacteria entering the embryo. Note the nuclei of the presumptive bacteriocytes (*black arrows*). Methylene blue, scale bar = 10 μm. (**c**) *Aphis craccae* (Aphidinae). The embryo at late blastula stage (longitudinal section). Note the newly formed bacteriocytes (*double black arrows*) and the opening at the posterior pole of the embryo (*white arrows*). *Black arrowhead* indicates flattened follicular cells surrounding the embryo. Methylene blue, scale bar = 10 μm. (**d**) *Cavariella theobaldi* (Aphidinae). The embryo contains bacteriocytes, which are packed with symbiotic bacteria (longitudinal section). Methylene blue, scale bar = 10 μm. (**e**) *Hamamelistes betulinus* (Hormaphidinae). Newly formed bacteriocyte filled with few bacteria *Buchnera aphidicola* (*white asterisks*). TEM, scale bar = 1 μm. (**f**) *Stomaphis quercus* (Lachninae). The dividing bacterium *Buchnera aphidicola*. *White arrowheads* indicate the external most "perisymbiotic membrane," which is derived from the host insect. TEM, scale bar = 1 μm. *bc* bacteriocyte, *bn* bacteriocyte nucleus, *e* embryo, *ov* ovary

body of the female and their embryos (Fig. 3.3a). At the time the embryos are at the blastula stage, the symbionts commence the invasion of their posterior poles (Fig. 3.3b, c). The embryo is surrounded by a single layer of flattened follicular cells (Fig. 3.3c), and embryo's interior is filled with a syncytial mass of nuclei of presumptive bacteriocytes (Fig. 3.3b). As the released symbionts accumulate around the posterior pole of the embryo, the follicular cells, as well as cells of the embryo, separate from each other to form a wide opening that facilitates the entry of microorganisms (Fig. 3.3b, c). The bacteriocytes, as well as the ovaries, are one of the first cells/organs to differentiate during embryonic development (Fig. 3.3c, d) (Braendle et al. 2003; Buchner 1965; Koga et al. 2012; Miura et al. 2003; Pyka-Fosciak and Szklarzewicz 2008; Wegierek et al. 2017; Wilkinson et al. 2003). During further development, bacteriocytes undergo polyploidization and are gradually colonized by symbionts (Fig. 3.3d, e).

Because the symbiotic bacteria enter the cytoplasm of bacteriocytes, follicular cells, and oocytes using endocytic vesicle pathway, they remain surrounded by an additional, host vesicle-derived membrane, termed the perisymbiotic membrane (Fig. 3.3f).

3.3.2 *Infestation of Undifferentiated Germ Cells (Cystocytes) and Young Oocytes*

As a rule, symbionts invade the ovaries of older females containing full-grown oocytes (for the classification and characterization of insect ovaries, see Biliński 1998; Büning 1994); however, in several groups of insects (some scale insects, some planthoppers, some heteropterans, some hymenopterans, mallophagans, termites, cockroaches), the larval ovaries containing undifferentiated germ cells (cystocytes) or ovaries of young adult females with early previtellogenic oocytes become infested (Buchner 1965; Fukatsu et al. 2007; Kuechler et al. 2010, 2011; Kupper et al. 2016; Matsuura et al. 2012; Niżnik and Szklarzewicz 2007; Sacchi et al. 1988, 2000; Schneider 1940; Swiatoniowska et al. 2013). Interestingly, in some insects possessing two or more types of symbionts (e.g., in the cottony cushion scale, *Icerya purchasi*, or the planthopper *Cixius nervosus*), one of the symbionts invades the larval ovary, whereas the others infest the older (vitellogenic or choriogenic) oocytes in the ovary of the adult female (Niżnik and Szklarzewicz 2007). In the second larval instar of *Icerya purchasi*, the rod-shaped bacteria (i.e., the secondary symbionts occupying the bacteriocyte epithelium; see above) infect the cystocytes (Fig. 3.4a, b). During further ovary development, cystocytes differentiate into oocytes and trophocytes (nurse cells), and both contain the symbionts (Niżnik and Szklarzewicz 2007). Bacteria present in the trophocytes migrate through the trophic core and nutritive cord into developing oocytes (Fig. 3.4c). The primary symbionts of members of other families of scale insects such as *Puto*

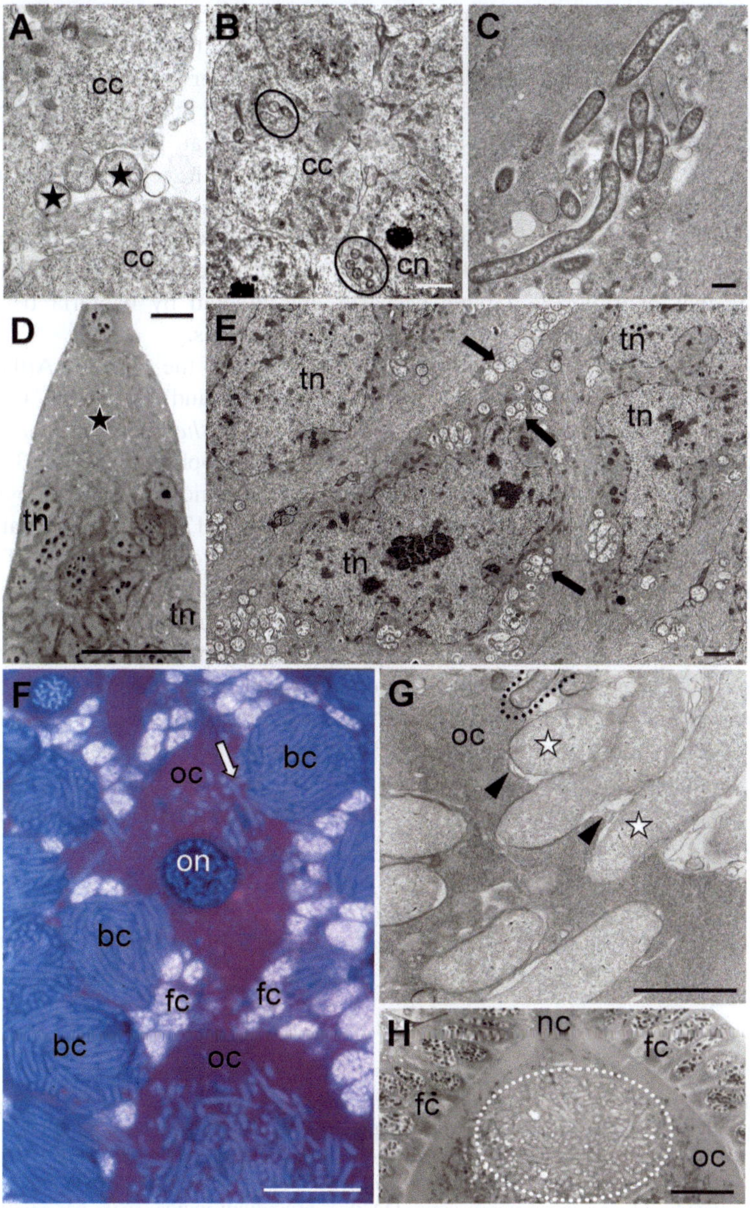

Fig. 3.4 Infestation of undifferentiated germ cells (cystocytes) and young oocytes by symbiotic microorganisms (**a**) *Icerya purchasi* (Hemiptera, Coccomorpha, Monophlebidae). Rod-shaped bacteria (*black asterisks*) invade the cystocyte. TEM, scale bar = 1 μm. (**b**) *Icerya purchasi* (Hemiptera, Coccomorpha, Monophlebidae). Fragment of the cystocyte cluster constituting the larval ovary. In the cystocyte cytoplasm, rod-shaped bacteria are present (encircled by a *black line*). TEM, scale bar = 1 μm. (**c**) *Marchalina hellenica* (Hemiptera, Coccomorpha, Marchalinidae). Bacteria migrating through the nutritive cord. TEM, scale bar = 1 μm. (**d**) *Cixius nervosus* (Hemiptera, Fulgoromorpha, Cixiidae). The anterior region of the tropharium (longitudinal section) filled with numerous rod-shaped bacteria (*black asterisk*). Methylene blue, scale bar = 20 μm. (**e**) *Cixius nervosus* (Hemiptera, Fulgoromorpha, Cixiidae). Fragment of the

albicans (Putoidae) (Szklarzewicz et al. 2010) and *Marchalina hellenica* (Marchalinidae) (Szklarzewicz et al. 2013) are also transported this way.

A similar situation to that of *Icerya purchasi* was observed in the bedbug *Cimex lectularius* (Heteroptera, Cimicidae) (Buchner 1965) and in the planthopper *Cixius nervosus* (Fulgoromorpha, Cixiidae) (Szklarzewicz et al. 2007). Ultrastructural observations have demonstrated that the tropharium of the telotrophic ovariole of *Cixius nervosus* is tightly packed with rod-shaped bacteria (Fig. 3.4d). These bacteria were also observed in nutritive cords and in arrested oocytes. The presence of symbionts in the tropharium (Fig. 3.4e), in the nutritive cords, and in the oocytes suggests that ovarioles of *Cixius nervosus* are infected by bacteria before the differentiation of cystocytes into oocytes and trophocytes.

In several species of heteropteran bugs belonging to the families Artheneidae (*Chilacis typhae*), Blissidae (*Ischnodemus sabuleti*), and Lygaeidae (*Arocatus longiceps*, *Kleidocerys resedae*, *Nysius ericae*, and *Nithecus jacobaeae*), which possess the ring-shaped "infection zone" within the tropharium (Figs. 3.2a, b, c and 3.4f), the symbiotic bacteria infect the previtellogenic oocytes occupying this part of the tropharium (Kuechler et al. 2010, 2011, 2012; Matsuura et al. 2012; Schneider 1940; Swiatoniowska et al. 2013). The mode of transmission of symbionts to the next generation has been thoroughly studied in two species belonging to Lygaeidae: Orsillinae—*Nysius ericae* and *Nithecus jacobaeae* only (Swiatoniowska et al. 2013). In these species, the cytoplasm of the bacteriocytes contains numerous, large, elongated gammaproteobacteria (Fig. 3.2b, c). In addition, *Nysius ericae* contains small, rod-shaped bacteria of an unknown identity and function (Fig. 3.2c). In adult females, the symbionts leave the bacteriocytes and enter the previtellogenic oocytes (Fig. 3.4f, g). Initially, i.e., in previtellogenic oocytes within the infection zone, the bacteria are dispersed throughout the entire cytoplasm (Fig. 3.4f); however, in developing oocytes in the vitellarium, they gather in the anterior pole of the oocyte, forming a "symbiont ball" (Fig. 3.4h).

In the carpenter ant, *Camponotus floridanus*, the symbiotic gammaproteobacteria *Blochmannia floridanus* infect the follicular cells surrounding the lower part of the germarium of the polytrophic ovarioles (Kupper et al.

Fig. 3.4 (continued) tropharium containing rod-shaped bacteria (*black arrows*) in the vicinity of the trophocyte nuclei. TEM, scale bar = 2 μm. (**f**) *Nysius ericae* (Hemiptera, Heteroptera, Lygaeidae: Orsilinae). Fragment of the ovariole containing the infection zone (longitudinal section). *White arrow* indicates the symbiotic bacteria that leave the bacteriocyte cytoplasm and enter the oocyte. Fluorescence microscope, DAPI + propidium iodide, scale bar = 20 μm. (**g**) *Nysius ericae* (Hemiptera, Heteroptera, Lygaeidae: Orsilinae). Fragment of the infection zone. The large elongated bacteria (*white asterisks*) and small rod-shaped bacteria (encircled by a *black dotted line*) leave the bacteriocyte and enter the oocyte. *Black arrowheads* indicate the oolemma. TEM, scale bar = 2 μm. (**h**) *Nithecus jacobaeae* (Hemiptera, Heteroptera, Lygaeidae: Orsilinae). Fragment of the early vitellogenic oocyte (longitudinal section). Note the "symbiont ball" (encircled by a *white dotted line*) at the anterior pole of the oocyte. Methylene blue, scale bar = 20 μm. *bc* bacteriocyte, *cc* cystocyte, *cn* cystocyte nucleus, *fc* follicular cells, *nc* nutritive cord, *oc* oocyte, *on* oocyte nucleus, *tn* trophocyte nucleus

2016). At the time the cystocytes differentiate into oocytes and trophocytes, the symbionts leave the follicular cells and begin to invade early previtellogenic oocytes via an endocytic pathway. Thus, in these insects, bacteria never infect the trophocytes. In contrast to *Camponotus floridanus*, in which germaria are infected, in some mallophagans (e.g., in a slender pigeon louse *Columbicola columbae*), the symbiotic bacteria after leaving the ovarial ampullae begin to invade the posterior ends of polytrophic ovarioles containing previtellogenic oocytes (Buchner 1965; Fukatsu et al. 2007). The infestation of young, previtellogenic oocytes has also been observed in the panoistic ovarioles of termites and cockroaches (Buchner 1965; Sacchi et al. 1988, 2000).

3.3.3 Modes of Infestation of the Older Oocytes

In most insects, symbionts infect the posterior pole of the terminal oocytes; however, there are significant differences in the course of this process (Buchner 1965; Cheung and Purcell 1999; Kobiałka et al. 2015, 2016; Kot et al. 2012, 2014; Michalik et al. 2009, 2013, 2014a; Niżnik and Szklarzewicz 2007; Pyka-Fosciak and Szklarzewicz 2008; Sacchi et al. 2008; Szklarzewicz and Moskal 2001; Szklarzewicz et al. 2006, 2013, 2016).

In the oviparous females of aphids (i.e., in all generations of the Adelgoidea and oviparous generations of Aphidoidea), the bacteriocytes begin to adhere closely to the choriogenic oocytes located terminally within the ovariole (Fig. 3.5a). Subsequently, symbionts leave the bacteriocytes and start to invade the ovarioles. As a rule, symbionts (both P-symbionts and S-symbionts) migrate to the perivitelline space (i.e., space between the follicular epithelium and oocyte) through the wide spaces between neighboring follicular cells (Fig. 3.5b). In some species ((*Adelges viridis* (Adelgoidea: Adelgidae), *Clethrobius comes* (Aphidoidea: Aphididae: Drepanosiphinae), *Schizolachnus pineti* (Aphidoidea: Aphididae: Lachninae)), however, symbionts also pass through the cytoplasm of the follicular cells (Fig. 3.5c) via an endocytic/exocytic pathway (Michalik et al. 2013, 2014a). After passing through the follicular epithelium and perivitelline space (Fig. 3.5b), symbionts enter the ooplasm eventually forming the characteristic "symbiont ball" at the posterior pole of the oocyte (Fig. 3.5d, e). It is worth mentioning that the time of the infestation of the oocyte by bacteria correlates with the process of choriogenesis (Michalik et al. 2013, 2014a; Pyka-Fosciak and Szklarzewicz 2008). During the migration of the bacteria from the body cavity to the ovarioles, the egg envelopes are incomplete: apical and lateral regions of the oocyte are surrounded by egg envelopes, whereas the oocyte posterior pole remains free of envelopes, enabling symbionts to enter the oocyte. The follicular cells begin to synthesize precursors of missing egg envelopes only after the symbionts have entered the oocyte (Michalik et al. 2013, 2014a; Pyka-Fosciak and Szklarzewicz 2008). It should be stressed that symbiont transmission that relies on internalization of symbionts in the ooplasm of the posterior pole of older (i.e., choriogenic) oocytes is very rare among insects and,

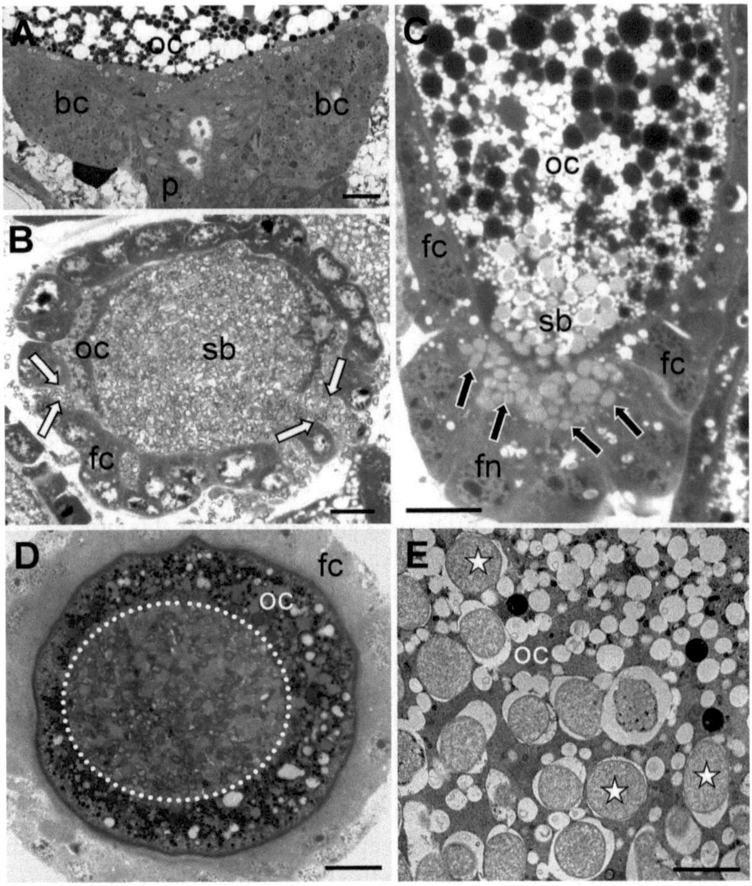

Fig. 3.5 Transovarial transmission of symbionts in oviparous aphids (Hemiptera, Aphidomorpha). (**a**) *Prociphilus fraxini* (Aphidoidea, Aphididae: Eriosomatinae). Bacteriocytes with *Buchnera aphidicola* bacterium gathered around the posterior pole of the terminal oocyte (longitudinal section). Methylene blue, scale bar = 10 μm. (**b**) *Sipha* (*Rungsia*) *maydis* (Aphidoidea, Aphididae: Drepanosiphinae). Posterior pole of oocyte (cross section). *White arrows* indicate symbiotic microorganisms, which migrate to the oocyte through the spaces between neighboring follicular cells. Methylene blue, scale bar = 10 μm. (**c**) *Adelges* (*Sacchiphantes*) *viridis* (Adelgoidea, Adelgidae). Posterior pole of the oocyte (longitudinal section). *Black arrows* indicate symbiotic microorganisms, which migrate to the oocyte through the cytoplasm of follicular cells and through the spaces between neighboring follicular cells. Methylene blue, scale bar = 10 μm. (**d**) *Eulachnus rileyi* (Aphidoidea, Aphididae: Lachninae). Symbiotic microorganisms at the posterior pole of the terminal oocyte in the form of the "symbiont ball" (encircled by a *white dotted line*) (cross section). Methylene blue, scale bar = 10 μm. (**e**) *Adelges (Sacchiphantes) viridis* (Adelgoidea, Adelgidae). Fragment of the "symbiont ball." Note symbiotic bacteria (*white asterisks*) in the oocyte cytoplasm. TEM, scale bar = 10 μm. *bc* bacteriocyte, *fc* follicular cells, *fn* follicular cell nucleus, *oc* oocyte, *p* ovariolar stalk (pedicel), *sb* "symbiont ball"

to our knowledge, apart from oviparous aphids, this mode of transmission has not been reported for other insects.

In psyllids, some scale insects (e.g., *Icerya purchasi* and *Palaeococcus fuscipennis* (both from Monophlebidae family) and auchenorrhynchans, the initial stages of symbiont transmission are similar to that in oviparous aphids (i.e., symbiotic microorganisms leave the bacteriocytes, begin to migrate toward ovaries, and gather around the posterior pole of the terminal oocyte) (Buchner 1965; Kobiałka et al. 2015, 2016; Kot et al. 2012, 2014; Niżnik and Szklarzewicz 2007; Sacchi et al. 2008; Szklarzewicz et al. 2006). There are, however, significant differences in the manner of the infestation of the ovariole in aphids and other insects: (1) in aphids, symbionts migrate mainly through the wide spaces between follicular cells (Fig. 3.5b), whereas in most insects, this migration takes place through the follicular cells (via endocytic/exocytic pathway) (Fig. 3.6b, c, d, e), and (2) in aphids, after passing through the follicular epithelium and perivitelline space, the microorganisms immediately enter the ooplasm (Fig. 3.5b, c, d, e), whereas in most insects, they gather in the deep invagination of the oolemma, forming a "symbiont ball" outside of the oocyte (Fig. 3.6f, g). The microorganisms residing in the perivitelline space closely adhere to one another (Fig. 3.6g, h). The symbionts stay in the perivitelline space until the end of oocyte growth.

It should be noted that in psyllids, scale insects, and auchenorrhynchans, prior to leaving the bacteriocytes, the pleomorphic bacteria change their shape into an almost spherical (Fig. 3.6a). Simultaneously, they begin to stain very intensely with methylene blue (Fig. 3.6a). It seems that the above changes are connected with the reorganization of bacterial cytoskeleton during the migration into the ovaries.

In a few families of scale insects (Pseudococcidae, Eriococcidae, Putoidae), the symbiotic bacteria infest the anterior region of the telotrophic ovariole (Buchner 1955; von Dohlen et al. 2001; Michalik et al. 2016; Szklarzewicz et al. 2017). At the time the ovarioles contain oocytes in the stage of late vitellogenesis, the bacteria (or whole bacteriocytes in *Puto superbus*; see below) begin to surround the neck region of the ovariole (i.e., the region between the tropharium and vitellarium). During symbiont/bacteriocyte migration, the posterior and lateral regions of the oocyte are surrounded by egg coverings. The only region of the oocyte surface that is devoid of eggshells is its anterior pole, which is attached to the nutritive cord (Fig. 3.7a). The bacteria may migrate either through the follicular cells (via endocytic/exocytic pathway) or through the spaces between follicular cells (Fig. 3.7b), whereas the bacteriocytes crawl between neighboring follicular cells (not shown). After reaching the perivitelline space (Fig. 3.7c), the bacteria/bacteriocytes accumulate in the depression of the oolemma at the anterior pole of the oocyte (Fig. 3.7d). At that time, in Eriococcidae family, the nutritive cord degenerates, and the bacteria start to enter the oocyte (Fig. 3.7f), whereas in Pseudococcidae and Putoidae, the bacteria/bacteriocytes remain in the invagination of oolemma until the end of oocyte growth (Figs. 3.7e and 3.8e). It should be stressed that although the symbiotic association between *Gossyparia spuria* and *Acanthococcus aceris* represents an evolutionary young condition (Michalik et al.

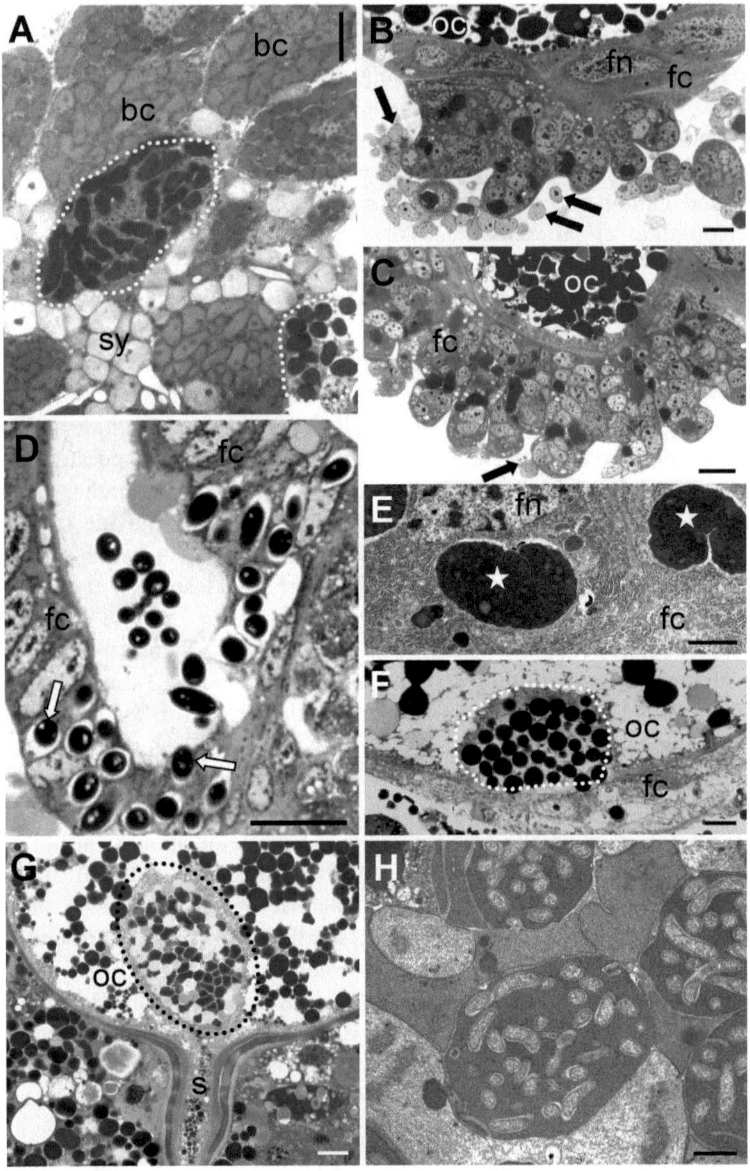

Fig. 3.6 Infestation of the terminal pole of the choriogenic oocyte by symbiotic bacteria. (**a**) *Cacopsylla mali* (Hemiptera, Psyllomorpha, Psyllidae). Fragment of the bacteriome. The symbiotic bacteria (encircled by a *white dotted line*) before migration to the ovaries stain intensively with methylene blue and change shape to almost spherical. Methylene blue, scale bar = 10 μm. (**b, c**) *Evacanthus interruptus* (Hemiptera, Cicadomorpha, Cicadellidae: Evacanthinae). Posterior pole of the terminal oocyte (*b*, longitudinal section; *c*, cross section). Consecutive stages of migration of symbionts into the perivitelline space through the cytoplasm of follicular cells. *Black arrows* indicate the bacteria that invade the projections of follicular cells. Methylene blue, scale bar = 20 μm. (**d**) *Conomelus anceps* (Hemiptera, Fulgoromorpha, Delphacidae). Posterior end of the ovariole (longitudinal section). *White arrows* indicate the yeast-like symbionts which invade the ovariole through the follicular epithelium. Methylene blue, scale bar = 20 μm. (**e**) *Cacopsylla ulmi* (Hemiptera,

2016), these scale insects have already developed and established a stable mode of symbiont transmission.

In insects, as a rule, before the migration to the ovaries, the bacteria are released from the bacteriocyte cytoplasm; however, in some insects such as cockroaches (Sacchi et al. 1988), mallophagans (Buchner 1965), whiteflies (Buchner 1965; Costa et al. 1993; Szklarzewicz and Moskal 2001), and scale insect *Puto superbus* (Putoidae) (Buchner 1965; Szklarzewicz et al. 2017), the whole and intact bacteriocytes penetrate the ovaries. In the cockroach *Blattella germanica*, bacteriocytes enter the ovaries; however, shortly before entry into the ovarioles, the symbionts leave bacteriocyte and individually migrate through the ovariole sheath and follicular epithelium to the perivitelline space (Sacchi et al. 1988). In the mallophagan genus *Goniodes*, the bacteriocytes traverse the epithelium of the ovariolar stalk (pedicel); however, during migration, they degenerate and release the symbionts (Buchner 1965). In consequence, just as in *Blattella germanica*, the oocytes of *Goniodes* are infected by individual bacteria. In whiteflies, in the last larval instar (puparium), bacteriocytes containing both primary and secondary symbionts separate from the bacteriomes and gather around the posterior ends of ovarioles (Fig. 3.8a). In adult females, which possess early vitellogenic oocytes, whole and intact bacteriocytes migrate through the spaces between follicular cells surrounding the posterior pole of early vitellogenic oocytes into the perivitelline space (Fig. 3.8b, c). During this migration, bacteria become elongated and stain more intensely with methylene blue (Fig. 3.8c). In the perivitelline space, the bacteriocytes accumulate in the deep invagination of the oolemma over the hydropylar stalk (Fig. 3.8d). In the whitefly *Aleurochiton aceris* and in the scale insect *Puto superbus* (Fig. 3.8e) (see above), the bacteria do not leave the bacteriocyte cytoplasm until the end of oocyte growth.

Fig. 3.6 (continued) Psyllomorpha, Psyllidae). Fragment of the follicular cell with migrating bacteria (*white asterisks*). TEM, scale bar = 2 μm. (**f**) *Palaeococcus fuscipennis* (Hemiptera, Coccomorpha, Monophlebidae). The posterior pole of the oocyte (longitudinal section). Pleomorphic bacteria (encircled by a *white dotted line*) accumulate in the deep depression of the oolemma. Methylene blue, scale bar = 10 μm. (**g**) *Aphalara polygoni* (Hemiptera, Psyllomorpha, Aphalaridae). Concentration of symbionts ("symbiont ball") (encircled by a *black dotted line*) in the deep invagination of oolemma over the stalk which will serve to anchor the egg to the host plant (longitudinal section). Methylene blue, scale bar = 1 μm (**a**, **e**, and **g**—courtesy of Marta Kot, Jagiellonian University, Kraków, Poland). (**h**) *Macrosteles laevis* (Hemiptera, Cicadomorpha, Cicadellidae: Deltocephalinae). The bacteria forming the "symbiont ball" closely adhere to each other. TEM, scale bar = 2 μm. *bc* bacteriocyte, *fc* follicular cells, *fn* follicular cell nucleus, *oc* oocyte, *s* stalk of the oocyte, *sy* syncytial region of the bacteriome containing the secondary symbionts

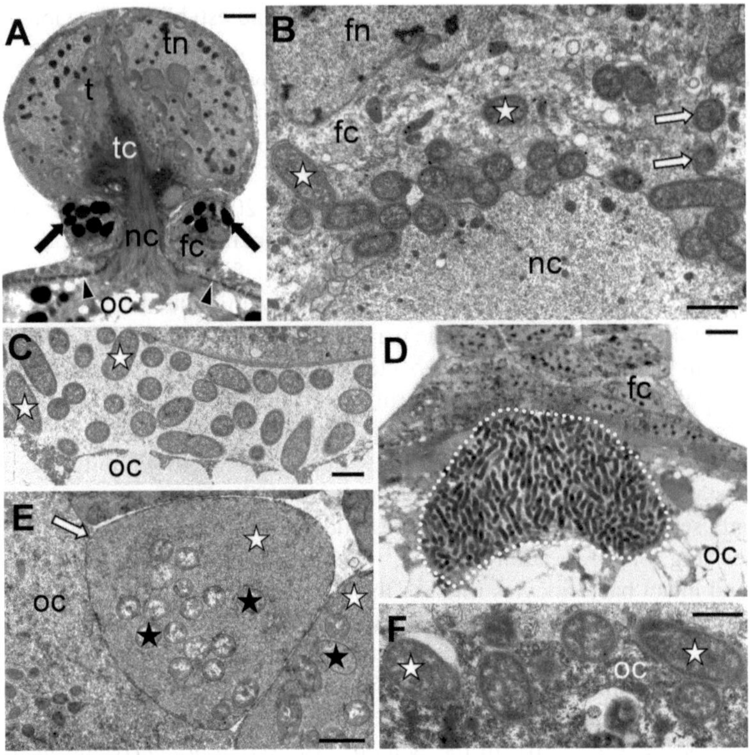

Fig. 3.7 Infestation of the anterior pole of the oocyte by symbiotic microorganisms in scale insects (Hemiptera, Coccomorpha). (**a**) *Phenacoccus aceris* (Pseudococcidae: Phenacoccinae). Anterior end of the ovariole (longitudinal section). Symbiotic bacteria (*black arrows*) inside follicular cells surrounding the neck region of the ovariole. *Black arrowheads* indicate egg envelopes. Methylene blue, scale bar = 10 μm. (**b**) *Gossyparia spuria* (Eriococcidae). Symbiotic bacteria migrate through the cytoplasm of follicular cells (*white asterisks*) and between neighboring cells (*white arrows*). TEM, scale bar = 1 μm. (**c**) *Acanthococcus aceris* (Eriococcidae). Bacteria (*white asterisks*) begin to accumulate in the perivitelline space. TEM, scale bar = 1 μm. (**d**) *Acanthococcus aceris* (Eriococcidae). The neck region of the ovariole (longitudinal section). The accumulation of bacteria (encircled by a *white dotted line*) in the perivitelline space. Methylene blue, scale bar = 10 μm. (**e**) *Pseudococcus longispinus* (Pseudococcidae: Pseudococcinae). Pleomorphic bacteria (*white asterisks*) containing rod-shaped bacteria (*black asterisks*) localized in the deep depression of the oolemma. *White arrow* indicates the oolemma. TEM, scale bar = 1 μm. (**f**) *Gossyparia spuria* (Eriococcidae). Symbiotic bacteria (*white asterisks*) enter the oocyte. TEM, scale bar = 1 μm. *fc* follicular cells, *fn* follicular cell nucleus, *nc* nutritive cord, *oc* oocyte, *t* trophocyte, *tn* trophocyte nucleus, *tc* trophic core

3.4 Future Perspectives

Earlier histological studies and more recent ultrastructural observations have revealed that different insect taxa developed typical (i.e., uniform among all the representatives of the taxon) mode of symbiont transmission. The one exception to

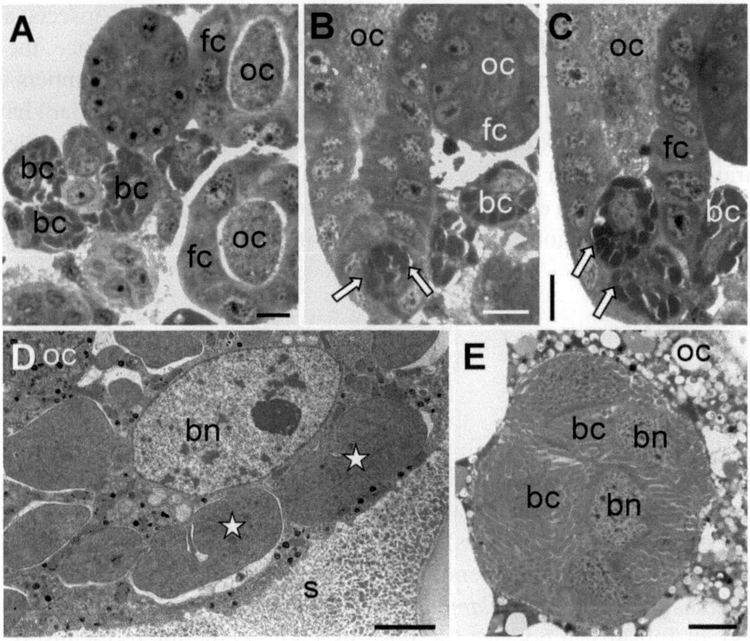

Fig. 3.8 Infestation of the ovariole by bacteriocytes which are whole and intact. (**a**) *Aleurochiton aceris* (Hemiptera, Aleyrodomorpha, Aleyrodidae). Bacteriocytes gather among the posterior ends of ovarioles (cross section) of the last instar before imago (termed puparium). Methylene blue, scale bar = 10 μm. (**b, c**) *Aleurochiton aceris* (Hemiptera, Aleyrodomorpha, Aleyrodidae). Consecutive stages of migration of the whole intact bacteriocytes (*white arrows*) into the perivitelline space in the adult female (longitudinal section). Methylene blue, scale bar = 10 μm. (**d**) *Aleurochiton aceris* (Hemiptera, Aleyrodomorpha, Aleyrodidae). Bacteriocyte with pleomorphic bacteria (*white asterisks*) in the deep invagination of the oolemma over the hydropylar stalk (longitudinal section). TEM, scale bar = 2 μm. (**e**) *Puto superbus* (Hemiptera, Coccomorpha, Putoidae). Bacteriocytes localized at the anterior pole of the full-grown oocyte ("symbiont ball") (cross section). Methylene blue, scale bar = 10 μm. *bc* bacteriocyte, *bn* bacteriocyte nucleus, *fc* follicular cells, *oc* oocyte, *s* hydropylar stalk

this rule is observed in scale insects, which are characterized by very diverse symbiotic systems in terms of the type of symbionts, their localization in the host insect body, and the mode of transmission to the next generation. Moreover, it has been shown that even close relatives—members of the same family (e.g., Eriococcidae) or related families (e.g., Monophlebidae and Marchalinidae), may have different symbionts, which are inherited in a different manner (Buchner 1965; Matsuura et al. 2009; Michalik et al. 2016; Niżnik and Szklarzewicz 2007; Szklarzewicz et al. 2006, 2013). According to the Polish coccidologist Jan Koteja (1985), this large diversity of symbionts, associated with scale insects, may result from the permanent contact of these insects with microorganisms present in the primary habitat (forest leaf litter). Koteja (1985) hypothesized that scale insects living in the litter were saprophagic but during further evolution, the different groups of scale insects changed their feeding behavior at different times and became plant sap feeders. Changing the feeding habit, scale insects acquired microorganisms, which then became their symbionts. In consequence, the symbionts of scale insects are polyphyletic. It is worth noting that the members of some scale insect families (e.g., Steingeliidae, Matsucoccidae, Dactylopiidae) have been regarded as asymbiotic (Tremblay 1977); however, recent, more detailed ultrastructural and molecular studies have revealed that they also contain symbiotic microorganisms (Koteja et al. 2003; Ramirez-Puebla et al. 2010; Szklarzewicz et al. 2014). Thus, the symbiotic systems of scale insects should be more extensively studied.

3.5 Conclusions

Insects, as hosts to obligate symbiotic microorganisms, developed different modes of transmission from one generation to the next: (1) the symbionts may invade dividing germ cells (cystocytes) (e.g., in some scale insects, bedbug *Cimex lectularius*, some planthoppers), early previtellogenic oocytes (in some heteropterans, carpenter ant *Camponotus floridanus*, mallophagans, cockroaches, termites), or full-grown oocytes (in most insects); (2) the microorganisms may be released from the bacteriocytes (in most insects), or whole and intact bacteriocytes may enter the ovarioles (in whiteflies, scale insect *Puto superbus*); (3) the symbionts may invade the anterior (in Pseudococcidae and Eriococcidae scale insects) or posterior pole of the oocyte (in most insects); and (4) the symbionts may enter the ooplasm (in aphids and some Eriococcidae scale insects) or may gather in the perivitelline space in the deep invagination of the oolemma (in most insects). The above observations strongly indicate that the course of transmission of symbionts between generations is related to the course of oogenesis of host insect and correlates with the stage of oocyte development.

References

Abt MC, Artis D (2009) The intestinal microbiota in health and disease: the influence of microbial products on immune cell homeostasis. Curr Opin Gastroenterol 25:496–502

Baumann P (1998) Symbiotic associations involving microorganisms. A special issue devoted to some less well known symbiotic associations. BioScience 48:254–255

Baumann P (2005) Biology of bacteriocyte-associated endosymbionts of plant sup-sucking insects. Annu Rev Microbiol 59:155–189

Baumann P, Baumann L, Lai CY, Rouhbakksh D, Moran NA, Clark MA (1995) Genetics, physiology and evolutionary relationships of the genus *Buchnera*: intracellular symbionts of aphids. Annu Rev Microbiol 49:55–94

Biliński S (1998) Introductory remarks. Folia Histochem Cytobiol 3:143–145

Braendle C, Miura T, Bickel R, Shingleton AW, Kambhampati S, Stern DL (2003) Developmental origin and evolution of bacteriocytes in the aphid – *Buchnera* symbiosis. PLoS Biol 1:70–76

Bright M, Bulgheresi S (2010) A complex journey: transmission of microbial symbionts. Nat Rev Microbiol 8:218–230

Buchner P (1955) Endosymbiosestudien an Schildläusen. III. Macrocerococcus und Puto zwei primitive Pseudococcinen. Z Morph Ökol Tiere 43:523–577

Buchner P (1965) Endosymbiosis of animals with plant microorganisms. Interscience Publishers, New York

Büning J (1994) The insect ovary: ultrastructure, previtellogenic growth and evolution. Chapman and Hall, London

Burke GR, Normark BB, Favret C, Moran NA (2009) Evolution and diversity of facultative symbionts from the aphid subfamily Lachninae. Appl Environ Microbiol 75:5328–5335

Cheng DJ, Hou RF (2001) Histological observations on transovarial transmission of a yeast-like symbiote in *Nilaparvata lugens* Stal (Homoptera, Delphacidae). Tissue Cell 33:273–279

Cheung W, Purcell AH (1999) Invasion of bacteroids and BEV bacterium into oocytes of the leafhopper *Euscelidius variegatus* Kirschbaum (Homoptera: Cicadellidae): an electron microscopic study. Zool Stud 38:69–75

Costa HS, Westcot DM, Ullman DE, Johnson MW (1993) Ultrastructure of the endosymbionts of the whitefly, *Bemisia tabaci* and *Trialeurodes vaporariorum*. Protoplasma 176:106–115

Dale C, Moran NA (2006) Molecular interactions between bacterial symbionts and their hosts. Cell 126:453–465

Dasch AG, Weiss E, Chang KP (1984) Endosymbionts of insects. In: Krieg NR (ed) Bergey's manual of systematic bacteriology, vol 1. Williams and Wilkins, Baltimore, pp 811–833

Dixon AFG (1985) Structure of aphid populations. Annu Rev Entomol 30:155–174

Douglas AE (1989) Mycetocyte symbiosis in insects. Biol Rev 64:409–434

Douglas AE (1998) Nutritional interactions in insect – microbial symbioses: aphids and their symbiotic bacteria *Buchnera*. Annu Rev Entomol 43:17–37

Douglas AE (2009) The microbial dimension in insect nutritional ecology. Funct Ecol 23:38–47

Eberle MW, McLean DL (1982) Initiation and orientation of the symbiote migration in the human body louse *Pediculus humanus* L. J Insect Physiol 28:417–422

Eberle MW, McLean DL (1983) Observation on symbiote migration in human body lice with scanning and transmission electron microscopy. Can J Microbiol 29:755–762

Fukatsu T, Hosokawa T (2002) Capsule-transmitted gut symbiotic bacterium of the Japanese common plataspid stinkbug, *Megacopta punctatissima*. Appl Environ Microbiol 68:389–396

Fukatsu T, Ishikawa H (1992) A novel eukaryotic extracellular symbiont in an aphid, *Astegopteryx styraci* (Homoptera, Aphididae, Hormaphidinae). J Insect Physiol 38:765–773

Fukatsu T, Ishikawa H (1996) Phylogenetic position of yeast-like symbiont of *Hamiltonaphis styraci* (Homoptera, Aphididae) based on 18S rDNA sequence. Insect Biochem Mol Biol 26:383–388

Fukatsu T, Nikoh N (1998) Two intracellular symbiotic bacteria from mulberry psyllid *Anomoneura mori* (Insecta, Homoptera). Appl Environ Microbiol 66:2748–2758

Fukatsu T, Nikoh N, Kawai R, Koga R (2000) The secondary endosymbiotic bacterium of the pea aphid, *Acyrthosiphon pisum* (Insecta: Homoptera). Appl Environ Microbiol 66:2748–2758

Fukatsu T, Koga R, Smith WA, Tanaka K, Nikoh N, Sasaki-Fukatsu K, Yoshizawa K, Dale C, Clayton DH (2007) Bacterial endosymbiont of the slender pigeon louse, *Columbicola columbae*, allied to endosymbionts of grain weevils and tsetse flies. Appl Environ Microbiol 73:6660–6668

Gatehouse LN, Sutherland P, Forgie SA, Kaji R, Christeller JT (2011) Molecular and histological characterization of primary (Betaproteobacteria) and secondary (Gammaproteobacteria) endosymbionts of three mealybug species. Appl Environ Microbiol 78:1187–1197

Gottlieb Y, Ghanim M, Gueguen G, Kontsedalov S, Vavre F, Fleury F, Zchori-Fein E (2008) Inherited intracellular ecosystem: symbiotic bacteria share bacteriocytes in whiteflies. FASEB J 22:2591–2599

Hongoh Y, Ishikawa H (2000) Evolutionary studies on uricases of fungal endosymbionts of aphids and planthoppers. Mol Evol 51:265–277

Hosokawa T, Kikuchi Y, Meng XY, Fukatsu T (2005) The making of symbiont capsule in the plataspid stinkbug *Megacopta punctatissima*. FEMS Microbiol Ecol 54:471–477

Husnik F, McCutcheon JP (2016) Repeated replacement of an intrabacterial symbiont in the tripartite nested mealybug symbiosis. PNAS 113:E5416–E5424

Husnik FN, Nikoh R, Koga R, Ross L, Duncan RP, Fujie M, Tanaka M, Satoh N, Bachtrog D, Wilson ACC, von Dohlen CD, Fukatsu T, McCutcheon JP (2013) Horizontal gene transfer from diverse bacteria to an insect genome enables a tripartite nested mealybug symbiosis. Cell 153:1567–1578

Ishikawa H (1989) Biochemical and molecular aspects of endosymbiosis in insects. Int Rev Cytol 116:1–45

Ishikawa H (2003) Insect symbiosis: an introduction. In: Bourtzis K, Miller TA (eds) Insect symbiosis. CRC Press, Boca Raton, pp 1–21

Kikuchi Y (2009) Endosymbiotic bacteria in insects: their diversity and culturability. Microbes Environ 24:195–204

Kikuchi Y, Fukatsu T (2003) Insect-bacterium mutualism without vertical transmission. In: Bourtzis K, Miller TA (eds) Insect symbiosis, vol 3. CRC Press, Boca Raton, pp 143–161

Kikuchi Y, Hosokawa T, Fukatsu T (2007) Insect-microbe mutualism without vertical transmission: a stinkbug acquires a beneficial gut symbiont from the environment every generation. Appl Environ Microbiol 73:4308–4316

Kindlmann P, Dixon AFG (1989) Developmental constraints in the evolution of reproductive strategies: telescoping of generations in parthenogenetic aphids. Funct Ecol 3:531–537

Kobiałka M, Michalik A, Walczak M, Junkiert Ł, Szklarzewicz T (2015) Symbiotic microorganisms of the leafhopper *Deltocephalus pulicaris* (Fallén, 1806) (Insecta, Hemiptera, Cicadellidae: Deltocephalinae): Molecular characterization, ultrastructure and transovarial transmission. Pol J Entomol 84:289–304

Kobiałka M, Michalik A, Walczak M, Junkiert Ł, Szklarzewicz T (2016) *Sulcia* symbiont of the leafhopper *Macrosteles laevis* (Ribaut, 1927) (Insecta, Hemiptera, Cicadellidae: Deltocephalinae) harbors *Arsenophonus* bacteria. Protoplasma 253:903–912

Koga R, Tsuchida T, Fukatsu T (2003) Changing partners in an obligate symbiosis: a facultative endosymbiont can compensate for loss of the essential endosymbiont *Buchnera* in an aphid. Proc R Soc London B 270:2543–2550

Koga R, Meng X-Y, Tsuchida T, Fukatsu T (2012) Cellular mechanism for selective vertical transmission of an obligate insect symbiont at the bacteriocyte-embryo interface. PNAS 109: E1230–E1237

Koga R, Bennett GM, Cryan JR, Moran NA (2013) Evolutionary replacement of symbionts in an ancient and diverse insect lineage. Environ Microbiol 15:2073–2081

Kono M, Koga R, Shimada M, Fukatsu T (2008) Infection dynamics of coexisting β and γ-proteobacteria in the nested endosymbiotic system of mealybugs. Appl Environ Microbiol 74:4175–4184

Kot M, Szklarzewicz T, Drohojowska J (2012) Ultrastructural studies on bacterial endosymbionts in psyllids (Insecta, Hemiptera: Psylloidea). Acta Biol Crac Ser Bot 54(Suppl 1):61

Kot M, Michalik A, Szklarzewicz T (2014) Primary and secondary endosymbionts of psyllids (Insecta, Hemiptera: Psylloidea). Acta Biol Crac Ser Bot 56(Suppl 1):66

Koteja J (1985) Essay on the prehistory of the scale insects (Homoptera, Coccinea). Ann Zool 38:461–503

Koteja J, Pyka-Fosciak G, Vogelgesang M, Szklarzewicz T (2003) Structure of the ovary in *Steingelia* (Sternorrhyncha: Coccinea), and its phylogenetic implications. Arthropod Struct Dev 32:247–256

Kuechler SM, Dettner K, Kehl S (2010) Molecular characterization and localization of the obligate endosymbiotic bacterium in the birch catkin bug *Kleidocerys resedae* (Heteroptera: Lygaeidae, Ischnorhynchinae). FEMS Microbiol Ecol 73:408–418

Kuechler SM, Dettner K, Kehl S (2011) Characterization of an obligate intracellular bacterium in the midgut epithelium of the bulrush bug *Chilacis typhae* (Heteroptera, Lygaeidae, Artheneinae). Appl Environ Microbiol 77:2869–2876

Kuechler SM, Renz P, Dettner K, Kehl S (2012) Diversity of symbiotic organs and bacterial endosymbionts of lygaeoid bugs of the families Blissidae and Lygaeidae (Hemiptera: Heteroptera: Lygaeoidea). Appl Environ Microbiol 78:2648–2659

Kupper M, Stigloher C, Feldhaar H, Gross R (2016) Distribution of the obligate endosymbiont *Blochmannia floridanus* and expression analysis of putative immune genes in ovaries of the carpenter ant *Camponotus floridanus*. Arthropod Struct Dev 45:475–487

Lukasik P, Guo H, Van Asch M, Ferrari J, Godfray HCJ (2013) Protection against a fungal pathogen conferred by the aphid facultative endosymbionts *Rickettsia* and *Spiroplasma* is expressed in multiple host genotypes and species and is not influenced by co-infection with another symbiont. J Evol Biol 26:2654–2661

Matsuura Y, Koga R, Nikoh N, Meng XY, Hanada S, Fukatsu T (2009) Huge symbiotic organs in giant scale insects of the genus *Drosicha* (Coccoidea: Monophlebidae) harbor flavobacterial and enterobacterial endosymbionts. Zoolog Sci 26:448–456

Matsuura Y, Kikuchi Y, Hosokawa T, Koga R, Meng X-Y, Kamagata Y, Nikoh N, Fukatsu T (2012) Evolution of symbiotic organs and endosymbionts in lygaeid stinkbugs. ISME J 6:397–409

McCutcheon JP, Moran NA (2007) Parallel genomic evolution and metabolic interdependence in an ancient symbiosis. PNAS 104:19392–19397

McCutcheon JP, von Dohlen CD (2011) An interdependent metabolic patchwork in the nested symbiosis of mealybugs. Curr Biol 21:1366–1372

McCutcheon JP, McDonald BR, Moran NA (2009) Convergent evolution of metabolic roles in bacterial co-symbionts of insects. PNAS 106:15394–15399

Michalik A, Jankowska W, Szklarzewicz T (2009) Ultrastructure and transovarial transmission of endosymbiotic microorganisms in *Conomelus anceps* and *Metcalfa pruinosa* (Insecta, Hemiptera, Fulgoromorpha). Folia Biol (Kraków) 57:131–137

Michalik A, Gołas A, Kot M, Wieczorek K, Szklarzewicz T (2013) Endosymbiotic microorganisms in *Adelges* (*Sacchiphantes*) *viridis* (Insecta, Hemiptera, Adelgoidea: Adelgidae): molecular characterization, ultrastructure and transovarial transmission. Arthropod Struct Dev 42:531–538

Michalik A, Szklarzewicz T, Jankowska W, Wieczorek K (2014a) Endosymbiotic microorganisms of aphids (Hemiptera: Sternorrhyncha: Aphidoidea): ultrastructure, distribution and transovarial transmission. Eur J Entomol 111:91–104

Michalik A, Jankowska W, Kot M, Gołas A, Szklarzewicz T (2014b) Symbiosis in the green leafhopper, *Cicadella viridis* (Hemiptera, Cicadellidae). Association *in statu nascendi*? Arthropod Struct Dev 43:579–587

Michalik K, Szklarzewicz T, Kalandyk-Kołodziejczyk M, Jankowska W, Michalik A (2016) Bacteria belonging to the genus *Burkholderia* are obligatory symbionts of the eriococcids *Acanthococcus aceris* Signoret, 1875 and *Gossyparia spuria* (Modeer, 1778) (Insecta, Hemiptera, Coccoidea). Arthropod Struct Dev 45:265–272

Miura T, Braendle C, Shingleton A, Sisk G, Kambhampati S, Stern DL (2003) A comparison of parthenogenetic and sexual embryogenesis of the pea aphid *Acyrthosiphon pisum* (Hemiptera: Aphidoidea). J Exp Zool 295B:59–81

Montllor CB, Maxmen A, Purcell AH (2002) Facultative bacterial endosymbionts benefit pea aphids *Acyrthosiphon pisum* under heat stress. Ecol Entomol 27:189–195

Moran NA (1992) The evolution of aphid life cycles. Annu Rev Entomol 37:321–348

Moran NA, Baumann P (2000) Bacterial endosymbionts in animals. Curr Opin Microbiol 3:270–275

Moran NA, Telang A (1998) Bacteriocyte-associated symbionts of insects: a variety of insect groups harbor ancient prokaryotic endosymbionts. BioScience 48:295–304

Niżnik S, Szklarzewicz T (2007) Structure and development of hermaphroditic gonad in *Icerya purchasi* (Insecta, Hemiptera, Coccinea: Monophlebidae). Zool Polon 52:71–90

Nováková E, Hypša V, Klein J, Foottit RG, von Dohlen CD, Moran NA (2013) Reconstructing the phylogeny of aphids (Hemiptera: Aphididae) using DNA of the obligate symbiont *Buchnera aphidicola*. Mol Phylogenet Evol 68:42–54

Oliver KM, Russel JA, Moran NA, Hunter MS (2003) Facultative bacterial symbionts in aphids confer resistance to parasitic wasps. PNAS 100:1803–1807

Oliver KM, Moran NA, Hunter MS (2006) Cost and benefits of a superinfection of facultative symbionts in aphids. Proc R Soc B 273:1273–1280

Oliver KM, Degnan PH, Burke GR, Moran NA (2010) Facultative symbionts in aphids and the horizontal transfer of ecologically important traits. Annu Rev Entomol 55:247–266

Pyka-Fosciak G, Szklarzewicz T (2008) Germ cell cluster formation and ovariole structure in viviparous and oviparous generations of the aphid *Stomaphis quercus*. Int J Dev Biol 52:259–265

Ramirez-Puebla ST, Rosenblueth M, Chavez-Moreno CK, Catanho Pereira de Lyra MC, Tecante A, Martinez-Romero A (2010) Molecular phylogeny of the genus *Dactylopius* (Hemiptera: Dactylopiidae) and identification of the symbiotic bacteria. Environ Entomol 39:1178–1183

Ries E (1931) Die Symbiose der Läuse und Federlinge. Z Morphol Őkol Tiere 20:233–367

Sacchi L, Grigolo A, Mazzini M, Bigliardi E, Baccetti B, Laudani U (1988) Symbionts in the oocytes of *Blattella germanica* L. (Dictyoptera: Blattellidae): their mode of transmission. Int J Insect Morphol Embryol 17:437–446

Sacchi L, Nalepa CA, Lenz M, Bandi C, Corona S, Grigolo A, Bigliardi E (2000) Transovarial transmission of symbiotic bacteria in *Mastotermes darwiniensis* (Isoptera: Mastotermitidae): ultrastructural aspects and phylogenetic implications. Ann Entomol Soc Am 93:1308–1313

Sacchi L, Genchi M, Clementi E, Bigliardi E, Avanzatti AM, Pajoroi M, Negri I, Marzorati M, Gonella E, Alma A, Daffonchio D, Bandi C (2008) Multiple symbiosis in the leafhopper *Scaphoideus titanus* (Hemiptera: Cicadellidae): details of transovarial transmission of *Cardinium* sp. and yeast-like endosymbionts. Tissue Cell 40:231–242

Scarborough CL, Ferrari J, Godfray HCJ (2005) Aphid protected from pathogen by endosymbiont. Science 310:1781

Schneider G (1940) Beitrage zur Kenntnis der symbiontischen Einrichtungen der Heteropteren. Z Morphol Őkol Tiere 36:565–644

Spaulding AW, von Dohlen CD (1998) Phylogenetic characterization and molecular evolution of bacterial endosymbionts in psyllids (Hemiptera, Sternorrhyncha). Mol Biol Evol 15:1506–1513

Swiatoniowska M, Ogorzalek A, Golas A, Michalik A, Szklarzewicz T (2013) Ultrastructure, distribution and transovarial transmission of symbiotic microorganisms in *Nysius ericae* and *Nithecus jacobaeae* (Heteroptera: Lygaeidae: Orsillinae). Protoplasma 250:325–332

Szklarzewicz T, Moskal A (2001) Ultrastructure, distribution, and transmission of endosymbionts in the whitefly *Aleurochiton aceris* Modeer (Insecta, Hemiptera, Aleyrodinea). Protoplasma 218:45–53

Szklarzewicz T, Kędra K, Niżnik S (2006) Ultrastructure and transovarial transmission of endosymbiotic microorganisms in *Palaeococcus fuscipennis* (Burmeister) (Insecta, Hemiptera, Coccinea: Monophlebidae). Folia Biol (Kraków) 54:69–74

Szklarzewicz T, Jankowska W, Łukasiewicz K, Szymańska B (2007) Structure of the ovaries and oogenesis in *Cixius nervosus* (Cixiidae), *Javesella pellucida* and *Conomelus anceps* (Delphacidae) (Insecta, Hemiptera, Fulgoromorpha). Arthropod Struct Dev 36:199–207

Szklarzewicz T, Michalik A, Czaja A, Szydłowska S (2010) Germ cell cluster formation and ovariole structure in *Puto albicans* and *Crypticerya morrilli* (Hemiptera: Coccinea). Phylogenetic implications. Eur J Entomol 107:589–595

Szklarzewicz T, Kalandyk-Kolodziejczyk M, Kot M, Michalik A (2013) Ovary structure and transovarial transmission of endosymbiotic microorganisms in *Marchalina hellenica* (Insecta, Hemiptera, Coccomorpha: Marchalinidae). Acta Zool (Stockholm) 94:184–192

Szklarzewicz T, Michalik A, Kalandyk-Kołodziejczyk M, Kobiałka M, Simon E (2014) Ovary of *Matsucoccus pini* (Insecta, Hemiptera, Coccinea: Matsucoccidae). Morphology, ultrastructure and phylogenetic implications. Microsc Res Tech 77:327–334

Szklarzewicz T, Grzywacz B, Szwedo J, Michalik A (2016) Bacterial symbionts of the leafhopper *Evacanthus interruptus* (Linnaeus, 1758) (Insecta, Hemiptera, Cicadellidae: Evacanthinae). Protoplasma 253:379–391

Szklarzewicz T, Kalandyk-Kołodziejczyk K, Michalik K, Jankowska W, Michalik A (2017) Symbiotic microorganisms in *Puto superbus* (Leonardi, 1907) (Insecta, Hemiptera, Coccomorpha: Putoidae). Protoplasma. doi:10.1007/s00709-017-1135-7

Takiya DM, Tran P, Dietrich CH, Moran NA (2006) Co-cladogenesis spanning three phyla: leafhoppers (Insecta: Hemiptera: Cicadellidae) and their dual bacterial symbionts. Mol Ecol 15:4175–4191

Thao ML, Moran NA, Abbot P, Bernnan EB, Burckhardt DH, Baumann P (2000a) Cospeciation of psyllids and their prokaryotic endosymbionts. Appl Environ Microbiol 66:2898–2905

Thao ML, Clark MA, Baumann L, Brennan EB, Moran NA, Baumann P (2000b) Secondary endosymbionts of psyllids have been acquired multiple times. Curr Microbiol 41:300–304

Thao ML, Gullan PJ, Baumann P (2002) Secondary (γ-proteobacteria) endosymbionts infect the primary (β-proteobacteria) endosymbionts of mealybugs multiple times and coevolve with their host. Appl Environ Microbiol 68:3190–3197

Toenshoff ER, Gruber D, Horn M (2012) Co-evolution and symbiont replacement shaped the symbiosis between adelgids (Hemiptera: Adelgidae) and their bacterial symbionts. Environ Microbiol 14:1284–1295

Tremblay E (1977) Advances in endosymbiont studies in Coccoidea. VA Polytech Inst State Univ Res Div Bull 127:23–33

von Dohlen CD, Kohler S, Alsop ST, McManus WR (2001) Mealybug β-proteobacterial endosymbionts contain γ-proteobacterial symbionts. Nature 412:433–435

Vorburger C, Gehrer L, Rodriguez P (2010) A strain of the bacterial symbiont *Regiella insecticola* protects aphids against parasitoids. Biol Lett 6:109–111

Wegierek P, Michalik A, Wieczorek K, Kanturski M, Kobiałka M, Śliwa K, Szklarzewicz T (2017) Buchnera aphidicola of the birch blister aphid, *Hamamelistes betulinus* (Horváth, 1896) (Insecta, Hemiptera, Aphididae: Hormaphidinae): molecular characterization, transmission between generations and its geographic significance. Acta Zool (Stockholm). doi:10.1111/azo.12186

Wilkinson TL, Ishikawa H (2001) On the functional significance of symbiotic microorganisms in the Homoptera: a comparative study of *Acyrthosiphon pisum* and *Nilaparvata lugens*. Physiol Entom 26:86–93

Wilkinson TL, Fukatsu T, Ishikawa H (2003) Transmission of symbiotic bacteria *Buchnera* to parthenogenetic embryos in the aphid *Acyrthosiphon pisum* (Hemiptera: Aphidoidea). Arthropod Struct Dev 32:241–245

Wojciechowski W, Depa Ł, Kanturski M, Wegierek P, Wieczorek K (2015) An annotated checklist of the Aphids (Hemiptera: Aphidomorpha) of Poland. Pol J Entomol 84:383–420

Wu D, Daugherty SC, Van Aken SE, Pai GH, Watkins KL, Khouri H (2006) Metabolic complementarity and genomics of the dual symbiosis of sharpshooters. PLoS Biol 4:e188

Żelazowska M, Biliński SM (1999) Distribution and transmission of endosymbiotic microorganisms in the oocytes of the pig louse, *Haematopinus suis* (L.) (Insecta: Phthiraptera). Protoplasma 209:207–213

Part II
Oocyte Polarity: Molecular and Organellar Aspects and Developmental Consequences

Chapter 4
Acquisition of Oocyte Polarity

Mara Clapp and Florence L. Marlow

Abstract Acquisition of oocyte polarity involves complex translocation and aggregation of intracellular organelles, RNAs, and proteins, along with strict posttranscriptional regulation. While much is still unknown regarding the formation of the animal-vegetal axis, an early marker of polarity, animal models have contributed to our understanding of these early processes controlling normal oogenesis and embryo development. In recent years, it has become clear that proteins with self-assembling properties are involved in assembling discrete subcellular compartments or domains underlying subcellular asymmetries in the early mitotic and meiotic cells of the female germline. These include asymmetries in duplication of the centrioles and formation of centrosomes and assembly of the organelle and RNA-rich Balbiani body, which plays a critical role in oocyte polarity. Notably, at specific stages of germline development, these transient structures in oocytes are temporally coincident and align with asymmetries in the position and arrangement of nuclear components, such as the nuclear pore and the chromosomal bouquet and the centrioles and cytoskeleton in the cytoplasm. Formation of these critical, transient structures and arrangements involves microtubule pathways, intrinsically disordered proteins (proteins with domains that tend to be fluid or lack a rigid ordered three-dimensional structure ranging from random coils, globular domains, to completely unstructured proteins), and translational repressors and activators. This review aims to examine recent literature and key players in oocyte polarity.

M. Clapp
Department of Developmental and Molecular Biology, Albert Einstein College of Medicine of Yeshiva University, Bronx, NY, USA

F.L. Marlow (✉)
Department of Developmental and Molecular Biology, Albert Einstein College of Medicine of Yeshiva University, Bronx, NY, USA

Department of Neuroscience, Albert Einstein College of Medicine of Yeshiva University, Bronx, NY, USA

Department of Cell, Developmental and Regenerative Medicine, Icahn School of Medicine at Mount Sinai, One Gustave L. Levy Place, Box 1020, New York, NY 10029-6574, USA
e-mail: florence.marlow@mssm.edu

M. Kloc (ed.), *Oocytes*, Results and Problems in Cell Differentiation 63,
DOI 10.1007/978-3-319-60855-6_4

Abbreviations

γ-Tub	Gamma-tubulin
Ana1	Anastral spindle 1
AnVg	Animal-vegetal axis
Asl	Asterless
Bb	Balbiani body
Bcd	Bicoid
BRE	Bruno response elements
Bru	Bruno
Buc	Bucky ball
Cep135	Centrosomal protein of 135 kDa
Cnn	Centrosomin
Cp110	Centrosomal protein of 110 kDa
CPEB	Cytoplasmic polyadenylation element binding protein
Dazl	Deleted in azoospermia-like
Dplp	Pericentrin-like protein
Glo	Glorund
Grk	Gurken
hnRNP	Heterogeneous nuclear ribonucleoprotein
HTS	Huli-tai-shao
IDR	Intrinsically disordered proteins
INM	Inner nuclear membrane
KASH	Klarsicht, ANC-1, and Syne homology
KR rich	Lysine/arginine rich
LINC	Linker of the nucleoskeleton and cytoskeleton
Macf1	Microtubule actin cross-linking factor 1
METRO	Messenger transport organizer
Mgn	Magellan
MT	Microtubules
MTOC	Microtubule organizing center
NE	Nuclear envelope
Nos	Nanos
NP	Nuclear pore
NRE	Nanos response elements
ONM	Outer nuclear membrane
Orb	oo18 RNA binding protein
Osk	Oskar
P bodies	Processing bodies
PCM	Pericentriolar material
PGC	Primordial germ cells
piRNA	Piwi-interacting RNA
PLD	Prion-like domain
Plk4	Polo-like kinase 4

Plp	Pericentrin-like protein
Pum	Pumilio
Rbpms2	RNA binding protein with multiple splice isoforms 2
RNAbps	RNA binding proteins
RNP	Ribonucleoprotein
Sas-4	Something about silencing 4
Sas-5	Something about silencing 5
Sas-6	Something about silencing 6
Smg	Smaug
Spd-2	Spindle-defective protein 2
SRE	Smaug response elements
STIL	SCL/TAL1 interrupting locus
SUN	UNC-84
TCE	Translational control element
TGF-α	Transforming growth factor alpha
Xvelo	Vegetally localized

4.1 Introduction

Many cells and tissues that make up an individual rely on cellular polarity during development and later to carry out their various functions. In some tissues and cell types, the overt architecture betrays their polarized nature, for example, neurons or sperm. For other cells, their polarity is only revealed upon subcellular or molecular inspection. This cellular polarity manifests as asymmetrically distributed organelles, such as mitochondria, endoplasmic reticulum, or vesicles, as differential distribution of proteins on or near adjacent or opposing cellular membranes, or even as unique perinuclear cytoplasmic scaffolds or compartments formed by self-organizing proteins and their associated proteins and RNAs. Cellular polarity is required for normal oocyte development, successful fertilization, and subsequent embryo development. An illustration of this polarity is the oocyte animal-vegetal axis, which becomes the first embryonic axes in many vertebrates. The formation of the animal-vegetal axis occurs during early oogenesis and thus can be difficult to study, especially in mammals since the earliest stages of oocyte development are most prevalent in the developing embryo in utero. The anteroposterior and dorsoventral embryonic axes develop later during embryogenesis. Other vertebrate and invertebrate animal models have been developed to further explore this process of developmental acquisition of polarity, which has implications for fertility and early embryogenesis.

4.2 Centrosome Assembly and Polarized Division

The orientation of mitotic divisions determines whether they are symmetrical to generate two equivalent daughter cells or asymmetrical to generate daughters with different fates, as in the division of stem cells to produce one renewing daughter and one that will differentiate or as occurs in the early *C. elegans* embryo or later in *Drosophila* neuroblasts to generate daughters with distinct fates. Whether symmetrical or asymmetric, faithful chromosome segregation during mitotic division requires functional centrosomes. Centrosomes serve as the major site of microtubule nucleation and organization that drives spindle assembly during mitosis. Most cells have two centrosomes, comprised of linked but structurally distinct centrioles; the older "mother" centriole has more complex and robust microtubule-anchoring capacity compared to its "daughter" (Bornens 2012). Both centrioles are surrounded by an electron-dense matrix of pericentriolar material (PCM) that was once thought to lack organized structure. This view of the PCM has recently changed owing to super-resolution imaging studies in multiple species that captured a dynamic and evolutionarily conserved highly ordered arrangement of PCM proteins, including Pericentrin and Pericentrin-like protein (Plp) fibers radiating from the mother centriole during interphase (Lawo et al. 2012; Mennella et al. 2012; Sonnen et al. 2012; Fu and Glover 2012). During G1 phase of the cell cycle, the centrosome replicates via a mechanism mediated by self-assembling proteins (Fig. 4.1a–c). In *C. elegans*, the earliest step in centrosome duplication is recruitment of the WD40 repeat protein Something about silencing 7 (Sas-7), which binds to and recruits Spd-2 (Cep192 in humans) (Sugioka et al. 2017). There are not clear homologs for Sas-7 outside of nematodes, and the closest human proteins are not implicated in centriole duplication based on knockdown studies (Sugioka et al. 2017; Balestra et al. 2013) but are essential for basal body formation and ciliogenesis (Enjolras et al. 2012; Lee and Somlo 2014). Assembly of the daughter centriole continues with the asymmetric recruitment of Polo-like kinase 4 (Plk4)/Zyg-1 within the PCM of the mother centriole, which seeds recruitment of the self-assembling protein Something about silencing 6 (Sas-6) also known as Bld12p (O'Connell et al. 2001; Kitagawa et al. 2011; Van Breugel et al. 2011) (Fig. 4.1b). It has long been known that the centrosome expands as cells enter mitosis through a process known as "centrosome maturation." Centrosome maturation requires Polo-like kinase 1 (PLK1) (Lane and Nigg 1996) and Aurora A kinase (Hannak et al. 2001) and ensues by a mechanism in which a scaffold of oligomerizing proteins, including Pericentrin or Plp proteins, expands to form a matrix (Kim and Rhee 2014; Lee and Rhee 2011). This matrix is thought to serve as a scaffold for recruitment of PCM components and expansion of the PCM to promote robust spindle assembly. Expansion of the matrix involves phosphorylation and activation of Polo kinase, which once activated phosphorylates Centrosomin (Cnn) and thereby promotes a positive feedback loop for further recruitment of Spindle-defective protein 2 (Spd-2) (aka Cep192) and other PCM components, including gamma-tubulin and eventual spindle assembly (Fig. 4.1c) (Yang and Feldman

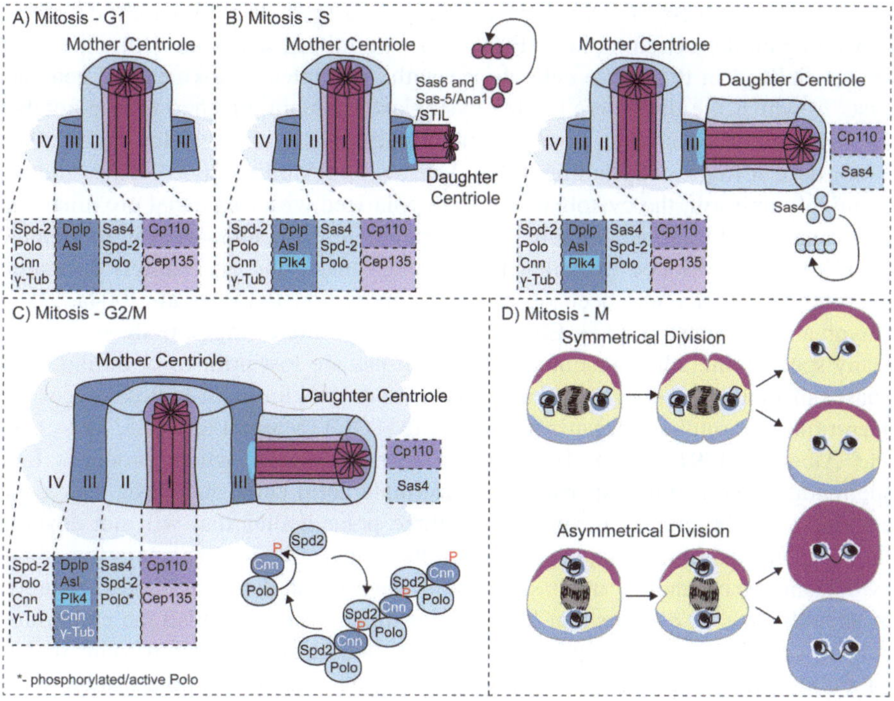

Fig. 4.1 Mitotic polarity requiring centrosomes. (**a**) Mitosis—G1: Mitotic division requires centrosomes to organize the microtubules and drive spindle formation. Centrosomes (usually two in each cell) are composed of centrioles and their associated pericentriolar material (PCM), which is assembled by highly ordered, self-assembling proteins. (**b**) Mitosis—S: The daughter centriole is seeded from the mother centriole, the more mature and complex centriole, via the asymmetric recruitment of Plk4 within the mother centriole's matrix (mPCM). This triggers polymerization of Sas6 and subsequent assembly of the protein Sas4 to form the daughter centriole's PCM (dPCM), which at this stage only contains a subset of the PCM scaffold proteins found in the mPCM matrix. (**c**) Mitosis—G2/M: Expansion of the mPCM matrix involves activation of Polo kinase and resulting phosphorylation of Cnn which in turns recruits more Spd-2 and γ-Tub via a positive feedback loop. The mPCM is responsible for coordinating spindle formation. The dPCM does not expand at this stage. (**d**) Mitosis—M: Depending upon the orientation of the centrosomes and resulting spindle with respect to localized content within a cell, division during mitosis can be symmetrical, yielding two identical daughters or asymmetrical, yielding daughters that inherit unique subsets of proteins, usually to specify distinct fates. Definitions of abbreviations in the figure: *PCM* pericentriolar material, *Spd-2* Spindle-defective protein 2, *Cnn* Centrosomin, *γ-Tub* gamma-tubulin, *Dplp* Pericentrin-like protein, *Asl* Asterless, *Sas-4* Something about silencing 4, *Cp110* Centrosomal protein of 110 kDa, *Cep135* Centrosomal protein of 135 kDa, *Plk4* Polo-like kinase 4, *Sas-6* Something about silencing 6, *Sas-5* Something about silencing 5, *Ana1* Anastral spindle 1, *STIL* SCL/TAL1 interrupting locus

2015; Wueseke et al. 2014; Giansanti et al. 2008; Dix and Raff 2007; Pelletier et al. 2004; Kemp et al. 2004; O'Connell et al. 2000, 2001).

The position of the centrosomes and the spindle determines the orientation of the division, which in cells with asymmetrically distributed content can either generate equivalent (symmetrical division) or unique (asymmetric division) daughters depending on the division plane (Fig. 4.1d). The mitotic divisions of the germline are specialized in that these cells divide with incomplete cytokinesis to generate cysts in which the daughters (called cystoblasts) remain attached to one another through structures known as cytoplasmic bridges or ring canals (Kloc et al. 2004a, 2008; de Cuevas and Spradling 1998; Marlow and Mullins 2008). Thus, though seemingly uniform, the cystoblast cells are polarized with ring canal proximal and ring canal distal regions. Notably, when examined, the centrosomes in the female germline are positioned proximal to the ring canals, and accordingly increased microtubule density has been observed in this region (Kloc et al. 2004a, 2008; Huynh and St Johnston 2004; Elkouby et al. 2016). Interestingly, in the oocytes of many species, including mammals, the centrioles are lost prior to or during early phases of meiosis. As a result, karyokinesis and cytokinesis in meiosis occur via acentriolar microtubule organizing centers (MTOC) (Maro et al. 1985, Messinger and Albertini 1991; Chap. 19). Unlike spermatogenesis, which produces four spermatids from a single spermatogonium (male germ cell), oogenesis yields one developmentally competent oocyte and three polar bodies that will not develop further. Due to these unequal divisions, the oocyte retains the majority of the cytoplasm and organelles and is significantly larger than the polar bodies. Notably, in animals, like *C. elegans* that break egg symmetry and establish the embryonic axis by a mechanism that requires fertilization and the paternal centrosome, maternal PCM components but not centriolar proteins are essential for polarization of the anterior posterior axis (Sugioka et al. 2017; Cowan and Hyman 2004). Similarly, in zebrafish, maternal *sas-6* is not required for oocyte or embryonic animal-vegetal axis formation but is required for centrosome duplication in the embryo (Yabe et al. 2007). Whether and which PCM proteins are involved in regulating oocyte polarity and axis determination in species that establish polarity prior to fertilization remains to be determined.

4.3 Asymmetry in the Nucleus of Meiotic Cells

Prior to the symmetric division in males and asymmetric reductive meiotic division of the female germline, conserved "nuclear asymmetries" are apparent in the position and the arrangement of the chromosomes within the nucleus in the germlines of both sexes. During the leptotene-zygotene transition of prophase I, the telomeres cluster in a prominent arrangement known as the chromosomal or zygotene bouquet. This clustering has been observed in meiotic cells (meiocytes) of all eukaryotes examined and tends to occur at the periphery of the nucleus that is proximal to the centrosome, when present, and corresponds to the site of organelle enrichment (e.g., Golgi and mitochondria) (Fig. 4.2) (Chikashige et al. 1994; Moens 1969). Telomere clustering has been proposed to facilitate chromosome pairing to promote synapsis. Consistent with such a role, yeast mutants that disrupt genes required for bouquet formation have compromised chromosome movement, homologous alignment, and meiotic

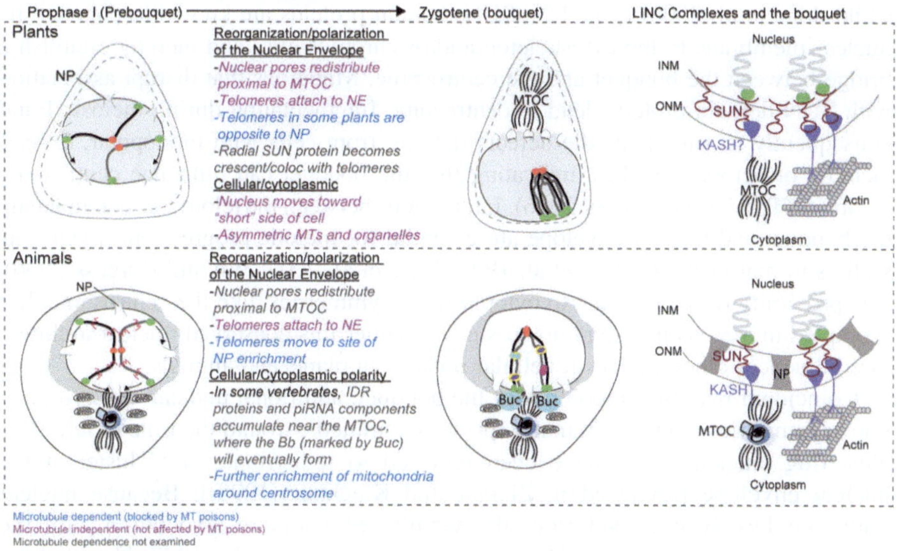

Fig. 4.2 Meiotic polarity, the chromosomal bouquet, and the nucleocytoplasmic axis. A prominent evolutionarily conserved feature of prophase I cells is formation of the chromosomal or zygotene bouquet. Prior to bouquet stage, the telomeres attach to the nuclear envelope (NE). At this stage, the nuclear pores are distributed around the NE of plants and animals. Prior to bouquet formation, asymmetries in cellular organelles, like mitochondria, are already apparent. In concert with, or more likely prior to clustering of the telomeres, the NE undergoes dramatic rearrangement as the nuclear pores cluster to one side of the NE, usually proximal to the MTOC. In some plants, the telomeres cluster opposite to the nuclear pores and cytoplasmic MTOC, whereas in animals the telomeres cluster near the clustered nuclear pores, proximal to the cytoplasmic centromeres when present, and aggregated organelles. This telomere bouquet is thought to facilitate chromosomal homologous pairing, thus contributing to genetic diversity, and occurs in meiotic cells of both sexes in plants and animals. Alignment of nuclear and cytoplasmic asymmetries is thought to occur via transmembrane LINC complexes that are composed of SUN on the inner nuclear membrane and KASH on the outer nuclear membrane. Microtubule-dependent (denoted in *blue*) and microtubule-independent (denoted in *purple*) mechanisms have been implicated in polarization of telomeres independent of nuclear pores and other cytoplasmic asymmetries in some plants, indicating that these events are not coupled. In zebrafish, telomere clustering and asymmetric enrichment of mitochondria requires microtubules, but whether NP clustering or cellular asymmetries like Bb formation are coupled has not been examined (denoted in *gray*). Definitions of abbreviations in the figure: *NP* nuclear pore, *MTOC* microtubule organizing center, *NE* nuclear envelope, *MT* microtubules, *INM* inner nuclear membrane, *ONM* outer nuclear membrane, *Bb* Balbiani body, *buc* bucky ball, *LINC* linker of the nucleoskeleton and cytoskeleton, *SUN* UNC-84, *KASH* Klarsicht, ANC-1, and Syne homology, *IDR* intrinsically disordered proteins, *piRNA* piwi-interacting RNA

recombination (Cooper et al. 1998; Conrad et al. 2008; Chikashige et al. 2006; Martin-Castellanos et al. 2005; Nimmo et al. 1998; Niwa et al. 2000). In yeast, the bouquet is connected to the centrosome, known as the spindle pole body, via a transmembrane bridge formed by interactions between bouquet proteins, telomeres, and SUN-KASH proteins (Yoshida et al. 2013; Ding et al. 2004; Cooper et al. 1998;

Chikashige et al. 2006) (Fig. 4.2). SUN domain proteins are known to connect the nuclear membrane to the cytoskeleton and/or chromosomes and thereby establish a bridge between the bouquet and the centrosome. Mutations that disrupt association with SUN domain proteins lead to centrosome fragmentation during meiosis I and consequently to varied spindle defects ranging from cells with monopolar, excess, lacking, or ectopic spindles, indicating that the bouquet can influence centrosome behavior (Tomita and Cooper 2006). Interactions between the telomeres that make up the bouquet and nuclear envelope are essential for meiotic progression in yeast, as well as in mammals (Conrad et al. 1997; Ding et al. 2007; Wu and Burgess 2006). The proximity of the bouquet to cytoplasmic asymmetries, like the centrosome, has prompted numerous investigators to study whether this conserved nuclear architecture influences other aspects of cellular polarity in plants and animals.

Coincident with the emergence of the bouquet, additional nuclear and cytoplasmic asymmetries manifest in many species examined. Around the time of telomere clustering, nuclear pores have been reported to redistribute and cluster at the nuclear envelope [reviewed in Zickler and Kleckner (1998)]. Because nuclear pores are largely excluded from the vicinity of telomere association with the nuclear envelope (Church 1976; Scherthan et al. 2000; Holm 1977; Hiraoka and Fuchikawa 1993), it is thought that modification of the nuclear envelope, likely via elimination or regulation of lamin, is important for bouquet stage rearrangements of the chromosomes and of the nuclear membrane. Evidence for this notion includes the expression of germline-specific lamina proteins, including lamins (Alsheimer and Benavente 1996; Furukawa and Hotta 1993; Smith and Benavente 1992; Vester et al. 1993). The relative position of the telomeres to the nuclear pores and their respective order/timing of clustering with respect to one another and other cytoplasmic asymmetries appear to vary somewhat in different species. For example, in some plants (Cowan et al. 2002) and animals (Elkouby et al. 2016), enrichment of microtubules and mitochondria within the cytoplasm occurs proximal to the clustered nuclear pores (marked by mAB414 or EM). As mentioned above, clustered nuclear pores have been observed proximal to the telomeres in some organisms and positioned on the opposing nuclear membrane in others (Fig. 4.2). Despite these differences, the alignment of these asymmetries has prompted the question of whether these nuclear and cellular events are coordinated and interdependent events or are coupled at all apart from their coincident emergence in meiotic cells.

If nuclear polarization and cytoplasmic polarization are interdependent events or if one directs the other, then disrupting clustering of telomeres or nuclear pores would be expected to perturb clustering of the other. On the other hand, if these were independent events, possibly organized by a common but unidentified polarity cue or coordinator, then disrupting bouquet formation would not affect nuclear pore clustering or microtubule enrichment. Evidence for the later scenario comes from studies in the rye grain in which bouquet formation was blocked with colchicine, an inhibitor of microtubule polymerization, and this disrupted the bouquet but not the nuclear pore clustering or nuclear displacement (Cowan et al. 2002). On the other hand, recent work in zebrafish oocytes examining the relationship between the bouquet and cytoplasmic asymmetries but using a different inhibitor of microtubule polymerization, nocodazole, showed the opposite result. In that study, both clustering of the telomeres and enrichment of mitochondria and ER marked by DiOC6 were disrupted (Elkouby et al. 2016), indicating that both events might be coordinated by microtubules. It is possible that this different finding reflects different requirements for microtubules in telomere clustering and cytoplasmic polarity between plants and animals at this stage or different degrees of microtubule inhibition in the two studies. It is also possible that the coordinating step is the clustering of or is proximal to clustering of the nuclear pores. In the Cowan et al. (2002) study, nuclear pore clustering remained intact when microtubules were inhibited as did cytoplasmic asymmetries, but in the Elkouby et al. (2016) study, the effects of microtubule inhibition on mAB414 clustering were not examined. Therefore, it remains possible that the apparent differences between gametes in these species may be explained in part by differences in the extent to which nuclear pore clustering occurs. So far, there is no evidence for a bouquet in *Drosophila* (Zickler and Kleckner 1998) or *C. elegans* (Mckee 2004), which would indicate that the bouquet itself is not essential for cellular polarity in those gametes. Notably, in meiotic cells of another insect, *Thermobia domestica*, a chromosomal bouquet has been detected and is associated with formation of nucleoplasmic asymmetry and concurrent aggregation of cytoplasmic organelles near the clustered telomeres by a mechanism proposed to involve centrosomal microtubules (Tworzydlo et al. 2014, 2017). Similarly, in mammalian (Rasmussen and Holm 1978) and zebrafish (Elkouby et al. 2016) gametes, organelles asymmetrically accumulate near the bouquet; however, clustering around the centrosome precedes the bouquet stage. Therefore, in these species, this aspect of cytoplasmic asymmetry is not established by the telomere clustering in the bouquet. It remains to be determined whether molecular asymmetries in the cytoplasm, including asymmetric localization of proteins and patterning mRNAs that spatially and temporally correlate with bouquet formation in zebrafish, are disrupted when telomere clustering is blocked (Heim et al. 2014; Elkouby et al. 2016).

4.4 Polarity Beyond the Nucleus

In addition to the asymmetries associated with meiotic recombination and division, such as the bouquet within the nucleus and later the position of the meiotic spindle, "subcellular asymmetries" are conserved features of oocytes, including asymmetric distribution of organelles (nucleus, mitochondria, endoplasmic reticulum, and Golgi), assembly of the cytoskeleton within the cytoplasm, and localization of RNAs and proteins within specialized non-membrane-enclosed structures or to specific regions of the cell cortex (animal or vegetal in vertebrates or anterior posterior in flies) at specific stages of oogenesis. Several membranous and non-membranous cellular structures have been shown to play a role in the acquisition of cellular polarity, including the centrosome discussed earlier, the Balbiani body (Bb), and the cytoskeleton and acentriolar organizing centers.

4.4.1 Balbiani Body

The Bb is an ancient asymmetric, transient collection of organelles found in vertebrate species, including humans that is one of the earliest recognizable hallmarks of oocyte polarity. This aggregate sometimes referred to as the mitochondrial cloud includes organelles (mitochondria, Golgi, endoplasmic reticulum) and various RNAs and proteins depending on the species examined. The cues leading to Bb development are not well understood. It has been hypothesized that the early germline cysts found within the ovary during mitotic proliferation develop a polarity that may be a precursor to the polarity later seen with establishment of the Bb (Kloc et al. 2004a). Germ cells' progeny in the ovary proliferate via mitosis before entering meiosis with some or all of the cells later becoming oocytes depending on the species. It is during this proliferation period that partial cytokinesis occurs and, as a result, germline cysts are created with interconnecting cells via intercellular bridges owing to incomplete cell division as discussed above (Marlow 2010). Some of these cystocytes will further progress and differentiate into oocytes (Kloc et al. 2004a). The transient Bb structure is present only in prophase I oocytes—it assembles next to the nucleus around zygotene stage and disassembles at the vegetal pole in animals like fish and frogs during diplotene (Fig. 4.3). Owing to its transient nature, the molecular factors regulating Bb development have remained elusive. A definitive function for the Bb has not been experimentally verified in most animals; however, several hypotheses regarding its possible role in early development have been put forth. The Bb houses numerous mitochondria, mRNAs, and RNA binding proteins (RNAbps) into an aggregate that first assembles adjacent to the nucleus and then moves toward the cell cortex (vegetal in vertebrates and posterior in flies). Because many of these same gene products later are found within the primordial germ cells (PGCs), the Bb has been thought to serve as depot for determinants of the future germ cells (Kloc et al. 2004b; Marlow 2010; Hartung and Marlow 2014). This role includes facilitating the transport of these germ cell products via specific RNA localization pathways in vertebrates and invertebrates (Kloc et al. 2004b; Kloc and Etkin 2005). In *Xenopus,* the early or messenger transport organizer (METRO) pathway and the microtubule-dependent, late, pathway are thought to be consecutive phases of Bb-mediated transport; whereby the early Bb cargos organize the cytoskeleton for transport of late cargos (Kloc and Etkin 2005). Entrapment of the germ plasm by the Bb may also serve a protective role preventing early degradation or early activation of gene products vital to germ cell development and embryo patterning [Chang et al. (2004) and reviewed in Marlow (2010)]. Consistent with this notion, several of the localized transcripts encode RNAbps, such as the conserved RNA binding protein deleted in azoospermia-like (*dazl*) and *nanos*, which have been implicated in germ cell translational activation (Collier et al. 2005) and migration/survival (Houston and King 2000; Draper et al. 2007). In insects, *Xenopus,* zebrafish, and mammals, the Bb may also play a role in selection of healthy mitochondria for future germ cells [Cox and Spradling (2003) and reviewed in Kloc et al. (2012)]. Consistent

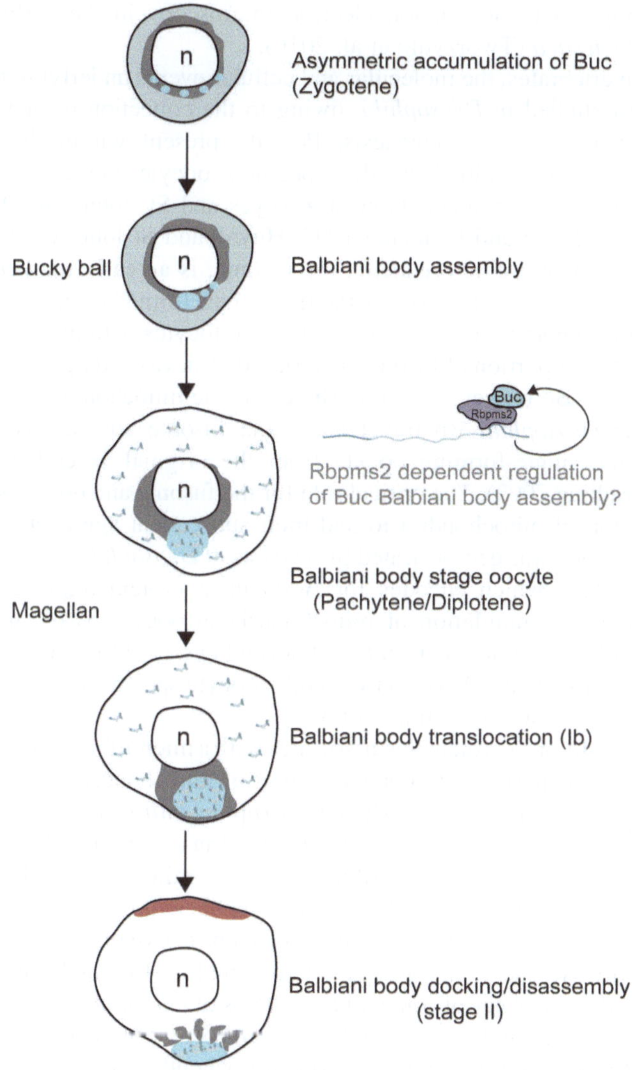

Fig. 4.3 Balbiani body. The Balbiani body starts to assemble after Buc accumulation near the perinuclear region during the zygotene stage of oogenesis. Translocation of the Balbiani body toward the vegetal cortex occurs via a mechanism dependent on Magellan. The transient nature of the Balbiani body is illustrated by its disassembly during the diplotene stage of oogenesis. Definitions of abbreviations in the figure: *Buc* Bucky ball, *rbpms2* RNA binding protein with multiple splice isoforms 2

with this notion, examination of indicators of mitochondrial respiration indicated that the most active and presumably healthiest mitochondria are localized to the Balbiani body (Wilding et al. 2001a, b; Zhang et al. 2008). Moreover, exclusion of deteriorating and dysfunctional mitochondria from the Balbiani body has been observed using 3D reconstruction electron microscopy in the early oocytes of *Thermobia domestica* (Tworzydlo et al. 2016).

Among invertebrates, the molecular and cellular events underlying formation of the Bb is best studied in *Drosophila,* owing to the collection of genetic mutants disrupting early events of oogenesis. Polarity present within the *Drosophila* germline cyst is transmitted to the specified oocyte via the fusome in a microtubule-dependent manner (Gonzalez-Reyes and St. Johnston 1994; Lin and Spradling 1995; Mach and Lehmann 1997; Huynh and St Johnston 2000; Bolivar et al. 2001). A cytoplasmic organelle, the fusome, is an early marker of polarity within the germline cysts. It traverses the ring canals formed during each cystocyte replication forming an asymmetric network of cystocytes with the eventual oocyte receiving the larger portion of fusome material (de Cuevas and Spradling 1998) and the centrosomes (Bolivar et al. 2001). The cystocyte mitochondria that eventually constitute the *Drosophila* Bb travel along the fusome and are asymmetrically enriched in the single-forming oocyte from the original 16-cell germline cyst (Cox and Spradling 2003). The critical role for the fusome and oocyte specification in the migration of mitochondria to and their subsequent aggregation within the mitochondria cloud was demonstrated by analysis of *Drosophila* mutants disrupting *huli-tai-shao* (*hts*), which encodes for Adducin, a protein required for fusome structure and for accumulation of mitochondria in oocytes (Cox and Spradling 2003). *Drosophila hts* mutants result in abnormal cysts with reduced cell numbers and no fusome (Lin et al. 1994); consequently, oocyte specification commonly fails in *hts* mutants (Yue and Spradling 1992).

Drosophila Milton, a protein that associates with a microtubule plus-end-directed molecular motor Kinesin I, promotes transport of mitochondria in the *Drosophila* oocyte (Cox and Spradling 2006). Mutants disrupting *milton* ($milt^{92}$) form a smaller mitochondrial cloud, due to diminished mitochondria transport to the oocyte, which results in a more symmetric distribution of the mitochondria throughout the germline cyst (Cox and Spradling 2006). Although there is only one Milton encoded in the *Drosophila* genome, there appears to be a corrective or compensatory mechanism to ensure the oocyte has adequate mitochondria (Cox and Spradling 2006). Interestingly, the discovery of some mutant alleles of *milton* ($milt^{EY01559}$, $milt^{k14514}$, *and* $milt^{k06704}$) that cause formation of larger than normal Bb with excess mitochondria demonstrated *milton's* role in regulating mitochondria content within the Bb of the oocyte (Cox and Spradling 2006). Although Kinesin I is required for localization of RNAs to the oocyte poles (Zimyanin et al. 2008; Brendza et al. 2000; Loiseau et al. 2010; Ghosh et al. 2012), Dynein complexes rather than Kinesin I are required for transport from the nurse cells to the oocyte (Jambor et al. 2014; Clark et al. 2007; Mische et al. 2007; Dienstbier et al. 2009). This raises the question of when the switch from Dynein- to Kinesin I-mediated transport occurs. Recently, *Dm*Tropomyosin1-I/C (TM1C), an atypical RNA-binding tropomyosin that is necessary for *osk* transport (Erdelyi et al. 1995), was shown through immunoprecipitation experiments to bind to Kinesin-1 (Gaspar et al. 2016; Veeranan-Karmegam et al. 2016) and to the *3′utr* of *osk* RNAs as they dimerize (Gaspar et al. 2016). Although ex vivo and in vivo localization assays indicated that

TM1C is required for the recruitment of Kinesin-1 (Gaspar et al. 2016), which travels with *osk* from the nurse cells to oocyte (Little et al. 2015), analysis of *osk* RNA localization in mutants disrupting *tm1c* showed that it is not necessary for this step of *osk* transport (Veeranan-Karmegam et al. 2016; Gaspar et al. 2016). Interestingly, Kinesin-I is inactive when it initially binds *osk* and later becomes activated during mid-oogenesis by a mechanism that depends on splicing of *osk* and assembly of the spliced localization complex and exon junction complex deposition (Gaspar et al. 2016), when it functions to mediate the localization of *osk* to the posterior pole of the oocyte (Gaspar et al. 2016; Veeranan-Karmegam et al. 2016). Interestingly, germ plasm and patterning RNAs transitioning to the posterior pole are only briefly associated with the Bb in *Drosophila* compared to other species (Cox and Spradling 2003). Strikingly, electron microscopy and analysis of Bb-associated molecules revealed that the *Drosophila* Bb is an ordered or polarized structure with the fusome products located more anteriorly and other RNAs or proteins (encoded by *orb, cup, and oskar*) located posteriorly (Cox and Spradling 2003). Studies of RNA and protein localization in *Xenopus* reveal a similar ordered arrangement of components within the Bb of this vertebrate (Kloc et al. 2002). These observations raise the question of how order is established within the Bb to support transport of organelles and RNA required for later oogenesis, germ plasm formation, and embryo development.

4.4.2 Polarization Via Intrinsically Disordered or Unstructured Proteins

Molecular regulation of Bb development is poorly understood, with only two known genes regulating Bb development in vertebrates. The *bucky ball* (*buc*) gene, also known as *vegetally localized (Xvelo1)*, is vertebrate specific (Bontems et al. 2009; Claussen and Pieler 2004). In *Xenopus*, *Xvelo1* was first identified based on localization of its RNA to the vegetal pole (Claussen and Pieler 2004), but its role in Bb development was not anticipated because the RNA was not detected in the Bb but instead localized via the late pathway (Claussen and Pieler 2004). It was a large-scale maternal-effect screen conducted in Mary Mullins' laboratory that identified a mutant disrupting a homolog of *Xvelo* that produced eggs without polarity, called *bucky ball,* that paved the way for discovery of this key Balbiani body regulator (Claussen and Pieler 2004; Bontems et al. 2009; Marlow and Mullins 2008; Dosch et al. 2004). In zebrafish, *bucky ball* is expressed solely in the female germline, localizes to the Bb, and has been shown to be critical for the assembly of the Bb (Bontems et al. 2009; Marlow and Mullins 2008). Consistent with its requirement for Bb formation in zebrafish, the subcellular localization of *buc* RNA and protein in early zebrafish oocytes is dynamic. Both initially are expressed in a pattern that reflects the phases of Bb development—their expression domains assemble in the perinuclear vicinity, expand toward vegetal as the Bb expands, and translocate to the vegetal pole where the Bb disassembles (Bontems et al. 2009; Heim et al. 2014). Interestingly, *buc* RNA and protein are later detected at the animal pole, possibly positioning the RNA for Buc function in germ plasm assembly at the cleavage furrows of the developing embryo (Bontems et al. 2009; Campbell et al. 2015b). *Buc* RNA expression persists through oogenesis and early embryogenesis until the midblastula transition (maternal to zygotic transition) (Bontems et al. 2009) when it is cleared along with other maternal RNAs by a *dicer*-dependent mechanism as evidenced by elevated *buc* in MZ*dicer* mutants (Le Good et al. 2005; Giraldez et al. 2006, 2005). This together with the strict maternal-effect phenotype of *buc* mutants indicates that functions of this *buc* gene are restricted to the germline. However, additional *buc-like* genes exist in human and other vertebrates, and a few have recently been described to be outside of the germline in zebrafish (Bontems et al. 2009; Skugor et al. 2016). The functions of these *buc-like* genes and whether they similarly promote the formation of subcellular structures await molecular and genetic analysis of their activity.

Initially, the molecular features of Buc/Velo provided no clue to its role in assembling the Bb or the RNAs, proteins, and organelles that localize there. Nonetheless, the failure to assemble the Bb based on electron microscopy, histological, and analysis of RNA localization in loss-of-function and gain-of-function *buc* contexts provided strong evidence that Bucky ball was crucial to the assembly of this ancient structure, recruitment of RNAs there, and subsequent patterning of the zebrafish embryo (Marlow and Mullins 2008; Heim et al. 2014; Bontems et al. 2009). The finding that Buc protein accumulates at zygotene prior to Balbiani body formation implicated the protein in the earliest stages of acquisition of oocyte polarity and suggested that oocyte polarity in zebrafish was established at zygotene or earlier (Heim et al. 2014). As a proxy for a lack of identifiable protein domains, transgenic-based rescue assays, and molecular approaches were pursued to provide mechanistic insight into Buc activities and to identify potential regulators of Balbiani body development and polarity. Attempts to identify domains of *buc* that could restore the Bb and polarity in mutants using only *buc* coding sequences did not reveal functional domains of Buc protein because expression of the full coding sequence led to ectopic smaller Balbiani bodies and polarity defects reminiscent of *buc* loss of function (Heim et al. 2014). Instead, this line of investigation revealed a requirement for *buc* introns and implicated posttranscriptional control mechanisms in *buc* regulation during oocyte polarization (Heim et al. 2014). However, the specific introns that confer proper regulation and the proteins involved in controlling posttranscriptional regulation of *buc* remain to be determined.

Candidate regulators of *buc* emerged from immunopurification approaches for Buc-interacting molecules (Heim et al. 2014; Campbell et al. 2015b) and proteomics-based approaches to catalog Balbiani body proteins (Boke et al. 2016). Among the proteins that interacted with the *bucky ball* gene products were resident Balbiani body RNAbps with canonical RNA recognition motifs, including deleted in azoospermia-like (*dazl)* and RNA binding protein with multiple splice isoforms 2 (*rbpms2),* both of which depend on *bucky ball* for their localization to the Balbiani body (Bontems et al. 2009; Heim et al. 2014; Maegawa et al. 2002). The identification of interactions between *buc* RNA and Buc protein and RNAbps in yeast two hybrid and RNAIPs (Heim et al. 2014; Nijjar and Woodland 2013) along with the failure to recruit RNAs to the Balbiani body in *buc* mutants and transgenics led to a model whereby accumulation of asymmetric Buc seeds assembly of the Balbiani body. Then through a positive feedback amplification mechanism that involves recruitment of more *buc* RNA to produce more Buc protein that through interactions with RNAbps, like *dazl* and *rbpms2*, recruits additional germ plasm and patterning RNAs to the asymmetric aggregate (Heim et al. 2014) (Figs. 4.3 and 4.4). Whether and how *dazl* and *rbpms2* RNAbps contribute to Bb development awaits experimental validation.

The interactions between Buc and RNAbps suggest that Buc may act as a scaffolding protein that stabilizes and recruits other proteins and RNAs. Evidence in support of this model comes from recent analysis of Xvelo/Buc protein motifs (Boke et al. 2016) (Fig. 4.4). Thioflavin T staining revealed the presence of β sheets in the *Xenopus* Bb, suggesting that the Bb organization may involve a network of amyloid-like fibers (Boke et al. 2016). The corresponding high concentration of Xvelo/Buc protein within the Bb of *Xenopus* (Boke et al. 2016) and zebrafish (Heim et al. 2014) suggests that it may provide a matrix on which the Bb develops and RNAs are recruited. Examination of the N-terminus of Xvelo identified a prion-like domain (PLD) that was suggestive of potential self-assembly and amyloid-like matrix formation (Boke et al. 2016). Wild-type Xvelo was demonstrated, using in vitro and oocyte assays to assemble an amyloid-like matrix that was competent to associate with mitochondria and RNAs (Boke et al. 2016) (Fig. 4.4). Initially, based on the lack of a canonical RNA recognition motif, it seemed likely that RNAs would be recruited indirectly via interactions between Buc protein and RNAbps that would engage the RNA. However, recent analysis of Xvelo/Buc identified protein motifs that suggest an enrichment of sequences with the potential to bind RNAs, raising the possibility that Buc might recruit a subset of RNAs via direct interaction (Boke et al. 2016) (Fig. 4.4). Using truncated versions of Xvelo in oocyte and in vitro assays, Boke and colleagues provided evidence for functionally distinct domains of Xvelo with respect to matrix assembly and RNA/organelle recruitment activities. Specifically, a mutant version of Xvelo disrupting the PLD was unable to self-assemble and failed to promote formation of the amyloid-like matrix that is presumably required for Bb formation in vivo, whereas a mutant lacking the KR-rich putative RNA binding region was sufficient for matrix assembly but failed to recruit RNAs and mitochondria (Boke et al. 2016) (Fig. 4.4). Notably, zebrafish mutants predicted to encode truncated versions of Buc that would include the PLD but lack the KR-rich region fail to assemble the Balbiani body (Bontems et al. 2009; Marlow and Mullins 2008). Consequently, RNAs and proteins that should localize to the Bb are not localized, and mitochondria as measured by DiOC6, MitoTracker, and EM are asymmetrically enriched in early prophase I but are not localized at diplotene when the Bb should be present (Marlow and Mullins 2008; Bontems et al. 2009; Elkouby et al. 2016). Whether recruitment

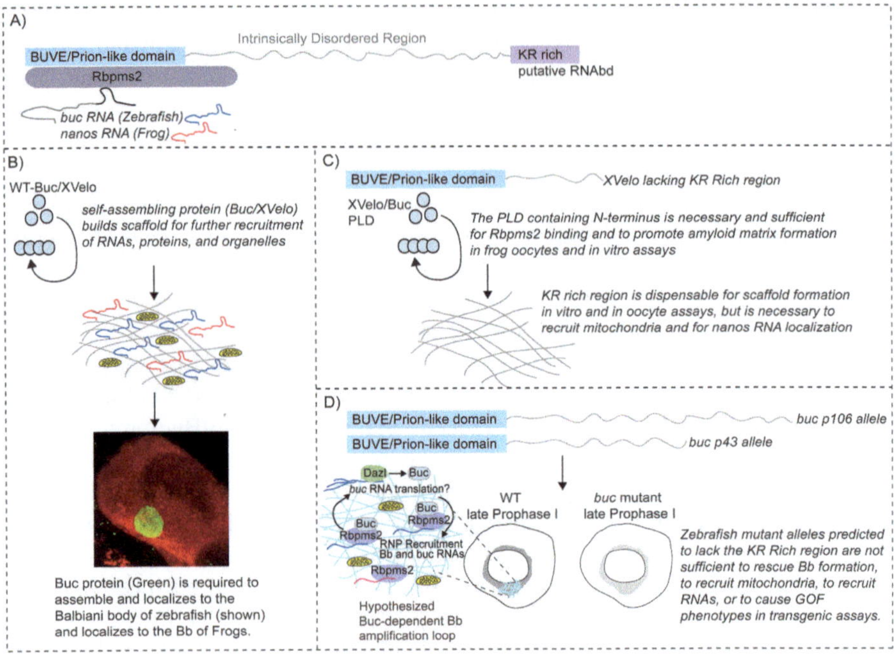

Fig. 4.4 Bucky ball/XVelo protein domains, molecular interactions, and functional regions. (**a**) Buc/XVelo is a disordered protein whose sequence includes a prion-like domain (PLD) at the N-terminus (conserved region also referred to as the BUVE domain) and a KR-rich domain at the C-terminus that is implicated in RNA binding. The N-terminus of Buc is required for its conserved interaction with the RNA binding protein Rbpms2 in zebrafish. In addition, RNAbps, such as Rbpms2 and Dazl, are known to bind *buc* mRNA. In *Xenopus,* Buc/Xvelo binds to *nanos* RNA. Buc's protein architecture and molecular interactions indicate that Buc could recruit RNAs by direct and/or indirect interactions with its associated RNAbps. (**b**) In WT zebrafish oocytes, Buc/Xvelo has been proposed to be a self-assembling protein that seeds assembly of a structure to which mitochondria, RNAs, and proteins, including *buc* RNA and protein, in zebrafish (*green* in confocal *Airyscan* projection) are recruited to form the Bb. This recruitment may involve translational activation of *buc* and a buc-dependent positive amplification feedback loop. Consistent with this model, Buc/Xvelo was recently shown to assemble a scaffold of amyloid-like β sheets that can recruit RNAs and mitochondria in oocyte and in vitro assays. (**c**) *XVelo* mutants lacking the KR-rich region can support scaffold formation; however, no mitochondria or RNAs are recruited there. (**d**) Mutations of zebrafish *buc* that are predicted to preserve the PLD but lacking the putative RNA binding region are not sufficient to support development or to localize RNAs and mitochondria in mutant oocytes. Definitions of abbreviations in the figure: *RNAbps* RNA binding proteins, *RNP* ribonucleoprotein, *buc* bucky ball, *dazl* deleted in azoospermia-like, *rbpms2* RNA binding protein with multiple splice isoforms 2, *Bb* Balbiani body, X*velo* vegetally localized, lysine/arginine rich (KR rich), *PLD* prion-like domain

of RNAs and organelles is mediated via direct interaction with Buc or indirectly via interactions with other proteins or RNAs remains to be determined.

Oskar, long known to interact with RNA binding proteins, was recently recognized as a noncanonical RNAbp in *Drosophila* (Yang et al. 2015; Jeske et al. 2015) and fulfills similar functions to *buc* in that both are self-assembling proteins and early regulators of germ plasm assembly and polarity (Marlow 2015). Flies produce two functionally distinct isoforms of Oskar, Long and Short Oskar. Short Oskar is the more abundant isoform and is essential for posterior patterning, whereas the Long isoform aids in affixing the mRNA and Short-Osk to the posterior but is not sufficient to rescue mutant defects (Markussen et al. 1995). Short Oskar was thought to seed and form a scaffold for germ plasm assembly and posterior patterning through interactions with RNA binding proteins and recruitment and binding of germ plasm-specific RNAs, including *(nanos) nos*, *(polar granule component) pgc*, *(germ cell-less) gcl*, and the RNA helicase *vasa*. Recently, two groups independently elucidated the crystal structure of Oskar and demonstrated using binding activity assays that although the N-terminal LOTUS domain has properties suggestive of nucleotide binding (e.g., winged helix structure), this domain instead mediates homodimerization of Oskar (Yang et al. 2015) and binding to the RNA helicase Vasa (Jeske et al. 2015). Surprisingly, both groups discovered that a C-terminal hydrolase-like (OSK) domain lacks residues necessary for enzymatic activity but can bind to *nanos* and *osk* RNAs (Jeske et al. 2015; Yang et al. 2015). Defining the crystal structure of Oskar has provided insight into its interactions with RNAbps and opened up new potential mechanisms for its contribution to oocyte polarity. More recently, Long Oskar has been shown to facilitate actin-dependent attachment of mitochondria to the oocyte posterior pole, thereby delegating these Long-Osk-selected mitochondria for enrichment in the future primordial germ cells (Hurd et al. 2016).

4.4.3 Microtubule Cross-Linking and Oocyte Polarity

As mentioned, the Bb structure is a transient structure, and genetic evidence from zebrafish indicates that its disassembly is as crucial for late stages of polarity as its assembly, because mutants disrupting *magellan (mgn)* cause similar egg polarity phenotypes to *buc* loss of function (Dosch et al. 2004). The *mgn* mutant phenotype is due to a deletion in the *microtubule actin cross-linking factor 1 (macf1)* gene (Gupta et al. 2010). Unlike *buc* mutants, initial Bb formation is intact; however, later, RNAs and Bb cargo are not delivered to the vegetal cortex. Instead, the Bb becomes abnormally large and persists near the nucleus, which in *mgn* mutant oocytes is proximal to the cortex rather than in the center of the cell (Gupta et al. 2010). This finding indicated that *mgn* contributes to a process that controls the size, translocation, and disassembly of the Bb. Mgn/Macf1 is known to contain domains for interaction with both the actin and microtubule cytoskeleton (Fig. 4.5) (Gregory and Brown 1998; Strumpf and Volk 1998; Chen et al. 2006; Applewhite et al. 2013; Sanchez-Soriano et al. 2009; Lee et al. 2000); thus, Bb translocation likely requires coordinated or sequential regulation of either or both cytoskeletal networks. The predicted protein, if stable, generated by the *mgn* mutant allele would lack the microtubule-binding domain but retain the domains that mediate interaction with

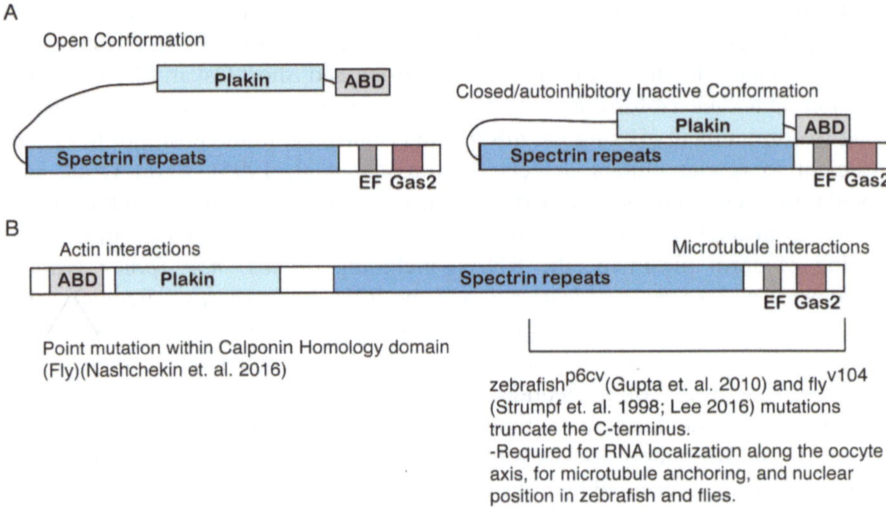

Fig. 4.5 Microtubule actin cross-linking protein contribution to polarity. *Magellan/microtubule actin cross-linking factor 1 (mgn/macf1)* or *shot* in *Drosophila* encodes proteins that interact with actin and microtubules. (**a**) MacF1 activity is thought to be regulated by intramolecular interactions between the amino and carboxy terminus of the protein, which render the protein inactive. (**b**) Generalized schematic of MacF1 protein domains. Distinct domains, including the plakin and spectrin repeats, of Mgn contribute to its biologic activity and interactions with the cytoskeleton. Mutations within each of the indicated domains cause polarity defects in zebrafish and flies. Definitions of abbreviations in the figure: *mgn* Magellan, *macf1* microtubule actin cross-linking factor 1

actin (Fig. 4.5) (Gupta et al. 2010). Consistent with this notion, examination of the cytoskeleton in the *mgn* zebrafish mutant revealed that actin filaments were preserved; however, microtubules failed to reach the oocyte cortex (Gupta et al. 2010). This finding provided evidence that a lack of microtubule tracks connecting to the cell cortex might underlie the failure to deliver the Bb cargo, but how this is regulated and which functions of *mgn/macf1* are essential to Bb development will require further structure function analysis. Nonetheless, it is clear that both the formation and dissociation of the Bb helps establish oocyte polarity vital for normal embryo development.

As discussed above, in *Drosophila*, only a single oocyte is specified among the 16 interconnected cystocytes. The remaining 15 daughters transfer most of their contents including mitochondria, proteins, RNAs, and centrosomes likely via a combination of microtubule-dependent and microtubule-independent mechanisms, because microtubule poisons blocked transport of some content but not others. Dynein is required to transport all of these components, including the centrosomes, to the oocyte (Li et al. 1994; Mcgrail and Hays 1997; Navarro et al. 2004; Serbus et al. 2005; Swan et al. 1999; Vaccari and Ephrussi 2002). The *mgn* counterpart in *Drosophila*, *shot* (Roper and Brown 2004), is a spectraplakin serving as a component of the fusome. Analyses of mutant clones for *shot* null alleles revealed a requirement for shot in oocyte specification and anchoring of microtubules to the fusome (Roper and Brown 2004). Moreover, *shot* null mutant clones exhibit decreased concentration of centrosomes and disrupt the polarized arrangement of microtubules in a minus-end orientation around the fusome (Roper and Brown 2004). In this context, *shot* was proposed to protect a subset of microtubule poison-resistant microtubules to promote centrosome migration. Recently two groups independently isolated new alleles of *shot*: a point mutation within the Calponin homology domain (Nashchekin et al. 2016) and an apparent dominant allele that eliminates the C-terminal domain (Lee et al. 2016) that mediates binding to EB1 (Fig. 4.5) (Strumpf and Volk 1998). Consistent with a dominant negative function for the latter allele, previous studies have shown the interactions between the C-terminal and N-terminal regions of Shot that render the protein inactive (Fig. 4.5) (Applewhite et al. 2013).

In contrast to the previously characterized alleles, the new alleles provide sufficient function for oocyte specification and revealed new *shot* functions in patterning the *Drosophila* oocyte (Nashchekin et al. 2016; Lee et al. 2016). Similar to zebrafish *mgn* mutants (Gupta et al. 2010), initial polarity is intact, as indicated by oocyte specification and an intact actin cytoskeleton in the *shot* mutant oocytes (Nashchekin et al. 2016; Lee et al. 2016). However, later microtubules are reduced in the anterior pole of *Drosophila shot* mutants, and consequently the localization of RNAs that should localize along the anterior posterior axis is impaired (Nashchekin et al. 2016; Lee et al. 2016). Consistent with its essential role in oocyte polarity, fluorescent reporters revealed that *Drosophila* Shot protein becomes specifically localized to anterior lateral regions and is excluded from the posterior pole by a complex mechanism that appears to involve local translation of localized *shot* RNA (Lee et al. 2016), tethering of Shot to the cortex (Lee et al. 2016) via residues that mediate actin binding (Nashchekin et al. 2016). This mechanism is Par1 dependent, as Shot protein is not excluded from the posterior in *par1* mutants (Nashchekin et al. 2016; Lee et al. 2016). It remains to be determined whether Par1 directly acts on Shot, but it seems more likely, based on the localization of *shot* RNA (Lee et al. 2016) and lack of detectable evidence for Shot phosphorylation (Nashchekin et al. 2016), that Par1 modifies the posterior cortex or proteins that localize there. Localization of Shot to the anterior pole in later oogenesis, where the microtubule minus ends and the microtubule minus-end binding protein Patronin are enriched in a *shot*-dependent manner, and isolation of Shot/Patronin complexes from oocytes provided evidence that a Shot/Patronin complex promotes polarity by creating a platform for localized MT assembly via MTOCs (Nashchekin et al. 2016). Accordingly, in *shot* mutants, failure to anchor the MTOCs results in disorganized microtubules that disrupt oocyte polarity (Nashchekin et al. 2016; Lee et al. 2016). Consistent with this model, Shot localizes to the ectopic MTOCs in *capu* gain of function alleles (Lee et al. 2016). Notably, the anchored MTOCs are acentrosomal based on the lack of γ-tubulin and PCM components associated with growing microtubules marked by an EB1-GFP reporter (Nashchekin et al. 2016). Interestingly, the microtubule-severing protein, Katanin, also localizes to the anterior cortex (Nashchekin et al. 2016), thus providing a localized mechanism to provide substrates for Patronin stabilization and new microtubule assembly and thereby eliminating the need for γ-Trc nucleation in oocytes, which like oocytes of other animals undergo the later stages of their development without centrioles. In addition to the similar defects in microtubule anchoring, disruption of microtubules but not actin, failure to deliver localized cargos in later stages of oogenesis, zebrafish *magellan* mutants, and the dominant negative *shot* allele lacking the C-terminal domain CTD all show defects in nuclear position and accumulation of RNAs and proteins proximal to the mislocalized nucleus and persisting Bb in zebrafish and "sponge body" in flies (Fig. 4.5) (Lee et al. 2016; Gupta et al. 2010). The similarities between these phenotypes are striking and suggest that *shot/macf1* functions in oogenesis may be conserved between invertebrates and vertebrates.

4.5 Posttranscriptional Control of Patterning: Lessons from *Drosophila*

The activity of several genes in *Drosophila* that are vital to establishing early oocyte polarity are regulated by posttranscriptional mechanisms, including spatial and temporal control of localization of the RNAs that they encode, including *oskar (osk)*, *gurken (grk)*, *bicoid (bcd)*, and *nanos (nos)* (reviewed in Weil 2014). Based on the common spatial and temporal localization patterns of several of these RNAs and in some cases molecular and genetic evidence pointing to shared translational regulators (discussed below), it seems reasonable to assume that similar posttranscriptional mechanisms would be utilized to silence these maternal RNAs or allow for their translation. However, recently, the increased utilization of and improved resolution of labeling and imaging methods, including in vivo probes for RNAs, have allowed investigators to visualize individual RNAs and the RNP complexes that they associate with as they transit through the oocyte. Application of these methods and sophisticated quantification with analysis of RNA localization in the *Drosophila* ovary (Little et al. 2015; Trcek et al. 2015) and in other contexts (e.g., the zebrafish germline) (Gross-Thebing et al. 2014; Campbell et al. 2015a) have provided compelling evidence that individual RNAs, even those headed to the same subcellular destination, can belong to distinct RNPs. The concept that these maternal RNAs, produced mostly in the nurse cells of *Drosophila* egg chambers, were maintained in a translationally silenced or "masked" state during transport into the oocyte has been appreciated for some time. According to this model, nascent RNAs would associate with repressive RNAbps in the nurse cells and in the oocyte until reaching their site of action, at which point the RNAs would encounter new factors that might regulate the length of their polyA tail and/or replace the repressive factor. Indeed, such mechanisms were uncovered and are discussed below, but recent findings suggest that just as there are distinct RNPs, it seems there is more than one way to silence and later activate a localized RNA not through repressors but instead by limiting and localizing translational activators. Evidence in support of mechanisms that allow precise temporal control of gene activity to generate patterning is discussed below.

4.5.1 *Oskar*

In the *Drosophila* female germline, prior to Bb development, *oskar* RNA is produced in the nurse cells, associates with the fusome, and translocates through the region where the Bb forms and eventually arrives to the posterior pole of the *Drosophila* oocyte in a microtubule and exon junction complex-dependent manner (Cox and Spradling 2003; Ghosh et al. 2012; Zimyanin et al. 2008). Although *oskar* mRNA is produced by the nurse cells early in oogenesis, Osk protein only accumulates later in oogenesis at the posterior pole of the oocyte where the germ plasm (known as pole plasm) assembles (Markussen et al. 1995). As loss of, misexpression of, or precocious translation of *oskar* causes anterior posterior polarity phenotypes in *Drosophila* (Smith et al. 1992), research has been directed toward understanding what controls *oskar* translation. Among the positive regulators of *osk* translation is oo18 RNA binding protein (Orb) [homolog of cytoplasmic polyadenylation element binding protein (CPEB)], which was implicated in translation of *osk* based on the lack of detectable Osk in strong *orb* mutants and reduced abundance of Osk in *orb* hypomorphic mutants (Chang et al. 1999; Markussen et al. 1995). That Orb might function in translational control via regulation of polyadenylation was first suggested based on its homology to *Xenopus* CPEB, which was previously shown to affect polyA tail length and translation (Hake and Richter 1994). Consistent with this expectation, a long poly(A) had previously been shown to be required for *osk* translation (Castagnetti and Ephrussi 2003), and the *oskar* RNA species from *orb* mutants had shorter PolyA tails (Chang et al. 1999). These findings implicated Orb as one activator and potential "unmasker" of *oskar* RNA.

Later, distinct roles for the RNA binding protein, Staufen, were identified as a mediator of *osk* localization (translationally repressed) and a localization-dependent translational activator of *osk* (Micklem et al. 2000). The RRM containing RNA binding protein Bruno (Bru), which binds to BRE (Bruno response elements) within the *osk 3′utr* (Kim-Ha et al. 1995; Snee et al. 2008), has been identified along with the later-recognized DEAD box protein Me31B (Nakamura et al. 2001) as repressors or "maskers" of *oskar* RNA based on increased Osk in mutant egg chambers. Me31 colocalized with *oskar* and other posterior group RNAs and was postulated to act by repressing *osk* by packaging the RNA into silencing RNPs (Nakamura et al. 2001). Similarly, one model for Bru-mediated repression suggested that Bru might force *osk* RNA into an oligomerized and silent state (Chekulaeva et al. 2006), consistent with this notion that Bru was recently shown to dimerize (Kim et al. 2015). In the second, nonmutually exclusive model, Bru represses *oskar* via association with Cup, which engages the 5′ cap-associated translation initiation factor eIF4E and blocks interaction with the translation initiation factor eIF4G, and thus translation (Nakamura et al. 2004; Wilhelm et al. 2003). Recent analyses using in vivo tethering assays and fluorescent reporters derived from phage have confirmed BRE-dependent repressor function of Bru but have also provided evidence for BRE-dependent region-specific translational activation (Kim et al. 2015). Using reporters with intact or mutated BREs, and lacking posterior localization elements of the *osk 3′utr*, the authors uncoupled this loss of translational repression from localization of the RNA (Kim et al. 2015). Based on previous evidence that Bru protein and activity is subject to posttranslational modification (Reveal et al. 2010; Yoshida et al. 2004) and the observed spatial and temporal switch from repressor to activator (Kim et al. 2015), both suggest that posttranslational modification might be involved. Identification and mutagenesis of phosphorylated residues in the reporter assays provided evidence that Bru dimerization, but not repression, was dependent on phosphorylation (Kim et al. 2015). Notably, replacement of wild-type Bru with versions defective in dimerization and, based on in vivo reporter assays, translational activation showed no defects in *oskar* regulation or patterning, indicating that this mode of regulation may be redundant to other modes of activation. Very interestingly, as discussed above, Osk has been recently shown to oligomerize and to bind to its own RNA (Yang et al. 2015); however, it is not known whether Osk binding promotes silencing or translation of the RNA. Given the noncoding function for *osk* RNA (Jenny et al. 2006) that Oskar protein only accumulates to significant levels at the posterior pole and previous evidence that Osk can bind to other repressors, like Smaug discussed below, it is tempting to speculate that Osk binding to the RNA would promote its translation, thus supporting Osk accumulation at the posterior pole.

As discussed, recent evidence provides support for a mechanism involving both local activation and restriction of Oskar activity to ensure that Oskar activity is tightly controlled both spatially and temporally and thereby allow activation of Oskar effectors in a similarly confined manner [for a comprehensive review focused on Oskar, see Lehmann (2016)]. In turn, the mechanism by which Oskar contributes to posterior asymmetry involves its effects on translational regulation of RNAs at the posterior pole such as the RNA binding protein *nanos (nos)*, which is also critical for patterning and germline development (Jeske et al. 2011; Zaessinger et al. 2006; Wang and Lehmann 1991; Gavis and Lehmann 1994; Gavis and Lehmann 1992). Smaug (smg), an RNA binding protein, is a key regulator in posttranscriptional regulation that remains inactive during oogenesis; however, during early embryo development, it accumulates and is responsible for maternal mRNA degradation [as reviewed in Laver et al. (2015)]. In the embryo, Smaug binds to the Smaug response elements (SRE) within the 3′utr of *nanos* and mediates deadenylation of the RNA via the CCR4-NOT deadenylase complex and thus promotes translational repression and degradation of *nos* RNA (Zaessinger et al. 2006). In vitro assays indicated that *nanos* repression is dependent on the SRE and can be alleviated by Short-Osk (Jeske et al. 2011). Pull-down and translation assays and examination of in vivo expression suggested that Oskar binding to Smaug prevents it from binding to the SRE elements of *nos* and thereby blocks deadenylation and alleviates translational repression (Zaessinger et al. 2006; Jeske et al. 2011; Dahanukar et al. 1999). Genetic analysis provided evidence that this activity is required in vivo to prevent ectopic Nos and abnormal patterning (head skeleton defects), as Nanos expression expands in mutants lacking *smaug*, *ccr4*, or when *oskar* is overexpressed (Zaessinger et al. 2006). Similar to *osk* and *nos*, *bicoid* (*bcd*), which is required for anterior patterning, is regulated by translational repression until localized to the anterior. In this case, the repressor Pumilio (Pum), which also binds to *nanos*, binds to Nanos response elements (NRE) within the *bcd* 3′UTR to silence the transcript by a mechanism that requires intact NREs. Accordingly, mutants disrupting *pum* or the NRE result in similar defects in anterior patterning (Gamberi et al. 2002). Thus, association with repressive factors that regulate RNA stability or translational state is a core mechanism to achieve spatial pattern within an individual cell or the early embryo.

4.5.2 Gurken

Gurken (grk) is a transforming growth factor alpha (TGF-α)-like protein and is an essential regulator of the prospective *Drosophila* embryonic axes whose activity is similarly regulated in part by localization of its mRNA in a microtubule and Dynein-dependent manner (Macdougall et al. 2003; Davidson et al. 2016; Thio et al. 2000; Neuman-Silberberg and Schupbach 1993, 1996; Gonzalez-Reyes et al. 1995; Roth et al. 1995). During translocation in the oocyte, *grk* mRNA localizes to and is translationally silenced within cytoplasmic compartments that regulate posttranscriptional events called processing bodies (P bodies), until it associates with the edge of the P body where it is translated (Weil et al. 2012). Although several positive regulators of *grk* translation have been identified, including roles for Orb and the poly(A) polymerase Wispy (Wisp), in polyadenylation and translation of *grk* (Norvell et al. 2015), the repressors have been elusive. Similar to Osk, the abundance of Grk is reduced in orb^{mel}, and the null allele (orb^{343}) lacks Grk expression (Chang et al. 2001). Like *oskar*, *gurken* is initially translationally silent. In earlier studies of *osk* regulation, the phenotypes caused by expression of regions of *osk 3′utr* led to models whereby limiting repressors would be overwhelmed or sequestered by excess target, thereby allowing an endogenous RNA to escape repression (Kim-Ha et al. 1995). Generally, loss of function of repressors is expected to lead to ectopic expression proteins encoded by target RNAs. However, although *grk* RNA contains repressor element sequences, like BREs that associate with the repressor Bru (Reveal et al. 2010), ectopic expression of Grk in nurse cells did not occur when potent translational repressors of *osk* (Me31B and Bru) or when a known *grk* repressor in oocytes, *squid* (*hnRNP*), was mutated (Davidson et al. 2016). Therefore, it seemed likely that another repressor or another mechanism contributes to *gurken* silencing. Notably, *orb* RNA is localized, and its translation is regulated by an Orb- and Wispy-dependent mechanism that leads to high levels of localized Orb in oocytes (Wong et al. 2011; Wong and Schedl 2011). Hints that another mechanism rather than another repressor might be involved came from the observation that misexpression of this normally spatially restricted activator in nurse cells was sufficient to promote recruitment and translation of *grk* to the P bodies of nurse cells, which are normally devoid of Orb and *grk* RNA (Davidson et al. 2016). Thus, in the case of *grk*, repression is achieved in part by localizing and limiting access to its translational activators.

Although localized RNAs are known in vertebrate oocytes, as discussed above, the mechanisms regulating their translational activation and repression during transit are not well understood. For example, RNAs, like *vasa* and *nanos*, are localized to the Balbiani body and later to various regions of the zebrafish oocyte cortex as RNAs but not as proteins (Kosaka et al. 2007), and *buc*, which shows complex regulation reminiscent of *oskar* and *gurken* in that each depends on regulation involving intronic sequences, as well as *3′utr*-mediated interactions with RNA binding proteins to ensure precise regulation of the timing and spatial control of activation (reviewed in Marlow (2015). Future studies are needed to determine if other mechanisms exist and whether these mechanisms are broadly used or are specific to distinct subsets of RNAs that must be spatially and temporally constrained to generate polarity and pattern in the developing *Drosophila* oocyte and prospective embryo.

4.6 Conclusion

Development of a new organism begins with the oocyte. This cell, like many of the cells that it will give rise to, has an underlying polarity as delineated above. Recent evidence indicates that this cellular polarity is orchestrated via mechanisms that involve disordered proteins, aggregates of organelles (Balbiani body), and cytoskeletal elements that impact cellular morphology and even polarity in mitotic and meiotic divisions. Regulation of the very few identified polarity factors and acquisition of polarity require posttranscriptional control mechanisms, involving key regulators that repress or activate translation in a temporal- and location-dependent fashion. As discussed above, only a limited number of factors that are essential for polarity have been identified, particularly in vertebrates. Much of what we understand has come from forward mutagenesis screens in genetic systems. Recent advances in genome editing and reverse genetics approaches and the sophisticated in vivo labeling and imaging approaches discussed have significantly improved access to this fundamental process. Increasingly, a combined approach reliant on genome engineering, imaging, and biochemical and biophysical approaches has clarified our understanding of the depth of complexity underlying polarization of this single cell. Going forward these technological advances that are only beginning to be applied today provide a platform to systematically test candidate factors identified through large-scale proteomics, RNAseq, and interaction partner screens to decipher their contribution to oocyte polarity. Significantly, the majority of human's first trimester miscarriages can be traced back to an aneuploidy etiology.

Given that meiosis in females is a highly polarized division, this prompts the question of whether or not a loss in oocyte polarity leads to an increased risk of aneuploidy in the developing embryo. Nonetheless, understanding the fundamental biology that underlies developmental competence of the oocyte has substantial implications for reproduction and fertility.

Acknowledgments Alexa Clapp is supported by an Einstein REI fellowship. Research on oocyte polarity and the germline in the Marlow lab is supported by NIHR01GM089979 and start-up funds to FLM. We are grateful to Odelya Kaufman for the fun discussions and critical reading and comments.

References

Alsheimer M, Benavente R (1996) Change of karyoskeleton during mammalian spermatogenesis: expression pattern of nuclear lamin C2 and its regulation. Exp Cell Res 228:181–188

Applewhite DA, Grode KD, Duncan MC, Rogers SL (2013) The actin-microtubule cross-linking activity of Drosophila Short stop is regulated by intramolecular inhibition. Mol Biol Cell 24:2885–2893

Balestra FR, Strnad P, Fluckiger I, Gonczy P (2013) Discovering regulators of centriole biogenesis through siRNA-based functional genomics in human cells. Dev Cell 25:555–571

Boke E, Ruer M, Wuhr M, Coughlin M, Lemaitre R, Gygi SP, Alberti S, Drechsel D, Hyman AA, Mitchison TJ (2016) Amyloid-like self-assembly of a cellular compartment. Cell 166:637–650

Bolivar J, Huynh JR, Lopez-Schier H, Gonzalez C, St Johnston D, Gonzalez-Reyes A (2001) Centrosome migration into the Drosophila oocyte is independent of BicD and egl, and of the organisation of the microtubule cytoskeleton. Development 128:1889–1897

Bontems F, Stein A, Marlow F, Lyautey J, Gupta T, Mullins MC, Dosch R (2009) Bucky ball organizes germ plasm assembly in zebrafish. Curr Biol 19:414–422

Bornens M (2012) The centrosome in cells and organisms. Science 335:422–426

Brendza RP, Serbus LR, Duffy JB, Saxton WM (2000) A function for kinesin I in the posterior transport of oskar mRNA and Staufen protein. Science 289:2120–2122

Campbell PD, Chao JA, Singer RH, Marlow FL (2015a) Dynamic visualization of transcription and RNA subcellular localization in zebrafish. Development 142:1368–1374

Campbell PD, Heim AE, Smith MZ, Marlow FL (2015b) Kinesin-1 interacts with Bucky ball to form germ cells and is required to pattern the zebrafish body axis. Development 142:2996–3008

Castagnetti S, Ephrussi A (2003) Orb and a long poly(A) tail are required for efficient oskar translation at the posterior pole of the Drosophila oocyte. Development 130:835–843

Chang JS, Tan L, Schedl P (1999) The Drosophila CPEB homolog, orb, is required for oskar protein expression in oocytes. Dev Biol 215:91–106

Chang JS, Tan L, Wolf MR, Schedl P (2001) Functioning of the Drosophila orb gene in gurken mRNA localization and translation. Development 128:3169–3177

Chang P, Torres J, Lewis RA, Mowry KL, Houliston E, King ML (2004) Localization of RNAs to the mitochondrial cloud in Xenopus oocytes through entrapment and association with endoplasmic reticulum. Mol Biol Cell 15:4669–4681

Chekulaeva M, Hentze MW, Ephrussi A (2006) Bruno acts as a dual repressor of oskar translation, promoting mRNA oligomerization and formation of silencing particles. Cell 124:521–533

Chen HJ, Lin CM, Lin CS, Perez-Olle R, Leung CL, Liem RK (2006) The role of microtubule actin cross-linking factor 1 (MACF1) in the Wnt signaling pathway. Genes Dev 20:1933–1945

Chikashige Y, Ding DQ, Funabiki H, Haraguchi T, Mashiko S, Yanagida M, Hiraoka Y (1994) Telomere-led premeiotic chromosome movement in fission yeast. Science 264:270–273

Chikashige Y, Tsutsumi C, Yamane M, Okamasa K, Haraguchi T, Hiraoka Y (2006) Meiotic proteins bqt1 and bqt2 tether telomeres to form the bouquet arrangement of chromosomes. Cell 125:59–69

Church K (1976) Arrangement of chromosome ends and axial core formation during early meiotic prophase in the male grasshopper Brachystola magna by 3D, E.M. reconstruction. Chromosoma 58:365–376

Clark A, Meignin C, Davis I (2007) A Dynein-dependent shortcut rapidly delivers axis determination transcripts into the Drosophila oocyte. Development 134:1955–1965

Claussen M, Pieler T (2004) Xvelo1 uses a novel 75-nucleotide signal sequence that drives vegetal localization along the late pathway in Xenopus oocytes. Dev Biol 266:270–284

Collier B, Gorgoni B, Loveridge C, Cooke HJ, Gray NK (2005) The DAZL family proteins are PABP-binding proteins that regulate translation in germ cells. EMBO J 24:2656–2666

Conrad MN, Dominguez AM, Dresser ME (1997) Ndj1p, a meiotic telomere protein required for normal chromosome synapsis and segregation in yeast. Science 276:1252–1255

Conrad MN, Lee CY, Chao G, Shinohara M, Kosaka H, Shinohara A, Conchello JA, Dresser ME (2008) Rapid telomere movement in meiotic prophase is promoted by NDJ1, MPS3, and CSM4 and is modulated by recombination. Cell 133:1175–1187

Cooper JP, Watanabe Y, Nurse P (1998) Fission yeast Taz1 protein is required for meiotic telomere clustering and recombination. Nature 392:828–831

Cowan CR, Carlton PM, Cande WZ (2002) Reorganization and polarization of the meiotic bouquet-stage cell can be uncoupled from telomere clustering. J Cell Sci 115:3757–3766

Cowan CR, Hyman AA (2004) Centrosomes direct cell polarity independently of microtubule assembly in *C. elegans* embryos. Nature 431:92–96

Cox RT, Spradling AC (2003) A Balbiani body and the fusome mediate mitochondrial inheritance during Drosophila oogenesis. Development 130:1579–1590

Cox RT, Spradling AC (2006) Milton controls the early acquisition of mitochondria by Drosophila oocytes. Development 133:3371–3377

Dahanukar A, Walker JA, Wharton RP (1999) Smaug, a novel RNA-binding protein that operates a translational switch in Drosophila. Mol Cell 4:209–218

Davidson A, Parton RM, Rabouille C, Weil TT, Davis I (2016) Localized translation of gurken/TGF-alpha mRNA during axis specification is controlled by access to Orb/CPEB on processing bodies. Cell Rep 14:2451–2462

de Cuevas M, Spradling AC (1998) Morphogenesis of the Drosophila fusome and its implications for oocyte specification. Development 125:2781–2789

Dienstbier M, Boehl F, Li X, Bullock SL (2009) Egalitarian is a selective RNA-binding protein linking mRNA localization signals to the dynein motor. Genes Dev 23:1546–1558

Ding DQ, Yamamoto A, Haraguchi T, Hiraoka Y (2004) Dynamics of homologous chromosome pairing during meiotic prophase in fission yeast. Dev Cell 6:329–341

Ding X, Xu R, Yu J, Xu T, Zhuang Y, Han M (2007) SUN1 is required for telomere attachment to nuclear envelope and gametogenesis in mice. Dev Cell 12:863–872

Dix CI, Raff JW (2007) Drosophila Spd-2 recruits PCM to the sperm centriole, but is dispensable for centriole duplication. Curr Biol 17:1759–1764

Dosch R, Wagner DS, Mintzer KA, Runke G, Wiemelt AP, Mullins MC (2004) Maternal control of vertebrate development before the midblastula transition: mutants from the zebrafish I. Dev Cell 6:771–780

Draper BW, Mccallum CM, Moens CB (2007) nanos1 is required to maintain oocyte production in adult zebrafish. Dev Biol 305:589–598

Elkouby YM, Jamieson-Lucy A, Mullins MC (2016) Oocyte polarization is coupled to the chromosomal bouquet, a conserved polarized nuclear configuration in meiosis. PLoS Biol 14:e1002335

Enjolras C, Thomas J, Chhin B, Cortier E, Duteyrat JL, Soulavie F, Kernan MJ, Laurencon A, Durand B (2012) Drosophila chibby is required for basal body formation and ciliogenesis but not for Wg signaling. J Cell Biol 197:313–325
Erdelyi M, Michon AM, Guichet A, Glotzer JB, Ephrussi A (1995) Requirement for Drosophila cytoplasmic tropomyosin in oskar mRNA localization. Nature 377:524–527
Fu J, Glover DM (2012) Structured illumination of the interface between centriole and peri-centriolar material. Open Biol 2:120104
Furukawa K, Hotta Y (1993) cDNA cloning of a germ cell specific lamin B3 from mouse spermatocytes and analysis of its function by ectopic expression in somatic cells. EMBO J 12:97–106
Gamberi C, Peterson DS, He L, Gottlieb E (2002) An anterior function for the Drosophila posterior determinant Pumilio. Development 129:2699–2710
Gaspar I, Sysoev V, Komissarov A, Ephrussi A (2016) An RNA-binding atypical tropomyosin recruits kinesin-1 dynamically to oskar mRNPs. EMBO J 36(3):319–333
Gavis ER, Lehmann R (1992) Localization of nanos RNA controls embryonic polarity. Cell 71:301–313
Gavis ER, Lehmann R (1994) Translational regulation of nanos by RNA localization. Nature 369:315–318
Ghosh S, Marchand V, Gaspar I, Ephrussi A (2012) Control of RNP motility and localization by a splicing-dependent structure in oskar mRNA. Nat Struct Mol Biol 19:441–449
Giansanti MG, Bucciarelli E, Bonaccorsi S, Gatti M (2008) Drosophila SPD-2 is an essential centriole component required for PCM recruitment and astral-microtubule nucleation. Curr Biol 18:303–309
Giraldez AJ, Cinalli RM, Glasner ME, Enright AJ, Thomson JM, Baskerville S, Hammond SM, Bartel DP, Schier AF (2005) MicroRNAs regulate brain morphogenesis in zebrafish. Science 308:833–838
Giraldez AJ, Mishima Y, Rihel J, Grocock RJ, Van Dongen S, Inoue K, Enright AJ, Schier AF (2006) Zebrafish MiR-430 promotes deadenylation and clearance of maternal mRNAs. Science 312:75–79
Gonzalez-Reyes A, Elliott H, St Johnston D (1995) Polarization of both major body axes in Drosophila by gurken-torpedo signalling. Nature 375:654–658
Gonzalez-Reyes A, St. Johnston D (1994) Role of oocyte position in establishment of anterior-posterior polarity in Drosophila. Science 266:639–642
Gregory SL, Brown NH (1998) kakapo, a gene required for adhesion between and within cell layers in Drosophila, encodes a large cytoskeletal linker protein related to plectin and dystrophin. J Cell Biol 143:1271–1282
Gross-Thebing T, Paksa A, Raz E (2014) Simultaneous high-resolution detection of multiple transcripts combined with localization of proteins in whole-mount embryos. BMC Biol 12:55
Gupta T, Marlow FL, Ferriola D, Mackiewicz K, Dapprich J, Monos D, Mullins MC (2010) Microtubule actin crosslinking factor 1 regulates the Balbiani body and animal-vegetal polarity of the zebrafish oocyte. PLoS Genet 6:e1001073
Hake LE, Richter JD (1994) CPEB is a specificity factor that mediates cytoplasmic polyadenylation during Xenopus oocyte maturation. Cell 79:617–627
Hannak E, Kirkham M, Hyman AA, Oegema K (2001) Aurora-A kinase is required for centrosome maturation in Caenorhabditis elegans. J Cell Biol 155:1109–1116
Hartung O, Marlow F (2014) Get it together: how RNA-binding proteins assemble and regulate germ plasm in the oocyte and embryo. Nova Science, New York
Heim AE, Hartung O, Rothhamel S, Ferreira E, Jenny A, Marlow FL (2014) Oocyte polarity requires a Bucky ball-dependent feedback amplification loop. Development 141:842–854
Hiraoka T, Fuchikawa Y (1993) An electronmicroscopic and morphometric analysis on the cell polarity in the synaptic stage of meiosis. Cytologia 58:77–84
Holm PB (1977) Three-dimensional reconstruction of chromosome pairing during the zygotene stage of meiosis in Lilium longiflorum (thunb.) Carlsberg Res Commun 42:103

Houston DW, King ML (2000) A critical role for Xdazl, a germ plasm-localized RNA, in the differentiation of primordial germ cells in Xenopus. Development 127:447–456
Hurd TR, Herrmann B, Sauerwald J, Sanny J, Grosch M, Lehmann R (2016) Long Oskar controls mitochondrial inheritance in *Drosophila melanogaster*. Dev Cell 39:560–571
Huynh JR, St Johnston D (2000) The role of BicD, Egl, Orb and the microtubules in the restriction of meiosis to the Drosophila oocyte. Development 127:2785–2794
Huynh JR, St Johnston D (2004) The origin of asymmetry: early polarisation of the Drosophila germline cyst and oocyte. Curr Biol 14:R438–R449
Jambor H, Mueller S, Bullock SL, Ephrussi A (2014) A stem-loop structure directs oskar mRNA to microtubule minus ends. RNA 20:429–439
Jenny A, Hachet O, Zavorszky P, Cyrklaff A, Weston MD, Johnston DS, Erdelyi M, Ephrussi A (2006) A translation-independent role of oskar RNA in early Drosophila oogenesis. Development 133:2827–2833
Jeske M, Bordi M, Glatt S, Muller S, Rybin V, Muller CW, Ephrussi A (2015) The crystal structure of the Drosophila germline inducer Oskar identifies two domains with distinct vasa helicase- and RNA-binding activities. Cell Rep 12:587–598
Jeske M, Moritz B, Anders A, Wahle E (2011) Smaug assembles an ATP-dependent stable complex repressing nanos mRNA translation at multiple levels. EMBO J 30:90–103
Kemp CA, Kopish KR, Zipperlen P, Ahringer J, O'Connell KF (2004) Centrosome maturation and duplication in *C. elegans* require the coiled-coil protein SPD-2. Dev Cell 6:511–523
Kim G, Pai CI, Sato K, Person MD, Nakamura A, Macdonald PM (2015) Region-specific activation of oskar mRNA translation by inhibition of Bruno-mediated repression. PLoS Genet 11:e1004992
Kim S, Rhee K (2014) Importance of the CEP215-pericentrin interaction for centrosome maturation during mitosis. PLoS One 9:e87016
Kim-Ha J, Kerr K, Macdonald PM (1995) Translational regulation of oskar mRNA by bruno, an ovarian RNA-binding protein, is essential. Cell 81:403–412
Kitagawa D, Vakonakis I, Olieric N, Hilbert M, Keller D, Olieric V, Bortfeld M, Erat MC, Fluckiger I, Gonczy P, Steinmetz MO (2011) Structural basis of the 9-fold symmetry of centrioles. Cell 144:364–375
Kloc M, Bilinski S, Dougherty MT, Brey EM, Etkin LD (2004a) Formation, architecture and polarity of female germline cyst in Xenopus. Dev Biol 266:43–61
Kloc M, Bilinski S, Etkin LD (2004b) The Balbiani body and germ cell determinants: 150 years later. Curr Top Dev Biol 59:1–36
Kloc M, Dougherty MT, Bilinski S, Chan AP, Brey E, King ML, Patrick CW Jr, Etkin LD (2002) Three-dimensional ultrastructural analysis of RNA distribution within germinal granules of Xenopus. Dev Biol 241:79–93
Kloc M, Etkin LD (2005) RNA localization mechanisms in oocytes. J Cell Sci 118:269–282
Kloc M, Ghobrial RM, Borsuk E, Kubiak JZ (2012) Polarity and asymmetry during mouse oogenesis and oocyte maturation. Results Probl Cell Differ 55:23–44
Kloc M, Jaglarz M, Dougherty M, Stewart MD, Nel-Themaat L, Bilinski S (2008) Mouse early oocytes are transiently polar: three-dimensional and ultrastructural analysis. Exp Cell Res 314:3245–3254
Kosaka K, Kawakami K, Sakamoto H, Inoue K (2007) Spatiotemporal localization of germ plasm RNAs during zebrafish oogenesis. Mech Dev 124:279–289
Lane HA, Nigg EA (1996) Antibody microinjection reveals an essential role for human polo-like kinase 1 (Plk1) in the functional maturation of mitotic centrosomes. J Cell Biol 135:1701–1713
Laver JD, Marsolais AJ, Smibert CA, Lipshitz HD (2015) Regulation and function of maternal gene products during the maternal-to-zygotic transition in Drosophila. Curr Top Dev Biol 113:43–84
Lawo S, Hasegan M, Gupta GD, Pelletier L (2012) Subdiffraction imaging of centrosomes reveals higher-order organizational features of pericentriolar material. Nat Cell Biol 14:1148–1158

Le Good JA, Joubin K, Giraldez AJ, Ben-Haim N, Beck S, Chen Y, Schier AF, Constam DB (2005) Nodal stability determines signaling range. Curr Biol 15:31–36

Lee J, Lee S, Chen C, Shim H, Kim-Ha J (2016) shot regulates the microtubule reorganization required for localization of axis-determining mRNAs during oogenesis. FEBS Lett 590:431–444

Lee K, Rhee K (2011) PLK1 phosphorylation of pericentrin initiates centrosome maturation at the onset of mitosis. J Cell Biol 195:1093–1101

Lee S, Harris KL, Whitington PM, Kolodziej PA (2000) Short stop is allelic to kakapo, and encodes rod-like cytoskeletal-associated proteins required for axon extension. J Neurosci 20:1096–1108

Lee SH, Somlo S (2014) Cyst growth, polycystins, and primary cilia in autosomal dominant polycystic kidney disease. Kidney Res Clin Pract 33:73–78

Lehmann R (2016) Germ plasm biogenesis--an Oskar-centric perspective. Curr Top Dev Biol 116:679–707

Li M, Mcgrail M, Serr M, Hays TS (1994) Drosophila cytoplasmic dynein, a microtubule motor that is asymmetrically localized in the oocyte. J Cell Biol 126:1475–1494

Lin H, Spradling AC (1995) Fusome asymmetry and oocyte determination in Drosophila. Dev Genet 16:6–12

Lin H, Yue L, Spradling AC (1994) The Drosophila fusome, a germline-specific organelle, contains membrane skeletal proteins and functions in cyst formation. Development 120:947–956

Little SC, Sinsimer KS, Lee JJ, Wieschaus EF, Gavis ER (2015) Independent and coordinate trafficking of single Drosophila germ plasm mRNAs. Nat Cell Biol 17:558–568

Loiseau P, Davies T, Williams LS, Mishima M, Palacios IM (2010) Drosophila PAT1 is required for Kinesin-1 to transport cargo and to maximize its motility. Development 137:2763–2772

Macdougall N, Clark A, Macdougall E, Davis I (2003) Drosophila gurken (TGFalpha) mRNA localizes as particles that move within the oocyte in two dynein-dependent steps. Dev Cell 4:307–319

Mach JM, Lehmann R (1997) An Egalitarian-BicaudalD complex is essential for oocyte specification and axis determination in Drosophila. Genes Dev 11:423–435

Maegawa S, Yamashita M, Yasuda K, Inoue K (2002) Zebrafish DAZ-like protein controls translation via the sequence 'GUUC'. Genes Cells 7:971–984

Markussen FH, Michon AM, Breitwieser W, Ephrussi A (1995) Translational control of oskar generates short OSK, the isoform that induces pole plasma assembly. Development 121:3723–3732

Marlow F (2015) Primordial germ cell specification and migration. F1000Res 4

Marlow FL (2010) Maternal control of development in vertebrates: my mother made me do it! Morgan & Claypool Life Sciences, San Rafael

Marlow FL, Mullins MC (2008) Bucky ball functions in Balbiani body assembly and animal-vegetal polarity in the oocyte and follicle cell layer in zebrafish. Dev Biol 321:40–50

Maro B, Howlett SK, Webb M (1985) Non-spindle microtubule organizing centers in metaphase II-arrested mouse oocytes. J Cell Biol 101:1665–1672

Martin-Castellanos C, Blanco M, Rozalen AE, Perez-Hidalgo L, Garcia AI, Conde F, Mata J, Ellermeier C, Davis L, San-Segundo P, Smith GR, Moreno S (2005) A large-scale screen in *S. pombe* identifies seven novel genes required for critical meiotic events. Curr Biol 15:2056–2062

Mcgrail M, Hays TS (1997) The microtubule motor cytoplasmic dynein is required for spindle orientation during germline cell divisions and oocyte differentiation in Drosophila. Development 124:2409–2419

Mckee BD (2004) Homologous pairing and chromosome dynamics in meiosis and mitosis. Biochim Biophys Acta 1677:165–180

Mennella V, Keszthelyi B, Mcdonald KL, Chhun B, Kan F, Rogers GC, Huang B, Agard DA (2012) Subdiffraction-resolution fluorescence microscopy reveals a domain of the centrosome critical for pericentriolar material organization. Nat Cell Biol 14:1159–1168
Messinger SM, Albertini DF (1991) Centrosome and microtubule dynamics during meiotic progression in the mouse oocyte. J Cell Sci 100(Pt 2):289–298
Micklem DR, Adams J, Grunert S, St Johnston D (2000) Distinct roles of two conserved Staufen domains in oskar mRNA localization and translation. EMBO J 19:1366–1377
Mische S, Li M, Serr M, Hays TS (2007) Direct observation of regulated ribonucleoprotein transport across the nurse cell/oocyte boundary. Mol Biol Cell 18:2254–2263
Moens PB (1969) The fine structure of meiotic chromosome polarization and pairing in Locusta migratoria spermatocytes. Chromosoma 28:1–25
Nakamura A, Amikura R, Hanyu K, Kobayashi S (2001) Me31B silences translation of oocyte-localizing RNAs through the formation of cytoplasmic RNP complex during Drosophila oogenesis. Development 128:3233–3242
Nakamura A, Sato K, Hanyu-Nakamura K (2004) Drosophila cup is an eIF4E binding protein that associates with Bruno and regulates oskar mRNA translation in oogenesis. Dev Cell 6:69–78
Nashchekin D, Fernandes AR, St Johnston D (2016) Patronin/shot cortical foci assemble the noncentrosomal microtubule array that specifies the Drosophila anterior-posterior axis. Dev Cell 38:61–72
Navarro C, Puthalakath H, Adams JM, Strasser A, Lehmann R (2004) Egalitarian binds dynein light chain to establish oocyte polarity and maintain oocyte fate. Nat Cell Biol 6:427–435
Neuman-Silberberg FS, Schupbach T (1993) The Drosophila dorsoventral patterning gene gurken produces a dorsally localized RNA and encodes a TGF alpha-like protein. Cell 75:165–174
Neuman-Silberberg FS, Schupbach T (1996) The Drosophila TGF-alpha-like protein Gurken: expression and cellular localization during Drosophila oogenesis. Mech Dev 59:105–113
Nijjar S, Woodland HR (2013) Protein interactions in Xenopus germ plasm RNP particles. PLoS One 8:e80077
Nimmo ER, Pidoux AL, Perry PE, Allshire RC (1998) Defective meiosis in telomere-silencing mutants of Schizosaccharomyces pombe. Nature 392:825–828
Niwa O, Shimanuki M, Miki F (2000) Telomere-led bouquet formation facilitates homologous chromosome pairing and restricts ectopic interaction in fission yeast meiosis. EMBO J 19:3831–3840
Norvell A, Wong J, Randolph K, Thompson L (2015) Wispy and Orb cooperate in the cytoplasmic polyadenylation of localized gurken mRNA. Dev Dyn 244:1276–1285
O'Connell KF, Caron C, Kopish KR, Hurd DD, Kemphues KJ, Li Y, White JG (2001) The *C. elegans* zyg-1 gene encodes a regulator of centrosome duplication with distinct maternal and paternal roles in the embryo. Cell 105:547–558
O'Connell KF, Maxwell KN, White JG (2000) The spd-2 gene is required for polarization of the anteroposterior axis and formation of the sperm asters in the *Caenorhabditis elegans* zygote. Dev Biol 222:55–70
Pelletier L, Ozlu N, Hannak E, Cowan C, Habermann B, Ruer M, Muller-Reichert T, Hyman AA (2004) The *Caenorhabditis elegans* centrosomal protein SPD-2 is required for both pericentriolar material recruitment and centriole duplication. Curr Biol 14:863–873
Rasmussen SW, Holm PB (1978) Human meiosis II. Chromosome pairing and recombination nodules in human spermatocytes. Carlsb Res Commun 43:275
Reveal B, Yan N, Snee MJ, Pai CI, Gim Y, Macdonald PM (2010) BREs mediate both repression and activation of oskar mRNA translation and act in trans. Dev Cell 18:496–502
Roper K, Brown NH (2004) A spectraplakin is enriched on the fusome and organizes microtubules during oocyte specification in Drosophila. Curr Biol 14:99–110
Roth S, Neuman-Silberberg FS, Barcelo G, Schupbach T (1995) cornichon and the EGF receptor signaling process are necessary for both anterior-posterior and dorsal-ventral pattern formation in Drosophila. Cell 81:967–978

Sanchez-Soriano N, Travis M, Dajas-Bailador F, Goncalves-Pimentel C, Whitmarsh AJ, Prokop A (2009) Mouse ACF7 and drosophila short stop modulate filopodia formation and microtubule organisation during neuronal growth. J Cell Sci 122:2534–2542

Scherthan H, Jerratsch M, Li B, Smith S, Hulten M, Lock T, De Lange T (2000) Mammalian meiotic telomeres: protein composition and redistribution in relation to nuclear pores. Mol Biol Cell 11:4189–4203

Serbus LR, Cha BJ, Theurkauf WE, Saxton WM (2005) Dynein and the actin cytoskeleton control kinesin-driven cytoplasmic streaming in Drosophila oocytes. Development 132:3743–3752

Skugor A, Tveiten H, Johnsen H, Andersen O (2016) Multiplicity of Buc copies in Atlantic salmon contrasts with loss of the germ cell determinant in primates, rodents and axolotl. BMC Evol Biol 16:232

Smith A, Benavente R (1992) Identification of a short nuclear lamin protein selectively expressed during meiotic stages of rat spermatogenesis. Differentiation 52:55–60

Smith JL, Wilson JE, Macdonald PM (1992) Overexpression of oskar directs ectopic activation of nanos and presumptive pole cell formation in Drosophila embryos. Cell 70:849–859

Snee M, Benz D, Jen J, Macdonald PM (2008) Two distinct domains of Bruno bind specifically to the oskar mRNA. RNA Biol 5:1–9

Sonnen KF, Schermelleh L, Leonhardt H, Nigg EA (2012) 3D-structured illumination microscopy provides novel insight into architecture of human centrosomes. Biol Open 1:965–976

Strumpf D, Volk T (1998) Kakapo, a novel cytoskeletal-associated protein is essential for the restricted localization of the neuregulin-like factor, vein, at the muscle-tendon junction site. J Cell Biol 143:1259–1270

Sugioka K, Hamill DR, Lowry JB, Mcneely ME, Enrick M, Richter AC, Kiebler LE, Priess JR, Bowerman B (2017) Centriolar SAS-7 acts upstream of SPD-2 to regulate centriole assembly and pericentriolar material formation. eLife 6:e25358

Swan A, Nguyen T, Suter B (1999) Drosophila Lissencephaly-1 functions with Bic-D and dynein in oocyte determination and nuclear positioning. Nat Cell Biol 1:444–449

Thio GL, Ray RP, Barcelo G, Schupbach T (2000) Localization of gurken RNA in Drosophila oogenesis requires elements in the 5′ and 3′ regions of the transcript. Dev Biol 221:435–446

Tomita K, Cooper JP (2006) The meiotic chromosomal bouquet: SUN collects flowers. Cell 125(1):19–21

Trcek T, Grosch M, York A, Shroff H, Lionnet T, Lehmann R (2015) Drosophila germ granules are structured and contain homotypic mRNA clusters. Nat Commun 6:7962

Tworzydlo W, Kisiel E, Jankowska W, Bilinski SM (2014) Morphology and ultrastructure of the germarium in panoistic ovarioles of a basal "apterygotous" insect, Thermobia domestica. Zoology (Jena) 117:200–206

Tworzydlo W, Kisiel E, Jankowska W, Witwicka A, Bilinski SM (2016) Exclusion of dysfunctional mitochondria from Balbiani body during early oogenesis of Thermobia. Cell Tissue Res 366(1):191–201

Tworzydlo W, Marek M, Kisiel E, Bilinski SM (2017) Meiosis, Balbiani body and early asymmetry of Thermobia oocyte. Protoplasma 254:649–655

Vaccari T, Ephrussi A (2002) The fusome and microtubules enrich Par-1 in the oocyte, where it effects polarization in conjunction with Par-3, BicD, Egl, and dynein. Curr Biol 12:1524–1528

Van Breugel M, Hirono M, Andreeva A, Yanagisawa HA, Yamaguchi S, Nakazawa Y, Morgner N, Petrovich M, Ebong IO, Robinson CV, Johnson CM, Veprintsev D, Zuber B (2011) Structures of SAS-6 suggest its organization in centrioles. Science 331:1196–1199

Veeranan-Karmegam R, Boggupalli DP, Liu G, Gonsalvez GB (2016) A new isoform of Drosophila non-muscle Tropomyosin 1 interacts with Kinesin-1 and functions in oskar mRNA localization. J Cell Sci 129:4252–4264

Vester B, Smith A, Krohne G, Benavente R (1993) Presence of a nuclear lamina in pachytene spermatocytes of the rat. J Cell Sci 104:557–563

Wang C, Lehmann R (1991) Nanos is the localized posterior determinant in Drosophila. Cell 66:637–647

Weil TT (2014) mRNA localization in the Drosophila germline. RNA Biol 11:1010–1018
Weil TT, Parton RM, Herpers B, Soetaert J, Veenendaal T, Xanthakis D, Dobbie IM, Halstead JM, Hayashi R, Rabouille C, Davis I (2012) Drosophila patterning is established by differential association of mRNAs with P bodies. Nat Cell Biol 14:1305–1313
Wilding M, Carotenuto R, Infante V, Dale B, Marino M, Di Matteo L, Campanella C (2001a) Confocal microscopy analysis of the activity of mitochondria contained within the 'mitochondrial cloud' during oogenesis in Xenopus laevis. Zygote 9:347–352
Wilding M, Dale B, Marino M, Di Matteo L, Alviggi C, Pisaturo ML, Lombardi L, De Placido G (2001b) Mitochondrial aggregation patterns and activity in human oocytes and preimplantation embryos. Hum Reprod 16:909–917
Wilhelm JE, Hilton M, Amos Q, Henzel WJ (2003) Cup is an eIF4E binding protein required for both the translational repression of oskar and the recruitment of Barentsz. J Cell Biol 163:1197–1204
Wong LC, Costa A, Mcleod I, Sarkeshik A, Yates J, Kyin S, Perlman D, Schedl P (2011) The functioning of the Drosophila CPEB protein Orb is regulated by phosphorylation and requires casein kinase 2 activity. PLoS One 6:e24355
Wong LC, Schedl P (2011) Cup blocks the precocious activation of the orb autoregulatory loop. PLoS One 6:e28261
Wu HY, Burgess SM (2006) Ndj1, a telomere-associated protein, promotes meiotic recombination in budding yeast. Mol Cell Biol 26:3683–3694
Wueseke O, Bunkenborg J, Hein MY, Zinke A, Viscardi V, Woodruff JB, Oegema K, Mann M, Andersen JS, Hyman AA (2014) The *Caenorhabditis elegans* pericentriolar material components SPD-2 and SPD-5 are monomeric in the cytoplasm before incorporation into the PCM matrix. Mol Biol Cell 25:2984–2992
Yabe T, Ge X, Pelegri F (2007) The zebrafish maternal-effect gene cellular atoll encodes the centriolar component sas-6 and defects in its paternal function promote whole genome duplication. Dev Biol 312:44–60
Yang N, Yu Z, Hu M, Wang M, Lehmann R, Xu RM (2015) Structure of Drosophila Oskar reveals a novel RNA binding protein. Proc Natl Acad Sci USA 112:11541–11546
Yang R, Feldman JL (2015) SPD-2/CEP192 and CDK are limiting for microtubule-organizing center function at the centrosome. Curr Biol 25:1924–1931
Yoshida M, Katsuyama S, Tateho K, Nakamura H, Miyoshi J, Ohba T, Matsuhara H, Miki F, Okazaki K, Haraguchi T, Niwa O, Hiraoka Y, Yamamoto A (2013) Microtubule-organizing center formation at telomeres induces meiotic telomere clustering. J Cell Biol 200:385–395
Yoshida S, Muller HA, Wodarz A, Ephrussi A (2004) PKA-R1 spatially restricts Oskar expression for Drosophila embryonic patterning. Development 131:1401–1410
Yue L, Spradling AC (1992) hu-li tai shao, a gene required for ring canal formation during Drosophila oogenesis, encodes a homolog of adducin. Genes Dev 6:2443–2454
Zaessinger S, Busseau I, Simonelig M (2006) Oskar allows nanos mRNA translation in Drosophila embryos by preventing its deadenylation by Smaug/CCR4. Development 133:4573–4583
Zhang YZ, Ouyang YC, Hou Y, Schatten H, Chen DY, Sun QY (2008) Mitochondrial behavior during oogenesis in zebrafish: a confocal microscopy analysis. Develop Growth Differ 50:189–201
Zickler D, Kleckner N (1998) The leptotene-zygotene transition of meiosis. Annu Rev Genet 32:619–697
Zimyanin VL, Belaya K, Pecreaux J, Gilchrist MJ, Clark A, Davis I, St Johnston D (2008) In vivo imaging of oskar mRNA transport reveals the mechanism of posterior localization. Cell 134:843–853

Chapter 5
The Pole (Germ) Plasm in Insect Oocytes

Szczepan M. Bilinski, Mariusz K. Jaglarz, and Waclaw Tworzydlo

Abstract Animal germline cells are specified either through zygotic induction or cytoplasmic inheritance. Zygotic induction takes place in mid- or late embryogenesis and requires cell-to-cell signaling leading to the acquisition of germline fate de novo. In contrast, cytoplasmic inheritance involves formation of a specific, asymmetrically localized oocyte region, termed the germ (pole) plasm. This region contains maternally provided germline determinants (mRNAs, proteins) that are capable of inducing germline fate in a subset of embryonic cells. Recent data indicate that among insects, the zygotic induction represents an ancestral condition, while the cytoplasmic inheritance evolved at the base of Holometabola or in the last common ancestor of Holometabola and its sister taxon, Paraneoptera.

In this chapter, we first describe subsequent stages of morphogenesis of the pole plasm and polar granules in the model organism, *Drosophila melanogaster*. Then, we present an overview of morphology and cytoarchitecture of the pole plasm in various holometabolan and paraneopteran insect species. Finally, we focus on phylogenetic hypotheses explaining the known distribution of two different strategies of germline specification among insects.

5.1 Introduction

Animal germline cells are specified either through zygotic induction or cytoplasmic inheritance (see Extavour and Akam 2003; Extavour 2007; Quan and Lynch 2016 for a review). Zygotic induction takes place in mid- or late embryogenesis; it requires cell-to-cell signaling and leads to the acquisition of germline fate de novo. In contrast, cytoplasmic inheritance (called also maternal provision) involves formation of a specific, asymmetrically localized oocyte region, termed the germ plasm. This region contains maternally provided germline determinants (mRNAs, proteins) that are capable of inducing germline fate in a subset of embryonic cells. In insects, this region of the oocyte cytoplasm (ooplasm) is called the pole plasm or

---

S.M. Bilinski (✉) • M.K. Jaglarz • W. Tworzydlo
Department of Developmental Biology and Morphology of Invertebrates, Institute of Zoology and Biomedical Research, Jagiellonian University, Gronostajowa 9, 30-387, Krakow, Poland
e-mail: szczepan.bilinski@uj.edu.pl

M. Kloc (ed.), *Oocytes*, Results and Problems in Cell Differentiation 63,
DOI 10.1007/978-3-319-60855-6_5

oosome. Interpretation of published data in a phylogenetic context leads to the suggestion that zygotic induction is more common among animal taxa and most probably represents the ancestral mode (mechanism) of germline specification (ancestral condition), whereas cytoplasmic inheritance evolved secondarily (derived condition) and independently in several animal lineages (Extavour and Akam 2003; Quan and Lynch 2016).

The best understood example of cytoplasmic inheritance comes from the studies of the fruit fly, *Drosophila melanogaster*. In this model species, the maternally derived germline determinants become localized to the posterior pole where they participate in the formation of the pole plasm (see Sect. 5.3). During early embryogenesis, the embryonic (blastoderm) nuclei that interact with the pole plasm (germline determinants) become cellularized before the other blastoderm nuclei. These cells, called pole cells, give rise to primordial germ cells (PGCs). Classical experimental studies clearly indicate that the pole plasm and the pole cells are both necessary and sufficient for PGC specification (see Mahowald 2001 for a review). Recent data suggest that the mode of germline specification described in the fruit fly is not ancestral among insects. So far, the asymmetrically located pole plasm and pole cells have been described only in taxa nested within Holometabola (insects undergoing complete metamorphosis) and three species (the book louse, *Liposcelis divergens*; mullein thrips, *Haplothrips verbasci*; and pea aphid, *Acyrthosiphon pisum*; see Goss 1952; Heming 1979; Lynch et al. 2011; Ewen-Campen et al. 2013; Lin et al. 2014 for further details), belonging to Paraneoptera, the sister group of Holometabola (Kristensen 1981). These observations led to the idea that the cytoplasmic inheritance is a derived feature of Holometabola (Lynch et al. 2011) or Holometabola and its sister taxon, Paraneoptera (Ewen-Campen et al. 2013) (Fig. 5.1, triangle and diamond, respectively). Although this idea seems to be well established, the evolution of germline specification among insects must have been much more complicated, because in several holometabolous lineages, cytoplasmic inheritance has been secondarily lost (Fig. 5.1, taxon names marked with asterisk) and replaced (secondarily) by zygotic induction (discussed in Lynch et al. 2011; Quan and Lynch 2016).

Intriguingly, both classical histological and recent data additionally indicate that cytoplasmic inheritance is, as a rule, confined to those insect taxa that are characterized by meroistic ovaries. As evolution (anagenesis) of insect ovaries is rather complicated and often not fully understood, we will briefly review present knowledge of insect ovary types in the next section.

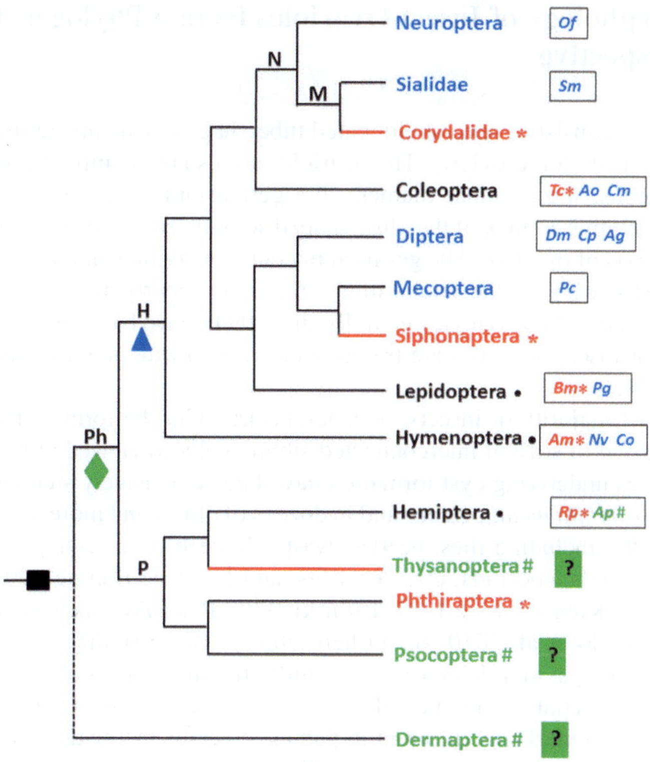

Fig. 5.1 Simplified cladogram of Paraneoptera + Holometabola (=Phalloneoptera) (based on hypotheses of Henning (1981) and Kristensen (1995). The distribution of ovary types is superimposed: *black box* indicates putative origin of meroistic ovary; *red thick lines* indicate reversion to neopanoism. Evolutionary origin of the cytoplasmic inheritance as suggested by Ewen-Campen et al. (2013) is indicated by a *green diamond*, while that proposed by Lynch et al. (2011) and Quan and Lynch (2016) by a *blue triangle*. Strategy of germline specification operating in a given taxon (order, family, or species) is indicated by a *color* of its name and marked with *asterisk*, *black dot*, or #. Taxon names in *blue* (not marked), cytoplasmic inheritance; names in *red* (marked with *asterisk*), zygotic induction; while those in *black* (marked with *black dot*), both strategies: in some species cytoplasmic inheritance in other zygotic induction. Names in *green* (marked with #)—cytoplasmic inheritance suggested or independently derived; *boxes* with ?—data not conclusive. Ph Phalloneoptera, P Paraneoptera, H Holometabola, N Neuropterida, M Megaloptera, Of *Osmylus fulvicephalus*, Sm *Sialis mitsuhashii*, Tc *Tribolium castaneum*, Ao *Acanthoscelides obtectus*, Cm *Callosobruchus maculatus*, Dm *Drosophila melanogaster*, Cp *Culex pipiens*, Ag *Anopheles gambiae*, Pc *Panorpa communis*, Bm *Bombyx mori*, Pg *Pectinophora gossypiella*, Am *Apis mellifera*, Nv *Nasonia vitripennis*, Co *Cosmoconus meridionator*, Rp *Rhodnius prolixus*, Ap *Acyrthosiphon pisum* (literature cited in the text)

5.2 Morphology of Insect Ovarioles from a Phylogenetic Perspective

Insect ovaries consist of several elongated tube-shaped elements, termed ovarioles (see Büning 1994 for a review). The ovarioles are usually composed of three well-defined regions: the terminal filament, the germarium, and the vitellarium. The terminal filament is a stack of flat, disk-shaped somatic cells, oriented perpendicular to the long axis of ovariole. The germarium contains dividing and/or differentiating oogonial cells, whereas the vitellarium consists of linearly arranged, sequentially growing ovarian follicles (reviewed in Büning 1994). Two basic morpho-functional categories of insect ovarioles are traditionally recognized, panoistic and meroistic (Brandt 1874).

In the vast majority of insects, oogenesis starts with the formation of germline cysts composed of several interconnected sibling cells (reviewed in Büning 1994). The processes underlying cyst formation have been extensively studied in a model species, *Drosophila melanogaster* and in dozens of closer and more distant relatives of the fruit fly, including flies, beetles, wasps, butterflies, lacewings, hemipterans (true bugs, aphids, coccoids, etc.), earwigs, and lice (see Büning 1994; Klag and Bilinski 1994; Kubrakiewicz 1997; Bilinski 1998; Pyka-Fosciak and Szklarzewicz 2008; Tworzydlo et al. 2010 for further details). These studies have shown that initial stages of cyst morphogenesis are similar (evolutionary stable) in all investigated species. In contrast, the fate of interconnected cyst cells in subsequent stages of oogenesis is remarkably different in panoistic versus meroistic ovarioles. In the panoistic ovarioles, the germline cysts split into functionally equivalent cells that become definitive oocytes (Pritsch and Büning 1989; Gottanka and Büning 1990). In the meroistic ovarioles, the cyst cells remain joined; one of them differentiates into the oocyte, and the remaining cells form the nurse cells. The main function of nurse cells is the synthesis and subsequent transport of the macromolecules and organelles to the oocyte cytoplasm, while the oocyte nucleus, as a rule, remains transcriptionally quiescent. Depending on spatial organization of the ovariole as well as relationships between the oocyte(s) and the nurse cells, two distinct categories of meroistic ovarioles are usually recognized: polytrophic and telotrophic. In the polytrophic ovariole, each oocyte is accompanied by its own group of nurse cells, whereas in the telotrophic ovariole, all nurse cells are retained in the germarium, thereby forming a trophic chamber.

Analysis of the distribution of the ovary/ovariole categories in various insect lineages in the light of phylogenetic hypotheses of Henning (1981) and Kristensen (1995) leads to the following conclusions:

- Panostic ovaries represent a plesiomorphic character (ancestral condition) inherited from the common hexapod ancestor. This ovarian category is characteristic for all basally branching taxa.

- Meroistic-polytrophic ovaries evolved, from the panoistic ones, in the common ancestor of Dermaptera and Phalloneoptera (Paraneoptera + Holometabola) (Fig. 5.1, black box).
- In some groups (Thysanoptera, Siphonaptera, Megaloptera: Corydalidae), meroistic ovaries have been reversed to the panoistic state (Fig. 5.1, red thick lines). These ovaries are characterized by the (secondary) loss of nurse cells; they are termed neopanoistic to indicate evolutionary distinctness from primary panoistic ones (see Stys and Bilinski 1990; Büning 1993, 1994 for further details).

Figure 5.1 shows a simplified cladogram of Paraneoptera + Holometabola (=Phalloneoptera). The distribution of ovariole types as well as the phylogenetic pattern of a single acquisition and multiple losses of the pole plasm and pole cells are superimposed on the cladogram. Two scenarios of the evolutionary origin of the cytoplasmic inheritance as suggested by Lynch et al. (2011) and Ewen-Campen et al. (2013) are indicated on the cladogram by triangle and diamond, respectively. Analysis of the superimposed events suggests that at least some of the multiple losses of the cytoplasmic inheritance (germ plasm) are associated with the loss of nurse cells, i.e., a reduction to the neopanoistic condition. Due to the sparse sampling, the proposed scenario/s are far from being complete and understood; therefore, we present new and unpublished data regarding ooplasm differentiation in the representatives of two non-holometabolous orders, Thysanoptera and Dermaptera (see Sects. 5.5 and 5.6).

5.3 The Origin and Assembly of Pole Plasm in *Drosophila*

It is well established that in *Drosophila*, pole plasm constituents have maternal origin and are synthesized during oogenesis, well ahead of embryogenesis (reviewed in Mahowald 2001; Becalska and Gavis 2009; Rangan et al. 2009; Lehmann 2016). The origin and interactions of the pole plasm components are both spatially and temporally regulated during oogenesis; therefore, a brief description of the fruit fly ovary architecture and functioning is necessary.

The *Drosophila* ovaries are composed of 12–14 canonical meroistic-polytrophic ovarioles (King 1970, see also Sect. 5.2) that comprise multicellular complexes called ovarian follicles or egg chambers. Each follicle is composed of a centrally located cyst of germline cells surrounded peripherally by somatic cells, the follicular cells. The germline cysts originate from four consecutive mitotic divisions of a cystoblast, a specialized stem cell-derived progenitor cell. Because these divisions are not followed by complete cytokineses, the resulting 16 cells are interconnected by intercellular bridges, also known as ring canals. Within each 16-cell cyst, one cell adopts an oocyte fate, while the remaining 15 differentiate into nurse cells (for a detailed description of the germline cyst formation and development, see de Cuevas et al. 1997; Bastock and St Johnston 2008). During egg chamber development, nurse cells become polyploid and highly transcriptionally active. In contrast, the oocyte is largely transcriptionally quiescent, enters meiosis, and ultimately becomes a haploid egg cell. During late oogenesis, the nurse cells undergo apoptosis and rapidly pass most of their cytoplasmic contents into the oocyte in a process called nurse cell dumping (Mahajan-Miklos and Cooley 1994; Cavaliere et al. 1998). Subsequently, the ooplasm is thoroughly mixed by microtubule-based streaming (reviewed in Mahajan-Miklos and Cooley 1994; Becalska and Gavis 2009).

Another important aspect of ovarian follicle development is specification of the anterior-posterior and dorsoventral polarity that is critical for the proper embryonic development (for review, see Bastock and St Johnston 2008; Roth and Lynch 2009). Within the ovarian follicle, nurse cells occupy invariably the anterior pole, while the oocyte resides at the posterior one. The establishment of the axes is aided by the reciprocal signal exchange between germline cells and the surrounding somatic (follicular) cells. Subsequently, the oocyte becomes polarized by asymmetric distribution of specific developmental factors in distinct regions of the ooplasm. The formation of a distinct pole (germ) plasm domain is a morphological manifestation of the early oocyte asymmetry.

5.3.1 Pole Plasm Contains Specific Ribonucleoprotein Complexes

Expression-based analyses have identified numerous factors localized to the pole plasm. The biochemical nature of many of these factors has been uncovered over the last three decades. It has been shown that a number of maternally provided proteins (e.g., Oskar, Vasa, Tudor, and Aubergine) and mRNAs (including *oskar*, *cyclinB*, *nanos*, *polar granule component*, and *germ cell-less*) are enriched in *Drosophila* pole plasm (Fig. 5.2a) and at least some of these factors are engaged in PGC specification, maintenance, and migration (Lecuyer et al. 2007; Frise et al. 2010; Gao and Arkov 2013; Jambor et al. 2015). In addition, certain pole plasm constituents (e.g., *nanos* mRNA) are necessary for the specification of the abdominal region of the future embryo (Wang and Lehmann 1991).

The pole plasm-specific mRNAs appear to share three important characteristics. First, they associate with proteins and are transported in the form of ribonucleoprotein (RNP) particles. Second, they are translationally repressed in the unlocalized form. Third, localized mRNAs are protected from premature degradation or decay until they fulfill their function/s (Gavis and Lehmann 1994; Kim-Ha et al. 1995; Markussen et al. 1995; Rongo et al. 1995; Smibert et al. 1996; Rangan et al. 2009). Despite these similarities, molecular analyses and live cell imaging revealed that various constituents of RNP particles are targeted to the oocyte posterior pole by distinct mechanisms. Four localization scenarios have been reported so far: (1) active transport along microtubules, as for *oskar* RNP particles

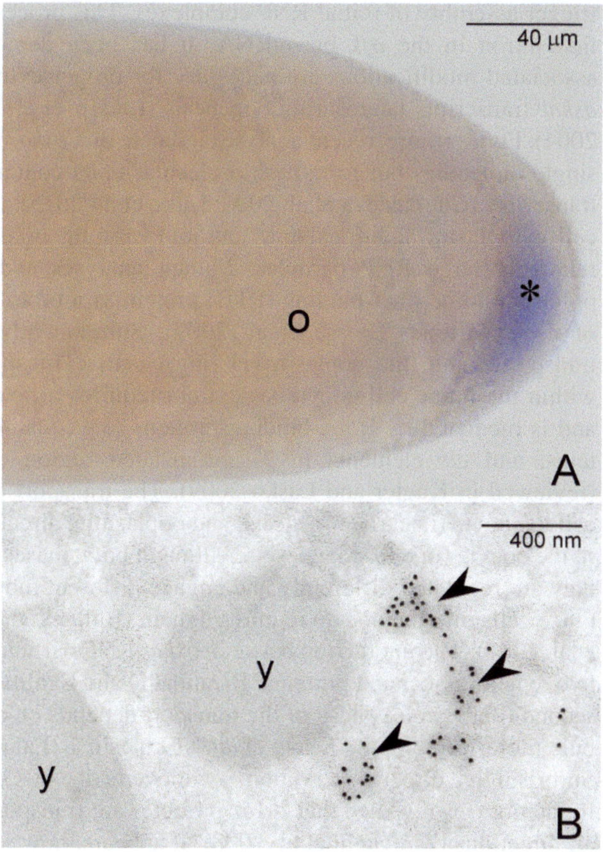

Fig. 5.2 Pole plasm in *Drosophila melanogaster*. (**a**) Fragment of the oocyte (o) with distinct pole plasm (*asterisk*) labeled by anti-Vasa antibody. Whole-mount preparation. (**b**) Pole plasm containing polar granule precursors (*arrowheads*) labeled by anti-Vasa antibody. *y* Yolk sphere. Immunogold EM

(Clark et al. 2007; Zimyanin et al. 2008); (2) localized translation within the posterior ooplasm and repression elsewhere, as for Osk (Kim-Ha et al. 1995); (3) passive diffusion within ooplasm and local entrapment at the posterior region, as for *cyclin B*, *germ cell-less*, and *nanos* mRNAs (Forrest and Gavis 2003); and (4) degradation of mRNA in the ooplasm except for the posterior pole, as for *nanos* mRNA (Zaessinger et al. 2006).

5.3.2 *oskar Is a Critical Player in Drosophila Pole Plasm Formation*

Among the factors targeted to the oocyte posterior pole, the transport and localization of *oskar* mRNA are probably best characterized. *oskar* is transcribed in the nurse cell nuclei during early stages of oogenesis. In the nucleoplasm, *cis*-regulatory elements in *oskar* transcripts are recognized by *trans*-acting proteins, which trigger assembly of initial RNP complexes. This process starts with splicing of the first intron in the *osk* pre-mRNA. It has been demonstrated that the splicing-associated modifications are necessary for proper cytoplasmic localization of the *oskar* transcripts later during oogenesis (Le Hir et al. 2001; Hachet and Ephrussi 2004). Furthermore, recent analyses indicate that *oskar* mRNAs are not regulated as single molecules but form multimolecular units containing many copies of *oskar* transcripts (Chekulaeva et al. 2006; Little et al. 2015). Following transport to nurse cell cytoplasm, additional proteins join existing *oskar* RNP complexes forming relatively large RNP particles. Mutant analyses and in vitro assays identified polypyrimidine tract-binding (PTB) protein as a promoting factor in the assembly of these particles (Besse et al. 2009). Subsequently, *oskar* RNP particles are transported into the ooplasm via ring canals. Transport of *oskar* RNP particles within the nurse cell-oocyte syncytium requires a polarized cytoskeletal network and is mediated by RNA-binding proteins (e.g., Staufen), appropriate motor proteins, and *cis* elements in 3′ untranslated region (3′UTR) of *oskar* mRNAs (reviewed in Kugler and Lasko 2009). The transport occurs in two phases: nurse cell to oocyte (in early oogenesis) and ooplasmic, directed toward the posterior pole of the oocyte (in mid-oogenesis). Although both phases are microtubule-dependent, they are regulated differently and engage different motor proteins. Transport from nurse cells into the oocyte requires dynein (Bullock and Ish-Horowicz 2001; Clark et al. 2007). The interaction between *oskar* RNP particles and dynein is mediated by two dynein-associated proteins: Bicaudal D and Egalitarian (Clark et al. 2007). The second (ooplasmic) phase of the transport depends on a plus end-directed microtubule motor protein, the heavy chain of kinesin 1 (Palacios and St Johnston 2002). Surprisingly, direct observations of movements of GFP-tagged *oskar* mRNA in living oocytes revealed that *oskar* particles are transported (along microtubules) in all directions (Zimyanin et al. 2008). It appears, however, that a slight bias toward the posterior oocyte pole is sufficient for the adequate posterior accumulation of *oskar* transcripts.

Previous studies suggested that the pole plasm components become anchored in the posterior ooplasm by an actin-based mechanism. More recent high-resolution live imaging revealed, however, rather unexpectedly, that the pole plasm anchoring is dynamic and requires constant trafficking of RNP particles at the posterior oocyte cortex (Sinsimer et al. 2013). According to a current model, motility of RNP particles depends on cortical microtubules, kinesin and dynein motor proteins, as well as interplay between the microtubular and actin cytoskeletons.

During transport, *oskar* mRNA is translationally repressed by the RNA-binding proteins. One such protein is Bruno. This *trans*-acting factor binds to dedicated sequences, called Bruno response elements, in 3′UTR of *oskar* mRNA (Kim-Ha et al. 1995; Chekulaeva et al. 2006). According to current hypotheses, Bruno interacts with Cup protein, which in turn binds to and inactivates the eukaryotic translation initiation factor-4E (Nakamura et al. 2004; Chekulaeva et al. 2006; Besse et al. 2009). It has also been proposed that formation of densely packed RNP particles in which *oskar* mRNAs become inaccessible to the translational machinery is also required for translational repression (Besse et al. 2009).

Once *oskar* mRNAs reach the posterior ooplasm and are properly localized, the binding of activators, such as Orb, releases the repression and translation starts (Chang et al. 1999; Kim et al. 2015). However, this process is not straightforward. Alternative translation initiation from two in-frame start codons in *oskar* mRNA generates two isoforms of Oskar protein: Short and Long, each with distinct properties (Markussen et al. 1995; Breitwieser et al. 1996). While the Long Oskar is responsible for anchoring *oskar* mRNA and Short Oskar isoform within the posterior cortical region of the oocyte, the more abundant Short Oskar interacts with several germline-associated mRNAs (e.g., *nanos*, *polar granule component*, and *germ cell-less*) and proteins (e.g., Vasa) and is necessary for nucleating the pole plasm as well as the proper patterning of the posterior region of the embryo (Markussen et al. 1995; Breitwieser et al. 1996; Vanzo and Ephrussi 2002). After fulfilling its functions, Oskar is phosphorylated by two kinases (Par-1 and GSK-3/Shaggy), and subsequently Short Oskar is targeted for proteosomal-dependent degradation by SCF(-Slimb) ubiquitin ligase (Morais-de-Sá et al. 2013). Apparently, the same pathway is used to remove mislocalized Oskar.

5.3.3 *The Assembly of the* Drosophila *Pole Plasm Occurs in a Stepwise Sequence*

Formation of pole plasm is initiated by *oskar* mRNA trafficking to the oocyte posterior cytoplasm during mid-oogenesis, followed by synthesis of Oskar protein. Expression of *oskar* in ectopic locations has indicated that it can attract other factors (mRNAs and proteins) to new locations and organize a functional pole plasm capable of inducing pole cell formation (Ephrussi and Lehmann 1992; Smith et al. 1992). Moreover, overexpression of *oskar* showed that its level of expression directly influences the size of the pole plasm and positively correlates with the number of PGCs (Ephrussi and Lehmann 1992; Smith et al. 1992). The mechanism by which Oskar recruits other pole plasm components was until recently unknown. However, two recent analyses of the high-resolution crystal structure of *Drosophila* Oskar have revealed that the protein contains two domains: C-terminal OSK domain and N-terminal LOTUS domain. The first domain binds RNA, whereas the latter directly interacts with Vasa, a DEAD-box RNA helicase (Jeske et al. 2015; Yang et al. 2015). Because the OSK domain binds the 3′UTRs of *oskar* and *nanos* mRNA in vitro, it has been suggested that Oskar may regulate certain aspects of RNA metabolism, e.g., stability, translation, and/or localization, through direct interaction with 3′UTRs (Yang et al. 2015).

The initial formation of the pole plasm domain is followed by next wave of localization of specific mRNAs (including *oskar*, *nanos*, *cyclin B*, and *polar granule component*) to the posterior ooplasm during advanced stages of oogenesis (Dalby and Glover 1992; Nakamura et al. 1996; Forrest and Gavis 2003; Sinsimer et al. 2011). Until recently, the dynamics of pole plasm formation have remained a mystery. However, quantitative single-molecule imaging demonstrated that *nanos*, *cyclin B*, and *polar granule component* transcripts are initially translocated as RNP particles containing single-mRNA molecules but merge into larger heterogeneous granules specifically at the posterior pole of the oocyte (Little et al. 2015). This mechanism effectively separates the germ cell-destined transcripts from others and ensures their proper segregation into the forming pole cells.

Aubergine is another important factor in the *Drosophila* germline development. It plays a role in early oogenesis by regulating *oskar* RNA translation, but it has also been implicated in pole cell specification, consistent with its presence in the pole plasm (Wilson et al. 1996; Harris and Macdonald 2001; Becalska et al. 2011). Recently, the results of cross-linking immunoprecipitation experiments have indicated that Aubergine acts as adhesive trap mediating retention of germ plasm-specific mRNAs within the posterior ooplasm (Vourekas et al. 2016). Aubergine is a member of the evolutionarily conserved Piwi family proteins, which associate with piwi-interacting RNAs (piRNAs) and facilitate silencing of transposomes in the germline (reviewed in Mani and Juliano 2013). It has therefore been suggested that Aubergine association with mRNAs may be mediated by low-specificity pairing between the piRNAs and the target mRNA. In a recently proposed model, Aubergine may facilitate packaging of germ cell-destined mRNAs into polar granules (see below). However, it is still not entirely clear how the selection of these mRNAs may occur. It is worth adding that Aubergine also functions in early embryonic development (Barckmann et al. 2015).

5.3.4 Polar (Germ) Granules

Ultrastructural studies revealed that the *Drosophila* pole plasm contains distinct large (up to 500 nm in diameter) roughly spherical organelles called polar or germ granules (for review, see Mahowald 2001). They are composed of electron-dense granulo-fibrillar material, which is not surrounded by membrane. Another characteristic feature of *Drosophila* polar granules is that they are frequently accompanied by mitochondria and, in activated eggs, by polysomes. Following fertilization, polar granules are segregated to the forming pole cells (Lerit and Gavis 2011).

Biochemical and molecular analyses established that the germ granules are aggregates of RNP particles (reviewed in Voronina et al. 2011; Gao and Arkov 2013; Lehmann 2016). Their constituents are synthesized during oogenesis and accumulate in the pole plasm of the oocytes/eggs and early embryos. Localization of various RNAs and proteins to the germ granules coincides with nurse cell dumping and the cytoplasm streaming within the oocyte. Although a list of polar granule components is still not complete, it is well documented that these organelles contain proteins such as Oskar, Vasa, Tudor, and Aubergine (Fig. 5.2b) and a number of distinct RNAs, both coding and noncoding (Hay et al. 1988, 1990; Lasko and Ashburner 1990; Dalby and Glover 1992; Nakamura et al. 1996; Forrest and Gavis 2003; Thomson and Lasko 2004; Jones and Macdonald 2007; Anne 2010;

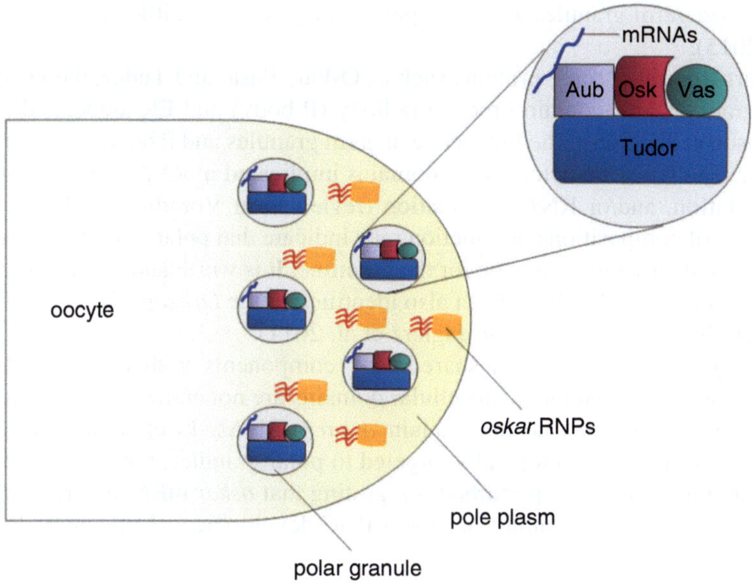

Fig. 5.3 Schematic representation of the distribution of molecular constituents in the pole plasm of *Drosophila melanogaster*. *Aub* Aubergine, *Osk* Oskar, *RNPs* ribonucleoprotein particles, *Vas* Vasa. See text for further details

Sinsimer et al. 2011). It has been also established that a critical step in polar granule formation is Oskar's direct interaction with Vasa followed by recruitment of Tudor, Piwi, and Aubergine (Breitwieser et al. 1996; Harris and Macdonald 2001; Arkov et al. 2006; Thomson et al. 2008; Kirino et al. 2009, 2010; Nishida et al. 2009; Anne 2010; Fig. 5.3). It has been suggested that Tudor serves as a scaffold for germ granule assembly, and its interaction with other germ granule components is mediated by 11 specific protein motifs, called Tudor domains. These domains bind to proteins that contain methylated arginine and lysine residues (Thomson and Lasko 2005; Arkov et al. 2006; Kirino et al. 2009, 2010; Arkov and Ramos 2010; Liu et al. 2010).

The inner architecture of the polar granules has been recently unveiled using super-resolution single-mRNA in situ fluorescence hybridization combined with structured illumination microscopy. This study revealed that the *Drosophila* polar granules are highly ordered entities. While proteins (Oskar, Vasa, Aubergine, and Tudor) are distributed evenly throughout the granules, mRNA molecules (*nanos*, *polar granule component*, and *germ cell-less*) form homotypic clusters and specifically concentrate either in the center or periphery of the granules (Trcek et al. 2015). Interestingly, such a spatial distribution is maintained during embryogenesis, indicating its functional significance. In addition, recent high-resolution analyses also indicate, contrary to previous speculations, that the

Drosophila germ granules are heterogeneous aggregates (Little et al. 2015; Trcek et al. 2015).

Apart from the "core" proteins such as Oskar, Vasa, and Tudor, the *Drosophila* polar granules also contain processing body (P body) and ER-associated proteins (Thomson et al. 2008). The link between germ granules and P bodies is of particular interest. The latter are cytoplasmic domains implicated in RNA storage, repression of translation, and/or RNA degradation (reviewed in Voronina et al. 2011). The similarity of composition and function may indicate that polar granules are in fact P body-related structures, specific for the germline. It is worth adding, in this context, that canonical P bodies have been also identified in the *Drosophila* nurse cells (Liu and Gall 2007; Lim et al. 2009; Jaglarz et al. 2011).

Drosophila polar granules share many components with the pole plasm. It appears, however, that these subcellular domains are not entirely equivalent. Significantly, polar granules do not contain *oskar* mRNAs. In experiments in which *oskar* transcripts were ectopically targeted to polar granules, both Vasa expression and PGC formation were perturbed, suggesting that *oskar* mRNAs separation from the polar granules is essential for proper PGC development (Little et al. 2015).

5.4 Pole Plasm in "Non-*Drosophila*" Holometabolous Insects

Drosophila-like modes of germline specification have been found in several representatives of most major lineages of holometabolous insects including Diptera (Zissler and Sander 1982; Gutzeit 1985), Mecoptera (Ando 1973, this chapter), Coleoptera (Lynch et al. 2011), Neuroptera (this chapter), Megaloptera (Suzuki et al. 1981), Lepidoptera (Berg and Gassner 1978), and Hymenoptera (Meng 1968; Zissler and Sander 1982; Bilinski 1991; Klag and Bilinski 1993; Lynch et al. 2011). Genetic analyses have indicated that in certain dipterans (*Anopheles*, *Aedes*, and *Culex*, Juhn and James 2006; Juhn et al. 2008), coleopterans (*Callosobruchus*, Benton et al. 2016), and hymenopterans (*Nasonia*, Lynch et al. 2011), the induction of the pole plasm (and pole cells) relies, as in *Drosophila*, on *oskar* orthologs. It has been additionally shown that regulatory interactions upstream and downstream of *Nasonia oskar* are largely conserved with *Drosophila*, suggesting that these two regulatory networks are homologous and have a common origin (Lynch et al. 2011; Quan and Lynch 2016). On the other hand, the absence of *oskar* from the genomes of such insects as *Bombyx mori* (Xia et al. 2004), *Apis mellifera* (Wienstock et al. 2006), and *Tribolium castaneum* (Richards et al. 2006) clearly coincides with the lack of the pole plasm and pole cells in these species. All these results led to the suggestion that *oskar* represents a key factor in the evolution of the cytoplasmic inheritance in Holometabola (Lynch et al. 2011; Quan and Lynch 2016). It has also been speculated that *oskar* originated at the base of Holometabola and that this gene has been subsequently lost several times in independent holometabolous lineages (Lynch et al. 2011; Quan and Lynch 2016). Finally, Ewen-Campen and colleagues (2012) have demonstrated that *oskar* first arose much earlier, in the common ancestor of Holometabola and Orthoptera, and that its well-known role in germline specification is a derived condition of holometabolans (Ewen-Campen et al. 2012). To this evolutionary scenario, we may add one remark: the secondary loss of the cytoplasmic inheritance is apparently related not only to the loss of *oskar* gene per se but also to the disappearance of the nurse cells, where *oskar* is, as a rule, transcribed. Such a situation most probably happened during the evolution of groups characterized by the neopanoistic ovaries, such as Megaloptera: Corydalidae (dobsonflies) (Fig. 5.4c) and Siphonaptera (fleas) (Simiczyjew and Margas 2001) (Fig. 5.1).

Morphologically, the pole plasm of holometabolans is similar to that of *Drosophila*. It occupies the posterior oocyte pole and contains characteristic particles (termed germ or polar granules) or electron-dense accumulations not organized in particles, termed nuage (Zissler and Sander 1982). Molecular composition and functions of the nuage are discussed in Jaglarz et al. (2011), Voronina et al. (2011), and Kloc et al. (2014). Usually, the pole plasm is clearly recognizable even at the level of a light microscope, as it is virtually devoid of any reserve materials, i.e., yolk spheres and lipid droplets (Fig. 5.4a, b, d, e). Subsequent stages of the morphogenesis of the pole plasm and polar granules have been described only in some representatives of Diptera (e.g., *Drosophila*, *Miastor*) and Hymenoptera (e.g., *Cosmoconus*, *Lissonota*, *Pimpla*) (Meng 1968; Mahowald and Stoiber 1974; Klag and Bilinski 1993; Mahowald 2001). In this context, we have analyzed formation of the polar plasm in the scorpionfly, *Panorpa communis*, a representative of the old and plesiomorphic insect order Mecoptera.

The ovaries of *Panorpa* are meroistic polytrophic, and each ovarian follicle consists of an oocyte and three nurse cells (Ramamurty 1964; Bilinski et al. 1998; Ando 1973). At the onset of previtellogenesis, numerous roughly spherical nuage aggregations arise in the cytoplasm of the nurse cells (Bilinski et al. 1998). As oogenesis progresses, these aggregations are transferred to the ooplasm (Fig. 5.5a, asterisks) and accumulate next to the posterior oocyte pole forming a disk-shaped pole plasm (oosome) (Ando 1973; Fig. 5.5b, arrows). The latter contains irregular accumulations of nuage material (Fig. 5.5c, arrowheads) that do not constitute defined particles or polar granules. Above results and previously published data suggest that in all holometabolans, constituents of the polar plasm/polar granules (germline determinants) are synthesized in the cytoplasm of the nurse cells. Subsequently, these macromolecules are transported to the oocyte cytoplasm (ooplasm) and are localized to the posterior pole.

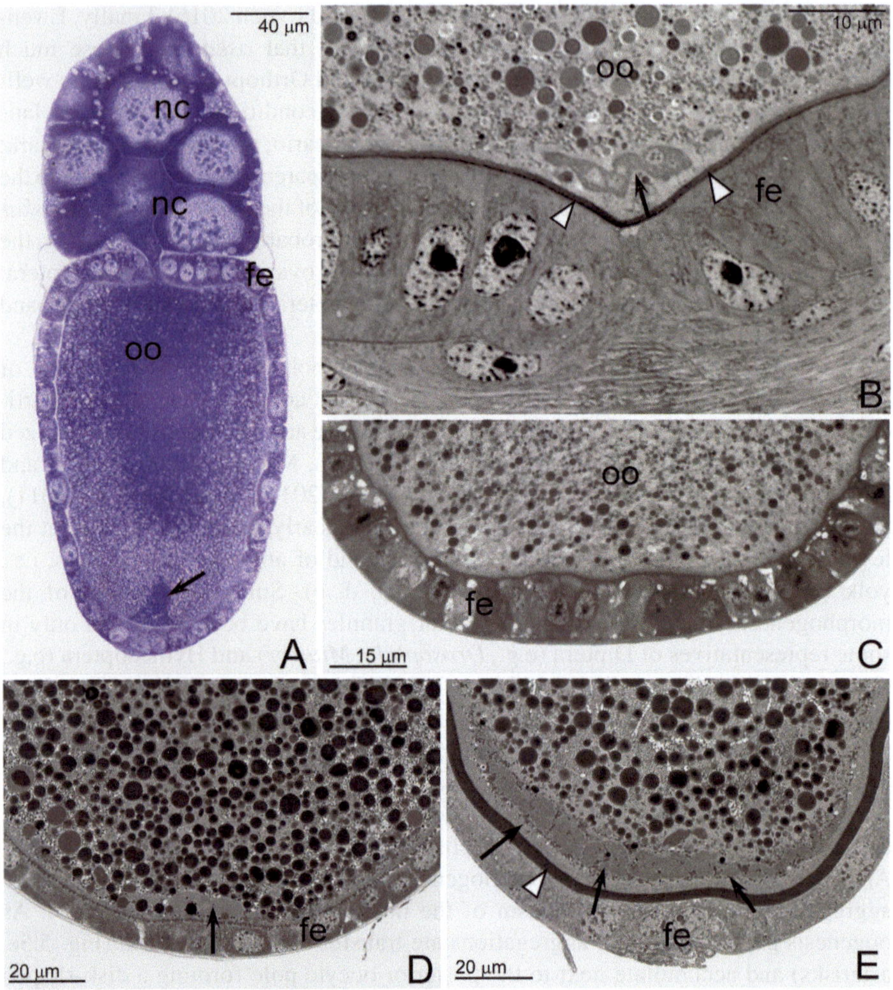

Fig. 5.4 The pole plasm (or its absence) in the oocytes of holometabolous insects. (**a**) Hymenoptera, *Cosmoconus meridionator* (early vitellogenesis); (**b**) Neuroptera, *Osmylus fulvicephalus* (postvitellogenesis; courtesy of Prof. Janusz Kubrakiewicz, Institute of Experimental Biology, Wroclaw University); (**c**) Megaloptera, Corydalidae, *Corydalus peruvianus* (early vitellogenesis); (**d**) Diptera, *Rhagio* sp. (early vitellogenesis); (**e**) Diptera, *Rhagio* sp. (postvitellogenesis). The pole plasm (*arrows*) is clearly visible at the posterior pole. Note lack of the pole plasm in *Corydalus* oocyte (**c**). *fe* Follicular epithelium, *nc* nurse cell, *oo* oocyte, egg envelope (*open arrowheads*). (**a**) semithin section stained with toluidine blue; (**b–e**) semithin sections stained with methylene blue

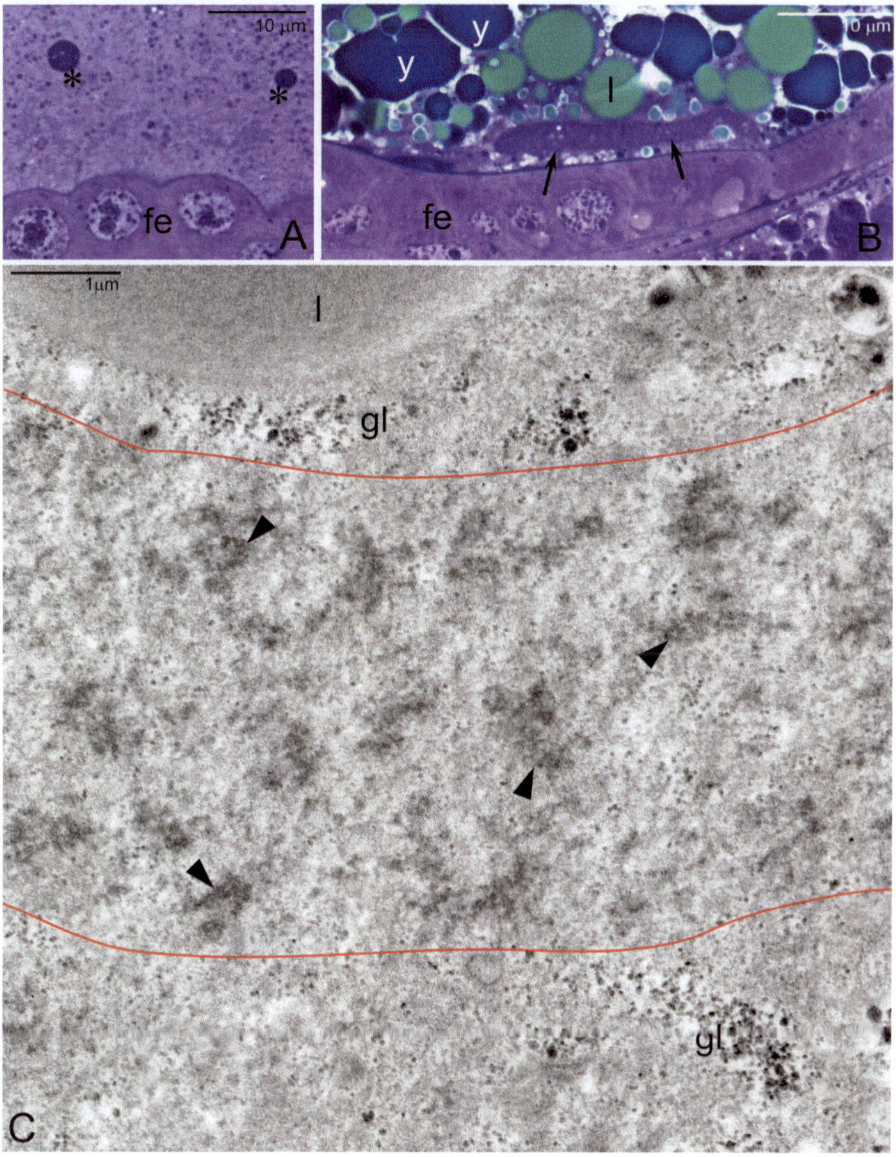

Fig. 5.5 The pole plasm in oocytes of *Panorpa communis* (Mecoptera). (**a**) previtellogenic oocyte; note spherical aggregations of nuage material (asterisks). (**b**) posterior pole of late vitellogenic oocyte; note the pole plasm (*arrows*). (**c**) posterior pole of late vitellogenic oocyte, TEM. The pole plasm (delimited with *red lines*) contains several accumulations of nuage material (*arrowheads*). *fe* Follicular epithelium, *gl* glycogen particles, *l* lipid droplets, *y* yolk granules. (**a**, **b**) semithin sections stained with methylene blue; (**c**) ultrathin section

5.5 Pole Plasm in Paraneoptera

The pole plasm and/or pole cells have been described only in three paraneopteran groups: Psocoptera (book lice), Thysanoptera (thrips), and Aphidoidea (aphids). As the processes found in these taxa are apparently different, we will discuss them separately.

5.5.1 Psocoptera

In *Liposcelis divergens*, the posteriorly located pole cells are specified relatively early, after formation of cellular blastoderm. Interestingly, the "preformed" pole plasm and polar granules have not been identified in fertilized eggs of this species (Goss 1952). Apparently, further studies or even reinvestigation of germline specification in this insect taxon is badly needed.

5.5.2 Thysanoptera

Classical histological studies have revealed that the pole plasm and polar granules are present at the posterior end of an unfertilized egg of *Haplothrips verbasci* (Heming 1979). This is quite astonishing because all thysanopterans are characterized by neopanoistic ovaries that most probably evolved by a secondary loss of nurse cells (Stys and Bilinski 1990; Büning 1993, 1994) (Fig. 5.1). In this context, we have followed the formation of the pole plasm in another representative of Thysanoptera, *Taeniothrips* sp., and found that the putative pole plasm of this species arises at the onset of vitellogenesis (Fig. 5.6a) and is clearly recognizable as oogenesis progresses (Fig. 5.6b). EM analysis has additionally indicated that the pole plasm is electron-transparent and does not contain either polar granules or electron-dense nuage-like material (not shown). The latter observation implies that germline specification should be reinvestigated also in this insect lineage.

5.5.3 Aphidoidea

Recent molecular analyses have allowed for identification of a preformed pole plasm in the pea aphid, *Acyrthosiphon pisum* (Lin et al. 2014). It contains protein ApVas1 encoded by an ortholog of *Drosophila* gene *vasa*. On the other hand, *oskar* (or its ortholog) is not present in the pea aphid genome, indicating that the aphid germ plasm evolved independently and is *not* homologous to the germ plasm of holometabolans (Lin et al. 2014).

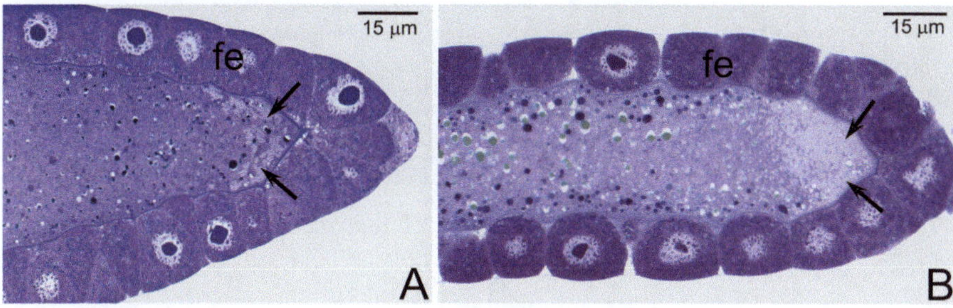

Fig. 5.6 The pseudo-pole plasm in oocytes of *Taeniothrips* sp. (Thysanoptera). (**a**) early vitellogenesis; (**b**) late vitellogenesis. Note that pseudo-pole plasm (*arrows*) is apparently more "transparent" than the rest of the ooplasm. *fe* Follicular epithelium. Semithin sections stained with methylene blue

5.6 Putative Pole Plasm in Dermaptera

Although dermapterans are one of the basally branched insect lineages, they have simple polytrophic ovarioles in which ovarian follicles consist of two cells only: an oocyte and a single nurse cell (Tworzydlo and Bilinski 2008; Tworzydlo et al. 2010). Therefore, the question arises: whether (or not) developing dermapteran oocytes/egg cells contain the pole plasm? To answer this question, we have analyzed advanced stages of oogenesis in three dermapteran species (*Forficula auricularia*, *Chaetospania borneensis*, and *Opisthocosmia silvestris*). To our surprise, the posterior oocyte pole of all three species is occupied by a morphologically recognizable yolk-free cytoplasm (Fig. 5.7). Initially, this cytoplasm region has a dome-like shape (Fig. 5.7a); later it flattens and firmly adheres to plasma membrane surrounding the posterior pole (Fig. 5.7b). At the EM level, this putative pole plasm contains elongated mitochondria and numerous vesicles of the rough endoplasmic reticulum (RER vesicles) filled with fine granular material (Fig. 5.7c). Morphologically identical RER vesicles have also been found in the cytoplasm of the nurse cells (Fig. 5.7d). Histochemical tests have shown that the RER vesicles (both present in the nurse cells and in putative pole plasm) are AgNOR positive (Fig. 5.7d, e), suggesting that the RER vesicles are formed in the nurse cell cytoplasm and (subsequently) are transported to the posterior oocyte pole. The role of the dermapteran (putative) pole plasm and its RER vesicles remains elusive; its implication in the germline specification will require molecular analyses. In this context, we should stress only that the RER vesicles found in dermapteran oocytes are morphologically clearly different from bona fide polar granules, as the latter are never membrane-bound.

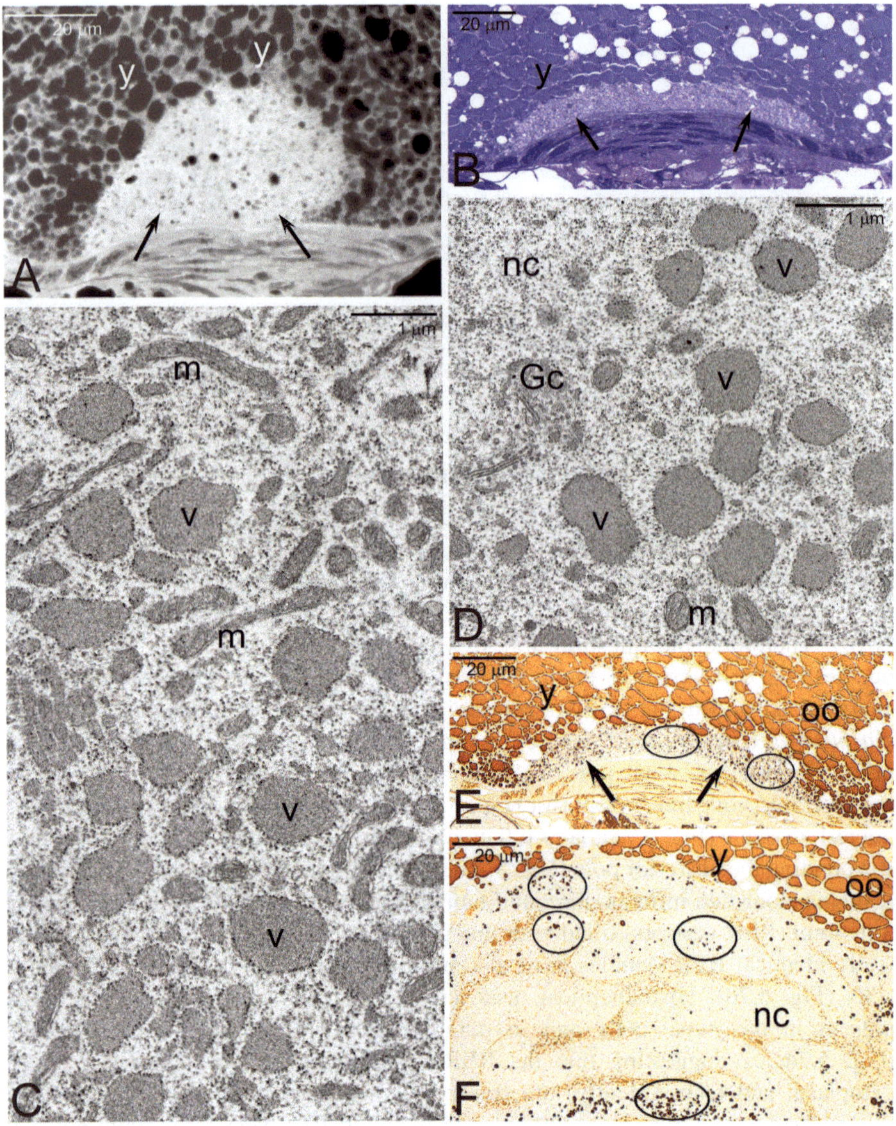

Fig. 5.7 The pseudo-pole plasm in dermapteran oocytes. (**a**) *Chaetospania borneensis* (early vitellogenesis); (**b–d**) *Opisthocosmia silvestris* (late vitellogenesis); (**e, f**) *Chaetospania borneensis* (late vitellogenesis). The pseudo-pole plasm (*arrows*) is devoid of reserve materials and contains RER vesicles filled with fine granular material (v in **c**). Similar vesicles are also present in the nurse cell cytoplasm (v in **d**). The vesicles are strongly AgNOR positive (**e, f**—*encircled*). *Gc* Golgi complex, *m* mitochondria, *nc* nurse cell, *oo* oocyte, *y* yolk spheres. (**a**) semithin histocryl section stained with propidium iodide; (**b**) semithin histocryl section stained with methylene blue; (**c, d**) ultrathin sections; (**e, f**) semithin histocryl sections stained with AgNOR technique

5.7 Conclusions

Despite relatively numerous studies, evolution of germline specification among insects is still debatable. There are at least two alternatives of phylogenetic interpretation of accumulated data.

1. The pole plasm and pole cells arose relatively early—at the base of Dermaptera and Phalloneoptera (Paraneoptera + Holometabola), i.e., simultaneously with the appearance of the nurse cells (Fig. 5.1, black box) or in the common ancestor of Paraneoptera + Holometabola (Fig. 5.1, diamond). Subsequently, the cytoplasmic inheritance has been lost independently in several insect lineages and replaced by the zygotic induction. Interestingly, all data clearly indicate that the putative loses have been much more frequent among Paraneoptera than within Holometabola. Such a situation is not surprising, as reversals to the pseudoprimitive condition (i.e., cytoplasmic inheritance → zygotic induction) are more likely to occur from a less advanced stage than from the more advanced one.
2. The pole plasm evolved much later—in the common ancestor of all holometabolans (Fig. 5.1, triangle). The morphologically recognizable "pole plasm" found in some paraneopteran groups, and in dermapterans, evolved independently.

In our opinion, accumulated data speak in favor of the second hypothesis. The following arguments are in support of this view:

- Although the gene *oskar* evolved as early as in the common ancestor of Orthoptera and Phalloneoptera, it gained its new germline function much later—at the base of Holometabola (Ewen-Campen et al. 2012).
- The germ plasm of aphids is not homologous to that of holometabolans; its specification does not rely on an *oskar* ortholog (Lin et al. 2014).
- The putative pole plasm of such non-holometabolous insects as Thysanoptera and Dermaptera does not contain either nuage accumulations or bona fide germinal granules; therefore, it cannot be interpreted as genuine pole plasm. For such morphologically distinct region of the oocyte, we propose tentatively the term "pseudo-pole plasm."

Acknowledgments We are grateful to Dr. Bozena Simiczyjew (Institute of Experimental Biology, Wroclaw University) for providing fixed *Panorpa* ovarian follicles. We also thank Ms. E. Kisiel and A. Jankowska for the technical assistance.

References

Ando H (1973) Old oocytes and newly laid eggs of scorpion-flies and hanging-flies (Mecoptera: Panorpidae and Bittacidae). Sci Rep Tokyo Kyoiku Daigaku Ser B 15:163–187

Anne J (2010) Targeting and anchoring Tudor in the pole plasm of the *Drosophila* oocyte. PLoS One 5(12):e14362

Arkov AL, Ramos A (2010) Building RNA-protein granules: insight from the germline. Trends Cell Biol 20:482–490

Arkov AL, Wang JY, Ramos A, Lehmann R (2006) The role of Tudor domains in germline development and polar granule architecture. Development 133:4053–4062

Barckmann B, Pierson S, Dufourt J, Papin C, Armenise C, Port F, Grentzinger T, Chambeyron S, Baronian G, Desvignes JP, Curk T, Simonelig M (2015) Aubergine iCLIP reveals piRNA-dependent decay of mRNAs involved in germ cell development in the early embryo. Cell Rep 12:1205–1216

Bastock R, St Johnston D (2008) *Drosophila* oogenesis. Curr Biol 18:1082–1087

Becalska AN, Gavis ER (2009) Lighting up mRNA localization in *Drosophila* oogenesis. Development 136:2493–2503

Becalska AN, Kim YR, Belletier NG, Lerit DA, Sinsimer KS, Gavis ER (2011) Aubergine is a component of a nanos mRNA localization complex. Dev Biol 349:46–52

Benton MA, Kenny NJ, Conrads KH, Roth S, Lynch JA (2016) Deep, staged transcriptomic resources for the novel coleopteran models *Atrachya menetriesi* and *Callosobruchus maculatus*. bioRxiv. doi:10.1101/035998

Berg GJ, Gassner G (1978) Fine structure of the blastoderm embryo in the pink bollworm, *Pectinophora gossypiella* (Saunders) (Lepidoptera: Gelechiidae). Int J Insect Morphol Embryol 7:81–105

Besse F, Lopez de Quinto S, Marchand V, Trucco A, Ephrussi A (2009) *Drosophila* PTB promotes formation of high-order RNP particles and represses oskar translation. Genes Dev 23:195–207

Bilinski SM (1991) Morphological markers of anteroposterior and dorsoventral polarity in developing oocytes of the hymenopteran *Cosmoconus meridionator* (Ichneumonidae). Roux Arch Dev Biol 200:330–335

Bilinski SM (1998) Introductory remarks. Folia Histochem Cytobiol 36:143–145

Bilinski SM, Büning J, Simiczyjew B (1998) The ovaries of Mecoptera: basic similarities and one exception to the rule. Folia Histochem Cytobiol 36:189–195

Brandt A (1874) Ueber die Eiroehren der *Blatta orientalis (Periplaneta)*. Mem Acad Imp Sci 21: 1–30

Breitwieser W, Markussen FH, Horstmann H, Ephrussi A (1996) Oskar protein interaction with Vasa represents an essential step in polar granule assembly. Genes Dev 10:2179–2188

Bullock SL, Ish-Horowicz D (2001) Conserved signals and machinery for RNA transport in *Drosophila* oogenesis and embryogenesis. Nature 414:611–616

Büning J (1993) Germ cell cluster formation in insect oocytes. Int J Insect Morphol Embryol 22: 237–253

Büning J (1994) The insect ovary. Ultrastructure, previtellogenic growth and evolution. Chapman & Hall, London

Cavaliere V, Taddei C, Gargiulo G (1998) Apoptosis of nurse cells at the late stages of oogenesis of *Drosophila melanogaster*. Dev Genes Evol 208:106–112

Chang JS, Tan L, Schedl P (1999) The *Drosophila* CPEB homolog, orb, is required for oskar protein expression in oocytes. Dev Biol 215:91–106

Chekulaeva M, Hentze MW, Ephrussi A (2006) Bruno acts as a dual repressor of *oskar* translation, promoting mRNA oligomerization and formation of silencing particles. Cell 124:521–533

Clark A, Meignin C, Davis I (2007) A dynein-dependent shortcut rapidly delivers axis determination transcripts into the *Drosophila* oocyte. Development 134:1955–1965

Dalby B, Glover DM (1992) 3′non-translated sequences in *Drosophila* cyclin B transcripts direct posterior pole accumulation late in oogenesis and peri-nuclear association in syncytial embryos. Development 115:989–997
de Cuevas M, Lilly M, Spradling AC (1997) Germline cyst formation in *Drosophila*. Annu Rev Genet 31:405–428
Ephrussi A, Lehmann R (1992) Induction of germ cell formation by *oskar*. Nature 358:387–392
Ewen-Campen B, Srouji JR, Schwager EE, Extavour CG (2012) *oskar* predates the evolution of germ plasm in insects. Curr Biol 22:2278–2283
Ewen-Campen B, Donoughe S, Clarke DN, Extavour CG (2013) Germ cell specification requires mechanisms rather than germ plasm in a basally branching insect. Curr Biol 23:835–842
Extavour CG (2007) Evolution of the bilaterian germ line: lineage origin and modulation of specification mechanisms. Integr Comp Biol 47:770–785
Extavour CG, Akam M (2003) Mechanisms of germ cell specification across the metazoans: epigenesis and preformation. Development 130:5869–5884
Forrest KM, Gavis ER (2003) Live imaging of endogenous RNA reveals a diffusion and entrapment mechanism for *nanos* mRNA localization in *Drosophila*. Curr Biol 13:1159–1168
Frise E, Hammonds AS, Celniker SE (2010) Systematic image-driven analysis of the spatial *Drosophila* embryonic expression landscape. Mol Syst Biol 6:345
Gao M, Arkov AL (2013) Next generation organelles: structure and role of germ granules in the germline. Mol Reprod Dev 80:610–623
Gavis ER, Lehmann R (1994) Translational regulation of *nanos* by RNA localization. Nature 369: 315–318
Goss RJ (1952) The early embryology of the book louse, *Liposcelis divergens* Badonnel (Psocoptera; Liposcelidae). J Morphol 91:135–167
Gottanka J, Büning J (1990) Oocytes develop from interconnected cystocytes in the panoistic ovaries of *Nemoura* sp. (Pictet) (Plecoptera: Nemouridae). Int J Insect Morphol Embryol 19:219–225
Gutzeit HO (1985) Oosome formation during *in vitro* oogenesis in *Bradysia tritici* (syn. *Sciara ocellaris*). Roux Arch Dev Biol 195:173–181
Hachet O, Ephrussi A (2004) Splicing of *oskar* RNA in the nucleus is coupled to its cytoplasmic localization. Nature 428:959–963
Harris AN, Macdonald PM (2001) Aubergine encodes a *Drosophila* polar granule component required for pole cell formation and related to eIF2C. Development 128:2823–2832
Hay B, Jan LY, Jan YN (1988) A protein component of *Drosophila* polar granules is encoded by *vasa* and has extensive sequence similarity to ATP-dependent helicases. Cell 55:577–587
Hay B, Jan LY, Jan YN (1990) Localization of Vasa, a component of *Drosophila* polar granules, in maternal-effect mutants that alter embryonic anteroposterior polarity. Development 109: 425–433
Heming BS (1979) Origin and fate of germ cells in male and female embryos of *Haplothrips verbasci* (Osborn) (Insecta, Thysanoptera, Phlaeothripidae). J Morphol 160:323–344
Henning W (1981) Insect phylogeny. Wiley, Chichester
Jaglarz MK, Kloc M, Jankowska W, Szymanska B, Bilinski SM (2011) Nuage morphogenesis becomes more complex: two translocation pathways and two forms of nuage coexist in *Drosophila* germline syncytia. Cell Tissue Res 344:169–181
Jambor H, Surendranath V, Kalinka AT, Mejstrik P, Saalfeld S, Tomancak P (2015) Systematic imaging reveals features and changing localization of mRNAs in *Drosophila* development. eLife 4:e05003
Jeske M, Bordi M, Glatt S, Müller S, Rybin V, Müller CW, Ephrussi A (2015) The crystal structure of the *Drosophila* germline inducer Oskar identifies two domains with distinct Vasa helicase- and RNA-binding activities. Cell Rep 12:587–598
Jones JR, Macdonald PM (2007) Oskar controls morphology of polar granules and nuclear bodies in *Drosophila*. Development 134:233–236

Juhn J, James AA (2006) oskar gene expression in the vector mosquitoes *Anopheles gambiae* and *Aedes aegypti*. Insect Mol Biol 15:363–372
Juhn J, Marinotti O, Calvo E, James AA (2008) Gene structure and expression of nanos (nos) and oskar (osk) orthologues of the vector mosquito, *Culex quinquefasciatus*. Insect Mol Biol 17: 545–552
Kim G, Pai C-I, Sato K, Person MD, Nakamura A, Macdonald PM (2015) Region-specific activation of *oskar* mRNA translation by inhibition of Bruno-mediated repression. PLoS Genet 11:e1004992
Kim-Ha J, Kerr K, Macdonald PM (1995) Translational regulation of *oskar* mRNA by Bruno, an ovarian RNA-binding protein, is essential. Cell 81:403–412
King RC (1970) Ovarian development in *Drosophila melanogaster*. Academic, New York
Kirino Y, Kim N, de Planell-Saguer M, Khandros E, Chiorean S, Klein PS, Rigoutsos I, Jongens TA, Mourelatos Z (2009) Arginine methylation of Piwi proteins catalysed by dPRMT5 is required for Ago3 and Aub stability. Nat Cell Biol 11:652–658
Kirino Y, Vourekas A, Sayed N, de Lima Alves F, Thomson T, Lasko P, Rappsilber J, Jongens TA, Mourelatos Z (2010) Arginine methylation of Aubergine mediates Tudor binding and germ plasm localization. RNA 16:70–78
Klag J, Bilinski S (1993) Oosome formation in two ichneumonid wasps. Tissue Cell 25:121–128
Klag J, Bilinski S (1994) Germ cell cluster formation and oogenesis in the hymenopteran *Coleocentrotus soldanskii*. Tissue Cell 26:699–706
Kloc M, Jedrzejowska I, Tworzydlo W, Bilinski SM (2014) Balbiani body, nuage and sponge bodies – the germ plasm pathway players. Arthropod Struct Dev 43:341–348
Kristensen NP (1981) Phylogeny of insect orders. Annu Rev Entomol 26:135–157
Kristensen NP (1995) Forty years' insect phylogenetic systematics. Zool Beitr N F 36:83–124
Kubrakiewicz J (1997) Germ cells cluster organization in polytrophic ovaries of Neuroptera. Tissue Cell 29:221–228
Kugler J-M, Lasko P (2009) Localization, anchoring and translational control of *oskar*, *gurken*, *bicoid* and *nanos* mRNA during *Drosophila* oogenesis. Fly 3:15–28
Lasko PF, Ashburner M (1990) Posterior localization of vasa protein correlates with, but is not sufficient for, pole cell development. Genes Dev 4:905–921
Le Hir H, Gatfield D, Braun IC, Forler D, Izaurralde E (2001) The protein Mago provides a link between splicing and mRNA localization. EMBO Rep 2:1119–1124
Lecuyer E, Yoshida H, Parthasarathy N, Alm C, Babak T, Cerovina T, Hughes TR, Tomancak P, Krause HM (2007) Global analysis of mRNA localization reveals a prominent role in organizing cellular architecture and function. Cell 131:174–187
Lehmann R (2016) Germ plasm biogenesis – an oskar-centric perspective. Curr Top Dev Biol 116: 679–707
Lerit DA, Gavis ER (2011) Transport of germ plasm on astral microtubules directs germ cell development in *Drosophila*. Curr Biol 21:439–448
Lim AK, Tao L, Kai T (2009) piRNAs mediate posttranscriptional retroelement silencing and localization to pi-bodies in the *Drosophila* germline. J Cell Biol 186:333–342
Lin GW, Cook CE, Miura T, Chang CC (2014) Posterior localization of ApVas1 positions the preformed germ plasm in the sexual oviparous pea aphid *Acyrthosiphon pisum*. EvoDevo 5:18
Little SC, Sinsimer KS, Lee JJ, Wieschaus EF, Gavis ER (2015) Independent and coordinate trafficking of single *Drosophila* germ plasm mRNAs. Nat Cell Biol 17:558–568
Liu JL, Gall JG (2007) U bodies are cytoplasmic structures that contain uridine-rich small nuclear ribonucleoproteins and associate with P bodies. Proc Natl Acad Sci 104:11655–11659
Liu H, Wang JY, Huang Y, Li Z, Gong W, Lehmann R, RM X (2010) Structural basis for methylarginine-dependent recognition of Aubergine by Tudor. Genes Dev 24:1876–1881
Lynch JA, Ozüak O, Khila A, Abouheif E, Desplan C, Roth S (2011) The phylogenetic origin of oskar coincided with the origin of maternally provisioned germ plasm and pole cells at the base of the Holometabola. PLoS Genet 7:e1002029

Mahajan-Miklos S, Cooley L (1994) Intercellular cytoplasm transport during *Drosophila* oogenesis. Dev Biol 165:336–351

Mahowald AP (2001) Assembly of the *Drosophila* germ plasm. Int Rev Cytol 203:187–213

Mahowald AP, Stoiber D (1974) The origin of the nurse chamber in ovaries of *Miastor* (Diptera: Cecidomyiidae). Roux Archiv 176:159–166

Mani SR, Juliano CE (2013) Untangling the web: the diverse functions of the PIWI/piRNA pathway. Mol Reprod Dev 80:632–664

Markussen FH, Michon AM, Breitwieser W, Ephrussi A (1995) Translational control of oskar generates short OSK, the isoform that induces pole plasma assembly. Development 121:3723–3732

Meng C (1968) Strukturwandel und histochemische Befunde insbesondere am Ooson während der Oogenese und nach der Ablage des Eies von *Pimpla turrionallae* L. (Hymenoptera, Ichneumonidae). Roux Arch Dev Biol 161:162–208

Morais-de-Sá E, Vega-Rioja A, Trovisco V, St Johnston D (2013) Oskar is targeted for degradation by the sequential action of Par-1, GSK-3, and the SCF(-Slimb) ubiquitin ligase. Dev Cell 26:303–314

Nakamura A, Amikura R, Mukai M, Kobayashi S, Lasko P (1996) Requirement for a noncoding RNA in *Drosophila* polar granules for germ cell establishment. Science 274:2075–2079

Nakamura A, Sato K, Hanyu-Nakamura K (2004) *Drosophila* cup is an eIF4E binding protein that associates with Bruno and regulates *oskar* mRNA translation in oogenesis. Dev Cell 6:69–78

Nishida KM, Okada TN, Kawamura T, Mituyama T, Kawamura Y, Inagaki S, Huang H, Chen D, Kodama T, Siomi H et al (2009) Functional involvement of Tudor and dPRMT5 in the piRNA processing pathway in *Drosophila* germlines. EMBO J 28:3820–3831

Palacios IM, St Johnston D (2002) Kinesin light chain-independent function of the Kinesin heavy chain in cytoplasmic streaming and posterior localisation in the *Drosophila* oocyte. Development 129:5473–5485

Pritsch M, Büning J (1989) Germ cell clusters in panoistic ovary of Thysanoptera (Insecta). Zoomorphology 108:309–313

Pyka-Fosciak G, Szklarzewicz T (2008) Germ cell cluster formation and ovariole structure in viviparous and oviparous generations of the aphid *Stomaphis quercus*. Int J Dev Biol 52: 259–265

Quan H, Lynch J (2016) The evolution of insect germline specification strategies. Curr Opin Insect Sci 13:99–105

Ramamurty PS (1964) Distribution of RNA in the ooplasm of the scorpion fly. Sci Cult 30: 459–461

Rangan P, DeGennaro M, Jaime-Bustamante K, Coux RX, Martinho RG, Lehmann R (2009) Temporal and spatial control of germ-plasm RNAs. Curr Biol 19:72–77

Richards S, Gibbs RA, Weinstock GM, Brown SJ, Denell R et al (2006) The genome of the model beetle and pest *Tribolium castaneum*. Nature 452:949–955

Rongo C, Gavis ER, Lehmann R (1995) Localization of *oskar* RNA regulates oskar translation and requires Oskar protein. Development 121:2737–2746

Roth S, Lynch JA (2009) Symmetry breaking during *Drosophila* oogenesis. Cold Spring Harb Perspect Biol 1:a001891

Simiczyjew B, Margas W (2001) Ovary structure in the bat flea *Ischnopsyllus* spp. (Siphonaptera: Ischnopsyllidae). Phylogenetic implications. Zool Pol 46:5–14

Sinsimer KS, Jain RA, Chatterjee S, Gavis ER (2011) A late phase of germ plasm accumulation during *Drosophila* oogenesis requires Lost and Rumpelstiltskin. Development 138:3431–3440

Sinsimer KS, Lee JJ, Thiberge SY, Gavis ER (2013) Germ plasm anchoring is a dynamic state that requires persistent trafficking. Cell Rep 5:1169–1177

Smibert CA, Wilson JE, Kerr K, Macdonald PM (1996) Smaug protein represses translation of unlocalized *nanos* mRNA in the *Drosophila* embryo. Genes Dev 10:2600–2609

Smith JL, Wilson JE, Macdonald PM (1992) Overexpression of oskar directs ectopic activation of *nanos* and presumptive pole cell formation in *Drosophila* embryos. Cell 70:849–859

Stys P, Bilinski S (1990) Ovariole types and phylogeny of hexapods. Biol Rev 65:401–429

Suzuki N, Shimizu S, Ando H (1981) Early embryology of the alderfly, *Sialis mitsuhashii* okamoto. Int J Insect Morphol Embryol 10:409–418

Thomson T, Lasko P (2004) *Drosophila tudor* is essential for polar granule assembly and pole cell specification, but not for posterior patterning. Genesis 40:164–170

Thomson T, Lasko P (2005) Tudor and its domains: germ cell formation from a Tudor perspective. Cell Res 15:281–291

Thomson T, Liu N, Arkov A, Lehmann R, Lasko P (2008) Isolation of new polar granule components in *Drosophila* reveals P body and ER associated proteins. Mech Dev 125:865–873

Trcek T, Grosch M, York A, Shroff H, Lionnet T, Lehmnann R (2015) *Drosophila* germ granules are structured and contain homotypic mRNA clusters. Nat Commun 6:7962

Tworzydlo W, Bilinski SM (2008) Structure of ovaries and oogenesis in dermapterans. I. Origin and functioning of the ovarian follicles. Arthropod Struct Dev 37:310–320

Tworzydlo W, Bilinski SM, Kočárek P, Haas F (2010) Ovaries and germline cysts and their evolution in Dermaptera (Insecta). Arthropod Struct Dev 39:360–368

Vanzo NF, Ephrussi A (2002) Oskar anchoring restricts pole plasm formation to the posterior of the *Drosophila* oocyte. Development 129:3705–3714

Voronina E, Seydoux G, Sassone-Corsi P, Nagamori I (2011) RNA granules in germ cells. Cold Spring Harb Perspect Biol 3:a002774

Vourekas A, Alexiou P, Vrettos N, Maragkakis M, Mourelatos Z (2016) Sequence-dependent but not sequence-specific piRNA adhesion traps mRNAs to the germ plasm. Nature 531:390–394

Wang C, Lehmann R (1991) Nanos is the localized posterior determinant in Drosophila. Cell 66: 637–647

Wienstock GM, Robinson GE, Gibbs RA, Worley KC, Evans JD et al (2006) Insight into social insects from the genome of the honeybee, *Apis mellifera*. Nature 443:931–949

Wilson JE, Connell JE, Macdonald PM (1996) *aubergine* enhances *oskar* translation in the *Drosophila* ovary. Development 122:1631–1639

Xia QY, Zhou ZY, Lu C, Cheng DJ, Dai FY et al (2004) A draft sequence for the genome of the domesticated silkworm (*Bombyx mori*). Science 306:1837–1940

Yang N, Yu Z, Hu M, Wang M, Lehmann R, Xu RM (2015) Structure of *Drosophila* Oskar reveals a novel RNA binding protein. Proc Natl Acad Sci USA 112:11541–11546

Zaessinger S, Busseau I, Simonelig M (2006) Oskar allows *nanos* mRNA translation in *Drosophila* embryos by preventing its deadenylation by Smaug/CCR4. Development 133:4573–4583

Zimyanin VL, Belaya K, Pecreaux J, Gilchrist MJ, Clark A, Davis I et al (2008) In vivo imaging of *oskar* mRNA transport reveals the mechanism of posterior localization. Cell 134:843–853

Zissler D, Sander K (1982) The cytoplasmic architecture of the insect egg cell. In: King RC, Akai H (eds) Insect ultrastructure. Plenum Publishing, New York

Chapter 6
Multiple Functions of the DEAD-Box Helicase Vasa in *Drosophila* Oogenesis

Mehrnoush Dehghani and Paul Lasko

Abstract The DEAD-box helicase Vasa (Vas) has been most extensively studied in the fruit fly, *Drosophila melanogaster*, and numerous roles for it in germline development have been discovered. Here, we summarize the present state of knowledge about processes during oogenesis that involve Vas, as well as functions of Vas as a maternal determinant of embryonic spatial patterning and germ cell specification. We review literature that implicates Vas in Piwi-interacting RNA (piRNA) biogenesis in germline cells and in regulating mitosis in germline stem cells (GSCs). We describe the functions of Vas in translational activation of two mRNAs, *gurken* (*grk*) and *mei-P26*, which encode proteins that are important regulators of developmental processes, as Grk specifies both the dorsal-ventral and the anterior-posterior axis of the embryo and Mei-P26 promotes GSC differentiation. The role of Vas in assembly of polar granules, ribonucleoprotein particles that accumulate in the posterior pole plasm of the oocyte and are essential for germ cell specification and posterior embryonic patterning, is also described.

6.1 A Brief Summary of *Drosophila* Oogenesis

The DEAD-box helicase Vas functions at many steps during oogenesis. In order to discuss its roles, it is first necessary to outline in overall terms how oogenesis proceeds in *Drosophila*. More detailed descriptions of this developmental process are available (King 1970; Bastock and St Johnston 2008; McLaughlin and Bratu 2015).

Each ovary is composed of approximately 12–14 ovarioles, which can be considered as assembly lines for egg production. At the anterior tip of each ovariole, there is a morphologically distinct region called the germarium, and 2–4 germline stem cells (GSCs) are located near its anterior end. These GSCs are maintained by contact with cap cells, somatic cells that are immediately adjacent at the anterior of the GSCs. Asymmetric division of GSCs results in two daughter cells, one of which is another GSC and the other, which no longer contacts the cap cells, differentiates into a

M. Dehghani • P. Lasko (✉)
Department of Biology, McGill University, 3649 Promenade Sir William Osler, Montréal, Québec, Canada, H3G 0B1
e-mail: paul.lasko@mcgill.ca

M. Kloc (ed.), *Oocytes*, Results and Problems in Cell Differentiation 63,
DOI 10.1007/978-3-319-60855-6_6

cystoblast (CB). Each CB undergoes four mitotic divisions to produce a cyst of 16 interconnected germ cells, in which one cell at the posterior differentiates into the oocyte and the others become nurse cells. Until late oogenesis, nurse cells remain interconnected among themselves and with the oocyte through cytoplasmic bridges called ring canals. Germline cysts that leave the germarium at this stage are called egg chambers. Meiosis in ovaries starts in the germline cysts, where 2 cells among 16 start meiosis but only 1 differentiates into an oocyte and remains in prophase I until late oogenesis, while the other follows the nurse cell fate. The oocyte then proceeds to metaphase I, where it is arrested again until fertilization (Ables 2015; Lehmann 2012). Egg chamber polarity, which includes posterior position of oocyte and differentiation of epithelial follicle cells, is established via reciprocal interactions between germline and somatic follicle cells (Assa-Kunik et al. 2007; Roth and Lynch 2009; Torres et al. 2003). The oocyte itself is also polarized through a process that originally depends on the cyst structure and becomes more pronounced as the egg chambers develop through a series of 14 defined stages (King 1970) that ultimately produce the mature egg. Oocyte polarity is internally linked to repositioning of the microtubule-organizing center (Grieder et al. 2000; Steinhauer and Kalderon 2006; Theurkauf et al. 1992). Anterior posterior (AP) axis establishment in the oocyte is associated with asymmetric localization of key mRNAs, such as *bicoid* (*bcd*) and *oskar* (*osk*) to the anterior and posterior poles, respectively (Berleth et al. 1988; Ephrussi et al. 1991). The dorsal-ventral (DV) axis is determined by a TGF-α-like protein, gurken (Grk), that accumulates along with its mRNA in the future dorsal-anterior corner of the oocyte, tightly associated with the oocyte nucleus (Neuman-Silberberg and Schüpbach 1993). Oocyte patterning determines polarization of the overlaying epithelial follicle cells, which is critical for proper development of egg shell structures such as respiratory appendages and the micropyle in the anterior-dorsal side of the egg (Berg 2005). During stages 10B and 11, nurse cells transport their cytoplasmic content to the oocyte and undergo apoptosis (Cavaliere et al. 1998). The mature oocyte (stage 14) becomes activated and completes meiosis as it passes through the oviduct (Horner and Wolfner 2008; Mahowald et al. 1983).

6.2 DEAD-Box Family of RNA Helicases

Vasa (Vas) is among the founding members of the DEAD-box protein family, which represents the largest group of nucleic acid helicases (Linder et al. 1989; Linder and Fuller-Pace 2013). DEAD-box helicases unwind double-stranded RNAs in a non-processive manner, an activity requiring ATP binding and hydrolysis.

All DEAD-box helicases contain a highly conserved core of approximately 400 amino acids that is composed of two recombinase A (RecA)-like domains and contains 12 characteristic motifs, which are involved in RNA binding, ATP binding, and inter-domain interactions (Fig. 6.1a). The name of this family originates from motif II, the Asp-Glu-Ala-Asp (D-E-A-D) motif. A similar mechanism of duplex separation is utilized by all DEAD-box proteins, whereby concomitant with ATP binding the

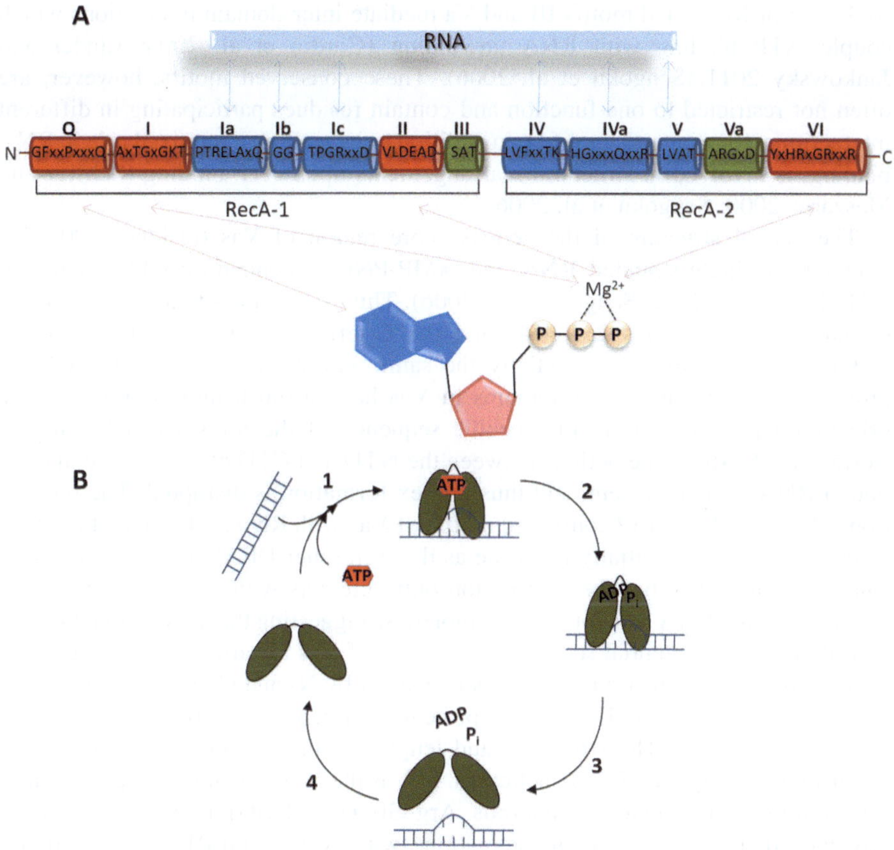

Fig. 6.1 Conserved structure and mechanism of function of DEAD-box proteins. (**a**) DEAD-box proteins share 12 conserved motifs in their core region, which are involved in RNA binding (*blue*), ATP binding (*red*), and inter-domain communications (*green*). These motifs are arranged within two RecA domains (Linder and Jankowsky 2011). (**b**) In DEAD-box proteins, ATP binding, which occurs concomitant with RNA binding, induces a closed conformation in the protein and results in RNA duplex separation (step 1). ATP hydrolysis and product (ADP and P_i) release (steps 2, 3) are subsequently required for dissociation of unwound RNA and to reset the enzyme for further catalytic cycles (step 4) (Chen et al. 2008)

helicase core undergoes a conformation change, closing the cleft between the two RecA domains (Chen et al. 2008; Linder and Fuller-Pace 2013; Rudolph and Klostermeier 2015). The closed state imposes a sharp bend in the RNA strand, bound opposite to the ATP-binding site, and causes local duplex destabilization and RNA unwinding. ATP hydrolysis occurs subsequently to restore open conformation and release RNA before the helicase can start another catalytic cycle (Fig. 6.1b).

Motifs Q, I (also known as the Walker A motif), II (also known as the Walker B motif), and VI are involved in ATP binding and hydrolysis; motifs Ia, Ib, Ic, IV, IVa, and V mainly contain residues which interact with the sugar phosphate backbone of RNA; and motifs III and Va mediate inter-domain interactions which couple ATP binding with RNA unwinding (Cordin et al. 2006; Linder and Jankowsky 2011; Sengoku et al. 2006). These conserved motifs, however, are often not restricted to one function and contain residues participating in different interactions; for example, motif Ia, in addition to its well-characterized role in RNA binding, is involved in structural rearrangements upon ATP binding (Schwer and Meszaros 2000; Sengoku et al. 2006).

The crystal structure of the helicase core region of Vas (residues 200–623) bound to a single-stranded RNA and AMP-PNP, a nonhydrolyzable analog of ATP, has been solved (Sengoku et al. 2006). This work shows that the N-terminal domain (NTD, residues 233–454) and the C-terminal domain (CTD, residues 463–621) of Vas fold in essentially the same way as for the other DEAD-box proteins, although these two domains in Vas have a much more closed relative orientation possibly due to the specific sequence of the 8-residue linker region (residues 455–462). Interactions between the NTD and CTD of Vas ensure that the bound RNA is sharply bent, and thus duplex formation is disrupted. The interactions of the residues in 12 conserved motifs of Vas with RNA, ATP, and other parts of the protein are essentially the same as those in other DEAD-box proteins. This study also indicates that the core region only interacts with the phosphate-ribose backbone of the RNA and not the base moieties, suggesting that this region does not contribute to any potential sequence specificity of Vas toward its RNA targets.

The conserved helicase core of Vas is flanked by N- and C-terminal sequences, as is the case for most DEAD-box proteins (Hilbert et al. 2009; Rudolph and Klostermeier 2015). The structures and lengths of these flanking domains differ extensively among DEAD-box helicases. In Vas the N-terminal flanking domain of 200 amino acids contains numerous Arg-Gly-Gly (RGG) motifs and a short sequence that is implicated in interactions with two Cullin-RING E3 ubiquitin ligase specificity receptors (Kugler et al. 2010; Dehghani and Lasko 2015). The C-terminal flanking domain of Vas composed of 40 amino acids ends with a series of acidic residues punctuated by a single Trp in the penultimate position. The flanking regions of Vas, with the exception of the stretch of acidic residues and the Trp, evolve very rapidly; for example, in the closely related species, *D. simulans*, the predicted Vas protein (GD23992; Hu et al. 2013) is 51 amino acids longer and only 80% identical to its counterpart in *D. melanogaster*, with most of the divergence in the N-terminal flanking domain.

6.3 Vas and the Biogenesis of piRNAs

Maintaining genome integrity is particularly important in the germline, because deleterious mutations in these cells could be transmitted to the next generation. Active transposable elements (TEs) impose a constant threat to the host cells through their mobility and insertion into essential genes. Therefore, the germline has evolved defense mechanisms to protect its genome against these internal enemies. Most notably the Piwi-interacting RNA (piRNA) pathway, which is highly conserved among animals, targets TE-derived transcripts for degradation and thus blocks TE replication (Grimson et al. 2008). Increasing evidence shows that Vas is a critical component of piRNA biogenesis.

6.3.1 piRNA Pathway Protects Genome Integrity in the Germline

Transposable elements, which were first identified over 50 years ago, are DNA sequences that can move from one location in the genome to another (Pray 2008). TEs comprise a significant portion of all eukaryotic genomes; about 4% of the *Drosophila* genome (and 50% of the human genome) is occupied by transposons and transposon-like repetitive elements (Kaminker et al. 2002; Mills et al. 2007). Transposons are classified based on their mode of propagation. Retrotransposons employ a "copy and paste" mechanism by which the original insertion remains in the genome and a new copy is generated by reverse transcription, resulting in amplification of the element. In contrast DNA transposons move through a "cut and paste" mechanism, which includes excision of the original copy and subsequent repair of the resulting gap in the DNA (Pray 2008). Transposon mobility often has deleterious effects by causing mutations or altering expression of neighboring genes, although it can be also evolutionarily beneficial. The germline is a battlefield for the race between transposons and the host, as TE-mediated changes in the genome could be transmitted to the new generation when they affect germ cells. TEs have evolved mechanisms to ensure their vertical propagation across generations. For example, expression of the active P element transposase in *Drosophila* is restricted to the germline (Seleme et al. 1999). In parallel germ cells have evolved special strategies to tame transposons and protect their genome integrity. One highly conserved mechanism to do this is mediated by piRNAs, a large class of small non-coding RNAs, which target and cleave TE-derived RNA molecules (reviewed by Haase 2016).

piRNAs derive from single-stranded RNA precursors and associate with members of the Piwi clade, which is named after *Drosophila* Piwi (P-element induced wimpy testis) and includes Piwi, Aub, and Ago3 (Meister 2013). Over 90% of piRNAs derive from discrete genomic loci, called piRNA clusters (Fig. 6.2, Brennecke et al. 2007). These loci are mainly found in pericentromeric and telomeric heterochromatin and act as a registry, which contains inactive copies or truncated fragments of all transposons hosted in the genome. Numerous genetic studies have provided a functional link between piRNA clusters and transposon silencing. For example, mutations in the *flamenco*/*COM* locus lead to upregulation of the retrotransposons *gypsy*, *Idefix*, and *ZAM* (Desset et al. 2003; Pelisson et al. 1994). Little is known about piRNA cluster transcription from these, otherwise silenced, heterochromatic regions. The heterochromatin protein 1 (HP1) variant,

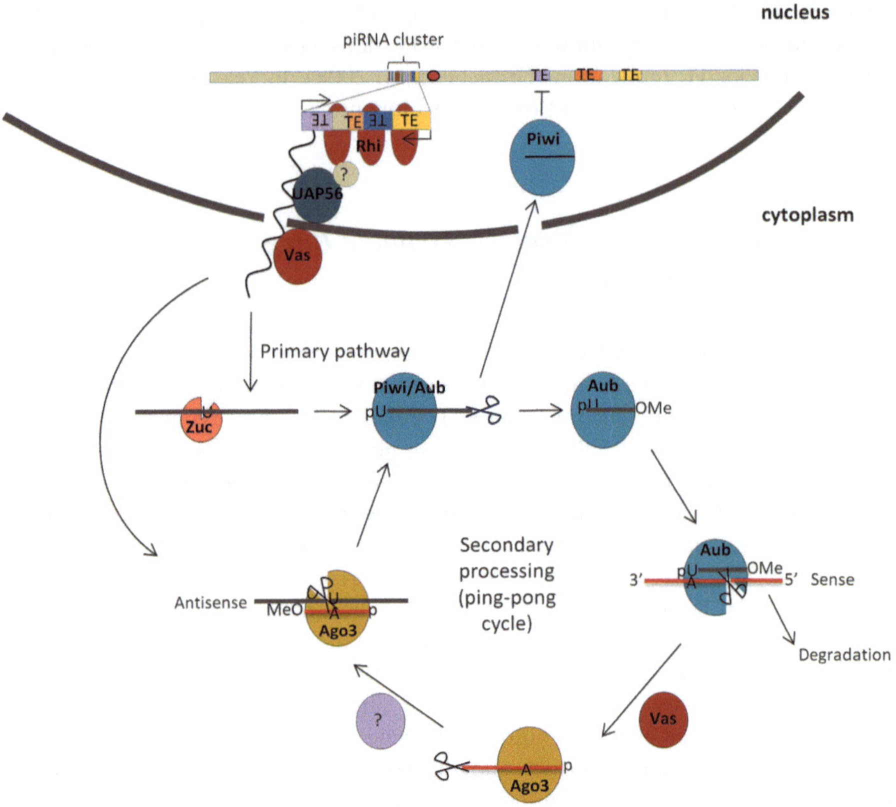

Fig. 6.2 piRNA transcripts are derived from transposon-rich heterochromatin, called piRNA clusters, in the pericentromeric or subtelomeric regions (Brennecke et al. 2007). The *Drosophila* HP1 variant, Rhino (Rhi), binds to piRNA cluster sequences and colocalizes with the DEAD-box protein UAP56 adjacent to the nuclear pore (Klattenhoff et al. 2009; Zhang et al. 2012). The other DEAD-box protein, Vas, is localized to the cytoplasmic face of the nuclear envelope and functions with UAP56 to transfer cluster transcripts to the cytoplasm. During primary processing, the long piRNA transcripts (antisense) are cleaved upstream of the 5′ uridine (U) bias by the nuclease Zucchini (Zuc) and loaded on aubergine (Aub) or Piwi (Nishimasu et al. 2012). The 3′ end of primary piRNA is trimmed by an unknown nuclease and is 2′-O-methylated (Ome). The Piwi-piRNA complex is imported to the nucleus to exert transposon silencing (Huang et al. 2013), whereas Aub-piRNA complex targets transposon transcripts (sense) for cleavage. The 3′ cleavage product is transferred to Ago3 with the help of Vas and undergoes further 3′ end trimming (Xiol et al. 2014). Ago3-piRNA complex then targets antisense transcripts to produce piRNA intermediate fragments, which are delivered to Aub for starting a new cycle (Hirakata and Siomi 2016). TE, transposable element

Rhino (Rhi), is required for production of piRNA precursors from the 42AB dual-strand cluster (Klattenhoff et al. 2009). A recent study shows that an interaction between Rhi and the transcription termination factor Cutoff (Cuff), via the adaptor protein Deadlock, is required for piRNA production, through noncanonical transcription that involves preventing RNA polymerase II from termination and protecting the 5′ end of the nascent transcripts (Mohn et al. 2014). The long single-stranded transcripts that result are exported to the cytoplasm for processing into small RNAs, through a mechanism that involves the DEAD-box protein UAP56 (Zhang et al. 2012). On piRNA clusters, close to the inner surface of the nuclear membrane, UAP56 colocalizes with Rhino (Fig. 6.2). UAP56 foci are located directly opposite Vas-containing foci on cytoplasmic side of nuclear pores. This together with biochemical and genetic evidence suggests that UAP56 and Vas interact to transfer piRNA transcripts across the nuclear envelope.

Once transcripts reach the perinuclear region of the cytoplasm, they undergo primary processing, which requires activity of the ssRNA-specific endonuclease, Zucchini (Zuc), to generate the monophosphorylated 5′ end (Nishimasu et al. 2012). Subsequently, the pre-piRNA is loaded on Aub or Piwi and the 3′ end is trimmed, by an unknown mechanism, to produce mature piRNAs (Haase 2016; Meister 2013). Piwi-piRNA complexes are translocated to the nucleus to bind complementary sequences in the genome and, by recruiting epigenetic factors such as HP1, silence gene expression (Huang et al. 2013). piRNA-guided epigenetic silencing by Piwi is also an alternative mode for transposon regulation (Saito 2013).

The primary piRNA loaded on Aub targets complementary sequences in transposon transcripts, for their cleavage by Aub slicer activity. Simultaneously collaboration between Aub and Ago3 produces secondary piRNAs through an amplification cycle referred to as the "ping-pong loop."

6.3.2 *Vas Function in piRNA Pathway*

The role of Vas in piRNA biogenesis was first discovered by studying Mvh (mouse Vasa homolog)-deficient mice (Kuramochi-Miyagawa et al. 2010). Defective spermatogenesis in these mutants was associated with elevated levels of several retrotransposons and impaired de novo DNA methylation of TE regulatory regions, resembling phenotypes observed in Mili- or Miwi2 (mouse homologs of Piwi)-deficient mice (Deng and Lin 2002; Kuramochi-Miyagawa et al. 2004). Further analyses indicated that Miwi2-bound piRNAs are significantly decreased in Mvh mutants and suggested a crucial role for Mvh in the ping-pong cycle (Kuramochi-Miyagawa et al. 2010).

In *Drosophila vas*-null females have aberrant ovarioles with the majority of oocytes not reaching vitellogenic stages of development (Styhler et al. 1998; Tomancak et al. 1998). The phenotype of *vas*-null mutants resembles those of mutants affecting the piRNA pathway, and indeed the activity of the ping-pong pathway is greatly reduced in *vas*-null ovaries (Zhang et al. 2012). Vas is localized to, and is required for, the assembly of the nuage, a specialized perinuclear structure that is most prominent in the large polyploid nurse cells (Hay et al. 1990; Lasko and Ashburner 1990; Liang et al. 1994). Many piRNA components including Aub and Ago3 and the Tud domain proteins Tej and Kumo also localize to the nuage, supporting the idea that nuage functions as a processing site for the amplification cycle of the piRNA pathway (Anand and Kai 2012; Harris and Macdonald 2001;

Patil and Kai 2010; Webster et al. 2015). Vas-containing nuage structures are often localized at or near the cytoplasmic face of nuclear pores, adjacent to and dependent on localization of Rhi and UAP56 foci on the nuclear side (Zhang et al. 2012). As for *rhi* and *uap56* mutants, several transposon families are overexpressed in *vas*-null ovaries. These observations, together with co-immunoprecipitation of piRNA cluster transcripts with Vas and the reduced levels of germline-specific piRNAs in *vas*-null ovaries, suggest that Vas and UAP56 cooperate on opposite sides of nuclear pores for the export of cluster transcripts to cytoplasm.

The role of Vas in the piRNA amplification cycle remained obscure for a long time partly due to the dynamic and transient nature of Vas interactions with RNA, as is the case for other RNA helicases (Linder and Jankowsky 2011). ATP hydrolysis by DEAD-box proteins and product dissociation are required for progression of repeated RNA-unwinding events; thus, mutations in residues, such as those in motif II, that are specifically involved in ATPase activity might be expected to create an RNA clamp and stabilize transient complexes. Therefore, such a mutant allele of *vas* (vas^{DQAD}) was tested in the BmN4 cell line, derived from the silkworm *Bombyx mori*, which is the only cell culture model with active piRNA biogenesis pathways (Kawaoka et al. 2009; Xiol et al. 2014). This study provided several important results that clarified the mechanism of Vas function in the ping-pong cycle. Fluorescence recovery after photobleaching (FRAP) experiments indicated that in contrast to eGFP-BmVas (*B. mori* Vas) in nuage granules, which is rapidly replaced from surrounding cytoplasm, eGFP-BmVas^{DQAD} fails to recover after photobleaching (Xiol et al. 2014). Furthermore, the mutation stabilized Vas in a complex that contains a number of proteins involved in the piRNA amplification cycle such as Siwi (the *Bombyx* ortholog of Aub), Ago3, and Qin/Kumo. Deep sequencing of small RNA associated with Vas^{DQAD} shows that these have characteristics of piRNAs and, in addition, a majority of them corresponded to the Siwi-bound antisense piRNAs. The Vas-Siwi-Ago3-Qin amplifier complex also contained poly(A) piRNA precursors corresponding to transposon transcripts, suggesting that Vas cooperates with Siwi for processing sense strands either by regulating the slicer activity of Siwi or by transferring the sense piRNA intermediates from Siwi to Ago3. Further experiments, using a construct expressing an artificial secondary piRNA precursor (sense), supported the latter function by indicating that 100% of reporter-derived fragments associated with Vas^{DQAD} have 5′ monophosphate ends already generated by Siwi slicer cleavage activity. Together these observations support a model, whereby Vas in its RNA-bound closed conformation provides a platform for the assembly of a piRNA amplifier complex and promotes transfer of the sliced precursor fragments from Siwi/Aub to Ago3 (Xiol et al. 2014).

Co-immunoprecipitation experiments in a subsequent study isolated Siwi from a wild-type BmVas complex, albeit at much lower levels than from the BmVas^{DQAD} complex, which ruled out the possibility that the mutation led to an artifactual association (Nishida et al. 2015). This study also shows that FLAG-BmVas purified from BmN4 cells could liberate RNA from Siwi-piRISC in vitro. The Siwi/BmVas complex, which is required for production of BmAgo3-secondary piRNAs, is found distinct from a Siwi/BmSpn-E/BmQin complex, involved in primary piRNA production, thus excluding Vas from the primary phase of piRNA processing.

6.4 The Posterior Pole Plasm of the *Drosophila* Oocyte Specifies the Germline in the Developing Embryo

In many organisms, including *Drosophila*, *C. elegans*, *Xenopus*, and zebrafish, there is a physical continuity of germline across generations, meaning that a specific cytoplasmic region in the oocyte, termed germ plasm, contains maternally inherited mRNA-protein complexes that segregate into subsequently developing primordial germ cells (PGCs, Lesch and Page 2012). In the *Drosophila* embryo, the maternally derived germ plasm (called pole plasm), which is located at the posterior pole, is sufficient to induce germ cell fate. This was proven by demonstrating that transplantation of posterior cytoplasm from an early embryo or unfertilized egg to an ectopic site of another embryo induces germ cells at this site, and further these germ cells are functional if transplanted to the posterior pole of a host embryo (Illmensee and Mahowald 1974).

Since *Drosophila* relies so heavily on maternally supplied proteins and mRNA for early development, there is elaborate posttranscriptional regulation of maternal mRNAs mainly at the levels of mRNA localization and translation (Becalska and Gavis 2009; Lasko 2012). Nurse cell-produced materials including germ cell determinants are unidirectionally transported to the oocyte via polarized microtubules (MTs), which in stages 2–6 oocytes emanate from a microtubule-organizing center (MTOC) at the posterior region and extend their plus ends through the ring canals into the nurse cells (Harrison and Huebner 1997; Roth and Lynch 2009; Weil 2014; Goldman and Gonsalvez 2017). In stages 7–10a (mid-oogenesis), MTs reorganize, presumably in response to signals from the posterior follicle cells, and emanate from the lateral and anterior cortex with a slightly biased polarity toward the posterior (Parton et al. 2011; Steinhauer and Kalderon 2006; Zimyanin et al. 2008). mRNAs and proteins involved in germ cell specification and embryonic patterning then become localized to the pole plasm at the posterior region of the oocyte. Different mechanisms have been described for localization of germ cell determinants (Meignin and Davis 2010): (1) degradation in bulk cytoplasm and stabilization at the posterior pole, as observed for *nanos* (*nos*) mRNA (Zaessinger et al. 2006); (2) diffusion and anchoring, for example, reported for *cyclin B*, *germ cell-less* (*gcl*), and *nos* (Forrest and Gavis 2003); (3) active transport mediated by motor proteins and *cis* elements in 3′ UTR of mRNAs such as *oskar* (*osk*) (Clark et al. 2007; Zimyanin et al. 2008); and (4) localized translation in the pole plasm and repression elsewhere, as for Osk (Kim-Ha et al. 1995). Following fertilization and during syncytial divisions, pole plasm complexes are recruited by posterior nuclei through dynein-mediated transport on astral microtubules that emanate from centrosomes (Lerit and Gavis 2011). By this mechanism germ plasm becomes segregated into the pole buds, which pinch off from the posterior somatic region at mitotic cycle 9 (stage 3, Campos-Ortega and Hartenstein 1985) and undergo two more divisions before the embryo undergoes cellularization. These nuclear divisions are asynchronous and lag behind the synchronous somatic divisions. Germ cell precursors then enter a prolonged quiescence until arriving in the presumptive gonads in late embryogenesis (Su et al. 1998). The mitotic quiescence is partly due to translational repression of *cyclin B* by RNA-binding proteins Nos and Pumilio (Pum) (Asaoka-Taguchi et al. 1999).

6.4.1 Vas Is a Component of the Polar Granules, RNPs that Assemble in the Drosophila *Pole Plasm*

The critical role of pole plasm in *Drosophila* germ cell formation was first demonstrated by UV-irradiating the posterior pole of fertilized eggs, which resulted in sterile flies (Geigy 1931). Germ cell formation can be restored to irradiated embryos by transplantation of a new pole plasm (Okada et al. 1974). As mentioned in the introduction, transplantation of pole plasm to an ectopic site results in ectopic formation of pole cells, suggesting that posterior cytoplasm contains all the necessary components for germ cell formation (Illmensee and Mahowald 1974). Ultrastructural analyses show that the main components of germ plasm, the polar granules, which first appear in the posterior region at stage 9, are composed of a non-membrane-bound fibrous meshwork. Polar granules appear as non-membrane-bound fibrous material and become associated with the surface of mitochondria in later-stage oocytes and with the ribosomes in the activated egg (Mahowald 1968; Mahowald et al. 1983). The full composition of polar granules is unknown, but several mRNAs and proteins, including Vas, associate with them.

Vas and other polar granule components expressed during oogenesis are essential for germ cell specification and posterior patterning in the embryos that develop after fertilization of the eggs. In fact, *vas* was first identified in screens for maternal-effect lethal mutations that impair embryonic patterning (Schüpbach and Wieschaus 1986a, b). Homozygous females for *vas* hypomorphic alleles produce embryos which lack polar granules, fail to form germ cells, and exhibit defects in abdominal segmentation (Schüpbach and Wieschaus 1986b). The *vas* gene was subsequently cloned and found to encode a DEAD-box helicase (Hay et al. 1988b; Lasko and Ashburner 1988; Linder et al. 1989). Further immunostaining experiments showed that Vas localizes to the pole plasm and is present in the pole cells and their germ cell derivatives in all stages of development (Hay et al. 1990; Lasko and Ashburner 1990).

Vas localization to the posterior pole plasm depends upon two genes that also encode pole plasm components, *osk* and *stau*, as females mutant for either of these genes do not display posterior localization of Vas. However, Vas localization is initially unaffected in mutants of most other pole plasm components, including *nos*, *tud*, *valois* (*vls*), *gcl*, and *pgc* (Jongens et al. 1992; Lasko and Ashburner 1990;

Nakamura et al. 1996). Yeast two-hybrid experiments indicate direct interaction between Vas and Osk, which is considered an essential step for initiating pole plasm assembly (Breitwieser et al. 1996). This interaction persists upon deleting residues 321–661 of Vas. Recent work further demonstrates that deletion of amino acids 3–200 of Vas, but not 3–139, severely reduces its localization to the pole plasm and to the nuage (Dehghani and Lasko 2015). As the latter occurs independently of Osk, the interval 140–200 must be involved in additional interactions between Vas and its protein partners that promote Vas localization. Furthermore, this work shows that a non-conservative point mutation in T546, an RNA-binding residue, abolishes Vas localization. Since mutations in other RNA-binding motifs do not have similar effects, T546 most likely plays a role in protein-protein interactions required for Vas localization (Dehghani and Lasko 2015; Sengoku et al. 2006). Precise localization of Vas at the posterior region, followed by other components of pole plasm, depends on the C-terminal motif, 636–646, which is conserved among closely related *Drosophila* species. When only a form of Vas deleted for this motif is present, the concentration of pole plasm at the posterior region is insufficient for its function, resulting in a phenotype where most pole buds fail to develop into germ cells and instead adopt a somatic fate (Dehghani and Lasko 2015).

Posterior localization of Vas becomes detectable at stage 10a (Lasko and Ashburner 1990). Live imaging of GFP-tagged Vas and Aub indicates that nuage particles become displaced to the cytoplasm of nurse cells and then enter the oocyte through ring canals (Snee and Macdonald 2004). This study argues against the previous model suggesting that pole plasm is directly formed from the nuage fragments, by providing evidence that Vas becomes dissociated from these precursors and is later entrapped in the pole plasm, after its de novo nucleation by Osk. The direct interaction between Vas and Osk is essential for pole plasm assembly, as long Osk, which has significantly lower affinity to Vas compared to short Osk, is not able to induce ectopic Vas localization and germ cell formation in the anterior region (Breitwieser et al. 1996). Furthermore, persistent trafficking of pole plasm components, such as Vas, on cortical microtubules, mediated by dynein, is required for their retention at the posterior cortex (Sinsimer et al. 2013).

There is evidence that regulation of Vas stability through mechanisms that affect its ubiquitination state contributes to maintenance of its posterior localization. Vas can be co-purified from *Drosophila* ovaries and embryos with the ubiquitin-specific protease Fat facets (Faf), which also localizes to the pole plasm (Huang et al. 1995; Liu et al. 2003). Embryos from *faf* mutant females form fewer pole cells than wild type, consistent with decreased levels of Vas and its posterior localization in these mutants. In addition, Vas purified from *faf* ovaries is highly ubiquitinated, suggesting that Vas is stabilized in pole plasm through its Faf-dependent deubiquitination. Vas ubiquitination in pole plasm is also modulated by the two ubiquitin Cullin-RING E3 ligase specificity receptors, Fsn and Gustavus (Gus), which both bind to the same motif (DINNN) on Vas (Kugler et al. 2010; Styhler et al. 2002). *fsn* mutants show a precocious accumulation of Vas in pole plasm, whereas *gus* mutants additionally lacking one copy of *cullin-5* exhibit delay in posterior localization of Vas. While Fsn clearly acts as a negative regulator of Vas stability, the function of Gus is more complex, as its overexpression also reduces levels of Vas in the ovaries.

In *vas* mutants, posterior localization of *nos* mRNA is abolished, consistent with defects in abdominal segmentation in these mutants (Gavis and Lehmann 1992). Nos, together with another RNA-binding protein Pum, blocks translation of the maternal mRNA, *hunchback* (*hb*), thereby creating a gradient of Hb protein across the egg, which opposes the steep posterior-anterior gradient of Nos (Murata and Wharton 1995; Tautz 1988). Hb acts as a transcriptional activator of the genes essential for development of anterior structures; thus, its repression in the posterior region is critical for abdominal segmentation (Lehmann and Nüsslein-Volhard 1987).

6.5 Vas Is a Translational Activator of *gurken* and *mei-P26* mRNAs

The EGFR ligand, gurken (Grk), is secreted by the oocyte and induces dorsal fate in the overlying epithelial follicle cells, through its interaction with the EGFR receptor, Torpedo (Top) (González-Reyes et al. 1995; Roth et al. 1995). During oogenesis *grk* mRNA is expressed by the nurse cells and transported to the oocyte via a mechanism depending on dynein and its cofactors, BicD and Egl (Mach and Lehmann 1997; Neuman-Silberberg and Schüpbach 1996). Throughout oogenesis *grk* mRNA remains tightly associated with the oocyte nucleus, which in early stages is positioned at the posterior. Starting at stage 8, growing microtubules push the nucleus to the anterodorsal corner of the oocyte, and accordingly *grk* mRNA forms a crescent between nucleus and cell cortex in this region (Steinhauer and Kalderon 2006; Zhao et al. 2012). The exact mechanisms of *grk* localization are unknown, but this process is mediated at different stages by dispersed elements in the 5′ UTR, coding region, and 3′ UTR of the *grk* transcript (Thio et al. 2000).

Genetic interactions suggest that translation of unlocalized *grk* mRNA is inhibited through a complex including Cup and Squid (Sqd) and that another component of this complex, PABP55B, through its interaction with Encore (Enc), facilitates translation of localized *grk* by a yet unknown mechanism (Clouse et al. 2008). *grk* mRNA is not efficiently translated in *orb* mutant ovaries, suggesting that polyadenylation mediates translation of localized *grk*, a mechanism through which PABP55B is also proposed to exert its regulatory effect on Grk translation (Chang and Matzuk 2001; Clouse et al. 2008). Furthermore, Grk protein level, but not the level of its mRNA, is severely reduced in *vas*-null mutants, whereas mRNA localization remains normal (Styhler et al. 1998). A direct interaction between Vas and the translation factor, eIF5B, required for 60S ribosomal subunit joining at the initiation step (Pestova et al. 2000), was found through a yeast two-hybrid screen against *Drosophila* ovarian cDNAs (Carrera et al. 2000). A mutant form of Vas, Vas$^{\Delta617}$, does not interact with eIF5B and once expressed in the ovaries does not support Grk translation, suggesting that Vas directly activates Grk expression by recruiting eIF5B to *grk* transcripts (Johnstone and Lasko 2004).

Grk translation is blocked as a response to meiotic checkpoints that in turn are activated by unrepaired dsDNA breaks (DSBs, Ghabrial and Schupbach 1999). Thus, mutations in genes such as *okra* and *spindle-B*, which are involved in DSB repair, cause defects in dorsoventral patterning. *vas* mutants produce similar nuclear morphology phenotypes as spindle-class mutants, suggesting that meiotic checkpoints act through downregulation of Vas (Styhler et al. 1998; Tomancak et al. 1998). Vas expression is not affected in spindle-class mutants; however, band shift assays indicate that the mobility of Vas protein from ovarian lysates prepared from these mutants is decreased compared to wild type and is restored by an additional mutation in *mei-41*, a gene involved in checkpoint pathway. These observations support a model, whereby meiotic checkpoints suppress Vas activity in Grk translation through posttranslation modifications (Abdu et al. 2002; Ghabrial and Schupbach 1999). Surprisingly, reducing the genetic dosage of the translation factor eIF1A suppresses this effect in *spnB* mutants, increasing both Grk level and the amount of grk mRNA associated with polysomes (Li et al. 2014). eIF1A facilitates recruitment of Met-tRNAi-eIF2-GTP ternary complex (TC) to the 40S ribosomal subunit and thus promotes translation initiation (Hinnebusch 2014). The unexpected inhibitory effect of eIF1A on Grk translation is thought to be associated with three upstream open reading frames (uORFs) in the *grk* 5′ UTR (Li et al. 2014). uORFs often reduce translation from the main ORF, by attracting pre-initiation complex and creating a block on the mRNA as the ribosome stalls on premature stop codons (Barbosa et al. 2013). eIF1A mutations could result in a larger fraction of ribosomes passing the uORFs before TC recruitment, enabling more efficient translation from the main ORF (Fekete et al. 2005; Li et al. 2014).

In addition to its likely role in Grk translation, Vas has been implicated in translation of *mei-P26*, which promotes GSC differentiation through its function in the microRNA (miRNA) pathway (Liu et al. 2009; Neumuller et al. 2008; Page et al. 2000). Vas-mRNA complexes were purified from early embryos through a tandem immunoprecipitation approach with anti-Vas antibody (Liu et al. 2009). In this manner 221 mRNAs co-purified with Vas, 13.3% of which were enriched in the pole cells. One of the recovered mRNAs was *mei-P26*, and subsequent work showed that Vas directly interacts with 3′ UTR of *mei-P26*, depending on a stretch of ten consecutive U residues, and promotes Mei-P26 expression (Liu et al. 2009). Polyuridine tracts are not overrepresented among the mRNAs that co-purified with Vas; however, germ cell-enriched mRNAs appear more frequently when the Vas-bound mRNAs are filtered for the presence of a (U)-rich motif, an approach which may identify other direct targets of Vas. Preferential binding of Vas to the polyuridine motif compared to the neighboring motifs depends on sequences in its N- or C-terminal regions, as the Vas core protein (VCP, residues 200–623) does not exhibit such discriminatory affinity. Mei-P26 protein level is significantly decreased in $vas^{\Delta 617}$ ovaries, suggesting that Vas interaction with eIF5B is required for translational activation of Mei-P26 in GSCs.

A direct role for Vas in activating translation of one or more pole plasm RNAs has long been assumed but has never been convincingly demonstrated. There is evidence that the translational repressor Bruno (Bru), which represses translation of *osk* outside of the pole plasm, interacts with Vas. This raises the possibility of Vas being involved in translational activation of *osk* in the pole plasm by removing Bru (Webster et al. 1997).

6.6 Vas Function in Regulating Mitosis

In *vas*-null (vas^{PH165}) ovaries, germarium atrophy is associated with a decrease in GSCs, CBs and dividing cystocyte clusters, which are recognized based on their specific structures, spectrosomes, and fusomes (Styhler et al. 1998). This phenotype resembles, although is less severe than, that observed in null *nos* mutants. Nos is required for GSC proliferation (Forbes and Lehmann 1998); however, evidence to support that Vas function in regulating GSC divisions is mediated by *nos* is still lacking. More recently it was shown that concomitant with a delay in mitotic progression, a higher percentage of GSCs and CBs in *vas* null, compared to wild type, do not progress to metaphase and remain in prometaphase (Pek and Kai 2011a). Consistent with this Vas becomes localized at pericentromeric piRNA clusters in GSCs and CBs during prometaphase/metaphase, dependent on its interaction with two piRNA pathway components Aub and Spn-E. Co-IP from GSC-/CB-enriched *bag-of-marbles* (*bam*) mutant ovaries indicates that Vas directly interacts with the condensin I components, Barren (Barr) and CAP-D2. Localization of Barr to chromosomes, which is required for chromosome condensation and segregation, is abolished in *vas*-null GSCs and CBs and is restored by transgenically expressed Vas (Bhat et al. 1996; Pek and Kai 2011a; Somma et al. 2003). Interestingly $vas^{\Delta 617}$, which does not interact with eIF5B, could still rescue Barr recruitment to the chromosomes in *vas* null, suggesting that *vas* role in mitotic progression of germline stem cells does not require its function in translation regulation. Recent work also shows that the fecundity of *vas*-null females is surprisingly restored by expressing Vas proteins that carry mutations in the conserved core domains and are very likely to be enzymatically inactive. This suggests that an early function of Vas, possibly associated with mitotic division of GSCs, is independent of its catalytic activity and perhaps requires Vas only as a scaffolding protein (Dehghani and Lasko 2015).

Bel, the *Drosophila* ortholog of mammalian DDX3, has the closest sequence homology to Vas and is expressed both in somatic cells and in germ cells (Johnstone et al. 2005). Bel, which is required for larval growth and adult fertility, can functionally substitute for yeast Ded1p, a protein implicated in translation (Chuang et al. 1997), suggesting that the essential role of Bel in development is through regulation of translation. In addition, Bel has been shown to function in gene silencing through RNA interference (RNAi) (Ulvila et al. 2006; Zhou et al. 2008). Similar to Vas, which regulates mitotic progression in GSCs via its interaction with piRNA components, Bel functions, through endogenous (endo)-siRNA pathway, to promote chromosome segregation in somatic cells (Pek and Kai 2011a, b). This function is through a physical interaction between Bel and the condensin I components, Barr and CAP-D2, while localization of Bel to the pericentric region of chromosomes is mediated by endo-siRNA pathway components, Ago2 and Dicer-2. The C-terminal end of Vas is distinct from Bel and other DEAD-box proteins and contains several acidic residues. Replacement of the seven C-terminal residues of Vas with those of Bel ($Vas^{c. bel}$) most profoundly affects early progression of oogenesis and piRNA biogenesis. Oocytes that escape this early block, when $Vas^{c. bel}$ is examined in a *vas* hypomorphic background, are however less severely affected than for most *vas* mutants, and some give rise to viable, fertile embryos (Dehghani and Lasko 2015, 2016). This suggests that the specific interactions of Vas with components of piRNA pathway are dependent on its unique acidic C-terminus. The C-terminal motif also contains a tryptophan residue (Trp660), which is very highly conserved among Vas orthologues. Mutating this residue to glutamic acid abolishes Vas function in germ cell formation and embryonic viability and dramatically decreases its function in piRNA pathway and activating Grk translation; yet it does not significantly affect female fecundity. Thus, these two elements in the Vas C-terminal end, Trp660 and the acidic residues, are both required for the full activity of Vas in *Drosophila* oogenesis and early embryogenesis (Dehghani and Lasko 2016).

6.7 Conclusion

Nearly 30 years of intensive study has revealed a large number of roles for Vas in germline development in *Drosophila*. This work has implicated Vas in a variety of RNA-dependent processes that underlie key developmental events in oogenesis and early embryogenesis. A great deal has been learned about exactly how Vas helps facilitate piRNA biogenesis and translational activation, but mechanistic detail is lacking about some of its other functions. Future work aimed at understanding the composition of Vas-containing ribonucleoprotein complexes will lead to further insights into how Vas functions during transit-amplifying divisions of germ cells and in the pole plasm.

References

Abdu U, Brodsky M, Schupbach T (2002) Activation of a meiotic checkpoint during *Drosophila* oogenesis regulates the translation of *gurken* through Chk2/Mnk. Curr Biol 12:1645–1651

Ables ET (2015) Drosophila oocytes as a model for understanding meiosis: an educational primer to accompany "Corolla is a novel protein that contributes to the architecture of the synaptonemal complex of Drosophila". Genetics 199:17–23

Anand A, Kai T (2012) The tudor domain protein Kumo is required to assemble the nuage and to generate germline piRNAs in *Drosophila*. EMBO J 31:870–882

Asaoka-Taguchi M, Yamada M, Nakamura A, Hanyu K, Kobayashi S (1999) Maternal Pumilio acts together with Nanos in germline development in *Drosophila* embryos. Nat Cell Biol 1: 431–437

Assa-Kunik E, Torres IL, Schejter ED, St Johnston D, Shilo BZ (2007) *Drosophila* follicle cells are patterned by multiple levels of Notch signaling and antagonism between the Notch and JAK/STAT pathways. Development 134:1161–1169

Barbosa C, Peixeiro I, Romao L (2013) Gene expression regulation by upstream open reading frames and human disease. PLoS Genet 9:e1003529

Bastock R, St Johnston D (2008) *Drosophila* oogenesis. Curr Biol 18:R1082–R1087

Becalska AN, Gavis ER (2009) Lighting up mRNA localization in *Drosophila* oogenesis. Development 136:2493–2503

Berg CA (2005) The *Drosophila* shell game: patterning genes and morphological change. Trends Genet 21:346–355

Berleth T, Burri M, Thoma G, Bopp D, Richstein S, Frigerio G, Noll M, Nüsslein-Volhard C (1988) The role of localization of bicoid RNA in organizing the anterior pattern of the *Drosophila* embryos. EMBO J 7:1749–1756

Bhat MA, Philp AV, Glover DM, Bellen HJ (1996) Chromatid segregation at anaphase requires the barren product, a novel chromosome-associated protein that interacts with topoisomerase II. Cell 87:1103–1114

Breitwieser W, Markussen FH, Horstmann H, Ephrussi A (1996) Oskar protein interaction with Vasa represents an essential step in polar granule assembly. Genes Dev 10:2179–2188

Brennecke J, Aravin AA, Stark A, Dus M, Kellis M, Sachidanandam R, Hannon GJ (2007) Discrete small RNA-generating loci as master regulators of transposon activity in *Drosophila*. Cell 128:1089–1103

Campos-Ortega JA, Hartenstein V (1985) The embryonic development of *Drosophila melanogaster*. Springer, Berlin

Carrera P, Johnstone O, Nakamura A, Casanova J, Jäckle H, Lasko P (2000) VASA mediates translation through interaction with a *Drosophila* yIF2 homolog. Mol Cell 5:181–187

Cavaliere V, Taddei C, Gargiulo G (1998). Apoptosis of nurse cells at the late stages of oogenesis of *Drosophila melanogaster*. Dev Genes Evol 208:106–112

Chang H, Matzuk MM (2001) Smad5 is required for mouse primordial germ cell development. Mech Dev 104:61–67

Chen Y, Potratz JP, Tijerina P, Del Campo M, Lambowitz AM, Russell R (2008) DEAD-box proteins can completely separate an RNA duplex using a single ATP. Proc Natl Acad Sci USA 105:20203–20208

Chuang RY, Weaver PL, Liu Z, Chang TH (1997) Requirement of the DEAD-box protein Ded1p for messenger RNA translation. Science 275:1468–1471

Clark A, Meignin C, Davis I (2007) A Dynein-dependent shortcut rapidly delivers axis determination transcripts into the *Drosophila* oocyte. Development 134:1955–1965

Clouse KN, Ferguson SB, Schupbach T (2008) Squid, Cup, and PABP55B function together to regulate *gurken* translation in *Drosophila*. Dev Biol 313:713–724

Cordin O, Banroques J, Tanner NK, Linder P (2006) The DEAD-box protein family of RNA helicases. Gene 367:17–37

Dehghani M, Lasko P (2015) *In vivo* mapping of the functional regions of the DEAD-box helicase Vasa. Biol Open 4:450–462

Dehghani M, Lasko P (2016) C-terminal residues specific to Vasa among DEAD-box helicases are required for its functions in piRNA biogenesis and embryonic patterning. Dev Genes Evol 226(6):401–412

Deng W, Lin HF (2002) *miwi*, a murine homolog of *piwi*, encodes a cytoplasmic protein essential for spermatogenesis. Dev Cell 2:819–830

Desset S, Meignin C, Dastugue B, Vaury C (2003) COM, a heterochromatic locus governing the control of independent endogenous retroviruses from *Drosophila melanogaster*. Genetics 164: 501–509

Ephrussi A, Dickinson LK, Lehmann R (1991) Oskar organizes the germ plasm and directs localization of the posterior determinant *nanos*. Cell 66:37–50

Fekete CA, Applefield DJ, Blakely SA, Shirokikh N, Pestova T, Lorsch JR, Hinnebusch AG (2005) The eIF1A C-terminal domain promotes initiation complex assembly, scanning and AUG selection *in vivo*. EMBO J 24:3588–3601

Forbes A, Lehmann R (1998) Nanos and Pumilio have critical roles in the development and function of *Drosophila* germline stem cells. Development 125:679–690

Forrest KM, Gavis ER (2003) Live imaging of endogenous RNA reveals a diffusion and entrapment mechanism for *nanos* mRNA localization in *Drosophila*. Curr Biol 13:1159–1168

Gavis ER, Lehmann R (1992) Localization of *nanos* RNA controls embryonic polarity. Cell 71: 301–313

Geigy R (1931) Action de l'ultra-violet sur le pole germinal dans l'oeuf de *Drosophila melanogaster*. Rev Suisse Zool 38:187–288

Ghabrial A, Schupbach T (1999) Activation of a meiotic checkpoint regulates translation of Gurken during *Drosophila* oogenesis. Nat Cell Biol 1:354–357

Goldman CH, Gonsalvez GB (2017) The role of microtubule motors in mRNA localization and patterning within the Drosophila oocyte. In: Kloc M (ed) Oocytes. Springer, Heidelberg

González-Reyes A, Elliott H, Stjohnston D (1995) Polarization of both major body axes in *Drosophila* by Gurken-Torpedo signaling. Nature 375:654–658

Grieder NC, de Cuevas M, Spradling AC (2000) The fusome organizes the microtubule network during oocyte differentiation in *Drosophila*. Development 127:4253–4264

Grimson A, Srivastava M, Fahey B, Woodcroft BJ, Chiang HR, King N, Degnan BM, Rokhsar DS, Bartel DP (2008) Early origins and evolution of microRNAs and Piwi-interacting RNAs in animals. Nature 455:1193–1115

Haase AD (2016) A small RNA-based immune system defends germ cells against mobile genetic elements. Stem Cells Int 2016:7595791

Harris AN, Macdonald PM (2001) *aubergine* encodes a *Drosophila* polar granule component required for pole cell formation and related to eIF2C. Development 128:2823–2832

Harrison RE, Huebner E (1997) Unipolar microtubule array is directly involved in nurse cell-oocyte transport. Cell Motil Cytoskeleton 36:355–362

Hay B, Jan LY, Jan YN (1988b) A protein component of *Drosophila* polar granules is encoded by *vasa* and has extensive sequence similarity to ATP-dependent helicases. Cell 55:577–587

Hay B, Jan LY, Jan YN (1990) Localization of Vasa, a component of *Drosophila* polar granules, in maternal-effect mutants that alter embryonic anteroposterior polarity. Development 109: 425–433

Hilbert M, Karow AR, Klostermeier D (2009) The mechanism of ATP-dependent RNA unwinding by DEAD-box proteins. Biol Chem 390:1237–1250

Hinnebusch AG (2014) The scanning mechanism of eukaryotic translation initiation. Annu Rev Biochem 83:779–812

Hirakata S, Siomi MC (2016) piRNA biogenesis in the germline: from transcription of piRNA genomic sources to piRNA maturation. Biochim Biophys Acta 1859:82–92

Horner VL, Wolfner MF (2008) Transitioning from egg to embryo: triggers and mechanisms of egg activation. Dev Dyn 237:527–544

Huang YZ, Baker RT, Fischervize JA (1995) Control of cell fate by a deubiquitinating enzyme encoded by the *fat facets* gene. Science 270:1828–1831

Huang XA, Yin H, Sweeney S, Raha D, Snyder M, Lin H (2013) A major epigenetic programming mechanism guided by piRNAs. Dev Cell 24:502–516

Hu TT, Eisen MB, Thornton KR, Andolfatto P (2013) A second-generation assembly of the *Drosophila simulans* genome provides new insights into patterns of lineage-specific divergence. Genome Res 23:89–98

Illmensee K, Mahowald AP (1974) Transplantation of posterior polar plasm in *Drosophila* – induction of germ cells at anterior pole of egg. Proc Natl Acad Sci USA 71:1016–1020

Johnstone O, Lasko P (2004) Interaction with eIF5B is essential for Vasa function during development. Development 131:4167–4178

Johnstone O, Deuring R, Bock R, Linder P, Fuller MT, Lasko P (2005) Belle is a *Drosophila* DEAD-box protein required for viability and in the germ line. Dev Biol 277:92–101

Jongens TA, Hay B, Jan LY, Jan YN (1992) The *germ cell-less* gene product – a posterior localized component necessary for germ-cell development in *Drosophila*. Cell 70:569–584

Kaminker JS, Bergman CM, Kronmiller B, Carlson J, Svirskas R, Patel S, Frise E, Wheeler DA, Lewis SE, Rubin GM, Ashburner M, Celniker SE (2002) The transposable elements of the *Drosophila melanogaster* euchromatin: a genomics perspective. Genome Biol 3: RESEARCH0084

Kawaoka S, Hayashi N, Suzuki Y, Abe H, Sugano S, Tomari Y, Shimada T, Katsuma S (2009) The *Bombyx* ovary-derived cell line endogenously expresses Piwi/Piwi-interacting RNA complexes. RNA 15:1258–1264

Kim-Ha J, Kerr K, Macdonald PM (1995) Translational regulation of oskar messenger RNA by Bruno, an ovarian RNA-binding protein, is essential. Cell 81:403–412

King RC (1970) Ovarian development in *Drosophila melanogaster*. Academic Press, New York

Klattenhoff C, Xi H, Li C, Lee S, Xu J, Khurana JS, Zhang F, Schultz N, Koppetsch BS, Nowosielska A et al (2009) The *Drosophila* HP1 homolog Rhino is required for transposon silencing and piRNA production by dual-strand clusters. Cell 138:1137–1149

Kugler JM, Woo JS, Oh BH, Lasko P (2010) Regulation of *Drosophila* Vasa *in vivo* through paralogous cullin-RING E3 ligase specificity receptors. Mol Cell Biol 30:1769–1782

Kuramochi-Miyagawa S, Kimura T, Ijiri TW, Isobe T, Asada N, Fujita Y, Ikawa M, Iwai N, Okabe M, Deng W et al (2004) *mili*, a mammalian member of *piwi* family gene, is essential for spermatogenesis. Development 131:839–849

Kuramochi-Miyagawa S, Watanabe T, Gotoh K, Takamatsu K, Chuma S, Kojima-Kita K, Shiromoto Y, Asada N, Toyoda A, Fujiyama A et al (2010) MVH in piRNA processing and gene silencing of retrotransposons. Genes Dev 24:887–892

Lasko PF (2012) mRNA localization and translational control in *Drosophila* oogenesis. Cold Spring Harb Perspect Biol 4:pii: a012294

Lasko PF, Ashburner M (1988) The product of the *Drosophila* gene *vasa* is very similar to eukaryotic initiation factor-4A. Nature 335:611–617

Lasko PF, Ashburner M (1990) Posterior localization of Vasa protein correlates with, but is not sufficient for, pole cell development. Genes Dev 4:905–921

Lehmann R, Nüsslein-Volhard C (1987) *hunchback*, a gene required for segmentation of an anterior and posterior region of the *Drosophila* embryo. Dev Biol 119:402–417

Lehmann R (2012) Germline stem cells: origin and destiny. Cell Stem Cell 10:729–739

Lerit DA, Gavis ER (2011) Transport of germ plasm on astral microtubules directs germ cell development in *Drosophila*. Curr Biol 21:439–448

Lesch BJ, Page DC (2012) Genetics of germ cell development. Nat Rev Genetics 13:781–794

Li W, Klovstad M, Schuepbach T (2014) Repression of Gurken translation by a meiotic checkpoint in *Drosophila* oogenesis is suppressed by a reduction in the dose of eIF1A. Development 141:3910–3921

Liang L, Diehljones W, Lasko P (1994) Localization of Vasa protein to the *Drosophila* pole plasm is independent of its RNA-binding and helicase activities. Development 120:1201–1211

Linder P, Fuller-Pace FV (2013) Looking back on the birth of DEAD-box RNA helicases. Biochim Biophys Acta 1829:750–755

Linder P, Jankowsky E (2011) From unwinding to clamping – the DEAD box RNA helicase family. Nat Rev Mol Cell Biol 12:505–516

Linder P, Lasko PF, Ashburner M, Leroy P, Nielsen PJ, Nishi K, Schnier J, Slonimski PP (1989) Birth of the D-E-A-D box. Nature 337:121–122

Liu NK, Dansereau DA, Lasko P (2003) Fat facets interacts with Vasa in the *Drosophila* pole plasm and protects it from degradation. Curr Biol 13:1905–1909

Liu NK, Han H, Lasko P (2009) Vasa promotes *Drosophila* germline stem cell differentiation by activating mei-P26 translation by directly interacting with a (U)-rich motif in its 3′ UTR. Genes Dev 23:2742–2752

Mach JM, Lehmann R (1997) An Egalitarian-BicaudalD complex is essential for oocyte specification and axis determination in *Drosophila*. Genes Dev 11:423–435

Mahowald AP (1968) Polar granules in *Drosophila*: II. Ultrastructural changes during early embryogenesis. J Exp Zool 167:237–261

Mahowald AP, Goralski TJ, Caulton JH (1983) *In vitro* activation of *Drosophila* eggs. Dev Biol 98:437–445

McLaughlin JM, Bratu DP (2015) *Drosophila melanogaster* oogenesis: an overview. Methods Mol Biol 1328:1–20

Meignin C, Davis I (2010) Transmitting the message: intracellular mRNA localization. Curr Opin Cell Biol 22:112–119

Meister G (2013) Argonaute proteins: functional insights and emerging roles. Nat Rev Genet 14: 447–459

Mills RE, Bennett EA, Iskow RC, Devine SE (2007) Which transposable elements are active in the human genome? Trends Genet 23:183–191

Mohn F, Sienski G, Handler D, Brennecke J (2014) The Rhino-Deadlock-Cutoff complex Licenses noncanonical transcription of dual-strand piRNA clusters in *Drosophila*. Cell 157:1364–1379

Murata Y, Wharton RP (1995) Binding of Pumilio to maternal Hunchback messenger RNA is required for posterior patterning in *Drosophila* embryos. Cell 80:747–756

Nakamura A, Amikura R, Mukai M, Kobayashi S, Lasko PF (1996) Requirement for a noncoding RNA in *Drosophila* polar granules for germ cell establishment. Science 274:2075–2079

Neuman-Silberberg FS, Schupbach T (1993) The *Drosophila* dorsoventral patterning gene *gurken* produces a dorsally localized RNA and encodes a TGF-alpha-like protein. Cell 75:165–174

Neuman-Silberberg FS, Schupbach T (1996) The *Drosophila* TGF-alpha-like protein Gurken: expression and cellular localization during *Drosophila* oogenesis. Mech Dev 59:105–113

Neumuller RA, Betschinger J, Fischer A, Bushati N, Poernbacher I, Mechtler K, Cohen SM, Knoblich JA (2008) Mei-P26 regulates microRNAs and cell growth in the *Drosophila* ovarian stem cell lineage. Nature 454:241–245

Nishida KM, Iwasaki YW, Murota Y, Nagao A, Mannen T, Kato Y, Siomi H, Siomi MC (2015) Respective functions of two distinct Siwi complexes assembled during Piwi-interacting RNA biogenesis in *Bombyx* germ cells. Cell Rep. 10:193–203

Nishimasu H, Ishizu H, Saito K, Fukuhara S, Kamatani MK, Bonnefond L, Matsumoto N, Nishizawa T, Nakanaga K, Aoki J et al (2012) Structure and function of Zucchini endoribonuclease in piRNA biogenesis. Nature 491:284–287

Okada M, Kleinman IA, Schneide HA (1974) Restoration of fertility in sterilized *Drosophila* eggs by transplantation of polar cytoplasm. Dev Biol 37:43–54

Page SL, McKim KS, Deneen B, Van Hook TL, Hawley RS (2000) Genetic studies of mei-P26 reveal a link between the processes that control germ cell proliferation in both sexes and those that control meiotic exchange in *Drosophila*. Genetics 155:1757–1772

Parton RM, Hamilton RS, Ball G, Yang L, Cullen CF, Lu W, Ohkura H, Davis I (2011) A PAR-1-dependent orientation gradient of dynamic microtubules directs posterior cargo transport in the *Drosophila* oocyte. J Cell Biol 194:121–135

Patil VS, Kai T (2010) Repression of retroelements in *Drosophila* germline via piRNA pathway by the Tudor domain protein Tejas. Curr Biol 20:724–730

Pek JW, Kai T (2011a) A Role for Vasa in regulating mitotic chromosome condensation in *Drosophila*. Curr Biol 21:39–44

Pek JW, Kai T (2011b) DEAD-box RNA helicase Belle/DDX3 and the RNA interference pathway promote mitotic chromosome segregation. Proc Natl Acad Sci USA 108:12007–12012

Pelisson A, Song SU, Prudhomme N, Smith PA, Bucheton A, Corces VG (1994) Gypsy transposition correlates with the production of a retroviral envelope-like protein under the tissue-specific control of the *Drosophila flamenco* gene. EMBO J 13:4401–4411

Pestova TV, Lomakin IB, Lee JH, Choi SK, Dever TE, Hellen CU (2000) The joining of ribosomal subunits in eukaryotes requires eIF5B. Nature 403:332–335
Pray LA (2008) Transposons: the jumping genes. Nat Educ 1:204
Roth S, Lynch JA (2009) Symmetry breaking during *Drosophila* oogenesis. Cold Spring Harb Perspect Biol 1:a001891
Roth S, Neuman-Silberberg FS, Barcelo G, Schüpbach T (1995) *cornichon* and the EGF receptor signaling process are necessary for both anterior-posterior and dorsal-ventral pattern formation in *Drosophila*. Cell 81:967–978
Rudolph MG, Klostermeier D (2015) When core competence is not enough: functional interplay of the DEAD-box helicase core with ancillary domains and auxiliary factors in RNA binding and unwinding. Biol Chem 396:849–865
Saito K (2013) The epigenetic regulation of transposable elements by Piwi-interacting RNAs in *Drosophila*. Genes Genet Syst 88:9–17
Schupbach T, Wieschaus E (1986a) Germline autonomy of maternal-effect mutations altering the embryonic body pattern of *Drosophila*. Dev Biol 113:443–448
Schupbach T, Wieschaus E (1986b) Maternal-effect mutations altering the anterior-posterior pattern of the *Drosophila* embryo. Roux Arch Dev Biol 195:302–317
Schwer B, Meszaros T (2000) RNA helicase dynamics in pre-mRNA splicing. EMBO J 19: 6582–6591
Seleme MD, Busseau I, Malinsky S, Bucheton A, Teninges D (1999) High-frequency retrotransposition of a marked I factor in *Drosophila melanogaster* correlates with a dynamic expression pattern of the ORF1 protein in the cytoplasm of oocytes. Genetics 151:761–771
Sengoku T, Nureki O, Nakamura A, Satoru KI, Yokoyama S (2006) Structural basis for RNA unwinding by the DEAD-box protein *Drosophila* Vasa. Cell 125:287–300
Sinsimer KS, Lee JJ, Thiberge SY, Gavis ER (2013) Germ plasm anchoring is a dynamic state that requires persistent trafficking. Cell Rep 5:1169–1177
Snee MJ, Macdonald PM (2004) Live imaging of nuage and polar granules: evidence against a precursor-product relationship and a novel role for Oskar in stabilization of polar granule components. J Cell Sci 117:2109–2120
Somma MP, Fasulo B, Siriaco G, Cenci G (2003) Chromosome condensation defects in *barren* RNA-interfered *Drosophila* cells. Genetics 165:1607–1611
Steinhauer J, Kalderon D (2006) Microtubule polarity and axis formation in the *Drosophila* oocyte. Dev Dyn 235:1455–1468
Styhler S, Nakamura A, Swan A, Suter B, Lasko P (1998) Vasa is required for Gurken accumulation in the oocyte, and is involved in oocyte differentiation and germline cyst development. Development 125:1569–1578
Styhler S, Nakamura A, Lasko P (2002) Vasa localization requires the SPRY-domain and SOCS-box containing protein, Gustavus. Dev Cell 3:865–876
Su TT, Campbell SD, O'Farrell PH (1998) The cell cycle program in germ cells of the *Drosophila* embryo. Dev Biol 196:160–170
Tautz D (1988) Regulation of the *Drosophila* segmentation gene *hunchback* by 2 maternal morphogenetic centers. Nature 332:281–284
Theurkauf WE, Smiley S, Wong ML, Alberts BM (1992) Reorganization of the cytoskeleton during *Drosophila* oogenesis – implications for axis specification and intercellular transport. Development 115:923–936
Thio GL, Ray RP, Barcelo G, Schupbach T (2000) Localization of *gurken* RNA in *Drosophila* oogenesis requires elements in the 5′ and 3′ regions of the transcript. Dev Biol 221:435–446
Tomancak P, Guichet A, Zavorszky P, Ephrussi A (1998) Oocyte polarity depends on regulation of *gurken* by Vasa. Development 125:1723–1732
Torres IL, Lopez-Schier H, St Johnston D (2003) A notch/delta-dependent relay mechanism establishes anterior-posterior polarity in *Drosophila*. Dev Cell 5:547–558

Ulvila J, Parikka M, Kleino A, Sormunen R, Ezekowitz RA, Kocks C, Ramet M (2006) Double-stranded RNA is internalized by scavenger receptor-mediated endocytosis in *Drosophila* S2 cells. J Biol Chem 281:14370–14375

Webster PJ, Liang L, Berg CA, Lasko P, Macdonald PM (1997) Translational repressor Bruno plays multiple roles in development and is widely conserved. Genes Dev 11:2510–2521

Webster A, Li S, Hur JK, Wachsmuth M, Bois JS, Perkins EM, Patel DJ, Aravin AA (2015) Aub and Ago3 are recruited to nuage through two mechanisms to form a ping-pong complex assembled by Krimper. Mol Cell 59:564–575

Weil TT (2014) mRNA localization in the *Drosophila* germline. RNA Biol 11:1010–1018

Xiol J, Spinelli P, Laussmann MA, Homolka D, Yang Z, Cora E, Couté Y, Conn S, Kadlec J, Sachidanandam R, Kaksonen M, Cusack S, Ephrussi A, Pillai RS (2014) RNA clamping by Vasa assembles a piRNA amplifier complex on transposon transcripts. Cell 157:1698–1711

Zaessinger S, Busseau I, Simonelig M (2006) Oskar allows *nanos* mRNA translation in *Drosophila* embryos by preventing its deadenylation by Smaug/CCR4. Development 133:4573–4583

Zhang F, Wang J, Xu J, Zhang Z, Koppetsch BS, Schultz N, Vreven T, Meignin C, Davis I, Zamore PD, et al. (2012). UAP56 couples piRNA clusters to the perinuclear transposon silencing machinery. Cell 151: 871-884

Zhao T, Graham OS, Raposo A, St Johnston D (2012) Growing microtubules push the oocyte nucleus to polarize the *Drosophila* dorsal-ventral axis. Science 336:999–1003

Zhou R, Hotta I, Denli AM, Hong P, Perrimon N, Hannon GJ (2008) Comparative analysis of argonaute-dependent small RNA pathways in *Drosophila*. Mol Cell 32:592–599

Zimyanin VL, Belaya K, Pecreaux J, Gilchrist MJ, Clark A, Davis I, St Johnston D (2008) *In vivo* imaging of *oskar* mRNA transport reveals the mechanism of posterior localization. Cell 134: 843–853

Chapter 7
The Role of Microtubule Motors in mRNA Localization and Patterning Within the *Drosophila* Oocyte

Chandler H. Goldman and Graydon B. Gonsalvez

Abstract Messenger RNA (mRNA) localization is a powerful and prevalent mechanism of post-transcriptional gene regulation, enabling the cell to produce protein at the exact location at which it is needed. The phenomenon of mRNA localization has been observed in many types of cells in organisms ranging from yeast to man. Thus, the process appears to be widespread and highly conserved. Several model systems have been used to understand the mechanism by which mRNAs are localized. One such model, and the focus of this chapter, is the egg chamber of the female *Drosophila melanogaster*. The polarity of the developing *Drosophila* oocyte and resulting embryo relies on the specific localization of three critical mRNAs: *gurken*, *bicoid*, and *oskar*. If these mRNAs are not localized during oogenesis, the resulting progeny will not survive. The study of these mRNAs has served as a model for understanding the general mechanisms by which mRNAs are sorted. In this chapter, we will discuss how the localization of these mRNAs enables polarity establishment. We will also discuss the role of motor proteins in the localization pathway. Finally, we will consider potential mechanisms by which mRNAs can be anchored at their site of localization. It is likely that the lessons learned using the *Drosophila* oocyte model system will be applicable to mRNAs that are localized in other organisms as well.

Abbreviations

bcd	*bicoid*
dhc	*Dynein heavy chain*
egl	*Egalitarian*
grk	*gurken*
khc	*Kinesin heavy chain*

C.H. Goldman • G.B. Gonsalvez (✉)
Cellular Biology and Anatomy, Medical College of Georgia, Augusta University, 1459 Laney Walker Blvd., CB2917, Augusta, GA 30912, USA
e-mail: ggonsalvez@augusta.edu

M. Kloc (ed.), *Oocytes*, Results and Problems in Cell Differentiation 63,
DOI 10.1007/978-3-319-60855-6_7

klc	*Kinesin light chain*
osk	*oskar*
RNP	Ribonucleoprotein
tm1	*Tropomyosin1*

7.1 Introduction

Research over the past two decades has revealed that eukaryotic cells possess numerous mechanisms to post-transcriptionally regulate gene expression. A few examples include alternative splicing, regulated nuclear export, microRNAs, translational regulation, polyA tail lengthening, and regulation of mRNA stability. An additional mechanism used by many cell types in diverse organisms to modulate gene expression is mRNA localization (Buxbaum et al. 2015). In this scenario, the transcript that codes for a particular protein is selectively localized within the cell. Once localized, the mRNA is translated, thus resulting in a high local concentration of the protein (Martin and Ephrussi 2009). A significant benefit of this mechanism is that it enables the cell to spatially regulate gene expression.

One of the earliest examples of this phenomenon was the localization of β-actin mRNA to the leading edge of migrating fibroblast cells (Lawrence and Singer 1986). Since that time, examples of mRNA localization have been identified in *S. cerevisiae* (yeast), *Xenopus* oocytes (frog), zebra fish, and a variety of mammalian cell types (Oh and Houston 2017; Kloc et al. 2002; Martin and Ephrussi 2009; Medioni et al. 2012; Singer-Kruger and Jansen 2014). Until recently, however, these examples were thought to represent unique cases: the exception rather than the rule. Several large genome-wide screens have challenged this dogma. In one striking example, approximately 70% of examined mRNAs were shown to be specifically localized in *Drosophila* embryos (Lecuyer et al. 2007). Similar findings have also been described in neurons and migratory cells (Cajigas et al. 2012; Mili et al. 2008). Thus, the emerging picture suggests that mRNA localization is a more prevalent and potent mechanism of gene regulation than was previously recognized.

There are four main routes by which mRNAs are known to localize. The most commonly used mechanism, and the main focus of this chapter, is directed transport along cytoskeletal filaments. A second mechanism involves random diffusion of the mRNA coupled with selective anchoring at a particular cellular site. A third mechanism is the transport of mRNAs within a specialized oocyte-specific organelle known as the Balbiani body or the mitochondrial cloud. The final mechanism relies on degradation of the mRNA at most cellular sites accompanied with selective protection from degradation at the desired location. It should be noted that these mechanisms are not mutually exclusive. In fact, the complex localization pathways of many mRNAs involve deploying these mechanisms in combination in order to achieve the correct temporal and spatial localization of the transcript. A good example of this concept is the localization of *oskar* (*osk*) mRNA in *Drosophila* oocytes, a topic that will be covered later in this chapter.

Perhaps the best-characterized example of a localized transcript is the sorting of *ASH1* mRNA to the bud tip of dividing yeast cells (Gonsalvez et al. 2005). This sorting enables Ash1 protein (Ash1p) to selectively enter the nucleus of the daughter cell. Within the daughter cell nucleus, Ash1p inhibits transcription of *HO* endonuclease (Sil and Herskowitz 1996), a factor that is required for mating-type switching. Thus, the localization of *ASH1* mRNA ultimately restricts mating-type switching to mother cells, a situation that is advantageous for diploid formation. The localization of *ASH1* mRNA involves binding of the transcript by the RNA binding protein, She2p (Bohl et al. 2000; Long et al. 2000). In addition to binding the mRNA, She2p also directly binds the adaptor protein She3p (Bohl et al. 2000; Long et al. 2000; Takizawa and Vale 2000). She3p functions as an adaptor in this pathway by linking the *ASH1* mRNA/She2p complex to the myosin V motor, Myo4p. This enables the messenger ribonucleoprotein (RNP) complex to associate with actin filaments for transport to the daughter cell (Long et al. 1997; Takizawa et al. 1997). Thus, in the case of *ASH1* mRNA, a series of direct interactions have been identified that link the mRNA to the appropriate cytoskeletal filament for transport.

The benefit of the yeast model system is its exquisite simplicity. It is, after all, a unicellular organism that can be studied in the lab in a haploid state. By contrast, mammalian model systems such as the mouse offer a biology that is more similar to our own. However, with this similarity comes experimental complexity. Although there are several known examples of mRNAs that are localized in mammalian cells, understanding their mechanism of localization has proven challenging. In between both of these extremes lies the *Drosophila melanogaster* model system. Like yeast, *Drosophila* provides a genetically tractable model with the availability of sophisticated cell biological tools. In addition, with the *Drosophila* system, processes such as mRNA localization can be studied using native tissues instead of isolated cells. Thus, the insights gained from *Drosophila* will likely relate to processes that take place in our own cells and tissues. In support of this notion, factors that were first identified as being required for mRNA localization in the *Drosophila* egg chamber were later shown to perform a similar function in mammalian neurons (Giorgi et al. 2007; Heraud-Farlow and Kiebler 2014).

7.2 The *Drosophila* Egg Chamber as a Model for mRNA Localization

To a certain extent, the fate of the *Drosophila* embryo is sealed at birth. Even at the earliest stage of embryogenesis, the anterior-posterior polarity of the embryo is set. The critical events that are required for progressing through these stages of embryogenesis occur within the oocyte of the mother. If these processes fail within the female germline, the embryo will not survive.

Each ovary is composed of several ovarioles (Fig. 7.1). The germline stem cells are found within the germarium, at the anterior tip of the ovariole. Division of a

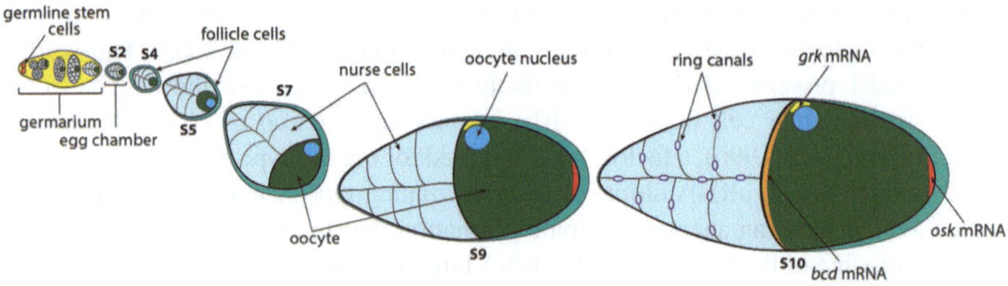

Fig. 7.1 Schematic of an ovariole. The germline stem cells reside in a region known as the germarium. Division of a germline stem cell gives rise to a cystoblast. This cell undergoes four rounds of cell division to produce a cyst containing 16 cells. One of these cells will differentiate to form the oocyte, whereas the remainder will become nurse cells. As the cyst exits the germarium, it becomes surrounded by a layer of somatic cells referred to as follicle cells. This structure is now called an egg chamber. Egg chambers undergo 14 stages of development before the oocyte is competent for fertilization. Stages 2, 4, 5, 7, 9, and 10 are indicated in this figure. The localization of *grk*, *bcd*, and *osk* mRNA in a stage 10 egg chamber is also indicated

germline stem cell results in a daughter cell referred to as a cystoblast (Spradling et al. 2001; Spradling 1993). This cell undergoes four rounds of cell division with incomplete cytokinesis to produce a cyst containing 16 germ cells. Fifteen of these cells will differentiate into a nurse cell lineage, whereas a single cell will be specified as the oocyte (Spradling 1993). As the cyst progresses through the germarium, it becomes surrounded by a layer of somatic epithelial cells called follicle cells (Spradling 1993). This layer of somatic cells and the underlying germline is collectively referred to as an egg chamber (Fig. 7.1). The egg chamber undergoes 14 stages of maturation before it is competent for fertilization (Spradling 1993).

The polarity of the oocyte and the resulting embryo depends on the precise localization of three critical mRNAs: *gurken* (*grk*), *oskar* (*osk*), and *bicoid* (*bcd*). These mRNAs, as well as the majority of mRNAs in the egg chamber, are transcribed by nurse cell nuclei. The oocyte is halted in prophase I of meiosis and is therefore thought to be transcriptionally quiescent with perhaps a few exceptions (Ables 2015; Saunders and Cohen 1999). Because the 16-cell egg chamber was produced as a consequence of cell division with incomplete cytokinesis, the nurse cells and oocyte share a common cytoplasm. Ring canals, which represent the contractile actomyosin ring formed during cytokinesis, persist and allow for transport of mRNAs and proteins between nurse cells and the oocyte.

7.3 Cytoskeletal Roadways and Their Associated Motors

Active transport of molecules within the cell requires at least two components, the cytoskeletal filaments that serve as tracks for transport and the motor proteins that translocate cargo along these tracks. The types of cytoskeletal filaments that are used for cargo transport are microtubules and actin. In the majority of cases, long-

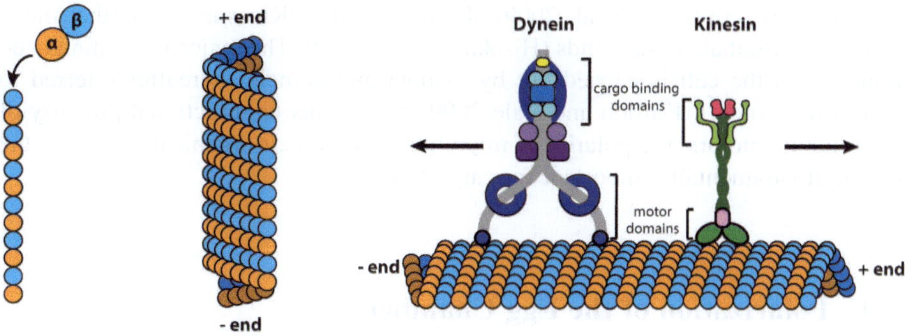

Fig. 7.2 Microtubules and motors. A microtubule is composed of repeating units of alpha- and beta-tubulin dimers. The end of the microtubule with an exposed beta-tubulin subunit is designated as the plus-end. The end with the exposed alpha-tubulin subunit is the minus-end. In general, Kinesin motors transport cargo toward microtubule plus-ends, whereas Dynein is involved in minus-end-directed transport

distance intracellular transport uses the microtubule cytoskeleton, whereas actin is more commonly used for short-range movement and cortical anchoring. There are exceptions to this, however. For instance, the localization of *ASH1* mRNA to the bud tip of dividing yeast cells involves actin and its associated myosin motor (Long et al. 1997; Takizawa et al. 1997).

Microtubules are hollow tubes consisting of 13 protofilaments held together by numerous interactions along their lateral surface. The protofilaments are composed of dimers of alpha- and beta-tubulin that stack in a precise and repeating orientation (Fig. 7.2). The net result of this is that the microtubule is inherently polarized. The end of the microtubule with an exposed beta-tubulin subunit is defined as the plus-end, whereas the end with an exposed alpha-tubulin subunit is the minus-end (Fig. 7.2) (Galjart 2010). In cells, however, the minus-end is rarely exposed, often being buried within a microtubule-organizing center (Luders and Stearns 2007).

Microtubule formation is a highly regulated process. In many cell types, microtubules originate from a centrosome. Found within the centrosome are numerous microtubule-organizing centers. Often, the centrosome resides adjacent to the nucleus. In these cells, the minus-ends of microtubules are buried within the centrosome, and the plus-ends extend toward the cell periphery. However, this is not always the case. For example, in the axonal process of most neurons, plus-ends extend away from the cell body. By contrast, microtubules within the dendrites of these same cells display a mixed orientation; some have plus-ends that extend away from the cell body, whereas others originate from the periphery of the process and grow toward the cell body (Kapitein and Hoogenraad 2015). The *Drosophila* oocyte contains a microtubule network that displays a complex organizational pattern (Theurkauf 1994). In addition, as discussed in the subsequent section, this pattern changes as the egg chamber matures.

Microtubule motors come in two general varieties. Kinesin motors, of which there are 45 in mammalian cells, in general move cargo toward the plus-end of microtubules (Hirokawa et al. 2009). There are a few Kinesin motors that move toward microtubule minus-ends (Hirokawa et al. 2009). The majority of minus-end transport in the cell is carried out by cytoplasmic Dynein (hereafter referred to simply as Dynein) (Kardon and Vale 2009). Given this unidirectional property of microtubule motors, the polarity of microtubules and their organization within the cell are fundamentally important for cargo transport.

7.4 Polarization of the Egg Chamber

The initiating event in oocyte polarization is the localization of *grk* mRNA and protein at the posterior of early-stage egg chambers (Fig. 7.3a). Between stages 6 and 7, Grk protein signals the overlying follicle cells to specify posterior fate (Gonzalez-Reyes et al. 1995; Neuman-Silberberg and Schupbach 1996; Roth et al. 1995). These cells in turn signal back to the oocyte. This follicle cell-based signal initiates a reorganization of microtubules within the oocyte. Prior to this stage, microtubule minus-ends are enriched at the posterior of the oocyte. Upon receiving the follicle cell-based signal, this organization changes, and minus-ends become enriched along the anterior and cortical regions of the oocyte, with plus-ends displaying a posterior bias (Parton et al. 2011; Sanghavi et al. 2012; Steinhauer and Kalderon 2006). The net result of this phenomenon is that the oocyte nucleus, which was localized at the posterior pole in stage 6 egg chambers, migrates to the dorsal-anterior corner (Gonzalez-Reyes et al. 1995; Roth et al. 1995). Another consequence of this reorganization is that *grk* mRNA becomes localized at the dorsal-anterior cortex of the oocyte, adjacent to the oocyte nucleus. Grk protein that is produced at this site is secreted and signals the overlying follicle cells to adopt dorsal fate (Gonzalez-Reyes et al. 1995; Roth et al. 1995) (Fig. 7.3b). This signaling event establishes the dorsoventral polarity of the oocyte.

The anterior-posterior polarity of the oocyte and future embryo is established by the combined action of *osk* and *bcd* mRNAs. As the egg chamber matures, *osk* mRNA becomes localized at the posterior pole (Ephrussi et al. 1991; Kim-Ha et al. 1991) (Fig. 7.3c). This localization pattern coupled with translational regulation results in the specific accumulation of Osk protein exclusively at the oocyte posterior (Kim-Ha et al. 1995). Osk is a potent morphogen and is responsible for establishing the germ plasm (Bilinski et al. 2017; Ephrussi and Lehmann 1992;

Fig. 7.3 mRNA localization in *Drosophila* oocytes. (**a**, **b**) Localization of *grk* mRNA in stage 6 (**a**) and stage 9 (**b**) egg chambers. (**c**) Localization of *osk* mRNA in a stage 10 egg chamber. (**d**) Localization of *bcd* mRNA in a stage 10 egg chamber

Smith et al. 1992). Germ plasm instructs the formation of posterior structures and primordial germ cells in the resulting embryo (Gavis and Lehmann 1992). In contrast to *osk*, *bcd mRNA* is localized at the anterior margin of the oocyte where it remains in a translationally quiescent state until embryogenesis (Berleth et al. 1988) (Fig. 7.3d). Once translated, Bcd is responsible for formation of anterior structures in the embryo (Berleth et al. 1988).

The three polarity-determining mRNAs, *grk*, *bcd*, and *osk*, all use an active microtubule-/motor-based mechanism for their localization. However, other models of localization also exist in *Drosophila*. For instance, localization of *nanos* (*nos*) mRNA at the posterior pole involves a diffusion and entrapment mechanism (Forrest and Gavis 2003). In addition, *nos* mRNA molecules present in the anterior portion of the embryo are actively degraded, whereas those that localize at the posterior pole are protected from degradation (Zaessinger et al. 2006). A similar degradation/selective protection mechanism contributes to the asymmetric localization of *hsp83* mRNA (Bashirullah et al. 2001). This chapter will focus on the mechanism by which *grk*, *bcd*, and *osk* mRNAs are localized and the important role played by microtubule motors in this process.

7.4.1 *Localization of* grk *mRNA*

Although the oocyte nucleus is generally thought to be transcriptionally quiescent, it appears to transcribe *grk* mRNA in mid-stage egg chambers (Saunders and Cohen 1999). In early stage egg chambers, however, *grk* mRNA is transcribed primarily by nurse cells and transported into the oocyte (Caceres and Nilson 2005).

The transport of *grk* mRNA into the oocyte appears to be aided by the Dynein motor. This was shown by injecting fluorescently labeled *grk* into the nurse cell cytoplasm. The injected mRNA moved in a rapid and directional manner, first toward nurse cell ring canals and subsequently translocated through the ring canals into the oocyte (Clark et al. 2007). Clark and colleagues further demonstrated that this movement was Dynein dependent; the mRNA particles moved significantly slower in hypomorphic *dynein heavy chain* (*dhc*) mutants (Clark et al. 2007). Dhc is the motor subunit of the Dynein complex. In addition, Egalitarian (Egl), an RNA-binding protein that was shown to function as a Dynein adaptor, was able to bind localization sequences present in *grk* mRNA (Dienstbier et al. 2009). Depletion of Egl resulted in a reduction of *grk* mRNA and protein within the oocyte (Sanghavi et al. 2016). Thus, the available evidence suggests that although *grk* mRNA can passively enter the oocyte, the Dynein motor and its cofactors ensure a robust oocyte enrichment of this transcript.

In mid-stage egg chambers, *grk* mRNA specifically localizes at the dorsal-anterior corner of the oocyte (Fig. 7.3b) (Neuman-Silberberg and Schupbach 1993). By following the transport of injected fluorescent *grk* mRNA, MacDougall and colleagues showed that this localization occurred in two distinct steps (MacDougall et al. 2003). First, the mRNA moves toward the anterior of the oocyte.

Next, the anterior-localized mRNA particles move toward the dorsal surface (MacDougall et al. 2003).

Several lines of evidence suggest that the primary motor involved in the transport of *grk* mRNA within the oocyte is Dynein. First, Dynein is a minus-end directed motor, and minus-ends of microtubules are concentrated at the anterior of the oocyte (Clark et al. 1997). Second, injection of function-blocking antibodies against Dhc significantly inhibited the localization of injected *grk* mRNA (MacDougall et al. 2003). The same result was obtained by genetically inhibiting the activity of Dynein (MacDougall et al. 2003). Consistent with these results, ultrastructural analyses revealed that *grk* mRNA was present on microtubules in a complex with Dynein and its cofactors, Egl and BicD (Delanoue et al. 2007). Finally, Dynein light chain (Dlc) was shown to directly bind localization sequences present within *grk* mRNA and mutation in *dlc* compromised *grk* localization (Rom et al. 2007). A similar phenotype was observed upon depletion of the Dynein adaptor, Egl (Sanghavi et al. 2016). Collectively, these data indicate that the movement of *grk* mRNA toward the anterior of the oocyte is highly dependent on Dynein.

The second step in *grk* localization involves the specific movement of the mRNA toward the dorsal-anterior corner of the oocyte (MacDougall et al. 2003). This localization pattern is not shared by other Dynein-dependent cargoes. Thus, the mechanism by which this step occurs is much less clear. One hypothesis that has been put forth is that *grk* particles translocate toward the dorsal corner using a specialized subset of microtubules that, for whatever reason, is not used by other Dynein-dependent cargo. Consistent with this hypothesis, MacDougall and colleagues demonstrated the presence of a dense array of microtubules that surrounds the oocyte nucleus, forming a basket or cage-like structure (MacDougall et al. 2003). It is likely that this population of microtubules contribute to the unique localization pattern of *grk* mRNA. Presumably, some component of the *grk* messenger ribonucleoprotein (mRNP) complex enables *grk* particles to translocate along these oocyte nucleus-anchored microtubules.

Once *grk* mRNA has been delivered to its site of localization, it has to be stably anchored. Interestingly, Dynein appears to play a role in this process as well. Unexpectedly, however, the role of Dynein in this process is not as a motor but rather as a static anchor. Using ultrastructural analysis, Delanoue and colleagues demonstrated that *grk* mRNA resides within non-membrane-bound structures at the dorsal-anterior corner of the oocyte termed "sponge bodies." They hypothesized that sponge bodies represented sites of *grk* mRNA anchoring (Delanoue et al. 2007). Dynein was also found in sponge bodies and inhibition of Dynein activity correlated with rapid dissolution of sponge bodies and loss of *grk* mRNA anchoring (Delanoue et al. 2007). By contrast, inhibition of BicD was not associated with sponge body loss or defecting anchoring of *grk* mRNA (Delanoue et al. 2007). Thus, whereas Dynein functions to transport and anchor *grk* mRNA, Dynein cofactors such as Egl and BicD are specifically required for the transport step. An RNA-binding protein called Squid (Sqd) also appears to function in sponge body maintenance and *grk* anchoring (Delanoue et al. 2007). The current model suggests that Sqd is involved in converting *grk* transport particles into anchored sponge bodies (Delanoue et al. 2007).

A somewhat paradoxical finding was that loss of Kinesin heavy chain (Khc), the motor subunit of Kinesin-1, also resulted in *grk* mRNA delocalization (Brendza et al. 2002; Duncan and Warrior 2002; Januschke et al. 2002). This result was surprising because Kinesin-1 is a plus-end-directed microtubule motor and plus-ends are enriched toward the posterior pole of the oocyte (Clark et al. 1994; Parton et al. 2011; Sanghavi et al. 2012). One interpretation of this result is that Kinesin-1 is directly involved in *grk* mRNA transport and localization. An alternative interpretation, and one that is generally accepted by the field, is that Kinesin-1 transports Dynein toward the posterior of the oocyte (Brendza et al. 2002; Duncan and Warrior 2002; Januschke et al. 2002). This enables the loading of Dynein onto microtubule plus-ends in order for this motor to mediate minus-end-directed runs. Thus, loss of Khc affects the localization of Dynein, and this in turn affects the localization of *grk* mRNA. However, experiments that would definitely distinguish between these interpretations are lacking. For instance, if acute inhibition of Khc using function blocking antibodies impaired the transport of injected fluorescent *grk* mRNA, it would suggest a more direct role for this motor in *grk* mRNA localization. Alternative, if acute inhibition of Khc had little effect on *grk* mRNA transport, it would suggest that Khc participates more indirectly in this process.

7.4.2 Localization of bcd *mRNA*

In contrast to *grk* mRNA, there is no evidence that *bcd* mRNA is transcribed by the oocyte. Thus, it is very likely that most, if not all, molecules of *bcd* mRNA are transcribed in the nurse cells and transported into the oocyte. The motor responsible for this transport step appears to be Dynein.

Mische and colleagues examined the transport of GFP-tagged Exuperentia (Exu), an RNA-binding protein that is essential for *bcd* mRNA localization (Cha et al. 2001; Mische et al. 2007; Theurkauf and Hazelrigg 1998). Presumably, Exu associates with *bcd* mRNA in vivo, and, therefore, movement of Exu was used as proxy for the motility of *bcd* mRNA in these early studies. GFP-Exu was found to move from the nurse cells into the oocyte in a microtubule- and Dynein-dependent manner (Mische et al. 2007). The Dynein dependence of this process was shown using a hypomorphic mutant combination of *dhc* alleles, as well as by overexpressing a dominant negative version of p150-Glued, a core component of the Dynactin complex (Mische et al. 2007). Almost all aspects of in vivo Dynein function rely on Dynactin (Kardon and Vale 2009).

Interestingly, although the motility of GFP-Exu particles was dramatically reduced when Dynein was compromised, the motility of these particles was enhanced in null alleles of *khc*, the motor subunit of Kinesin-1 (Mische et al. 2007). As noted previously, Kinesin-1 is a plus-end-directed microtubule motor (Hirokawa et al. 2009). One interpretation of this result is that Khc somehow restrains the movement of GFP-Exu particles. Consistent with this notion, research over the past decade has revealed that many types of cargo are simultaneously associated with opposite polarity motors (Jolly and Gelfand 2011). In the case of some cargoes, their motility depends on both motors. For others, the predominant motor is restrained by the activity of the opposite polarity motor. The analogy that is often used to describe this behavior is a tug-of-war, where both motors compete for movement of the same cargo (Kural et al. 2005). When one of these motors is removed from the equation, the other motor moves the cargo more frequently and more rapidly. In the case of GFP-Exu and *bcd* mRNA by extension, the predominant motor is likely to be Dynein. When Dynein was inhibited, all movement was compromised (Mische et al. 2007). This hypothesis suggests that when Khc is absent, Dynein is freed up to move these particles faster and more frequently.

Once in the oocyte, *bcd* mRNA localizes along the anterior margin (Berleth et al. 1988). In stage 9 egg chambers, *bcd* mRNA is enriched along the dorsal and ventral corners. However, by stage 10b, this localization shifts, and the mRNA is seen in a disklike pattern at the center of the anterior cortex (Fig. 7.4a–c) (Berleth et al. 1988). Consistent with this shift in localization, the mechanism by which *bcd* is localized at these stages appears to be quite different.

In a recent paper, Trovisco and colleagues directly visualized *bcd* mRNA in vivo using a fluorescent protein-tethering strategy (Trovisco et al. 2016). This approach was first developed for monitoring mRNA localization in yeast (Bertrand et al. 1998). In this strategy, the mRNA to be examined is modified to contain multiple MS2 stem-loops. The MS2 stem-loops are specifically bound by the MS2 coat protein (MCP). Thus, if a fluorescently tagged MCP is used, the MS2 stem-loop containing mRNA is now fluorescent labeled in vivo and its motility can be analyzed in live tissue (Fig. 7.4d).

Using this approach, Trovisco and colleagues demonstrated that *bcd* mRNA particles at the anterior of the oocyte underwent rapid motility (Trovisco et al. 2016). However, the particles appeared to move in a random orientation; some

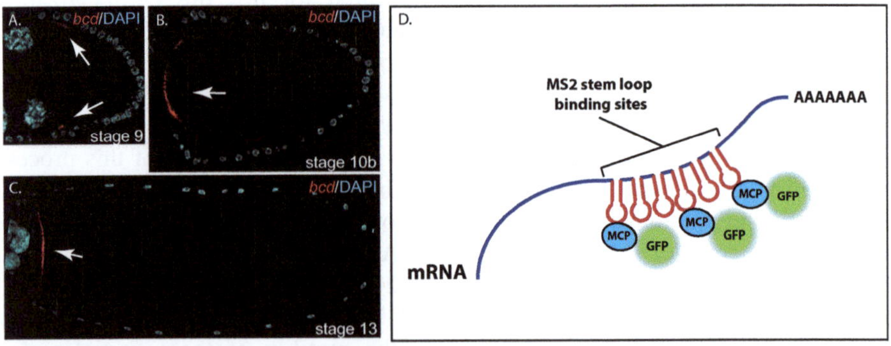

Fig. 7.4 *bcd* mRNA localization and in vivo mRNP visualization. (**a–c**) Localization of *bcd* mRNA in stage 9 (**a**), stage 10b (**b**), and stage 13 (**c**) egg chambers. (**d**) In order to visualize the movement of mRNA in live tissue, the mRNA of interest is fused to several MS2 stem-loop sequences. The MS2 stem-loops are specifically bound by the MCP protein. Thus, if the MCP protein is tagged with a fluorescent protein, this fusion will bind to the target mRNA, enabling the mRNA to be visualized and tracked in live tissues

moved toward the anterior of the oocyte, whereas others moved in the opposite direction (Trovisco et al. 2016). Despite this random movement, the majority of *bcd* mRNA particles were retained at the anterior cortex, presumably anchored there by some unknown component. When these particles were examined in a hypomorphic *dhc* mutant, their motility was compromised. Fewer particles moved, and those that did moved with reduced velocity (Trovisco et al. 2016). Thus, the primary motor involved in transporting these particles is Dynein. This is consistent with the observation that microtubule minus-ends are abundant at the anterior of the oocyte (Clark et al. 1997). This is also consistent with the finding that genetic disruption or depletion of Dynein results in delocalization of *bcd* mRNA (Duncan and Warrior 2002; Januschke et al. 2002; Sanghavi et al. 2013).

Reminiscent of what was observed for *grk* mRNA and the nurse cell to oocyte transport properties of GFP-Exu, loss of Khc also compromised the localization of *bcd* mRNA (Trovisco et al. 2016). In live egg chambers, the number of motile *bcd* particles and their average velocity were increased in the absence of Khc (Trovisco et al. 2016).

As before, several scenarios can be envisioned to explain this phenotype. One possibility is that Khc also associates with *bcd* mRNA and acts to restrain the motility of Dynein (the tug-of-war model described previously). Another possibility is that Khc is required for localizing Dynein at the posterior pole and a defect in this process indirectly affects *bcd* particle motility (Brendza et al. 2002). In order to distinguish between these hypotheses, Trovisco and colleagues examined the movement of *bcd* particles in a hypomorphic *khc* mutant (Trovisco et al. 2016). Khc levels are unaffected in the mutant, but the speed with which Khc can move cargo is reduced (Brendza et al. 1999). Importantly, this slow-mutant version of Khc is still able to localize Dynein at the posterior pole (Trovisco et al. 2016). Interestingly, the percentage of motile particles remained the same in these mutants in comparison to wild type. These findings suggest that when Khc is absent, the mislocalized Dynein at the anterior of the oocyte transports some percentage of *bcd* mRNA particles away from the anterior, resulting in a more diffuse localization of the transcript. Thus, the correct localization of *bcd* mRNA depends on the precise amount of Dynein at the oocyte anterior. Too little and the mRNA is delocalized; too much and the motor inappropriately transports the mRNA away from the anterior cortex.

In later-stage egg chambers, *bcd* mRNA relocalizes to a central disk at the anterior margin of the oocyte (Fig. 7.4b, c). Using a similar MS2 tagging strategy, Weil and colleagues demonstrated that in late-stage egg chambers, *bcd* mRNA particles are continuously transported by Dynein toward the anterior cortex (Weil et al. 2006). Thus, in contrast to stage 9 egg chambers, where transport is random, transport of *bcd* particles in later stages is biased toward the anterior cortex. A further difference is that maintenance of *bcd* mRNA at the anterior of more mature egg chambers relies less on anchoring than on active transport (Weil et al. 2006). The one constant between these stages, however, is the involvement of Dynein (Trovisco et al. 2016; Weil et al. 2006).

At first glance, the mechanism of *bcd* mRNA localization seems needlessly complex. Why would distinct mechanisms be required to localize *bcd* mRNA at these various stages? In order to answer this question, one needs to realize that many different processes are occurring simultaneously in egg chambers. With respect to *bcd* mRNA, the organization of oocyte microtubules changes between stages 9 and 10b. At stage 9, microtubules at the anterior of the oocyte are not organized in a polarized manner, some have their plus-ends pointed toward the anterior cortex; others have their plus-ends pointed away (Parton et al. 2011; Sanghavi et al. 2012; Zimyanin et al. 2008). As the egg matures, the microtubules become reorganized in order to facilitate a process called ooplasmic streaming (discussed later in the chapter). In essence, ooplasmic streaming mixes the cytoplasm of the oocyte much like a washing machine. Microtubules become bundled along the lateral cortex to facilitate ooplasmic streaming, and a microtubule organizing center is formed within a central area along the anterior cortex (Schnorrer et al. 2002; Theurkauf 1994). The net effect of this reorganization is that *bcd* mRNA particles are continuously moved by Dynein toward this central area, lest they get swept up and delocalized by the fast cytoplasmic flows that accompany streaming.

7.4.3 Localization of osk *mRNA*

Whereas *grk* and *bcd* mRNAs are targeted to the anterior of the oocyte, *osk* mRNA is localized at the posterior pole (Ephrussi et al. 1991; Kim-Ha et al. 1991) (Fig. 7.3c). However, as with *bcd* and *grk*, *osk* appears to be transported from the nurse cells into the oocyte by Dynein. Overexpression of Dynamitin (Dmn), a component of the Dynactin complex, has been shown to compromise Dynein function in multiple cell types (Burkhardt et al. 1997; Presley et al. 1997). In egg chambers, Dmn overexpression results in lower levels of *osk* mRNA in the oocyte as well as accumulation of *osk* in nurse cells (Januschke et al. 2002). Similarly, depletion of the Dynein adaptor, Egl, results in delocalization of *osk* mRNA in the oocyte and higher levels of transcript in the cytoplasm of nurse cells (Sanghavi et al. 2016). Lastly, Jambor and colleagues identified a 67-nucleotide stem-loop structure within *osk* mRNA that targets it for minus-end transport (Jambor et al. 2014). The presence of this element, referred to as the oocyte-entry sequence or OES, is required for enrichment of *osk* mRNA within the oocyte (Jambor et al. 2014). Thus, the current model suggests that Dynein associates with *osk*, likely via the OES, and transports the mRNA from nurse cells into the oocyte (Jambor et al. 2014).

The next phase of *osk* mRNA localization involves its transport toward the posterior pole. The motor that is most directly involved in this process is the plus-end motor, Kinesin-1. *osk* mRNA is completely delocalized in null mutants of *kinesin heavy chain* (*khc*) (Brendza et al. 2000). Consistent with this finding, live imaging of *osk* particles using the MS2-MCP tagging strategy revealed that motility was almost completely abrogated in the absence of Khc (Zimyanin et al. 2008). Zimyanin and colleagues further demonstrated that *osk* mRNA is transported toward the posterior pole using a "biased random walk" mechanism (Zimyanin et al. 2008). In other words, *osk* particles could be detected moving toward both the anterior and posterior poles. However, when net directionality of the entire population was considered, there was a slight bias in movements toward the oocyte posterior. This bias, if coupled with an anchoring mechanism once particles are delivered to the posterior pole, can account for the dramatic posterior enrichment that is observed for *osk* mRNA.

Interestingly, one of the factors involved in anchoring *osk* mRNA at the posterior pole is Osk protein itself. *osk* mRNA is capable of producing two distinct protein isoforms, a short and long version (Markussen et al. 1995). This is accomplished via translation initiating from two distinct start codons. Long-Osk is translated from the first ATG, whereas Short-Osk is translated from a downstream ATG (Markussen et al. 1995). Despite the fact that all of the amino acids present in Short-Osk are also present in Long-Osk, the two isoforms appear to possess different activities in vivo. Short-Osk is involved in formation of germ plasm, whereas Long-Osk functions as an atypical RNA-binding protein and anchors its own mRNA at the posterior pole (Jeske et al. 2015; Markussen et al. 1995; Vanzo and Ephrussi 2002). In egg chambers that are specifically lacking Long-Osk, *osk* mRNA is still enriched toward the posterior region of the oocyte, but it is not anchored in a tight cortical crescent (Vanzo and Ephrussi 2002).

In addition to Long-Osk, the actin cytoskeleton also appears to be required for anchoring *osk* mRNA. Disruption of the actin cytoskeleton using pharmacological agents compromises the anchoring of *osk* mRNA (Sanghavi et al. 2012). A similar phenotype is also observed in mutants of certain actin-binding proteins (Babu et al. 2004; Jankovics et al. 2002; Polesello et al. 2002). A final factor that appears to play a role in *osk* anchoring is the endocytic pathway (Sanghavi et al. 2012; Tanaka and Nakamura 2008; Vanzo et al. 2007). Endocytosis within the oocyte occurs at a higher level at the posterior cortex, and this appears to promote the formation of long filamentous actin (F-actin) fibers (Tanaka and Nakamura 2008; Vanzo et al. 2007). It is likely that these F-actin fibers act as anchoring sites for *osk* mRNA.

Most cargoes that are transported by Kinesin-1 are linked to the Khc motor subunit via the canonical adaptor, Kinesin light chain (Klc) (Hirokawa et al. 2009). In this regard, *osk* mRNA appears to be somewhat unique. Whereas *osk* is completely delocalized in *khc* mutants, the mRNA remains localized in null alleles of *klc* (Palacios and St Johnston 2002). The light chain is not completely dispensable, however. Klc along with a related protein, Pat1, appears to have a stimulatory role in Kinesin-1-based motility (Loiseau et al. 2010). When Klc and Pat1 are both absent, the localization of *osk* mRNA is compromised (Loiseau et al. 2010). Nevertheless, it appears that Khc is linked to *osk* mRNA via an alternative adaptor.

The identity of this alternative adaptor was recently codiscovered by two groups. Using a proteomic approach, Veeranan-Karmegam and colleagues identified a novel isoform of Tropomyosin1, referred to as Tm1C/I, as a direct binding partner of Khc (Veeranan-Karmegam et al. 2016). This finding was somewhat surprising because Tropomyosins are considered to be actin-biding proteins (Gunning et al. 2015). Tm1C/I is the first example of a Tropomyosin that is capable of binding a microtubule motor. Veeranan-Karmegam and colleagues went on to show that loss of Tm1C/I produced the same *osk* delocalization phenotype as loss of Khc

(Veeranan-Karmegam et al. 2016). Interestingly, Tm1C/I appears to function specifically in the *osk* localization pathway. Whereas loss of Khc results in *grk* and *bcd* mRNA delocalization, as well as delocalization of the oocyte nucleus, these cargoes were unaffected in Tm1C/I nulls (Veeranan-Karmegam et al. 2016).

In a concurrent study, Gaspar and colleagues identified the same Tropomyosin isoform as being required for Khc-based *osk* motility (Gaspar et al. 2016). Gaspar and colleagues demonstrated that the expression of Tm1C/I was specially downregulated in a previously characterized allele of *tm1* (Erdelyi et al. 1995; Gaspar et al. 2016). *osk* mRNA was delocalized in this mutant, and expression of transgenic Tm1C/I was able to completely rescue the phenotype (Gaspar et al. 2016). Importantly, Gaspar and colleagues demonstrated that this unique isoform of Tm1C/I was able to directly and specifically associate with *osk* mRNA (Gaspar et al. 2016). From a mechanistic standpoint, Tm1C/I appears to recruit Khc to *osk* mRNA and to activate the motor for plus-end transport (Gaspar et al. 2016). Collectively, these two studies provided the first demonstration of a direct link between *osk* mRNA and the microtubule cytoskeleton.

While Kinesin-1 is credited as the primary microtubule motor involved in *osk* transport within the oocyte, Dynein also appears to play a role. Initial studies detailed earlier in this section suggested that Dynein was important for nurse cell to oocyte transport of *osk* mRNA. Recent studies indicate that a role for Dynein might also extend to transport of *osk* within the oocyte. Dhc co-localizes with *osk* mRNA and Khc at the posterior pole in wild-type egg chambers and within a central focus in polarity mutants (McGrail and Hays 1997; Sanghavi et al. 2013). This, along with biochemical data, suggests that Dynein is a component of the *osk* mRNP in vivo (Sanghavi et al. 2013). In addition, depletion of Dhc and its adaptor protein Egl results in delocalization of *osk* mRNA within the oocyte, not just a retention of the mRNA in nurse cells (Sanghavi et al. 2013, 2016). A similar phenotype was observed upon depletion of Dynein regulators such as components of the Dynactin complex and Lis1 (Liu et al. 2015; Swan et al. 1999). However, the mechanism by which Dynein contributes to the posterior localization of *osk* mRNA is not known. One possibility is that Dynein acts as a restraint on Kinesin activity and that this, in turn, is required for efficient *osk* localization. Consistent with this notion, mutants in which Khc is hyperactive result in *osk* delocalization (Krauss et al. 2009). Also consistent with hypothesis, *osk* particles displayed increased velocity in hypomorphic *dhc* mutants (Zimyanin et al. 2008). Further experiments will be needed to determine the precise role of Dynein in this process.

7.5 Motors in Cytoplasmic Streaming

The mature *Drosophila* oocyte is a very large cell with a diameter around 500 microns. By contrast, the commonly used HeLa cell has a diameter around 50 microns. Thus, diffusion is insufficient to uniformly distribute proteins and mRNAs within late-stage oocytes. Rather, an active, motor-driven process called cytoplasmic streaming is required (Quinlan 2016). The cytoplasm of stage 10b oocytes is mixed rapidly in a swirling motion. Microtubules and the Kinesin-1 motor are essential for this process (Palacios and St Johnston 2002; Quinlan 2016). Interestingly, however, much like the localization of *osk* mRNA, Khc is essential for streaming but Klc is not (Palacios and St Johnston 2002).

The mechanism by which Khc facilitates cytoplasmic streaming was recently solved by Lu and colleagues (Lu et al. 2016). These authors discovered that a region within Khc that had been shown to bind microtubules in an ATP-independent fashion was required for streaming (Lu et al. 2016). In other systems, Khc has been shown to slide microtubules using this domain (Jolly and Gelfand 2010; Jolly et al. 2010). Put simply, the N-terminal motor domain of Khc is bound to one microtubule in an ATP-dependent manner, and this short region within the C-terminal tail binds a different parallel microtubule in an ATP-independent manner. In this scenario, the motor is using one of the microtubules as a track and the other microtubule as cargo. The force that is generated causes the microtubules to slide apart. Lu and colleagues demonstrated the presence of cortically anchored and static microtubules (Lu et al. 2016). Deeper in the oocyte, the microtubules were much more dynamic (Lu et al. 2016). The working model suggests that Khc binds to cortical microtubules as tracks and to subcortical microtubules as cargo (Lu et al. 2016). The sliding of these microtubules generates forces that result in streaming.

Interestingly, it appears that in addition to mixing the contents of the oocyte, cytoplasmic streaming also contributes to the localization of *osk* mRNA. Injection of fluorescently labeled *osk* mRNA in stage 10 oocytes resulted in the mRNA being swept up by streaming forces and delivered to the posterior pole where it was anchored (Glotzer et al. 1997). Consistent with this finding, *osk* mRNA localization was partially defective in *khc* mutants that were compromised for streaming (Lu et al. 2016). Thus, the localization of *osk* involves direct movement by Khc and also a streaming-facilitated movement toward the posterior pole.

In contrast to Khc, Dynein appears to inhibit streaming. The rate of streaming was increased in oocytes that were injected with function-blocking antibodies against Dhc (Serbus et al. 2005). It is therefore possible that Dhc antagonizes the activity of Khc, acting as a restraint on this motor. Another restraint of Kinesin-driven cytoplasmic streaming is the actin cytoskeleton. Mid-stage egg chambers were shown to contain an actin mesh (Dahlgaard et al. 2007; Quinlan et al. 2007). At stage 10b, this actin mesh is disassembled and streaming ensues (Quinlan 2016). Mutants in actin-binding proteins such as Cappuccino and Spire fail to establish this actin mesh (Dahlgaard et al. 2007). Consequently, fast cytoplasmic streaming can be observed in these mutants in earlier-stage egg chambers. This is correlated with defects in mRNA localization and cell polarity establishment (Dahlgaard et al. 2007). Thus, although streaming is required for mixing the contents of the oocyte, it must occur at the correct stage of egg chamber maturation.

7.6 The Road Ahead

The study of *grk*, *bcd*, and *osk* localization has revealed some common mechanisms. All three mRNAs appear to depend on active, microtubule-based transport for their localization. In addition, minus-end and plus-end motors seem to cooperate to ensure the correct localization of these mRNAs. These studies have also revealed significant differences. *grk* mRNA anchoring appears to require the Dynein motor to function as a static anchor, whereas *osk* utilizes its own protein product and the actin cytoskeleton for anchoring. However, these mRNA are not alone in the cytoplasmic milieu. Recent studies have revealed that hundreds, if not thousands, of mRNAs are specifically localized in *Drosophila* and mammalian systems (Jambor et al. 2015; Mili et al. 2008; Wilk et al. 2016). What is the mechanism by which these mRNAs are localized? Do they bear similarity to the localization of *grk*, *bcd*, and *osk*? Questions such as these will drive research in the near future. Although we have learned a great deal regarding mRNA localization and motor-based transport over the past two decades, the story is far from complete. The road ahead, regardless of where it leads, promises to be filled with exciting new discoveries. We look forward to the ride!

Acknowledgments Thanks to members of the Gonsalvez lab for critical reading and input on this manuscript. Projects on mRNA localization and microtubule motors in the Gonsalvez lab are supported by a grant from the National Institutes of Health (R01GM100088).

References

Ables ET (2015) Drosophila oocytes as a model for understanding meiosis: an educational primer to accompany "corolla is a novel protein that contributes to the architecture of the synaptonemal complex of Drosophila". Genetics 199:17–23

Babu K, Cai Y, Bahri S, Yang X, Chia W (2004) Roles of Bifocal, Homer, and F-actin in anchoring Oskar to the posterior cortex of Drosophila oocytes. Genes Dev 18:138–143

Bashirullah A, Cooperstock RL, Lipshitz HD (2001) Spatial and temporal control of RNA stability. Proc Natl Acad Sci USA 98:7025–7028

Berleth T, Burri M, Thoma G, Bopp D, Richstein S, Frigerio G, Noll M, Nusslein-Volhard C (1988) The role of localization of bicoid RNA in organizing the anterior pattern of the Drosophila embryo. EMBO J 7:1749–1756

Bertrand E, Chartrand P, Schaefer M, Shenoy SM, Singer RH, Long RM (1998) Localization of ASH1 mRNA particles in living yeast. Mol Cell 2:437–445

Bilinski SM, Jaglarz MK, Tworzydlo W (2017) The pole (germ) plasm in insect oocytes. In: Kloc M (ed) Oocytes. Springer, Heidelberg

Bohl F, Kruse C, Frank A, Ferring D, Jansen RP (2000) She2p, a novel RNA-binding protein tethers ASH1 mRNA to the Myo4p myosin motor via She3p. EMBO J 19:5514–5524

Brendza KM, Rose DJ, Gilbert SP, Saxton WM (1999) Lethal kinesin mutations reveal amino acids important for ATPase activation and structural coupling. J Biol Chem 274:31506–31514

Brendza RP, Serbus LR, Duffy JB, Saxton WM (2000) A function for kinesin I in the posterior transport of oskar mRNA and Staufen protein. Science 289:2120–2122

Brendza RP, Serbus LR, Saxton WM, Duffy JB (2002) Posterior localization of dynein and dorsal-ventral axis formation depend on kinesin in Drosophila oocytes. Curr Biol 12:1541–1545

Burkhardt JK, Echeverri CJ, Nilsson T, Vallee RB (1997) Overexpression of the dynamitin (p50) subunit of the dynactin complex disrupts dynein-dependent maintenance of membrane organelle distribution. J Cell Biol 139:469–484

Buxbaum AR, Haimovich G, Singer RH (2015) In the right place at the right time: visualizing and understanding mRNA localization. Nat Rev Mol Cell Biol 16:95–109

Caceres L, Nilson LA (2005) Production of gurken in the nurse cells is sufficient for axis determination in the Drosophila oocyte. Development 132:2345–2353

Cajigas IJ, Tushev G, Will TJ, tom Dieck S, Fuerst N, Schuman EM (2012) The local transcriptome in the synaptic neuropil revealed by deep sequencing and high-resolution imaging. Neuron 74:453–466

Cha BJ, Koppetsch BS, Theurkauf WE (2001) In vivo analysis of Drosophila bicoid mRNA localization reveals a novel microtubule-dependent axis specification pathway. Cell 106:35–46

Clark I, Giniger E, Ruohola-Baker H, Jan LY, Jan YN (1994) Transient posterior localization of a kinesin fusion protein reflects anteroposterior polarity of the Drosophila oocyte. Curr Biol 4: 289–300

Clark IE, Jan LY, Jan YN (1997) Reciprocal localization of Nod and kinesin fusion proteins indicates microtubule polarity in the Drosophila oocyte, epithelium, neuron and muscle. Development 124:461–470

Clark A, Meignin C, Davis I (2007) A Dynein-dependent shortcut rapidly delivers axis determination transcripts into the Drosophila oocyte. Development 134:1955–1965

Dahlgaard K, Raposo AA, Niccoli T, St Johnston D (2007) Capu and Spire assemble a cytoplasmic actin mesh that maintains microtubule organization in the Drosophila oocyte. Dev Cell 13: 539–553

Delanoue R, Herpers B, Soetaert J, Davis I, Rabouille C (2007) Drosophila Squid/hnRNP helps Dynein switch from a gurken mRNA transport motor to an ultrastructural static anchor in sponge bodies. Dev Cell 13:523–538

Dienstbier M, Boehl F, Li X, Bullock SL (2009) Egalitarian is a selective RNA-binding protein linking mRNA localization signals to the dynein motor. Genes Dev 23:1546–1558

Duncan JE, Warrior R (2002) The cytoplasmic dynein and kinesin motors have interdependent roles in patterning the Drosophila oocyte. Curr Biol 12:1982–1991

Ephrussi A, Lehmann R (1992) Induction of germ cell formation by oskar. Nature 358:387–392

Ephrussi A, Dickinson LK, Lehmann R (1991) Oskar organizes the germ plasm and directs localization of the posterior determinant nanos. Cell 66:37–50

Erdelyi M, Michon AM, Guichet A, Glotzer JB, Ephrussi A (1995) Requirement for Drosophila cytoplasmic tropomyosin in oskar mRNA localization. Nature 377:524–527

Forrest KM, Gavis ER (2003) Live imaging of endogenous RNA reveals a diffusion and entrapment mechanism for nanos mRNA localization in Drosophila. Curr Biol 13:1159–1168

Galjart N (2010) Plus-end-tracking proteins and their interactions at microtubule ends. Curr Biol 20:R528–R537

Gaspar I, Sysoev V, Komissarov A, Ephrussi A (2016) An RNA-binding atypical tropomyosin recruits kinesin-1 dynamically to oskar mRNPs. EMBO J 36(3):319–333

Gavis ER, Lehmann R (1992) Localization of nanos RNA controls embryonic polarity. Cell 71: 301–313

Giorgi C, Yeo GW, Stone ME, Katz DB, Burge C, Turrigiano G, Moore MJ (2007) The EJC factor eIF4AIII modulates synaptic strength and neuronal protein expression. Cell 130:179–191

Glotzer JB, Saffrich R, Glotzer M, Ephrussi A (1997) Cytoplasmic flows localize injected oskar RNA in Drosophila oocytes. Curr Biol 7:326–337

Gonsalvez GB, Urbinati CR, Long RM (2005) RNA localization in yeast: moving towards a mechanism. Biol Cell 97:75–86

Gonzalez-Reyes A, Elliott H, St Johnston D (1995) Polarization of both major body axes in Drosophila by gurken-torpedo signalling. Nature 375:654–658

Gunning PW, Hardeman EC, Lappalainen P, Mulvihill DP (2015) Tropomyosin - master regulator of actin filament function in the cytoskeleton. J Cell Sci 128:2965–2974
Heraud-Farlow JE, Kiebler MA (2014) The multifunctional Staufen proteins: conserved roles from neurogenesis to synaptic plasticity. Trends Neurosci 37:470–479
Hirokawa N, Noda Y, Tanaka Y, Niwa S (2009) Kinesin superfamily motor proteins and intracellular transport. Nat Rev Mol Cell Biol 10:682–696
Jambor H, Mueller S, Bullock SL, Ephrussi A (2014) A stem-loop structure directs oskar mRNA to microtubule minus ends. RNA 20:429–439
Jambor H, Surendranath V, Kalinka AT, Mejstrik P, Saalfeld S, Tomancak P (2015) Systematic imaging reveals features and changing localization of mRNAs in Drosophila development. Elife 4. doi:10.7554/eLife.05003
Jankovics F, Sinka R, Lukacsovich T, Erdelyi M (2002) MOESIN crosslinks actin and cell membrane in Drosophila oocytes and is required for OSKAR anchoring. Curr Biol 12:2060–2065
Januschke J, Gervais L, Dass S, Kaltschmidt JA, Lopez-Schier H, St Johnston D, Brand AH, Roth S, Guichet A (2002) Polar transport in the Drosophila oocyte requires Dynein and Kinesin I cooperation. Curr Biol 12:1971–1981
Jeske M, Bordi M, Glatt S, Muller S, Rybin V, Muller CW, Ephrussi A (2015) The crystal structure of the Drosophila germline inducer oskar identifies two Domains with distinct Vasa helicase- and RNA-binding activities. Cell Rep 12:587–598
Jolly AL, Gelfand VI (2010) Cytoplasmic microtubule sliding: an unconventional function of conventional kinesin. Commun Integr Biol 3:589–591
Jolly AL, Gelfand VI (2011) Bidirectional intracellular transport: utility and mechanism. Biochem Soc Trans 39:1126–1130
Jolly AL, Kim H, Srinivasan D, Lakonishok M, Larson AG, Gelfand VI (2010) Kinesin-1 heavy chain mediates microtubule sliding to drive changes in cell shape. Proc Natl Acad Sci USA 107:12151–12156
Kapitein LC, Hoogenraad CC (2015) Building the neuronal microtubule cytoskeleton. Neuron 87: 492–506
Kardon JR, Vale RD (2009) Regulators of the cytoplasmic dynein motor. Nat Rev Mol Cell Biol 10:854–865
Kim-Ha J, Smith JL, Macdonald PM (1991) oskar mRNA is localized to the posterior pole of the Drosophila oocyte. Cell 66:23–35
Kim-Ha J, Kerr K, Macdonald PM (1995) Translational regulation of oskar mRNA by bruno, an ovarian RNA-binding protein, is essential. Cell 81:403–412
Kloc M, Zearfoss NR, Etkin LD (2002) Mechanisms of subcellular mRNA localization. Cell 108: 533–544
Krauss J, Lopez de Quinto S, Nusslein-Volhard C, Ephrussi A (2009) Myosin-V regulates oskar mRNA localization in the Drosophila oocyte. Curr Biol 19:1058–1063
Kural C, Kim H, Syed S, Goshima G, Gelfand VI, Selvin PR (2005) Kinesin and dynein move a peroxisome in vivo: a tug-of-war or coordinated movement? Science 308:1469–1472
Lawrence JB, Singer RH (1986) Intracellular localization of messenger RNAs for cytoskeletal proteins. Cell 45:407–415
Lecuyer E, Yoshida H, Parthasarathy N, Alm C, Babak T, Cerovina T, Hughes TR, Tomancak P, Krause HM (2007) Global analysis of mRNA localization reveals a prominent role in organizing cellular architecture and function. Cell 131:174–187
Liu G, Sanghavi P, Bollinger KE, Perry L, Marshall B, Roon P, Tanaka T, Nakamura A, Gonsalvez GB (2015) Efficient endocytic uptake and maturation in Drosophila oocytes requires Dynamitin/p50. Genetics 201:631–649
Loiseau P, Davies T, Williams LS, Mishima M, Palacios IM (2010) Drosophila PAT1 is required for Kinesin-1 to transport cargo and to maximize its motility. Development 137:2763–2772
Long RM, Singer RH, Meng X, Gonzalez I, Nasmyth K, Jansen RP (1997) Mating type switching in yeast controlled by asymmetric localization of ASH1 mRNA. Science 277:383–387

Long RM, Gu W, Lorimer E, Singer RH, Chartrand P (2000) She2p is a novel RNA-binding protein that recruits the Myo4p-She3p complex to ASH1 mRNA. EMBO J 19:6592–6601

Lu W, Winding M, Lakonishok M, Wildonger J, Gelfand VI (2016) Microtubule-microtubule sliding by kinesin-1 is essential for normal cytoplasmic streaming in Drosophila oocytes. Proc Natl Acad Sci USA 113:E4995–E5004

Luders J, Stearns T (2007) Microtubule-organizing centres: a re-evaluation. Nat Rev Mol Cell Biol 8:161–167

MacDougall N, Clark A, MacDougall E, Davis I (2003) Drosophila gurken (TGFalpha) mRNA localizes as particles that move within the oocyte in two dynein-dependent steps. Dev Cell 4: 307–319

Markussen FH, Michon AM, Breitwieser W, Ephrussi A (1995) Translational control of oskar generates short OSK, the isoform that induces pole plasma assembly. Development 121: 3723–3732

Martin KC, Ephrussi A (2009) mRNA localization: gene expression in the spatial dimension. Cell 136:719–730

McGrail M, Hays TS (1997) The microtubule motor cytoplasmic dynein is required for spindle orientation during germline cell divisions and oocyte differentiation in Drosophila. Development 124:2409–2419

Medioni C, Mowry K, Besse F (2012) Principles and roles of mRNA localization in animal development. Development 139:3263–3276

Mili S, Moissoglu K, Macara IG (2008) Genome-wide screen reveals APC-associated RNAs enriched in cell protrusions. Nature 453:115–119

Mische S, Li M, Serr M, Hays TS (2007) Direct observation of regulated ribonucleoprotein transport across the nurse cell/oocyte boundary. Mol Biol Cell 18:2254–2263

Neuman-Silberberg FS, Schupbach T (1993) The Drosophila dorsoventral patterning gene gurken produces a dorsally localized RNA and encodes a TGF alpha-like protein. Cell 75:165–174

Neuman-Silberberg FS, Schupbach T (1996) The Drosophila TGF-alpha-like protein Gurken: expression and cellular localization during Drosophila oogenesis. Mech Dev 59:105–113

Oh D, Houston DW (2017) RNA localization in the vertebrate oocyte: establishment of oocyte polarity and localized mRNA assemblages. In: Kloc M (ed) Oocytes. Springer, Heidelberg

Palacios IM, St Johnston D (2002) Kinesin light chain-independent function of the Kinesin heavy chain in cytoplasmic streaming and posterior localisation in the Drosophila oocyte. Development 129:5473–5485

Parton RM, Hamilton RS, Ball G, Yang L, Cullen CF, Lu W, Ohkura H, Davis I (2011) A PAR-1-dependent orientation gradient of dynamic microtubules directs posterior cargo transport in the Drosophila oocyte. J Cell Biol 194:121–135

Polesello C, Delon I, Valenti P, Ferrer P, Payre F (2002) Dmoesin controls actin-based cell shape and polarity during Drosophila melanogaster oogenesis. Nat Cell Biol 4:782–789

Presley JF, Cole NB, Schroer TA, Hirschberg K, Zaal KJ, Lippincott Schwartz J (1997) ER to Golgi transport visualized in living cells. Nature 389:81–85

Quinlan ME (2016) Cytoplasmic streaming in the Drosophila oocyte. Annu Rev Cell Dev Biol 32: 173–195

Quinlan ME, Hilgert S, Bedrossian A, Mullins RD, Kerkhoff E (2007) Regulatory interactions between two actin nucleators, Spire and Cappuccino. J Cell Biol 179:117–128

Rom I, Faicevici A, Almog O, Neuman-Silberberg FS (2007) Drosophila Dynein light chain (DDLC1) binds to gurken mRNA and is required for its localization. Biochim Biophys Acta 1773:1526–1533

Roth S, Neuman-Silberberg FS, Barcelo G, Schupbach T (1995) cornichon and the EGF receptor signaling process are necessary for both anterior-posterior and dorsal-ventral pattern formation in Drosophila. Cell 81:967–978

Sanghavi P, Lu S, Gonsalvez GB (2012) A functional link between localized Oskar, dynamic microtubules, and endocytosis. Dev Biol 367:66–77

Sanghavi P, Laxani S, Li X, Bullock SL, Gonsalvez GB (2013) Dynein associates with oskar mRNPs and is required for their efficient net plus-end localization in Drosophila oocytes. PLoS One 8:e80605

Sanghavi P, Liu G, Veeranan-Karmegam R, Navarro C, Gonsalvez GB (2016) Multiple roles for Egalitarian in polarization of the Drosophila egg chamber. Genetics 203(1):415–432

Saunders C, Cohen RS (1999) The role of oocyte transcription, the 5′UTR, and translation repression and derepression in Drosophila gurken mRNA and protein localization. Mol Cell 3:43–54

Schnorrer F, Luschnig S, Koch I, Nusslein-Volhard C (2002) Gamma-tubulin37C and gamma-tubulin ring complex protein 75 are essential for bicoid RNA localization during drosophila oogenesis. Dev Cell 3:685–696

Serbus LR, Cha BJ, Theurkauf WE, Saxton WM (2005) Dynein and the actin cytoskeleton control kinesin-driven cytoplasmic streaming in Drosophila oocytes. Development 132:3743–3752

Sil A, Herskowitz I (1996) Identification of asymmetrically localized determinant, Ash1p, required for lineage-specific transcription of the yeast HO gene. Cell 84:711–722

Singer-Kruger B, Jansen RP (2014) Here, there, everywhere. mRNA localization in budding yeast. RNA Biol 11:1031–1039

Smith JL, Wilson JE, Macdonald PM (1992) Overexpression of oskar directs ectopic activation of nanos and presumptive pole cell formation in Drosophila embryos. Cell 70:849–859

Spradling AC (1993) Developmental genetics of oogenesis. In: Michael Bate AMA (ed) The development of Drosophila melanogaster. Cold Spring Harbor Laboratory Press, New York

Spradling A, Drummond-Barbosa D, Kai T (2001) Stem cells find their niche. Nature 414:98–104

Steinhauer J, Kalderon D (2006) Microtubule polarity and axis formation in the Drosophila oocyte. Dev Dyn 235:1455–1468

Swan A, Nguyen T, Suter B (1999) Drosophila Lissencephaly-1 functions with Bic-D and dynein in oocyte determination and nuclear positioning. Nat Cell Biol 1:444–449

Takizawa PA, Vale RD (2000) The myosin motor, Myo4p, binds Ash1 mRNA via the adapter protein, She3p. Proc Natl Acad Sci USA 97:5273–5278

Takizawa PA, Sil A, Swedlow JR, Herskowitz I, Vale RD (1997) Actin-dependent localization of an RNA encoding a cell-fate determinant in yeast. Nature 389:90–93

Tanaka T, Nakamura A (2008) The endocytic pathway acts downstream of Oskar in Drosophila germ plasm assembly. Development 135:1107–1117

Theurkauf WE (1994) Microtubules and cytoplasm organization during Drosophila oogenesis. Dev Biol 165:352–360

Theurkauf WE, Hazelrigg TI (1998) In vivo analyses of cytoplasmic transport and cytoskeletal organization during Drosophila oogenesis: characterization of a multi-step anterior localization pathway. Development 125:3655–3666

Trovisco V, Belaya K, Nashchekin D, Irion U, Sirinakis G, Butler R, Lee JJ, Gavis ER, St Johnston D (2016) bicoid mRNA localises to the Drosophila oocyte anterior by random Dynein-mediated transport and anchoring. elife 5:e17537

Vanzo NF, Ephrussi A (2002) Oskar anchoring restricts pole plasm formation to the posterior of the Drosophila oocyte. Development 129:3705–3714

Vanzo N, Oprins A, Xanthakis D, Ephrussi A, Rabouille C (2007) Stimulation of endocytosis and actin dynamics by Oskar polarizes the Drosophila oocyte. Dev Cell 12:543–555

Veeranan-Karmegam R, Boggupalli DP, Liu G, Gonsalvez GB (2016) A new isoform of Drosophila non-muscle Tropomyosin 1 interacts with Kinesin-1 and functions in oskar mRNA localization. J Cell Sci 129:4252–4264

Weil TT, Forrest KM, Gavis ER (2006) Localization of bicoid mRNA in late oocytes is maintained by continual active transport. Dev Cell 11:251–262

Wilk R, Hu J, Blotsky D, Krause HM (2016) Diverse and pervasive subcellular distributions for both coding and long noncoding RNAs. Genes Dev 30:594–609

Zaessinger S, Busseau I, Simonelig M (2006) Oskar allows nanos mRNA translation in Drosophila embryos by preventing its deadenylation by Smaug/CCR4. Development 133:4573–4583

Zimyanin VL, Belaya K, Pecreaux J, Gilchrist MJ, Clark A, Davis I, St Johnston D (2008) In vivo imaging of oskar mRNA transport reveals the mechanism of posterior localization. Cell 134: 843–853

Chapter 8
Phosphoinositides and Cell Polarity in the *Drosophila* Egg Chamber

Julie Jouette, Sandra Claret, and Antoine Guichet

Abstract Phosphatidylinositol phosphates (PIPs) are essential membrane components. They are localized at distinct membrane domains and recruit distinct effectors; they play an important role in the maintenance of membrane identity. They are essential for many cellular functions that include membrane trafficking, cytoskeletal organization, cell polarity and tissue morphogenesis. Cell polarity is also controlled by a set of polarity proteins, the PAR proteins, well conserved among bilaterians. These proteins are part of two dynamic networks that are engaged in a mutual negative-feedback regulation. PAR proteins control cell polarity by regulating cytoskeletal organization, asymmetric distributions of cellular components and directional transport through the cells. They share common activities with the PIPs in the control of intracellular polarity. Therefore, the analysis of potential cross talks between polarity proteins and PIPs is particularly important. The *Drosophila* egg chamber provides a very good model system to study the processes that control cell polarity. It includes the oocyte, a large cell in which asymmetric transport is very easy to monitor. Furthermore, the oocyte is surrounded by a follicular epithelium that allows the study of cross talks between polarity and tissue morphogenesis. This review focuses on the polarization of *Drosophila* egg chamber and our understanding of PIPs requirement during *Drosophila* oogenesis and discusses the relationship between PIPs and polarity proteins.

8.1 Organization and Development of the Fly Egg Chamber

In *Drosophila*, the oogenesis process fulfils two essential tasks. First, it provides the formation of a haploid oocyte after meiosis completion. Second, it ensures the asymmetric localization in the oocyte of several mRNAs that encode polarity axis determinants controlling the establishment of the anteroposterior and the dorsoventral

J. Jouette • S. Claret (✉) • A. Guichet (✉)
Institut Jacques Monod, UMR 7592, CNRS, Université Paris Diderot, 15 rue Hélène Brion, Bât Buffon, 75205 Paris, France
e-mail: julie.jouette@ijm.fr; sandra.claret@ijm.fr; antoine.guichet@ijm.fr

M. Kloc (ed.), *Oocytes*, Results and Problems in Cell Differentiation 63,
DOI 10.1007/978-3-319-60855-6_8

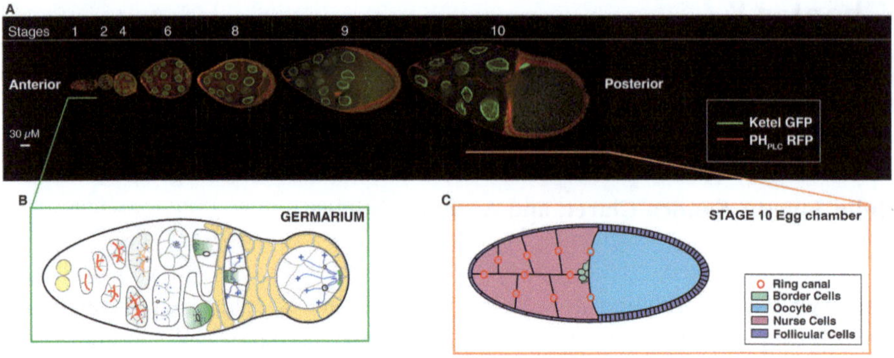

Fig. 8.1 An ovariole is organized in a succession of developing egg chambers. (**a**) An ovariole expressing the GFP-tagged importin-beta Ketel that highlights the nuclear envelope (Villanyi et al. 2008) and the RFP-tagged PH domain of the phospholipase gamma that highlights PI(4,5)P2 at the plasma membrane (Box 8.1) (Claret et al. 2014). The development of egg chambers is illustrated from the germarium (*left*) to the stage 10 (*right*). (**b**) Illustration of the germarium (adapted from [Huynh and St Johnston 2004)] showing the germline stem cells (*yellow*) that divide four times to produce a 16-cell cyst. These cells divide incompletely and remain joined by actin-rich ring canals and a fibrous structure named fusome (*red*) containing both cytoskeletal and membrane elements. As the cyst matures, it travels along the germarium and is surrounded by somatic follicle cells (*orange*). At that stage within the germline cyst, the oocyte has been selected, and it is the only cell that remains in meiosis. Then, the centrosomes (*blue point*) and the components of the Balbiani body, vesicles, mitochondria, mRNAs and associated proteins (*green*), progressively accumulate in the oocyte from the nurse cells. Before exiting the germarium, the egg chamber is established with a follicular epithelium (*orange*) surrounding a germline cyst in which the oocyte is positioned at the posterior. Microtubules (*blue*) are organized in the cyst with their minus ends at the posterior of the oocyte and their plus ends in the adjacent nurse cells. (**c**) Illustration of a stage 10 egg chamber. The nurse cells and the oocyte (shown in *pink* and *blue*, respectively) are surrounded by the follicular cells (*violet*). The ring canals connecting the nurse cells and the oocyte are shown as *red circles*. The border cells (*green*) that have migrated between the nurse cells are adjacent to the oocyte

axis of the embryo and future adult fly. Like in other metazoans, oocyte development occurs in ovaries. The ovaries are composed of about 16 ovarioles, which contain sequentially developing egg chambers (Fig. 8.1a). Each ovariole is tipped by the germarium, a structure comprising germline and somatic stem cells whose progeny will subsequently shape the egg chambers (Fig. 8.1b). Egg chambers are connected by a subset of follicle cells forming a stalk. The egg chamber development takes about 5 days and has been divided into 14 stages based on morphological criteria. Stage one corresponds to the budding of the egg chamber from the germarium and stage 14 is the mature egg.

In the germarium, germline stem cells undergo four rounds of divisions with an incomplete cytokinesis giving rise to a cyst of 16 cells interconnected by cytoplasmic bridges or ring canals (Fig. 8.1b). One of the two cells, connected by four ring canals, will differentiate into the oocyte, whereas the 15 remaining cells exit meiosis, undergo DNA replication without mitosis and become polyploid nurse cells [Fig. 8.1b; reviewed in (Huynh and St Johnston 2004)]. The oocyte progresses through meiosis but arrests its cell cycle in the G2 phase of meiosis I before exiting the germarium until the mature egg is laid and activated (Mirouse et al. 2006; Xiang et al. 2007). During oogenesis, oocyte transcription is largely silenced. In order to allow rapid and dramatic oocyte growth, the nurse cells synthesize nutrients and cytoplasmic components including proteins, mRNAs and membrane vesicles, which are transported to the oocyte through the ring canals (Spradling 1993).

The development of the oocyte and its associated nurse cells is coordinated with the growth and patterning of the somatic follicular epithelium. As the germ cell cyst matures and moves along the germarium, migrating somatic follicle cells surround it [Fig. 8.1b; reviewed in (Huynh and St Johnston 2004; Bastock and St Johnston 2008)]. At the posterior of the germarium, the egg chamber is established with a follicular epithelium surrounding a germline cyst in which the oocyte is positioned at the posterior. Subsequently the follicle cells proliferate until reaching about 1000 cells at stage 6. At that stage, they cease proliferation and switch to an endoreplicative cycle [review in Horne-Badovinac and Bilder (2005)]. At the same stage, communication between the posterior follicle cells and the oocyte triggers an important reorganization of the oocyte microtubules, which leads to the establishment of a complex microtubule array that directs the asymmetric transport of the polarity axis determinants for the future embryo (Gonzalez-Reyes et al. 1995; Roth et al. 1995). From stage 7 to 9, the local translation in the oocyte beneath the antero-dorsal follicle cells of *gurken* mRNA, which encodes a TGFα, mediates EGF signalling events between the oocyte and the follicle cells. This signalling will direct the establishment of the dorsoventral polarity within the follicular epithelium through differentiation programmes [reviewed in (Cheung et al. 2011)]. From stage 10, follicle cells deposit vitelline membrane above the oocyte plasma membrane and eventually secrete the components necessary for the eggshell (Fig. 8.2a) that will cover the oocyte. In parallel, tissue morphogenesis within the follicular epithelium will organize the eggshell secondary structures that include two dorsal appendages for embryonic respiration, an operculum for larval exit and a micropyle for sperm entry (Fig. 8.2e) [reviewed in (Spradling 1993; Berg 2005)].

8.2 The Egg Chamber: A Host of Polarity

In the egg chamber, several types of polarities cohabit to ensure proper oocyte development. The follicle cells that surround the oocyte exhibit a classical epithelial polarity with the apical surface directed toward the germ cells including the nurse cells and the oocyte and the basal surface in contact with the basement membrane of the extracellular matrix (Fig. 8.2d). The apical side of the follicle cells is the focus of signalling events and exchanges with the oocyte (Cheung et al. 2011). The basal side of the follicle cells interacts with the extracellular matrix (Fig. 8.2e). In particular it controls the establishment and the maintenance of a basal planar polarity with the basement membrane that drives the elongation of the egg

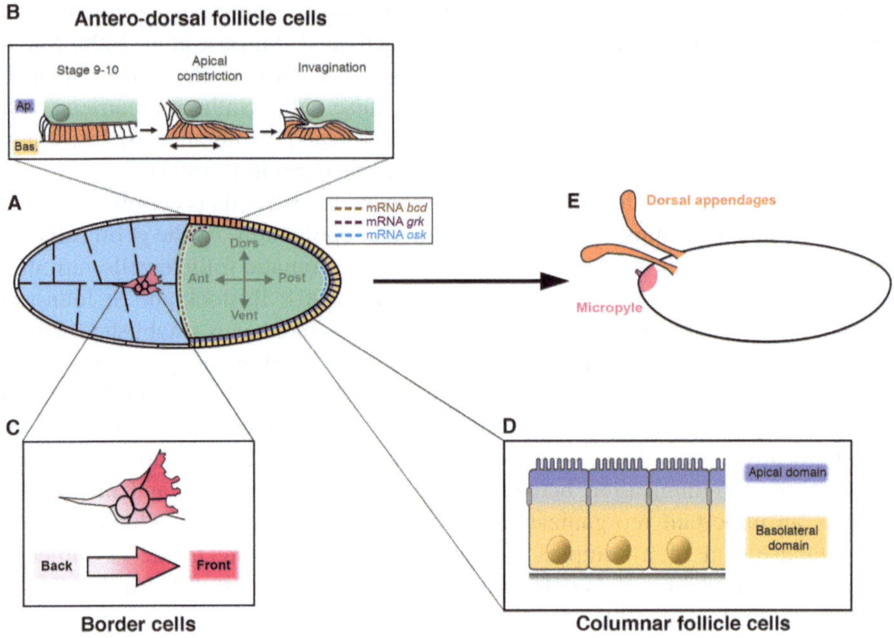

Fig. 8.2 The *Drosophila* egg chamber: a host of polarity. (**a**) Schematic representation of a stages 9–10 egg chamber. The nurse cells and the oocyte are in *blue* and *green*, respectively. The three mRNAs *bicoid* (*bcd*) in brown, *oskar* (*osk*) in pale *blue* and *gurken* (*grk*) in *violet* are localized in the oocyte at the anterior, at the posterior and above the nucleus (*green circle*) at the antero-dorsal corner, respectively. Different cell types within the follicular epithelium are illustrated: the migratory border cells (*pink*) which will form the micropyle, the antero-dorsal follicle cells (*brown*) that will form the dorsal appendages through tubulogenesis and the columnar follicle cells (*orange* and *violet*). (**b**) Illustration of the stepwise morphological process resulting in tube formation that will constitute the dorsal appendages. (**c**) Illustration of the polarity within the migrating border cells cluster. (**d**) Illustration of the apico-basal polarity within the columnar follicle cells. The apical domain (*violet*) is facing the oocyte (**a**). The basal domain (*orange*) is facing the basement membrane and the extracellular matrix (*black thick line*). The adherens junctions that maintain the cells within the epithelium are shown in *grey*. (**e**) Schematic representation of the deposited egg at the end of oogenesis showing dorsal appendages (*orange*) and the micropyle (*pink*)

chamber during oogenesis [reviewed in (Horne-Badovinac and Bilder 2005)]. Hence, the diversity of follicle cell communication with their neighbouring environment implies a tight control of polarized vesicular trafficking directed either to the apical microvilli-enriched apex [reviewed in (Cheung et al. 2011)] or to the basal domain of the plasma membrane (Devergne et al. 2014, 2017). Furthermore, at stage 9, some follicle cells named anterior polar cells recruit neighbouring cells to form the border cell cluster and start to migrate between the nurse cells until they reach the oocyte [Fig. 8.2c; reviewed in (Rorth 2002; Montell et al. 2012)]. This collective migration implicates a reorientation of the apico-basal polarity in favour of a front-rear polarity (Fig. 8.2c). Finally at stage 11, a group of antero-dorsal follicle cells reorganizes switching from a 2D epithelium to a 3D tubular epithelial structure that will constitute the future dorsal appendages [Fig. 8.2a; reviewed in (Berg 2005)]. During the early stages of this tubulogenesis process (Fig. 8.2a), the follicle cells elongate and modify the distribution of polarity proteins along the apico-basal axis in order to apically constrict and eventually to invaginate (Claret et al. 2014; Peters and Berg 2016).

The *Drosophila* oocyte is a highly polarized giant cell whose anteroposterior and dorsoventral axes are set up before fertilization [reviewed in (Riechmann and Ephrussi 2001; St Johnston 2005; St Johnston and Ahringer 2010)].

The polarity determinants of the future embryo are transcribed in the nurse cells and localized as mRNAs to specific and distinct regions in the oocyte. Importantly, translation of these mRNAs does not take place during their localization and is only released once they are localized. The main mRNA determinants are the *bicoid* (*bcd*), *oskar* (*osk*) and *gurken* (*grk*) mRNAs (Fig. 8.2a). They encode a transcription factor, a scaffold RNA-binding protein and TGFα protein, respectively [reviewed in (Riechmann and Ephrussi 2001)]. The *bcd* mRNA is localized along the anterior plasma membrane of the oocyte, and the Bicoid protein will specify the anterior structures of the embryo (Berleth et al. 1988). The *osk* mRNA is localized at the posterior of the oocyte, and the Oskar protein will specify the posterior structures and the germ cells of the embryo (Ephrussi et al. 1991). The *grk* mRNA is localized above the nucleus near the intersection between the anterior and the dorsal plasma membrane. The Gurken protein will specify the dorsoventral axis of the follicular epithelium and of the future embryo (Neuman-Silberberg and Schuepbach 1993). The localization of these three mRNA determinants relies on the microtubule cytoskeleton and on associated molecular motors [reviewed in (St Johnston 2005; Meignin and Davis 2010; Goldman and Gonzalves 2017)]. Hence, in the oocyte, the organization of the microtubule network is particularly important for the asymmetric transport of mRNAs.

8.3 PAR Proteins and Egg Chamber Polarization

PAR proteins constitute core modules of polarity protein complexes essential to build and sustain intracellular asymmetry [reviewed in (Rodriguez-Boulan and Macara 2014)]. They were originally identified in *C. elegans* in a search for genes that are required for establishment of anteroposterior polarity during the first steps of embryogenesis, and they are conserved among bilaterians [reviewed in (St Johnston and Ahringer 2010)]. They are involved in polarity control in all cell types from migrating mesenchymal cells to neurons [reviewed in (Goldstein and Macara 2007)]. They function through complementary protein networks that include GTPases, protein kinases, ubiquitin ligase, scaffold and adapter proteins [reviewed in (St Johnston and Ahringer 2010; Tepass 2012)]. In the oocyte, two mutually exclusive protein networks specify the anteroposterior axis [Fig. 8.3a; reviewed in (St Johnston and Ahringer 2010)]. The module that includes the

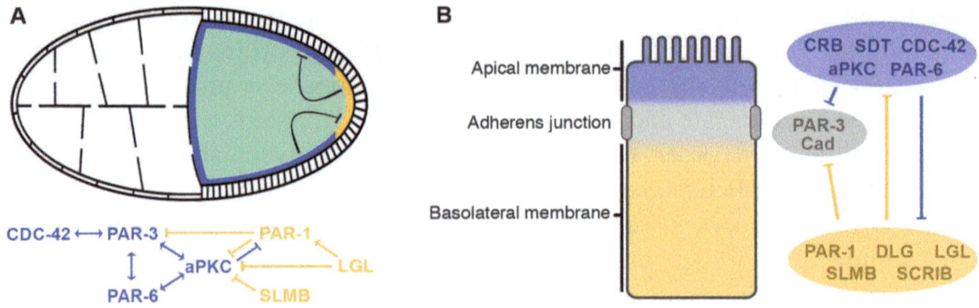

Fig. 8.3 Polarity proteins within the egg chamber. (**a**) Schematic distribution of the PAR proteins in the oocyte at stages 9–10. In the oocyte (*green*), the two PAR protein modules, anterior (*mauve*) and posterior (*orange*), are maintained through a mutual exclusion (*black negative interaction arrows*). The PAR proteins required for the anterior polarity of the oocyte (CDC-42, PAR-3, PAR-6 and aPKC) are shown in *violet*, and the PAR proteins required for the posterior polarity of the oocyte (PAR-1, LGL and SLMB) are shown in *orange*. The positive and negative interactions between the different proteins are shown. (**b**) Idealized schematic of a columnar epithelial follicle cells. The *left* panel shows three regions along the plasma membrane of an epithelial cell. The right panel shows the apical polarity proteins (*violet*), which include CRB, STD, CDC-42, aPKc and PAR-6. The PAR-3 protein clusters the adherens junctions (*grey*) by interacting with the cadherin-catenin complex (CAD). The basolateral polarity proteins (*orange*) include PAR-1, DLG, LGL, SLMB and SCRIB. The apical and basolateral polarity proteins restrict the localization of PAR-3, which in turn positions the adherens junction cluster (CAD). The apical-basal polarity axis of the cell is maintained through the mutual exclusion between the apical polarity proteins (*blue*) and the basolateral proteins (*orange*)

CDC42, PAR-3 (also named Bazooka in *Drosophila*), PAR-6 and aPKC proteins specifies the anterior domain, whereas the one including PAR-1, PAR-5 (14.3.3), Lethal (2) giant larvae (LGL) and Supernumerary limbs (SLMB) proteins specifies the posterior domain [reviewed in (St Johnston and Ahringer 2010; Morais-de-Sa et al. 2014)]. The PAR proteins are required very early during oocyte development. In the germarium, once the oocyte is specified, the PAR proteins are required in the oocyte for the anterior to posterior migration of the centrosomes and the Balbiani body, a structure containing vesicles, mitochondria, mRNAs and associated proteins [Fig. 8.1b; reviewed in (Huynh and St Johnston 2004)]. The posterior positioning of this structure is required to maintain oocyte fate and later to specify the germ cells in the future embryo. However, this is not the final polarized organization within the oocyte, since a signalling dialogue between the oocyte and the posterior follicle cells triggers at stage 6 the ultimate anteroposterior oocyte polarization [reviewed in (Bastock and St Johnston 2008; Roth and Lynch 2009)]. The two polarity modules mentioned above are crucial for this polarization step [reviewed in (St Johnston and Ahringer 2010)]. In particular, they control the correct anteroposterior organization of the microtubule network and the positioning of the nucleus. PAR-3, CDC42, PAR-6 and aPKC act together first to position and maintain the nucleus in an antero-dorsal position, second to sustain the minus-end microtubule bias toward the anterior and third to restrict to the posterior the activity of the other PAR proteins associated with PAR-1 (Fig. 8.3a). In turn, PAR-1, LGL,

PAR-5 (14-3-3) and the E3 ubiquitin ligase SLMB act together first to inhibit the positioning of microtubule minus ends at the posterior and sustain the plus-end microtubule bias at the posterior and second to restrict to the anterior the activity of the other PAR proteins associated with PAR-3 [reviewed in (St Johnston and Ahringer 2010; Morais-de-Sa et al. 2014)].

In follicle cells as in other epithelial cells, the PAR proteins are essential for establishment and maintenance of the apico-basal polarity [reviewed in (St Johnston and Ahringer 2010)]. In this case, the polarity modules are slightly more complex. Different sets of polarity proteins are involved for the specification of at least three domains within the follicle cells, the apical domain, the adherens junction cluster and the basolateral domain (Fig. 8.3b). The apical domain is controlled by a module that includes the transmembrane protein Crumbs (CRB) and its associated scaffold proteins, Stardust (STD) and PATJ; they work together with the CDC42, PAR-6 and aPKC complex (Fig. 8.3b). The association of CRB, STD, PAR6 and aPKC leads to the exclusion of PAR-3 just underneath this region. In turn, PAR-3 will set at this position the location of the adherens junction cluster by interacting with cadherin-catenin complexes (Fig. 8.3b). The basolateral domain is controlled by a set of polarity proteins that includes the scaffold proteins Disc Large (DLG), Scribbled, LGL, the protein kinase PAR-1 and the E3 ubiquitin ligase SLMB. Like in the oocyte, the positioning of the different polarity modules is maintained by a mutual exclusion process through phosphorylated-mediated repression [Fig. 8.3b; reviewed in (St Johnston and Ahringer 2010; Tepass 2012; Morais-de-Sa et al. 2014)]. PAR proteins are also required for border cell migration (Fig. 8.2c). PAR-1 is required to promote border cell detachment from the epithelium (Majumder et al. 2012), while PAR-3 together with PAR-6 and aPKC maintains their cohesion during their migration (Pinheiro and Montell 2004). However, further studies will be needed to better understand the involvement of different polarity modules in the establishment of these different polarities.

8.4 Phosphatidylinositol Phosphates

Membrane lipid phosphatidylinositols, known as phosphoinositides (PI), also play an important role in the control of cell polarity [reviewed in (Rodriguez-Boulan and Macara 2014; Schink et al. 2016)]. PI can be phosphorylated by PI kinases on the 3-, 4- and 5- hydroxyl groups of the inositol head group singularly or in various combinations to generate seven species of PI phosphorylated derivatives (PIP) (see Fig. 8.4 and Table 8.1). Each PIP, by acting like a zip code or "PIP" code, specifies the identity of the associated membrane. The differential distribution of PIPs in cells (Box 8.1) has an essential function in membrane identity and trafficking through recognition by specific effectors [reviewed in (Shewan et al. 2011; Balla 2013; Jackson et al. 2016)]. For instance, PI4P is enriched in the Golgi apparatus, but it is also weakly present at the plasma membrane. PI3P is associated with early endosomes and multivesicular bodies (MVB), whereas PI(3,5)P2 associates with

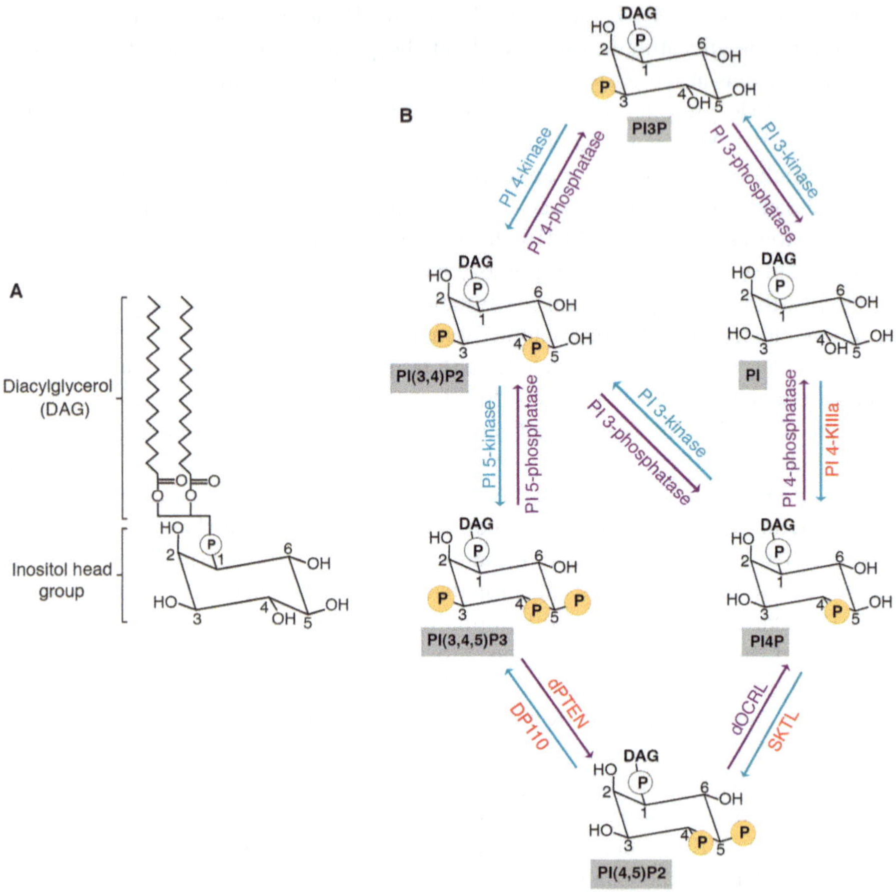

Fig. 8.4 Structure of phosphatidylinositol and six phosphoinositides (PIPs). (**a**) Representation of a phosphatidylinositol. The two acyl chains esterified to glycerol are inserted in the hydrophobic region of the cytoplasmic leaflet membranes. P highlights a phosphate group. (**b**) Catalytic conversion between phosphoinositides by kinases (*blue arrows*) and phosphatases (*green arrows*). Enzymes whose mutation leads to a phenotype in the *Drosophila* egg chamber are indicated in *red*

late endosomes and lysosomes. The PI(4,5)P2, the most abundant phosphoinositide, is found primarily at the plasma membrane [reviewed in (Hilgemann 2007; Balla 2013)]. PI(3,4,5)P3 and PI(3,4)P2 are also associated with the plasma membrane, but they represent a minor fraction of PIs (<10% of PI(4,5)P2) [reviewed in (Gassama-Diagne and Payrastre 2009)]. It is worth mentioning that recent studies have revealed a more complex picture with minor populations of PIs found occasionally outside their signature membrane [reviewed in (Tan et al. 2015; Schink et al. 2016)]. In the oocyte, PI(4,5)P2 can also be detected in endosomes (unpublished result, J Jouette, A Guichet, S Claret). The distribution of PI(4,5)P2 and PI(3,4,5)P3 at the plasma membrane is not random but is strongly related to the

Table 8.1 The phosphoinositides phosphate, localizations and functions

Phosphoinositide phosphate	Major subcellular localization	Functions	Produced by	
			Kinase	Phosphatase
PI3P	Early endosomes and MVB	Endosome sorting Membrane trafficking pathway to the lysosome (Pendaries et al. 2005)	PIK359F/ vps34 CG2699/ PI3K21B	PI4-phos-phatase/ Fig. 8.4
PI4P	Golgi apparatus and plasma membrane	Protein recruitment to the Golgi (D'Angelo et al. 2008) Constitutive secretion from the late Golgi (Pendaries et al. 2005)	**PI4KIIIα** CG7004/ Fwd	CG6562/ Synj CG3573/ dOcrl INPP5E
PI5P	–	Hypothetical role: Trafficking of late endosome to the plasma membrane (Lecompte et al. 2008) Local decrease of cytoskeletal membrane adhesion (Pendaries et al. 2005)	–	PI4-phosphatase
PI(3,5)P2	Late endosome and lysosome	Recycling of vacuole membrane (Odorizzi et al. 2000)	PI3P5K/ Fab1	–
PI(4,5)P2	Plasma membrane and endosome	Endocytosis (vesicle formation and scission) (Bethoney et al. 2009) Exocytosis (exocyst localization) (Liu et al. 2007) Endosomal recycling pathway (Tan et al. 2015) Actin nucleation (Zhang et al. 2012) Junctional stability (Goni et al. 2014) Directed cell migration Cell adhesion (Saarikangas et al. 2010)	**PIP5K/ SKTL** CG3682/ PIP5K59B CG17471/ PIP4K	**INPP3/ dPTEN**
PI(3,4)P2	Plasma membrane	–	–	INPP5E (?) (Echard 2012)
PI(3,4,5)P2	Plasma membrane	Endocytose Cell growth (Hawkins et al. 2006) Cytoskeletal organization (Oikawa et al. 2004) Cell mobility (Saarikangas et al. 2010)	**PI3K92E/ DP110**	–

The kinases and the phosphatase involved in egg chamber development are indicated in bold organization of cell polarity [reviewed in (Shewan et al. 2011; Rodriguez-Boulan and Macara 2014)]. Therefore these two PIs and their relations with polarity proteins form the focus of this review. PI(4,5)P2 and PI(3,4,5)P3 control numerous processes such as cell growth and division, cell migration and cell adhesion, vesicular transport and cytoskeletal organization (Table 8.1). They assure these functions either by direct binding and recruitment of target proteins or through hydrolysis to yield second messengers, such as diacylglycerol (DAG) and inositol triphosphate (IP3), which further promote cell signalling and tissue homeostasis [reviewed in (Balla 2013)].

Box 8.1 In Situ Detection of PIPs in Cells is Commonly Achieved by Fusing a Protein Module Containing a Specific PIP-binding Domain with a Fluorescent Protein

PI4P can be detected by tagging the PH domain of Osh2 or the P4M domain of the bacterial protein SidM. Various PH domains bind PI4P, often along with a protein in a coincidence detection mechanism. P4M is a bacterial PI4P-binding domain. PI3P can be detected by tagging the FYVE motif. PI(4,5)P2 is detected by tagging the PH domain of the phospholipase gamma and PI (3,4,5)P3 by tagging the PH domain of GRP1 [reviewed in (Idevall-Hagren and De Camilli 2015)]. PIP4 and PI(4,5)P2 can also be detected by specific antibodies (Tan et al. 2015). It is worth noting that the classical methods of PIP visualization do not permit the identification individual PIP microdomains. The very small size (few tens of nanometers) of microdomains and the impossibility to fix lipids using regular chemical fixatives make their detection very difficult. Studies using quick-freezing electron microscopy (EM), an approach that preserves the intact PM structure of live cells much better than chemical fixation, may improve detection of the nanoscale PIPs.

In the egg chamber, PI(4,5)P2 and PI(3,4,5)P3 are present at the plasma membrane, but their distribution is different in the oocyte and in the follicle cells. The follicular epithelium cells exhibit a clear apico-basal polarity [Fig. 8.5b, c; (Claret et al. 2014)]. The PI(4,5)P2 is mostly detected along the apical membrane with a slight localization along the lateral membrane, but it is excluded from the basal plasma membrane (Fig. 8.5b). PI(3,4,5)P3 is detected both at the lateral and basal plasma membrane but is excluded from the apical side (Fig. 8.5b). In follicle cells, the localization of those two PIPs is similar to that observed in epithelial cells in other model systems [reviewed in (Shewan et al. 2011; Rodriguez-Boulan and Macara 2014)]. As such, they are likely to play an important role in recruiting different cargos according to their "PIP" code, which will function either at the apical or the basal compartments. In columnar follicle cells, the asymmetry between PI(4,5)P2 and PI(3,4,5)P3 is stable and appears related to the position of the adherens junctions [Fig. 8.5c, (Claret et al. 2014)]. However, at the onset of the tubular morphogenesis process that gives rise to the dorsal appendages, the positioning of these two PIPs is inverted (Claret et al. 2014). PI(3,4,5)P3 becomes enriched at the apex, while PI(4,5)P2 is excluded from the apical side and accumulates more laterally. This inversion contributes to the apical constriction of the anterior-dorsal cell placode before invagination of the epithelium. In the oocyte the situation is different; both PI(4,5)P2 and PI(3,4,5)P3 are isotropically distributed and seem to overlap at the plasma membrane [Fig. 8.5a, (Gervais et al. 2008)]. It is worth noting that in the oocyte, adherens junction clusters are only found within the ring canals that connect the oocyte to the nurse cells (Gonzalez-Reyes and St Johnston 1998; Fichelson et al. 2010). Furthermore cadherin removal does not affect oocyte polarity (Fichelson et al. 2010; Loyer et al. 2015). Although PI(4,5)P2 and PI(3,4,5)P3 appear to overlap in the oocyte, it is possible that they could also segregate, like in mammalian cells, into distinct nanoscale regions within the plasma membrane where they could carry out specific functions (Wang and Richards 2012). In the migrating border cells, the distribution of the PIPs has not been well investigated yet. In other models of migrating cell, PI(3,4,5)P3 is present at the leading edge, while PI(4,5)P2 accumulates preferentially at the trailing edge [reviewed in (Schink et al. 2016)].

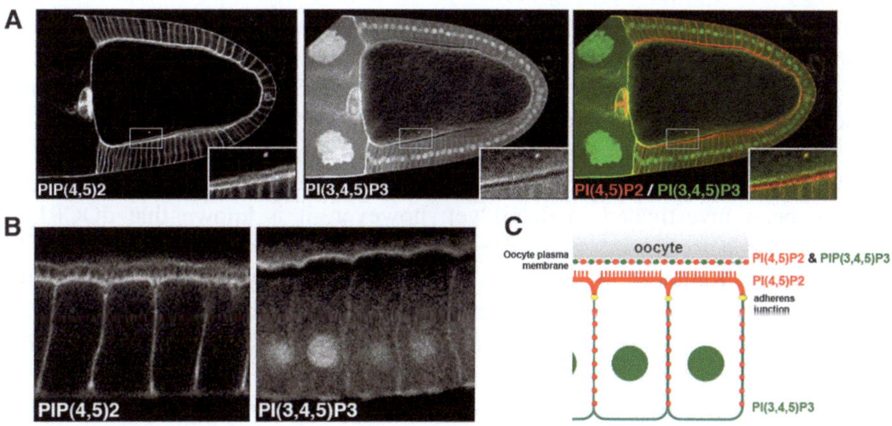

Fig. 8.5 Distribution of PI(4,5)P2 and PI(3,4,5)P3 within the egg chamber. (**a**) PI(4,5)P2 (*red*) and PI(3,4,5)P3 (*green*) overlap along the plasma membrane of the oocyte. In follicle cells, PI(4,5)P2 (*red*) and PI(3,4,5)P3 (*green*) have a distinct distribution, which is magnified in (**b**). (**b**) Enlarged distribution of PI(4,5)P2 and PI(3,4,5)P3 in columnar follicle cells. PI(4,5)P2 is mainly enriched at the apical side and absent from the basal side. PI(3,4,5)P3 is present along the basolateral part of the plasma membrane and is absent from the apical part. A fraction of PI(3,4,5)P3 is also detected in the nucleus. PI(4,5)P2 and PI(3,4,5)P3 are detected with a PH domain of the phospholipase gamma fused to RFP and a PH domain of GRP1 fused to GFP, respectively (for further details, see Box 8.1). (**c**) Schematic representation of PI(4,5)P2 and PI(3,4,5)P3 distribution along the plasma membrane in the oocyte and in columnar follicle cells

8.5 PIP Kinases and PIP Phosphatases Regulate PIP Homeostasis

PI(4,5)P2 can be generated by the activity of two classes of phosphatidylinositol phosphate kinase (PIPK) enzymes, designated PIP5K and PIP4K (Fig. 8.4; Table 8.1). PIP5K phosphorylates PI4P at position 5 of the inositol ring, whereas PIP4K phosphorylates PI5P at position 4 [reviewed in (Schink et al. 2016)]. Two PIP5Ks and one PIP4K are present in the *Drosophila* genome, but only the PIP5K Skittles (SKTL) is required for cell polarity and is essential for PI(4,5)P2 synthesis in the egg chamber (Gervais et al. 2008; Gupta et al. 2013; Claret et al. 2014; Chakrabarti et al. 2015). PI4P is generated by the activity of PI 4-kinases (PI4Ks) (Fig. 8.4; Table 8.1). Two such enzymes are present in the *Drosophila* genome. PI4KIIIalpha is mainly required for the synthesis of PI4P in the egg chamber. In the absence of PI4KIIIalpha, the levels of PI4P and PI(4,5)P2 are strongly reduced (Tan et al. 2014). It is interesting to note that PI(4,5)P2 is mostly produced by PI4P phosphorylation. Accordingly, the inactivation of PI4KIIIalpha recapitulates the egg chamber defects observed in PIP5K SKTL mutants (Gervais et al. 2008; Claret et al. 2014; Tan et al. 2014). PI(3,4,5)P3 can be generated by the activity of PI3 kinases (PI3Ks) (Fig. 8.4b; Table 8.1). In *Drosophila* a single PI3K composed of a catalytic subunit encoded by the PI3K92E gene (DP110) and a regulatory subunit encoded by the PI3K21B gene (DP60) is present. Both proteins show high homology to the corresponding human proteins (Leevers et al. 1996). In the egg chamber, DP110 inactivation leads to the strong reduction of PI(3,4,5)P3 (Hsu et al. 2008; Jagut et al. 2013; Claret et al. 2014). The PIP phosphatases participate also in PIP homeostasis [reviewed in (Schink et al. 2016)]. Inositol polyphosphate 5-phosphatase (INPP5) converts PI(4,5)P2 into PI4P (Fig. 8.4b; Table 8.1). At least two INPP5 are expressed in the *Drosophila* egg chamber, dOCRL (Ben El Kadhi et al. 2011) and dINPP5E (Park et al. 2015). Their function during oogenesis has not been investigated in detail yet; however, it is known that dOCRL is important for actomyosin contractile activity and cytokinesis in *Drosophila* S2 cells (Ben El Kadhi et al. 2011), whereas dINPP5E, by controlling PI(4,5)P2 homeostasis, regulates ciliary membrane trafficking in chordotonal neuron cilia (Park et al. 2015). Inositol polyphosphate 3-phosphatase (INPP3) converts PI(3,4,5)P3 into PI(4,5)P2 (Fig. 8.4; Table 8.1). The INPP3 PTEN is expressed in the egg chamber (von Stein et al. 2005), and its inactivation leads to an increase of PI(3,4,5)P3 (Claret et al. 2014).

8.6 Relationship Between PI(4,5)P2, PI(3,4,5)P and Cell Polarity

In many cell types such as epithelia, PI(4,5)P2 and PI(3,4,5)P3 exhibit a reciprocal distribution along the plasma membrane. Studies in polarized Madin-Darby canine kidney (MDCK) cells have shown that both PI(4,5)P2 and PI(3,4,5)P3 play an important role in cell polarity [reviewed in (Gassama-Diagne and Payrastre 2009; Shewan et al. 2011; Rodriguez-Boulan and Macara 2014)]. Modification of PI (3,4,5)P3 homeostasis level in MDCK cells triggers major polarity changes with new basolateral domains when PI(3,4,5)P3 is ectopically inserted in the apical surface and a reduction of lateral surfaces when PI(3,4,5)P3 level is reduced (Gassama-Diagne et al. 2006). On the other hand, the ectopic insertion of PI(4,5) P2 at the basolateral surface triggers the relocalization of apical proteins to the basolateral membrane (Martin-Belmonte et al. 2007). The role of PI(4,5)P2 and PI (3,4,5)P3 in cell polarity and tissue morphogenesis has been investigated in vivo in *Drosophila* egg chamber and its follicular epithelium (Claret et al. 2014). The restricted number of PIP kinases in *Drosophila* has facilitated this analysis. As described above, PI(4,5)P2 synthesis is controlled by the PIP5K SKTL and synthesis of PI(3,4,5)P3 by the PI3K DP110. SKTL inactivation in follicle cells once the epithelium has been specified leads to cell morphology changes and ultimately cell extrusion (Claret et al. 2014). This event occurs through an apical constriction associated with the modification of the actin cytoskeleton and the disorganization of the adherens junctions between neighbouring cells, a process reminiscent of that observed during epithelial-mesenchymal transition (EMT). In the follicular epithelium, a group of antero-dorsal cells undergoes a stepwise morphological process that includes length extension along the apical-basal axis, formation of thickened placode, apical cell constriction and invagination (Fig. 8.2b), resulting in tube formation that will constitute the dorsal appendages of the eggshell [Fig. 8.2e, (Berg 2005)]. The modification of PI(4,5)P2 level strongly impairs this tubulogenesis process; both PI(4,5)P2 reduction and PI(4,5)P2 increase dramatically affect the formation of dorsal appendages. However, in follicle cells, the inactivation of DP110, which strongly reduces PI(3,4,5)P3 level, does not trigger any defect in epithelial organization. Likewise, the increase of PI(3,4,5)P3 level in those cells does not affect the epithelium integrity. Accordingly, the modification of PI(3,4,5)P3 homeostasis does not affect the tubulogenesis process controlling the formation of the dorsal appendages. Hence, these results suggest that in vivo during follicle cells morphogenesis, PI(4,5)P2 but not PI(3,4,5)P3 is important for cell polarity (Claret et al. 2014). However, the tight control of PI(3,4,5)P3 level is important for the control of cell size and cell cycle progression (Goberdhan et al. 1999).

In oocyte, like in follicle cells, PI(4,5)P2 is important for polarity. In *SKTL* mutant oocytes, the organization of the actin cytoskeleton and of the microtubule network is affected. The polarized distribution of the *bcd*, *osk* and *grk* mRNAs is compromised (Fig. 8.2a), leading to fully ventralized egg (Gervais et al. 2008). On the other hand, PI(3,4,5)P3 does not seem to be essential for cell polarity in the oocyte, as is the case for follicle cells.

Inactivation of DP110 in germ cells induces growth defects that reduce the fecundity but with no sign of polarity impairment (Orme et al. 2006; Hsu et al. 2008; Jagut et al. 2013). Little is known about the putative involvement of PIPs in border cell migration. However, SKTL inactivation impairs border cell migration; cells dissociate from the cluster and migrate to incorrect destinations (unpublished results, J Jouette, A Guichet, S Claret). This suggests that PI(4,5)P2 is important for the correct migration of border cells although further investigation will be required to clarify this point. Taken together, these data highlight the multiple roles of PI (4,5)P2 in processes determining and maintaining cell polarity in the egg chamber, while PI(3,4,5)P3 is not essential in these processes.

8.7 Relationship Between PIPs and Polarity Proteins

PAR protein modules are crucial for the establishment and maintenance of cell polarity. Interestingly some interactions have been found between PAR proteins and PIPs. PI(4,5)P2 and PI(3,4,5)P3 can interact in vitro with PAR-3 (Wu et al. 2007; Krahn et al. 2010). This is further strengthened by the fact that a PAR-3 gene knockdown can be rescued by a chimeric PAR-3 protein in which the C-terminal region that interacts with the PIPs has been substituted with the PH domain of the phospholipase-C gamma, which specifically binds PI(4,5)P2 (Krahn et al. 2010). These data suggest that the anchorage of PAR-3 to the plasma membrane is controlled by PI(4,5)P2. In the oocyte, PI(4,5)P2 depletion induced by SKTL knockdown leads to the delocalization of several PAR proteins such as PAR-3, PAR-1 and LGL (Gervais et al. 2008). In that case, the polarity of the oocyte is also severely affected, the organization of both the actin and the microtubule cytoskeleton is disturbed and the asymmetric transport of the polarity axis determinants controlling the establishment of the anteroposterior and the dorsoventral axis of the embryo is disorganized (Gervais et al. 2008). Similar polarity defects are observed when the level of PI4P is reduced by PIKIIIα knockdown (Tan et al. 2014). However, when DP110 is knocked down in the oocyte, PI(3,4,5)P3 reduction does not seem to affect cell polarity although the development of the germline cyst is compromised (Jagut et al. 2013). It is difficult to assess the effect of the absence of PI(4,5)P2 on the localization of the PAR proteins in the oocyte because this leads to the delocalization of several PAR proteins (Gervais et al. 2008) which are themselves engaged in mutual negative-feedback regulation that affect their repartition within the cell [Fig. 8.3a; reviewed in (St Johnston and Ahringer 2010)]. In follicular cells, the situation is different. PI(4,5)P2 depletion once the epithelium is established leads to cell delamination through a process reminiscent of EMT (Claret et al. 2014). In this case, PAR-3 and aPKC are delocalized, but the lateral distribution of PAR-1 and DLG is not affected. Moreover, in the follicle cells, PAR-3 inactivation, once the epithelium is established, leads to cell delamination, whereas PAR-1 inactivation does not affect epithelial integrity (Claret et al. 2014). Furthermore, the expression in follicle cells of a mutated form of PAR-3 that cannot be phosphorylated by PAR-1 and spreads laterally along the plasma membrane mimics the morphological changes when either PAR-3 or SKTL is mutated (Claret et al. 2014). Taken together, this indicates that the polarity defects observed in a SKTL mutant are linked to PAR-3 and to the anchorage of PAR-3 to the plasma membrane by PI(4,5)P2 in particular.

Given the importance of PI(4,5)P2 for PAR-3 localization, one can propose that in turn Par-3 plays an important role in the control of PI(4,5)P2 homeostasis. Interestingly the INNPP3 PTEN that converts PI(3,4,5)P3 into PI(4,5)P2 has been identified as a PAR-3 interacting partner in *Drosophila* (von Stein et al. 2005). Furthermore, studies on MDCK cells have shown that PTEN mediates the apical enrichment of PI(4,5)P2 (Martin-Belmonte et al. 2007). An attractive model would be that PAR-3, by recruiting PTEN, controls the fine tuning of PI(4,5)P2 and thereby maintains its apical localization. However, this seems unlikely, at least in the *Drosophila* egg chamber, since PTEN knockdown either in the oocyte or in the follicle cells does not cause any cell polarity defect reminiscent of that of a PAR-3 knockdown (von Stein et al. 2005; Claret et al. 2014). In fact, PTEN knockdown does not affect PI(4,5)P2 but does affect PI(3,4,5)P3 level. In the absence of PTEN, the level of PI(3,4,5)P3 increases, and mutant cells exhibit defects in cell size and cell cycle progression (von Stein et al. 2005; Claret et al. 2014). It is possible that the link between PAR-3 and PI(4,5)P2 homeostasis is rather at the level of the PIP5K SKTL, because recent results indicate that SKTL can be immunoprecipitated with PAR-3 (unpublished results, J Jouette, A Guichet, S Claret).

8.8 Concluding Remarks

An important challenge for the future is to understand in more details how the PIPs, which play key roles in cell function, control of membrane identity and vesicular trafficking in particular, modulate cell polarity processes. Recent results suggest that PAR proteins are also involved in the organization of membrane compartments. It would be particularly interesting to further study the relationship between PAR proteins and PIPs during membrane trafficking.

Acknowledgments We thank Catherine Jackson and Jean-Antoine Lepesant for comments on the manuscript and the Association pour la Recherche sur le Cancer (grant PJA 20161204931) for funding our research on the relationship between PIPs and polarity proteins.

References

Balla T (2013) Phosphoinositides: tiny lipids with giant impact on cell regulation. Physiol Rev 93(3):1019–1137

Bastock R, St Johnston D (2008) Drosophila oogenesis. Curr Biol 18(23):R1082–R1087

Ben El Kadhi K, Roubinet C, Solinet S, Emery G, Carreno S (2011) The inositol 5-phosphatase dOCRL controls PI(4,5)P2 homeostasis and is necessary for cytokinesis. Curr Biol 21(12): 1074–1079

Berg CA (2005) The Drosophila shell game: patterning genes and morphological change. Trends Genet 21(6):346–355

Berleth T, Burri M, Thoma G, Bopp D, Richstein S, Frigerio G, Noll M, Nusslein-Volhard C (1988) The role of localization of bicoid RNA in organizing the anterior pattern of the Drosophila embryo. EMBO J 7(6):1749–1756

Bethoney KA, King MC, Hinshaw JE, Ostap EM, Lemmon MA (2009) A possible effector role for the pleckstrin homology (PH) domain of dynamin. Proc Natl Acad Sci USA 106(32): 13359–13364

Chakrabarti P, Kolay S, Yadav S, Kumari K, Nair A, Trivedi D, Raghu P (2015) A dPIP5K dependent pool of phosphatidylinositol 4,5 bisphosphate (PIP2) is required for G-protein coupled signal transduction in Drosophila photoreceptors. PLoS Genet 11(1):e1004948

Cheung LS, Schupbach T, Shvartsman SY (2011) Pattern formation by receptor tyrosine kinases: analysis of the Gurken gradient in Drosophila oogenesis. Curr Opin Genet Dev 21(6):719–725

Claret S, Jouette J, Benoit B, Legent K, Guichet A (2014) PI(4,5)P2 produced by the PI4P5K SKTL controls apical size by tethering PAR-3 in Drosophila epithelial cells. Curr Biol 24(10): 1071–1079

D'Angelo G, Vicinanza M, De Matteis MA (2008) Lipid-transfer proteins in biosynthetic pathways. Curr Opin Cell Biol 20(4):360–370

Devergne O, Tsung K, Barcelo G, Schupbach T (2014) Polarized deposition of basement membrane proteins depends on Phosphatidylinositol synthase and the levels of Phosphatidylinositol 4,5-bisphosphate. Proc Natl Acad Sci USA 111(21):7689–7694

Devergne O, Sun GH, Schupbach T (2017) Stratum, a homolog of the human GEF Mss4, partnered with Rab8, controls the basal restriction of basement membrane proteins in epithelial cells. Cell Rep 18(8):1831–1839

Echard A (2012) Phosphoinositides and cytokinesis: the "PIP" of the iceberg. Cytoskeleton (Hoboken) 69(11):893–912

Ephrussi A, Dickinson LK, Lehmann R (1991) Oskar organizes the germ plasma and directs localization of the posterior determinant nanos. Cell 66(1):37–50

Fichelson P, Jagut M, Lepanse S, Lepesant JA, Huynh JR (2010) lethal giant larvae is required with the par genes for the early polarization of the Drosophila oocyte. Development 137(5): 815–824

Gassama-Diagne A, Payrastre B (2009) Phosphoinositide signaling pathways: promising role as builders of epithelial cell polarity. Int Rev Cell Mol Biol 273:313–343

Gassama-Diagne A, Yu W, ter Beest M, Martin-Belmonte F, Kierbel A, Engel J, Mostov K (2006) Phosphatidylinositol-3,4,5-trisphosphate regulates the formation of the basolateral plasma membrane in epithelial cells. Nat Cell Biol 8(9):963–970

Gervais L, Claret S, Januschke J, Roth S, Guichet A (2008) PIP5K-dependent production of PIP2 sustains microtubule organization to establish polarized transport in the Drosophila oocyte. Development 135(23):3829–3838

Goberdhan DC, Paricio N, Goodman EC, Mlodzik M, Wilson C (1999) Drosophila tumor suppressor PTEN controls cell size and number by antagonizing the Chico/PI3-kinase signaling pathway. Genes Dev 13(24):3244–3258

Goldman CH, Gonsalvez GB (2017) The role of microtubule motors in mRNA localization and patterning within the Drosophila oocyte. In: Kloc M (ed) Oocytes. Springer, Heidelberg

Goldstein B, Macara IG (2007) The PAR proteins: fundamental players in animal cell polarization. Dev Cell 13(5):609–622

Goni GM, Epifano C, Boskovic J, Camacho-Artacho M, Zhou J, Bronowska A, Martin MT, Eck MJ, Kremer L, Grater F et al (2014) Phosphatidylinositol 4,5-bisphosphate triggers activation of focal adhesion kinase by inducing clustering and conformational changes. Proc Natl Acad Sci USA 111(31):E3177–E3186

Gonzalez-Reyes A, St Johnston D (1998) The Drosophila AP axis is polarised by the cadherin-mediated positioning of the oocyte. Development 125(18):3635–3644

Gonzalez-Reyes A, Elliott H, St Johnston D (1995) Polarization of both major body axes in Drosophila by gurken-torpedo signalling. Nature 375(6533):654–658

Gupta A, Toscano S, Trivedi D, Jones DR, Mathre S, Clarke JH, Divecha N, Raghu P (2013) Phosphatidylinositol 5-phosphate 4-kinase (PIP4K) regulates TOR signaling and cell growth during Drosophila development. Proc Natl Acad Sci USA 110(15):5963–5968

Hawkins PT, Anderson KE, Davidson K, Stephens LR (2006) Signalling through Class I PI3Ks in mammalian cells. Biochem Soc Trans 34(Pt 5):647–662

Hilgemann DW (2007) Local PIP(2) signals: when, where, and how? Pflugers Arch 455(1):55–67

Horne-Badovinac S, Bilder D (2005) Mass transit: epithelial morphogenesis in the Drosophila egg chamber. Dev Dyn 232(3):559–574

Hsu HJ, LaFever L, Drummond-Barbosa D (2008) Diet controls normal and tumorous germline stem cells via insulin-dependent and -independent mechanisms in Drosophila. Dev Biol 313(2):700–712

Huynh JR, St Johnston D (2004) The origin of asymmetry: early polarisation of the Drosophila germline cyst and oocyte. Curr Biol 14(11):R438–R449

Idevall-Hagren O, De Camilli P (2015) Detection and manipulation of phosphoinositides. Biochim Biophys Acta 1851(6):736–745

Jackson CL, Walch L, Verbavatz JM (2016) Lipids and their trafficking: an integral part of cellular organization. Dev Cell 39(2):139–153

Jagut M, Mihaila-Bodart L, Molla-Herman A, Alin MF, Lepesant JA, Huynh JR (2013) A mosaic genetic screen for genes involved in the early steps of Drosophila oogenesis. G3 (Bethesda) 3(3):409–425

Krahn MP, Klopfenstein DR, Fischer N, Wodarz A (2010) Membrane targeting of Bazooka/PAR-3 is mediated by direct binding to phosphoinositide lipids. Curr Biol 20(7):636–642

Lecompte O, Poch O, Laporte J (2008) PtdIns5P regulation through evolution: roles in membrane trafficking? Trends Biochem Sci 33(10):453–460

Leevers SJ, Weinkove D, MacDougall LK, Hafen E, Waterfield MD (1996) The Drosophila phosphoinositide 3-kinase Dp110 promotes cell growth. EMBO J 15(23):6584–6594

Liu J, Zuo X, Yue P, Guo W (2007) Phosphatidylinositol 4,5-bisphosphate mediates the targeting of the exocyst to the plasma membrane for exocytosis in mammalian cells. Mol Biol Cell 18(11):4483–4492

Loyer N, Kolotuev I, Pinot M, Le Borgne R (2015) Drosophila E-cadherin is required for the maintenance of ring canals anchoring to mechanically withstand tissue growth. Proc Natl Acad Sci USA 112(41):12717–12722

Majumder P, Aranjuez G, Amick J, McDonald JA (2012) Par-1 controls myosin-II activity through myosin phosphatase to regulate border cell migration. Curr Biol 22(5):363–372

Martin-Belmonte F, Gassama A, Datta A, Yu W, Rescher U, Gerke V, Mostov K (2007) PTEN-mediated apical segregation of phosphoinositides controls epithelial morphogenesis through Cdc42. Cell 128(2):383–397

Meignin C, Davis I (2010) Transmitting the message: intracellular mRNA localization. Curr Opin Cell Biol 22(1):112–119

Mirouse V, Formstecher E, Couderc JL (2006) Interaction between Polo and BicD proteins links oocyte determination and meiosis control in Drosophila. Development 133(20):4005–4013

Montell DJ, Yoon WH, Starz-Gaiano M (2012) Group choreography: mechanisms orchestrating the collective movement of border cells. Nat Rev Mol Cell Biol 13(10):631–645

Morais-de-Sa E, Mukherjee A, Lowe N, St Johnston D (2014) Slmb antagonises the aPKC/Par-6 complex to control oocyte and epithelial polarity. Development 141(15):2984–2992

Neuman-Silberberg FS, Schuepbach T (1993) The Drosophila dorsoventral patterning gene gurken produces a dorsally localized RNA and encodes a TGF-alpha-like protein. Cell 75(1):165–174

Odorizzi G, Babst M, Emr SD (2000) Phosphoinositide signaling and the regulation of membrane trafficking in yeast. Trends Biochem Sci 25(5):229–235

Oikawa T, Yamaguchi H, Itoh T, Kato M, Ijuin T, Yamazaki D, Suetsugu S, Takenawa T (2004) PtdIns(3,4,5)P3 binding is necessary for WAVE2-induced formation of lamellipodia. Nat Cell Biol 6(5):420–426

Orme MH, Alrubaie S, Bradley GL, Walker CD, Leevers SJ (2006) Input from Ras is required for maximal PI(3)K signalling in Drosophila. Nat Cell Biol 8(11):1298–1302

Park J, Lee N, Kavoussi A, Seo JT, Kim CH, Moon SJ (2015) Ciliary Phosphoinositide regulates ciliary protein trafficking in Drosophila. Cell Rep 13(12):2808–2816

Pendaries C, Tronchere H, Racaud-Sultan C, Gaits-Iacovoni F, Coronas S, Manenti S, Gratacap MP, Plantavid M, Payrastre B (2005) Emerging roles of phosphatidylinositol monophosphates in cellular signaling and trafficking. Adv Enzyme Regul 45:201–214

Peters NC, Berg CA (2016) Dynamin-mediated endocytosis is required for tube closure, cell intercalation, and biased apical expansion during epithelial tubulogenesis in the Drosophila ovary. Dev Biol 409(1):39–54

Pinheiro EM, Montell DJ (2004) Requirement for Par-6 and Bazooka in Drosophila border cell migration. Development 131(21):5243–5251

Riechmann V, Ephrussi A (2001) Axis formation during Drosophila oogenesis. Curr Opin Genet Dev 11(4):374–383

Rodriguez-Boulan E, Macara IG (2014) Organization and execution of the epithelial polarity programme. Nat Rev Mol Cell Biol 15(4):225–242

Rorth P (2002) Initiating and guiding migration: lessons from border cells. Trends Cell Biol 12(7): 325–331

Roth S, Lynch JA (2009) Symmetry breaking during Drosophila oogenesis. Cold Spring Harb Perspect Biol 1(2):a001891

Roth S, Neuman-Silberberg FS, Barcelo G, Schupbach T (1995) Cornichon and the EGF receptor signaling process are necessary for both anterior-posterior and dorsal-ventral pattern formation in Drosophila. Cell 81(6):967–978

Saarikangas J, Zhao H, Lappalainen P (2010) Regulation of the actin cytoskeleton-plasma membrane interplay by phosphoinositides. Physiol Rev 90(1):259–289

Schink KO, Tan KW, Stenmark H (2016) Phosphoinositides in control of membrane dynamics. Annu Rev Cell Dev Biol 32:143–171

Shewan A, Eastburn DJ, Mostov K (2011) Phosphoinositides in cell architecture. Cold Spring Harb Perspect Biol 3(8):a004796

Spradling A (1993) Developmental genetics of oogenesis. In: Martinez-Arias B (ed) The development of Drosophila melanogaster. Cold Spring Harbor Laboratory Press, New York, pp 1–70

St Johnston D (2005) Moving messages: the intracellular localization of mRNAs. Nat Rev Mol Cell Biol 6(5):363–375

St Johnston D, Ahringer J (2010) Cell polarity in eggs and epithelia: parallels and diversity. Cell 141(5):757–774

Tan J, Oh K, Burgess J, Hipfner DR, Brill JA (2014) PI4KIIIalpha is required for cortical integrity and cell polarity during Drosophila oogenesis. J Cell Sci 127(Pt 5):954–966

Tan X, Thapa N, Choi S, Anderson RA (2015) Emerging roles of PtdIns(4,5)P2—beyond the plasma membrane. J Cell Sci 128(22):4047–4056

Tepass U (2012) The apical polarity protein network in Drosophila epithelial cells: regulation of polarity, junctions, morphogenesis, cell growth, and survival. Annu Rev Cell Dev Biol 28: 655–685

Villanyi Z, Debec A, Timinszky G, Tirian L, Szabad J (2008) Long persistence of importin-beta explains extended survival of cells and zygotes that lack the encoding gene. Mech Dev 125(3-4):196–206

von Stein W, Ramrath A, Grimm A, Muller-Borg M, Wodarz A (2005) Direct association of Bazooka/PAR-3 with the lipid phosphatase PTEN reveals a link between the PAR/aPKC complex and phosphoinositide signaling. Development 132(7):1675–1686

Wang J, Richards DA (2012) Segregation of PIP2 and PIP3 into distinct nanoscale regions within the plasma membrane. Biol Open 1(9):857–862

Wu H, Feng W, Chen J, Chan LN, Huang S, Zhang M (2007) PDZ domains of Par-3 as potential phosphoinositide signaling integrators. Mol Cell 28(5):886–898

Xiang Y, Takeo S, Florens L, Hughes SE, Huo LJ, Gilliland WD, Swanson SK, Teeter K, Schwartz JW, Washburn MP et al (2007) The inhibition of polo kinase by matrimony maintains G2 arrest in the meiotic cell cycle. PLoS Biol 5(12):e323

Zhang L, Mao YS, Janmey PA, Yin HL (2012) Phosphatidylinositol 4, 5 bisphosphate and the actin cytoskeleton. Subcell Biochem 59:177–215

Chapter 9
RNA Localization in the Vertebrate Oocyte: Establishment of Oocyte Polarity and Localized mRNA Assemblages

Denise Oh and Douglas W. Houston

Abstract RNA localization is a fundamental mechanism for controlling cell structure and function. Early development in fish and amphibians requires the localization of specific mRNAs to establish the initial differences in cell fates prior to the onset of zygotic genome activation. RNA localization in these oocytes (e.g., *Xenopus* and zebrafish) requires that animal-vegetal polarity be established early in oogenesis, mediated by formation of the Balbiani body/mitochondrial cloud. This structure serves as a platform for assembly and transport of germline determinants to the future vegetal pole and also sets up the machinery for the localization of non-germline transcripts later in oogenesis. Understanding these polarization and localization mechanisms is critical for understanding the basis for early embryonic development in these organisms and also for understanding the role of RNA compartmentalization in animal gametogenesis. Here we outline recent advances in elucidating the molecular basis for the establishment of oocyte polarity at the level of Balbiani body assembly as well as the formation of RNP assemblies for early and late pathway mRNA localization in the oocyte.

9.1 Introduction

The eggs of many animals are highly polarized, often leading to different cell fates arising from different regions of the egg. The amphibian egg is a long-studied example of a polarized egg: the pigmented animal pole region, from which the polar body is extruded, forms the nervous system and epidermis, whereas the gut derivatives are derived from the vegetal pole. Additionally, in anuran amphibians such as *Xenopus*, the future germline arises from vegetal cells through the inheritance of germ plasm, a cytologically distinct collection of mitochondria and other organelles, along with dense germ granules. Classic amphibian embryological experiments have demonstrated the general importance of regional vegetal factors

D. Oh • D.W. Houston (✉)
Department of Biology, The University of Iowa, 257 BB, Iowa City, IA 52242, USA
e-mail: douglas-houston@uiowa.edu

M. Kloc (ed.), *Oocytes*, Results and Problems in Cell Differentiation 63,
DOI 10.1007/978-3-319-60855-6_9

in critical aspects of development, including dorsal axis formation, germ layer establishment, and primordial germ cell (PGC) fate. Relative displacement of the vegetal cortex was implicated in dorsoventral axis patterning (Ancel and Vintembenger 1948), and vegetal cells induce mesoderm fate in more equatorial and animal cells (Nieuwkoop 1969; Sudarwati and Nieuwkoop 1971). Additionally, ultraviolet irradiation of the vegetal pole (but not the animal pole) leads to loss of PGCs and sterility that is rescuable by transplantation of vegetal cytoplasm (Bounoure 1937; Bounoure et al. 1954; Smith 1966; Ikenishi et al. 1974; Züst and Dixon 1975; Wakahara 1977).

Since these seminal experiments, a number of equally important studies have established primary roles for individual maternal messenger RNAs (mRNAs) localized during oogenesis in mediating germ layer and PGC fates in *Xenopus* development. Since the cloning of *gdf1 (alias vg1)*, the first localized mRNA to be identified in any organism (Rebagliati et al. 1985; Weeks and Melton 1987), improvements in technology and the sequencing of *Xenopus* genomes (Hellsten et al. 2010; Session et al. 2016) have allowed the large-scale characterization and annotation of vegetally localized mRNAs. Subtractive hybridization, different microarray strategies, and RNA sequencing have all been used and have generated quite consistent and progressively more complete results (reviewed in Houston 2013). Estimates of the total number of vegetally localized RNAs, around 400 (out of 10,000 maternally expressed mRNAs), are satisfyingly close to the original estimates of vegetal RNA localization obtained from kinetic hybridization experiments [~4% of total poly(A) RNA (Carpenter and Klein 1982; Cuykendall and Houston 2010)]. Additionally, specific pathways mediating the localization of transcripts to the vegetal pole have been characterized, involving regulation of the cytoskeleton and specific RNA-binding proteins. In some cases, homologous proteins mediate RNA localization in somatic cells, suggesting the existence of a conserved core mechanism for RNA transport in cells (Kloc et al. 2002b; King et al. 2005; Houston 2013).

In *Xenopus* and zebrafish, animal-vegetal polarity is established early in oogenesis, in conjunction with the formation of the conspicuous Balbiani body/mitochondrial cloud. This structure serves as a hub for localization for the so-called early pathway/METRO RNAs and is the main pathway for localization to the germ plasm (Forristall et al. 1995; Kloc and Etkin 1995). During the vitellogenic stages, this mitochondrial cloud breaks down, and the fragments localize to the future vegetal pole cortex, establishing a definitive structural polarity to the oocyte. This fragmenting cloud sets up the path for "late pathway" mRNAs to localize to the cortex (reviewed in Kloc et al. 2002b). These mRNAs generally lack strict commonality of function but rather seem to act broadly in germ layer induction and patterning in the vegetal hemisphere. A subset of mRNAs can localize to the germ plasm but also more widely form a little understood "intermediate pathway" (reviewed in Houston 2013). This review will highlight recent advances in understanding the establishment of animal-vegetal polarity and formation of the Balbiani body in both *Xenopus* and zebrafish as well as the formation of ribonucleoprotein (RNP) granules involved in RNA localization and their roles in the various localization pathways.

9.2 The Origin of Animal-Vegetal Polarity During Oogenesis

9.2.1 Oocyte Formation and Polarization

Oogenesis in vertebrates begins when PGCs reach the genital ridge and form germline cysts, with oogonia remaining connected by ring canals after incomplete cell divisions (Pepling et al. 1999). These ring canals are intercellular bridges thought to be important in distributing nutrients and maintaining synchronous development signals in the premeiotic oogonia. In *Xenopus* and also zebrafish, oogenesis is a continuous process in the adult, with cysts arising from germline stem cells at different time points and resulting in a collection of asynchronous stages in the mature female ovary (Pelegri 2003; Rasar and Hammes 2006). Cysts form through a series of four mitotic divisions in the nest providing 16 cystocytes all of which will eventually form oocytes (Kloc et al. 2004a). When the ring canals are broken down, the cystocytes are disconnected and enter pre-stage I of oogenesis where prophase of meiosis I begins (al-Mukhtar and Webb 1971; Coggins 1973).

Subsequently, the Balbiani body develops from one of many “premitochondrial clouds” (Heasman et al. 1984; Kloc et al. 1996) adjacent to the germinal vesicle (GV; i.e., the nucleus in oocytes) and becomes positioned toward the future vegetal pole. This marks the first incidence of asymmetry and the orientation of AV polarity in a cell that must expand at least 100 times in size (Dumont 1972; Selman et al. 1993). As the oocyte concludes stage I and transitions into stage II, the oocyte is arrested in diplotene of meiosis I. During the subsequent progression through oogenesis (Dumont 1972; Selman 1993), there is a massive growth phase to accommodate the accumulation of yolk, synthesis of proteins and RNAs, as well as a drastic rearrangement of cytoskeleton.

The formation and function of the Balbiani body,[1] or mitochondrial cloud, in early oocytes have remained largely mysterious since its discovery in myriapods and arachnids (reviewed in Wilson 1928; Kloc et al. 2004b). The Balbiani body appears in early oocytes of many animals, including the mouse, which was thought to be a notable exception until recently (Guraya 1979; Kloc et al. 2004b, 2008; Pepling et al. 2007). In organisms with maternal inheritance of germ plasm, the Balbiani body serves as the source for this material (Heasman et al. 1984), whereas the structure uniformly disperses in mammals (Pepling et al. 2007; Kloc et al. 2008) and presumably other organisms as well. The ultrastructure of the Balbiani body is similar to other intracellular “bodies” including the chromatoid body in spermatids, the germ plasm in oocytes, and other RNA-processing structures (e.g., P bodies, stress

[1]The Balbiani body was named in honor of Édouard Gérard Balbiani, a Haitian-born French naturalist and embryologist, by his student Louis Félix Henneguy (*la vesicule de Balbiani/le corps vitellin de Balbiani*). Balbiani studied the structure, which was first identified by von Wittich and Carus in 1845 and 1850, respectively, and termed the “yolk/vitellin nucleus”, *Dotterkern*, or *corps vitellin* in older literature (Wilson 1928; Kloc et al. 2004b).

granules). It is likely that the assembly of such a structure in germ cells is essential for gametogenesis. And, this structure may have been adapted to localize and store mRNAs and contribute to germ plasm in diverse animal lineages where it has evolved (reviewed in Extavour and Akam 2003). Additionally, the Balbiani body may act to select and concentrate the more active (i.e., "healthy") mitochondria (Kogo et al. 2011; Boke et al. 2016), a feature that would also be useful for future PGCs inheriting germ plasm, since early embryos typically do not generate new mitochondria [*Xenopus*: (El Meziane et al. 1989); mouse: (Shoubridge and Wai 2007)].

9.2.2 Molecular Regulation of Balbiani Body Assembly

Maternal effect genetic screens in zebrafish have provided critical insight into Balbiani body assembly (Dosch et al. 2004; Wagner et al. 2004). Mutants in the *buckyball* (*buc*) locus showed a loss of AV polarity during oogenesis, failure of Balbiani body formation, and concomitant loss of germ plasm RNA localization and animal pole cytoplasmic streaming and blastodisc formation (Dosch et al. 2004; Marlow and Mullins 2008; Bontems et al. 2009). Buc encodes a rapidly evolving novel vertebrate protein with homology to Velo1, the product of a localized mRNA in *Xenopus* [(Claussen and Pieler 2004; Bontems et al. 2009); Velo1 hereafter]. In fish embryos, overexpressed Velo1 can induce Balbiani body-like structures competent to recruit localized mRNAs (Bontems et al. 2009), suggesting that Velo1 may function analogously to *Drosophila oskar* in assembling the germ plasm, although there is no sequence or structural homology. Recent data suggest that Velo1 contributes to Balbiani body assembly by forming amyloid-like fibrils (Boke et al. 2016; see below).

Zebrafish *magellan* (*mgn*) mutants also exhibit Balbiani body defects (Dosch et al. 2004; Gupta et al. 2010), although in this case the Balbiani body is enlarged and mispositioned. Germ plasm RNAs can localize but are not distributed vegetally. The *mgn* locus maps to *microtubule-actin cross-linking factor 1* (*macf1*; (Gupta et al. 2010), which encodes a large and conserved spectraplakin-related protein, distantly related to the Shortstop protein in *Drosophila*, a component of the organellar fusome involved in germline cyst formation (Bottenberg et al. 2009). Macf1 has several functional motifs and can interact with F-actin, microtubules, and intermediate filaments to regulate cytoskeletal organization and organelle trafficking (Suozzi et al. 2012). Recent data suggest that Macf1 is required for basal body positioning and maintenance of polarity in retinal photoreceptors, suggesting similar mechanisms may mediate oocyte polarity and apicobasal polarity. Fusome material (likely containing Macf1) is clustered with the centrosome and premitochondrial cloud aggregates in *Xenopus* germline cysts (Kloc et al. 2004a), providing a site of possible functional interaction that might explain the apparent opposing effects for Macf1 and Velo1. Also, the Balbiani body becomes surrounded by a "net" of cytokeratin (Heasman et al. 1984; Gard et al. 1997), which could interact with Macf1 during its localization or disassembly.

Evidence suggests that Balbiani body assembly involves a feedback localization loop with *velo1* mRNA and protein and self-organizing properties of Velo1 protein itself. In zebrafish, Velo1 protein is localized asymmetrically and perinuclearly in premitochondrial cloud-stage oocytes [zygotene; prior to *velo1* mRNA localization; (Heim et al. 2014)]. Also, Velo1 protein binds to putative translation-promoting RNA-binding proteins Dazl and Rbpms2 (*alias* Hermes), which themselves bind *velo1* mRNA (as well as other mRNAs, including possibly their own) (Heim et al. 2014)—thus, a local stochastic activation of Velo1 translation would be sufficient to recruit additional Velo1 and amplify its own expression and localization. However, *velo1* RNA is not localized to the mitochondrial cloud in *Xenopus*, so it is possible that recruitment of the protein is more critical.

Interestingly, Elkouby et al. (2016) show that the Balbiani body in zebrafish arises subsequent to the formation of the "chromosome bouquet," a polarized arrangement of telomeres at the nuclear envelope, during zygotene of meiosis I (Fig. 9.1). The bouquet is formed during the early stages of prophase and is thought to function in homologous chromosome pairing (Zickler and Kleckner 1998; Scherthan 2001; Harper et al. 2004). Both the bouquet and Balbiani body are conserved in animal meiosis but had not previously been linked together. Co-localization studies showed that Balbiani body/germ plasm components localize to the bouquet at the nuclear envelope during zygotene, along with mitochondria, the centrosome, and nuclear pore proteins (Elkouby et al. 2016). These germ plasm components include *dazl* mRNA (see above), Velo1 and Ddx4 (*alias* Vasa) proteins, as well as Piwi pathway proteins (Piwil1/2, *alias* Ziwi/Zili). Analysis of *buc* mutant zebrafish showed that bouquet formation and initial polarization are normal in *buc* mutants and that AV polarity is eventually lost in the absence of Balbiani body assembly. Importantly, inhibition of microtubule polymerization (nocodazole) blocked the clustering of telomeres, centrosomes and Balbiani components,

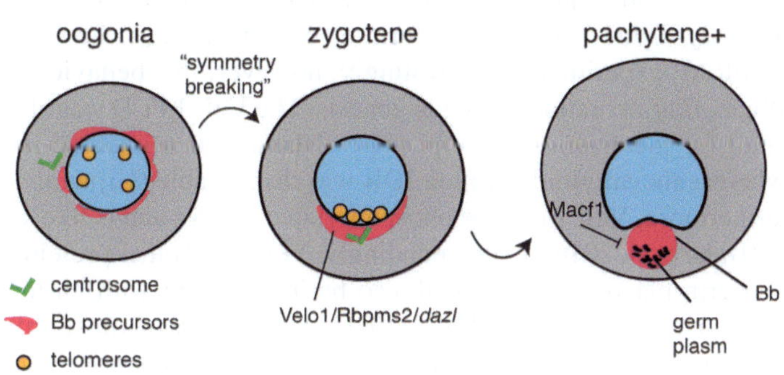

Fig. 9.1 Generalized stages of Balbiani body assembly (Bb). Stages of oogenesis are indicated above each panel. Early oogonia are not polarized and undergo a symmetry-breaking event upon meiotic entry, leading to chromosomal bouquet formation, clustering of telomeres, and alignment with the centrosome. Subsequently, in zygotene of meiosis I, Bb precursors are polarized, including Velo1 protein and RNA, Rbpms2, and *dazl* RNA. Velo1 self-organizes into an amyloid scaffold for the Bb, and localized RNAs are recruited and assembled into germ plasm/germ granules in a subregion of the Bb. Macf1 plays an inhibitory role in Bb assembly

suggesting that bouquet microtubules are the critical connection between nuclear asymmetry and the assembly of the Balbiani body and establishment of the future vegetal pole (Elkouby et al. 2016). The role of microtubules in this process remains unclear, as does the ultimate source of asymmetry, although Elkouby et al. (2016) hint that the final cell division plane in the germline cyst may be involved.

9.2.3 Structural Role of Velo1 in the Balbiani Body and Germ Plasm

Recent data suggest that Velo1 likely plays a key structural role in the formation and possibly the dispersal of the Balbiani body and germ plasm. Full-length Velo1 protein is highly enriched in the Balbiani body (Nijjar and Woodland 2013a; Heim et al. 2014; Boke et al. 2016). In *Xenopus* there is also a shorter variant lacking a region in the C-terminal intrinsically disordered region (IDR) that does not localize to the Balbiani body but accumulates in later-stage germ plasm islands (Nijjar and Woodland 2013a). Depletion of this short *velo1* isoform RNA with antisense oligos (but not the full-length RNA) resulted in dispersal of germ plasm islands in cultured stage VI oocytes (Nijjar and Woodland 2013a). Additionally, Balbiani bodies bind Thioflavin T, a fluorescent probe specific for amyloid fibrils (Boke et al. 2016), suggesting a role for these structures in the Balbiani body. The N-terminus of Velo1 contains a prion-like domain (PLD), a sequence involved in promoting conformational changes to form amyloid fibers (Alberti et al. 2009). Tagged Velo1 can localize to the Balbiani body in a manner dependent on its specific PLD (Nijjar and Woodland 2013a; Boke et al. 2016). Also, Velo1 protein can form amyloid fibrils in vitro and can also directly mediate clustering of mitochondria and RNA. The latter activity was non-sequence-specific however (Boke et al. 2016).

Balbiani body Velo1 is thought to form a stable matrix, as shown through low mobility in FRAP experiments. Interestingly, however, this behavior may change during Balbiani fragmentation during oogenesis. The Velo1 PLD was also necessary and sufficient for localization to germ plasm islands when injected into stage VI oocytes, whereas the amyloid-forming IDR was dispensable (Nijjar and Woodland 2013a). Furthermore, Velo1 may be phosphorylated in eggs and is likely not present as amyloid (Boke et al. 2016). Correspondingly, Velo1 and other proteins are mobile in stage VI germ plasm, which would also be inconsistent with an inert amyloid structure (Nijjar and Woodland 2013a). The significance of these differences in Velo1 behavior are not clear, but it has been known for some time, based on ultrastructural and mRNA localization studies, that germ plasm assembles in a specific vegetal subregion of the Balbiani body (Kloc and Etkin 1995; Kloc et al. 2002b). Thus, the formation of *bona fide* germ plasm may involve different structural forms of Velo1, likely in addition to specific protein-protein interactions and the formation of germ plasm RNP granules [i.e., see (Nijjar and Woodland 2013a; Aguero et al. 2016)]. These mechanisms will be briefly reviewed in the following section. It is interesting to note however that these RNPs are thought to form via structural transition states involving disordered protein domains, although involving hydrogel states as opposed to amyloid.

Taken together, these studies suggest that the Balbiani body is a semi-inert mass of amyloid and other proteins, mitochondria, and RNAs that is localized by nuclear asymmetry in early meiosis and formed through localized self-organization. The Balbiani body may contribute to general early oogenesis and also serve as the site for RNA localization and germ plasm assembly in animals that possess it.

9.3 RNA Localization Mechanisms

9.3.1 RNA Localization Pathways in Xenopus

Two main general mechanisms for RNA localization in *Xenopus* oocytes have been identified (Forristall et al. 1995; Kloc and Etkin 1995): an early pathway in stage I oocytes which localizes RNAs to the germ plasm for germline specification and axis formation and a late pathway occurring around stage III that localizes RNAs via microtubules to the cortex throughout the vegetal hemisphere for somatic fate (reviewed in Houston 2013; summarized in Fig. 9.2). A number of RNA sequence motifs within larger localization elements (LE) have been identified in addition to required RNA-binding proteins; however, the same *cis-* and *trans*-acting components appear to be used by both localization pathways. These elements generally include superclusters of UUCAC (VM1) and UUUCU (E2) motifs, which are bound by Igf2bp3 and Ptbp1 proteins, respectively (reviewed in Houston 2013). However, these motifs may only be a requirement for a distinct population of RNAs, including the well-studied *gdf1* and *vegt*. Additional localized RNAs including *dndl*, *velol*, and *plin2* do not possess clustered UUCAC and UUUCU motifs, yet they continue to localize to the vegetal pole (Chan et al. 1999; Claussen et al. 2004; Horvay et al. 2006), indicating that much remains unknown about the ultimate determinants of how RNAs are recognized for transport.

One of the main roles of mRNA localization is to spatially and temporally regulate translation of transcripts to restrict protein expression. Although long associated with the germline, it is becoming increasingly apparent that the compartmentalization of various aspects of RNA metabolism into large-scale macroscopic bodies or assemblages is critical for normal cell structure and function (reviewed in Voronina et al. 2011; Toretsky and Wright 2014; Courchaine et al. 2016). Additionally, these bodies lack surrounding membranes and are thought to form "droplets" by what is termed liquid-liquid phase separation (Courchaine et al. 2016). Such bodies include in the nucleus, the nucleoli, Cajal bodies, and nuclear speckles, and in the cytoplasm, the P bodies, stress granules, and germ granules (Voronina et al. 2011). Centrosomes likely also exhibit this behavior (Zwicker et al. 2014). Liquid phase transition behavior was first seen for nematode P granules, homologues of the germ granules (Brangwynne et al. 2009).

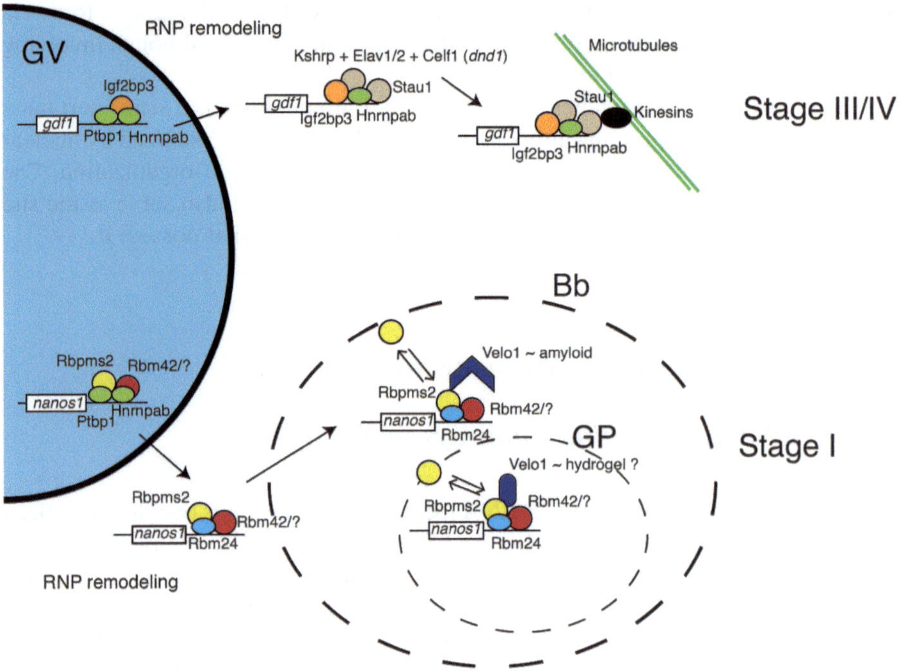

Fig. 9.2 Schematic of RNA localization pathways in the *Xenopus* oocyte. For the late localization pathway (*stage III/IV*), Ptbp1 and Hnrnpab associate with target RNAs in the germinal vesicle (GV, e.g., *gdf1*) along with Igf2bp3. Upon nuclear export, RNPs are remodeled to remove Ptbp1and allow binding of Igf2bp3 to RNA. Other factors also associate with the complex, including Kshrp and Elav proteins and/or Celf1 in the case of *dnd1* mRNA localization. Also, Stau1 binds late localizing RNAs in the cytoplasm and couples them to the microtubule cytoskeleton. For the early pathway (*stage I*), RNPs are likely nucleated in the GV by Ptbp1 and Hnrnpab, in common with the late pathway, along with Rbpms2/Hermes and an undefined RNA-binding protein partner (e.g., Rbm42 and others). In the cytoplasm, RNPs are remodeled to remove Ptbp1and Hnrnpab and incorporate other RNA-binding proteins (e.g., Rbm24). These RNPs then associate with the Bb and then the germ plasm (GP) by diffusion-entrapment, likely mediated by Rbpms2 interaction with different conformational aggregate forms of the Velo1 protein. *Double arrows* indicate that Rbpms2 is likely in equilibrium in these complexes

This transition has since been shown to depend on low complexity/intrinsically disordered protein regions (LC/IDRs) present in many RNA-binding proteins, with droplet formation occurring at high local concentrations of proteins and RNAs (Kato et al. 2012; Han et al. 2012). Additionally, because these structures are dynamic, their regulated phase transitions would be a compelling mechanism for the packaging and unpacking of cell fate determinants. Table 9.1 summarizes the RNA-binding proteins discussed in this review and identifies their propensity to oligomerize.

Table 9.1 RNA localization proteins

Protein	Family	RNA-binding domains	Oligomerization	References
Celf1	ELAV/Hu like	3 RRMs	Yes	Bauermeister et al. (2015), Arthur et al. (2009)
Dazl	RRM containing	1 RRM	Yes	Heim et al. (2014), Fu et al. (2015)
Ddx4 (Vasa)	DEAD box RNA helicase	DEAD domain	Yes	Nott et al. (2015)
Dnd1	RRM containing	2 RRMs	No	Kedde et al. (2007), Koebernick et al. (2010)
Elav1/2	ELAV/Hu like	3 RRMs	Yes	Cosson et al. (2006), Devaux et al. (2006)
Hnrnpab	hnRNP/RRM	2 RRMs	Yes	Xiang et al. (2015), Aguero et al. (2016)
Igf2bp3 (Vera)	IMP/VICKZ	KH + RRM	No	Deshler et al. (1997), Havin et al. (1998)
Ptbp1	hnRNP/RRM	4 RRMs	No	Monie et al. (2005)
Rbm24b	RRM containing	1 RRM	Yes	Nijjar and Woodland (2013a)
Rbm42b	RRM containing	1 RRM	Yes	Nijjar and Woodland (2013a)
Rbpms2 (Hermes)	RRM containing	1 RRM	Yes	Heim et al. (2014), Sagnol et al. (2014)
Staufen1	Double-stranded RNA binding	DSRM	Yes	Martel et al. (2010)
Velo1 (Buc)	YppG like	Predicted RNA binding C-terminal (K/R rich)	Yes	Boke et al. (2016)

Characteristics of RNA-binding proteins involved in RNA localization and discussed in this review. The general mode of RNA binding is noted, as is the ability for self-interaction (oligomerization). All of these proteins contain IDRs either at the N- and C-termini and/or interspersed between the RNA-binding domains (our unpublished observations)

9.3.2 Germ Plasm RNA Localization

RNA localization to the germ plasm and germ granules is thought to occur by a diffusion/entrapment-mediated mechanism (Zhou and King 1996; Kloc et al. 1996; Chang et al. 2004), but the details of this mechanism is unclear. Cytoskeletal reorganization is thought to be dispensable for direct RNA localization to the germ plasm (Chang et al. 2004), but there may be a requirement indirectly through organelle trafficking (Heinrich and Deshler 2009). Structure-function analysis of the *nanos1* 3′UTR identified distinct motifs required for localization to the mitochondrial cloud/Balbiani body and to germ granules (Zhou and King 1996; Kloc et al. 2000, 2002a; Chang et al. 2004). However, these motifs are bound by similar proteins involved in late pathway RNAs [i.e., Igf2bp3 and Ptbp1 (Chang et al. 2004)], suggesting that other proteins might provide the germ plasm specificity.

A series of recent studies have implicated the RNA-binding protein Rbpms2 (*alias* Hermes) in germ plasm RNA localization and germ granule formation. Rbpms2 is an RRM-containing RNA-binding protein initially identified in heart development but also as a vegetally and Balbiani body localized mRNA in *Xenopus* (Gerber et al. 2002; Zearfoss et al. 2004) and zebrafish (Kosaka et al. 2007; Marlow and Mullins 2008). In addition to having likely roles in controlling the cell cycle, Rbpms2 binds to *nanos1* mRNA (likely with other RNA-binding protein partners; see below) and localizes to germ granules in forming and maturing Balbiani bodies, as shown by immunoelectron microscopy (Song et al. 2007). Rbpms2 and *nanos1* are co-localized in stage VI oocyte germ plasm islands, along with another germ plasm RNA, *pgat* (*alias xpat*) (Nijjar and Woodland 2013b). These structures likely represent germ granules, since this localization has been demonstrated by electron microscopy, along with *ddx25* (Kloc et al. 2002a). Interestingly, Rbpms2 showed rapid recovery in germ plasm in FRAP experiments, whereas *nanos1* did not, suggesting that Rbpms2 may shuttle *nanos1* into more stable germ granules (Nijjar and Woodland 2013b). Relatedly, Rbpms2 can bind to *velo1* RNA and protein, and this interaction may facilitate recruitment to the Balbiani body to initiate germ plasm assembly (Heim et al. 2014).

Rbpms2 also binds other RNA-binding proteins, Rbm24b and Rbm42b, as well as the Velo1 isoforms. Bimolecular fluorescence complementation analysis suggests that these interact cytoplasmically within the germ plasm and that Rbpms2 and Rbm42b interact in the germinal vesicle (Nijjar and Woodland 2013b). Additionally, Rbpms2 likely initiates *nanos1* RNP granule formation in the germinal vesicle. Rbpms2 requires other oocyte factors for RNA binding and self-multimerizes upon association with RNA (Gerber et al. 2002). One of these requisite oocyte proteins could be Rbm42b (see above). Recent data show that Rbpms2 can bind *nanos1* in the germinal vesicle and can also interact with Ptbp1 (*alias* hnRNPI), likely also in the GV (Aguero et al. 2016). Ptbp1 has been implicated in nuclear RNP formation and remodeling in the context of *gdf1* localization (Lewis et al. 2008). It is tempting to speculate that differential remodeling might occur depending on the ultimate localization pathway, with Rbpms2 serving as a key Balbiani body determinant in the vertebrate oocyte. Also, many of these proteins are predicted to undergo liquid–liquid phase separation (Courchaine et al. 2016), although biochemical studies are lacking in the *Xenopus* system.

Recent studies in cultured cells have shown that the DEAD-box helicase 4 (Ddx4), a conserved component of germ plasm-related granules, undergoes similar liquid phase separation events (Nott et al. 2015). The N-terminus of Ddx4 is an intrinsically disordered region (IDR) and mediates the phase transition.

Interestingly, these Ddx4 droplets are less stable than and distinct from amyloid and are dependent on electrostatic interactions (as opposed to hydrogen bonding between amyloid beta-sheets). Arginine methylation and other perturbations could destabilize the droplets, suggesting a mechanism for dynamic regulation. Additionally, the droplets excluded double-stranded DNA and only incorporated RNAs, consistent with the formation of RNP-rich germ granules (Nott et al. 2015).

Overall, these observations suggest that germ plasm RNAs are nucleated into granules in the germinal vesicle, possibly by association of the Rbpms2 protein. These proto-germ plasm granules would then diffuse until anchored to the Balbiani body and ultimately incorporated into the germ plasm. It is unclear however how these would be distinguished from other localized mRNA RNP granules or what would mediate their organization within the proposed amyloid Balbiani body. Differential use of Velo1 isoforms or Velo1 phosphorylation may be involved (Nijjar and Woodland 2013a; Boke et al. 2016).

9.3.3 *Late Pathway RNAs and RNP Formation*

Late pathway RNA localization differs from the early pathway mainly in that localization occurs after translocation of the Balbiani body vegetally and requires microtubule transport. Indeed, these RNAs likely follow in the wake of the localizing Balbiani body, possibly using transport apparatus assembled for that purpose. Numerous studies have identified a core set of proteins that bind vegetally localized mRNAs, although in many cases, these proteins do not distinguish between different pathways (reviewed in Houston 2013). Recent work by Snedden et al. (2013) has demonstrated that these proteins, which include Igf2bp3, Ptbp1, Hnrnpab, Prrp, Khsrp, and possibly Elav family proteins, bind RNAs that are targeted to both the animal and vegetal hemispheres. Intriguingly, *phax* (*phosphorylated adaptor for RNA export*), an animally localized RNA, was found to contain vegetal localization motifs (Snedden et al. 2013), suggesting that these are not the main determinant for vegetal localization. However, the presence of Staufen1 (Stau1), a ds-RNA-binding protein and a kinesin motor protein linker (Yoon and Mowry 2004; Allison et al. 2004), was implicated in late pathway localization, as cross-linking experiments demonstrated the absence of Stau1 from complexes bound to early pathway RNAs and animal hemisphere transcripts (Snedden et al. 2013). Stau1 thus might provide the link to kinesin-based transport and is thought to associate with *gdf1* RNPs upon nuclear export (Kress et al. 2004). The specificity of Stau1 association with late localizing transcripts is not well understood, and coupling with nuclear export may not be required, since cytoplasmically injected RNAs are still able to achieve proper localization.

Recent work studying the localization of *dnd1*, whose 3′UTR localization element lacks obvious similarity to that of *gdf1*, identified novel candidate localization proteins of the Elav family, including Elavl1/2 (ELAV like neuron-specific RNA-binding proteins 1 and 2; (Arthur et al. 2009) and Celf1 [CUGBP, Elav-like family member 1; (Bauermeister et al. 2015)]. These proteins are conserved RRM-type RNA-binding proteins involved in RNA processing at various levels, including mRNA stability, splicing, polyadenylation, and translational control (reviewed in (Hinman and Lou 2008; Dasgupta and Ladd 2012). Celf1 also has roles in regulating mRNA deadenylation in *Xenopus* embryos through U(A/G)-rich EDEN (embryo deadenylation) elements (Paillard et al. 1998). These proteins bind to a central region of the *dnd1* 3′UTR enriched in uracil and guanine, although no specific sequences were identified, and both Elavl1/2 and Celf1 could co-immunoprecipitate with each other and with core RNP components Stau1, Igf2bp3, and Hnrnpab, likely in an RNA-dependent manner (Arthur et al. 2009; Bauermeister et al. 2015).

RNA binding of Celf1 was suggested to be in low affinity (Bauermeister et al. 2015) and Celf1 bound a subset, but not all vegetally-localizing RNAs, including *vegt*, *trim36*, and *spire1*. All of these RNAs appear to be able to localize through the late pathway (Zhang and King 1996; Le Goff et al. 2006; Cuykendall and Houston 2009), although *dnd1*, *trim36*, and *spire1* likely belong to the "intermediate" class of localized RNAs since they also preferentially accumulate in the germ plasm (Horvay et al. 2006; Le Goff et al. 2006; Cuykendall and Houston 2009). Interestingly, *vegt* mRNA can also accumulate somewhat in the germ plasm (Sudou et al. 2016), suggesting that many localizing RNAs may have this property, possibly through nonspecific RNP droplet association. Curiously, *vegt* undergoes a "relocalization" during the cleavage stages and becomes enriched dorsoanimally and accumulates around vegetal nuclei (Sudou et al. 2016), although the specificity and significance of this behavior is unclear.

Elavl2 and Celf1 are able to oligomerize (Devaux et al. 2006; Cosson et al. 2006), suggesting these proteins may be involved in forming RNP granules through liquid–liquid phase transitions. These proteins are also required for localization of the RNAs they bind, but not unbound ones (Arthur et al. 2009; Bauermeister et al. 2015), and their mutual interactions suggest that these proteins may form heterogeneous RNP granules for the localization of a subset of vegetal transcripts. Oligomerization has been shown to be required for Celf1-mediated deadenylation (Cosson et al. 2006) and may be true for RNA localization. Celf1 along with Igf2bp3 may also have a role in anchoring at the vegetal cortex (Git et al. 2009; Bauermeister et al. 2015). Elavl1/2 along with Dnd1 itself are thought to protect *dnd1* and other RNAs from miRNA-mediated degradation in germ plasm-containing cells (Kedde et al. 2007; Koebernick et al. 2010). In general, RNA localization is correlated with translation repression and stabilization, but the interplay of the various RNA-binding proteins involved in these processes and their regulation spatiotemporally are not well understood mechanistically.

9.4 Conclusions and Future Directions

Localized determinants in the egg play critical roles in development. Recent years have seen a great deal of progress in not only identifying many of these determinants but also in discovering how cell polarity is established in the oocyte and key determinants are localized in the first place.

The Balbiani body has garnered recent attention in this regard. This structure is conspicuous in the oocytes of animals and its role in serving as the source of germ plasm (Heasman et al. 1984), and a hub for localized RNAs (Forristall et al. 1995; Kloc and Etkin 1995) had been well studied. However, not all animals use germ plasm for specification of the germline yet still develop Balbiani bodies, including myriapods (where the Balbiani body was first observed) (Extavour and Akam 2003) and mouse (Pepling et al. 2007). It has been speculated that the Balbiani body sequesters high-quality mitochondria and suppresses respiration, thus limiting potential oxidative damage during oocyte dormancy and passage to the next generation (Kogo et al. 2011; Boke et al. 2016). However, there is a similar structure formed in sperm, the chromatoid body, which does not pass on mitochondria. Thus, it is likely the Balbiani body minimally serves a more general function in gametogenesis, regulating cytoplasmic RNA processing and storage and transposon silencing. And, this body could then serve as a scaffold allowing for the evolution of novel functions, including the germ plasm in multiple lineages where this occurs.

The demonstration that the Balbiani body forms in association with the chromosomal bouquet in meiosis shows that nuclear asymmetry can drive oocyte polarity (Elkouby et al. 2016). However, it remains unclear to what extent bouquet formation occurs stochastically along the nuclear envelope or is itself directed by existing polarity determinants in the oogonia, such as the centrosome or mitochondrial aggregates. If this were the case, then vegetal polarity would be carried over from the polarity in the progenitor cells. In either case, the mechanisms linking the chromosomal bouquet to localizing Velo1 and nascent Balbiani body components need to be elucidated; these could include the SUN-KASH domain proteins involved in homolog pairing in meiosis (Starr and Fridolfsson 2010). Additionally, the mechanisms controlling *dazl* mRNA localization to the bouquet would be interesting to determine, since Dazl plays numerous roles during germline development (Fu et al. 2015). Rbpms2 may also have a critical role in this early localization, given its role in later RNA localization events.

An association of the Balbiani body and the chromosomal bouquet was also seen in a primitive insect *Thermobia* (Tworzydlo et al. 2016; Bilinski et al. 2017), suggesting its conservation in animals. The significance of coupling bouquet formation with Balbiani assembly is not known, but linking a meiotic event with Balbiani assembly may ensure that its formation is limited to germ cells. Interestingly, Balbiani body assembly occurs within a "nuclear cleft" (Elkouby et al. 2016) which is likely similar to the classically characterized "nuclear bay" concavity that typically surrounds the centrosome (Wilson 1928).[2] Recently these nuclear bays have been shown to accumulate proteins destined for degradation in the pericentriolar material (Fuentealba et al. 2008), potentially implicating these structures in protein turnover during oocyte polarity and Balbiani body formation. Additionally, the ability of Velo1 to assemble into amyloid aggregates and co-aggregate mitochondria and RNA (Boke et al. 2016) suggests that the process is largely self-organizing and does not require specific directed transport and assembly.

The Balbiani body serves as the site for germ plasm assembly, which also likely occurs through self-organization. RNAs have been shown to localize to germ plasm through diffusion-entrapment, and recent data suggest a key role for Rbpms2 as a shuttling protein. Although different pathways for RNA localization have been known for some time, involving specific *cis*-RNA sequence elements, identifying specific *trans*-acting proteins has been more problematic. Definitive germ plasm assembly may involve phase transitions of Velo1 from the amyloid structure used in Balbiani body assembly to hydrogel or other forms. Velo1 can associate with RNAs but nonspecifically and may thus provide an initial low-affinity step in RNA localization, after which more specific interactions mediate the localization into germ plasm/germ granules. This latter process is not well understood; only one RNA element and no *trans*-acting factors have been identified.

Rbpms2 may be a key determinant for the early pathway since it can interact with *nanos1* RNA and Velo1 protein, although how it recognizes specific RNAs is unclear. Similarly, StauI appears specific for late pathway RNAs, and the specificity of its association is similarly not understood. Since StauI is a dsRNA-binding protein, it is tempting to speculate that RNP assembly and remodeling create a double-stranded region recognized by StauI and coupled to the transport machinery.

The number of LE studied mechanistically has been increasing but is still limited and has identified divergent LE using a set of overlapping general RNA-binding proteins. Unifying these mechanisms as well as identifying distinct cellular "addresses" in the oocyte for various localized RNAs will be one of the future challenges. RNA localization research suffers several inherent technical limitations, including the promiscuous nature of many RNA-protein interactions and the propensity of RNPs to interact in liquid phase granules. Also, many studies have relied on overexpressed proteins and RNAs, which, while informative in general, may fail to uncover subtle features of localization.

Increasingly sophisticated technological advancements are becoming available and could shed light on RNA localization mechanisms in the oocyte. These methods include high-throughput analysis of RNAs bound to specific proteins using various "Clip-omics" (cross-linking and immunoprecipitation) approaches [i.e., PAR-Clip, HITS-Clip, iClip; e.g., (Ascano et al. 2011)], similar proteomics strategies to

[2]Curiously, Wilson, a leading figure in the early years of cell and developmental biology, considered the term centrosome "undesirable", apparently preferring "central body" instead (Wilson 1928, p. 30).

identify interacting proteins, and computational prediction of RNA-binding protein binding sites (Li et al. 2013). Additionally, genome editing should allow more robust labeling and tagging of localized RNAs in vivo to gain a more precise picture of endogenous RNA localization mechanisms [e.g., (Nelles et al. 2016)]. Last, the increasing applicability of single-molecule analysis and biochemistry is likely to enable unforeseen detail of molecular interactions in the assembly of localization granules. *Xenopus* has proven especially suitable for studying RNA localization, owing to the large size and accessibility of oocytes and abundant material for biochemical or physical fractionation. Many of these newer methods are only now being applied to oocyte RNA localization, and given the large number of localized RNAs identified along different pathways in *Xenopus*, there is considerable opportunity to elucidate many novel basic aspects of RNA localization.

Acknowledgments This work was supported by the University of Iowa and by NIH grant GM083999 (DWH).

References

Aguero T, Zhou Y, Kloc M et al (2016) Hermes (Rbpms) is a critical component of RNP complexes that sequester germline RNAs during oogenesis. J Dev Biol 4:2. doi:10.3390/jdb4010002

Alberti S, Halfmann R, King O et al (2009) A systematic survey identifies prions and illuminates sequence features of prionogenic proteins. Cell 137:146–158. doi:10.1016/j.cell.2009.02.044

Allison R, Czaplinski K, Git A et al (2004) Two distinct Staufen isoforms in *Xenopus* are vegetally localized during oogenesis. RNA 10:1751–1763. doi:10.1261/rna.7450204

al-Mukhtar K, Webb A (1971) An ultrastructural study of primordial germ cells, oogonia and early oocytes in *Xenopus laevis*. J Embryol Exp Morphol 26:195–217

Ancel P, Vintembenger P (1948) Recherches sur le déterminisme de la symétrie bilatérale dans l'œuf des amphibiens. Bull Biol Fr Belg 31(Suppl):1–182

Arthur PK, Claussen M, Koch S et al (2009) Participation of *Xenopus* Elr-type proteins in vegetal mRNA localization during oogenesis. J Biol Chem 284:19982–19992. doi:10.1074/jbc.M109.009928

Ascano M, Hafner M, Cekan P et al (2011) Identification of RNA-protein interaction networks using PAR-CLIP. WIREs RNA 3:159–177. doi:10.1002/wrna.1103

Bauermeister D, Claussen M, Pieler T (2015) A novel role for Celf1 in vegetal RNA localization during *Xenopus* oogenesis. Dev Biol 405:214–224. doi:10.1016/j.ydbio.2015.07.005

Bilinski SM, Jaglarz MK, Tworzydlo W (2017) The pole (germ) plasm in insect oocytes. In: Kloc M, Kubiak JZ (eds) Oocytes: maternal information and functions, Results and problems in cell differentiation. Springer, Cham

Boke E, Ruer M, Wühr M et al (2016) Amyloid-like self-assembly of a cellular compartment. Cell 166:637–650. doi:10.1016/j.cell.2016.06.051

Bontems F, Stein A, Marlow F et al (2009) Bucky ball organizes germ plasm assembly in zebrafish. Curr Biol 19:414–422. doi:10.1016/j.cub.2009.01.038

Bottenberg W, Sanchez-Soriano N, Alves-Silva J et al (2009) Context-specific requirements of functional domains of the Spectraplakin short stop in vivo. Mech Dev 126:489–502. doi:10.1016/j.mod.2009.04.004

Bounoure L (1937) Le sort de la lignée germinale chez la Grenouille rousse après l'action des rayons ultra-violets sur le pôle inférieur de l'oeuf. CR Acad Sci Paris 204:1837

Bounoure L, Aubry R, Huck ML (1954) Nouvelles recherches experimentales sur les origines de la lignee reproductrice chez la Grenouille rousse. J Embryol Exp Morph 2:245–263

Brangwynne CP, Eckmann CR, Courson DS et al (2009) Germline P granules are liquid droplets that localize by controlled dissolution/condensation. Science 324:1729–1732. doi:10.1126/science.1172046

Carpenter CD, Klein WH (1982) A gradient of poly(A)+ RNA sequences in *Xenopus laevis* eggs and embryos. Dev Biol 91:43–49

Chan AP, Kloc M, Etkin LD (1999) fatvg encodes a new localized RNA that uses a 25-nucleotide element (FVLE1) to localize to the vegetal cortex of *Xenopus* oocytes. Development 126: 4943–4953

Chang P, Torres J, Lewis RA et al (2004) Localization of RNAs to the mitochondrial cloud in *Xenopus* oocytes through entrapment and association with endoplasmic reticulum. Mol Biol Cell 15:4669–4681. doi:10.1091/mbc.E04-03-0265

Claussen M, Pieler T (2004) Xvelo1 uses a novel 75-nucleotide signal sequence that drives vegetal localization along the late pathway in *Xenopus* oocytes. Dev Biol 266:270–284

Claussen M, Horvay K, Pieler T (2004) Evidence for overlapping, but not identical, protein machineries operating in vegetal RNA localization along early and late pathways in *Xenopus* oocytes. Development 131:4263–4273. doi:10.1242/dev.01283

Coggins LW (1973) An ultrastructural and radioautographic study of early oogenesis in the toad *Xenopus laevis*. J Cell Sci 12:71–93

Cosson B, Gautier-Courteille C, Maniey D et al (2006) Oligomerization of EDEN-BP is required for specific mRNA deadenylation and binding. Biol Cell 98:653–665. doi:10.1042/BC20060054

Courchaine EM, Lu A, Neugebauer KM (2016) Droplet organelles? EMBO J 35:1603–1612. doi:10.15252/embj.201593517

Cuykendall TN, Houston DW (2009) Vegetally localized *Xenopus* trim36 regulates cortical rotation and dorsal axis formation. Development 136:3057–3065. doi:10.1242/dev.036855

Cuykendall TN, Houston DW (2010) Identification of germ plasm-associated transcripts by microarray analysis of *Xenopus* vegetal cortex RNA. Dev Dyn 239:1838–1848. doi:10.1002/dvdy.22304

Dasgupta T, Ladd AN (2012) The importance of CELF control: molecular and biological roles of the CUG-BP, Elav-like family of RNA-binding proteins. WIREs RNA 3:104–121. doi:10.1002/wrna.107

Deshler JO, Highett MI, Schnapp BJ (1997) Localization of *Xenopus* Vg1 mRNA by Vera protein and the endoplasmic reticulum. Science 276:1128–1131

Devaux A, Colegrove-Otero LJ, Standart N (2006) *Xenopus* ElrB, but not ElrA, binds RNA as an oligomer: possible role of the linker. FEBS Lett 580:4947–4952. doi:10.1016/j.febslet.2006.08.012

Dosch R, Wagner DS, Mintzer KA et al (2004) Maternal control of vertebrate development before the midblastula transition: mutants from the zebrafish I. Dev Cell 6:771–780. doi:10.1016/j.devcel.2004.05.002

Dumont JN (1972) Oogenesis in *Xenopus laevis* (Daudin). I. Stages of oocyte development in laboratory maintained animals. J Morphol 136:153–179. doi:10.1002/jmor.1051360203

El Meziane A, Callen JC, Mounolou JC (1989) Mitochondrial gene expression during *Xenopus laevis* development: a molecular study. EMBO J. doi:10.1371/journal.pgen.1005019.s006

Elkouby YM, Jamieson-Lucy A, Mullins MC (2016) Oocyte polarization is coupled to the chromosomal bouquet, a conserved polarized nuclear configuration in meiosis. PLoS Biol 14:e1002335. doi:10.1371/journal.pbio.1002335

Extavour CG, Akam M (2003) Mechanisms of germ cell specification across the metazoans: epigenesis and preformation. Development 130:5869–5884. doi:10.1242/dev.00804

Forristall C, Pondel M, Chen L, King M (1995) Patterns of localization and cytoskeletal association of two vegetally localized RNAs, Vg1 and Xcat-2. Development 121:201–208

Fu X-F, Cheng S-F, Wang L-Q et al (2015) DAZ family proteins, key players for germ cell development. Int J Biol Sci 11:1226–1235. doi:10.7150/ijbs.11536

Fuentealba LC, Eivers E, Geissert D et al (2008) Asymmetric mitosis: unequal segregation of proteins destined for degradation. Proc Natl Acad Sci 105:7732–7737. doi:10.1073/pnas.0803027105

Gard DL, Cha BJ, King E (1997) The organization and animal-vegetal asymmetry of cytokeratin filaments in stage VI *Xenopus* oocytes is dependent upon F-actin and microtubules. Dev Biol 184:95–114. doi:10.1006/dbio.1997.8508

Gerber WV, Vokes SA, Zearfoss NR, Krieg PA (2002) A role for the RNA-binding protein, hermes, in the regulation of heart development. Dev Biol 247:116–126. doi:10.1006/dbio.2002.0678

Git A, Allison R, Perdiguero E et al (2009) Vg1RBP phosphorylation by Erk2 MAP kinase correlates with the cortical release of Vg1 mRNA during meiotic maturation of *Xenopus* oocytes. RNA 15:1121–1133. doi:10.1261/rna.1195709

Gupta T, Marlow FL, Ferriola D et al (2010) Microtubule actin crosslinking factor 1 regulates the Balbiani body and animal-vegetal polarity of the zebrafish oocyte. PLoS Genet 6:e1001073. doi:10.1371/journal.pgen.1001073

Guraya SS (1979) Recent advances in the morphology, cytochemistry, and function of Balbiani's vitelline body in animal oocytes. Int Rev Cytol 59:249–321

Han TW, Kato M, Xie S et al (2012) Cell-free formation of RNA granules: bound RNAs identify features and components of cellular assemblies. Cell 149:768–779. doi:10.1016/j.cell.2012.04.016

Harper L, Golubovskaya I, Cande WZ (2004) A bouquet of chromosomes. J Cell Sci 117:4025–4032. doi:10.1242/jcs.01363

Havin L, Git A, Elisha Z et al (1998) RNA-binding protein conserved in both microtubule- and microfilament-based RNA localization. Genes Dev 12:1593–1598

Heasman J, Quarmby J, Wylie CC (1984) The mitochondrial cloud of *Xenopus* oocytes: the source of germinal granule material. Dev Biol 105:458–469

Heim AE, Hartung O, Rothhämel S et al (2014) Oocyte polarity requires a Bucky ball-dependent feedback amplification loop. Development 141:842–854. doi:10.1242/dev.090449

Heinrich B, Deshler JO (2009) RNA localization to the Balbiani body in *Xenopus* oocytes is regulated by the energy state of the cell and is facilitated by kinesin II. RNA 15:524–536. doi:10.1261/rna.975309

Hellsten U, Harland RM, Gilchrist MJ et al (2010) The genome of the Western clawed frog *Xenopus* tropicalis. Science 328:633–636. doi:10.1126/science.1183670

Hinman MN, Lou H (2008) Diverse molecular functions of Hu proteins. Cell Mol Life Sci 65:3168–3181. doi:10.1007/s00018-008-8252-6

Horvay K, Claussen M, Katzer M et al (2006) *Xenopus* Dead end mRNA is a localized maternal determinant that serves a conserved function in germ cell development. Dev Biol 291:1–11. doi:10.1016/j.ydbio.2005.06.013

Houston DW (2013) Regulation of cell polarity and RNA localization in vertebrate oocytes. Int Rev Cell Mol Biol 306:127–185. doi:10.1016/B978-0-12-407694-5.00004-3

Ikenishi K, Kotani M, Tanabe K (1974) Ultrastructural changes associated with UV irradiation in the "germinal plasm" of *Xenopus laevis*. Dev Biol 36:155–168

Kato M, Han TW, Xie S et al (2012) Cell-free formation of RNA granules: low complexity sequence domains form dynamic fibers within hydrogels. Cell 149:753–767. doi:10.1016/j.cell.2012.04.017

Kedde M, Strasser MJ, Boldajipour B et al (2007) RNA-binding protein Dnd1 inhibits MicroRNA access to target mRNA. Cell 131:1273–1286. doi:10.1016/j.cell.2007.11.034

King ML, Messitt TJ, Mowry KL (2005) Putting RNAs in the right place at the right time: RNA localization in the frog oocyte. Biol Cell 97:19–33. doi:10.1042/BC20040067

Kloc M, Etkin LD (1995) Two distinct pathways for the localization of RNAs at the vegetal cortex in *Xenopus* oocytes. Development 121:287–297

Kloc M, Larabell C, Etkin LD (1996) Elaboration of the messenger transport organizer pathway for localization of RNA to the vegetal cortex of *Xenopus* oocytes. Dev Biol 180:119–130. doi:10.1006/dbio.1996.0289

Kloc M, Bilinski S, Pui-Yee Chan A, Etkin LD (2000) The targeting of Xcat2 mRNA to the germinal granules depends on a cis-acting germinal granule localization element within the 3′UTR. Dev Biol 217:221–229. doi:10.1006/dbio.1999.9554

Kloc M, Dougherty MT, Bilinski S et al (2002a) Three-dimensional ultrastructural analysis of RNA distribution within germinal granules of *Xenopus*. Dev Biol 241:79–93. doi:10.1006/dbio.2001.0488

Kloc M, Zearfoss NR, Etkin LD (2002b) Mechanisms of subcellular mRNA localization. Cell 108:533–544

Kloc M, Bilinski S, Dougherty MT et al (2004a) Formation, architecture and polarity of female germline cyst in *Xenopus*. Dev Biol 266:43–61

Kloc M, Bilinski S, Etkin LD (2004b) The Balbiani body and germ cell determinants: 150 years later. Curr Top Dev Biol 59:1–36. doi:10.1016/S0070-2153(04)59001-4

Kloc M, Jaglarz M, Dougherty M et al (2008) Mouse early oocytes are transiently polar: three-dimensional and ultrastructural analysis. Exp Cell Res 314:3245–3254. doi:10.1016/j.yexcr.2008.07.007

Koebernick K, Loeber J, Arthur PK et al (2010) Elr-type proteins protect *Xenopus* Dead end mRNA from miR-18-mediated clearance in the soma. Proc Natl Acad Sci 107:16148–16153. doi:10.1073/pnas.1004401107

Kogo N, Tazaki A, Kashino Y et al (2011) Germ-line mitochondria exhibit suppressed respiratory activity to support their accurate transmission to the next generation. Dev Biol 349:462–469. doi:10.1016/j.ydbio.2010.11.021

Kosaka K, Kawakami K, Sakamoto H, Inoue K (2007) Spatiotemporal localization of germ plasm RNAs during zebrafish oogenesis. Mech Dev 124:279–289. doi:10.1016/j.mod.2007.01.003

Kress TL, Yoon YJ, Mowry KL (2004) Nuclear RNP complex assembly initiates cytoplasmic RNA localization. J Cell Biol 165:203–211. doi:10.1083/jcb.200309145

Le Goff C, Laurent V, Le Bon K et al (2006) pEg6, a spire family member, is a maternal gene encoding a vegetally localized mRNA in *Xenopus* embryos. Biol Cell 98:697–708. doi:10.1042/BC20050095

Lewis RA, Gagnon JA, Mowry KL (2008) PTB/hnRNP I is required for RNP remodeling during RNA localization in *Xenopus* oocytes. Mol Cell Biol 28:678–686. doi:10.1128/MCB.00999-07

Li X, Kazan H, Lipshitz HD, Morris QD (2013) Finding the target sites of RNA-binding proteins. WIREs RNA 5:111–130. doi:10.1002/wrna.1201

Marlow FL, Mullins MC (2008) Bucky ball functions in Balbiani body assembly and animal-vegetal polarity in the oocyte and follicle cell layer in zebrafish. Dev Biol 321:40–50. doi:10.1016/j.ydbio.2008.05.557

Martel C, Dugré-Brisson S, Boulay K et al (2010) Multimerization of Staufen1 in live cells. RNA 16:585–597. doi:10.1261/rna.1664210

Monie TP, Hernandez H, Robinson CV et al (2005) The polypyrimidine tract binding protein is a monomer. RNA 11:1803–1808. doi:10.1261/rna.2214405

Nelles DA, Fang MY, O'Connell MR et al (2016) Programmable RNA tracking in live cells with CRISPR/Cas9. Cell 165:488–496. doi:10.1016/j.cell.2016.02.054

Nieuwkoop PD (1969) The formation of mesoderm in Urodelean amphibians. I. Induction by the endoderm. Wilhelm Roux' Arch EntwMech Org 162:341–373

Nijjar S, Woodland HR (2013a) Protein interactions in *Xenopus* germ plasm RNP particles. PLoS One 8:e80077

Nijjar S, Woodland HR (2013b) Localisation of RNAs into the germ plasm of vitellogenic *Xenopus* oocytes. PLoS One 8:e61847. doi:10.1371/journal.pone.0061847.t002

Nott TJ, Petsalaki E, Farber P et al (2015) Phase transition of a disordered nuage protein generates environmentally responsive membraneless organelles. Mol Cell 57:936–947. doi:10.1016/j.molcel.2015.01.013

Paillard L, Omilli F, Legagneux V et al (1998) EDEN and EDEN-BP, a cis element and an associated factor that mediate sequence-specific mRNA deadenylation in *Xenopus* embryos. EMBO J 17:278–287. doi:10.1093/emboj/17.1.278

Pelegri F (2003) Maternal factors in zebrafish development. Dev Dyn 228:535–554. doi:10.1002/dvdy.10390

Pepling ME, de Cuevas M, Spradling AC (1999) Germline cysts: a conserved phase of germ cell development? Trends Cell Biol 9:257–262

Pepling ME, Wilhelm JE, O'Hara AL et al (2007) Mouse oocytes within germ cell cysts and primordial follicles contain a Balbiani body. Proc Natl Acad Sci 104:187–192. doi:10.1073/pnas.0609923104

Rasar MA, Hammes SR (2006) The physiology of the *Xenopus laevis* ovary. Methods Mol Biol 322:17–30. doi:10.1007/978-1-59745-000-3_2

Rebagliati MR, Weeks DL, Harvey RP, Melton DA (1985) Identification and cloning of localized maternal RNAs from *Xenopus* eggs. Cell 42:769–777

Sagnol S, Yang Y, Bessin Y et al (2014) Homodimerization of RBPMS2 through a new RRM-interaction motif is necessary to control smooth muscle plasticity. Nucleic Acids Res 42:10173–10184. doi:10.1093/nar/gku692

Scherthan H (2001) A bouquet makes ends meet. Nat Rev Mol Cell Biol. doi:10.1038/35085086

Selman K, Wallace RA, Sarka A, Qi X (1993) Stages of oocyte development in the zebrafish, *Brachydanio rerio*. J Morphol 218:203–224

Session AM, Uno Y, Kwon T et al (2016) Genome evolution in the allotetraploid frog *Xenopus laevis*. Nature 538:336–343. doi:10.1038/nature19840

Shoubridge EA, Wai T (2007) Mitochondrial DNA and the mammalian oocyte. Curr Top Dev Biol 77:87–111. doi:10.1016/S0070-2153(06)77004-1

Smith LD (1966) The role of a "germinal plasm" in the formation of primordial germ cells in Rana pipiens. Dev Biol 14:330–347

Snedden DD, Bertke MM, Vernon D, Huber PW (2013) RNA localization in *Xenopus* oocytes uses a core group of trans-acting factors irrespective of destination. RNA 19:889–895. doi:10.1261/rna.038232.113

Song H-W, Cauffman K, Chan AP et al (2007) Hermes RNA-binding protein targets RNAs-encoding proteins involved in meiotic maturation, early cleavage, and germline development. Differentiation 75:519–528. doi:10.1111/j.1432-0436.2006.00155.x

Starr DA, Fridolfsson HN (2010) Interactions between nuclei and the cytoskeleton are mediated by SUN-KASH nuclear-envelope bridges. Annu Rev Cell Dev Biol 26:421–444. doi:10.1146/annurev-cellbio-100109-104037

Sudarwati S, Nieuwkoop PD (1971) Mesoderm formation in the anuran *Xenopus laevis* (Daudin). Dev Genes Evol 166:189–204. doi:10.1007/BF00650029

Sudou N, Garcés-Vásconez A, López-Latorre MA et al (2016) Transcription factors Mix1 and VegT, relocalization of vegt mRNA, and conserved endoderm and dorsal specification in frogs. Proc Natl Acad Sci 113:5628–5633. doi:10.1073/pnas.1605547113

Suozzi KC, Wu X, Fuchs E (2012) Spectraplakins: master orchestrators of cytoskeletal dynamics. J Cell Biol 197:465–475. doi:10.1083/jcb.201112034

Toretsky JA, Wright PE (2014) Assemblages: functional units formed by cellular phase separation. J Cell Biol 206:579–588. doi:10.1083/jcb.201404124

Tworzydlo W, Marek M, Kisiel E, Bilinski SM (2016) Meiosis, Balbiani body and early asymmetry of Thermobia oocyte. Protoplasma. doi:10.1007/s00709-016-0978-7

Voronina E, Seydoux G, Sassone-Corsi P, Nagamori I (2011) RNA granules in germ cells. Cold Spring Harb Perspect Biol. doi:10.1101/cshperspect.a002774

Wagner DS, Dosch R, Mintzer KA et al (2004) Maternal control of development at the midblastula transition and beyond: mutants from the zebrafish II. Dev Cell 6:781–790. doi:10.1016/j.devcel.2004.04.001

Wakahara M (1977) Partial characterization of "primordial germ cell-forming activity" localized in vegetal pole cytoplasm in anuran eggs. Development 39:221–233

Weeks DL, Melton DA (1987) A maternal mRNA localized to the vegetal hemisphere in *Xenopus* eggs codes for a growth factor related to TGF-beta. Cell 51:861–867

Wilson E (1928) The cell in development and heredity, 3rd edn. Macmillian, New York

Xiang S, Kato M, Wu LC, Lin Y, Ding M, Zhang Y, Yu Y, McKnight SL (2015) The LC domain of hnRNPA2 adopts similar conformations in hydrogel polymers, liquid-like droplets, and nuclei. Cell 163:829–839. doi:10.1016/j.cell.2015.10.040

Yoon YJ, Mowry KL (2004) *Xenopus* Staufen is a component of a ribonucleoprotein complex containing Vg1 RNA and kinesin. Development 131:3035–3045. doi:10.1242/dev.01170

Zearfoss NR, Chan AP, CF W et al (2004) Hermes is a localized factor regulating cleavage of vegetal blastomeres in *Xenopus laevis*. Dev Biol 267:60–71. doi:10.1016/j.ydbio.2003.10.032

Zhang J, King ML (1996) *Xenopus* VegT RNA is localized to the vegetal cortex during oogenesis and encodes a novel T-box transcription factor involved in mesodermal patterning. Development 122:4119–4129

Zhou Y, King ML (1996) Localization of Xcat-2 RNA, a putative germ plasm component, to the mitochondrial cloud in *Xenopus* stage I oocytes. Development 122:2947–2953

Zickler D, Kleckner N (1998) The leptotene-zygotene transition of meiosis. Annu Rev Genet 32: 619–697. doi:10.1146/annurev.genet.32.1.619

Züst B, Dixon K (1975) The effect of u.v. irradiation of the vegetal pole of *Xenopus laevis* eggs on the presumptive primordial germ cells. J Embryol Exp Morph 34:209–220

Zwicker D, Decker M, Jaensch S et al (2014) Centrosomes are autocatalytic droplets of pericentriolar material organized by centrioles. Proc Natl Acad Sci 111:E2636–E2645. doi:10.1073/pnas.1404855111

Part III
Epigenetic, Transcriptional and Translational Regulation in Oocytes

Chapter 10
DNA Methyltransferases in Mammalian Oocytes

Fatma Uysal and Saffet Ozturk

Abstract Epigenetic mechanisms play important roles in properly occurring mammalian oogenesis. One of these mechanisms is DNA methylation adding a methyl group to the fifth carbon atom of the cytosine residues using S-adenosyl-L-methionine as a methyl donor. DNA methylation generally takes place at cytosine-phosphate-guanine (CpG) dinucleotide sites and rarely occurs at cytosine-phosphate-thymine (CpT), cytosine-phosphate-adenine (CpA), or cytosine-phosphate-cytosine sites, known as non-CpG sites. Basically, two different DNA methylation processes are identified: de novo methylation and maintenance methylation. While the de novo methylation functions in methylation of unmethylated DNA strands, maintenance methylation is capable of methylating hemi-methylated DNA strands following DNA replication. Both DNA methylation processes are catalyzed by special DNA methyltransferase (DNMT) enzymes. To date, five different DNMTs have been identified: DNMT1, DNMT3A, DNMT3B, DNMT3L, and DNMT2. In this chapter, we focus particularly on temporal and spatial expression of DNMTs in mammalian oocytes and granulosa cells.

10.1 Introduction

Genomic imprinting is briefly defined as some genes having maternal and paternal copies are expressed in a parent-specific manner (Barlow and Bartolomei 2014). In this process, epigenetic mechanisms such as histone modifications including methylation, acetylation, phosphorylation, sumoylation and ubiquitylation, and DNA methylation play important roles in proper imprinting of the target genes (Henikoff and Smith 2015). DNA methylation functions in transcriptional repression or activation of development-related genes such as *H19* (Bartolomei et al. 1993) and insulin-like growth factor 2 (*IGF2*) (Stoger et al. 1993) genes, X-chromosome inactivation (Heard et al. 1997), genomic imprinting (Reik and Dean 2001; Kaneda et al. 2004), cell differentiation (Agarwal and Rao 1998), aging (Richardson 2003),

F. Uysal • S. Ozturk (✉)
Department of Histology and Embryology, School of Medicine, Akdeniz University, Campus, 07070 Antalya, Turkey
e-mail: sozturk@akdeniz.edu.tr

M. Kloc (ed.), *Oocytes*, Results and Problems in Cell Differentiation 63,
DOI 10.1007/978-3-319-60855-6_10

tumorigenesis (Feinberg et al. 2002; Feinberg 2000), and other cellular processes (Robertson and Wolffe 2000). It is important to note that environmental factors (Virani et al. 2016), tobacco smoking (Ambatipudi et al. 2016), aging (Yue et al. 2012), diabetes (Ma et al. 2012), some assisted reproductive technologies such as superovulation (Liang et al. 2013), in vitro culture conditions (Fernandez-Gonzalez et al. 2007; Zaitseva et al. 2007; Deshmukh et al. 2011), and cryopreservation (Petrussa et al. 2014) are known to affect DNA methylation event.

In brief, DNA methylation adds a methyl group to the fifth carbon atom of the cytosine residues using S-adenosyl-L-methionine (AdoMet) as a methyl donor (Chen and Li 2004; Portela and Esteller 2010). DNA methylation commonly takes place in cytosine-phosphate-guanine (CpG) dinucleotide sites and rarely occurs in cytosine-phosphate-thymine (CpT), cytosine-phosphate-adenine (CpA), and cytosine-phosphate-cytosine sites, known as non-CpG sequences (Chen and Li 2004; Reik and Dean 2001). Two different DNA methylation mechanisms have been identified as de novo and maintenance methylation. De novo methylation is capable of methylating the unmethylated DNA sequences and plays important roles in establishing new genomic imprints in germ cells (Law and Jacobsen 2010). Maintenance methylation provides methylation of hemi-methylated DNA sequences following DNA replication. Each DNA methylation process is catalyzed by specific DNA methyltransferase (DNMT) enzymes: DNMT1, DNMT2, DNMT3A, DNMT3B, and DNMT3L (Bestor 2000). Importantly, dynamic expression of the DNMTs has been comprehensively evaluated in the female (Uysal et al. 2015) and male germ cells (Uysal et al. 2016).

10.2 Structure and Particular Functions of DNMTs

The DNMT enzyme was first isolated and cloned from the bacterium *Haemophilus haemolyticus* and is composed of 327 amino acid residues (Kumar et al. 1994). Generally, mammalian DNMTs consist of three structural regions: N-terminal regulatory domain, C-terminal catalytic domain, and a central linker region (Araujo et al. 2001). The N-terminal regulatory domain determines subcellular localization of the DNMTs and contains a proliferating cell nuclear antigen-binding domain (PBD), a nuclear localization signal (NLS), a cytosine-rich zinc finger DNA-binding motif (ATRX), a polybromo homology domain (PHD), and a PWWP tetrapeptide chromatin binding domain (Bestor 2000; Posfai et al. 1989). The C-terminal catalytic domain includes ten different characteristic motifs, and six of them (I, IV, VI, VIII, IX, and X) are evolutionally conserved in mammals. The motifs have specific functions: (1) the motif I has an AdoMet binding site, (2) the motif IV can associate with the substrate cytosine at the active site, (3) the motif VI consists of glutamyl residues serving as a donor, (4) the motif IX provides stability to the substrate binding site, and (5) the motif X creates an AdoMet binding site; however, the motif VIII's function remains unclear. The last region of DNMTs, the central linker domain, is composed of repeated lysine-glycine dipeptides and provides a connection between the N- and C-terminal domains. Although all known DNMTs exhibit almost similar structural features, they achieve unique missions as well as similar roles.

The well-known function of DNMT1 is to methylate hemi-methylated DNA strands during genomic replication (Cedar and Bergman 2012; Pradhan et al. 1999). It is important to note that DNMT1 also participates in de novo methylation (Lorincz et al. 2002). Consistently, lack of *Dnmt1* gene (*Dnmt1*$^{-/-}$) causes embryonic lethality, loss of imprinting, and global demethylation of the genome (Yoder et al. 1997; Kurihara et al. 2008; Okano et al. 1998a). To date, three functional DNMT1 isoforms have been identified: the pachytene spermatocyte form of DNMT1 (DNMT1p), the oocyte-specific form of DNMT1 (DNMT1o), and the somatic form of DNMT1 (DNMT1s) (Ko et al. 2005).

The DNMT3A enzyme, one of the de novo methyltransferases, plays a critical role in de novo methylation of unmethylated DNA strands so that new imprints are established (Okano et al. 1998a). To date, two RNA isoforms of *DNMT3A* have been identified (Ostler et al. 2007). DNMT3A isoforms are expressed from maternally stored mRNAs; therefore they are detected in mouse oocytes and early embryos until embryonic genome activation (EGA) (Watanabe et al. 2002). The *Dnmt3a* knockout (*Dnmt3a*$^{-/-}$) mouse models exhibit partial defective embryonic development, and some of these mice are found to be viable (Okano et al. 1999). Another de novo methyltransferase, DNMT3B, specifically functions in de novo DNA methylation (Okano et al. 1998a). DNMT3B specializes in methylation of the CpG dinucleotides at the repeated DNA sequences present in the pericentric satellite DNA region of eukaryotic chromosomes (Okano et al. 1999; Kato et al. 2007). Mutations in the *DNMT3B* gene lead to genetic immunodeficiency disorder in humans known as centromere instability and facial anomaly syndromes (ICF syndrome) (Gowher et al. 2005). More than 20 DNMT3B isoforms have been identified (Ostler et al. 2007); however, the role of these isoforms remains elusive. Lack of the *Dnmt3b* gene (*Dnmt3b*$^{-/-}$) or *Dnmt1*$^{-/-}$ causes embryonic lethality in mouse models (Okano et al. 1998a).

Interestingly, DNMT3L differs from other DNMTs because it lacks the catalytic motif and does not have any methyltransferase activity (Chedin et al. 2002; Turek-Plewa and Jagodzinski 2005). However, DNMT3L contributes to de novo methylation process via interacting with either DNMT3A or DNMT3B by binding to their carboxyl terminal domains (Neri et al. 2013). DNMT3L functions in establishment of maternal genomic imprints during oogenesis (Bourc'his et al. 2001). *Dnmt3l* knockout (*Dnmt3l*$^{-/-}$) leads to abnormal maternal imprinting and male infertility (Bourc'his et al. 2001; Bourc'his and Bestor 2004).

Lastly, DNMT2 is prominently different from all other DNMTs because instead of transferring methyl groups to the cytosine residues in genomic DNA, DNMT2 methylates cytosine 38 in the anticodon loop of aspartic acid transfer RNA (Goll et al. 2006). It is important to note that because DNMT2 does not have a N-terminal regulatory domain, it does not perform de novo or maintenance methylation (Turek-Plewa and Jagodzinski 2005). Lack of *Dnmt2* gene (*Dnmt2*$^{-/-}$) in mouse results in losing RNA methyltransferase activity, but it does not lead to any phenotypic abnormality (Okano et al. 1998b).

10.3 DNA Methylation During Oogenesis

Female germ cells originate from primordial germ cells (PGCs). The PGCs present in the yolk sac endoderm migrate from the yolk sac to the right and left sides of the posterior body wall. When PGCs reach the gonadal ridges, they differentiate into oogonia. Oogenesis is briefly defined as a process of developing oogonia into mature oocytes, which begins before birth and is completed after puberty. During prenatal development, in the developing ovary, the oogonia undergo successive mitotic divisions. Subsequently, oocytes enter meiotic division and become arrested at meiotic prophase I stage. These early oocytes are also resided in primordial follicles. After puberty, the immature germinal vesicle (GV) oocytes undergo nucleus GV breakdown to complete first meiotic division and, at ovulation, until fertilization, become arrested in metaphase II (MII) (Verlhac and Terret 2016).

During oogenesis, DNA methylation is a critical mechanism because expression of certain genes required for development, growth, and maturation of oocytes is regulated by this mechanism. Erasure of DNA methylation in PGCs is carried out during migration to genital ridge, and then de novo DNA methylation occurs for establishment of allele-specific imprints. Genome-wide DNA methylation is established at GV oocytes during early oogenesis and it is fulfilled before germinal vesicle breakdown takes place (Yan et al. 2014). When Kageyama et al. (2007) analyzed the global DNA methylation status in the growing oocytes from the 5-, 10-, and 15-day-old mouse ovaries, the DNA methylation levels is increased from 5-day to 15-day, respectively (Kageyama et al. 2007). This increase is compatible with creating genomic imprints in the growing oocytes between days 10 and 15 (Kageyama et al. 2007).

In oocytes, DNA methylation catalyzed by DNMTs establishes genomic imprints, and it reaches the highest levels in the MII oocytes when compared to GV oocytes (Yang et al. 2007) (Fig. 10.1). It is known that proper methylation establishment in the oocyte stages is very important for normal oogenesis and preimplantation embryo development. Recent studies have shown that ART procedures such as vitrification may alter DNA methylation in the oocytes, resulting in abnormal embryo development. Vitrification causes decreased global DNA methylation in vitrified oocytes in comparison to control counterparts (Liang et al. 2014). On the other hand, some investigators found that there is no significant difference in DNA methylation status between vitrified and control oocytes (Yan et al. 2014; Liu et al. 2016). More detailed studies are required to establish whether ART applications negatively affect development of the oocytes and early embryos.

DNA methylation is also crucial for somatic granulosa cells that surround oocytes during follicular development. DNA methylation may provide repression or activation of development-related genes in granulosa cells. In humans, DNA methylation is

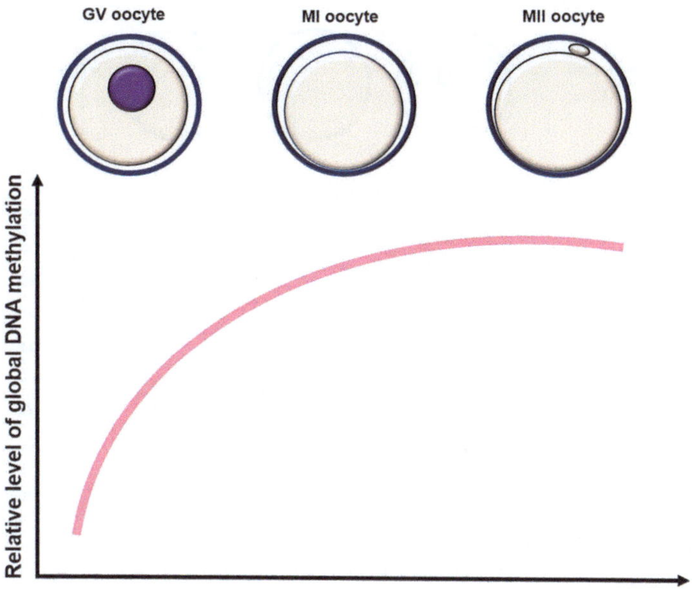

Fig. 10.1 A diagram summarizing the levels of global DNA methylation in the GV, MI, and MII oocytes. The global DNA methylation increases from GV to MII oocytes. *GV* germinal vesicle, *MI* metaphase I, *MII* metaphase II

initially detected in the granulosa cells obtained from developing follicles (Huntriss et al. 2004). A recent study revealed that global DNA methylation in the granulosa cells of old bovines is much lower than in the young ones (Goto et al. 2013). Similarly, granulosa cells obtained from aged donors had decreased DNA methylation levels compared with young age donors (Yu et al. 2015). Thus, the age-related decline of ovarian function, oocyte quality, and altered gene expression may be associated with abnormal DNA methylation in the granulosa cells.

10.3.1 DNMT1 Expression in Oocytes

The primary function of DNMT1 in mammalian oocytes is to maintain methylation in the hemi-methylated DNA strands after DNA methylation. In mouse, *Dnmt1* is expressed in both fetal (E17.5–E18.5) and postnatal ovaries (days 1, 5, 10, 15, 21, 35, 70) and in the oocytes collected from the same days (Ratnam et al. 2002). However, *Dnmt1o*, oocyte isoform of *Dnmt1*, is only transcribed in the postnatal ovaries and oocytes and, is undetectable levels in fetal ovaries and oocytes (Ratnam et al. 2002). In the MII oocytes, *DNMT1o* is expressed at higher levels than *Dnmt1* (Ratnam et al. 2002). In mouse, the DNMT1 protein is present in the cytoplasm of both GV and MII oocytes (Hirasawa et al. 2008) (Fig. 10.2). Notably, DNMT1 is concentrated in the cortex region of MII oocytes (Mertineit et al. 1998).

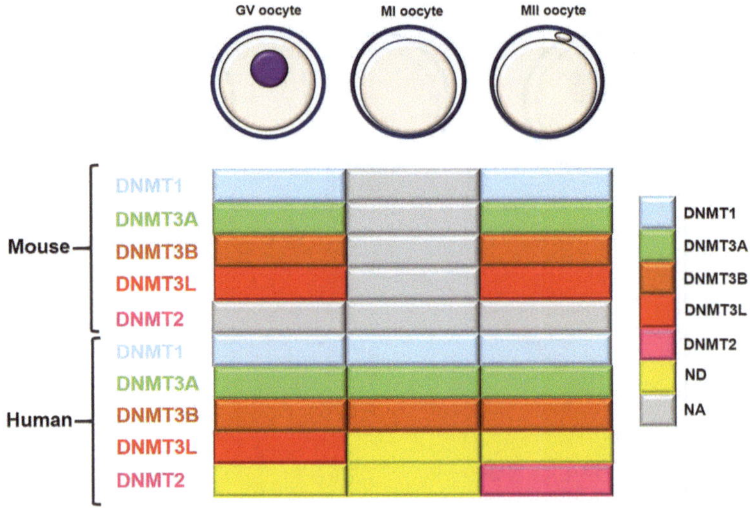

Fig. 10.2 Expression of DNMTs in mouse and human oocytes. Expression of each DNMT is depicted at different *colors*. *GV* germinal vesicle, *MI* metaphase I, *MII* metaphase II, *NA* not analyzed, *ND* not detected. There are expression differences in the GV, MI, and MII oocytes

In bovines, DNMT1 is found to be expressed in all GV stage (from GV0 to GV3) oocytes and is predominantly responsible for establishment of DNA methylation (Lodde et al. 2009a). Additionally, global DNA methylation increases from GV0 to GV3, which correlates with decreased *Dnmt1* transcriptional activity in the GV0 and GV1 oocytes (Lodde et al. 2009a). Lodde et al. (2009b) showed that DNMT1 is also expressed in in vitro matured cow MII oocytes (Lodde et al. 2009b). In humans, *DNMT1* is detected in the secondary follicles, granulosa cells, and GV and MII oocytes, but it is absent in the primordial and primary follicles (Huntriss et al. 2004). DNMT1 protein localizes in the nuclei of human GV oocytes and in the cytoplasm of the MI and MII oocytes (Fig. 10.2), and DNMT1s protein, one of the DNMT1 isoforms, exhibits similar expression pattern (Petrussa et al. 2014).

Because DNMT1 is critical for producing competent oocytes and early embryos, some factors being used in ART may alter the *Dnmt1* gene expression, resulting in abnormal DNA methylation, abnormal imprints, poor embryo quality, and abnormal oocyte maturation. DNMT1o deficiency in human oocytes obtained by ART leads to abnormal genome-wide DNA methylation and poor embryo quality (Whidden et al. 2016). The loss of DNMT1 expression during somatic cell nuclear transfer disrupts the global DNA methylation or maternal imprints in the early embryos (Wei et al. 2011). In vitro culture condition also changes *Dnmt1* expression: *Dnmt1* levels are lower in the GV oocytes produced from culture of fetal mouse germ cells isolated from 12.5 days postcoitum (dpc) mouse ovary than in in vivo-produced GV oocytes (Song et al. 2009). Lastly, aging leads to increased *Dnmt1* gene expression in adult sheep donors when compared to the prepubertal donors (Fang et al. 2016). This reveals that donor age seems to change expression profile of the *Dnmt1* gene during oocyte maturation and subsequently disrupts the embryonic development (Fang et al. 2016).

10.3.2 DNMT3A and DNMT3B Expression in Mammalian Oocytes

Both *Dnmt3a* and *Dnmt3b* mRNA expression have been detected in the mouse, cat, and bovine oocytes (Bourc'his et al. 2001; O'Doherty et al. 2012; Kaneda et al. 2004). In rhesus monkey oocytes, *DNMT3A* mRNA is expressed in all oocyte stages, and the expression levels progressively decrease from GV to MII oocytes (Vassena et al. 2005). On the other hand, another de novo methyltransferase, *DNMT3B*, is expressed at low levels in the GV and MII rhesus monkey oocytes (Vassena et al. 2005). In mouse, GV and MII oocytes express *Dnmt3b* at low levels (Vassena et al. 2005). In human, *DNMT3A* mRNA expression is detected in the ovarian follicles from primordial to secondary follicles, granulosa cells, and GV and MII oocytes; however, the primordial and early primary follicles do not express *DNMT3B* but express the *DNMT1* gene (Huntriss et al. 2004). And, there is *DNMT3B* expression in the late primary and secondary follicles, granulosa cells, and GV and MII oocytes (Huntriss et al. 2004) (Fig. 10.2).

In mouse, DNM3A protein is localized around the nuclei of GV oocytes and in the cytoplasm of the MII oocytes, but interestingly GV and MII oocytes do not produce the DNMT3B protein despite the presence of *Dnmt3b* mRNA in these oocyte stages (Hirasawa et al. 2008). In human, DNMT3A is expressed in the GV, MI, and MII oocytes at day 0 (2 h after oocyte pickup); however, DNMT3A is absent in the GV oocytes at day 1 (in vitro matured overnight) (Petrussa et al. 2014) (Fig. 10.2). In contrast, the GV, MI, and MII oocytes either from day 0 or 1 produce the DNMT3B protein (Petrussa et al. 2014) (Fig. 10.2). The expression of DNMT3A2, one of the DNMT3A isoforms, increases progressively during oocyte growth and is responsible for the establishment of methylation imprints (Hara et al. 2014). A recent study showed that DNMT3A2 is essential for creating genomic imprinting in the mouse oocytes (Ma et al. 2015). It is important to note that aging may play a role in changing *Dnmt3a* gene expression but not in *Dnmt3b* expression (Fang et al. 2016). Most likely, the change in the *Dnmt3a* gene expression causes impaired oocyte maturation in the GV and MII oocytes and subsequent abnormal embryonic development (Fang et al. 2016). In vitro culture also disrupts the *Dnmt3a* and *Dnmt3b* expression in the cultured GV oocytes compared with in vivo-produced counterparts (Song et al. 2009).

10.3.3 DNMT3L and DNMT2 Expression in Mammalian Oocytes

DNMT3L mRNA expression is detected at low levels in rhesus monkey GV and MII oocytes (Vassena et al. 2005). In mouse, DNMT3L protein expression is present in the fetal oocytes, but non-growing GV oocytes at the diameters between 40 and 49 μm do not express DNMT3L protein; however DNMT3L protein is strongly expressed in the fully grown oocytes between 50 and 59 μm (Hara et al. 2014)

(Fig. 10.2). And, DNMT3L expression decreases in the cultured mouse GV oocytes compared with in vivo GV oocytes (Song et al. 2009). In humans, Huntriss et al. (2004) reported that *DNMT3L* was not transcribed in any of follicular stages, GV and MII oocytes (Huntriss et al. 2004). Another study showed that DNMT3L is undetectable in human GV, MI, and MII oocytes (Petrussa et al. 2014) (Fig. 10.2).

The *Dnmt2* mRNA expression is detected at higher levels in rhesus monkey GV oocytes than in other oocyte stages (Vassena et al. 2005). In contrast, *DNMT2* mRNA expression is not detectable in any stage of human oocytes (Huntriss et al. 2004). Moreover, human follicles and granulosa cells do not transcribe the *DNMT2 gene* (Huntriss et al. 2004). It is important to note that in humans the DNMT2 protein is only expressed in MII oocytes (Petrussa et al. 2014) (Fig. 10.2).

10.4 Conclusion and Future Remarks

In conclusion, correct establishment and maintenance of global and de novo DNA methylation are largely carried out by DNMTs, and these processes are essential for achieving normal oogenesis in mammals. The spatial and temporal expression patterns and subcellular localization of the DNMT enzyme exhibit prominent differences in the mammalian oocytes and granulosa cells during folliculogenesis. Because granulosa cells are essential for producing competent oocytes, correctly established DNA methylation in the granulosa cells is also important for the regulation of expression of genes related to follicular development. Therefore, the lack of DNMTs leads to various types of defects such as impaired oocyte maturation, embryonic lethality, abnormal maternal imprinting, and altered gene expression. Recent studies reported that ART technology is a potential disruptor of global DNA methylation and DNMTs expression; therefore, it may lead to changing development-related gene expression in the oocytes. To understand the molecular background of the epigenetic disorders derived from ART applications, new studies are required to determine the intra- or intercellular mechanisms and signaling pathways regulating expression of DNMT genes.

References

Agarwal S, Rao A (1998) Modulation of chromatin structure regulates cytokine gene expression during T cell differentiation. Immunity 9(6):765–775

Ambatipudi S, Cuenin C, Hernandez-Vargas H, Ghantous A, Le Calvez-Kelm F, Kaaks R, Barrdahl M, Boeing H, Aleksandrova K, Trichopoulou A, Lagiou P, Naska A, Palli D, Krogh V, Polidoro S, Tumino R, Panico S, Bueno-de-Mesquita B, Peeters PH, Quiros JR, Navarro C, Ardanaz E, Dorronsoro M, Key T, Vineis P, Murphy N, Riboli E, Romieu I, Herceg Z (2016) Tobacco smoking-associated genome-wide DNA methylation changes in the EPIC study. Epigenomics 8(5):599–618. doi:10.2217/epi-2016-0001

Araujo FD, Croteau S, Slack AD, Milutinovic S, Bigey P, Price GB, Zannis-Hadjopoulos M, Szyf M (2001) The DNMT1 target recognition domain resides in the N terminus. J Biol Chem 276 (10):6930–6936. doi:10.1074/jbc.M009037200

Barlow DP, Bartolomei MS (2014) Genomic imprinting in mammals. Cold Spring Harb Perspect Biol 6(2). doi:10.1101/cshperspect.a018382

Bartolomei MS, Webber AL, Brunkow ME, Tilghman SM (1993) Epigenetic mechanisms underlying the imprinting of the mouse H19 gene. Genes Dev 7(9):1663–1673

Bestor TH (2000) The DNA methyltransferases of mammals. Hum Mol Genet 9(16):2395–2402

Bourc'his D, Bestor TH (2004) Meiotic catastrophe and retrotransposon reactivation in male germ cells lacking Dnmt3L. Nature 431(7004):96–99. doi:10.1038/nature02886

Bourc'his D, Xu GL, Lin CS, Bollman B, Bestor TH (2001) Dnmt3L and the establishment of maternal genomic imprints. Science 294(5551):2536–2539. doi:10.1126/science.1065848

Cedar H, Bergman Y (2012) Programming of DNA methylation patterns. Annu Rev Biochem 81:97–117. doi:10.1146/annurev-biochem-052610-091920

Chedin F, Lieber MR, Hsieh CL (2002) The DNA methyltransferase-like protein DNMT3L stimulates de novo methylation by Dnmt3a. Proc Natl Acad Sci USA 99(26):16916–16921. doi:10.1073/pnas.262443999

Chen T, Li E (2004) Structure and function of eukaryotic DNA methyltransferases. Curr Top Dev Biol 60:55–89. doi:10.1016/S0070-2153(04)60003-2

Deshmukh RS, Ostrup O, Ostrup E, Vejlsted M, Niemann H, Lucas-Hahn A, Petersen B, Li J, Callesen H, Hyttel P (2011) DNA methylation in porcine preimplantation embryos developed in vivo and produced by in vitro fertilization, parthenogenetic activation and somatic cell nuclear transfer. Epigenetics 6(2):177–187

Fang Y, Zhang X, Zhang J, Zhong R, Zhou D (2016) Global DNA methylation and related mRNA profiles in sheep oocytes and early embryos derived from pre-pubertal and adult donors. Anim Reprod Sci 164:144–151. doi:10.1016/j.anireprosci.2015.11.022

Feinberg AP (2000) DNA methylation, genomic imprinting and cancer. Curr Top Microbiol Immunol 249:87–99

Feinberg AP, Cui H, Ohlsson R (2002) DNA methylation and genomic imprinting: insights from cancer into epigenetic mechanisms. Semin Cancer Biol 12(5):389–398

Fernandez-Gonzalez R, Ramirez MA, Bilbao A, De Fonseca FR, Gutierrez-Adan A (2007) Suboptimal in vitro culture conditions: an epigenetic origin of long-term health effects. Mol Reprod Dev 74(9):1149–1156. doi:10.1002/mrd.20746

Goll MG, Kirpekar F, Maggert KA, Yoder JA, Hsieh CL, Zhang X, Golic KG, Jacobsen SE, Bestor TH (2006) Methylation of tRNAAsp by the DNA methyltransferase homolog Dnmt2. Science 311 (5759):395–398. doi:10.1126/science.1120976

Goto H, Iwata H, Takeo S, Nisinosono K, Murakami S, Monji Y, Kuwayama T (2013) Effect of bovine age on the proliferative activity, global DNA methylation, relative telomere length and telomerase activity of granulosa cells. Zygote 21(3):256–261. doi:10.1017/S0967199411000499

Gowher H, Liebert K, Hermann A, Xu G, Jeltsch A (2005) Mechanism of stimulation of catalytic activity of Dnmt3A and Dnmt3B DNA-(cytosine-C5)-methyltransferases by Dnmt3L. J Biol Chem 280(14):13341–13348. doi:10.1074/jbc.M413412200

Hara S, Takano T, Fujikawa T, Yamada M, Wakai T, Kono T, Obata Y (2014) Forced expression of DNA methyltransferases during oocyte growth accelerates the establishment of methylation imprints but not functional genomic imprinting. Hum Mol Genet 23(14):3853–3864. doi:10.1093/hmg/ddu100

Heard E, Clerc P, Avner P (1997) X-chromosome inactivation in mammals. Annu Rev Genet 31:571–610. doi:10.1146/annurev.genet.31.1.571

Henikoff S, Smith MM (2015) Histone variants and epigenetics. Cold Spring Harb Perspect Biol 7 (1):a019364. doi:10.1101/cshperspect.a019364

Hirasawa R, Chiba H, Kaneda M, Tajima S, Li E, Jaenisch R, Sasaki H (2008) Maternal and zygotic Dnmt1 are necessary and sufficient for the maintenance of DNA methylation imprints during preimplantation development. Genes Dev 22(12):1607–1616. doi:10.1101/gad.1667008

Huntriss J, Hinkins M, Oliver B, Harris SE, Beazley JC, Rutherford AJ, Gosden RG, Lanzendorf SE, Picton HM (2004) Expression of mRNAs for DNA methyltransferases and methyl-CpG-binding proteins in the human female germ line, preimplantation embryos, and embryonic stem cells. Mol Reprod Dev 67(3):323–336. doi:10.1002/mrd.20030

Kageyama S, Liu H, Kaneko N, Ooga M, Nagata M, Aoki F (2007) Alterations in epigenetic modifications during oocyte growth in mice. Reproduction 133(1):85–94. doi:10.1530/REP-06-0025

Kaneda M, Okano M, Hata K, Sado T, Tsujimoto N, Li E, Sasaki H (2004) Essential role for de novo DNA methyltransferase Dnmt3a in paternal and maternal imprinting. Nature 429 (6994):900–903. doi:10.1038/nature02633

Kato Y, Kaneda M, Hata K, Kumaki K, Hisano M, Kohara Y, Okano M, Li E, Nozaki M, Sasaki H (2007) Role of the Dnmt3 family in de novo methylation of imprinted and repetitive sequences during male germ cell development in the mouse. Hum Mol Genet 16(19):2272–2280. doi:10.1093/hmg/ddm179

Ko YG, Nishino K, Hattori N, Arai Y, Tanaka S, Shiota K (2005) Stage-by-stage change in DNA methylation status of Dnmt1 locus during mouse early development. J Biol Chem 280 (10):9627–9634. doi:10.1074/jbc.M413822200

Kumar S, Cheng X, Klimasauskas S, Mi S, Posfai J, Roberts RJ, Wilson GG (1994) The DNA (cytosine-5) methyltransferases. Nucleic Acids Res 22(1):1–10

Kurihara Y, Kawamura Y, Uchijima Y, Amamo T, Kobayashi H, Asano T, Kurihara H (2008) Maintenance of genomic methylation patterns during preimplantation development requires the somatic form of DNA methyltransferase 1. Dev Biol 313(1):335–346. doi:10.1016/j.ydbio.2007.10.033

Law JA, Jacobsen SE (2010) Establishing, maintaining and modifying DNA methylation patterns in plants and animals. Nat Rev Genet 11(3):204–220. doi:10.1038/nrg2719

Liang XW, Cui XS, Sun SC, Jin YX, Heo YT, Namgoong S, Kim NH (2013) Superovulation induces defective methylation in line-1 retrotransposon elements in blastocyst. Reprod Biol Endocrinol 11:69. doi:10.1186/1477-7827-11-69

Liang Y, Fu XW, Li JJ, Yuan DS, Zhu SE (2014) DNA methylation pattern in mouse oocytes and their in vitro fertilized early embryos: effect of oocyte vitrification. Zygote 22(2):138–145. doi:10.1017/S0967199412000512

Liu MH, Zhou WH, Chu DP, Fu L, Sha W, Li Y (2016) Ultrastructural changes and methylation of human oocytes vitrified at the germinal vesicle stage and matured in vitro after thawing. Gynecol Obstet Invest. doi:10.1159/000448143

Lodde V, Modina SC, Franciosi F, Zuccari E, Tessaro I, Luciano AM (2009a) Localization of DNA methyltransferase-1 during oocyte differentiation, in vitro maturation and early embryonic development in cow. Eur J Histochem 53(4):199–207

Lodde V, Modina SC, Franciosi F, Zuccari E, Tessaro I, Luciano AM (2009b) Localization of DNA methyltransferase-1 during oocyte differentiation, in vitro maturation and early embryonic development in cow. Eur J Histochem 53(4):e24. doi:10.4081/ejh.2009.e24

Lorincz MC, Schubeler D, Hutchinson SR, Dickerson DR, Groudine M (2002) DNA methylation density influences the stability of an epigenetic imprint and Dnmt3a/b-independent de novo methylation. Mol Cell Biol 22(21):7572–7580

Ma JY, Li M, Ge ZJ, Luo Y, Ou XH, Song S, Tian D, Yang J, Zhang B, Ou-Yang YC, Hou Y, Liu Z, Schatten H, Sun QY (2012) Whole transcriptome analysis of the effects of type I diabetes on mouse oocytes. PLoS One 7(7):e41981. doi:10.1371/journal.pone.0041981

Ma P, de Waal E, Weaver JR, Bartolomei MS, Schultz RM (2015) A DNMT3A2-HDAC2 complex is essential for genomic imprinting and genome integrity in mouse oocytes. Cell Rep 13(8):1552–1560. doi:10.1016/j.celrep.2015.10.031

Mertineit C, Yoder JA, Taketo T, Laird DW, Trasler JM, Bestor TH (1998) Sex-specific exons control DNA methyltransferase in mammalian germ cells. Development 125(5):889–897

Neri F, Krepelova A, Incarnato D, Maldotti M, Parlato C, Galvagni F, Matarese F, Stunnenberg HG, Oliviero S (2013) Dnmt3L antagonizes DNA methylation at bivalent promoters and favors DNA methylation at gene bodies in ESCs. Cell 155(1):121–134. doi:10.1016/j.cell.2013.08.056
O'Doherty AM, O'Shea LC, Fair T (2012) Bovine DNA methylation imprints are established in an oocyte size-specific manner, which are coordinated with the expression of the DNMT3 family proteins. Biol Reprod 86(3):67. doi:10.1095/biolreprod.111.094946
Okano M, Xie S, Li E (1998a) Cloning and characterization of a family of novel mammalian DNA (cytosine-5) methyltransferases. Nat Genet 19(3):219–220. doi:10.1038/890
Okano M, Xie S, Li E (1998b) Dnmt2 is not required for de novo and maintenance methylation of viral DNA in embryonic stem cells. Nucleic Acids Res 26(11):2536–2540
Okano M, Bell DW, Haber DA, Li E (1999) DNA methyltransferases Dnmt3a and Dnmt3b are essential for de novo methylation and mammalian development. Cell 99(3):247–257
Ostler KR, Davis EM, Payne SL, Gosalia BB, Exposito-Cespedes J, Le Beau MM, Godley LA (2007) Cancer cells express aberrant DNMT3B transcripts encoding truncated proteins. Oncogene 26(38):5553–5563. doi:10.1038/sj.onc.1210351
Petrussa L, Van de Velde H, De Rycke M (2014) Dynamic regulation of DNA methyltransferases in human oocytes and preimplantation embryos after assisted reproductive technologies. Mol Hum Reprod 20(9):861–874. doi:10.1093/molehr/gau049
Portela A, Esteller M (2010) Epigenetic modifications and human disease. Nat Biotechnol 28 (10):1057–1068. doi:10.1038/nbt.1685
Posfai J, Bhagwat AS, Posfai G, Roberts RJ (1989) Predictive motifs derived from cytosine methyltransferases. Nucleic Acids Res 17(7):2421–2435
Pradhan S, Bacolla A, Wells RD, Roberts RJ (1999) Recombinant human DNA (cytosine-5) methyltransferase. I. Expression, purification, and comparison of de novo and maintenance methylation. J Biol Chem 274(46):33002–33010
Ratnam S, Mertineit C, Ding F, Howell CY, Clarke HJ, Bestor TH, Chaillet JR, Trasler JM (2002) Dynamics of Dnmt1 methyltransferase expression and intracellular localization during oogenesis and preimplantation development. Dev Biol 245(2):304–314. doi:10.1006/dbio.2002.0628
Reik W, Dean W (2001) DNA methylation and mammalian epigenetics. Electrophoresis 22 (14):2838–2843. doi:10.1002/1522-2683
Richardson B (2003) Impact of aging on DNA methylation. Ageing Res Rev 2(3):245–261
Robertson KD, Wolffe AP (2000) DNA methylation in health and disease. Nat Rev Genet 1 (1):11–19. doi:10.1038/35049533
Song Z, Min L, Pan Q, Shi Q, Shen W (2009) Maternal imprinting during mouse oocyte growth in vivo and in vitro. Biochem Biophys Res Commun 387(4):800–805. doi:10.1016/j.bbrc.2009.07.131
Stoger R, Kubicka P, Liu CG, Kafri T, Razin A, Cedar H, Barlow DP (1993) Maternal-specific methylation of the imprinted mouse Igf2r locus identifies the expressed locus as carrying the imprinting signal. Cell 73(1):61–71
Turek-Plewa J, Jagodzinski PP (2005) The role of mammalian DNA methyltransferases in the regulation of gene expression. Cell Mol Biol Lett 10(4):631–647
Uysal F, Akkoyunlu G, Ozturk S (2015) Dynamic expression of DNA methyltransferases (DNMTs) in oocytes and early embryos. Biochimie 116:103–113. doi:10.1016/j.biochi.2015.06.019
Uysal F, Akkoyunlu G, Ozturk S (2016) DNA methyltransferases exhibit dynamic expression during spermatogenesis. Reprod Biomed Online 33(6):690–702. doi:10.1016/j.rbmo.2016.08.022
Vassena R, Dee Schramm R, Latham KE (2005) Species-dependent expression patterns of DNA methyltransferase genes in mammalian oocytes and preimplantation embryos. Mol Reprod Dev 72(4):430–436. doi:10.1002/mrd.20375
Verlhac MH, Terret ME (2016) Oocyte maturation and development. F1000Res 5. doi:10.12688/f1000research.7892.1
Virani S, Rentschler KM, Nishijo M, Ruangyuttikarn W, Swaddiwudhipong W, Basu N, Rozek LS (2016) DNA methylation is differentially associated with environmental cadmium exposure based on sex and smoking status. Chemosphere 145:284–290. doi:10.1016/j.chemosphere.2015.10.123

Watanabe D, Suetake I, Tada T, Tajima S (2002) Stage- and cell-specific expression of Dnmt3a and Dnmt3b during embryogenesis. Mech Dev 118(1–2):187–190

Wei Y, Huan Y, Shi Y, Liu Z, Bou G, Luo Y, Zhang L, Yang C, Kong Q, Tian J, Xia P, Sun QY, Liu Z (2011) Unfaithful maintenance of methylation imprints due to loss of maternal nuclear Dnmt1 during somatic cell nuclear transfer. PLoS One 6(5):e20154. doi:10.1371/journal.pone.0020154

Whidden L, Martel J, Rahimi S, Richard Chaillet J, Chan D, Trasler JM (2016) Compromised oocyte quality and assisted reproduction contribute to sex-specific effects on offspring outcomes and epigenetic patterning. Hum Mol Genet. doi:10.1093/hmg/ddw293

Yan J, Zhang L, Wang T, Li R, Liu P, Yan L, Qiao J (2014) Effect of vitrification at the germinal vesicle stage on the global methylation status in mouse oocytes subsequently matured in vitro. Chin Med J 127(23):4019–4024

Yang X, Smith SL, Tian XC, Lewin HA, Renard JP, Wakayama T (2007) Nuclear reprogramming of cloned embryos and its implications for therapeutic cloning. Nat Genet 39(3):295–302. doi:10.1038/ng1973

Yoder JA, Soman NS, Verdine GL, Bestor TH (1997) DNA (cytosine-5)-methyltransferases in mouse cells and tissues. Studies with a mechanism-based probe. J Mol Biol 270(3):385–395. doi:10.1006/jmbi.1997.1125

Yu B, Russanova VR, Gravina S, Hartley S, Mullikin JC, Ignezweski A, Graham J, Segars JH, DeCherney AH, Howard BH (2015) DNA methylome and transcriptome sequencing in human ovarian granulosa cells links age-related changes in gene expression to gene body methylation and 3′-end GC density. Oncotarget 6(6):3627–3643. doi:10.18632/oncotarget.2875

Yue MX, Fu XW, Zhou GB, Hou YP, Du M, Wang L, Zhu SE (2012) Abnormal DNA methylation in oocytes could be associated with a decrease in reproductive potential in old mice. J Assist Reprod Genet 29(7):643–650. doi:10.1007/s10815-012-9780-4

Zaitseva I, Zaitsev S, Alenina N, Bader M, Krivokharchenko A (2007) Dynamics of DNA-demethylation in early mouse and rat embryos developed in vivo and in vitro. Mol Reprod Dev 74(10):1255–1261. doi:10.1002/mrd.20704

Chapter 11
Accumulation of Chromatin Remodelling Enzyme and Histone Transcripts in Bovine Oocytes

V. Lodde, A.M. Luciano, F. Franciosi, R. Labrecque, and M.A. Sirard

Abstract During growth, the oocyte accumulates mRNAs that will be required in the later stages of oogenesis and early embryogenesis until the activation of the embryonic genome. Each of these developmental stages is controlled by multiple regulatory mechanisms that ensure proper protein production. Thus mRNAs are stabilized, stored, recruited, polyadenylated, translated and/or degraded over a period of several days. As a consequence, understanding the biological significance of changes in the abundance of transcripts during oocyte growth and differentiation is rather complex. Nevertheless the availability of transcriptomic platforms applicable to scarce samples such as oocytes has generated large amounts of data that depict the transcriptome of oocytes under different conditions. Despite several technical constrains related to protein determination in oocytes that still limit the possibility to verify certain hypothesis, it is now possible to use mRNA levels to start building plausible scenarios. To start deciphering the changes in the level of specific mRNAs involved in chromatin remodelling, we have performed a meta-analysis of existing microarray datasets from germinal vesicle (GV) stage bovine oocytes during the final stages of oocyte differentiation. We then analysed the expression profiles of histone and histone-remodelling enzyme mRNAs and correlated these with the major histone modifications known to occur at the same period, based on data available in the literature. We believe that this approach could reveal the function of specific enzymes in the oocyte. In turn, this information will be useful in future studies, which final ambitious goal is to decipher the 'oocyte-specific histone code'.

The online version of this chapter (doi:10.1007/978-3-319-60855-6_11) contains supplementary material, which is available to authorized users.

V. Lodde • A.M. Luciano • F. Franciosi
Reproductive and Developmental Biology Laboratory, Department of Health, Animal Sciences and Food Safety, University of Milan, Milan, Italy

R. Labrecque • M.A. Sirard (✉)
Centre de Recherche en Biologie de la Reproduction, Faculté des Sciences de l'Agriculture et de l'Alimentation, Département des Sciences Animales, Pavillon INAF, Université Laval, Québec, QC, Canada, G1V 0A6
e-mail: Marc-Andre.Sirard@fsaa.ulaval.ca

M. Kloc (ed.), *Oocytes*, Results and Problems in Cell Differentiation 63,
DOI 10.1007/978-3-319-60855-6_11

11.1 Premise[1]

Mammalian oocytes enter meiotic prophase I during foetal life. In the foetus, the meiotic cell cycle arrests at the diplotene stage of prophase I until after puberty. When meiotic cell cycle arrests, a single layer of flattened granulosa cells surrounds the oocytes, thus forming the primordial follicles. These follicles are quiescent and constitute a pool from where they will be recruited for growth (follicle activation) and cyclically selected for ovulation during reproductive life. During subsequent follicle development (see below), the growth phase of the oocytes occurs, and their diameters increase considerably. Once the growth phase is concluded, the sole 'fully grown' oocytes selected for ovulation resume meiosis in response to luteinizing hormone (LH), while subordinate follicles, enclosing oocytes that have not been selected for ovulation, undergo degeneration through atresia (Scaramuzzi et al. 2011; Gougeon 1986). Upon meiotic resumption, which occurs only after puberty is reached, meiosis continues to the metaphase of meiosis II (MII stage), occurring when the oocytes are released from the follicles. In mono-ovulatory species such as cattle, one MII stage per reproductive cycle is released, whereas the remaining follicles undergo atresia. On the contrary in poly-ovulatory species, such as mouse, multiple oocytes are released. Oogenesis and folliculogenesis are linked with the oocyte growing and developing in an intimate and mutually dependent relationship with the somatic cells of the follicle. Folliculogenesis progresses through sequential stages, leading to the formation and development of a fluid-filled cavity called the antrum (antral follicle stage), which also increases in size.

The nucleus of oocytes arrested at the diplotene I stage of meiosis is surrounded by an intact nuclear envelope and is named germinal vesicle (GV). Thus the expression 'GV stage' is often used to indicate when the oocyte is collected from the ovarian follicle before meiotic resumption occurs. During meiotic arrest, and particularly during the final oocyte growth phase leading to the formation of the fully grown and differentiated oocyte, the chromatin enclosed within the GV is subjected profound morphological, structural and functional remodelling (De La Fuente 2006; Luciano and Lodde 2013). Thus, GV stage oocytes can be further subdivided according to the level of chromatin compaction, which can be observed under fluorescence microscopy. This is biologically relevant since chromatin configuration is considered a marker of oocyte differentiation and developmental competence in all the species studied so far (De La Fuente 2006; Luciano and Lodde 2013).

As summarized in Fig. 11.1, in cow four GV stage oocytes have been characterized (from GV0 to GV3), with increasing level of chromatin compaction

[1]Please also refer to (Leung and Adashi 2003) and (Hyttel 2010) as general bibliographic references for the premise.

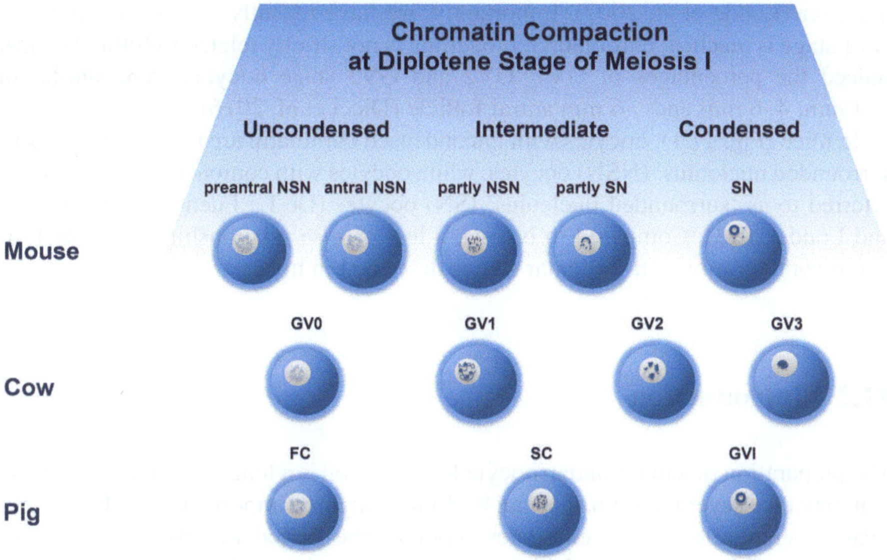

Fig. 11.1 Graphic representation of chromatin compaction in the oocyte nucleus (germinal vesicle, GV), occurring during meiotic arrest at the diplotene stage. Morphologically, the chromosomes seem to lose their individuality forming a loose chromatin mass, which in turn undergoes profound and dynamic rearrangements within the GV before the meiotic resumption. Although different terminology has been used in different mammals, chromatin compaction is a common process in which the oocyte chromatin condenses and rearranges progressively passing through intermediate configurations. In mouse, the chromatin is initially found dispersed throughout the nucleoplasm and appears mainly decondensed in a configuration termed 'non-surrounded nucleolus' (NSN). Thereafter, chromatin condensation increases, and few chromatin foci start to associate with the nucleolar periphery until a complete rim of heterochromatin is formed in close apposition with the nucleolus (surrounded nucleolus, SN configuration). Notably, NSN-type configuration is typical of oocytes collected from primordial to growing preantral follicles; nonetheless, NSN oocytes are also found in antral follicles that indeed enclose both NSN and SN oocytes; the origin of the oocyte should always be carefully considered when studying mouse oocytes with NSN chromatin configuration (Mattson and Albertini 1990; Zuccotti et al. 1995; Bouniol-Baly et al. 1999). In cow, the oocytes with uncondensed chromatin (GV0 oocytes) are typically found in early antral follicles (0.5–2 mm), while they are absent in larger follicles. In contrast, oocytes with increasing levels of compaction (GV1, GV2 and GV3) are collected from medium antral follicle (2–8 mm diameter) (Lodde et al. 2007), and their distribution does not depend on follicular size within this category (see text) (Dieci et al. 2016). In post-pubertal pig, oocytes with uncondensed chromatin (termed 'filamentous chromatin', FC configuration) represent the majority of oocytes collected from early antral follicle (0.5–2 mm), while oocytes with intermediate (stringy chromatin, SC configuration) and high level of compaction (GV1 configuration) are distributed in medium antral follicles (3–6 mm) (Dieci et al. 2013; Bui et al. 2007) [Modified from Luciano and Lodde (2013)]

(Lodde et al. 2007). GV0 oocytes, which typically show uncondensed chromatin, are found in early antral follicles (0.5–2 mm), while they are absent in antral follicle larger than 2 mm. In contrast, oocytes with increasing levels of compaction (GV1, GV2 and GV3) are typically collected from medium antral follicle >2 mm in diameter (Lodde et al. 2007). Importantly, we have recently shown that once the GV1 stage is reached, chromatin remodelling is not strictly related to follicular size. Indeed the percentages of GV1, GV2 and GV3 stage oocytes were similar in 2–4 mm, 4–6 mm and >6 mm antral follicle (Dieci et al. 2016).

In mice (Fig. 11.1), oocytes with uncondensed chromatin are referred to as 'non-surrounded nucleolus' (NSN) oocytes, while oocytes with compacted chromatin are referred to as 'surrounded nucleolus' (SN) oocytes (De La Fuente 2006; Luciano and Lodde 2013). Compared to bovine, a less precise relationship between chromatin configuration and follicular size is described in mice.

11.2 Introduction

The preparation of a mammalian oocyte for ovulation is a long and complex process that may take several months but can be divided in three important periods: growth, final differentiation and meiotic resumption. The period of growth starts with activation of the primordial follicle into an activated one. This is an irreversible process that will lead to atresia or ovulation in a period of several weeks in rodents or several months in large mammals. During growth, the oocyte transcription machinery is very active as the cell accumulates organelles, new structures such as the zona *pellucida* and RNAs that will be required when growth eventually ceases in the antral follicle. Timing of completion of the growth phase and of transcriptional inactivation in relation to follicular development is species dependent. Thus, for example, the growth of mouse oocytes is already completed in the early antral follicles (Sorensen and Wassarman 1976), while bovine oocytes are still in their growth phase (Dieci et al. 2016; Lodde et al. 2008). As follicular growth progresses into follicular differentiation, either in the process of atresia or in dominant and preovulatory follicles, the oocyte transcription capacity decreases and eventually stops few hours before ovulation (Hunter and Moor 1987) in preparation for meiotic resumption. In this last phase, which is strictly regulated by post-transcriptional events, stored mRNAs are the only source of information for translation.

Due to the transcriptional silencing, the oocyte relies on previously stored mRNAs for protein formation as the proteins have a limited half-life. The amount of mRNA that oocytes accumulate starts to increase significantly when follicles reach the multilayer phase of development (Fair et al. 1997). Normally, mRNA is translated within hours; therefore, the accumulation of mRNA in oocytes requires a protection mechanism that evolved in animals producing eggs. In *Xenopus*, one of the proposed mechanisms involves the storage of maternal RNA using a specific configuration where the mRNA is de-polyadenylated on the 3′ end and capped on the 5′ end. In this species, the protein MASKIN associates with the cytoplasmic polyadenylation element binding protein (CPEB) located in the 3′ untranslated region (UTR) on mRNAs that contain a cytoplasmic polyadenylation element (CPE) which represses translation through the inhibitory action on elF4E (Richter 2007). The recruitment of mRNA for translation is associated with longer polyA tails (more than 100 A), while mRNAs with shorter tails (around 20 A) are not recruited (Richter 2007). These mechanisms are conserved in rodents (Clarke 2012; Conti et al. 2015) as well as in large mammals. For instance, we recently surveyed bovine oocytes to identify mRNAs with very long or very short polyA tails as indicators of the timing of translation during meiotic resumption (Gohin et al. 2014). This study led to the confirmation of *cis*-motifs as regulators of re-polyadenylation of maternal RNAs and to the identification of new sequences potentially involved in embryonic genome activation (Gohin et al. 2014). This study also illustrated quite clearly that some RNAs are translated during maturation, while others remain in storage for the early embryonic period. In another study from our group, the analysis of polysomes from oocytes at the beginning versus the end of maturation confirmed the specific nature of recruitment and translation in bovine oocytes (Scantland et al. 2011). In addition we accumulated data from the analyses of bovine oocyte transcriptomic signatures in different physiological contexts, using the platform within the EmbryoGENE programme [http://emb-bioinfo.fsaa.ulaval.ca (Robert et al. 2011)] that offers the unique opportunity to conduct meta-analysis across different datasets.

The availability of all the above-mentioned information becomes a key factor to assess the role of different histone transcripts or histone-modifying enzymes in the oocyte. However, since oocytes are available in limited supply, the capacity to analyse protein amounts is limited, and most studies relied on the amplification of mRNA through primers bearing a T7 promoter, or reverse transcription followed by PCR, to obtain information of the oocyte transcriptome. If this process is quite reliable in somatic cells, it becomes problematic in oocytes not only because of the limited supply of material but also because RNA is stabilized, stored, polyadenylated, recruited and translated or degraded over a period of several days from the transcriptional arrest to the embryonic genome activation, which occurs at the eight-cell stage in bovine (Barnes and First 1991). Because mRNA is stored and de-polyadenylated, the length of the polyA tail is an issue for measurements. If the tail is short, the extraction protocols that use the capturing ability of poly T columns or filters are not efficient. Then, if the primers used for the reverse transcription include a poly T sequence, it will not include the stored RNA, and finally, if a T7 polymerase-based approach is used for RNA amplification (such as in microarray experiments), the presence of a polyA tail will exclude some non-polyadenylated RNAs. Therefore, careful attention must be given to the interpretation of RNA levels obtained with these techniques prior to the eight-cell stage. These considerations have been clearly demonstrated in the past (Gilbert et al. 2009), and we will not use terms such as up-regulation or overexpression in this review unless the protein level has been verified by other means.

To start deciphering the changes in the levels of specific mRNAs involved in chromatin remodelling, and to try to overcome the above-mentioned limitations, we have analysed the expression profiles of histone and histone-remodelling enzyme mRNAs in bovine GV oocytes during the final stages of oocyte differentiation (which is a critical step in the acquisition of the competence to sustain embryonic development) and correlated these with the major histone modifications (mainly methylation and acetylation) known to occur at the same period, based on data available in the literature. In particular, the list of genes that we considered in this review was generated by conducting a meta-analysis of two datasets from previous published studies: one from oocytes isolated from small (0.5–2 mm) and middle (2–8 mm) antral follicles and selected according to the degree of large-scale chromatin configuration (Labrecque et al. 2015) and one from oocytes isolated from follicles of different sizes (<3 mm, 3–5 mm, 5–8 mm, >8 mm) (Labrecque et al. 2016). From the list of genes obtained by the meta-analysis, which basically represent the transcripts that are not randomly expressed during competence acquisition, and thus potentially important before meiotic resumption and ovulation, we arbitrarily selected the histone and histone-related genes. Then, the dynamic expression profile of each of the selected targets was evaluated separately in the original microarray datasets in each of the different experimental settings (chromatin configuration or follicular size). These data are presented and discussed in the following paragraphs.

11.3 Identification of Histone and Histone-Related mRNA Targets

To identify specific mRNAs related to histones and histone-modifying enzymes, we first performed one meta-analysis using NetworkAnalyst (http://www.networkanalyst.ca/) (Xia et al. 2015) by combining two datasets derived from previous microarray-based transcriptomic analysis of oocytes with different chromatin configurations (GV0-1-2-3) (Labrecque et al. 2015) or collected from follicles of different dimensions (<3 mm, 3–5 mm, 5–8 mm, >8 mm) (Labrecque et al. 2016). All datasets were generated using the EmbryoGENE bovine transcriptome microarray (Robert et al. 2011). Microarray gene expression data from these earlier studies were retrieved from the ELMA database (http://emb-bioinfo.fsaa.ulaval.ca/Home/index.html) and pooled together for meta-analysis. The analysis focused on a specific cell type within the ovarian follicle: the germinal vesicle (GV) stage oocyte. For the first dataset, we contrasted GV0 (characterized by uncondensed chromatin configuration) versus GV1, GV2 and GV3 (which are characterized by increasing levels of chromatin compaction). For the second group, GV oocytes from follicles 3 to 5 mm, 5 to 8 mm and larger than 8 mm were compared to the <3 mm group. Data processing involved conversion of all gene probes to common 'Entrez ID' according to the EmbryoGENE platform annotation file. In cases where multiple probes or transcripts are mapped to the same gene, NetworkAnalyst summarized them into a single value for the corresponding gene performing an average of the values. All the datasets were already log2-transformed and normalized-within, so 'quantile normalization' was performed in NetworkAnalyst.

First of all, the programme analyses individual microarray datasets separately and subsequently performs meta-analysis. Differential expression analysis of each dataset was performed individually using a moderated t test based on the Limma algorithm and a false discovery rate (FDR) of 0.05. This step was for reference purpose only. We performed meta-analysis using Fisher's method with a significance level of $p < 0.05$. Fisher's method of meta-analysis combines p-values from the multiple datasets independently of the sample size within each study (Xia et al. 2015).

The list of genes generated by the meta-analysis was used to arbitrarily select histone- or histone-modifying enzyme-related targets. Then, each of these targets was searched in the original microarray dataset that is accessible through an interactive web interface (Khan et al. 2016), which has been updated to include data from oocytes and cumulus cells (http://emb-bioinfo.fsaa.ulaval.ca/IMAGE/). Finally, the expression profile of each target was extracted from the interactive web interface, which allowed the evaluation of both constitutive and isoform (where present) for each target. Thus, the dynamic expression profile of each of the selected transcripts (both constitutive and isoforms) was evaluated separately in each of the different experimental settings (chromatin configuration or follicular size).

11.4 Significance of Histone-Remodelling Enzyme and Histone Transcript Dynamics in Bovine Oocytes

The meta-analysis resulted in a list of 4340 targets affected (i.e. nonrandomly expressed) by the combination of chromatin status and follicular size (Supplemental Table 1). This list is substantially different from the combination of the two lists from each analysis (GV status and follicular size) that would have been obtained by simply generating a Venn diagram. From this list we arbitrarily selected the genes related to histones or histone-modifying enzymes, for which we then extracted the profiles according to the original microarray data (http://emb-bioinfo.fsaa.ulaval.ca/IMAGE/). These candidate targets, grouped according to known families and their expression profiles in each experimental setting (chromatin configuration and follicle size), are represented on graphs (see below) and discussed in the following paragraphs.

The first observation is that most of the changes are occurring between the GV0 stage and the more condensed chromatin statuses (GV1, GV2 and GV3). This is not surprising as the process of chromatin compaction is accompanied by progressive changes to the chromatin, which require different histones (e.g. the oocyte-specific linker histone H1oo) (Labrecque et al. 2016) or histone modifications [reviewed in (Canovas and Ross 2016; Clarke and Vieux 2015; Pan et al. 2012)].

It is worth mentioning here that while GV0 oocytes can only be found in small antral follicles (0.5–2 mm), once the chromatin starts the compaction process (GV0 to GV1 transition), the follicular size has no clear influence on the GV status (1–2 or 3), since it has been recently demonstrated that GV1, GV2 and GV3 are equally distributed in follicles of different sizes (i.e. follicles between 2–4, 4–6 and >6 mm in diameter) (Dieci et al. 2016). Thus, when looking at the profiles of individual targets derived from microarray data in each experimental setting, it is not surprising that one gene can be found significant in one experimental setting but not in the other. Similarly, it is not surprising to observe different increasing or decreasing trends for a specific transcript in the two experimental settings.

11.4.1 Transcripts Encoding Histone-Modifying Enzymes

Understanding the biological significance of changes in transcript abundance during chromatin remodelling is not an easy task. As already mentioned, one of the limiting factors is scarcity of material and the lack of specific antibodies that make it very difficult to properly track the encoded protein. In addition, during the transition from uncondensed (GV0) to more compacted configuration (GV1-2-3), a major transcriptional silencing occurs in mouse and bovine oocytes (Bouniol-Baly et al. 1999; Lodde et al. 2008); [reviewed in (Luciano and Lodde 2013)], and increased or decreased levels of specific transcripts may be due to the specific methodological approach used and have multiple and diverse significances as summarized in Table 11.1. Moreover, while a major silencing occurs at the beginning of the chromatin compaction process in bovine and mouse oocytes (reviewed in (Luciano and Lodde 2013), a low level of transcription could still be detected in bovine oocytes with intermediate configurations (GV1 and GV2), while only oocytes with fully compacted chromatin (GV3) were found to be completely silent (Lodde et al. 2008).

Therefore, in an effort to interpret the major findings of the present meta-analysis, wherever possible, we compared the changes in the abundance of transcripts encoding for histone-modifying enzymes with the expected phenotype (i.e. the change in the histone modification associated with the specific enzyme) whenever these data were available. Moreover, since the bovine data are lacking for most of the histone modifications, we have critically reviewed the literature and built, taking into consideration the species-specific differences in ovarian physiology, a 'consensus table' within and across species to have a global view of histone modification changes as they occur during chromatin compaction in mammals (Tables 11.2 and 11.3). This is possible since, as summarized in Fig. 11.1, the process of large-scale chromatin remodelling, in which the chromatin of prophase I-arrested mammalian oocytes condenses and progressively rearranges passing through intermediate configurations, is a well-conserved process in (and not limited to) mammalian species [reviewed in (Luciano and Lodde 2013)].

Discussion of original and published data relative to several histone modifications and related enzymes is reported below. The main interpretative key being the following: if changes in mRNA level for a particular transcript match with the occurrence of the corresponding histone modification, then the encoded protein is translated and most likely is functioning in the oocyte. In contrast, if the change in

Table 11.1 Possible significances of mRNA level changes in oocytes

Change of transcript level assessed during chromatin compaction	Meaning
Decrease	– Less transcription – Shortening of polyA tail and storage for future use – Degradation – Translation and subsequent degradation
Increase	– More transcription for storage for future use – More transcription for protein that are required at the time

the mRNA level does not correlate with any changes of the specific histone modification, then the mRNA is probably coding for proteins that are needed to processes that will occur later, during maturation or early embryonic development.

Since the precise description of level of histone modification marks in bovine oocytes isolated from antral follicles of different diameter is currently not available, most of the discussion is based on the comparison between histone modification and the corresponding enzyme mRNA as they occur in oocytes with different chromatin configurations, where a match between mRNA level and corresponding phenotype is possible. However, where relevant, we also took into consideration some of the mRNA variations observed during follicular growth. Unless otherwise indicated, the function of each gene refers to data available on gene card (http://www.genecards.org) and/or UniProt (http://www.uniprot.org) databases.

11.4.1.1 Transcripts Related to Histone H3 Lysine 27 Trimethylation (H3K27me3)

Among the differentially expressed transcripts that encode for lysine demethylases (KDMs) during chromatin compaction (Fig. 11.2a), *KDM6A* (also known as ubiquitously transcribed tetratricopeptide repeat X, UTX) is the only one that specifically demethylates H3K27 (Agger et al. 2007; Lan et al. 2007). Based on our analysis, no transcripts encoding H3K27-specific methylase change significantly in the conditions studied. Immunofluorescence studies revealed that the levels of H3K27me3 were high in fully grown bovine oocytes, decreased slightly during oocyte maturation (Table 11.2), steadily decreased after fertilization to reach a nadir at the eight-cell stage and then increased at the blastocyst stage (after embryonic genome activation) (Ross et al. 2008). However, no data are available for the earlier stages of oogenesis in this species. Nevertheless, in mouse oocytes, no major changes occur during chromatin compaction as H3K27me3 is always detected in both centromeric chromatin and in the rest of chromatin during the transition from the uncondensed 'non-surrounded nucleolus' (NSN) configuration to the 'compacted surrounded nucleolus' (SN) stage, with only a little increase in

Table 11.2 Overview of histones H3 and H4 methylation during chromatin compaction and meiotic maturation in mammals

Modification	Species	Uncondensed	Intermediate		Compacted	Fully grown	MII (in vivo/ in vitro)	References
H3Arg17me	Bovine	na	na	na	na	na	na/na	Sarmento et al. (2004)
	Mouse	na	na		na	+	−/na	
	Porcine	na	na		na	na	na/na	
H3Argme3	Bovine	na	na	na	na	na	na/na	Sarmento et al. (2004)
	Mouse	na	na		na	+	−/na	
	Porcine	na	na		na	na	na/na	
H3K27me1	Bovine	na	na	na	na	na	na/na	Park et al. (2009)
	Mouse	na	na		na	na	na/na	
	Porcine	na	na		na	+	na/+	
H3K27me3	Bovine	na	na	na	na	++	na/+/−	Bonnet-Garnier et al. (2012), Ross et al. (2008), Park et al. (2009)
	Mouse	− (p-Ch); + (c-Ch and r-Ch)	+/− (p-Ch); + (cc-Ch and r-Ch)		+/− (p-Ch); + (c-Ch and r-Ch)	+	na/na	
	Porcine	na	na		na	+	na/+	
H3K4me2	Bovine	na	na	na	na	na	na/na	Kageyama et al. (2007), Lin et al. (2016)
	Mouse	+	+		++	+	na/na	
	Porcine	na	na		na	na	na/na	
H3K4me3	Bovine	na	na	na	na	na	na/na	Kageyama et al. (2007), Bonnet-Garnier et al. (2012), Fulka (2008)
	Mouse	+ (excl. from p-Ch and c-Ch)	+ (excl. from p-Ch and c-Ch)		++ (excl. from p-Ch and c-Ch)	+	+/na	
	Porcine	na	na		na	na	na/na	

H3K9me1	Bovine	na	na	na	na	+	na/+	Wee et al. (2010)
	Mouse	na	na		na	na	na/na	
	Porcine	na	na		na	na	na/na	
H3K9me2	Bovine	na	na	na	na	+	na/+	Kageyama et al. (2007), Meglicki et al. (2008), Lin et al. (2016), Wee et al. (2010), Racedo et al. (2009), Bui et al. (2007)
	Mouse	+ (pCh and r-Ch)	+ (p-Ch and r-Ch)		++ (p-Ch and r-Ch)	+	na/+	
	Porcine	na	na		na	na	na/na	
H3K9me3	Bovine	na	na	na	na	+	na/+/−	Kageyama et al. (2007), Endo et al. (2005), Meglicki et al. (2008), Bonnet-Garnier et al. (2012), Zhang et al. (2016b), Wee et al. (2010), Bui et al. (2007), Russo et al. (2013)
	Mouse	+ (higher in p-Ch)	+ (higher in p-Ch)		+ (higher in p-Ch)	+	na/+	
	Porcine	+/−	+		+	+	na/+	
	Sheep	+	na		+	na	na/na	
H4K20me3	Bovine	na	na	na	na	na	na/na	Bonnet-Garnier et al. (2012)
	Mouse	+ (p-Ch only)	+ (p-Ch only)		+ (p-Ch only)	+	+/na	
	Porcine	na	na		na	na	na/na	

Data available in the literature have been critically revised to build an overview of patterns of H3 and H4 methylation within and across species. *Uncondensed* category includes NSN configuration in mouse (Note that only data of NSN collected from antral follicles are reported; for data on early growing NSN, please refer to the specific reference), GV0 in cow and F in pig. *Intermediate* category includes pNSN and pSN in mouse, GV1 and GV2 in cow and SC in pig. *Condensed* configuration includes SN in mouse, GV3 in cow and GV1 in pig. *p-Ch* pericentromeric chromatin, *c-Ch* centromeric chromatin, *r-Ch* rest of the chromatin, *na* not assessed, (+), present; (+/−), weak; (−), absent

Table 11.3 Overview of histones H3 and H4 acetylation during chromatin compaction and meiotic maturation in mammals

Modification	Species	Uncondensed	Intermediate		Compacted	Fully grown	MII (in vivo/ in vitro)	References
H3K14ac	Bovine	na	na	na	na	+	na/+/–	Meglicki et al. (2008), Lin et al. (2016), Wang et al. (2006, 2011), Bui et al. (2007)
	Mouse	+ (excl. from p-Ch)	+ (excl. from p-Ch)		++ (excl. from p-Ch)	+	na/–	
	Porcine	+	+		+	+	na/?	
H3K18ac	Bovine	na	na	na	na	na	na/na	Kageyama et al. (2007), Bui et al. (2007)
	Mouse	na	na		na	+	na/na	
	Porcine	+	+		+	+	na/–	
H3K4ac	Bovine	na	na		na	na	na/na	Endo et al. (2005)
	Mouse	na	na		na	na	na/na	
	Porcine	na	na		na	+	na/–	
H3K9,K1ac	Bovine	na	na		na	na	na/na	Nagashima et al. (2007), Meglicki et al. (2008)
	Mouse	+ (excl. from p-Ch)	+ (excl. from p-Ch)		+ (excl. from p-Ch)	+	na/–	
	Porcine	na	na		na	na	na/na	
H3K9ac	Bovine	na	na	na	na	+	na/–	Kageyama et al. (2007), Endo et al. (2005), Fulka (2008), Wang et al. (2006), Wee et al. (2010), Bui et al. (2007)
	Mouse	na	na		na	+	–/na	
	Porcine	+	na		+	+	na/–	
H4hyperac	Bovine	na	na	na	na	na	na/na	Sarmento et al. (2004), Fulka (2008)
	Mouse	na	na		na	+	–/na	
	Porcine	na	na		na	na	na/na	
H4K12ac	Bovine	+/–	+	+	+	+	na/–	This paper; Kageyama et al. (2007), Kim et al. (2003), Akiyama et al. (2006), Endo et al. (2005), Maalouf et al. (2008), Lin et al. (2016), Franciosi et al. (2012), Fulka (2008), Wang et al. (2006), Racedo et al. (2009)
	Mouse	+	+		++	++	–/–	
	Porcine	na	na		na	+	na/?	
	Horse	+++	++		+	+	–/–	

H4K16ac	Bovine	na	na	na	na	+	na/−	Kim et al. (2003), Akiyama et al. (2006), Maalouf et al. (2008), Lin et al. (2016), Franciosi et al. (2012), Wang et al. (2006)
	Mouse	+	+		++	+	−/−	
	Porcine	na	na		na	+	na/−	
	Horse	+	+		++	+	+/−	
H4K5ac	Bovine	−	+/−	+	+	+	na/−	This paper; Kageyama et al. (2007), Kim et al. (2003), Nagashima et al. (2007), Endo et al. (2005), Maalouf et al. (2008), Bonnet-Garnier et al. (2012), Wang et al. (2006), Wee et al. (2010)
	Mouse	+ (excl. from p-Ch and c-Ch)	++ (excl. from p-Ch and c-Ch)		++ (excl. from p-Ch and c-Ch)	++	na/−	
	Porcine	na	na		na	+	na/−	
H4K8ac	Bovine	na	na	na	na	+	na/+/−	Kim et al. (2003), Akiyama et al. (2006), Endo et al. (2005), Maalouf et al. (2008), Franciosi et al. (2012), Wang et al. (2006)
	Mouse	na	na		na	+	+/−/+/−	
	Porcine	na	na		na	na	na/?	
	Horse	+++	++		+	+	−/−	

Data available in the literature have been critically revised to build an overview of histones H3 and H4 acetylation within and across species. *Uncondensed* category includes NSN configuration in mouse (*Note* that only data of NSN collected from antral follicles are reported; for data on early growing NSN, please refer to the specific reference), GV0 in cow and F in pig. *Intermediate* category includes pNSN and pSN in mouse, GV1 and GV2 in cow, and SC in pig. *Condensed* configuration includes SN in mouse, GV3 in cow and GV1 in pig. *p-Ch* pericentromeric chromatin, *c-Ch* centromeric chromatin, *r-Ch* rest of the chromatin, *na* not assessed, (+), present; (+/−), weak; (−), absent; (?), discordant data in the literature within the same species

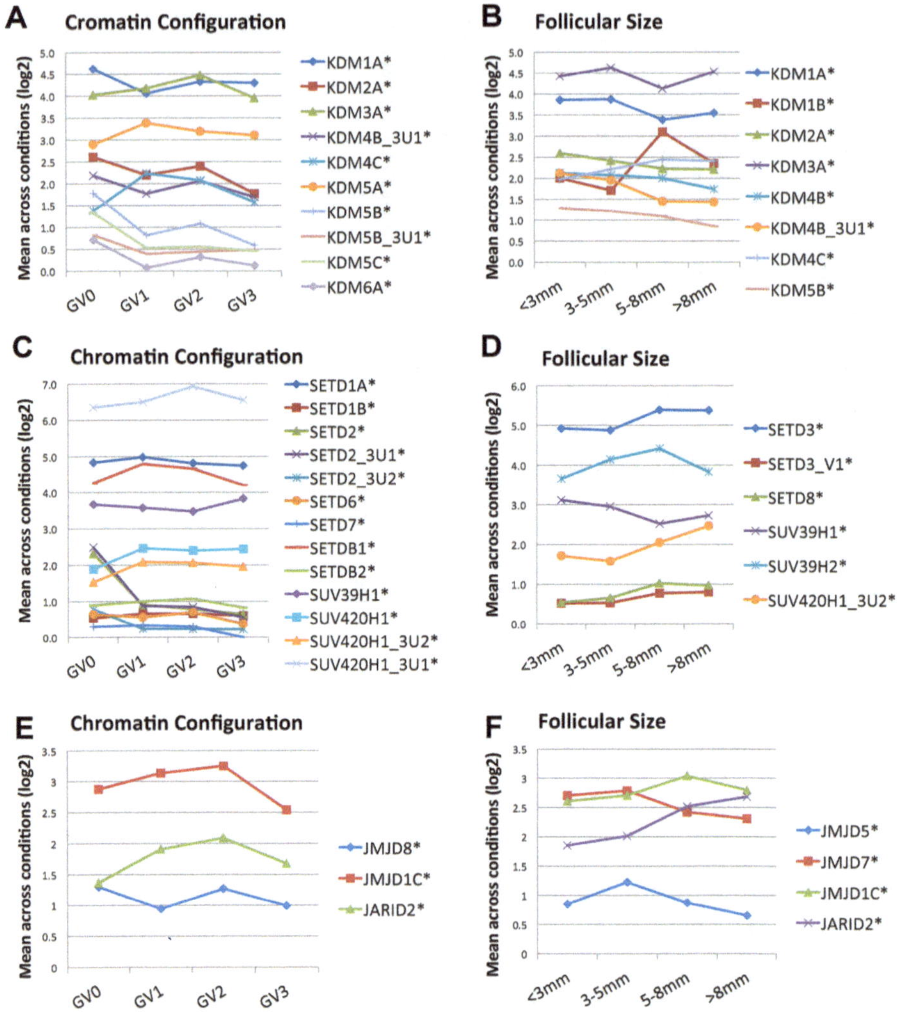

Fig. 11.2 Changes of mRNAs encoding for histone remodelling enzymes according to chromatin configuration (*left panel*) and follicular growth (*right panel*). The data represent the profiles in the original microarray analysis and are obtained/imported from the EmbryoGENE profiler website (http://emb-bioinfo.fsaa.ulaval.ca/IMAGE/)

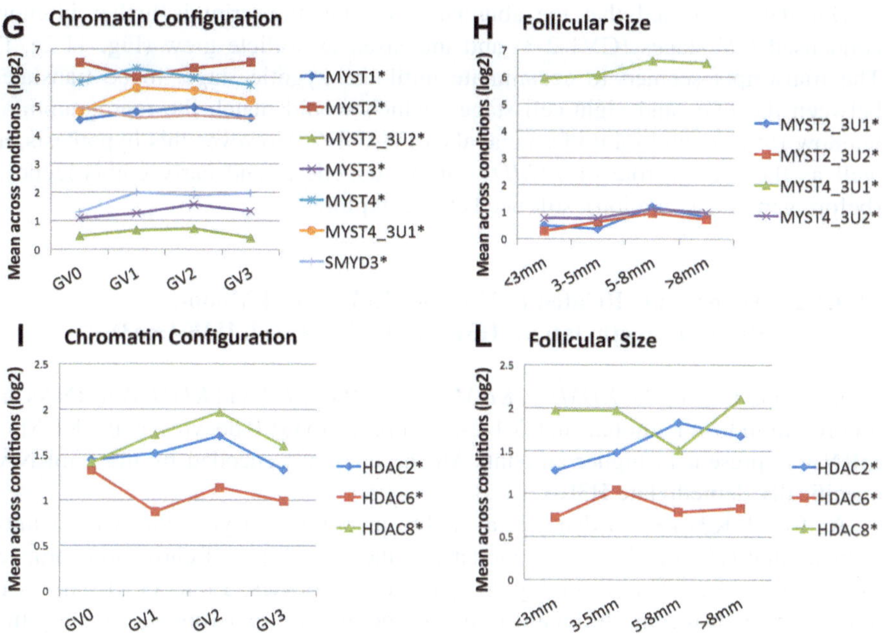

Fig. 11.2 (continued)

the pericentromeric chromatin (Table 11.2). This is also in agreement with more recent findings in mice using highly sensitive chromatin immunoprecipitation-based techniques that revealed widespread deposition of H3K27me3 in early growing oocytes (before establishment of the DNA methylome) and profound changes after fertilization but not during the later stages of oocyte differentiation (i.e. comparing oocytes from 2- to 8-week-old mice) (Zheng et al. 2016). Thus the lower amount of *KDM6A* in GV1-2-3 oocytes (when compared to GV0) might indicate RNA degradation rather than translation and protein production since H3K27me3 is maintained at this stage. De-adenylation can be excluded since *KDM6A* mRNA could not be detected in fully grown bovine oocytes (and up to the blastocyst stage) using either random primers or oligo-dT for RT-PCR (Canovas et al. 2012) or RNA-seq approaches (http://emb-bioinfo.fsaa.ulaval.ca/IMAGE/). This supports the idea that *KDM6A* mRNA degradation is associated with chromatin compaction in bovine GV oocytes, and it is not required during early development.

The Jarid2/Jumonji (*JARID2*) transcript encodes a protein that is a component of the polycomb repressive complex 2 (PRC2), which is essential for the development of multiple organs in mice, and is required for embryonic stem cell (ESC) differentiation (Landeira and Fisher 2011). It was recently demonstrated that PRC2 methylates JARID2 and, in turn, JARID2 methylation regulates H3K27me3 deposition during mouse ESC differentiation (Sanulli et al. 2015).

Our data indicated that the abundance of this transcript is higher in more condensed GV stages (GV1-2-3) and increased as follicle grew (Fig. 11.2e, f). The transcripts seemed to accumulate until the zygotic stage, to be translated between the two- and eight-cell stages, which would match the reprogramming window (http://emb-bioinfo.fsaa.ulaval.ca/IMAGE/). However this hypothesis, as well as the precise role of *JARID2* during oogenesis and early embryogenesis (before genome activation), still needs to be explored.

11.4.1.2 Transcripts Related to Histone H3 Lysine 4 Mono-, Bi- or Trimethylation (H3K4me1, H3K4me2, H3K4me3)

As shown in Fig. 11.2a, *KDM5C*, *KDM5B*, *KDM5B_3U1* and *KDM1A* mRNAs are all present in lower amounts in GV1-2-3 compared to GV0 oocytes, while KDM5A mRNA is present in higher amount. All the enzymes encoded by these mRNAs specifically demethylate H3K4.

As for H3K27me, no data are available in bovine oocytes, but data in mice indicate that H3K4me2 and me3 increase in the later stages of chromatin compaction (from intermediate to highly compacted SN oocytes, Table 11.2). This is in agreement with recent chromatin immunoprecipitation experiments showing that H3K4me3 is present in a pattern that the authors indicated as 'noncanonical' in fully grown and metaphase II (MII) mouse oocytes, when compared to late-stage embryos (after embryonic genome activation) and a somatic tissue (cerebral cortex) (Zhang et al. 2016a). Specifically, the DNA sequences in which H3K4me3 is present in fully grown and MII oocytes differ from the one in which is normally found in somatic tissues (Zhang et al. 2016a). Importantly, the same study has shown that H3K4me3 remains largely in a canonical pattern in growing oocytes (collected from 7- to 15-day-old mice) before adopting a noncanonical pattern in fully grown oocytes from 8-week-old mice. Moreover, the noncanonical H3K4me3 pattern in oocytes overlaps almost exclusively with partially methylated DNA domains (Zhang et al. 2016a), which further confirms emerging evidence that the modification state and sequence of DNA can affect the methylation states of accompanying histones in chromatin and vice versa (Rose and Klose 2014). Notably, the global occurrence of noncanonical H3K4me3 coincided with genome silencing (Zhang et al. 2016a). To see H3K4me3, which is normally considered a permissive epigenetic mark in somatic cells, so well distributed in a transcriptionally silent cell could mean openness for reprogramming in a context where Pol-II is not active, but this hypothesis requires further investigation.

All together these data suggest that the H3K4me pattern is remodelled during final oocyte differentiation and that, on a global scale, its level increases, which implies that some methylation must be removed while some other must be established at specific DNA sites. Therefore for certain transcripts such as *KDM5B*, the lower abundance in GV1-2-3 could be related to translation and subsequent degradation and to degradation without translation. The increase of *KDM5A* mRNA could be related to the activity of the protein in oocytes with higher compacted chromatin and/or increased accumulation for later use. Accordingly, KDM5B overexpression in fully grown mouse oocytes with SN configuration (which are transcriptionally inactive) led to removal of H3K4me3 and transcription reoccurrence, while overexpression of KDM5A did not affect H3K4me3 and transcription (Zhang et al. 2016a). Accordingly, *KDM5B* mRNA was not detected in fully grown bovine oocytes up to the genome activation at the eight-cell stage, while *KDM5A* mRNA was detected even before genome activation (http://emb-bioinfo.fsaa.ulaval.ca/IMAGE/).

For some other transcripts such as *KDM1A*, reduced levels in GV1-2-3 are potentially linked to de-adenylation for storage for later use (KDM1A is associated with H3K4me1, H3K4me2). In mouse oocytes, KDM1A of maternal origin was essential for chromatin modification and genome activation (Ancelin et al. 2016). Accordingly, in bovine, *KDM1A* mRNA was detected by RNA-seq at all stages up to the embryonic genome activation: it dropped at the early eight-cell stage and was up-regulated after genome activation (http://emb-bioinfo.fsaa.ulaval.ca/IMAGE/).

Notably, the global H3K4me increase observed in mice did not correspond to major changes in mRNA encoding H3K4me-specific methylases during chromatin compaction in bovine. Indeed, both the constitutive *SETD3*, whose encoded protein methylates H3K4 and K36, and the variant form mRNA (*SETD3_V*) were not significantly affected by chromatin configuration changes (Fig. 11.2c). However these transcripts accumulated in a follicle size-dependent manner (Fig. 11.2d), which could be related to changes occurring in preparation for ovulation and potentially reprogramming as suggested above.

Interestingly, *KDM1B* mRNA, which is also associated with H3K4me1 and H3K4me2 demethylation and doesn't change significantly in relation to chromatin configuration, was notably increased in follicles larger than 5 mm where the capacity to develop is slightly higher (Fig. 11.2b). In mice, this gene was highly expressed in growing oocytes at the time where genomic imprints are established (Ciccone et al. 2009). More recent studies showed that KDM1B is the primary H3K4me2 demethylase required for imprinted CpG islands (CGI) and the methylation of imprinted germline differentially methylated regions (gDMRs) in the oocyte (Stewart et al. 2015). Knock-out of the gene encoding KDM1B had no effect on mouse development and oogenesis. However, when both alleles were affected, KDM1B-deficient oocytes had higher levels of H3K4 methylation, and embryos derived from these oocytes died before mid-gestation (Ciccone et al. 2009). Therefore we could speculate that proper imprinting may occur in bovine oocytes coincident to the rise of *KDM1B* levels and could have an impact on embryo survival.

11.4.1.3 Transcripts Related to Histone H3 Lysine 9 Mono-, Bi- or Trimethylation (H3K9me1, H3K9me2, H3K9me3)

Some H3K9 methylase and demethylase transcripts were present in different amounts during chromatin compaction. Among the transcripts encoding demethylases (Fig. 11.2a), *KDM4C* (also known as JMJD2C), whose encoded protein demethylates H3K9me3 and H3K36me3 residues, was higher in GV1-2-3 compared to GV0. The constitutive *KDM4B* (H3K9me3specific) did not change significantly, while amount of the alternative 3′UTR (*KDM4B_3U1*) was lower in GV1-2-3. Among the transcripts encoding methylases, only *SETDB1*, a histone methyltransferase that specifically trimethylates H3K9, was significantly affected in GV1-2-3 (Fig. 11.2c). However, interpretation of the biological significance of these changes is complicated. In fact, although H3K9 methylation is generally considered a marker of pericentromeric heterochromatin forming chromocentres, all the data available on H3K9me1, H3K9me2 and H3K9me3 are based on immunofluorescence, and precise localization of H3K9me deposition during oocyte and embryonic development is lacking. Moreover, data in the literature are sometimes controversial. Nevertheless, both H3K9me2 and H3K9me3 were present in oocytes with less compacted chromatin with some increase during chromatin compaction, with H3K9me3 more clearly marking pericentromeric heterochromatin (but both also marking nonpericentromeric chromatin). Both marks (H3K9me2 and H3K9me3) seemed to be retained in MII oocytes, with some decrease observed in bovine (Table 11.2). Moreover, in bovine as in the mouse [reviewed in Canovas and Ross (2016)], some extent of H3K9me2 and me3 remodelling was reported from the zygote to the eight-cell-stage embryo (Pichugin et al. 2010; Wu et al. 2011; Santos et al. 2003). Therefore, *KDM4C* higher abundance in bovine oocytes with more compacted chromatin could be associated with an increased requirement for partial demethylation at the MII stage and/or before genome activation, which is in accordance with the expression profile identified by RNA-seq (http://emb-bioinfo.fsaa.ulaval.ca/IMAGE/). On the other hand, KDM4C (also known as JMJD2) was dispensable for female fertility and embryo development in a knock-out mouse model (Pedersen et al. 2014). Interestingly, the mRNA levels of *KDM4C* are also increasing as the follicle grows (Fig. 11.2b), either supporting the progressive shift towards condensed chromatin or the progressive accumulation for later embryonic use as the oocyte gets closer to ovulation.

Moreover both RNA-seq- (http://emb-bioinfo.fsaa.ulaval.ca/IMAGE/) and siRNA-mediated gene silencing experiments (Golding et al. 2015) supported the idea that *SETDB1* is required during bovine preimplantation development, and, therefore, higher amounts in GV1-2 could mean increased storage for future use. The slight decrease observed in GV3 is an interesting feature that remains to be studied. In mice, maternal deletion of SETDB1 results in defects in meiotic progression and preimplantation development (Eymery et al. 2016; Kim et al. 2016).

Interestingly *SUV39H1* and *SUV39H2* transcripts, which encode for methyltransferases that specifically trimethylate H3K9 using H3K9me1 as substrate, do not seem to be much affected by chromatin configuration since only *SUV39H1* increased in GV3 (Fig. 11.2c). However, both transcripts were significantly affected by follicular size, which could be related to pericentromeric rearrangements occurring in preparation for meiotic resumption in a follicle size-dependent manner.

11.4.1.4 Transcripts Related to Histone H3 Lysine 36 Methylation (H3K36me)

Some H3K36 methyl transferases and demethylases were significantly affected during chromatin compaction and/or follicular growth, including *KDM2A* (Fig. 11.2a), *SETD3* (Fig. 11.2d) and *SETD2* (3′UTR variants and constitutive; Fig. 11.2c). However, we were unable to find any reference in the literature on the occurrence of H3K36me during oocyte growth. Therefore it is not possible to predict the significance of these changes based on the analysis of the phenotype. Moreover, with the exception of one study showing an important role for KDM2A during mouse embryonic development (Kawakami et al. 2015), very little is known on the specific role of these enzymes during oocyte and embryonic development. Thus these are interesting new targets for future studies.

11.4.1.5 Transcripts Related to Histone H4 Lysine 20 Trimethylation (H4K20me3)

H4K20me3 is considered a repressive mark and hallmark of pericentromeric heterochromatin (together with H3K9me3) (Souza et al. 2009). In the mouse, it is always clearly associated with pericentromeric chromatin during the NSN to SN transition (Table 11.2). In the present analysis, transcripts of both the constitutive form of the H4K20me3-specific methyltransferase (*SUV420H1*) and one 3′UTR variant were more abundant in GV1-2-3, compared to GV0. Since the global intensity of this mark did not change during chromatin compaction, we can speculate that these transcripts are accumulated for later use during maturation and/or embryonic development according to the RNA-seq profile (http://emb-bioinfo.fsaa.ulaval.ca/IMAGE/).

In mice, the combined siRNA-mediated downregulation of Suv420h1 and Suv420h2 resulted in attenuation of H4K20me3 and concomitant accumulation of H4K20me, which led to aberrant chromosome alignment on the MII plate, together with impaired polar body I extrusion (Xiong et al. 2013). On the other hand, we cannot exclude that methyltransferase activity may also be required during chromatin compaction for the maintenance of the H4K20me3 state during oocyte differentiation.

As for the possible role in further embryonic development, H3K20me3 was not detected in mouse embryos and that ectopic expression of SUV420H1 at the one-cell stage increased H3K20me3 mark and impaired subsequent embryo development (Eid et al. 2016). However, since genomic activation occurs at different stages in mouse and bovine, these results do not help us in interpreting the specific role of maternally inherited *SUV420H1*. Moreover, to the best of our knowledge, no data are available on H4K20me3 during bovine preimplantation development. Therefore, the role of *SUV420H1* transcripts in bovine development remains to be fully elucidated by knockdown analysis.

11.4.1.6 Transcripts Related to Histone H3 and H4 Acetylation

Several changes in histone acetylation occur during oocyte chromatin compaction and subsequent meiotic resumption and maturation (Table 11.3). Despite this, relatively few transcripts related to histone acetylation/deacetylation activity were found to be significantly affected in our analysis, which might suggest that transcriptional regulation has limited relevance on histone acetylation processes at this stage, while other translational or post-translational regulatory mechanisms could have major roles.

We have assessed the levels of histone H4 lysine 12 and 5 acetylation (H4K12ac and H4K5ac) in cow oocytes. Overall, we observed an increased acetylation level of both K residues during chromatin compaction from GV0 to GV3 oocytes and a deacetylation following in vitro maturation (Fig. 11.3). However, acetylation of H4K12 and H4K5 seems to be stage dependent during oogenesis since H4K12ac occurred earlier than H4K5ac. As shown in Fig. 11.2g, transcripts for histone acetyltransferase 1 (*HAT1*), which encode protein that acetylates soluble (but not nucleosomal) H4K5 and H4K12, were slightly decreased in GV1-2-3 compared to GV0 oocyte. Moreover, both RNA-seq analysis (http://emb-bioinfo.fsaa.ulaval.ca/IMAGE/) and quantitative PCR using oligo-dT (McGraw 2002) detected relatively low levels of *HAT1* transcripts in in vitro matured MII oocytes and up to the blastocyst stage when its expression increased. Therefore the reduced amounts in GV1-2-3 could be associated with translation, which would somehow contribute to increased H4K12ac and H4K5ac during chromatin compaction, followed by transcript degradation.

Other interesting targets identified by the present analysis were several members of the MYST histone acetylase family. The levels of some transcripts of the MYST group were slightly affected by the chromatin status and/or the follicular size (Fig. 11.2g, h). However, since the precise histone residue target for each of the MYST enzymes has not yet been reported, it is difficult to match the increased/decreased mRNA levels in our contrasts with the observed phenotype (Table 11.3). Nevertheless, among the MYST transcripts, *MYST4* is probably the most promising. Although *MYST4* mRNAs are ubiquitous, immunohistochemistry performed on bovine ovarian sections and whole mount samples revealed that the MYST4 protein is present in oocytes and in the cytoplasm and nuclei of in vitro produced embryos (McGraw et al. 2007). Moreover, the MYST4 protein accumulated in the GV of oocytes and then concentrated in the vicinity of the meiotic spindle rather than on chromosomes when oocytes reached the MII stage. Given the global histone deacetylation in in vitro matured MII oocytes, it appears unlikely that MYST4 acetylates histones H3 and H4. However, a careful observation of immunofluorescent images of published studies in several species revealed that, although there is a general reduction of most acetylation marks, low levels of acetylation of some K residues were maintained in MII chromosomes (see, e.g. MII stage of Fig. 11.3a), and this could depend on the species and on the culture procedure. In horse, for example, H4K16 was deacetylated in chromosomes of in vitro matured oocytes,

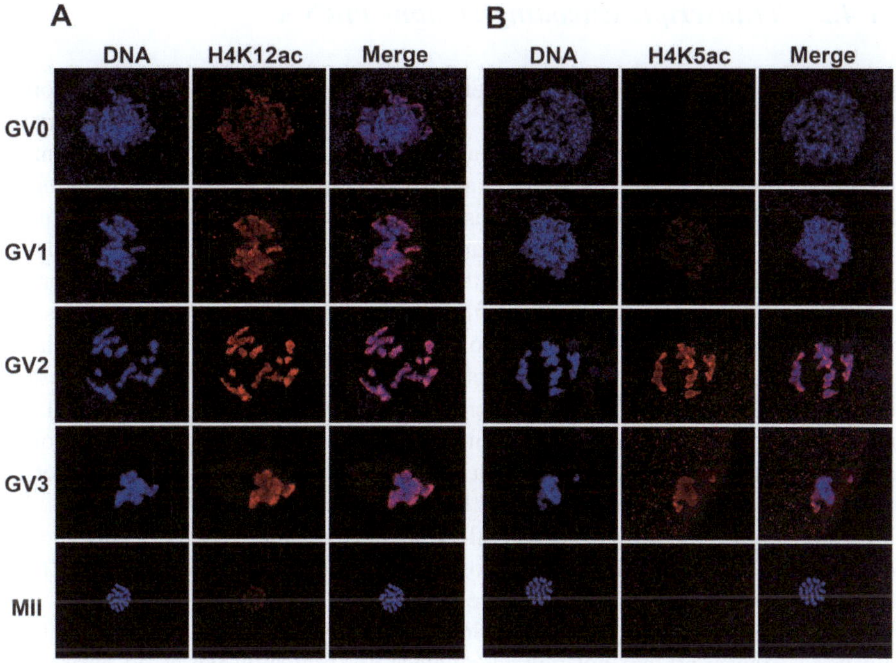

Fig. 11.3 Representative images showing H4K12 (**a**) and H4K5 (**b**) acetylation in GV0, GV1, GV2, GV3 and MII bovine oocytes. H4K12ac and H4K5ac levels were analysed by indirect immunofluorescence using polyclonal anti-AcH4K12 or anti-AcH4K5 antibodies (Upstate Biotechnologies, Inc., Lake Placid, NY, USA) and subsequent confocal microscopy. A total of 100 GV and 35 MII stage oocytes were analysed. Fluorescence intensities were arbitrarily classified as absent, weak or intense. H4K12 was highly acetylated in all the GV1, GV2 and GV3 oocytes and in 46.1% of GV0 oocytes, while the remaining 53.9% of GV0 oocytes showed none or weak H4K12 acetylation signal. H4K5 was absent in all the GV0 and in 20% of GV1 oocytes. Twenty percent of GV1 oocytes were weakly acetylated, while the remaining 60% were highly acetylated. Roughly half of the GV2 and GV3 oocytes were weakly acetylated, and the other half showed intense acetylation signals. All the images are the three-dimensional confocal reconstructions of each sample

while acetylation was retained in in vivo matured oocytes (Franciosi et al. 2012, 2015). So we cannot exclude that some processes responsible for the maintenance of histone acetylation marks may still function during oocyte maturation. This could explain why not all of the transcripts for members of the MYST family are changing in the same direction.

As for the histone deacetylase (HDAC) family, as shown in Fig. 11.2i, l, the levels of *HDAC2*, *HDAC6* and *HDAC8* transcripts were significantly affected by the chromatin status and the follicular size. However, since the precise histone residue targets for these enzymes have not been reported yet, it is difficult to match the increased/decreased mRNA levels with the observed phenotype.

11.4.2 Transcripts Encoding Histone mRNA

Histone mRNAs are particularly interesting in oocytes since histone isoforms (lately referred to as 'variants') are recognized as key players for chromatin function and are emerging as essential participants in mammalian development (Biterge and Schneider 2014; Maze et al. 2014). However, proper interpretation of histone mRNA level changes in oocytes is not an easy task, and the warnings made above for histone-modifying enzymes are even more relevant. We know indeed that histone transcripts are to be stored for the sperm nucleosome formation and for the first few cell divisions, which means that an increased level of the mRNA is not related to immediate translation. Moreover, taking into consideration data available in human and mouse for histone transcripts whose genes are present in clusters (*HIST1*, *HIST2* and *HIST3* cluster genes), and are typically not polyadenylated (Marzluff et al. 2002, 2008), we are limited to the ones that have a sufficient polyA stretch to generate an amplified product as previously discussed (Labrecque et al. 2015). Other histone transcripts might be present but invisible to our analysis. On the other hand, some histone variant mRNAs that are encoded by genes outside the canonical gene clusters are usually polyadenylated (Marzluff et al. 2002, 2008). Moreover, in mammalian somatic cells, histones can be further divided into replication-dependent and replication-independent histones: histone genes that are clustered (transcripts are not polyadenylated) are typically expressed during the S-phase and incorporated into chromatin in a DNA replication-dependent manner, while the others (transcripts are polyadenylated) are expressed throughout the cell cycle and incorporated into chromatin in a replication-independent manner (Marzluff et al. 2002, 2008).

Nevertheless, for the purpose of this review, we are currently limited by the still incomplete annotations for the histone-related genes in the bovine genome, by the lack of a consolidated unified nomenclature in the literature (Talbert et al. 2012) as well as by the scarcity of information on the pattern of nucleosome composition during oogenesis in mammals and particularly in bovine. For these reasons, a tentative match between changes of histone mRNA levels and relative phenotype is not possible. Therefore, discussion of changes in mRNA encoding for histone proteins will be mostly descriptive. Wherever possible, some hypotheses (sometime provocative) based on data presented in the literature are also presented; however, as for the discussion presented in the previous paragraphs, we are aware that all of the hypothesis presented here await appropriate validation. Changes of histone transcript levels are presented in Fig. 11.4 and organized according to gene family.

11.4.2.1 Histone Clusters 1 and 2 Family (HIST1H and HIST2H) Transcripts

As shown in Fig. 11.4a, the RNA levels of HIST1H family increase with the shift from GV0 to GV1-2-3. For the HIST2H family, the *HIST2H2AB* mRNA accumulates in a chromatin configuration-dependent manner, while the constitutive and the 3′UTR isoform of *HIST2H2BE* accumulate in relation to follicular growth, and *HIST2H3A* accumulates in both conditions.

Interestingly, a previous study has identified the activating transcription factor-2 (ATF-2) transcript, which is recognized as a histone acetyltransferase (Kawasaki et al. 2000), as an important quality factor in cleaving bovine embryos (Orozco-Lucero et al. 2014); moreover *ATF-2* is translated during bovine oocyte maturation (Gohin et al. 2014). EmbryoGENE dataset shows that the *ATF-2* constitutive transcript is significantly accumulating in oocytes from larger follicle size and in GV1 and 2 stage oocytes while decreasing in GV3 (http://emb-bioinfo.fsaa.ulaval.ca/IMAGE/). This enzyme interacts with H2B (Bruhat et al. 2007), and in our analysis, some genes encoding H2B such as *HIST1H2BA*, *HIST1H2BN* and *HIST2H2BE* transcripts mostly follow the same trend (Fig. 11.4a, b). This combination could create a more open context for DNA demethylation to occur, although no one has tested such hypothesis yet.

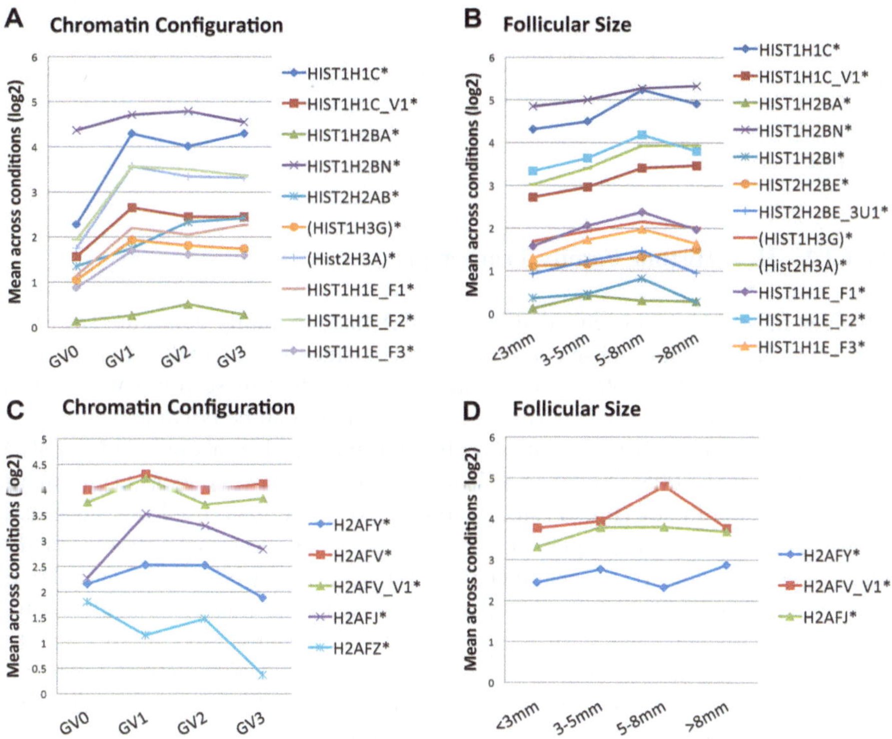

Fig. 11.4 Changes of mRNAs encoding for histone proteins according to chromatin configuration (*left panel*) and follicular growth (*right panel*). The data represent the profiles in the original microarray analysis and are obtained/imported from the EmbryoGENE profiler website (http://emb-bioinfo.fsaa.ulaval.ca/IMAGE/)

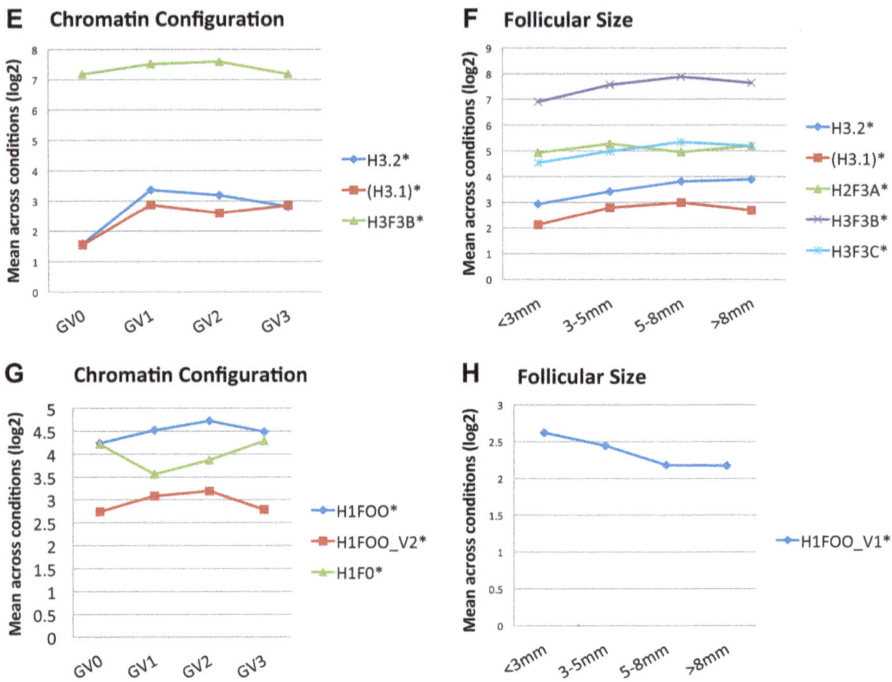

Fig. 11.4 (continued)

11.4.2.2 H2A Histone Family Transcripts

Changes in mRNA levels encoded by the H2A histone family genes are shown in Fig. 11.4c, d. In humans these genes are found outside the principal histone clusters (Marzluff et al. 2002, 2008), so it is not surprising that these mRNAs could be detected in our microarray experiments, as they should all be polyadenylated.

In particular, *H2AFJ* represents a replication-independent histone with a polyA tail (Nishida et al. 2005), and its level is high both in the more condensed chromatin and the larger follicles (Fig. 11.4c, d). For *H2AFZ*, we have used this gene as a stable gene reference in oocyte since it varies according to transcription capacity, and it is polyadenylated (Vigneault et al. 2007); accordingly it is significantly lower in GV1-GV2 and GV3 oocytes when compared to the transcriptionally highly active GV0 oocytes, but it does not change significantly (and is very stable) according to follicular size (http://emb-bioinfo.fsaa.ulaval.ca/IMAGE/cgi-bin/DoProfile.cgi?gene=H2AFZ&tissue=oocyte). *H2AFV*, which levels in our analysis change significantly according to chromatin compaction (and the splicing variant to both chromatin configuration and follicle size), is involved in DNA repair. Indeed, in *Drosophila*, both ataxia telangiectasia-mutated (ATM) and ataxia telangiectasia-related (ATR) kinases phosphorylate H2AV (γ-H2AV), and using this as a reporter for ATM/ATR activity, Joyce et al. (2011) found that γ-H2AV is continuously exchanged, requiring new phosphorylation at the break site until repair is completed (Joyce et al. 2011). Thus, they conclude that ATM is primarily required for the meiotic double-strand break (DSB) repair response, which includes functions in DNA damage repair and negative feedback control over the level of programmed DSBs during meiosis (Joyce et al. 2011).

To the best of our knowledge, still little is known about the function of each of the H2A protein variants in mammalian oocytes, and only recently they are starting to be studied. For example, some immunolocalization survey to detect their presence and localization changes during oogenesis, and early embryonic development, has been conducted (Wu et al. 2014). Interestingly, the above-mentioned H2AZ protein (H2AFZ in our dataset) was found in growing mouse oocytes (from 1- to 15-day-old mice) but not associated with the chromatin (it localizes in the cytoplasm). Then, H2AZ seems to be incorporated in the chromatin after priming of 28-day-old mice and stays in the chromatin during GVBD and MII. It is then absent from the nuclei of early embryos (from zygotes to morula) and then present in the nuclei of blastocyst (Wu et al. 2014). Moreover, H2AZ null mouse could not be generated since all the embryos died at day 7 post coitum (Faast et al. 2001). Up to that stage, embryo development looked normal, meaning a possible contribution of maternal H2AZ (Faast et al. 2001; Binda et al. 2013; Wu et al. 2014). In bovine, both RNA-seq (http://emb-bioinfo.fsaa.ulaval.ca/IMAGE/) and qRT-PCR with both random hexamers and oligo-dT (Vigneault et al. 2007) detected very low level of *H2AZ* transcript up to genome activation, excluding its possible involvement as a maternal transcript. However we cannot exclude that the maternal H2AZ protein could serve the role. Notably, SETD6 is responsible for H2AZ monomethylation (Binda et al. 2013). As shown in Fig. 11.2c, the level of a *SETD6* transcript's level is significantly affected by chromatin configuration changes, raising the hypothesis that H2AZ protein could be subjected to regulatory post-translational modification in preparation for embryonic development. However, we are aware that this hypothesis still needs to be tested.

11.4.2.3 H3 Histone Family Transcripts

For the transcripts encoded by the histone H3 family, some components show higher levels in the GV1-2-3 groups compared to GV0 (see Fig. 11.4e, f) as well as in oocytes from larger follicles compared to the <3 mm ones. Among these, *H3F3A* and *H3F3B* are particularly interesting. According to the current annotation, as in mouse (Wen et al. 2014), they both encode for the H3.3 protein. In our analysis the levels of these transcripts change significantly according to chromatin configuration and/or follicle size (Fig. 11.4e, f). Moreover previous data have shown that *H3F3A* is also associated with a long polyA tail and possess 3′UTR sequences associated with recruitment during oocyte maturation (Gohin et al. 2014). These transcripts and the encoded H3.3 protein could have a similar role to the one documented in the mouse.

In the mouse, indeed, different groups have demonstrated the importance of the histone variant H3.3 for chromatin reprogramming upon fertilization (Akiyama et al. 2011; Lin et al. 2013). In 2011, Akiyama and coworkers have shown that following fertilization, H3.3 is removed from the mouse female pronucleus, suggesting that the epigenetic marks carried by H3.3 in oocytes are erased, thus, suggesting that this process might participate in generating totipotency in early embryos (Akiyama et al. 2011). The same group also studied the distribution of H3.2, which is incorporated into the transcriptionally silent heterochromatin, whereas H3.1 and H3.3 occupy unusual heterochromatic and euchromatin locations, respectively (Akiyama et al. 2011). More recently, using both morpholino-mediated H3.3 knockdown in fully grown oocytes and genetic KO of the H3.3 chaperones Hira, Lin and coworkers have shown that Hira-mediated H3.3 incorporation is essential for parental genome reprogramming at fertilization (Lin et al. 2014). In another study, H3.3 siRNA-mediated knockdown in mouse MII oocytes compromised reprogramming and resulted in downregulation of key pluripotency genes; moreover maternally derived H3.3, and not H3.2 in the donor nucleus, was found to be essential for reprogramming of somatic cell nucleus (Wen et al. 2014).

Importantly, the depletion of the H3.3 chaperone Hira in developing mouse oocytes, and the consequent lack of continuous H3.3/H4 deposition, alters chromatin structure, increases DNAse I sensitivity and the accumulation of DNA damage, disrupts gene expression and leads to inefficient de novo DNA methylation and a severe fertility phenotype, thus unequivocally showing the importance of H3.3 continuous histone replacement for transcriptional regulation and normal developmental progression during oogenesis (Nashun et al. 2015). However, precise role of H3.3 protein and encoding mRNA change level still needs to be confirmed in bovine.

Changes of the other transcripts encoded by the H3 histone family, such as *H3F3C*, which rises with follicular size (Fig. 11.4f), await an interpretation. Interestingly, though, *SMYD3* (SET and MYND domain containing 3) is present in slightly higher levels in GV1-2-3 oocytes, and the encoded protein interacts with H3F3C (http://www.signalinggateway.org/molecule/query;jsessionid=d784ecaea9b654d77c593e053f32a7d52deb33c61364f64af4b5ff6d40c7a0b9?afcsid=A006027&type=interactioninfo&adv=latest&idx=26400349), whose mRNA is accumulated in a follicular size-dependent manner. Here again, one could speculate that the uptake and acetylation of H3F3C could be associated with the preparation for DNA access associated with postfertilization reprogramming. We are aware that all these interesting findings must be validated by specific KO experiment to validate the roles proposed by their expression patterns.

11.4.2.4 H1 Histone Family Transcripts

For the linker histones H1 family, the mRNA of the somatic form (*H1F0*) surprisingly goes up in the GV3 stage, while *H1FOO*, which so far is the only oocyte-specific H1 histone variant, is accumulated during chromatin compaction; only *H1FOO* mRNA level is affected by follicular size (Fig. 11.4g, h).

Interestingly we have found a 3′UTR sequence in the *H1FOO* that promotes translation during maturation and accordingly a longer polyA tail for this transcript between GV and MII (Gohin et al. 2014). Further possible significances of H1 histone family changes have been already discussed in our previous publications (Labrecque et al. 2015).

11.5 Conclusions

In the last decade, the availability of 'omic' technologies (mainly microarray- and RNA-seq-based transcriptomic platforms) applicable to scarce samples such as oocytes generated a large amount of data. However, these information are difficult to interpret not only for the limits imposed by the technical inaccurateness of protein determination in such samples, which precludes appropriate validation, but also for the peculiar physiology of the oocyte. Assessment of epigenetic modifications and in particular the histone-related ones is even more difficult. In fact, it is becoming more and more clear that oocytes escape common rules of the 'histone code' that applies to somatic cells (Jenuwein and Allis 2001; Suganuma and Workman 2011).

We have tried to overcome the present limitations by integrating data coming from different microarray analyses and critically combining changes of mRNA levels with the corresponding phenotype as well as changes occurring in the early embryo, which in turn enable the formulation of precise hypotheses. We believe that this approach could reveal the function of specific enzymes in the oocyte. In turn these information will be useful in future studies, in which the experimental disruption of these enzymes would finally reveal the biological significance of the corresponding histone modification. Clearly the ambitious goal would be deciphering of the oocyte-specific histone code. Finally, this approach could be particularly useful in large mammalian species, including humans, that most of the times do not allow the application of sophisticated genetic manipulation for the confirmation of experimental hypothesis.

Funding This work was supported by the Natural Sciences and Engineering Research Council of Canada (NSERC-CRSNG). V.L. acknowledges the receipt of a fellowship from the OECD Co-operative Research Programme: Biological Resource Management for Sustainable Agricultural Systems in 2016 (contract n. JA00091594). F.F. is supported by FP7-PEOPLE-2013-IOF GA 624874 MateRNA.

References

Agger K, Cloos PA, Christensen J, Pasini D, Rose S, Rappsilber J, Issaeva I, Canaani E, Salcini AE, Helin K (2007) UTX and JMJD3 are histone H3K27 demethylases involved in HOX gene regulation and development. Nature 449(7163):731–734. doi:10.1038/nature06145

Akiyama T, Nagata M, Aoki F (2006) Inadequate histone deacetylation during oocyte meiosis causes aneuploidy and embryo death in mice. Proc Natl Acad Sci USA 103(19):7339–7344

Akiyama T, Suzuki O, Matsuda J, Aoki F (2011) Dynamic replacement of histone H3 variants reprograms epigenetic marks in early mouse embryos. PLoS Genet 7(10):e1002279. doi:10.1371/journal.pgen.1002279

Ancelin K, Syx L, Borensztein M, Ranisavljevic N, Vassilev I, Briseno-Roa L, Liu T, Metzger E, Servant N, Barillot E, Chen CJ, Schule R, Heard E (2016) Maternal LSD1/KDM1A is an essential regulator of chromatin and transcription landscapes during zygotic genome activation. eLife 5. doi:10.7554/eLife.08851

Barnes FL, First NL (1991) Embryonic transcription in in vitro cultured bovine embryos. Mol Reprod Dev 29(2):117–123. doi:10.1002/mrd.1080290205

Binda O, Sevilla A, LeRoy G, Lemischka IR, Garcia BA, Richard S (2013) SETD6 monomethylates H2AZ on lysine 7 and is required for the maintenance of embryonic stem cell self-renewal. Epigenetics 8(2):177–183. doi:10.4161/epi.23416

Biterge B, Schneider R (2014) Histone variants: key players of chromatin. Cell Tissue Res 356 (3):457–466. doi:10.1007/s00441-014-1862-4

Bonnet-Garnier A, Feuerstein P, Chebrout M, Fleurot R, Jan HU, Debey P, Beaujean N (2012) Genome organization and epigenetic marks in mouse germinal vesicle oocytes. Int J Dev Biol 56(10–12):877–887. doi:10.1387/ijdb.120149ab

Bouniol-Baly C, Hamraoui L, Guibert J, Beaujean N, Szollosi MS, Debey P (1999) Differential transcriptional activity associated with chromatin configuration in fully grown mouse germinal vesicle oocytes. Biol Reprod 60(3):580–587

Bruhat A, Cherasse Y, Maurin AC, Breitwieser W, Parry L, Deval C, Jones N, Jousse C, Fafournoux P (2007) ATF2 is required for amino acid-regulated transcription by orchestrating specific histone acetylation. Nucleic Acids Res 35(4):1312–1321. doi:10.1093/nar/gkm038

Bui HT, Van Thuan N, Kishigami S, Wakayama S, Hikichi T, Ohta H, Mizutani E, Yamaoka E, Wakayama T, Miyano T (2007) Regulation of chromatin and chromosome morphology by histone H3 modifications in pig oocytes. Reproduction 133(2):371–382. doi:10.1530/REP-06-0099

Canovas S, Ross PJ (2016) Epigenetics in preimplantation mammalian development. Theriogenology 86(1):69–79. doi:10.1016/j.theriogenology.2016.04.020

Canovas S, Cibelli JB, Ross PJ (2012) Jumonji domain-containing protein 3 regulates histone 3 lysine 27 methylation during bovine preimplantation development. Proc Natl Acad Sci USA 109(7):2400–2405. doi:10.1073/pnas.1119112109

Ciccone DN, Su H, Hevi S, Gay F, Lei H, Bajko J, Xu G, Li E, Chen T (2009) KDM1B is a histone H3K4 demethylase required to establish maternal genomic imprints. Nature 461 (7262):415–418. doi:10.1038/nature08315

Clarke HJ (2012) Post-transcriptional control of gene expression during mouse oogenesis. Results Probl Cell Differ 55:1–21. doi:10.1007/978-3-642-30406-4_1

Clarke HJ, Vieux KF (2015) Epigenetic inheritance through the female germ-line: the known, the unknown, and the possible. Semin Cell Dev Biol 43:106–116. doi:10.1016/j.semcdb.2015.07.003

Conti M, Sousa Martins JP, Jin Han S, Franciosi F (2015) Translational control in the germ line. In: Menon KMJ, Goldstrohm AC (eds) Post-transcriptional mechanisms in endocrine regulation. Springer, Switzerland, pp 129–156. doi:10.1007/978-3-319-25124-0_7

De La Fuente R (2006) Chromatin modifications in the germinal vesicle (GV) of mammalian oocytes. Dev Biol 292(1):1–12

Dieci C, Lodde V, Franciosi F, Lagutina I, Tessaro I, Modina SC, Albertini DF, Lazzari G, Galli C, Luciano AM (2013) The effect of cilostamide on gap junction communication dynamics, chromatin remodeling, and competence acquisition in pig oocytes following parthenogenetic activation and nuclear transfer. Biol Reprod 89(3):68. doi:10.1095/biolreprod.113.110577

Dieci C, Lodde V, Labreque R, Dufort I, Tessaro I, Sirard MA, Luciano AM (2016) Differences in cumulus cell gene expression indicate the benefit of a pre-maturation step to improve in-vitro bovine embryo production. Mol Hum Reprod. doi:10.1093/molehr/gaw055

Eid A, Rodriguez-Terrones D, Burton A, Torres-Padilla ME (2016) SUV4-20 activity in the preimplantation mouse embryo controls timely replication. Genes and development 30 (22):2513–2526. doi:10.1101/gad.288969.116

Endo T, Naito K, Aoki F, Kume S, Tojo H (2005) Changes in histone modifications during in vitro maturation of porcine oocytes. Mol Reprod Dev 71(1):123–128

Eymery A, Liu Z, Ozonov EA, Stadler MB, Peters AH (2016) The methyltransferase Setdb1 is essential for meiosis and mitosis in mouse oocytes and early embryos. Development 143 (15):2767–2779. doi:10.1242/dev.132746

Faast R, Thonglairoam V, Schulz TC, Beall J, Wells JR, Taylor H, Matthaei K, Rathjen PD, Tremethick DJ, Lyons I (2001) Histone variant H2A.Z is required for early mammalian development. Curr Biol 11(15):1183–1187

Fair T, Hulshof SC, Hyttel P, Greve T, Boland M (1997) Nucleus ultrastructure and transcriptional activity of bovine oocytes in preantral and early antral follicles. Mol Reprod Dev 46 (2):208–215

Franciosi F, Lodde V, Goudet G, Duchamp G, Deleuze S, Douet C, Tessaro I, Luciano AM (2012) Changes in histone H4 acetylation during in vivo versus in vitro maturation of equine oocytes. Mol Hum Reprod 18(5):243–252. doi:10.1093/molehr/gar077

Franciosi F, Goudet G, Tessaro I, Papillier P, Dalbies-Tran R, Reigner F, Deleuze S, Douet C, Miclea I, Lodde V, Luciano AM (2015) In vitro maturation affects chromosome segregation, spindle morphology and acetylation of lysine 16 on histone H4 in horse oocytes. Reprod Fertil Dev. doi:10.1071/RD15350

Fulka H (2008) Changes in global histone acetylation pattern in somatic cell nuclei after their transfer into oocytes at different stages of maturation. Mol Reprod Dev 75(3):556–564. doi:10.1002/mrd.20840

Gilbert I, Scantland S, Sylvestre EL, Gravel C, Laflamme I, Sirard MA, Robert C (2009) The dynamics of gene products fluctuation during bovine pre-hatching development. Mol Reprod Dev 76(8):762–772. doi:10.1002/mrd.21030

Gohin M, Fournier E, Dufort I, Sirard MA (2014) Discovery, identification and sequence analysis of RNAs selected for very short or long poly A tail in immature bovine oocytes. Mol Hum Reprod 20(2):127–138. doi:10.1093/molehr/gat080

Golding MC, Snyder M, Williamson GL, Veazey KJ, Peoples M, Pryor JH, Westhusin ME, Long CR (2015) Histone-lysine N-methyltransferase SETDB1 is required for development of the bovine blastocyst. Theriogenology 84(8):1411–1422. doi:10.1016/j.theriogenology.2015.07.028

Gougeon A (1986) Dynamics of follicular growth in the human: a model from preliminary results. Hum Reprod 1(2):81–87

Hunter AG, Moor RM (1987) Stage-dependent effects of inhibiting ribonucleic acids and protein synthesis on meiotic maturation of bovine oocytes in vitro. J Dairy Sci 70(8):1646–1651. doi:10.3168/jds.S0022 0302(87)80192 3

Hyttel P (2010) Gametogenesis. In: Hyttel P, Sinowatz F, Vejlsted M (eds) Essentials of domestic animal embriology. Saunders/Elsevier, Edinburgh/New York, pp 32–56

Jenuwein T, Allis CD (2001) Translating the histone code. Science 293(5532):1074–1080. doi:10.1126/science.1063127

Joyce EF, Pedersen M, Tiong S, White-Brown SK, Paul A, Campbell SD, McKim KS (2011) Drosophila ATM and ATR have distinct activities in the regulation of meiotic DNA damage and repair. J Cell Biol 195(3):359–367. doi:10.1083/jcb.201104121

Kageyama S, Liu H, Kaneko N, Ooga M, Nagata M, Aoki F (2007) Alterations in epigenetic modifications during oocyte growth in mice. Reproduction 133(1):85–94

Kawakami E, Tokunaga A, Ozawa M, Sakamoto R, Yoshida N (2015) The histone demethylase Fbxl11/Kdm2a plays an essential role in embryonic development by repressing cell-cycle regulators. Mech Dev 135:31–42. doi:10.1016/j.mod.2014.10.001

Kawasaki H, Schiltz L, Chiu R, Itakura K, Taira K, Nakatani Y, Yokoyama KK (2000) ATF-2 has intrinsic histone acetyltransferase activity which is modulated by phosphorylation. Nature 405 (6783):195–200. doi:10.1038/35012097

Khan DR, Fournier E, Dufort I, Richard FJ, Singh J, Sirard MA (2016) Meta-analysis of gene expression profiles in granulosa cells during folliculogenesis. Reproduction 151(6):R103–R110. doi:10.1530/REP-15-0594

Kim JM, Liu H, Tazaki M, Nagata M, Aoki F (2003) Changes in histone acetylation during mouse oocyte meiosis. J Cell Biol 162(1):37–46

Kim J, Zhao H, Dan J, Kim S, Hardikar S, Hollowell D, Lin K, Lu Y, Takata Y, Shen J, Chen T (2016) Maternal Setdb1 is required for meiotic progression and preimplantation development in mouse. PLoS Genet 12(4):e1005970. doi:10.1371/journal.pgen.1005970

Labrecque R, Lodde V, Dieci C, Tessaro I, Luciano AM, Sirard MA (2015) Chromatin remodelling and histone m RNA accumulation in bovine germinal vesicle oocytes. Mol Reprod Dev 82(6):450–462. doi:10.1002/mrd.22494

Labrecque R, Fournier E, Sirard MA (2016) Transcriptome analysis of bovine oocytes from distinct follicle sizes: insights from correlation network analysis. Mol Reprod Dev 83 (6):558–569. doi:10.1002/mrd.22651

Lan F, Bayliss PE, Rinn JL, Whetstine JR, Wang JK, Chen S, Iwase S, Alpatov R, Issaeva I, Canaani E, Roberts TM, Chang HY, Shi Y (2007) A histone H3 lysine 27 demethylase regulates animal posterior development. Nature 449(7163):689–694. doi:10.1038/nature06192

Landeira D, Fisher AG (2011) Inactive yet indispensable: the tale of Jarid2. Trends Cell Biol 21 (2):74–80. doi:10.1016/j.tcb.2010.10.004

Leung PCK, Adashi EY (2003) The ovary. Elesevier, San Diego

Lin CJ, Conti M, Ramalho-Santos M (2013) Histone variant H3.3 maintains a decondensed chromatin state essential for mouse preimplantation development. Development 140 (17):3624–3634. doi:10.1242/dev.095513

Lin CJ, Koh FM, Wong P, Conti M, Ramalho-Santos M (2014) Hira-mediated H3.3 incorporation is required for DNA replication and ribosomal RNA transcription in the mouse zygote. Dev Cell 30(3):268–279. doi:10.1016/j.devcel.2014.06.022

Lin J, Chen F, Sun MJ, Zhu J, Li YW, Pan LZ, Zhang J, Tan JH (2016) The relationship between apoptosis, chromatin configuration, histone modification and competence of oocytes: a study using the mouse ovary-holding stress model. Sci Rep 6:28347. doi:10.1038/srep28347

Lodde V, Modina S, Galbusera C, Franciosi F, Luciano AM (2007) Large-scale chromatin remodeling in germinal vesicle bovine oocytes: interplay with gap junction functionality and developmental competence. Mol Reprod Dev 74(6):740–749. doi:10.1002/mrd.20639

Lodde V, Modina S, Maddox-Hyttel P, Franciosi F, Lauria A, Luciano AM (2008) Oocyte morphology and transcriptional silencing in relation to chromatin remodeling during the final phases of bovine oocyte growth. Mol Reprod Dev 75(5):915–924. doi:10.1002/mrd.20824

Luciano AM, Lodde V (2013) Changes of large-scale chromatin configuration during mammalian oocyte differentiation. In: Coticchio G, Albertini DF, De Santis L (eds) Oogenesis. Springer, London, pp 93–108. doi:10.1007/978-0-85729-826-3_7

Maalouf WE, Alberio R, Campbell KH (2008) Differential acetylation of histone H4 lysine during development of in vitro fertilized, cloned and parthenogenetically activated bovine embryos. Epigenetics 3(4):199–209

Marzluff WF, Gongidi P, Woods KR, Jin J, Maltais LJ (2002) The human and mouse replication-dependent histone genes. Genomics 80(5):487–498. doi:10.1006/geno.2002.6850

Marzluff WF, Wagner EJ, Duronio RJ (2008) Metabolism and regulation of canonical histone mRNAs: life without a poly(A) tail. Nat Rev Genet 9(11):843–854. doi:10.1038/nrg2438

Mattson BA, Albertini DF (1990) Oogenesis: chromatin and microtubule dynamics during meiotic prophase. Mol Reprod Dev 25(4):374–383

Maze I, Noh KM, Soshnev AA, Allis CD (2014) Every amino acid matters: essential contributions of histone variants to mammalian development and disease. Nat Rev Genet 15(4):259–271. doi:10.1038/nrg3673

McGraw S (2002) Quantification of histone acetyltransferase and histone deacetylase transcripts during early bovine embryo development. Biol Reprod 68(2):383–389. doi:10.1095/biolreprod.102.005991

McGraw S, Morin G, Vigneault C, Leclerc P, Sirard MA (2007) Investigation of MYST4 histone acetyltransferase and its involvement in mammalian gametogenesis. BMC Dev Biol 7:123. doi:10.1186/1471-213X-7-123

Meglicki M, Zientarski M, Borsuk E (2008) Constitutive heterochromatin during mouse oogenesis: the pattern of histone H3 modifications and localization of HP1alpha and HP1beta proteins. Mol Reprod Dev 75(2):414–428. doi:10.1002/mrd.20772

Nagashima T, Maruyama T, Furuya M, Kajitani T, Uchida H, Masuda H, Ono M, Arase T, Ozato K, Yoshimura Y (2007) Histone acetylation and subcellular localization of chromosomal protein BRD4 during mouse oocyte meiosis and mitosis. Mol Hum Reprod 13 (3):141–148

Nashun B, Hill PW, Smallwood SA, Dharmalingam G, Amouroux R, Clark SJ, Sharma V, Ndjetehe E, Pelczar P, Festenstein RJ, Kelsey G, Hajkova P (2015) Continuous histone replacement by Hira is essential for normal transcriptional regulation and de novo DNA methylation during mouse oogenesis. Mol Cell 60(4):611–625. doi:10.1016/j.molcel.2015.10.010

Nishida H, Suzuki T, Tomaru Y, Hayashizaki Y (2005) A novel replication-independent histone H2a gene in mouse. BMC Genet 6:10. doi:10.1186/1471-2156-6-10

Orozco-Lucero E, Dufort I, Robert C, Sirard MA (2014) Rapidly cleaving bovine two-cell embryos have better developmental potential and a distinctive mRNA pattern. Mol Reprod Dev 81(1):31–41. doi:10.1002/mrd.22278

Pan Z, Zhang J, Li Q, Li Y, Shi F, Xie Z, Liu H (2012) Current advances in epigenetic modification and alteration during mammalian ovarian folliculogenesis. J Genet Genomics 39(3):111–123. doi:10.1016/j.jgg.2012.02.004

Park KE, Magnani L, Cabot RA (2009) Differential remodeling of mono- and trimethylated H3K27 during porcine embryo development. Mol Reprod Dev 76(11):1033–1042. doi:10.1002/mrd.21061

Pedersen MT, Agger K, Laugesen A, Johansen JV, Cloos PA, Christensen J, Helin K (2014) The demethylase JMJD2C localizes to H3K4me3-positive transcription start sites and is dispensable for embryonic development. Mol Cell Biol 34(6):1031–1045. doi:10.1128/MCB.00864-13

Pichugin A, Le Bourhis D, Adenot P, Lehmann G, Audouard C, Renard JP, Vignon X, Beaujean N (2010) Dynamics of constitutive heterochromatin: two contrasted kinetics of genome restructuring in early cloned bovine embryos. Reproduction 139(1):129–137. doi:10.1530/REP-08-0435

Racedo SE, Branzini MC, Salamone D, Wojcik C, Rawe VY, Niemann H (2009) Dynamics of microtubules, motor proteins and 20S proteasomes during bovine oocyte IVM. Reprod Fertil Dev 21(2):304–312

Richter JD (2007) CPEB: a life in translation. Trends Biochem Sci 32(6):279–285. doi:10.1016/j.tibs.2007.04.004

Robert C, Nieminen J, Dufort I, Gagne D, Grant JR, Cagnone G, Plourde D, Nivet AL, Fournier E, Paquet E, Blazejczyk M, Rigault P, Juge N, Sirard MA (2011) Combining resources to obtain a comprehensive survey of the bovine embryo transcriptome through deep sequencing and microarrays. Mol Reprod Dev 78(9):651–664. doi:10.1002/mrd.21364

Rose NR, Klose RJ (2014) Understanding the relationship between DNA methylation and histone lysine methylation. Biochim Biophys Acta 1839(12):1362–1372. doi:10.1016/j.bbagrm.2014.02.007

Ross PJ, Ragina NP, Rodriguez RM, Iager AE, Siripattarapravat K, Lopez-Corrales N, Cibelli JB (2008) Polycomb gene expression and histone H3 lysine 27 trimethylation changes during bovine preimplantation development. Reproduction 136(6):777–785. doi:10.1530/REP-08-0045

Russo V, Bernabo N, Di Giacinto O, Martelli A, Mauro A, Berardinelli P, Curini V, Nardinocchi D, Mattioli M, Barboni B (2013) H3K9 trimethylation precedes DNA methylation during sheep oogenesis: HDAC1, SUV39H1, G9a, HP1, and Dnmts are involved in these epigenetic events. J Histochem Cytochem 61(1):75–89. doi:10.1369/0022155412463923

Santos F, Zakhartchenko V, Stojkovic M, Peters A, Jenuwein T, Wolf E, Reik W, Dean W (2003) Epigenetic marking correlates with developmental potential in cloned bovine preimplantation embryos. Curr Biol 13(13):1116–1121. doi:10.1016/s0960-9822(03)00419-6

Sanulli S, Justin N, Teissandier A, Ancelin K, Portoso M, Caron M, Michaud A, Lombard B, da Rocha ST, Offer J, Loew D, Servant N, Wassef M, Burlina F, Gamblin SJ, Heard E, Margueron R (2015) Jarid2 methylation via the PRC2 complex regulates H3K27me3 deposition during cell differentiation. Mol Cell 57(5):769–783. doi:10.1016/j.molcel.2014.12.020

Sarmento OF, Digilio LC, Wang Y, Perlin J, Herr JC, Allis CD, Coonrod SA (2004) Dynamic alterations of specific histone modifications during early murine development. J Cell Sci 117 (Pt 19):4449–4459

Scantland S, Grenon JP, Desrochers MH, Sirard MA, Khandjian EW, Robert C (2011) Method to isolate polyribosomal mRNA from scarce samples such as mammalian oocytes and early embryos. BMC Dev Biol 11:8. doi:10.1186/1471-213X-11-8

Scaramuzzi RJ, Baird DT, Campbell BK, Driancourt MA, Dupont J, Fortune JE, Gilchrist RB, Martin GB, McNatty KP, McNeilly AS, Monget P, Monniaux D, Vinoles C, Webb R (2011) Regulation of folliculogenesis and the determination of ovulation rate in ruminants. Reprod Fertil Dev 23(3):444–467. doi:10.1071/RD09161

Sorensen RA, Wassarman PM (1976) Relationship between growth and meiotic maturation of the mouse oocyte. Dev Biol 50(2):531–536

Souza PP, Volkel P, Trinel D, Vandamme J, Rosnoblet C, Heliot L, Angrand PO (2009) The histone methyltransferase SUV420H2 and heterochromatin proteins HP1 interact but show different dynamic behaviours. BMC Cell Biol 10:41. doi:10.1186/1471-2121-10-41

Stewart KR, Veselovska L, Kim J, Huang J, Saadeh H, Tomizawa S, Smallwood SA, Chen T, Kelsey G (2015) Dynamic changes in histone modifications precede de novo DNA methylation in oocytes. Genes Dev 29(23):2449–2462. doi:10.1101/gad.271353.115

Suganuma T, Workman JL (2011) Signals and combinatorial functions of histone modifications. Annu Rev Biochem 80:473–499. doi:10.1146/annurev-biochem-061809-175347

Talbert PB, Ahmad K, Almouzni G, Ausio J, Berger F, Bhalla PL, Bonner WM, Cande WZ, Chadwick BP, Chan SW, Cross GA, Cui L, Dimitrov SI, Doenecke D, Eirin-Lopez JM, Gorovsky MA, Hake SB, Hamkalo BA, Holec S, Jacobsen SE, Kamieniarz K, Khochbin S, Ladurner AG, Landsman D, Latham JA, Loppin B, Malik HS, Marzluff WF, Pehrson JR, Postberg J, Schneider R, Singh MB, Smith MM, Thompson E, Torres-Padilla ME, Tremethick DJ, Turner BM, Waterborg JH, Wollmann H, Yelagandula R, Zhu B, Henikoff S (2012) A unified phylogeny-based nomenclature for histone variants. Epigenetics Chromatin 5:7. doi:10.1186/1756-8935-5-7

Vigneault C, Gilbert I, Sirard MA, Robert C (2007) Using the histone H2a transcript as an endogenous standard to study relative transcript abundance during bovine early development. Mol Reprod Dev 74(6):703–715. doi:10.1002/mrd.20665

Wang Q, Yin S, Ai JS, Liang CG, Hou Y, Chen DY, Schatten H, Sun QY (2006) Histone deacetylation is required for orderly meiosis. Cell Cycle 5(7):766–774. doi:10.4161/cc.5.7.2627

Wang Z, Zhao T, Zhang P, Zhang S, Guan J, Ma X, Yin Y, Zhang J, Tang B, Li Z (2011) Histone deacetylase 1 down-regulation on developmental capability and histone acetylation in bovine oocytes and parthenogenetic embryos. Reprod Domest Anim 46(6):1022–1028. doi:10.1111/j.1439-0531.2011.01778.x

Wee G, Shin ST, Koo DB, Han YM (2010) Behaviors of ATP-dependent chromatin remodeling factors during maturation of bovine oocytes in vitro. Mol Reprod Dev 77(2):126–135. doi:10.1002/mrd.21113

Wen D, Banaszynski LA, Liu Y, Geng F, Noh KM, Xiang J, Elemento O, Rosenwaks Z, Allis CD, Rafii S (2014) Histone variant H3.3 is an essential maternal factor for oocyte reprogramming. Proc Natl Acad Sci USA 111(20):7325–7330. doi:10.1073/pnas.1406389111

Wu X, Li Y, Xue L, Wang L, Yue Y, Li K, Bou S, Li GP, Yu H (2011) Multiple histone site epigenetic modifications in nuclear transfer and in vitro fertilized bovine embryos. Zygote 19 (1):31–45. doi:10.1017/S0967199410000328

Wu BJ, Dong FL, Ma XS, Wang XG, Lin F, Liu HL (2014) Localization and expression of histone H2A variants during mouse oogenesis and preimplantation embryo development. Genet Mol Res 13(3):5929–5939. doi:10.4238/2014.August.7.8

Xia J, Gill EE, Hancock RE (2015) NetworkAnalyst for statistical, visual and network-based meta-analysis of gene expression data. Nat Protoc 10(6):823–844. doi:10.1038/nprot.2015.052

Xiong K, Wu W, Wang X, Ma X, Chen J, Liu H (2013) Mouse oocyte meiosis is disturbed by knockdown of Suv4-20h. Reprod Fertil Dev 25(3):503–510. doi:10.1071/RD12116

Zhang B, Zheng H, Huang B, Li W, Xiang Y, Peng X, Ming J, Wu X, Zhang Y, Xu Q, Liu W, Kou X, Zhao Y, He W, Li C, Chen B, Li Y, Wang Q, Ma J, Yin Q, Kee K, Meng A, Gao S, Xu F, Na J, Xie W (2016a) Allelic reprogramming of the histone modification H3K4me3 in early mammalian development. Nature 537(7621):553–557. doi:10.1038/nature19361

Zhang S, Wang F, Fan C, Tang B, Zhang X, Li Z (2016b) Dynamic changes of histone H3 lysine 9 following trimethylation in bovine oocytes and pre-implantation embryos. Biotechnol Lett 38 (3):395–402. doi:10.1007/s10529-015-2001-3

Zheng H, Huang B, Zhang B, Xiang Y, Du Z, Xu Q, Li Y, Wang Q, Ma J, Peng X, Xu F, Xie W (2016) Resetting epigenetic memory by reprogramming of histone modifications in mammals. Mol Cell 63(6):1066–1079. doi:10.1016/j.molcel.2016.08.032

Zuccotti M, Piccinelli A, Rossi PG, Garagna S, Redi CA (1995) Chromatin organization during mouse oocyte growth. Mol Reprod Dev 41(4):479–485

Chapter 12
Translational Regulation in the Mammalian Oocyte

Andrej Susor and Michal Kubelka

Abstract Fully grown oocytes arrest meiosis at prophase I and deposit maternal RNAs. A subset of maternal transcripts is stored in a dormant state in the oocyte, and the timely driven translation of specific mRNAs guides meiotic progression, the oocyte-embryo transition, and early embryo development. In the absence of transcription, the regulation of gene expression in oocytes is controlled almost exclusively at the level of transcriptome and proteome stabilization and at the level of protein synthesis.

This chapter focuses on the recent findings on RNA distribution related to the temporal and spatial translational control of the meiotic cycle progression in mammalian oocytes. We discuss the most relevant mechanisms involved in the organization of the oocyte's maternal transcriptome storage and localization, and the regulation of translation, in correlation with the regulation of oocyte meiotic progression.

12.1 Introduction

Translational control of specific mRNAs is a widespread mechanism of gene regulation and contributes to a range of diverse biological processes in many cell types. During the meiotic division of mammalian oocytes (the so-called oocyte maturation), protein synthesis plays an important role in controlling the progress of meiosis, since the regulation of gene expression on the level of transcription is ceased (De La Fuente et al. 2004). After a period of active transcription during growth, the nucleus (germinal vesicle, GV) of mammalian oocytes becomes transcriptionally inactive. Transcription is ceased at the onset of meiotic resumption and becomes fully activated in the zygotic genome activation (ZGA) stage (Schultz 1993; Matova and Cooley 2001), which occurs at the two-cell stage in mice and four- to eight-cell stage in cattle and humans (Flach et al. 1982; Clegg and Pikó 1982; Braude et al. 1988; Pavlok et al. 1993; Toralová et al. 2009).

A. Susor (✉) • M. Kubelka
Institute of Animal Physiology and Genetics, CAS, Rumburska 89, 277 21 Libechov, Czech Republic
e-mail: susor@iapg.cas.cz

M. Kloc (ed.), *Oocytes*, Results and Problems in Cell Differentiation 63,
DOI 10.1007/978-3-319-60855-6_12

Maternal RNAs are uniquely stable, with a half-life of 8 to 12 days (Bachvarova and De Leon 1980; Brower et al. 1981) and, in response to the appropriate stimulus, will drive reentry into meiosis and control the rate of mitotic cell divisions during the cleavage of the early embryo (Schultz 1993; Tadros and Lipshitz 2009). Therefore, the regulation of gene expression in oocytes and early embryos is controlled almost exclusively at the level of mRNA stabilization and translation and posttranslational modifications of proteins.

In order to facilitate the transition from maternal-to-embryonic control, oocytes need to store and then selectively utilize an abundance of compounds. Therefore, as opposed to somatic cells, which divide after doubling their volume, the mouse oocyte grows around 200–300 times in volume without dividing, and during these ~2.5 weeks of growth, there is a period of intense metabolic activity resulting in the synthesis of large amounts of mRNAs, proteins, and macromolecular structures (Yanagimachi 1994). It is no longer assumed that once an mRNA is made, it will be predictably translated into a protein. While some portion of this newly generated material is specifically required for oocyte function, a number of these mRNA molecules are likely to be endowed to the embryo after fertilization to facilitate early cleavage divisions. However, as opposed to mRNA from most somatic cells, oocyte mRNA is extremely stable. Also, it has been observed that the ratio of poly (A) to total RNA is higher in oocytes than in somatic cells (Sternlicht and Schultz 1981). New NGS methodologies have started to provide the first high-resolution, global views of poly(A) tail-length variants and their relevance to mRNA stability and translational efficiency in cells and tissues from a variety of species, including yeast, *Drosophila*, zebrafish, *Xenopus*, mouse, and human. These genome-wide studies suggest that the regulation of poly(A) tails in oocytes shapes the translatomic landscape (Lim et al. 2016).

In the absence of transcription, the completion of meiosis and early embryo development in mammals relies significantly on maternally synthesized RNAs and proteins (Curtis et al. 1995; Nothias et al. 1995; Brandhorst 1985).

At the onset of the first meiotic division, nuclear envelope breaks down [NEBD; G2/M transition, chromosomes condense, and a bipolar spindle forms from the microtubule-organizing centers (Schuh and Ellenberg 2007)]. During meiosis I, the spindle migrates from the center of the oocyte to the cortex, and the oocyte undergoes an asymmetric division resulting in a large egg competent for fertilization and a relatively small polar body. Correct positioning of the spindle during asymmetric cell division ensures the successful partitioning of the cellular determinants (Kusch et al. 2003). Stored RNA is used for cellular divisions and functions during the period between transcriptional arrest and the transcriptional resumption that occurs after the maternal-to-embryonic transition.

While protein synthesis is not required for NEBD in the fully grown oocyte (Schultz and Wassarman 1977; de Vantéry et al. 1996), it is a prerequisite for normal meiotic progression and early embryo development. Temporal translation of stored maternal messages, however, has been evolutionarily conserved as a mechanism to ensure a continuing supply of protein to the actively metabolizing oocyte, pacing the oocyte maturation and transition to the embryo. Stage-specific translation during transcriptional silence, coupled with the turnover and degradation of some proteins synthesized in the growing oocyte, can significantly alter the protein content of the maturing oocyte and early embryo development (Bachvarova 1981; Su et al. 2007).

12.2 RNA Localization in the Mammalian Oocyte

RNA localization is observed prior to the appearance of protein, consistent with the view that transcript localization generally predetermines protein distribution at most subcellular destinations. Subcellular mRNA trafficking has been demonstrated as a mechanism to control protein distribution, although it is generally believed that most protein localization occurs subsequent to translation. The localization of mRNA within a cell is an essential prerequisite for the correct transfer of genetic information, and it is also a very efficient way to orchestrate cellular processes. In many species, including *Drosophila* and *Xenopus*, synthesis of proteins is localized by compartmentalization of mRNAs (Dubowy and Macdonald 1998; King et al. 2005; Holt and Bullock 2009). This has been seen to be critical, e.g., for the determination of the animal and vegetal poles of *Xenopus* embryos, which requires the accurate and asymmetric distribution of several mRNAs (Nieuwkoop 1985).

In situ hybridization analysis of the subcellular distribution of thousands of mRNAs during *Drosophila* oogenesis and early embryogenesis has shown that the majority of mRNAs are subcellularly localized (Lécuyer et al. 2007; Jambor et al. 2015). A number of new and striking localization patterns have been observed. The expression of mRNA is restricted to subsets of blastomeres and mRNAs that are asymmetrically localized in the cytoplasm or to the nuclei of cells. Jambor et al. (2015) and Lécuyer et al. (2007) have discovered that dozens of detected mRNAs are limited to only a small number of subcellular domains, e.g., spindle, nucleus, and perinuclear region, or they are membrane associated. The variety of subcellular mRNA localized implies an equivalent variety of localization mechanisms. Tight correlations between mRNA distribution and subsequent protein localization and function indicate major roles for mRNA localization. Localization of mRNA linked to local translation is a fundamental mechanism for the successful passing on of genetic information to the next generation, as well as having an effective influence on cellular processes. Spatial segregation of protein synthesis in cells is crucially dependent on the correct positioning of mRNAs. This asymmetrical distribution of mRNA might be more thermodynamically efficient than transporting proteins, since fewer RNA molecules need to be moved. It is also possible that spatially controlled translation offers a more subtle control of local protein activity (Weatheritt et al. 2014) in comparison to other mechanisms. Furthermore, proteins synthesized locally are structurally and functionally distinct from the transported proteins: they are more likely to contain domains that promote protein-protein interactions and are subject of a tighter regulation of expression and of more extensive posttranslational modification (Weatheritt et al. 2014). The fully grown oocytes of many invertebrate and nonmammalian vertebrate species are not only

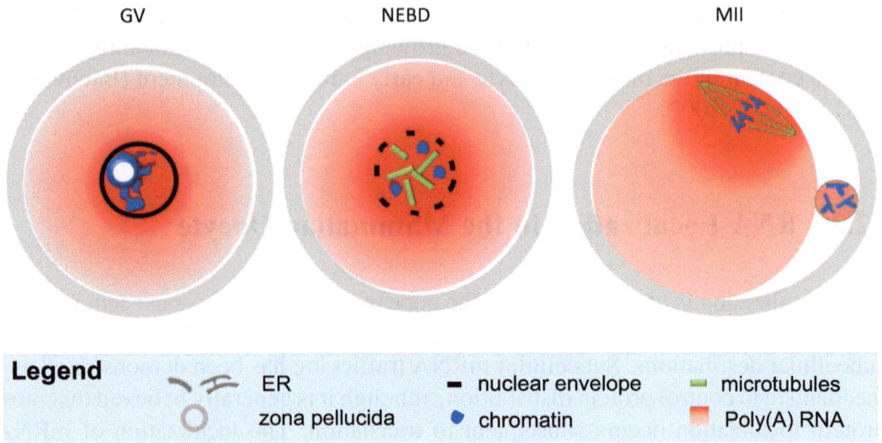

Fig. 12.1 Scheme of localization of poly(A) RNA population in mouse oocytes in the GV, NEBD, and MII stages. Nucleus of a fully grown oocyte shows the high poly(A) RNA signal that is retained in the chromosomal area and disappears as the oocyte proceeds through meiosis to the MII stage (adapted from Susor et al. 2016)

asymmetrical but also polar in the distribution of their localized RNAs, as well as their organelles and proteins. Oocyte polarity also dictates the patterning of the future embryo in lower organisms. Little, however, is known about this patterning in mammalian oocytes and early embryos through the localization of mRNAs, except for the reported accumulation of RNA in the cortex of the oocyte (Li et al. 2008; Flemr et al. 2010), active polysomes localized at the meiotic spindle (Romasko et al. 2013) and nucleus (Kopecný et al. 1995; Susor et al. 2015). Examples of known localization of global poly(A) and specific RNAs in the mouse oocyte are shown in Figs. 12.1, 12.2, and 12.4a.

12.2.1 Growing Oocyte Asymmetry: The Balbiani Body

In humans, as in the mouse, the Balbiani body (BB) is present as a perinuclear aggregate that includes mitochondria, ER, granulo-fibrillar material (GFM), and other organelles and is transiently present in a spherical conformation (Gondos 1987). The BB is polarly located within the oocytes of many species. In *Xenopus*, BBs are known to be associated with the germinal granules responsible for the determination of germ cell fate. BBs are also present in developing oocytes of mammals including humans (de Smedt et al. 2000; Pepling and Spradling 2001; Kloc et al. 2004; Pepling et al. 2007). A fully developed and easily recognizable BB is already present in the oocytes of E17.5 mouse embryos and persists postnatally till day 3. It disperses throughout the cytoplasm in the majority of oocytes from the cyst-stage germ cells to the newly forming primordial follicles (Jefferson et al. 2006; Kloc et al. 2008; Marlow 2010). A connection of BB to germinal

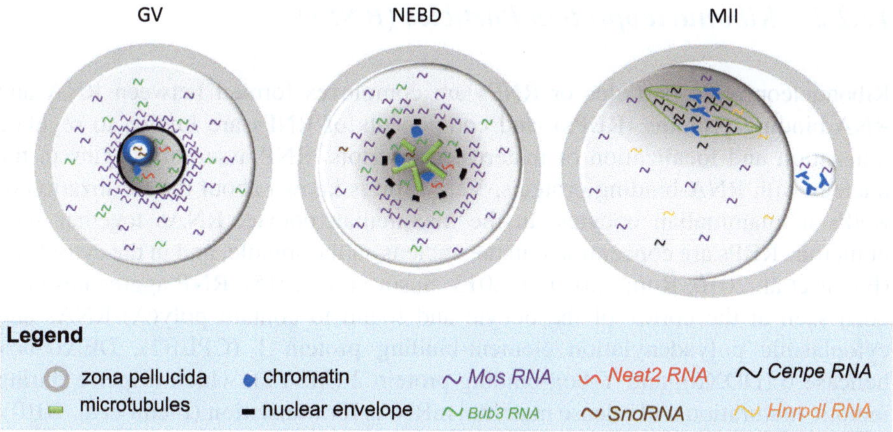

Fig. 12.2 Example of localization of the specific RNAs in the mouse oocyte through meiosis. The nucleus of a fully grown oocyte containing a pool of specific mRNAs and noncoding RNAs, such as *Neat2 (Malat1)* or *Bub3* mRNA; the latter is also present in the cytoplasm, while *Mos* mRNA is localized exclusively in cytoplasm. The oocyte shows enrichment of *Centromere protein E (Cenpe)* and *hnRNP (heterogeneous nuclear ribonucleoprotein)* mRNAs at the spindle area. Scheme is based on the studies of (VerMilyea et al. 2011; Romasko et al. 2013; Susor et al. 2015). Adapted from (Susor et al. 2016)

granules has been proven in *Xenopus*, where BB has been shown to contain germinal granule proteins and RNAs (Heasman et al. 1984; Kloc and Etkin 1995, 2005; Kloc et al. 2002a, b; Zhou and King 2004) that are incorporated into germ cells (Forristall et al. 1995; Kloc et al. 1998, 2002a). Mouse cyst-stage germ cell BB contains LSM14A protein, also known as Trailerhitch in *Drosophila* (Pepling et al. 2007), which is known to be involved in RNA metabolism. Despite the presence of RNPs in BBs, their fate remains elusive. Although the germ cell-specific mRNAs, such as *Vasa, Nanos, Gasz*, and *Dazl*, localize to the BB of lower vertebrates, BB or polarized localization of these mRNAs in the early oocytes of mammals has not been observed (Houston et al. 1998; Maegawa et al. 1999; Houston and King 2000; Knaut et al. 2000; Mita and Yamashita 2000, McNeilly et al. 2000, Kloc et al. 2001, 2002a, Yan et al. 2002, 2004; Kosaka et al. 2007). One explanation for the absence of localized germ plasm mRNAs in mammalian oocytes could be the different mode or developmental timing of primordial germ cell specification in mammals compared to *Xenopus* and zebrafish. Published studies provide evidence that mouse oocytes develop using molecular and developmental mechanisms widely conserved throughout the animal kingdom. The developmental requirement of the BB for the mammalian oocyte remains to be determined.

12.2.2 Ribonucleoprotein Particles (RNPs)

Ribonucleoprotein particles or RNPs are complexes formed between RNA and RNA-binding proteins (RBPs), and components of RNPs are known to regulate translation and localization of maternal transcripts. RNP assembly begins in the nucleus with RNA-binding proteins. Very little is known about the localization of RNPs in mammalian oocytes. In the mammalian oocyte, RNAs together with numerous RBPs are concentrated in the nucleus, at the spindle, and in the cytoplasm (Flemr et al. 2010; Romasko et al. 2013; Susor et al. 2015). RNP aggregates have been seen at the cortex of the oocyte and found to contain poly(A) RNAs and cytoplasmic polyadenylation element-binding protein 1 (CPEB1), DEAD-box helicase 6 (DDX6), and Y-box-binding protein 2 (MSY2), which disperse during meiotic maturation and release maternal mRNAs for translation (Flemr et al. 2010). The storage of ribosomal components is a prerequisite for the meiotic and developmental competence of the mouse oocyte (Monti et al. 2013). It has been proposed that storage of maternal RNPs might be organized in the cytoplasmic lattices (CPLs), a large supramolecular fibrillar complex composed of maternal RNAs, ribosomes, and microfilaments (McGaughey and Capco 1989; Capco et al. 1993). Several pieces of evidence suggest that inactive ribosomes are actually embedded in the oocyte CPLs. Biochemical data in the mouse indicate that 75–80% of ribosomes are not incorporated into polysomes at ovulation and do not engage in protein synthesis *in vitro* (Bachvarova and De Leon 1977). A remarkable feature of mammalian oocyte maturation is the significant degradation of rRNA and ribosomes (Clegg and Pikó 1982). Moreover, genome-wide transcriptome analysis has shown that mRNAs coding for ribosomal proteins are degraded during maturation and after fertilization (Zeng et al. 2004; Su et al. 2007). However, transcription in the mouse early embryo becomes effective only after the two-cell stage (Abe et al. 2015). Taken together, specific maternal ribosomal machineries are accumulated during oocyte development and most likely eliminated and replaced by embryonic ones after genome activation.

Kotani et al. (2013) have observed decreased *Cyclin B1 (Ccnb1)* RNA-FISH signals after the resumption of meiosis, without a decrease of *Ccnb1* mRNA, as measured by qRT-PCR. Dormant *Ccnb1* forms RNPs, which disassemble at the time of translational activation during mouse oocyte maturation. Disruption of *Ccnb1* RNA granules causes the acceleration of the timing of their translational activation after induction of maturation, whereas stabilization hinders translational activation. These authors suggest that RNA granule formation is critical for the regulation of timing of translational activation, when the transcript is loaded on the active polysomes, where mRNA-ribosome complex blocking probes hybridize with the target RNA. Moreover, RNPs might change physical state from solid to liquid and vice versa, which in turn might affect detection availability. RNA- and protein-rich germ line granules similarly undergo regulation of RNP transitions among diffuse, semiliquid, or solid states to modulate RNP translation in the cytoplasm of *the C. elegans* oocyte and early embryo. Such phase transitions may represent a fundamental physicochemical mechanism for structuring of the cytoplasm (Brangwynne et al. 2009). Therefore, large-scale organization of gene expression extends to the cytoplasm, where regulation of supramolecular states leads to specific patterns of RNP dynamics. Such phase structuring may represent a mechanism for functional assembly/disassembly of the specific RNPs in the mammalian oocytes and embryos.

12.2.3 Subcortical Maternal Complex (SCMC)

Populations of RNA-binding proteins (RBPs) and RNAs have been identified at the subcortical region of the cytoplasmic membrane of the mammalian oocyte (Ohsugi et al. 2008; Herr et al. 2008; Li et al. 2008, 2010, 2013; Flemr et al. 2010) and are believed to play a role in mRNA storage and metabolism. Subcortical complexes share components with P-bodies (processing bodies) including several RNA-binding proteins, such as DDX6, MSY2 (YBX2), and CPEB1 (Flemr et al. 2010). Consistent with their function as a storage compartment, they detach from the cortex upon resumption of meiosis, relocate toward the center of the oocyte, and dissolve. One of the components of the SCMC is the translational repressor DDX6 (Rck/p45) which is implicated in the control of maternal mRNAs in lower organisms (Weston and Sommerville 2006). DDX6 orthologues are found in a number of different RNA granules regulating mRNA stability and translation throughout female germ cell development and in the soma (Weston and Sommerville 2006). SCMC contains MATER (mater protein homolog), together with FILIA (KH domain containing 3 like, subcortical maternal complex member), FLOPED (HOEP19, oocyte-expressed protein), peptidyl arginine deiminase 6), and TLE6 (transducin-like enhancer of split 6) proteins, and this complex has been shown to be essential for preimplantation embryo development beyond the two-cell stage. The role of the SCMC complex in mRNA metabolism is not clear, although it is known that FLOPED is a putative RNA-binding protein (Pierre et al. 2007; Herr et al. 2008; Li et al. 2008). Absence of the SCMC significantly impedes development beyond the two-cell-stage embryo, and it seems likely that defects arise earlier, since the progression from one to two cells is delayed and initial cell division is often asymmetrical. This could result from abnormalities in syngamy, mitotic spindle formation (loss of FILIA causes defects in spindle assembly), cytokinesis, or cell cycle progression and could affect later events in embryonic development including axes formation, a subject of intense investigative interest (Rossant and Tam 2004; Louvet-Vallée et al. 2005; Motosugi et al. 2005; Kurotaki et al. 2007; Bischoff et al. 2008). Downregulation of oocyte- and embryo-restricted SCMC component PADI6 shows that maternal ribosomes are not able to properly store subsets of transcripts in the cytoplasmic lattice structures, which results in a reduction of total de novo protein synthesis at the two-cell stage (Yurttas et al. 2008).

12.2.4 Transzonal Projections and RNPs

SCMCs are localized directly at the oolemma, where cumulus cells have cell protrusions with gap junctions which facilitate direct intercellular communication between the oocyte and granulosa cells. This communication is bidirectional and involves the regulation of nutrient transfer and signaling between the two very different cell types. A number of findings have suggested that miRNAs from somatic cells participate in intrafollicular signaling and influence oocyte developmental potential. It has been shown that bovine cumulus cells and oocytes reciprocally affect the abundance of miRNAs in cellular compartments (Abd El Naby et al. 2013; Kloc and Kubiak 2017) and during different oocyte developmental stages. Age-associated changes in the human miRNA profile have also been suggested to influence the cumulus gene expression required for oocyte quality (Russell et al. 2016). Thus, SCMC might physically interact with this structure in the oocytes of various mammalian species. The SCMC might have a role in the RNA shuttling and harboring of imported RNPs (Macaulay et al. 2014, 2016) and other factors involved in the regulation of meiotic competence connected with the extracellular matrix.

Oocyte maturation is mediated by communication events with the cumulus cells that ensure oocyte quality for maturation, fertilization, and embryo development. Cumulus cells are in direct contact with the oolemma through cellular extensions known as transzonal projections (TZPs). TZPs and their potential role in the transfer of RNA to the oocyte have been previously described, especially for miRNA transition across gap junction channels (Katakowski et al. 2010; Zong et al. 2016). Valiunas et al. (2005) have proved that synthetic oligonucleotides (12-, 16-, and 24-mer) can selectively pass through gap junction channels. Larger RNA molecules produced from the cumulus cell nuclei have also been recorded traveling to the oocytes through the TZPs (Macaulay et al. 2014, 2016; Kloc and Kubiak 2017). Labeled RNP structures and vesicles have been observed at the junction of the two membranes. In addition, lncRNAs have also been identified in the cumulus cell TZPs and linked with bovine embryo quality (Caballero et al. 2014). This provision of RNPs from cumulus cells might also be utilized during the oocyte growth phase in other mammalian species. Similarly, RNPs shuttling between two cell types might be used during oocyte maturation, especially in species with a longer maturation time, e.g., human (35 h; Coticchio et al. 2012). These novel findings on the exchange of RNPs between the oocyte and its surrounding somatic cells, the cumulus cells, might represent another level of meiosis regulation and as such contribute to the subsequent developmental capacity of the oocyte.

The transfer of large structures like RNPs during oocyte growth and at the initiation of maturation may have significant consequences upon the quality of the oocyte. Taking into account that the proteome and the transcriptome of the oocyte may be modified by the cumulus cell transfer of RNAs and proteins, this could have a significant impact upon the way we view oocyte competence. It remains to be determined how these RNA transcripts in cumulus cells influence the developmental potential of the oocyte.

More study is needed directed to elucidation of the function of this RNA transfer, since the results obtained so far on the observed transfers of small and long RNAs have not described where these RNAs go and how, or if indeed, they are used. Clearly, the fully grown oocyte is able to produce an enormous amount of self-made RNA molecules; however, a qualitative contribution from transcriptionally active surrounding cells could improve the efficiency of the biological processes in the oocyte in the cell niche environment.

12.2.5 RNA Localization in the Nucleus

Recent studies have applied deep sequencing of transcriptome from cellular fractions to identify RNAs that are retained in the nucleus (Bhatt et al. 2012; Tilgner et al. 2012; Djebali et al. 2012; Pandya-Jones et al. 2013). These, however, have predominantly included long noncoding RNAs (lncRNA), such as *X inactive specific transcript (Xist)*, *nuclear paraspeckle assembly transcript 1 (Neat1)*, and *Neat2 (Malat1, metastasis-associated lung adenocarcinoma transcript 1)*, hyper-edited mRNA (Chen and Carmichael 2009), and incompletely spliced mRNA (Boutz et al. 2015), rather than mature protein-coding mRNAs. An earlier study has indicated that, in many cell types, a considerable fraction of poly(A) RNA (as much as 30%) is nuclear-retained and undetectable in the cytoplasm (Herman et al. 1976). More recently, RNA fluorescence in situ hybridization (RNA-FISH) analysis using an oligo dT probe to visualize poly(A) RNA population in mammalian cells has shown that a population of poly(A) RNA is enriched in nuclear speckles, also known as interchromatin granule clusters (Carter et al. 1991; Visa et al. 1993). The previous examples are from somatic cells in which nuclear RNA presence might be associated with active transcription. Developmentally competent fully grown mammalian oocytes are naturally transcriptionally silent; therefore, any presence of RNA in the nucleoplasm is most likely originating from a previous transcriptionally active period or imported from the cytoplasm.

Mammalian oocytes acquire meiotic/developmental competence to resume and complete meiosis; the chromosomes condense and cluster on the nuclear and surrounding nucleolar-like body (NLB, further as a nucleolus), which is manifested as a shift from NSN oocytes (non-surrounded nucleolus) to SN oocytes (surrounded nucleolus, fully grown oocyte). A general hypothesis has been proposed that only NSN oocytes actively synthesize rRNA, whereas the SN-type oocytes are transcriptionally silent (Bouniol-Baly et al. 1999; Pesty et al. 2007). The last period of transcriptional activity in growing mouse oocytes seems to be important for the full developmental competence of the oocytes, as indicated by analyses of these two categories of oocytes (Zuccotti et al. 2002; Inoue et al. 2008; Zuccotti et al. 2011).

Maternal nucleolus is important for the development of oocytes and embryos and may serve as a marker for the completion of oocyte growth and the normality of activation of the embryonic genome. The nucleolus, which is the most prominent nuclear organelle, is the site of 45S rRNA synthesis through the transcription of the 18S/5.8S/28S rRNA (ribosomal RNA) genes, as well as of the subsequent rRNA processing and RNP assembly. Indeed, the only active genes in the nucleolus are the rRNA genes, and the only RNA polymerase is RNA polymerase I (RNA pol I). rRNA functions as a scaffold for the dozens of ribosomal proteins composed of rRNA forming two unequal subunits of eukaryotic ribosomes, designated small subunit (40S) and large subunit (60S) according to their sedimentation coefficients. Nucleolus of a fully grown oocyte contains various RNA species (Kopecny et al. 1996a, b; Shishova et al. 2015a, b). Kopecny et al. (1995) detected RNPs in the nucleolus of porcine and rat oocytes containing m3G-capped snRNA. The maternal nucleolus contributes to the storage of maternal nucleoplasmic spliceosomal components, which remain to be concentrated in the nuclear area during the disappearance of the nucleolus at least for the first 2 h and become diluted into the cytoplasm in the MII oocytes (Vautier et al. 1994). These authors have also shown that nuclear RNPs, released post NEBD, reincorporate into both the male and female pronucleus. Aggregates of RNA inside nucleolus of fully grown mouse oocytes have been detected (Pochukalina et al. 1998; Pochukalina and Parfenov 2008) with the aggregates corresponding to local accumulations of ribosomes.

Retention of mRNA within the nucleoplasm has been described in plants and *Drosophila* (Lécuyer et al. 2007; Boothby and Wolniak 2011; Boothby et al. 2013; Göhring et al. 2014; Niedojadło et al. 2014; Bahar Halpern et al. 2015).

Genome-wide transcriptome analysis shows a substantial fraction (13.1% in liver cells and 30% in the MIN6 cells) of the lncRNAs and mRNAs in the nuclei of the cells (Bahar Halpern et al. 2015). It seems that nuclear retention of mRNA is a significant, previously underappreciated step in the mRNA life cycle. However, an example of this phenomenon has been provided in mammalian oocytes only recently, when endogenous poly(A) RNAs have been observed in the nucleus of a fully grown mouse GV oocyte (Figs. 12.1 and 12.4a, Susor et al. 2015). Poly(A) RNAs are distributed diffusely in the nucleus of growing oocytes, in comparison with fully grown oocytes, where the signal is clustered. Previous data has clearly shown that some transcripts, including *Bub3 (budding uninhibited by benzimidazoles 3*, mitotic checkpoint component), *Npml (nucleophosmin*, involved in ribosome biogenesis, centrosome duplication), *Survivin (baculoviral IAP repeat containing 5*, component of a chromosome passage protein complex), *Dazl (deleted In azoospermia like*, germ cell-specific RNA-binding protein), and *Pabpnl (poly(A)-binding protein, nuclear 1*; controls the size of the poly(A) tail of RNA) mRNAs, are equally distributed in both the nucleus and the cytoplasm, while other transcripts, such as *Mos (oocyte maturation factor Mos*), *Gapdh (glyceraldehyde-3-phosphate dehydrogenase*), *Tuba (tubulin alpha 3a*), *mTOR (mechanistic target of rapamycin*, component of mTORC1), *Eif4e (eukaryotic translation initiation factor 4E*, mRNA cap-binding protein), and *Camk2a (calcium-/calmodulin-dependent protein kinase II alpha*), are abundant in the cytoplasm and absent from the nucleus (Susor et al. 2015). Furthermore, other transcripts known to be localized in the nucleus in somatic cells, such as noncoding RNAs (*Malatl*, *U2*, and *U12*), as well as *Pabpnll* (Watson et al. 1992; König et al. 2007) mRNA, have been found in the nucleus of mouse oocytes. Apart from the specific mRNAs detected in the oocyte nucleus mentioned above, a pool of poly(A) mRNAs (as classified by Yang et al. (2011) according to the presence or absence of a poly (A) tail at their 3′ termini) has been also found to be localized in the nucleus; however, the identity of these mRNAs has yet to be determined.

After NEBD, a strong poly(A) RNA signal is present in the vicinity of the chromosomes, in the region of spindle formation, in contrast to the cytoplasm, where the signal is significantly lower (Figs. 12.1 and Fig. 12.4a, Susor et al. 2015).

What could be the role of nuclear retention of mature mRNAs in the transcriptionally silent cell? Nuclear RNA localization might serve as a means of a precise timing or coordination of protein expression in correlation to cellular processes occurring during mammalian oocyte meiosis I. Although this mechanism of translational control has not been documented before, it may prove to be a relatively common form of posttranscriptional gene regulation. The cell nucleus holds on to mRNA molecules that are not immediately needed to produce proteins but whose cytoplasmic presence is rapidly required upon specific cellular events such as NEBD. Observed nuclear localization of RNAs might represent a mechanism by which the cell ensures temporal and spatial translation of mRNAs important for the onset and progression of the dynamic processes of meiosis, especially spindle assembly after NEBD and asymmetric cytokinesis. The oocyte nucleus might serve as a reservoir of transcripts, which positively correlate with protein localization at the spindle forming area during the meiotic cell cycle (Amin et al. 2008; Li et al. 2009). This mechanism of action highlights an important paradigm for the essential role of the nucleus-retained stable transcripts in regulating gene expression (Prasanth et al. 2005; Bhatt et al. 2012; Carrieri et al. 2015).

The presence of mRNAs in the nucleus might be the result of the posttranscriptional retention of these mRNAs, or of a bidirectional process, when such mRNAs exported from the nucleus might be imported back again. In the *Drosophila* oocyte, Jambor et al. (2015) have shown that nuclear patterns of mRNA varied from a ring-like signal to dispersed foci or widespread distribution in the nucleoplasm and are not linked to the chromosomal position of the genes. The mechanisms responsible for either of these events are currently unknown in mammalian oocytes. The abovementioned transcripts are present in both cellular compartments, i.e., the nucleus and cytoplasm; however, we might expect to see specific mRNAs which occur exclusively in the nucleoplasm. This mRNA localization might lead to oocyte mRNA polarity similarly as in lower organisms (*Drosophila*, *Xenopus*). Nucleus-retained mRNAs might be provisionally excluded from translational machinery. Importantly, it has been shown that mRNA is not translated following injection into the *Xenopus* oocyte nucleus but is translated after NEBD (Braddock et al. 1994), which suggests a mechanism by which some transcripts are being sequestrated in the nucleus and become translated only after NEBD.

These examples suggest that nuclear mRNA retention may serve as a way to orchestrate protein expression in various cellular processes (Kambadur et al. 1998; Pilot et al. 2006; Brandt et al. 2006; Grosskortenhaus et al. 2006). This mechanism seems to be a relatively common form of posttranscriptional gene regulation. It is conceivable that in oocytes the RNA population retained in the nucleus might contribute to translation after NEBD at the dynamic area of the cell, where new spindle is assembled, enabling the segregation of meiotic chromosomes.

12.2.6 RNA Localization at the Spindle

Many of the most striking localization of RNA subcellular patterns observed in *Drosophila* oogenesis and early embryogenesis occur during cellular division (Lécuyer et al. 2007; Jambor et al. 2015). They show specific transcripts to be localized to spindle poles, centrosomes/centrioles, astral microtubules, around metaphase chromosomes, or along the mitotic spindles during anaphase and telophase. The genes in these categories show gene ontology term enrichments for cell division-related processes, implying an important role for localized mRNAs in the establishment, function, and regulation of cell division machineries. Very little is known about the programmed sorting of mRNAs, which produce characteristic and distinguishable transcriptomes on the spindle of the mammalian oocyte. Only a few studies have focused on MII mouse oocytes, such as the analysis of enrichment for maternal mRNAs encoding spindle-related and other proteins on the mouse oocyte metaphase II (MII) spindle (VerMilyea et al. 2011; Romasko et al. 2013). In order to elucidate possible enriched localization of mRNAs on the MII spindle in mouse oocytes, the authors have isolated samples of spindle chromosome complex by microsurgery followed by microarray analysis. The results show that the most prominently represented categories include proteins associated with the plasma membrane, chromatin/nuclear, signaling, and, as expected, spindle/cytoskeletal functions, followed by vesicle/endocytosis/protein transport, Golgi and endoplasmic reticulum, ubiquitination and protein degradation, and RNA binding. Spindle complexes are sites of targeted polysomal mRNA localization in diverse systems (Blower et al. 2007; Mili and Macara 2009), and spindles characteristically orchestrate key elements of mouse oocyte maturation (Brunet and Verlhac 2011). The presence and translation of mRNAs support spindle formation, maintenance, and function. Localized maternal mRNAs could be translated *in situ* to provide a local high concentration of proteins while minimizing potential deficiencies related to limitations in the speed or extent of protein accumulation from elsewhere within the ooplasm.

An attractive hypothesis suggests that the RNA pool in the nucleus remains in the "spindle formation area" post NEBD even after the disruption of the barriers between the nucleus and the cytoplasm, providing a local collection of very specific transcripts supporting various spindle functions. One can speculate that asymmetric targeting and translation of developmentally regulated mRNAs to one spindle pole could be a mechanism by which asymmetry is established in the mouse oocyte after reaching spindle bipolarity. Localization of specific mRNAs to the spindle might be evolutionarily conserved between mammals and other vertebrates, possibly serving as the mechanism for enhancing protein localization.

12.3 Translation in Mammalian Oocyte

Selective protein synthesis of germ cell determinants is governed temporally and spatially by both sequence-specific mRNA repression and translational activation. In order to control the proteome of a developing fertilizable oocyte, translation becomes an effective tool in the transcriptionally silent cell.

Temporal control of mRNA translation is an important mechanism for regulating cellular, neuronal, and developmental processes. However, the details of the mechanisms that coordinate the timing of translational activation remain largely unresolved. Fully grown oocytes arrest meiosis at prophase I and deposit dormant mRNAs. Translational control of mRNA translation relies on cis-acting elements, which provide binding sites for trans-acting factors. The cis-acting element-mediated cytoplasmic polyadenylation of dormant mRNAs drives the translational activation of the mRNAs (McGrew et al. 1989; Vassalli et al. 1989; Sheets et al. 1994; Stutz et al. 1998). To control cellular processes during cell cycle, the reprograming of translation to synthetize specific proteins might significantly contribute to the cellular physiology (Figs. 12.3 and 12.5).

12.3.1 Cap-Dependent Translation in Mammalian Oocyte

The control of translational initiation of specific cellular mRNAs occurs in a cap-dependent manner by the cap-binding protein complex eIF4F. This complex consists of three subunits: eIF4E, which specifically recognizes the cap structure of the mRNA; eIF4A, a helicase which unwinds mRNA secondary structures; and eIF4G1, a bridging protein responsible for the integrity of the eIF4F complex (Mader et al. 1995). The most important and limiting factor seems to be the cap-binding protein, eIF4E. Strictly eIF4E-dependent mRNAs, so-called eIF4E-sensitive transcripts, often possess long, highly structured 5′-untranslated regions (5′-UTR) or 5′-terminal oligopyrimidine (TOP) motifs (Dowling et al. 2010; Thoreen et al. 2012) and encode mostly regulatory proteins. Thoreen et al. (2012) show that mechanistic target of rapamycin complex 1 (mTORC1) almost entirely regulates the translation of transcripts with established 5′ terminal oligopyrimidine (TOP) motifs. 4E-BP1 inhibits translation initiation by interfering with the interaction between the cap-binding proteins eIF4E and eIF4G1. Loss of this interaction diminishes the capacity of eIF4E to bind TOP and TOP-like mRNAs much more than other mRNAs, explaining why mTOR (key component of mTORC1) inhibition selectively suppresses their translation. On the other hand, housekeeping proteins have short and unstructured 5′-UTRs and are effectively translated even

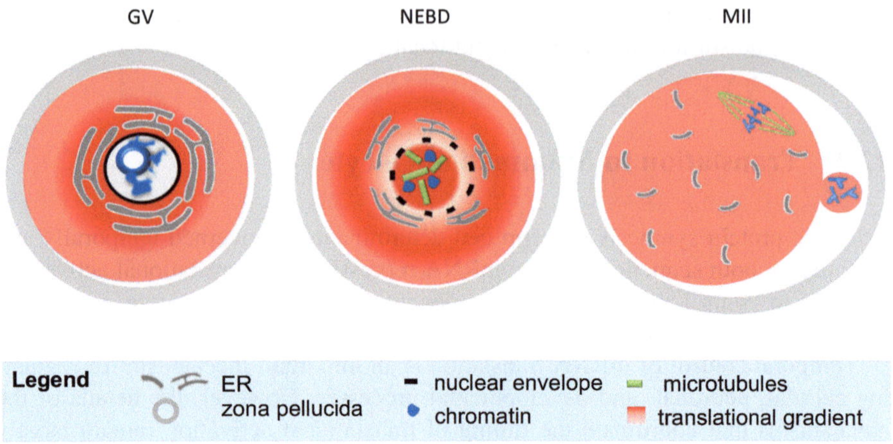

Fig. 12.3 Scheme of in situ translation in the GV, NEBD, and MII stages. A strong translational signal is present in the perinuclear region in GV stage; two translational hotspots (*CTA* chromosomal translational area; and *PTA* perispindular translational area) develop post NEBD that disappear after polar body extrusion in the MII stage (adapted from Susor et al. 2016)

when eIF4E is limited, for instance, by the action of repressive eIF4E-binding proteins (4E-BPs, see below).

EIF4E is a phosphoprotein with a single phosphorylation site at Ser209, regulated via MAPK and MNKs (Imataka et al. 1998; Ellederová et al. 2008). However, the impact of eIF4E phosphorylation on translational stimulation remains controversial. There is a growing body of evidence, however, that not phosphorylation per se but the turnover of the phosphorylation enhances translation (Scheper and Proud 2002; Tomek and Wollenhaupt 2012). The eIF4E engagement within the eIF4F complex is also controlled via the regulatory proteins binding to eIF4E, i.e., the 4E-BPs. Three isoforms (4E-BP1–4E-BP3) have been described in the somatic cell; however, only the 4E-BP1 protein is abundant in mouse and bovine oocytes (Mayer et al. 2014; Susor et al. 2015; Jansova et al. 2017). 4E-BP1 has to undergo phosphorylation to dissociate from eIF4E and in such a way to enable its coupling with eIF4G1 and the formation of the functional eIF4F complex (Gingras et al. 1999). The kinase responsible for phosphorylation of 4E-BP1 on several sites is mTOR, which itself is regulated by the PI3K/AKT signaling pathway (Scheper and Proud 2002).

Two different mTOR complexes have been described, which are associated with two different regulatory proteins: Raptor and Rictor. MTORC1 represents the complex of mTOR with Raptor, which is sensitive to rapamycin and is responsible for 4E-BP1 and ribosomal protein S6 kinase (S6K) phosphorylation. However, rapamycin-resistant phosphorylation of 4E-BP1 has been reported in several cell lines (Huynh et al. 2009; Dowling et al. 2010; Livingstone and Bidinosti 2012), and the effect of rapamycin treatment on translational control in bovine oocytes is currently not clear. Nevertheless, exposure of bovine oocytes to high doses of

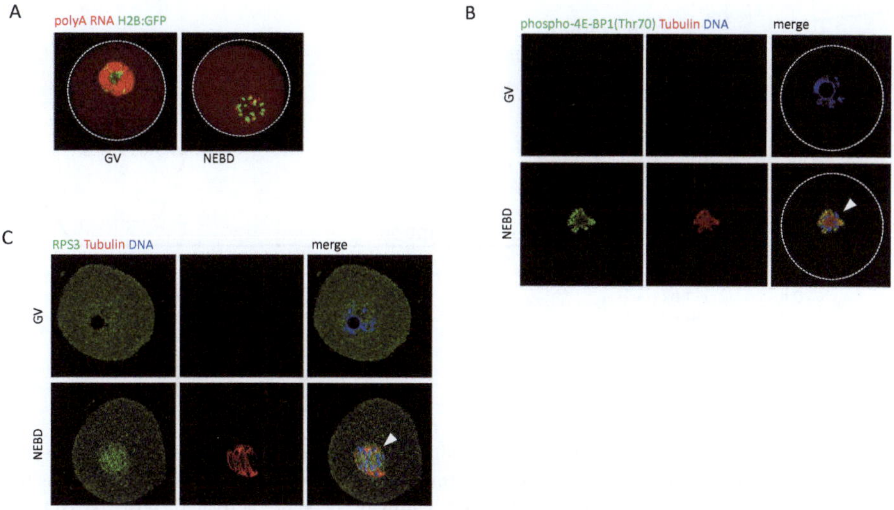

Fig. 12.4 Localization of the poly(A) RNA, inactive 4E-BP1, and ribosomal protein S3 in the mouse oocyte. (**a**) Live cell imaging shows a strong poly(A) RNA signal (molecular beacons *red*) present in the nucleus of the GV oocyte or in the vicinity of chromosomes (GFP *green*) in post-NEBD oocyte. (**b**) Phosphorylation of the suppressor of the cap-dependent translation 4E-BP1 leads to its inactivation at the newly forming spindle (tubulin *red*) depicted by arrowhead in the NEBD stage. In the GV oocyte, only non-phosphorylated 4E-BP1 is present. (**c**) Ribosomal protein S3 is localized evenly in the cytoplasm of the GV and NEBD oocytes with higher concentration at the newly forming spindle (depicted by *arrowhead*) in the NEBD oocyte (RPS3 *green*, tubulin *red*, and DNA *blue*). Interrupted *line* marks oocyte cortex.

rapamycin (Mayer et al. 2014) inhibits spindle migration and asymmetric division (in accordance to mouse oocytes, see Lee et al. 2012). Inhibition of mTOR has been found to prevent dissociation of 4E-BP1 from the initiation factor eIF4E and to correlatively suppress a burst of cap-dependent protein synthesis occurring at the resumption of meiosis (Figs. 12.4b and 12.5).

Although PLK1 is likely to phosphorylate 4E-BP1 during mitosis (Shang et al. 2012), and CDK1 kinase has been shown in mitosis to phosphorylate 4E-BP1 on Ser65 and Thr70 (Heesom et al. 2001; Jansova et al. 2017), phosphorylation of these sites requires priming phosphorylation on Thr37 and Thr46, which has also been reported to be mediated by mTOR (Gingras et al. 1999). These clear examples of involvement of the mTOR/4F axis in translational regulation during mitosis might be used as a model case for the meiotic cell. Increased phosphorylation of 4E-BP1 has also been seen during the meiotic progression of mammalian oocytes (Tomek et al. 2002; Ellederova et al. 2006), and different phosphorylated forms of 4E-BP1 have been shown to co-localize with the meiotic spindle in mouse oocytes (Romasko et al. 2013). In conclusion, mTOR appears to be of crucial importance in the formation of the active eIF4F complex, which stimulates the translation of eIF4E-sensitive mRNAs. Reported blocking of 4E-BP1 phosphorylation during

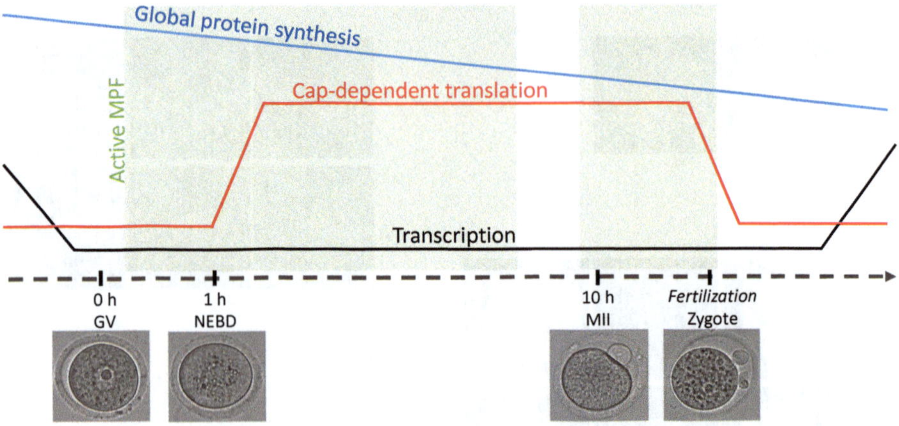

Fig. 12.5 Scheme of protein synthesis during meioses I and II in the mammalian oocyte. The fully grown oocyte is transcriptionally silent and becomes active after fertilization (*black line*). Positive regulators of the cap-dependent translational pathway become activated post NEBD and inactivated after fertilization (*red line*). In contrast, global protein synthesis gradually decreases during oocyte meiosis and early embryo development (*blue line*). Scheme is based on the studies of (Ellederova et al. 2006, 2008; Susor et al. 2008, 2015)

cultivation has resulted in the irreversible arrest of metaphase I in bovine oocytes (Mayer et al. 2014) and in abnormally formed spindles in MII mouse oocytes (Susor et al. 2015).

A similar study in *Drosophila* has shown a direct involvement of regulatory-associated protein of MTOR complex 1 (Raptor)and Ras homolog enriched in brain (RheB)which activates the protein kinase activity of mTORC1, two proteins of the mTOR complex, in chromosome misalignment (Goshima et al. 2007). Rapamycin treatment causes defects in meiotic spindle formation (mouse oocytes; Lee et al. 2012; Susor et al. 2015; Jansova et al. 2017), prevents second polar body emission (starfish, Lapasset et al. 2008), and induces chromosome mis-segregation in yeast and mammalian cells (Bonatti et al. 1998), demonstrating that a rapamycin-sensitive pathway is important for regulation of these important morphological processes during meiosis. mTOR pathway activity is enhanced during the M-phase of the mitotic and meiotic cell cycle (Kosaka et al. 2007; Vazquez-Martin et al. 2009; Yu et al. 2011; Susor et al. 2015; Shuda et al. 2015). M-phase-specific roles for mTOR signaling have been demonstrated by Yaba et al. (2008) and Gwinn et al. (2010). Indeed, an autophosphorylated form of mTOR has been localized to midzone microtubules and the midbody of the mitotic apparatus in cancer cells (Vazquez-Martin et al. 2009). mTOR and CDK1 are regulators of 4E-BP1 in the oocyte (Susor et al. 2015) and in the cumulus cells (Kogasaka et al. 2013; Jansova et al. 2017). Importantly, as the oocyte enters meiosis, levels of 4E-BP1 phosphorylation are higher than in any other part of the cell cycle. Meiotic phosphorylation of 4E-BP1 on Ser65 and Thr70 by mTOR acts to stimulate cap-dependent translation as the oocyte proceeds though meiosis (particularly after NEBD) and specific localization of the key cap-dependent translation regulatory factors has been observed (Guertin and Sabatini 2007; Romasko et al. 2013).

On the other hand, another 4E-BP1 regulator also emerges, CDK1 (Heesom et al. 2001; Shuda et al. 2015; Jansova et al. 2017). CDK1 is a key component in the initiation of NEBD at a time point for increasing cap-dependent translation (Fig. 12.5). Both kinases mTOR and CDK1 become highly active after the resumption of meiosis in mouse, human, and, also, bovine oocytes (Mayer et al. 2014; Susor et al. 2015; Hampl and Eppig 1995), which is similar to mitosis (Vazquez-Martin et al. 2009). In accordance with Heesom et al. (2001), Jansova et al. (2017) have shown that the main regulator of 4E-BP1 phosphorylation in mouse oocytes is mTOR. On the other hand, CDK1 activity is required in oocytes for full mTOR activation, rather than for direct 4E-BP1 phosphorylation. It is known that mTOR is phosphorylated and activated in mitotic cells by AKT (AKT serine/threonine kinase 1; Makker et al. 2014); however, according to Jansova et al. (2017), it seems that during mammalian meiosis this pathway is not sufficient for full mTOR activation, which is mediated by CDK1.

In mouse oocytes, the suppression of AKT activity delays the resumption of meiosis and is accompanied by a decrease in cyclin-dependent kinase 1 (CDK1) activity. CDK1 is the catalytic subunit of maturation-promoting factor (MPF) whose activity has been reported to be crucial for the meiotic maturation of oocytes (Kalous et al. 2006). AKT plays a substantial role during the progression of meiosis to the MI/MII stage (Kalous et al. 2009). Similar studies on bovine oocytes have demonstrated the involvement of AKT in the regulation of meiotic MI/MII transition (Tomek and Smiljakovic 2005). Kinase localization is a prerequisite for its effect on the substrate, and both kinases (mTOR, CDK1) and substrate 4E-BP1 are localized to the spindle (Susor et al. 2015; Jansova et al. 2017) in the oocyte, indicating local inactivation of the evenly distributed 4E-BP1 (Fig. 12.4b). In addition to translational initiation factors, ribosomal protein S3 (RPS3) and RNA-binding proteins are present at the mitotic (Jang et al. 2012) or newly forming meiotic spindle (Fig. 12.4c). RPS3 knockdown causes arrest in mitotic metaphase (Jang et al. 2012; Romasko et al. 2013), which resembles the effect of mTOR inhibition (Mayer et al. 2014) in the oocyte. The influence of known effector kinases in the inactivation of the translational repressor 4E-BP1 might be essential for the temporal and spatial translation of specific mRNAs at the spindle area to ensure errorless meiotic progression (Fig. 12.5).

12.3.2 Polyadenylation and Regulation of Translation

It is generally accepted that the length of the poly(A) tail of a given mRNA plays an important role in its translation rate, i.e., that mRNA with long poly(A) tail is translationally activated and, reciprocally, mRNA with short poly(A) tail is translationally repressed and can be potentially degraded. Nearly all pre-mRNAs undergo cleavage and polyadenylation in the nucleus, which is tightly connected with transcription and splicing (Martinson 2011; Proudfoot 2011; Nagaike and Manley 2011).

Certain cell types utilize de-adenylation of mRNAs as a tool for the regulation of gene expression. These mRNAs bind protein factors, which ensure their de-adenylation and thus prevent their translation. This is the case in postsynaptic regions of neuronal cells, in early embryonic cells and, especially, in germ cells (Colegrove-Otero et al. 2005; Bramham and Wells 2007; Radford et al. 2008). In oocytes, which are transcriptionally inactive, gene expression is regulated posttranscriptionally through the activation of stored transcripts for protein synthesis. A main mechanism by which translation is regulated in oocytes in a number of species is by the so-called cytoplasmic polyadenylation (CP) of selected transcripts, which enables time- and space-specific translation of the most important signaling molecules of the meiotic cell cycle, e.g., cyclins and Mos (Stebbins-Boaz et al. 1996). Messenger RNAs of such molecules contain in their 3-′untranslated region (3′UTR) at least two cis-factors: the cytoplasmic polyadenylation element (CPE) and the polyadenylation signal, hexanucleotide AAUAAA (which is also important for nuclear polyadenylation) (McGrew et al. 1989). Another cis-factor, which regulates mRNA translation, is the Pumilio-binding element (PBE) which is the site for the binding of regulatory proteins from the Pumilio family (PUF). Two Pumilio homologues have been identified in mammalian cells – PUM1 and PUM2 – and their binding to mRNA results in translational repression via the recruitment of CCR4-POP2-NOT deadenylase, which shortens the poly(A) tail of mRNAs. The motif recognized by the PUF proteins in 3′UTR has the sequence "UGUANAUA," where "N" can be any base (Zamore et al. 1997; Goldstrohm et al. 2006; Kadyrova et al. 2007; Suh et al. 2009). Piqué et al. (2008) have shown that the relative position of CPE vs PBE and vs polyadenylation signal in 3′UTR determines the rate and the timing of polyadenylation and the translation of the given mRNA.

The CPE element plays a specific role in cytoplasmic polyadenylation as it binds the CPE-binding protein, CPEB. So far four members of the CPEB family have been identified in vertebrates (CPEB1–CPEB4), and while CPEB2–CPEB3 share relatively high sequence homology, CPEB1 differs from the other members (Theis et al. 2003; Wang and Cooper 2010). The essential importance of CPEB1-dependent cytoplasmic polyadenylation for the development of germ cells has been documented by Tay and Richter (2001) using CPEB1 knockout mice. In these mice the development of mouse oocytes has been arrested in the pachytene stage by disrupting the synaptonemal complex. Two mRNAs coding for the synaptonemal complex proteins SCP1 and SCP3 contain CPEs in their 3′UTR and interact with CPEB1. In the knockout mice, these mRNAs possess shortened poly(A) tails, and their translation is impaired. The role of CPEB1 during meiotic maturation is less clear. Knockdown of CPEB1 by RNAi using a ZP3 promoter in diplotene-stage mouse oocytes does not block meiosis resumption in these oocytes, which have also been able to emit their first polar bodies, although a decrease of the follicle growth factor GDF9 and spindle abnormalities were observed (Racki and Richter 2006). On the other hand, the inhibition of CPEB1 function in porcine oocytes using microinjection of the dominant-negative CPEB1 mutant results in the inhibition of Cyclin B1 synthesis, ribosomal S6 kinase phosphorylation (an indicator of Mos activity), and meiotic resumption (Nishimura et al. 2010). These contradictory results might reflect the fact that while mouse oocytes do not require protein synthesis for their meiotic resumption (Fulka et al. 1986), protein synthesis is essential for meiotic resumption in porcine oocytes. It also suggests that specific differences in cytoplasmic polyadenylation and CPEB1 regulation may exist between different species.

The phosphorylation state of CPEB1 has a crucial importance for its function. Non-phosphorylated CPEB1 represses the translation initiation of CPE-containing mRNAs by recruiting other trans-acting factors such as Maskin and Pumilio (Stebbins-Boaz et al. 1999; Nakahata et al. 2001, 2003). In contrast, the CPEB1 phosphorylation at Ser174 activates the polyadenylation of a CPE-containing mRNA by excluding the poly(A)-specific ribonuclease (PARN) from its 3′UTR (Kim and Richter 2006). Phosphorylation of six additional CPEB sites occurs in *Xenopus* oocytes as a result of the rapid inducer of G2/M progression (RINGO)-mediated activation of CDK1 (Ferby et al. 1999; Mendez et al. 2002; Padmanabhan and Richter 2006). This allows the release of embryonic poly(A)-binding protein (ePAB) from the repression complex and enables its association with the nascent poly(A) tail synthesized by GLD2 polymerase (Kim and Richter 2006; Reyes and Ross 2016), which brings the translational machinery together in a "closed-loop" active configuration (Kahvejian et al. 2001; Brook et al. 2009).

The second wave of CPEB1 phosphorylation is mediated by polo-like kinase 1 (PLK1) and cyclin-dependent kinase 1 (CDK1), also aided by peptidylprolyl cis/trans isomerase, NIMA-interacting 1 (PIN1), which specifically binds to phosphorylated ser/thr-pro motifs to catalytically regulate the post-phosphorylation conformation of its substrates and leads to the degradation of CPEB1 by the ubiquitin-proteasome pathway (Reverte et al. 2001; Mendez et al. 2002; Setoyama et al. 2007; Karabinova et al. 2011; Nechama et al. 2013). The partial degradation of CPEB1 is thought to be necessary for the cytoplasmic polyadenylation of mRNAs, which contain two or more CPEs in their 3′UTR (Mendez et al. 2002; Piqué et al. 2008). CPEB1 also interacts with several other regulatory protein-/CP-associated factors, which are involved in the regulation of the poly(A) tail length and translation, such as symplekin, GLD2 (an atypical poly(A) polymerase), and cytoplasmic polyadenylation specificity factor (CPSF).

This model of CPEB1 function in the regulation of CP is based mainly on the results obtained from oocytes of *Xenopus laevis*. Although homologues of most of the abovementioned factors have been identified in mammals, some, e.g., symplekin, RINGO, and Maskin/TACC3, might play a different role in CP regulation in mammalian oocytes or might even regulate different processes in different cell types, suggesting that the precise regulation of CP in oocytes might be species specific.

Aurora kinase A (AURKA) has previously been thought to be the kinase responsible for CP-activating CPEB1 phosphorylation in *Xenopus* (Mendez et al. 2000) and also mouse oocytes (Hodgman et al. 2001). On the other hand, recently it has been shown that AURKA is not responsible for activatory CPEB1 phosphorylation in porcine oocytes (Komrskova et al. 2014). Even in *Xenopus* oocytes, CPEB1 activation by AURKA has been questioned by Keady et al. (2007), who have shown that in oocyte extracts, the phosphorylation of CPEB1 at Ser174 occurs despite the depletion of AURKA or the inhibition of its activity. They further suggest that ERK1/ERK2 MAP kinase might be involved in CPEB1 activation, although it does not phosphorylate CPEB1 on the activatory Ser174 site directly, which implies the involvement of another unknown kinase. Recently, Sha et al. (2016) also suggested an important role for ERK1/ERK2 MAP kinase in CPEB1 activation during the meiotic maturation of mouse oocytes. On the other hand, Ca^{2+}/calmodulin-dependent protein kinase II (CaMKII) has also been found to be responsible for CPEB1 activation in mammalian neurons (Atkins et al. 2004).

These proteins, the translation of which is regulated by CPEB1-mediated CP, are among the most important regulators of the meiotic and mitotic cell cycles of vertebrate oocytes and early embryos. Originally, *Mos* and *Ccnb1* mRNAs have been identified as being regulated by this mechanism; later, Eliscovich et al. (2008) have shown in *Xenopus* oocytes that a number of other mRNAs vital for meiotic progression are potential targets of CPEB-mediated translational regulation. These include mRNAs encoding *Tpx2* (microtubule nucleation factor), *Bub1* (mitotic checkpoint serine/threonine kinase), *Bub3* (mitotic checkpoint protein), *Eg5* (kinesin family member 11), *Mad1* and *Mad2* (mitotic spindle assembly checkpoint protein 1 and 2), *BubR1* (mitotic checkpoint serine/threonine kinase), *Xkid* (kinesin family member), *Cenp-E* (centromere protein), *Nek2B* (serine/threonine-protein kinase), and *Aurora A* and *Aurora B* (serine/threonine-protein kinases), all of which are integral in the formation and function of the meiotic spindle. These authors further show that CPEB1 at least partly localizes to the spindle and that the spindle-localized translational activation of CPE-regulated mRNAs, encoding for proteins with a known function in spindle assembly and chromosome segregation, is essential for the completion of the first meiotic division and for chromosome segregation in *Xenopus* oocytes. Similar results have been obtained in *Xenopus* early embryos, in which CPEB1 localizes at the animal pole of oocytes and later on at the embryonic spindles and centrosomes (Groisman et al. 2000). On the other hand, no specific localization of CPEB1 has been detected in bovine oocytes (Uzbekova et al. 2008).

From the above mentioned findings, it is clear that CPEB1 plays a role in the regulation of CP of important meiotic cell cycle regulators in vertebrate oocytes; however, much less is known about the function of the other members of the CPEB family, CPEB2–CPEB4. CPEB2 was first detected in spermatids and then later found in nearly all mouse tissues including ovaries (Kurihara et al. 2003), as well as in *Xenopus* oocytes (Li et al. 2009). CPEB2 binds to a similar 3′UTR sequence as CPE, i.e., "UUUUCAU" (Chen and Huang 2012), and up to now, translation of only one mRNA has been shown to be regulated by CPEB2, that is, hypoxia-inducible factor 1α (HIF-1α; Chen and Huang 2012). CPEB3 is predominantly expressed in neuronal cells (although it has also been detected in most mouse tissues including ovaries), and in the neuronal cells, its function has also been studied in depth (Huang et al. 2006; Chen and Huang 2012; Wang and Huang 2012). CPEB3 knockout mice display defects in memory processes, and their reproduction efficiency is also impaired (Chao et al. 2013). As for CPEB4, it is supposed that it has similar role as CPEB1, i.e., activation of CP. Igea and Méndez (2010) have described a role of CPEB4 in the regulation of CP during *Xenopus* oocyte meiotic maturation, especially during the second meiotic division when CPEB4 becomes accumulated. These authors have suggested a mechanism implying that CPEB4 replaces CPEB1 after its degradation post NEBD and regulates CP during the second meiotic division, i.e., CP of so-called "late" mRNAs. Although this hypothesis is supported by similar results by the same group on HeLa cells (Novoa et al. 2010), Huang et al. (2006) argue that CPEB4 does not bind to CPE and that it also requires a specific tertiary structure for its function. Furthermore, it has also been shown that the expression of CPEB4 does not change during the meiotic maturation of mouse and pig oocytes (Kracmarova and Kubelka, unpublished results). Recently, Giangarrà et al. (2015) have suggested that CPEB3 is the only member dispensable for mitotic cell division, whereas the other three members, CPEB1, CPEB2, and CPEB4, are essential for successful mitotic cell division. CPEB1 is supposed to be required for prophase entry, CPEB2 for metaphase, and CPEB4 for cytokinesis.

Altogether, these findings suggest that although cytoplasmic polyadenylation regulated by CPEB1 has originally been supposed to involve only several specific yet important processes occurring in a limited number of cell types (especially oocytes and early embryos), it may well represent a much more general mechanism of translational regulation, which might also play a role during mitosis in somatic cells (Novoa et al. 2010).

12.3.3 Translation at the Meiotic Spindle

Dramatic internal membrane restructuring occurs upon NEBD during metazoan meiosis and mitosis (Motlík and Kubelka 1990; Hetzer 2010; Puhka et al. 2012). The internal spatial organization of cells is critically important for cell physiology. This is illustrated by the characteristic positioning of organelles in many differentiated cell types and by the reorganization of the cellular interior post-G2/prophase of the cell cycle. Nuclear envelope dissolution and chromatin condensation are followed by spindle formation and the extrusion of the first polar body containing homologous chromosomes. Spindle formation requires protein synthesis (Hashimoto and Kishimoto 1988), indicating a possible role for translational control by localized mRNAs in this process. Although a fully grown mouse oocyte is able to resume meiosis without proteosynthesis, the formation of the spindle and the progression to metaphase II require active translation (Schultz and Wassarman 1977; Hashimoto and Kishimoto 1988; Kanmera et al. 1995). Antisense RNA experiments with fully grown oocytes have shown that the translation of a large variety of mRNAs is required for meiotic spindle organization (Svoboda et al. 2000; Terret et al. 2003; Stein et al. 2013; Oh et al. 2013), suggesting that the blockage of translation of specific proteins results in abnormalities in the first meiosis. These results indicate that the proteins synthesized shortly after NEBD play an important role in oocyte meiotic progression. Published studies also show that despite a decrease in overall protein synthesis in the mammalian oocyte during meiosis (Schultz et al. 1978; Tomek et al. 2002; Ellederova et al. 2006), there is a regulatory program that ensures temporal and spatial synthesis of specific proteins essential for meiotic progression and embryo development.

The global RNA population in the mouse oocyte is localized uniformly in the cytoplasm, although a strong signal of poly(A) mRNAs in the nucleus (Figs. 12.1 and 12.4a) has been reported (Susor et al. 2015). Among the processes which might be regulated by localization are the regulation of meiotic progression, spindle-dependent translation remodeling, and asymmetric division, typical for mammalian female meiosis (Vinot et al. 2004; Dumont et al. 2007).

The localization of maternal mRNAs at the spindle is evolutionarily conserved between mouse and lower organisms and is also seen in mitotic cells. Figure 12.2 shows example of known localization of maternal RNAs in the mouse oocyte. Studies from *Xenopus* have revealed enriched localization to spindle microtubules of mRNAs encoding spindle proteins (Blower et al. 2007). Groisman et al. (2000) have detected concentrations of *Bub3 (mitotic checkpoint protein)* and *Ccnb1 (Cyclin B1)* mRNAs on the mitotic apparatus in the animal pole of early *Xenopus* embryos. The discovery that *Ccnb1* mRNA and protein, together with CPEB1 and Maskin, are concentrated on spindles and centrosomes might suggest that these structures are the sites of translational control. Local *Ccnb1* and *Bub3* mRNA translation might be necessary to ensure that the protein product is delivered to the areas where it is needed. This mechanism might also be important in *Drosophila* embryos, where both *Ccnb1* mRNA and protein have been detected at the oocyte posterior and mitotic apparatus (Raff et al. 1990; Huang and Raff 1999). Localized maternal mRNAs could be translated in situ to provide a local high concentration of proteins while minimizing potential deficiencies related to limitations in the speed or extent of protein accumulation from elsewhere within the ooplasm. Spindle complexes are sites of targeted polysomal mRNA localization in diverse systems (Blower et al. 2007; Mili and Macara 2009), which is likely to be correlated with the fact that the spindles are preparing the chromosomes for division (Brunet and Verlhac 2011). The enrichment of various mRNAs on the meiotic spindle permits the localized translation of specific proteins in the area, where spindle is newly forming (Fig. 12.3; Susor et al. 2015). The localization of RNA population in the nucleus of a fully grown oocyte might be a predisposition for the specific localization of RNA at newly forming spindle post NEBD. Experiments using microsurgery (Romasko et al. 2013; VerMilyea et al. 2011) techniques to study the differential abundance of mRNA in various compartments of meiotically maturing mouse oocytes have revealed a significant abundance of specific mRNAs in the sample containing spindle. This suggests that by unknown mechanisms mRNAs in mammalian oocytes are not distributed equally within the whole-cell volume but rather from distinguishable transcriptomes, one of which being linked to the spindle. The rearrangements of the intracellular organelles post NEBD, such as the Golgi apparatus, ER, and other organelles, are less well understood. The results come from the studies of mitotic cells, where these organelles should be equally distributed to the newly divided daughter cells. However, cytokinesis in a fully grown oocyte represents an asymmetric division, where it is important to retain maternal RNAs, organelles, and other components (with the exception of reduced genomes) in the mature egg and not in the polar body. Moreover, it is unclear whether other organelles, such as ribosomes, undergo reorganization during mitosis and meiosis.

Dynamic interactions between membrane-bound and membraneless organelles and the microtubule cytoskeleton are crucial to establish, maintain, and remodel the internal organization of cells throughout the cell cycle. However, the molecular nature of these interactions remains poorly understood. During meiosis, chromosomes must be accurately segregated to opposite sides of the spindle by a microtubule-based structure known as a meiotic spindle. On the other hand, a microtubule-independent structure, the so-called spindle matrix, has been proposed to assist mitosis. Originally it has been envisioned as a static scaffold that could act as a tensile element mediating chromosome motion (Pickett-Heaps et al. 1984) and/or determining spindle length (Mitchison et al. 2005), although more recent findings suggest that the spindle matrix is rather a dynamic assembly of molecules confined to the vicinity of the spindle (Lince-Faria et al. 2009; Schweizer et al. 2014). However, the mechanism that underlies the microtubule-independent accumulation of proteins in the spindle region remains unknown. A membranous structure derived from the nuclear envelope and ER surrounds the mitotic/meiotic spindle (Hepler and Wolniak 1984; Susor et al. 2015), but its role in meiosis/mitosis remains poorly understood (NEBD stage in Figs. 12.1, 12.2, and 12.3). Two translationally active areas in the oocyte post NEBD are present, one in the vicinity of the chromosomes and second in the perispindular region, areas separated by a gap with low translational activity where LMN A/C (lamin A/C; component of nuclear lamina) and ER structures are present (Susor et al. 2015). We hypothesize that a gap with a low translational activity between the active areas is some sort of semipermeable membrane formed on the basis of lamins, ER, and possibly other constituents. A membranous "spindle envelope" has been suggested to work as a scaffold for the recruitment of proteins required for mitotic spindle assembly and function (Tsai et al. 2006). Supported by mathematical modeling, these findings have led to the proposal that lamin and possibly other components surrounding the spindle play a structural role in the orchestration of cell division, even in systems undergoing open mitosis (Liu and Zheng 2009; Civelekoglu-Scholey et al. 2010; Goodman et al. 2010; Capalbo et al. 2011). Results from well-studied mouse oocytes show a centrally localized nucleus surrounded by ER. ER continues to encompass the assembling and migrating bipolar spindle after NEBD and forms a semipermeable barrier, which might result in a localization of maternal RNAs, proteins, and organelles (e.g., chromosomes, ribosomes, centrosomes, and mitochondria). A semipermeable membrane formed on the basis of microfilaments, LMN, and possibly other constituents has been suggested (FitzHarris et al. 2007; Yu et al. 2011; Yi et al. 2013; Dalton and Carroll 2013; Schlaitz et al. 2013; Schweizer et al. 2015). Susor et al. (2015) and Sanfins et al. (2004) have determined that lamin A/C and lamin B structures surround newly forming spindle at least 3 h post NEBD, which might suggest the presence of another component of the semipermeable barrier in the oocyte.

12.3.4 Regulation of Translation in the Oocyte Through Exogenous Signaling

It has long been known that the maternal environment of the ovarian follicle nurtures the development of the oocyte and endows it with developmental competence. The oocyte directs the activity of its neighboring somatic cells via paracrine factors, such that oocyte and cumulus cell functions are mutually dependent (Kloc and Kubiak 2017). The importance of bidirectional communication between oocytes and follicular somatic cells for the acquisition of oocyte meiotic and developmental competence is well-established (Gilchrist et al. 2008; Su et al. 2009). The cumulus cells are directly attached to the oocyte during growth, and upon resumption of meiosis, the connections are lost. However, the oocyte is not a passive player in regulation of its own maturation, but rather actively regulates intrafollicular environment by directing the somatic cell function and that the oocyte plays a central role in the regulation of folliculogenesis and thereby its own development (Gilchrist et al. 2008). Chen et al. (2013) have reported that the translation of a subset of maternal mRNAs within the oocyte, possibly critical for embryo development, is regulated by signals from the somatic cell compartment under the control of EGF-like growth factors. In vitro data have shown an increase in oocyte mRNA translation through mTOR/4F axis upon exposure of oocyte cumulus complexes to amphiregulin, which utilizes the EGF receptor in cumulus cells (Fig. 12.6). Chen et al. (2013) also indicate that somatic cells, via EGF-like growth factor signaling, play an important role in regulating oocyte mRNA translation and, consequently, oocyte developmental competence. This mechanism works in a bidirectional way; Cakmak et al. (2016) have demonstrated that interleukin 7 (IL7) as a prototypic secreted factor is dependent on the activation of mRNA translation in synchrony with the cell cycle and that its translation is under the control of somatic cells. IL7 concentration in the follicular fluid correlates with the oocyte ability to reach the MII stage of meiotic maturation. Oocyte-derived IL7 could act on neighboring granulosa cells as a survival factor and promote the maturation of preovulatory oocytes through the activation of the PIK3/mTOR axis (Cheng et al. 2011). IL7 has previously been shown to suppress the apoptosis of cultured granulosa cells by activating the PI3K/AKT pathway (Cheng et al. 2011). These dynamic secretions are likely to be critical for promoting the final stages of maturation and oocyte developmental competence.

Therefore, mRNA translation depends on somatic cell cues that might be essential to promote developmental competence of mammalian oocyte. Further work is needed in order to gain a better understanding of the role of these signals

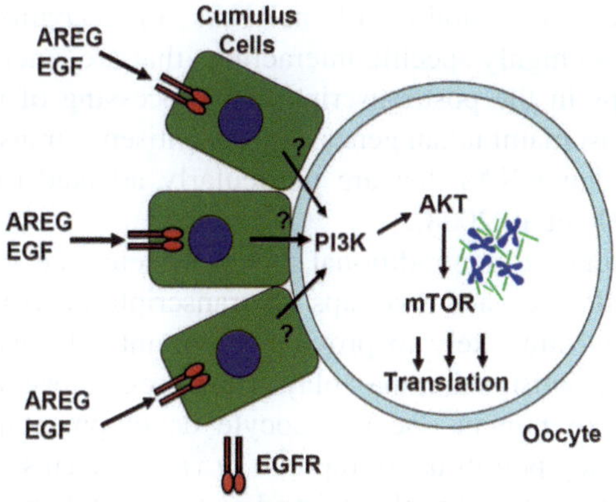

Fig. 12.6 Model of the signaling pathways involved in somatic cell control of oocyte translation. Cumulus cells are represented as *green squares* with a *blue* nucleus. An oocyte post NEBD is schematically represented with condensed chromosomes (*blue*) and newly forming spindle in the vicinity of chromosomes (*green*). EGFR dimers are represented by *red symbols*. EGF-like growth factor interacts with EGF receptors expressed on cumulus cells and cause activation of undefined signaling pathway/s through the cumulus-oocyte plasma membrane. These signals lead to PI3K activation and in activation of mTOR in the oocyte causing an increase in translation of a subset of maternal mRNAs (adapted from Chen et al. 2013)

in the regulation of the fate of mRNA molecules and hence the possibility that they are vital for the cumulus oocyte complex.

12.4 Future Directions

It is clear that the past decade has brought immense progress in the field of regulation of RNA metabolism in oocytes. However, the main body of our knowledge about the regulation of translation in connection with molecular processes in the female germ cell is garnered from lower organisms. Further work remains to determine the nature of RNP complexes in oocytes and their functional relationship to RNA stability, localization, and translation.

Although we live in the noncoding RNA (ncRNA) revolution era, up to now we have only a scarce knowledge about the ncRNA world, i.e., about their content and their role in oocytes and early embryos. It is known that noncoding RNAs, including the lncRNA, often function through interaction with proteins by forming RNPs (Rederstorff and Hüttenhofer 2011; Gong et al. 2012). lncRNA has recently been shown to be associated with the translation apparatus of developing and matured oocytes (Caballero et al. 2014); moreover it is known that lncRNAs influence translation of mRNAs (Mercer et al. 2009; Carrieri et al. 2012; Pircher et al. 2014). Despite the fact that the function of miRNA is globally suppressed during oocyte maturation and preimplantation development (Mercer et al. 2009; Suh et al. 2010), other classes of ncRNAs might contribute to regulation of translation in mammalian oocytes. The ability of ncRNAs to recognize complementary sequences also allows highly specific interactions that are amenable to the regulation of various steps in the posttranscriptional processing of mRNAs, including their translation. Most mammalian genes express antisense transcripts which might constitute a class of ncRNAs that are particularly adapted to the regulation of mRNA dynamics (He et al. 2008).

Comparative studies in nontraditional model systems are valuable in order to address the dissimilarities and overlaps in transcriptome composition between model organisms and are likely to provide important information regarding the components and mechanisms that may play critical regulatory roles in the physiology and pathology of human oocytes, oocyte developmental competence, and ovarian health and may permit us to improve *in vitro* systems which are currently being used for oocyte culture for clinical and experimental applications.

We therefore look forward to new discoveries in the near future of this terra incognita in this unique cell type.

Acknowledgments We thank Denisa Jansova for assistance with text and figures. This work was supported by GACR13-12291S, GACR15-22765S, EXCELLENCE CZ.02.1.01/0.0/0.0/15_003/0000460 OP RDE and Institutional Research Concept RVO67985904.

References

Abd El Naby WS, Hagos TH, Hossain MM, Salilew-Wondim D, Gad AY, Rings F, Cinar MU, Tholen E, Looft C, Schellander K, Hoelker M, Tesfaye D (2013) Expression analysis of regulatory microRNAs in bovine cumulus oocyte complex and preimplantation embryos. Zygote Camb Engl 21:31–51. doi:10.1017/S0967199411000566

Abe K-I, Yamamoto R, Franke V, Cao M, Suzuki Y, Suzuki MG, Vlahovicek K, Svoboda P, Schultz RM, Aoki F (2015) The first murine zygotic transcription is promiscuous and uncoupled from splicing and 3′ processing. EMBO J 34:1523–1537. doi:10.15252/embj.201490648

Amin MA, Matsunaga S, Uchiyama S, Fukui K (2008) Nucleophosmin is required for chromosome congression, proper mitotic spindle formation, and kinetochore-microtubule attachment in HeLa cells. FEBS Lett 582:3839–3844. doi:10.1016/j.febslet.2008.10.023

Atkins CM, Nozaki N, Shigeri Y, Soderling TR (2004) Cytoplasmic polyadenylation element binding protein-dependent protein synthesis is regulated by calcium/calmodulin-dependent protein kinase II. J Neurosci 24:5193–5201. doi:10.1523/JNEUROSCI.0854-04.2004

Bachvarova R (1981) Synthesis, turnover, and stability of heterogeneous RNA in growing mouse oocytes. Dev Biol 86:384–392

Bachvarova R, De Leon V (1977) Stored and polysomal ribosomes of mouse ova. Dev Biol 58:248–254

Bachvarova R, De Leon V (1980) Polyadenylated RNA of mouse ova and loss of maternal RNA in early development. Dev Biol 74:1–8

Bahar Halpern K, Caspi I, Lemze D, Levy M, Landen S, Elinav E, Ulitsky I, Itzkovitz S (2015) Nuclear Retention of mRNA in Mammalian Tissues. Cell Rep 13:2653–2662. doi:10.1016/j.celrep.2015.11.036

Bhatt DM, Pandya-Jones A, Tong A-J, Barozzi I, Lissner MM, Natoli G, Black DL, Smale ST (2012) Transcript dynamics of proinflammatory genes revealed by sequence analysis of subcellular RNA fractions. Cell 150:279–290. doi:10.1016/j.cell.2012.05.043
Bischoff M, Parfitt D-E, Zernicka-Goetz M (2008) Formation of the embryonic-abembryonic axis of the mouse blastocyst: relationships between orientation of early cleavage divisions and pattern of symmetric/asymmetric divisions. Development 135:953–962. doi:10.1242/dev.014316
Blower MD, Feric E, Weis K, Heald R (2007) Genome-wide analysis demonstrates conserved localization of messenger RNAs to mitotic microtubules. J Cell Biol 179:1365–1373. doi:10.1083/jcb.200705163
Bonatti S, Simili M, Galli A, Bagnato P, Pigullo S, Schiestl RH, Abbondandolo A (1998) Inhibition of the Mr 70,000 S6 kinase pathway by rapamycin results in chromosome malsegregation in yeast and mammalian cells. Chromosoma 107:498–506. doi:10.1007/s004120050335
Boothby TC, Wolniak SM (2011) Masked mRNA is stored with aggregated nuclear speckles and its asymmetric redistribution requires a homolog of mago nashi. BMC Cell Biol 12:45. doi:10.1186/1471-2121-12-45
Boothby TC, Zipper RS, van der Weele CM, Wolniak SM (2013) Removal of retained introns regulates translation in the rapidly developing gametophyte of Marsilea vestita. Dev Cell 24:517–529. doi:10.1016/j.devcel.2013.01.015
Bouniol-Baly C, Hamraoui L, Guibert J, Beaujean N, Szöllösi MS, Debey P (1999) Differential transcriptional activity associated with chromatin configuration in fully grown mouse germinal vesicle oocytes. Biol Reprod 60:580–587. doi:10.1095/biolreprod60.3.580
Boutz PL, Bhutkar A, Sharp PA (2015) Detained introns are a novel, widespread class of post-transcriptionally spliced introns. Genes Dev 29:63–80. doi:10.1101/gad.247361.114
Braddock M, Muckenthaler M, White MR, Thorburn AM, Sommerville J, Kingsman AJ, Kingsman SM (1994) Intron-less RNA injected into the nucleus of Xenopus oocytes accesses a regulated translation control pathway. Nucleic Acids Res 22:5255–5264
Bramham CR, Wells DG (2007) Dendritic mRNA: transport, translation and function. Nat Rev Neurosci 8:776–789. doi:10.1038/nrn2150
Brandhorst BP (1985) Informational content of the echinoderm egg. Dev Biol 1:525–576
Brandt A, Papagiannouli F, Wagner N, Wilsch-Bräuninger M, Braun M, Furlong EE, Loserth S, Wenzl C, Pilot F, Vogt N, Lecuit T, Krohne G, Großhans J (2006) Developmental control of nuclear size and shape by kugelkern and kurzkern. Curr Biol 16:543–552. doi:10.1016/j.cub.2006.01.051
Brangwynne CP, Eckmann CR, Courson DS, Rybarska A, Hoege C, Gharakhani J, Jülicher F, Hyman AA (2009) Germline P granules are liquid droplets that localize by controlled dissolution/condensation. Science 324:1729–1732. doi:10.1126/science.1172046
Braude P, Bolton V, Moore S (1988) Human gene expression first occurs between the four- and eight-cell stages of preimplantation development. Nature 332:459–461. doi:10.1038/332459a0
Brook M, Smith JWS, Gray NK (2009) The DAZL and PABP families: RNA-binding proteins with interrelated roles in translational control in oocytes. Reprod Camb Engl 137:595–617. doi:10.1530/REP-08-0524
Brower PT, Gizang E, Boreen SM, Schultz RM (1981) Biochemical studies of mammalian oogenesis: synthesis and stability of various classes of RNA during growth of the mouse oocyte in vitro. Dev Biol 86:373–383
Brunet S, Verlhac MH (2011) Positioning to get out of meiosis: the asymmetry of division. Hum Reprod Update 17:68–75. doi:10.1093/humupd/dmq044
Caballero J, Gilbert I, Fournier E, Gagné D, Scantland S, Macaulay A, Robert C (2014) Exploring the function of long non-coding RNA in the development of bovine early embryos. Reprod Fertil Dev 27:40–52. doi:10.1071/RD14338
Cakmak H, Franciosi F, Zamah AM, Cedars MI, Conti M (2016) Dynamic secretion during meiotic reentry integrates the function of the oocyte and cumulus cells. Proc Natl Acad Sci USA 113:2424–2429. doi:10.1073/pnas.1519990113

Capalbo L, D'Avino PP, Archambault V, Glover DM (2011) Rab5 GTPase controls chromosome alignment through Lamin disassembly and relocation of the NuMA-like protein Mud to the poles during mitosis. Proc Natl Acad Sci USA 108:17343–17348. doi:10.1073/pnas.1103720108

Capco DG, Gallicano GI, McGaughey RW, Downing KH, Larabell CA (1993) Cytoskeletal sheets of mammalian eggs and embryos: a lattice-like network of intermediate filaments. Cell Motil Cytoskeleton 24:85–99. doi:10.1002/cm.970240202

Carrieri C, Cimatti L, Biagioli M, Beugnet A, Zucchelli S, Fedele S, Pesce E, Ferrer I, Collavin L, Santoro C, Forrest ARR, Carninci P, Biffo S, Stupka E, Gustincich S (2012) Long non-coding antisense RNA controls Uchl1 translation through an embedded SINEB2 repeat. Nature 491:454–457. doi:10.1038/nature11508

Carrieri C, Forrest ARR, Santoro C, Persichetti F, Carninci P, Zucchelli S, Gustincich S (2015) Expression analysis of the long non-coding RNA antisense to Uchl1 (AS Uchl1) during dopaminergic cells' differentiation in vitro and in neurochemical models of Parkinson's disease. Front Cell Neurosci 9:114. doi:10.3389/fncel.2015.00114

Carter KC, Taneja KL, Lawrence JB (1991) Discrete nuclear domains of poly(A) RNA and their relationship to the functional organization of the nucleus. J Cell Biol 115:1191–1202. doi:10.1083/jcb.115.5.1191

Chao H-W, Tsai L-Y, Y-L L, Lin P-Y, Huang W-H, Chou H-J, W-H L, Lin H-C, Lee P-T, Huang Y-S (2013) Deletion of CPEB3 enhances hippocampus-dependent memory via increasing expressions of PSD95 and NMDA receptors. J Neurosci 33:17008–17022. doi:10.1523/JNEUROSCI.3043-13.2013

Chen L-L, Carmichael GG (2009) Altered nuclear retention of mRNAs containing inverted repeats in human embryonic stem cells: Functional role of a nuclear noncoding RNA. Mol Cell 35:467–478. doi:10.1016/j.molcel.2009.06.027

Chen P-J, Huang Y-S (2012) CPEB2–eEF2 interaction impedes HIF-1α RNA translation. EMBO J 31:959–971. doi:10.1038/emboj.2011.448

Chen J, Torcia S, Xie F, Lin C-J, Cakmak H, Franciosi F, Horner K, Onodera C, Song JS, Cedars MI, Ramalho-Santos M, Conti M (2013) Somatic cells regulate maternal mRNA translation and developmental competence of mouse oocytes. Nat Cell Biol 15:1415–1423. doi:10.1038/ncb2873

Cheng Y, Yata A, Klein C, Cho J-H, Deguchi M, Hsueh AJW (2011) Oocyte-expressed interleukin 7 suppresses granulosa cell apoptosis and promotes oocyte maturation in rats. Biol Reprod 84:707–714. doi:10.1095/biolreprod.110.086504

Civelekoglu-Scholey G, Tao L, Brust-Mascher I, Wollman R, Scholey JM (2010) Prometaphase spindle maintenance by an antagonistic motor-dependent force balance made robust by a disassembling lamin-B envelope. J Cell Biol 188:49–68. doi:10.1083/jcb.200908150

Clegg KB, Pikó L (1982) RNA synthesis and cytoplasmic polyadenylation in the one-cell mouse embryo. Nature 295:342–345. doi:10.1038/295342a0

Colegrove-Otero LJ, Minshall N, Standart N (2005) RNA-binding proteins in early development. Crit Rev Biochem Mol Biol 40:21–73. doi:10.1080/10409230590918612

Coticchio G, Dal-Canto M, Guglielmo M-C, Mignini-Renzini M, Fadini R (2012) Human oocyte maturation in vitro. Int J Dev Biol 56:909–918. doi:10.1387/ijdb.120135gv

Curtis D, Lehmann R, Zamore PD (1995) Translational regulation in development. Cell 81:171–178

Dalton CM, Carroll J (2013) Biased inheritance of mitochondria during asymmetric cell division in the mouse oocyte. J Cell Sci 126(Pt 13):2955–2964. doi:10.1242/jcs.128744 (Epub 2013 May 9)

De La Fuente R, Viveiros MM, Burns KH, Adashi EY, Matzuk MM, Eppig JJ (2004) Major chromatin remodeling in the germinal vesicle (GV) of mammalian oocytes is dispensable for global transcriptional silencing but required for centromeric heterochromatin function. Dev Biol 275:447–458. doi:10.1016/j.ydbio.2004.08.028

de Smedt V, Szöllösi D, Kloc M (2000) The balbiani body: asymmetry in the mammalian oocyte. Genesis 26:208–212 de Vantéry C, Gavin AC, Vassalli JD, Schorderet-Slatkine S (1996) An Accumulation of p34cdc2at the end of mouse oocyte growth correlates with the acquisition of meiotic competence. Dev Biol 174:335–344. doi:10.1006/dbio.1996.0078

Djebali S, Davis CA, Merkel A, Dobin A, Lassmann T, Mortazavi A, Tanzer A, Lagarde J,Lin W, Schlesinger F, Xue C, Marinov GK, Khatun J, Williams BA, Zaleski C, Rozowsky J, Röder M, Kokocinski F, Abdelhamid RF, Alioto T, Antoshechkin I, Baer MT, Bar NS, Batut P, Bell K, Bell I, Chakrabortty S, Chen X, Chrast J, Curado J, Derrien T, Drenkow J, Dumais E, Dumais J, Duttagupta R, Falconnet E, Fastuca M, Fejes-Toth K, Ferreira P, Foissac S, Fullwood MJ, Gao H, Gonzalez D, Gordon A, Gunawardena H, Howald C, Jha S, Johnson R, Kapranov P, King B, Kingswood C, Luo OJ, Park E, Persaud K, Preall JB, Ribeca P, Risk B, Robyr D, Sammeth M, Schaffer L, See L-H, Shahab A, Skancke J, Suzuki AM, Takahashi H, Tilgner H, Trout D, Walters N, Wang H, Wrobel J, Yu Y, Ruan X, Hayashizaki Y, Harrow J, Gerstein M, Hubbard T, Reymond A, Antonarakis SE, Hannon G, Giddings MC, Ruan Y, Wold B, Carninci P, Guigó R, Gingeras TR (2012) Landscape of transcription in human cells. Nature 489:101–108. doi:10.1038/nature11233

Dowling RJO, Topisirovic I, Alain T, Bidinosti M, Fonseca BD, Petroulakis E, Wang X, Larsson O, Selvaraj A, Liu Y, Kozma SC, Thomas G, Sonenberg N (2010) mTORC1-mediated cell proliferation, but not cell growth, controlled by the 4E-BPs. Science 328:1172–1176. doi:10.1126/science.1187532

Dubowy J, Macdonald PM (1998) Localization of mRNAs to the oocyte is common in Drosophila ovaries. Mech Dev 70:193–195

Dumont J, Petri S, Pellegrin F, Terret M-E, Bohnsack MT, Rassinier P, Georget V, Kalab P, Gruss OJ, Verlhac M-H (2007) A centriole- and RanGTP-independent spindle assembly pathway in meiosis I of vertebrate oocytes. J Cell Biol 176:295–305. doi:10.1083/jcb.200605199

Eliscovich C, Peset I, Vernos I, Méndez R (2008) Spindle-localized CPE-mediated translation controls meiotic chromosome segregation. Nat Cell Biol 10:858–865. doi:10.1038/ncb1746

Ellederova Z, Kovarova H, Melo-Sterza F, Livingstone M, Tomek W, Kubelka M (2006) Suppression of translation during in vitro maturation of pig oocytes despite enhanced formation of cap-binding protein complex eIF4F and 4E-BP1 hyperphosphorylation. Mol Reprod Dev 73:68–76. doi:10.1002/mrd.20368

Ellederová Z, Cais O, Susor A, Uhlírová K, Kovárová H, Jelínková L, Tomek W, Kubelka M (2008) ERK1/2 map kinase metabolic pathway is responsible for phosphorylation of translation initiation factor eIF4E during in vitro maturation of pig oocytes. Mol Reprod Dev 75:309–317. doi:10.1002/mrd.20690

Ferby I, Blazquez M, Palmer A, Eritja R, Nebreda AR (1999) A novel p34cdc2-binding and activating protein that is necessary and sufficient to trigger G2/M progression in Xenopus oocytes. Genes Dev 13:2177–2189

FitzHarris G, Marangos P, Carroll J (2007) Changes in endoplasmic reticulum structure during mouse oocyte maturation are controlled by the cytoskeleton and cytoplasmic dynein. Dev Biol 305(1):133–144. (Epub 2007 Feb 14)

Flach G, Johnson MH, Braude PR, Taylor RA, Bolton VN (1982) The transition from maternal to embryonic control in the 2-cell mouse embryo. EMBO J 1:681–686

Flemr M, Ma J, Schultz RM, Svoboda P (2010) P-body loss is concomitant with formation of a messenger RNA storage domain in mouse oocytes. Biol Reprod 82:1008–1017. doi:10.1095/biolreprod.109.082057

Forristall C, Pondel M, Chen L, King ML (1995) Patterns of localization and cytoskeletal association of two vegetally localized RNAs, Vg1 and Xcat-2. Development 121:201–208

Fulka J, Motlík J, Fulka J, Jílek F (1986) Effect of cycloheximide on nuclear maturation of pig and mouse oocytes. J Reprod Fertil 77:281–285

Giangarrà V, Igea A, Castellazzi CL, Bava F-A, Mendez R (2015) Global analysis of CPEBs reveals sequential and non-redundant functions in mitotic cell cycle. PloS One 10:e0138794. doi:10.1371/journal.pone.0138794

Gilchrist RB, Lane M, Thompson JG (2008) Oocyte-secreted factors: regulators of cumulus cell function and oocyte quality. Hum Reprod Update 14:159–177. doi:10.1093/humupd/dmm040

Gingras AC, Gygi SP, Raught B, Polakiewicz RD, Abraham RT, Hoekstra MF, Aebersold R, Sonenberg N (1999) Regulation of 4E-BP1 phosphorylation: a novel two-step mechanism. Genes Dev 13:1422–1437
Göhring J, Jacak J, Barta A (2014) Imaging of endogenous messenger RNA splice variants in living cells reveals nuclear retention of transcripts inaccessible to nonsense-mediated decay in Arabidopsis[C][W][OPEN]. Plant Cell 26:754–764. doi:10.1105/tpc.113.118075
Goldstrohm AC, Hook BA, Seay DJ, Wickens M (2006) PUF proteins bind Pop2p to regulate messenger RNAs. Nat Struct Mol Biol 13:533–539. doi:10.1038/nsmb1100
Gondos B (1987) Comparative studies of normal and neoplastic ovarian germ cells: 2. Ultrastructure and pathogenesis of dysgerminoma. Int J Gynecol Pathol 6:124–131
Gong C, Popp MW-L, Maquat LE (2012) Biochemical analysis of long non-coding RNA-containing ribonucleoprotein complexes. Methods 58:88–93. doi:10.1016/j.ymeth.2012.06.020
Goodman B, Channels W, Qiu M, Iglesias P, Yang G, Zheng Y (2010) Lamin B counteracts the kinesin Eg5 to restrain spindle pole separation during spindle assembly. J Biol Chem 285:35238–35244. doi:10.1074/jbc.M110.140749
Goshima G, Wollman R, Goodwin SS, Zhang N, Scholey JM, Vale RD, Stuurman N (2007) Genes required for mitotic spindle assembly in Drosophila S2 cells. Science 316:417–421. doi:10.1126/science.1141314
Groisman I, Huang Y-S, Mendez R, Cao Q, Theurkauf W, Richter JD (2000) CPEB, Maskin, and Cyclin B1 mRNA at the mitotic apparatus: implications for local translational control of cell division. Cell 103:435–447. doi:10.1016/S0092-8674(00)00135-5
Grosskortenhaus R, Robinson KJ, Doe CQ (2006) Pdm and Castor specify late-born motor neuron identity in the NB7-1 lineage. Genes Dev 20:2618–2627. doi:10.1101/gad.1445306
Guertin DA, Sabatini DM (2007) Defining the role of mTOR in cancer. Cancer Cell 12:9–22. doi:10.1016/j.ccr.2007.05.008
Gwinn DM, Asara JM, Shaw RJ (2010) Raptor is phosphorylated by cdc2 during mitosis. PloS One 5:e9197. doi:10.1371/journal.pone.0009197
Hampl A, Eppig JJ (1995) Analysis of the mechanism(s) of metaphase I arrest in maturing mouse oocytes. Development 121:925–933
Hashimoto N, Kishimoto T (1988) Regulation of meiotic metaphase by a cytoplasmic maturation-promoting factor during mouse oocyte maturation. Dev Biol 126:242–252
He Y, Vogelstein B, Velculescu VE, Papadopoulos N, Kinzler KW (2008) The antisense transcriptomes of human cells. Science 322:1855–1857. doi:10.1126/science.1163853
Heasman J, Quarmby J, Wylie CC (1984) The mitochondrial cloud of Xenopus oocytes: the source of germinal granule material. Dev Biol 105:458–469
Heesom KJ, Gampel A, Mellor H, Denton RM (2001) Cell cycle-dependent phosphorylation of the translational repressor eIF-4E binding protein-1 (4E-BP1). Curr Biol 11:1374–1379. doi:10.1016/S0960-9822(01)00422-5
Hepler PK, Wolniak SM (1984) Membranes in the mitotic apparatus: their structure and function. Int Rev Cytol 90:169–238
Herman RC, Williams JG, Penman S (1976) Message and non-message sequences adjacent to poly (A) in steady state heterogeneous nuclear RNA of HeLa cells. Cell 7:429–437
Herr JC, Chertihin O, Digilio L, Jha KN, Vemuganti S, Flickinger CJ (2008) Distribution of RNA binding protein MOEP19 in the oocyte cortex and early embryo indicates pre-patterning related to blastomere polarity and trophectoderm specification. Dev Biol 314:300–316. doi:10.1016/j.ydbio.2007.11.027
Hetzer MW (2010) The nuclear envelope. Cold Spring Harb Perspect Biol 2:a000539. doi:10.1101/cshperspect.a000539
Hodgman R, Tay J, Mendez R, Richter JD (2001) CPEB phosphorylation and cytoplasmic polyadenylation are catalyzed by the kinase IAK1/Eg2 in maturing mouse oocytes. Development 128:2815–2822
Holt CE, Bullock SL (2009) Subcellular mrna localization in animal cells and why it matters. Science 326:1212–1216. doi:10.1126/science.1176488
Houston DW, King ML (2000) A critical role for Xdazl, a germ plasm-localized RNA, in the differentiation of primordial germ cells in Xenopus. Development 127:447–456

Houston DW, Zhang J, Maines JZ, Wasserman SA, King ML (1998) A Xenopus DAZ-like gene encodes an RNA component of germ plasm and is a functional homologue of Drosophila boule. Development 125:171–180

Huang J, Raff JW (1999) The disappearance of cyclin B at the end of mitosis is regulated spatially in Drosophila cells. EMBO J 18:2184–2195. doi:10.1093/emboj/18.8.2184

Huang Y-S, Kan M-C, Lin C-L, Richter JD (2006) CPEB3 and CPEB4 in neurons: analysis of RNA-binding specificity and translational control of AMPA receptor GluR2 mRNA. EMBO J 25:4865–4876. doi:10.1038/sj.emboj.7601322

Huynh H, Ngo VC, Koong HN, Poon D, Choo SP, Thng CH, Chow P, Ong HS, Chung A, Soo KC (2009) Sorafenib and rapamycin induce growth suppression in mouse models of hepatocellular carcinoma. J Cell Mol Med 13:2673–2683. doi:10.1111/j.1582-4934.2009.00692.x

Igea A, Méndez R (2010) Meiosis requires a translational positive loop where CPEB1 ensues its replacement by CPEB4. EMBO J 29:2182–2193. doi:10.1038/emboj.2010.111

Imataka H, Gradi A, Sonenberg N (1998) A newly identified N-terminal amino acid sequence of human eIF4G binds poly(A)-binding protein and functions in poly(A)-dependent translation. EMBO J 17:7480–7489. doi:10.1093/emboj/17.24.7480

Inoue A, Nakajima R, Nagata M, Aoki F (2008) Contribution of the oocyte nucleus and cytoplasm to the determination of meiotic and developmental competence in mice. Hum Reprod 23:1377–1384. doi:10.1093/humrep/den096

Jambor H, Surendranath V, Kalinka AT, Mejstrik P, Saalfeld S, Tomancak P (2015) Systematic imaging reveals features and changing localization of mRNAs in Drosophila development. eLife. doi:10.7554/eLife.05003

Jang C-Y, Kim HD, Zhang X, Chang J-S, Kim J (2012) Ribosomal protein S3 localizes on the mitotic spindle and functions as a microtubule associated protein in mitosis. Biochem Biophys Res Commun 429:57–62. doi:10.1016/j.bbrc.2012.10.093

Jansova D, Koncicka M, Tetkova A, Cerna R, Malik R, Del Llano E, Kubelka M, Susor A (2017) Regulation of 4E-BP1 activity in the mammalian oocyte. Cell Cycle 16(10):927–939. doi:10.1080/15384101.2017.1295178 (Epub 2017 Mar 8)

Jefferson W, Newbold R, Padilla-Banks E, Pepling M (2006) Neonatal genistein treatment alters ovarian differentiation in the mouse: inhibition of oocyte nest breakdown and increased oocyte survival. Biol Reprod 74:161–168. doi:10.1095/biolreprod.105.045724

Kadyrova LY, Habara Y, Lee TH, Wharton RP (2007) Translational control of maternal Cyclin B mRNA by Nanos in the Drosophila germline. Development 134:1519–1527. doi:10.1242/dev.002212

Kahvejian A, Roy G, Sonenberg N (2001) The mRNA closed-loop model: the function of PABP and PABP-interacting proteins in mRNA translation. Cold Spring Harb Symp Quant Biol 66:293–300

Kalous J, Solc P, Baran V, Kubelka M, Schultz RM, Motlik J (2006) PKB/AKT is involved in resumption of meiosis in mouse oocytes. Biol Cell 98:111–123. doi:10.1042/BC20050020

Kalous J, Kubelka M, Solc P, Susor A, Motlík J (2009) AKT (protein kinase B) is implicated in meiotic maturation of porcine oocytes. Reproduction 138:645–654. doi:10.1530/REP-08-0461

Kambadur R, Koizumi K, Stivers C, Nagle J, Poole SJ, Odenwald WF (1998) Regulation of POU genes by castor and hunchback establishes layered compartments in the Drosophila CNS. Genes Dev 12:246–260. doi:10.1101/gad.12.2.246

Kanmera S, Sakakibara R, Ishiguro M (1995) Inhibition of polar body formation in mouse denuded oocytes cultured in vitro by protein synthesis inhibitors. Biol Pharm Bull 18:1255–1258

Karabinova P, Kubelka M, Susor A (2011) Proteasomal degradation of ubiquitinated proteins in oocyte meiosis and fertilization in mammals. Cell Tissue Res 346:1–9. doi:10.1007/s00441-011-1235-1

Katakowski M, Buller B, Wang X, Rogers T, Chopp M (2010) Functional microRNA is transferred between glioma cells. Cancer Res 70:8259–8263. doi:10.1158/0008-5472.CAN-10-0604

Keady BT, Kuo P, Martínez SE, Yuan L, Hake LE (2007) MAPK interacts with XGef and is required for CPEB activation during meiosis in Xenopus oocytes. J Cell Sci 120:1093–1103. doi:10.1242/jcs.03416

Kim JH, Richter JD (2006) Opposing polymerase-deadenylase activities regulate cytoplasmic polyadenylation. Mol Cell 24:173–183. doi:10.1016/j.molcel.2006.08.016

King ML, Messitt TJ, Mowry KL (2005) Putting RNAs in the right place at the right time: RNA localization in the frog oocyte. Biol Cell 97:19–33. doi:10.1042/BC20040067

Kloc M, Etkin LD (1995) Two distinct pathways for the localization of RNAs at the vegetal cortex in Xenopus oocytes. Development 121:287–297

Kloc M, Etkin LD (2005) RNA localization mechanisms in oocytes. J Cell Sci 118:269–282. doi:10.1242/jcs.01637

Kloc M, Kubiak JZ (2017) Exogenous molecules and organelles delivery in oogenesis. In: Kloc M (ed) Oocytes: maternal information and functions. Springer, Cham

Kloc M, Larabell C, Chan AP, Etkin LD (1998) Contribution of METRO pathway localized molecules to the organization of the germ cell lineage. Mech Dev 75:81–93

Kloc M, Bilinski S, Chan AP, Allen LH, Zearfoss NR, Etkin LD (2001) RNA localization and germ cell determination in Xenopus. Int Rev Cytol 203:63–91. Academic Press, ScienceDirect

Kloc M, Dougherty MT, Bilinski S, Chan AP, Brey E, King ML, Patrick CW, Etkin LD (2002a) Three-dimensional ultrastructural analysis of RNA distribution within germinal granules of Xenopus. Dev Biol 241:79–93. doi:10.1006/dbio.2001.0488

Kloc M, Zearfoss NR, Etkin LD (2002b) Mechanisms of subcellular mRNA localization. Cell 108:533–544

Kloc M, Bilinski S, Etkin LD (2004) The Balbiani body and germ cell determinants: 150 years later. Curr Top Dev Biol 59:1–36. doi:10.1016/S0070-2153(04)59001-4

Kloc M, Jaglarz M, Dougherty M, Stewart MD, Nel-Themaat L, Bilinski S (2008) Mouse early oocytes are transiently polar: three-dimensional and ultrastructural analysis. Exp Cell Res 314:3245–3254. doi:10.1016/j.yexcr.2008.07.007

Knaut H, Pelegri F, Bohmann K, Schwarz H, Nüsslein-Volhard C (2000) Zebrafish vasa RNA but not its protein is a component of the germ plasm and segregates asymmetrically before germline specification. J Cell Biol 149:875–888

Kogasaka Y, Hoshino Y, Hiradate Y, Tanemura K, Sato E (2013) Distribution and association of mTOR with its cofactors, raptor and rictor, in cumulus cells and oocytes during meiotic maturation in mice. Mol Reprod Dev 80:334–348. doi:10.1002/mrd.22166

Komrskova P, Susor A, Malik R, Prochazkova B, Liskova L, Supolikova J, Hladky S, Kubelka M (2014) Aurora kinase A is not involved in CPEB1 phosphorylation and cyclin B1 mRNA polyadenylation during meiotic maturation of porcine oocytes. PloS One 9:e101222. doi:10.1371/journal.pone.0101222

König H, Matter N, Bader R, Thiele W, Müller F (2007) Splicing segregation: the minor spliceosome acts outside the nucleus and controls cell proliferation. Cell 131:718–729. doi:10.1016/j.cell.2007.09.043

Kopecny V, Landa V, Pavlok A (1995) Localization of nucleic acids in the nucleoli of oocytes and early embryos of mouse and hamster: an autoradiographic study. Mol Reprod Dev 41:449–458. doi:10.1002/mrd.1080410407

Kopecny V, Biggiogera M, Laurincik J, Pivko J, Grafenau P, Martin TE, XD F, Fakan S (1996a) a Fine structural cytochemical and immunocytochemical analysis of nucleic acids and ribonucleoprotein distribution in nuclei of pig oocytes and early preimplantation embryos. Chromosoma 104:561–574

Kopecny V, Landa V, Malatesta M, Martin TE, Fakan S (1996b) b Immunoelectron microscope analyses of rat germinal vesicle-stage oocyte nucleolus-like bodies. Reprod Nutr Dev 36:667–679

Kosaka K, Kawakami K, Sakamoto H, Inoue K (2007) Spatiotemporal localization of germ plasm RNAs during zebrafish oogenesis. Mech Dev 124:279–289. doi:10.1016/j.mod.2007.01.003

Kotani T, Yasuda K, Ota R, Yamashita M (2013) Cyclin B1 mRNA translation is temporally controlled through formation and disassembly of RNA granules. J Cell Biol 202:1041–1055. doi:10.1083/jcb.201302139

Kurihara Y, Tokuriki M, Myojin R, Hori T, Kuroiwa A, Matsuda Y, Sakurai T, Kimura M, Hecht NB, Uesugi S (2003) CPEB2, a novel putative translational regulator in mouse haploid germ cells. Biol Reprod 69:261–268. doi:10.1095/biolreprod.103.015677

Kurotaki Y, Hatta K, Nakao K, Nabeshima Y-I, Fujimori T (2007) Blastocyst axis is specified independently of early cell lineage but aligns with the ZP shape. Science 316:719–723. doi:10.1126/science.1138591
Kusch J, Liakopoulos D, Barral Y (2003) Spindle asymmetry: a compass for the cell. Trends Cell Biol 13:562–569
Lapasset L, Pradet-Balade B, Vergé V, Lozano J-C, Oulhen N, Cormier P, Peaucellier G (2008) Cyclin B synthesis and rapamycin-sensitive regulation of protein synthesis during starfish oocyte meiotic divisions. Mol Reprod Dev 75:1617–1626. doi:10.1002/mrd.20905
Lécuyer E, Yoshida H, Parthasarathy N, Alm C, Babak T, Cerovina T, Hughes TR, Tomancak P, Krause HM (2007) Global analysis of mRNA localization reveals a prominent role in organizing cellular architecture and function. Cell 131:174–187. doi:10.1016/j.cell.2007.08.003
Lee S-E, Sun S-C, Choi H-Y, Uhm S-J, Kim N-H (2012) mTOR is required for asymmetric division through small GTPases in mouse oocytes. Mol Reprod Dev 79:356–366. doi:10.1002/mrd.22035
Li L, Baibakov B, Dean J (2008) A subcortical maternal complex essential for pre-implantation mouse embryogenesis. Dev Cell 15:416–425. doi:10.1016/j.devcel.2008.07.010
Li M, Li S, Yuan J, Wang Z-B, Sun S-C, Schatten H, Sun Q-Y (2009) Bub3 is a spindle assembly checkpoint protein regulating chromosome segregation during mouse oocyte meiosis. PloS One 4:e7701. doi:10.1371/journal.pone.0007701
Li L, Lu X, Dean J (2013) The maternal to zygotic transition in mammals. Mol Aspects Med 34:919–938. doi:10.1016/j.mam.2013.01.003
Li L, Zheng P, Dean J (2010) Maternal control of early mouse development. Development 137:859–870. doi:10.1242/dev.039487
Lim J, Lee M, Son A, Chang H, Kim VN (2016) mTAIL-seq reveals dynamic poly(A) tail regulation in oocyte-to-embryo development. Genes Dev. doi:10.1101/gad.284802.116
Lince-Faria M, Maffini S, Orr B, Ding Y, null CF, Sunkel CE, Tavares A, Johansen J, Johansen KM, Maiato H (2009) Spatiotemporal control of mitosis by the conserved spindle matrix protein Megator. J Cell Biol 184:647–657. doi:10.1083/jcb.200811012
Liu Z, Zheng Y (2009) A requirement for epsin in mitotic membrane and spindle organization. J Cell Biol 186:473–480. doi:10.1083/jcb.200902071
Livingstone M, Bidinosti M (2012) Rapamycin-insensitive mTORC1 activity controls eIF4E:4E-BP1 binding. F1000Res 1:4. doi:10.12688/f1000research.1-4.v1
Louvet-Vallée S, Vinot S, Maro B (2005) Mitotic spindles and cleavage planes are oriented randomly in the two-cell mouse embryo. Curr Biol 15:464–469. doi:10.1016/j.cub.2004.12.078
Macaulay AD, Gilbert I, Caballero J, Barreto R, Fournier E, Tossou P, Sirard M-A, Clarke HJ, Khandjian ÉW, Richard FJ, Hyttel P, Robert C (2014) The gametic synapse: RNA transfer to the bovine oocyte. Biol Reprod 91:90. doi:10.1095/biolreprod.114.119867
Macaulay AD, Gilbert I, Scantland S, Fournier E, Ashkar F, Bastien A, Saadi HAS, Gagné D, Sirard M-A, Khandjian ÉW, Richard FJ, Hyttel P, Robert C (2016) Cumulus cell transcripts transit to the bovine oocyte in preparation for maturation. Biol Reprod 94:16. doi:10.1095/biolreprod.114.127571
Mader S, Lee H, Pause A, Sonenberg N (1995) The translation initiation factor eIF-4E binds to a common motif shared by the translation factor eIF-4 gamma and the translational repressors 4E-binding proteins. Mol Cell Biol 15:4990–4997
Maegawa S, Yasuda K, Inoue K (1999) Maternal mRNA localization of zebrafish DAZ-like gene. Mech Dev 81:223–226. doi:10.1016/S0925-4773(98)00242-1
Makker A, Goel MM, Mahdi AA (2014) PI3K/PTEN/Akt and TSC/mTOR signaling pathways, ovarian dysfunction, and infertility: an update. J Mol Endocrinol 53:R103–R118. doi:10.1530/JME-14-0220
Marlow FL (2010) Maternal control of development in vertebrates: my mother made me do it! Morgan & Claypool Life Sciences, San Rafael, CA
Martinson HG (2011) An active role for splicing in 3'-end formation. Wiley Interdiscip Rev RNA 2:459–470. doi:10.1002/wrna.68

Matova N, Cooley L (2001) Comparative aspects of animal oogenesis. Dev Biol 231:291–320. doi:10.1006/dbio.2000.0120

Mayer S, Wrenzycki C, Tomek W (2014) Inactivation of mTor arrests bovine oocytes in the metaphase-I stage, despite reversible inhibition of 4E-BP1 phosphorylation. Mol Reprod Dev 81:363–375. doi:10.1002/mrd.22305

McGaughey RW, Capco DG (1989) Specialized cytoskeletal elements in mammalian eggs: structural and biochemical evidence for their composition. Cell Motil Cytoskeleton 13:104–111. doi:10.1002/cm.970130205

McGrew LL, Dworkin-Rastl E, Dworkin MB, Richter JD (1989) Poly(A) elongation during Xenopus oocyte maturation is required for translational recruitment and is mediated by a short sequence element. Genes Dev 3:803–815. doi:10.1101/gad.3.6.803

McNeilly JR, Saunders PT, Taggart M, Cranfield M, Cooke HJ, McNeilly AS (2000) Loss of oocytes in Dazl knockout mice results in maintained ovarian steroidogenic function but altered gonadotropin secretion in adult animals. Endocrinology 141:4284–4294. doi:10.1210/endo.141.11.7764

Mendez R, Hake LE, Andresson T, Littlepage LE, Ruderman JV, Richter JD (2000) Phosphorylation of CPE binding factor by Eg2 regulates translation of c-mos mRNA. Nature 404:302–307. doi:10.1038/35005126

Mendez R, Barnard D, Richter JD (2002) Differential mRNA translation and meiotic progression require Cdc2-mediated CPEB destruction. EMBO J 21:1833–1844. doi:10.1093/emboj/21.7.1833

Mercer TR, Dinger ME, Mattick JS (2009) Long non-coding RNAs: insights into functions. Nat Rev Genet 10:155–159. doi:10.1038/nrg2521

Mili S, Macara IG (2009) RNA localization and polarity: from A(PC) to Z(BP). Trends Cell Biol 19:156–164. doi:10.1016/j.tcb.2009.02.001

Mita K, Yamashita M (2000) Expression of Xenopus Daz-like protein during gametogenesis and embryogenesis. Mech Dev 94:251–255

Mitchison TJ, Maddox P, Gaetz J, Groen A, Shirasu M, Desai A, Salmon ED, Kapoor TM (2005) Roles of polymerization dynamics, opposed motors, and a tensile element in governing the length of Xenopus extract meiotic spindles. Mol Biol Cell 16:3064–3076. doi:10.1091/mbc.E05-02-0174

Monti M, Zanoni M, Calligaro A, Ko MSH, Mauri P, Redi CA (2013) Developmental Arrest and Mouse Antral Not-Surrounded Nucleolus Oocytes. Biol Reprod 88:2. doi:10.1095/biolreprod.112.103887

Motlík J, Kubelka M (1990) Cell-cycle aspects of growth and maturation of mammalian oocytes. Mol Reprod Dev 27:366–375. doi:10.1002/mrd.1080270411

Motosugi N, Bauer T, Polanski Z, Solter D, Hiiragi T (2005) Polarity of the mouse embryo is established at blastocyst and is not prepatterned. Genes Dev 19:1081–1092. doi:10.1101/gad.1304805

Nagaike T, Manley JL (2011) Transcriptional activators enhance polyadenylation of mRNA precursors. RNA Biol 8:964–967. doi:10.4161/rna.8.6.17210

Nakahata S, Katsu Y, Mita K, Inoue K, Nagahama Y, Yamashita M (2001) Biochemical identification of Xenopus Pumilio as a sequence-specific cyclin B1 mRNA-binding protein that physically interacts with a Nanos homolog, Xcat-2, and a cytoplasmic polyadenylation element-binding protein. J Biol Chem 276:20945–20953. doi:10.1074/jbc.M010528200

Nakahata S, Kotani T, Mita K, Kawasaki T, Katsu Y, Nagahama Y, Yamashita M (2003) Involvement of Xenopus Pumilio in the translational regulation that is specific to cyclin B1 mRNA during oocyte maturation. Mech Dev 120:865–880

Nechama M, Lin C-L, Richter JD (2013) An unusual two-step control of CPEB destruction by Pin1. Mol Cell Biol 33:48–58. doi:10.1128/MCB.00904-12

Niedojadło J, Kubicka E, Kalich B, Smoliński DJ (2014) Poly(A) RNAs including coding proteins RNAs occur in plant Cajal bodies. PLoS ONE. doi:10.1371/journal.pone.0111780

Nieuwkoop PD (1985) Inductive interactions in early amphibian development and their general nature. J Embryol Exp Morphol 89(Suppl):333–347

Nishimura Y, Kano K, Naito K (2010) Porcine CPEB1 is involved in Cyclin B translation and meiotic resumption in porcine oocytes. Anim Sci J 81:444–452. doi:10.1111/j.1740-0929.2010.00755.x

Nothias JY, Majumder S, Kaneko KJ, DePamphilis ML (1995) Regulation of gene expression at the beginning of mammalian development. J Biol Chem 270:22077–22080

Novoa I, Gallego J, Ferreira PG, Mendez R (2010) Mitotic cell-cycle progression is regulated by CPEB1 and CPEB4-dependent translational control. Nat Cell Biol 12:447–456. doi:10.1038/ncb2046

Oh JS, Susor A, Schindler K, Schultz RM, Conti M (2013) Cdc25A activity is required for the metaphase II arrest in mouse oocytes. J Cell Sci 126:1081–1085. doi:10.1242/jcs.115592

Ohsugi M, Zheng P, Baibakov B, Li L, Dean J (2008) Maternally derived FILIA-MATER complex localizes asymmetrically in cleavage-stage mouse embryos. Development 135:259–269. doi:10.1242/dev.011445

Padmanabhan K, Richter JD (2006) Regulated Pumilio-2 binding controls RINGO/Spy mRNA translation and CPEB activation. Genes Dev 20:199–209. doi:10.1101/gad.1383106

Pandya-Jones A, Bhatt DM, Lin C-H, Tong A-J, Smale ST, Black DL (2013) Splicing kinetics and transcript release from the chromatin compartment limit the rate of Lipid A-induced gene expression. RNA 19:811–827. doi:10.1261/rna.039081.113

Pavlok A, Kopečný V, Lucas-Hahn A, Niemann H (1993) Transcriptional activity and nuclear ultrastructure of 8-cell bovine embryos developed by in vitro maturation and fertilization of oocytes from different growth categories of antral follicles. Mol Reprod Dev 35:233–243. doi:10.1002/mrd.1080350304

Pepling ME, Spradling AC (2001) Mouse ovarian germ cell cysts undergo programmed breakdown to form primordial follicles. Dev Biol 234:339–351. doi:10.1006/dbio.2001.0269

Pepling ME, Wilhelm JE, O'Hara AL, Gephardt GW, Spradling AC (2007) Mouse oocytes within germ cell cysts and primordial follicles contain a Balbiani body. Proc Natl Acad Sci USA 104:187–192. doi:10.1073/pnas.0609923104

Pesty A, Miyara F, Debey P, Lefevre B, Poirot C (2007) Multiparameter assessment of mouse oogenesis during follicular growth in vitro. Mol Hum Reprod 13:3–9. doi:10.1093/molehr/gal089

Pickett-Heaps J, Spurck T, Tippit D (1984) Chromosome motion and the spindle matrix. J Cell Biol 99:137s–143s

Pierre A, Gautier M, Callebaut I, Bontoux M, Jeanpierre E, Pontarotti P, Monget P (2007) Atypical structure and phylogenomic evolution of the new eutherian oocyte- and embryo-expressed KHDC1/DPPA5/ECAT1/OOEP gene family. Genomics 90:583–594. doi:10.1016/j.ygeno.2007.06.003

Pilot F, Philippe J-M, Lemmers C, Chauvin J-P, Lecuit T (2006) Developmental control of nuclear morphogenesis and anchoring by charleston, identified in a functional genomic screen of Drosophila cellularisation. Development 133:711–723. doi:10.1242/dev.02251

Piqué M, López JM, Foissac S, Guigó R, Méndez R (2008) A combinatorial code for CPE-mediated translational control. Cell 132:434–448. doi:10.1016/j.cell.2007.12.038

Pircher A, Bakowska-Zywicka K, Schneider L, Zywicki M, Polacek N (2014) An mRNA-derived noncoding RNA targets and regulates the ribosome. Mol Cell 54:147–155. doi:10.1016/j.molcel.2014.02.024

Pochukalina GN, Parfenov VN (2008) Nucleolus transformation in oocytes of mouse antral follicles. Revealing of coilin and RNA polymerase I complex components. Tsitologiia 50:671–680

Pochukalina GN, Davis DS, Kostiuchek DF, Murti KG, Parfenov VN (1998) Splicing factors in oocyte nuclei from human antral follicles. Tsitologiia 40:239–247

Prasanth KV, Prasanth SG, Xuan Z, Hearn S, Freier SM, Bennett CF, Zhang MQ, Spector DL (2005) Regulating gene expression through RNA nuclear retention. Cell 123:249–263. doi:10.1016/j.cell.2005.08.033

Proudfoot NJ (2011) Ending the message: poly(A) signals then and now. Genes Dev 25:1770–1782. doi:10.1101/gad.17268411

Puhka M, Joensuu M, Vihinen H, Belevich I, Jokitalo E (2012) Progressive sheet-to-tubule transformation is a general mechanism for endoplasmic reticulum partitioning in dividing mammalian cells. Mol Biol Cell 23:2424–2432. doi:10.1091/mbc.E10-12-0950

Racki WJ, Richter JD (2006) CPEB controls oocyte growth and follicle development in the mouse. Development 133:4527–4537. doi:10.1242/dev.02651

Radford HE, Meijer HA, de Moor CH (2008) Translational control by cytoplasmic polyadenylation in Xenopus oocytes. Biochim Biophys Acta 1779:217–229. doi:10.1016/j.bbagrm.2008.02.002

Raff JW, Whitfield WG, Glover DM (1990) Two distinct mechanisms localise cyclin B transcripts in syncytial Drosophila embryos. Development 110:1249–1261

Rederstorff M, Hüttenhofer A (2011) cDNA library generation from ribonucleoprotein particles. Nat Protoc 6:166–174. doi:10.1038/nprot.2010.186

Reverte CG, Ahearn MD, Hake LE (2001) CPEB degradation during Xenopus oocyte maturation requires a PEST domain and the 26S proteasome. Dev Biol 231:447–458. doi:10.1006/dbio.2001.0153

Reyes JM, Ross PJ (2016) Cytoplasmic polyadenylation in mammalian oocyte maturation. Wiley Interdiscip Rev RNA 7:71–89. doi:10.1002/wrna.1316

Romasko EJ, Amarnath D, Midic U, Latham KE (2013) Association of maternal mRNA and phosphorylated EIF4EBP1 variants with the spindle in mouse oocytes: localized translational control supporting female meiosis in mammals. Genetics 195:349–358. doi:10.1534/genetics.113.154005

Rossant J, Tam PPL (2004) Emerging asymmetry and embryonic patterning in early mouse development. Dev Cell 7:155–164. doi:10.1016/j.devcel.2004.07.012

Russell DL, Gilchrist RB, Brown HM, Thompson JG (2016) Bidirectional communication between cumulus cells and the oocyte: old hands and new players? Theriogenology 86:62–68. doi:10.1016/j.theriogenology.2016.04.019

Sanfins A, Plancha CE, Overstrom EW, Albertini DF (2004) Meiotic spindle morphogenesis in in vivo and in vitro matured mouse oocytes: insights into the relationship between nuclear and cytoplasmic quality. Hum Reprod 19:2889–2899. doi:10.1093/humrep/deh528

Scheper GC, Proud CG (2002) Does phosphorylation of the cap-binding protein eIF4E play a role in translation initiation? Eur J Biochem 269:5350–5359

Schlaitz AL, Thompson J, Wong CC, Yates JR 3rd, Heald R (2013) REEP3/4 ensure endoplasmic reticulum clearance from metaphase chromatin and proper nuclear envelope architecture. Dev Cell 26(3):315–323. doi:10.1016/j.devcel.2013.06.016 (Epub 2013 Aug 1)

Schuh M, Ellenberg J (2007) Self-organization of MTOCs replaces centrosome function during Acentrosomal spindle assembly in live mouse oocytes. Cell 130:484–498. doi:10.1016/j.cell.2007.06.025

Schultz RM (1993) Regulation of zygotic gene activation in the mouse. Bioessays 15:531–538. doi:10.1002/bies.950150806

Schultz RM, Wassarman PM (1977) Biochemical studies of mammalian oogenesis: protein synthesis during oocyte growth and meiotic maturation in the mouse. J Cell Sci 24:167–194

Schultz RM, LaMarca MJ, Wassarman PM (1978) Absolute rates of protein synthesis during meiotic maturation of mammalian oocytes in vitro. Proc Natl Acad Sci 75:4160–4164

Schweizer N, Pawar N, Weiss M, Maiato H (2015) An organelle-exclusion envelope assists mitosis and underlies distinct molecular crowding in the spindle region. J Cell Biol 210 (5):695–704. doi:10.1083/jcb.201506107 (Epub 2015 Aug 24)

Schweizer N, Weiss M, Maiato H (2014) The dynamic spindle matrix. Curr Opin Cell Biol 28:1–7. doi:10.1016/j.ceb.2014.01.002

Setoyama D, Yamashita M, Sagata N (2007) Mechanism of degradation of CPEB during Xenopus oocyte maturation. Proc Natl Acad Sci USA 104:18001–18006. doi:10.1073/pnas.0706952104

Sha Q-Q, Dai X-X, Dang Y, Tang F, Liu J, Zhang Y-L, Fan H-Y (2016) MAPK cascade couples maternal mRNA translation and degradation to meiotic cell cycle progression in mouse oocyte. Development. doi:10.1242/dev.144410

Shang ZF, Yu L, Li B, Tu WZ, Wang Y, Liu XD, Guan H, Huang B, Rang WQ, Zhou PK (2012) 4E-BP1 participates in maintaining spindle integrity and genomic stability via interacting with PLK1. Cell Cycle 11:3463–3471. doi:10.4161/cc.21770

Sheets MD, Fox CA, Hunt T, Woude GV, Wickens M (1994) The 3'-untranslated regions of c-mos and cyclin mRNAs stimulate translation by regulating cytoplasmic polyadenylation. Genes Dev 8:926–938. doi:10.1101/gad.8.8.926

Shishova KV, Khodarovich YM, Lavrentyeva EA, Zatsepina OV (2015a) High-resolution microscopy of active ribosomal genes and key members of the rRNA processing machinery inside nucleolus-like bodies of fully-grown mouse oocytes. Exp Cell Res 337:208–218. doi:10.1016/j.yexcr.2015.07.024

Shishova KV, Lavrentyeva EA, Dobrucki JW, Zatsepina OV (2015b) Nucleolus-like bodies of fully-grown mouse oocytes contain key nucleolar proteins but are impoverished for rRNA. Dev Biol 397:267–281. doi:10.1016/j.ydbio.2014.11.022

Shuda M, Velásquez C, Cheng E, Cordek DG, Kwun HJ, Chang Y, Moore PS (2015) CDK1 substitutes for mTOR kinase to activate mitotic cap-dependent protein translation. Proc Natl Acad Sci USA 112:5875–5882. doi:10.1073/pnas.1505787112

Stebbins-Boaz B, Hake LE, Richter JD (1996) CPEB controls the cytoplasmic polyadenylation of cyclin, Cdk2 and c-mos mRNAs and is necessary for oocyte maturation in Xenopus. EMBO J 15:2582–2592

Stebbins-Boaz B, Cao Q, de Moor CH, Mendez R, Richter JD (1999) Maskin is a CPEB-associated factor that transiently interacts with elF-4E. Mol Cell 4:1017–1027

Stein P, Svoboda P, Schultz RM (2013) RNAi-based methods for gene silencing in mouse oocytes. Methods Mol Biol 957:135–151. doi:10.1007/978-1-62703-191-2_9

Sternlicht AL, Schultz RM (1981) Biochemical studies of mammalian oogenesis: kinetics of accumulation of total and poly(A)-containing RNA during growth of the mouse oocyte. J Exp Zool 215:191–200. doi:10.1002/jez.1402150209

Stutz A, Conne B, Huarte J, Gubler P, Völkel V, Flandin P, Vassalli J-D (1998) Masking, unmasking, and regulated polyadenylation cooperate in the translational control of a dormant mRNA in mouse oocytes. Genes Dev 12:2535–2548

Su Y-Q, Sugiura K, Woo Y, Wigglesworth K, Kamdar S, Affourtit J, Eppig JJ (2007) Selective degradation of transcripts during meiotic maturation of mouse oocytes. Dev Biol 302:104–117. doi:10.1016/j.ydbio.2006.09.008

Su Y-Q, Sugiura K, Eppig JJ (2009) Mouse oocyte control of granulosa cell development and function: paracrine regulation of cumulus cell metabolism. Semin Reprod Med 27:32–42. doi:10.1055/s-0028-1108008

Suh N, Crittenden SL, Goldstrohm A, Hook B, Thompson B, Wickens M, Kimble J (2009) FBF and its dual control of gld-1 expression in the Caenorhabditis elegans germline. Genetics 181:1249–1260. doi:10.1534/genetics.108.099440

Suh N, Baehner L, Moltzahn F, Melton C, Shenoy A, Chen J, Blelloch R (2010) MicroRNA function is globally suppressed in mouse oocytes and early embryos. Curr Biol 20:271–277. doi:10.1016/j.cub.2009.12.044

Susor A, Jelínková L, Karabínová P, Torner H, Tomek W, Kovárová H, Kubelka M (2008) Regulation of cap-dependent translation initiation in the early stage porcine parthenotes. Mol Reprod Dev 75:1716–1725. doi:10.1002/mrd.20913

Susor A, Jansova D, Cerna R, Danylevska A, Anger M, Toralova T, Malik R, Supolikova J, Cook MS, JS O, Kubelka M (2015) Temporal and spatial regulation of translation in the mammalian oocyte via the mTOR-eIF4F pathway. Nat Commun 6:6078. doi:10.1038/ncomms7078

Susor A, Jansova D, Anger M, Kubelka M (2016) Translation in the mammalian oocyte in space and time. Cell Tissue Res 363:69–84. doi:10.1007/s00441-015-2269-6

Svoboda P, Stein P, Hayashi H, Schultz RM (2000) Selective reduction of dormant maternal mRNAs in mouse oocytes by RNA interference. Development 127:4147–4156

Tadros W, Lipshitz HD (2009) The maternal-to-zygotic transition: a play in two acts. Development 136:3033–3042. doi:10.1242/dev.033183

Tay J, Richter JD (2001) Germ cell differentiation and synaptonemal complex formation are disrupted in CPEB knockout mice. Dev Cell 1:201–213

Terret ME, Lefebvre C, Djiane A, Rassinier P, Moreau J, Maro B, Verlhac M-H (2003) DOC1R: a MAP kinase substrate that control microtubule organization of metaphase II mouse oocytes. Development 130:5169–5177. doi:10.1242/dev.00731

Theis M, Si K, Kandel ER (2003) Two previously undescribed members of the mouse CPEB family of genes and their inducible expression in the principal cell layers of the hippocampus. Proc Natl Acad Sci USA 100:9602–9607. doi:10.1073/pnas.1133424100

Thoreen CC, Chantranupong L, Keys HR, Wang T, Gray NS, Sabatini DM (2012) A unifying model for mTORC1-mediated regulation of mRNA translation. Nature 485:109–113. doi:10.1038/nature11083

Tilgner H, Knowles DG, Johnson R, Davis CA, Chakrabortty S, Djebali S, Curado J, Snyder M, Gingeras TR, Guigó R (2012) Deep sequencing of subcellular RNA fractions shows splicing to be predominantly co-transcriptional in the human genome but inefficient for lncRNAs. Genome Res 22:1616–1625. doi:10.1101/gr.134445.111

Tomek W, Smiljakovic T (2005) Activation of Akt (protein kinase B) stimulates metaphase I to metaphase II transition in bovine oocytes. Reprod Camb Engl 130:423–430. doi:10.1530/rep.1.00754

Tomek W, Wollenhaupt K (2012) The "closed loop model" in controlling mRNA translation during development. Anim Reprod Sci 134:2–8. doi:10.1016/j.anireprosci.2012.08.005

Tomek W, Torner H, Kanitz W (2002) Comparative analysis of protein synthesis, transcription and cytoplasmic polyadenylation of mRNA during maturation of bovine oocytes in vitro. Reprod Domest Anim 37:86–91

Toralová T, Susor A, Nemcová L, Kepková K, Kanka J (2009) Silencing CENPF in bovine preimplantation embryo induces arrest at 8-cell stage. Reproduction 138:783–791. doi:10.1530/REP-09-0234

Tsai M-Y, Wang S, Heidinger JM, Shumaker DK, Adam SA, Goldman RD, Zheng Y (2006) A mitotic lamin B matrix induced by RanGTP required for spindle assembly. Science 311:1887–1893. doi:10.1126/science.1122771

Uzbekova S, Arlot-Bonnemains Y, Dupont J, Dalbiès-Tran R, Papillier P, Pennetier S, Thélie A, Perreau C, Mermillod P, Prigent C, Uzbekov R (2008) Spatio-temporal expression patterns of aurora kinases a, B, and C and cytoplasmic polyadenylation-element-binding protein in bovine oocytes during meiotic maturation. Biol Reprod 78:218–233. doi:10.1095/biolreprod.107.061036

Valiunas V, Polosina Y, Miller H, Potapova I, Valiuniene L, Doronin S, Mathias R, Robinson R, Rosen M, Cohen I, Brink P (2005) Connexin-specific cell-to-cell transfer of short interfering RNA by gap junctions. J Physiol 568:459–468. doi:10.1113/jphysiol.2005.090985

Vassalli JD, Huarte J, Belin D, Gubler P, Vassalli A, O'Connell ML, Parton LA, Rickles RJ, Strickland S (1989) Regulated polyadenylation controls mRNA translation during meiotic maturation of mouse oocytes. Genes Dev 3:2163–2171. doi:10.1101/gad.3.12b.2163

Vautier D, Besombes D, Chassoux D, Aubry F, Debey P (1994) Redistribution of nuclear antigens linked to cell proliferation and RNA processing in mouse oocytes and early embryos. Mol Reprod Dev 38:119–130. doi:10.1002/mrd.1080380202

Vazquez-Martin A, Oliveras-Ferraros C, Bernadó L, López-Bonet E, Menendez JA (2009) The serine 2481-autophosphorylated form of mammalian Target Of Rapamycin (mTOR) is localized to midzone and midbody in dividing cancer cells. Biochem Biophys Res Commun 380:638–643. doi:10.1016/j.bbrc.2009.01.153

VerMilyea MD, Maneck M, Yoshida N, Blochberger I, Suzuki E, Suzuki T, Spang R, Klein CA, Perry ACF (2011) Transcriptome asymmetry within mouse zygotes but not between early embryonic sister blastomeres. EMBO J 30:1841–1851. doi:10.1038/emboj.2011.92

Vinot S, Le T, Maro B, Louvet-Vallée S (2004) Two PAR6 proteins become asymmetrically localized during establishment of polarity in mouse oocytes. Curr Biol 14:520–525. doi:10.1016/j.cub.2004.02.061

Visa N, Puvion-Dutilleul F, Harper F, Bachellerie J-P, Puvion E (1993) Intranuclear distribution of poly(A) RNA determined by electron microscope in situ hybridization. Exp Cell Res 208:19–34

Wang X-P, Cooper NGF (2010) Comparative in silico analyses of cpeb1-4 with functional predictions. Bioinforma Biol Insights 4:61–83

Wang C-F, Huang Y-S (2012) Calpain 2 activated through N-methyl-D-aspartic acid receptor signaling cleaves CPEB3 and abrogates CPEB3-repressed translation in neurons. Mol Cell Biol 32:3321–3332. doi:10.1128/MCB.00296-12

Watson AJ, Wiemer KE, Arcellana-Panlilio M, Schultz GA (1992) U2 small nuclear RNA localization and expression during bovine preimplantation development. Mol Reprod Dev 31:231–240. doi:10.1002/mrd.1080310402

Weatheritt RJ, Gibson TJ, Babu MM (2014) Asymmetric mRNA localization contributes to fidelity and sensitivity of spatially localized systems. Nat Struct Mol Biol 21:833–839. doi:10.1038/nsmb.2876

Weston A, Sommerville J (2006) Xp54 and related (DDX6-like) RNA helicases: roles in messenger RNP assembly, translation regulation and RNA degradation. Nucleic Acids Res 34:3082–3094. doi:10.1093/nar/gkl409

Yaba A, Bianchi V, Borini A, Johnson J (2008) A putative mitotic checkpoint dependent on mTOR function controls cell proliferation and survival in ovarian granulosa cells. Reprod Sci 15:128–138. doi:10.1177/1933719107312037

Yan W, Rajkovic A, Viveiros MM, Burns KH, Eppig JJ, Matzuk MM (2002) Identification of Gasz, an evolutionarily conserved gene expressed exclusively in germ cells and encoding a protein with four ankyrin repeats, a sterile-alpha motif, and a basic leucine zipper. Mol Endocrinol 16:1168–1184. doi:10.1210/mend.16.6.0864

Yan W, Ma L, Zilinski CA, Matzuk MM (2004) Identification and characterization of evolutionarily conserved pufferfish, zebrafish, and frog orthologs of GASZ. Biol Reprod 70:1619–1625. doi:10.1095/biolreprod.103.024778

Yanagimachi R (1994) Mammalian fertilization, In Physiology of reproduction. The Ravens, New York, pp 189–317

Yang L, Duff MO, Graveley BR, Carmichael GG, Chen LL (2011) Genomewide characterization of non-polyadenylated RNAs. Genome Biol 12(2):R16. doi:10.1186/gb-2011-12-2-r16 (Epub 2011 Feb 16)

Yi K, Rubinstein B, Unruh JR, Guo F, Slaughter BD, Li R (2013) Sequential actin-based pushing forces drive meiosis I chromosome migration and symmetry breaking in oocytes. J Cell Biol 200(5):567–576. doi:10.1083/jcb.201211068 (Epub 2013 Feb 25)

Yu J, Yaba A, Kasiman C, Thomson T, Johnson J (2011) mTOR controls ovarian follicle growth by regulating granulosa cell proliferation. PloS One 6:e21415. doi:10.1371/journal.pone.0021415

Yurttas P, Vitale AM, Fitzhenry RJ, Cohen-Gould L, Wu W, Gossen JA, Coonrod SA (2008) Role for PADI6 and the cytoplasmic lattices in ribosomal storage in oocytes and translational control in the early mouse embryo. Development 135:2627–2636. doi:10.1242/dev.016329

Zamore PD, Williamson JR, Lehmann R (1997) The Pumilio protein binds RNA through a conserved domain that defines a new class of RNA-binding proteins. RNA 3:1421–1433

Zeng F, Baldwin DA, Schultz RM (2004) Transcript profiling during preimplantation mouse development. Dev Biol 272:483–496. doi:10.1016/j.ydbio.2004.05.018

Zhou Y, King ML (2004) Sending RNAs into the future: RNA localization and germ cell fate. IUBMB Life 56:19–27. doi:10.1080/15216540310001658886

Zong L, Zhu Y, Liang R, Zhao H-B (2016) Gap junction mediated miRNA intercellular transfer and gene regulation: a novel mechanism for intercellular genetic communication. Sci Rep. doi:10.1038/srep19884

Zuccotti M, Ponce RH, Boiani M, Guizzardi S, Govoni P, Scandroglio R, Garagna S, Redi CA (2002) The analysis of chromatin organisation allows selection of mouse antral oocytes competent for development to blastocyst. Zygote 10:73–78

Zuccotti M, Merico V, Cecconi S, Redi CA, Garagna S (2011) What does it take to make a developmentally competent mammalian egg? Hum Reprod Update 17:525–540. doi:10.1093/humupd/dmr009

Chapter 13
Regulation of Translationally Repressed mRNAs in Zebrafish and Mouse Oocytes

Tomoya Kotani, Kaori Maehata, and Natsumi Takei

Abstract From the beginning of oogenesis, oocytes accumulate tens of thousands of mRNAs for promoting oocyte growth and development. A large number of these mRNAs are translationally repressed and localized within the oocyte cytoplasm. Translational activation of these dormant mRNAs at specific sites and timings plays central roles in driving progression of the meiotic cell cycle, axis formation, mitotic cleavages, transcriptional initiation, and morphogenesis. Regulation of the localization and temporal translation of these mRNAs has been shown to rely on *cis*-acting elements in the mRNAs and *trans*-acting factors recognizing and binding to the elements. Recently, using model vertebrate zebrafish, localization itself and formation of physiological structures such as RNA granules have been shown to coordinate the accurate timings of translational activation of dormant mRNAs. This subcellular regulation of mRNAs is also utilized in other animals including mouse. In this chapter, we review fundamental roles of temporal regulation of mRNA translation in oogenesis and early development and then focus on the mechanisms of mRNA regulation in the oocyte cytoplasm by which the activation of dormant mRNAs at specific timings is achieved.

Abbreviations

APC/C	Anaphase-promoting complex/cyclosome
CPE	Cytoplasmic polyadenylation element
CPEB	Cytoplasmic polyadenylation element-binding protein
CPSF	Polyadenylation specificity factor
CSF	Cytostatic factor
DAZ	Deleted in Azoospermia
EJC	Exon-exon junction complex
Emi2	Early mitotic inhibitor 2
GV	Germinal vesicle

T. Kotani (✉) • K. Maehata • N. Takei
Department of Biological Sciences, Faculty of Science, Hokkaido University, North 10 West 8, Sapporo, Hokkaido 060-0810, Japan
e-mail: tkotani@sci.hokudai.ac.jp

M. Kloc (ed.), *Oocytes*, Results and Problems in Cell Differentiation 63,
DOI 10.1007/978-3-319-60855-6_13

GVBD	Germinal vesicle breakdown
MAPK	Mitogen-activated protein kinase
MBE	Musashi-binding element
METRO	Messenger transport organizer
MPF	Maturation/M-phase-promoting factor
PABP	Poly(A)-binding protein
PAP	Poly(A) polymerase
PBE	Pumilio-binding element
P-body	Processing body
PGC	Primordial germ cell
UTR	Untranslated region
TACC	Transforming acidic coiled-coil domain protein
4E-T	eIF4E-Transporter

13.1 Introduction

13.1.1 Oogenesis and Early Development of Vertebrates

In most vertebrates, oogonia differentiate into oocytes after the number of cells has increased through several mitotic divisions. Subsequently, the oocytes replicate DNA and enter prophase of meiosis I progressing until the diplotene stage, in which many transcripts are actively transcribed from four copies of each chromosome (Bachvarova et al. 1982; Hill and Macgregor 1980; Moore et al. 1974; Rosbash and Ford 1974). Oocytes remain arrested at the diplotene stage while growing. Fully grown oocytes in this stage are still unable to be fertilized. These oocytes are termed immature oocytes (Fig. 13.1). In response to specific cues such as hormones, oocytes resume meiotic cell cycles (Masui and Clarke 1979). An obvious morphological change that occurs after resumption of meiosis is germinal vesicle breakdown (GVBD), i.e., the breakdown of oocyte nuclear envelope, followed by spindle formation, chromosome segregation, and polar body extrusion. After entry into meiosis II, oocytes of most vertebrates are arrested again at metaphase II with a spindle and chromosomes aligned at the equatorial plane. The process from resumption of meiosis to arrest at metaphase II is termed oocyte maturation, through which oocytes acquire fertility. The oocytes arrested at metaphase II are termed mature oocytes (Fig. 13.1). Following maturation, oocytes are ovulated and fertilized with sperm.

In some vertebrate species, shortly after initiation of prophase of meiosis I, oocytes form the animal-vegetal axis through assembly and disassembly of the Balbiani body (also known as mitochondrial cloud) (Gupta et al. 2010; Heim et al. 2014; Marlow and Mullins 2008), a transient and universal structure consisting of mitochondria, Golgi apparatus, endoplasmic reticulum, and certain RNAs (Kloc et al. 2014). Oocytes of most vertebrates hold this axis during oogenesis until completion of oocyte maturation. Based on the animal-vegetal axis, the dorsal-ventral axis of the

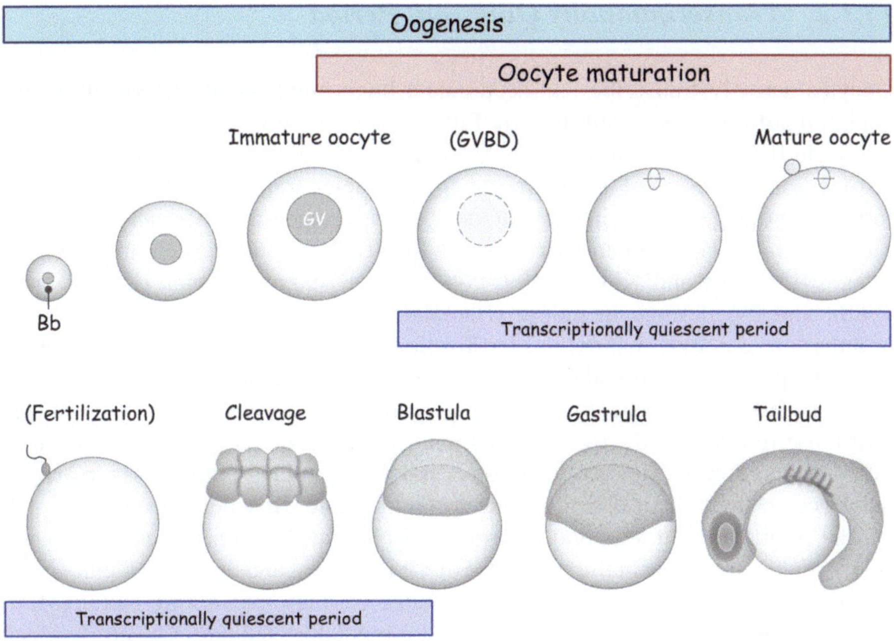

Fig. 13.1 Oogenesis, oocyte maturation, fertilization, and development. Diagrams refer to those of zebrafish. Oocytes are arrested at prophase I of meiosis I and actively transcribe huge numbers of mRNAs during growth. Fully grown oocytes in this stage are unable to be fertilized and are called immature oocytes. In response to specific signals such as hormones, oocytes begin oocyte maturation and undergo germinal vesicle breakdown (GVBD), followed by spindle formation and polar body extrusion. Oocytes skip an intervening period of DNA replication in the transition from meiosis I to meiosis II, which is important for generating a haploid egg. Most vertebrate oocytes are arrested at metaphase II, resulting in fertilizable mature oocytes. Following maturation, oocytes are ovulated and fertilized. Embryos initiate mitotic cleavages, followed by differentiation and morphogenesis. After the initiation of oocyte maturation, transcription becomes quiescent until mitotic cleavage stages. Although the exact time at which zygotic transcription is initiated varies from species to species, it is at the 1000-cell stage in zebrafish embryos. *Bb* Balbiani body, *GV* germinal vesicle

embryo is established shortly after fertilization in fish and frogs (Jesuthasan and Stahle 1997; Lu et al. 2011; Nojima et al. 2010; Tao et al. 2005; Tran et al. 2012; Vincent et al. 1986). At the same time, fertilized eggs complete meiosis, begin mitotic cleavages, and increase the number of cells (Fig. 13.1). When thousands of cells are produced, embryos begin morphogenesis. Gastrulation is an important process that generates the three germ layers (ectoderm, mesoderm, and endoderm) and the patterning of the dorsal-ventral and anterior-posterior regions.

13.1.2 Transcriptionally Quiescent Period

Oocytes actively transcribe and accumulate huge numbers of mRNAs from the beginning of oogenesis until they are fully grown. The transcription then becomes quiescent from the initiation of oocyte maturation (Susor and Kubelka 2017). After fertilization and fusion of the sperm and egg pronuclei, transcription from the newly formed zygotic genome is initiated at the 4000-cell stage in the African clawed frog *Xenopus laevis* (Newport and Kirschner 1982), 1000-cell stage in zebrafish (Harvey et al. 2013; Kane and Kimmel 1993), 4- to 8-cell stage in human (Braude et al. 1988), and 2-cell stage in mouse (Hamatani et al. 2004). This means that the progression of oocyte maturation, fertilization, and mitotic cleavage is thoroughly driven by posttranscriptional regulations in all animals. In addition, in zebrafish and frogs, dorsal-ventral axis formation and germ line development are regulated posttranscriptionally. Of these regulations, translational activation of dormant mRNAs stored in oocytes at specific sites and timings plays central roles in driving the processes that occur during oocyte maturation and early development.

The significance of temporal control of mRNA translation has been most extensively studied in oocyte maturation. One of the important and universal factors in the progression of oocyte maturation is maturation/M-phase-promoting factor (MPF) consisting of a catalytic subunit, Cdc2, and a regulatory subunit, Cyclin B, which phosphorylates a large number of substrates to regulate meiotic cell cycles as well as mitotic division in all eukaryotes studied so far (Nurse 1990). In contrast to the presence of Cdc2 protein, Cyclin B protein is absent in immature oocytes of the most of studied mammals, amphibians, and fish except *Xenopus* and mouse (German et al. 2015; Ihara et al. 1998; Kanayama et al. 2002; Kondo et al. 1997; Kotani and Yamashita 2002; Kotani et al. 2001; Levesque and Sirard 1996; Naito et al. 1995; Sakamoto et al. 1998; Tanaka and Yamashita 1995). Therefore, synthesis of Cyclin B protein from stored mRNAs is prerequisite for promoting oocyte maturation in these species. Although *Xenopus* and mouse contain Cyclin B protein in immature oocytes, newly synthesized Cyclin B protein is necessary for the transition between meiosis I and meiosis II, which occurs without an intervening period of DNA replication (Hochegger et al. 2001; Ledan et al. 2001). Mos, another universal factor, is synthesized from stored mRNAs in the early period prior to or at the same time as synthesis of Cyclin B protein in all vertebrates studied so far, and it promotes activation of MPF (Sagata et al. 1989). Attenuation of Mos protein synthesis causes a delay in GVBD, pronuclear formation after completion of meiosis I, and failure in arrest at metaphase II, through reduction in MPF activity (Araki et al. 1996; Choi et al. 1996; Dupre et al. 2002; Furuno et al. 1994; Gross et al. 2000; Kotani and Yamashita 2002). In the late period of oocyte maturation after entry into meiosis II, Wee1 protein is synthesized from stored mRNAs and acts as an inhibitory kinase for Cdc2 after fertilization (Murakami et al. 1999). Precocious synthesis of Wee1 protein in meiosis I causes DNA replication after completion of meiosis I by inactivating MPF (Nakajo et al. 2000). Therefore, Mos, Cyclin B, and Wee1 proteins must be synthesized in this order to correctly drive the progression of oocyte maturation.

13.1.3 A Model Vertebrate Zebrafish

Mechanisms of translational regulation of dormant mRNAs have been extensively studied in *Xenopus laevis* oocytes (Radford et al. 2008). Biochemical analyses have shown that the translational repression and activation of mRNAs depend on *cis*-acting elements typically located in the untranslated regions (UTRs) and *trans*-acting factors directly binding to the mRNAs. Analyses of reporter mRNAs injected into *Xenopus* oocytes have provided well-proven mechanisms of translational regulation of the mRNAs stored in oocytes (de Moor et al. 2005; MacNicol et al. 2015; Mendez and Richter 2001; Radford et al. 2008). However, recent studies have shown that mRNAs injected into the oocyte cytoplasm do not precisely recapitulate the regulation of endogenous mRNAs in *Drosophila* and zebrafish (Cha et al. 2001; Yasuda et al. 2010). These findings suggest that certain processes occurring during transcription in the nucleus are required for precise regulation of mRNAs in the cytoplasm.

Zebrafish is emerging as an excellent model vertebrate for biological research in the field of oogenesis and early development. Zebrafish can produce a large number of eggs and has short (2–3 months) generation time. These features facilitate genetic analysis including mutant screens and production of transgenic lines. Many maternal-effect mutants have been isolated and analyzed (Dosch et al. 2004; Kishimoto et al. 2004; Kotani and Kawakami 2008; Pelegri et al. 1999, 2004; Wagner et al. 2004). Fertilization and embryonic development in zebrafish proceed externally, and the embryos are transparent. These properties enable us to easily manipulate embryos and visualize particular cells during embryogenesis. In addition to genetic and developmental analyses, zebrafish is applicable to biochemical and cell biological analyses. Many localized mRNAs have been identified in zebrafish oocytes. Reporter mRNAs transcribed from transgenes can recapitulate the experiences of endogenous mRNAs in the nucleus, which are necessary for precise regulation of cytoplasmic localization and translation. In this chapter, we first describe accumulation and localization of some mRNAs during zebrafish oocyte growth.

13.2 Translationally Regulated mRNAs in Oocytes

13.2.1 Accumulation of Dormant mRNAs in Zebrafish Oocytes

During the course of oocyte growth, numerous mRNAs are transcribed from four copies of each chromosome. After being transcribed, some of the mRNAs are accumulated and stored in the oocyte cytoplasm as a translationally repressed form with apparent localizations. Zebrafish *cyclin B1* mRNA, for instance, is a prime example of such mRNAs that are transcribed, accumulated, and stored temporarily at a specific location (Kondo et al. 2001; Yasuda et al. 2010). Figure 13.2 shows diagrams depicting the physical movements of *cyclinB1* mRNA and some of the other mRNAs present at oocyte developmental stages, from the beginning of oogenesis to oocyte maturation.

Zebrafish oogenesis has been divided into five stages according to its morphological and physiological changes (Selman et al. 1993). At the very beginning of oogenesis, there is still no sign of *cyclin B1* mRNA localization (Fig. 13.2). In stage I of a primary growth stage, *cyclin B1* mRNAs are dispersed throughout the cytoplasm almost evenly. In stage II of a cortical alveolus stage, *cyclin B1* mRNAs start to move to the animal pole. Oocytes undergo vitellogenesis in stage III and resume meiosis after being fully grown in stage IV. In stage V, oocytes are arrested at metaphase II. In stages III–IV, there is already a clearly recognizable localization of mRNAs at the animal polar cytoplasm beneath a structure called the micropyle, which functions as a site of sperm entry and is located at the top of the animal pole. These mRNAs are inherited in the blastoderm formed at the animal pole of eggs after fertilization (see Fig. 13.2, lower panel). Translational repression of *cyclin B1* mRNA is released after resumption of meiosis (Kondo et al. 1997, 2001; Kotani et al. 2013; Yasuda et al. 2010), and the protein product, Cyclin B1, functions as the regulatory subunit of MPF.

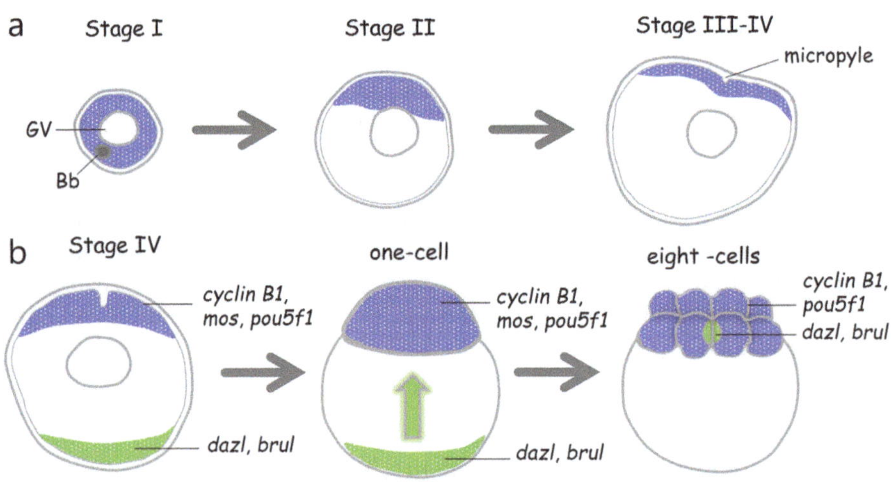

Fig. 13.2 Accumulation and localization of mRNAs during oogenesis and mitotic cleavages in zebrafish. (**a**) From the beginning of oogenesis, *cyclin B1* mRNA is transcribed and accumulated in oocytes (*blue*). *cyclin B1* mRNA is uniformly distributed within the cytoplasm except for the Balbiani body (Bb) at stage I and is transported to the animal pole of oocytes from stage II. The localization is restricted at the animal polar cytoplasm beneath the micropyle at stages III–IV. (**b**) In a region similar to that of *cyclin B1* mRNA, *mos* and *pou5f1* mRNAs are localized in stage IV oocytes (*blue*). These mRNAs are inherited in the blastoderm formed at the animal pole of fertilized eggs and are distributed within the cytoplasm. In contrast, *dazl* and *brul* mRNAs are localized at the vegetal pole of stage IV oocytes (*green*). These mRNAs move toward the animal pole after fertilization through cytoplasmic streaming (*arrow*). Four large aggregates containing *dazl* and *brul* mRNAs form at the distal ends of the cleavage furrows by the four-cell stage

Some other mRNAs, such as *mos* and *pou5f1*, are known to follow similar or even the same path as that followed by *cyclin B1* mRNA (see stage IV) (Horie and Kotani 2016; Howley and Ho 2000; Suzuki et al. 2009). *mos* mRNA is translated in the early period during oocyte maturation (Horie and Kotani 2016; Kajiura-Kobayashi et al. 2000). Mos protein is one of the factors promoting oocyte maturation and functions mainly through activating the mitogen-activated protein kinase (MAPK) signaling pathway. Pou5f1, also known as Pou2, has multiple roles after fertilization. It is one of the essential transcription factors that activate zygotic gene transcription during maternal-to-zygotic transition (Lee et al. 2013; Leichsenring et al. 2013).

In contrast to the examples mentioned above, *DAZ-like* (*dazl*) and *bruno-like* (*brul*) mRNAs are localized at the vegetal polar cytoplasm of the zebrafish oocyte (Kosaka et al. 2007; Maegawa et al. 1999; Suzuki et al. 2000). They then move slowly toward the animal pole after fertilization, running through the center of the egg by cytoplasmic streaming (shown in the diagram of Fig. 13.2 as a green arrow directed upward), and the journey ends eventually when they reach the blastoderm formed at the animal pole of the egg. At the four-cell stage, four large aggregates containing *dazl* and *brul* mRNAs form at the distal ends of the cleavage furrows (Hashimoto et al. 2004). *dazl* mRNA encodes a zebrafish homolog of Deleted in Azoospermia (DAZ), an RNA-binding protein that is known to be involved in primordial germ cell (PGC) development in *Xenopus* (Houston and King 2000) and spermatogenesis in mouse (Reijo et al. 1995; Ruggiu et al. 1997). Bruno-like belongs to a subgroup of the Elav-type RNA-binding proteins, and it is known to be involved in body patterning, PGC development, and gametogenesis in *Drosophila* (Kim-Ha et al. 1995; Webster et al. 1997). The cytoplasm containing the four large aggregates including localized *dazl* and *brul* mRNAs has been shown to function as germplasm containing the germ cell determinants, because depletion of the cytoplasm causes loss of PGC in zebrafish embryos (Hashimoto et al. 2004). However, how components assembled in these regions specify the germinal lineage remains to be investigated.

It is generally assumed that translational regulation of localized mRNAs can be divided into two types. One is spatial regulation (short-term repression), in which translation is repressed until the mRNAs reach their final destinations and is activated at the sites of mRNA localization. The other is temporal regulation (long-term repression), in which translation remains repressed even after the mRNAs have reached their final destinations and is activated at timings specific to distinct mRNAs, in most cases coupling with the time when protein products are required (Besse and Ephrussi 2008). These mRNAs described above are examples of temporal regulation of mRNA translation.

13.2.2 *Translational Activation of Dormant mRNAs During Oogenesis and Development*

From comprehensive analysis of transcription during zebrafish development by RNA sequencing, it was estimated that 18,000 out of 19,000 protein-coding transcripts are accumulated in eggs (Harvey et al. 2013). Although the exact number remains unknown, a large number of them including those described above are stored as a translationally repressed form in the oocyte cytoplasm. These dormant mRNAs are translated at timings specific to distinct mRNAs to properly drive the processes that occur during oogenesis and development. However, when and where the mRNAs are translated remains largely unresolved. We focus on the translation of some mRNAs, which is known to be essential for the progression of oocyte maturation and development.

13.2.2.1 Oocyte Maturation

During the period of oocyte maturation, thousands of dormant mRNAs are translated in a temporally ordered manner, which drives most of the processes that occur during oocyte maturation. Of these mRNAs, translational activation of *cyclin B1* mRNA after resumption of oocyte maturation is extremely important for promoting meiotic cell cycles and both nuclear and cytoplasmic maturation necessary for oocytes to acquire fertility in all animals. Cyclin B1 protein synthesized from stored mRNA forms a complex termed MPF with preexisting Cdc2 kinase and immediately activates the Cdc2 (de Moor and Richter 1999; Yamashita et al. 1995). MPF phosphorylates hundreds of substrates to properly promote meiotic cell cycles as well as cytoplasmic and nuclear maturation (Masui and Clarke 1979). For example, activated MPF translocates into the germinal vesicle and induces GVBD and chromosome condensation, followed by spindle formation with aligned chromosomes during the normal course of oocyte maturation. Precocious activation of MPF by the synthesis of Cyclin B1 from injected mRNA causes an abnormal organization of microtubules after GVBD in the Japanese brown frog *Rana japonica* oocytes and an error in segregation of chromosomes due to abnormal microtubule-kinetochore attachments in mouse oocytes (Davydenko et al. 2013; Kotani and Yamashita 2002). Therefore, in addition to the translational activation of *cyclin B1* mRNA, the accurate timing of activation is important to properly promote oocyte maturation.

At the end of meiosis I, Cyclin B1 is partially degraded by the anaphase-promoting complex/cyclosome (APC/C), leading to chromosome segregation and subsequent polar body extrusion. Continuous synthesis of Cyclin B1 from stored mRNAs is important to quickly elevate levels of MPF activity, which promotes entry into meiosis II without an intervening DNA replication, a process that is important for generating a haploid egg (Hochegger et al. 2001; Iwabuchi et al. 2000; Ledan et al. 2001). After entry into meiosis II, *cyclin B1* mRNA is successively translated, and the high activity of MPF maintains metaphase II arrest. Overall, translation of *cyclin B1* mRNAs plays a central role in the promotion of oocyte maturation.

The *mos* mRNA is translated in the early phase of oocyte maturation in all animals so far studied. The synthesized Mos protein immediately activates the MAPK pathway and is required for arrest at metaphase II as a component of cytostatic factor (CSF). Deficiency of Mos and inhibition of MAPK cause a delay in the occurrence of GVBD, pronuclear formation after completion of meiosis I, and failure in the arrest at metaphase II in frogs and mouse (Araki et al. 1996; Choi et al. 1996; Dupre et al. 2002; Furuno et al. 1994; Gross et al. 2000; Kotani and Yamashita 2002). Since MPF activity is reduced in these oocytes, Mos functions to promote activation of MPF most likely through stabilizing Cyclin B1. Independent of the MPF stabilization, the Mos-MAPK pathway induces microtubule reorganization prior to the activation of MPF in frog (Kotani and Yamashita 2002) or simultaneously with activation of MPF in mice (Verlhac et al. 1996). As in the case of *cyclin B1*, the timing of translational activation of *mos* mRNA is important for the progression of oocyte maturation, particularly for microtubule organization into a normal bipolar spindle and thus subsequent chromosome segregation.

The *emi2* mRNA encodes early mitotic inhibitor 2 (Emi2), an APC/C inhibitor, and is translated at the end of meiosis I during *Xenopus* and mouse oocyte maturation (Madgwick et al. 2006; Ohe et al. 2007). The synthesized Emi2 protein prevents complete degradation of Cyclin B1, which is required for the transition between meiosis I and meiosis II that occurs without an intervening period of DNA replication. After entry into meiosis II, Emi2 induces arrest at metaphase II as a component of CSF. Precocious synthesis of Emi2 in an early period of oocyte maturation at a level similar to that in meiosis II causes arrest of the meiotic cell cycle at metaphase I by most likely preventing the partial degradation of Cyclin B1 that is required for homologous chromosome segregation (Madgwick et al. 2006; Ohe et al. 2007; Tung et al. 2007). These findings indicate that the temporal regulation of *emi2* mRNA translation, that is, stable repression during the early period of meiosis I and activation at the end of meiosis I, is critical for proper progression of meiotic cell cycles and arrest at metaphase II, which is important for preventing parthenogenesis.

At the onset of meiosis II, *wee1* mRNA is translated in *Xenopus* oocytes (Charlesworth et al. 2000; Murakami and Vande Woude 1998; Nakajo et al. 2000). Wee1 protein is a kinase negatively regulating MPF activity by phosphorylation of Cdc2. Although Wee1 kinase seems to be unnecessary for the progression of meiosis, the synthesized Wee1 protein is required for extending the interphase of first embryonic cell cycle after fertilization, which may be necessary for fusion of the sperm and egg pronuclei (Murakami et al. 1999). Precocious synthesis of Wee1 in meiosis I causes a significant reduction in MPF activity after completion of meiosis I, resulting in DNA replication before entering into meiosis II (Nakajo et al. 2000). The stable repression of *wee1* mRNA during meiosis I and the translational activation after completion of meiosis I are therefore important for the normal progression of oocyte maturation and of the first embryonic cell cycle after fertilization.

13.2.2.2 Development

In addition to the period of oocyte maturation, certain numbers of dormant mRNAs are translated after fertilization, resulting in the promotion of various processes that occur not only by the time of transcriptional initiation but also later in embryogenesis. In regard to transcription, it has been shown that newly synthesized proteins from mRNAs stored in oocytes are required for initiation of zygotic gene transcription, because translational inhibition of eggs after fertilization causes a significant decrease in zygotic gene transcription in mouse (Hamatani et al. 2004; Wang and Latham 1997) and zebrafish (Harvey et al. 2013). Some transcriptional factors synthesized from stored mRNAs after fertilization have been isolated using mouse cleavage-stage embryos (Wang and Latham 2000). However, whether these factors direct the initial transcription of zygotic genes in mice remains to be elucidated.

Pou5f1 (also termed Oct4) was identified as an initiator of zygotic gene transcription using zebrafish embryos (Lee et al. 2013; Leichsenring et al. 2013). During the cleavage stages, *pou5f1* mRNA is actively translated, and the synthesized Pou5f1 protein triggers the initial transcription of zygotic genes through binding to the promoter sequences of distinct genes. Nanog and Sox19b are also actively synthesized from stored mRNAs in cleavage stages and promote the transcription of most initial zygotic genes cooperatively with Pou5f1. Interestingly, Pou5f1-deficient zebrafish embryos generated by maternal and zygotic mutation in the *pou5f1* coding sequence show severe defects similar to those caused by transcriptional inhibition when Nanog and Sox19b are simultaneously depleted, in which morphogenesis and differentiation are stopped before the beginning of gastrulation (Lee et al. 2013). These findings suggest that the initial zygotic transcription triggered by Pou5f1, Nanog, and Sox19b is required for subsequent transcriptional activation of numerous genes, which promote morphogenesis, differentiation, and gastrulation. Since Pou5f1/Oct4, Nanog, and Sox19b are regulatory factors for pluripotency in in vitro cultured stem cells, these findings provide a link between transcriptional activation in embryos and control of pluripotency in cells.

The Pou5f1-deficient zebrafish embryos derived from maternal and zygotic *pou5f1* mutants generate ectoderm and mesoderm layers during gastrulation but fail to form an endoderm layer (Lunde et al. 2004; Reim et al. 2004). In addition, cell movement during gastrulation is defective in the Pou5f1-deficient zebrafish embryos (Lachnit et al. 2008; Reim and Brand 2006). Furthermore, these embryos show an expansion of dorsal tissues to the ventral side (Belting et al. 2011; Reim and Brand 2006). Given that the zygotic *pou5f1* mutants seem to develop almost normally until the tail bud stage (Belting et al. 2001), translational activation of maternally provided *pou5f1* mRNA after fertilization is critical for lineage decision, morphogenesis, and patterning of embryo during development in zebrafish.

In *Xenopus*, the dorsal-ventral axis of embryos appears to be established shortly after fertilization by translocation of dormant *wnt11* mRNA, which is translated at the site of destination of the mRNA (Tao et al. 2005). *wnt11* mRNA is accumulated during oogenesis and localized at the cortex of the vegetal pole of fully grown oocytes (Ku and Melton 1993). Fertilization induces cortical rotation of the egg, by which *wnt11* mRNA translocates at the opposite side from the sperm entry point. Subsequent translation of *wnt11* mRNA during cleavage stages at the final destination results in activation of the canonical Wnt signaling pathway and induction of the transcription of organizer genes such as *chordin* by the late blastula stage (Schroeder et al. 1999; Tao et al. 2005). Protein products of these genes antagonize BMP signaling and specify the dorsal cell fate during the period of gastrulation.

Accumulated evidences from different sources have shown the significance of mRNA regulation at various levels including localization and temporal translation in order to promote oogenesis and development. This poses the question how do oocytes achieve the precise regulation of these mRNAs in a spatially and temporally ordered manner?

13.3 Mechanisms of mRNA Regulation in Oocytes

13.3.1 Mechanisms of the Localization of mRNAs Stored in Oocytes

Mechanisms of mRNA localization are generally important for regulating the sites where protein products are provided. The mechanisms of mRNA localization in the oocyte cytoplasm have been extensively analyzed in invertebrates (Martin and Ephrussi 2009; St Johnston 2005), because the nematode, *C. elegance*, and the fruit fly, *Drosophila*, are applicable for genetic analyses. In contrast to invertebrates, there are relatively few cases in which mechanisms of mRNA localization have been investigated in detail in vertebrate oocytes (Oh and Houston 2017).

Xcat2 mRNA encodes a *Xenopus* homolog of Nanos, and its localization changes dynamically during oogenesis. At the beginning of oogenesis, *Xcat2* mRNA is localized at the Balbiani body (mitochondrial cloud), and it is then incorporated into the germinal granules and finally localizes at the vegetal cortex in later stages. This pathway is called the messenger transport organizer (METRO) pathway (Kloc and Etkin 1995). Reporter mRNA injected into the nucleus of early-stage oocytes is localized at the Balbiani body within 22 h after injection (Zhou and King 1996). Using this system, Zhou and King (1996) showed that localization of *Xcat2* mRNA at the Balbiani body depends on the sequences of *Xcat2* 3′ UTR, particularly within 250 nucleotides proximal to the coding region. By using a similar system, Kloc et al. (2000) showed that the distal region of the 3′ UTR consisting of 164 nucleotides is responsible for the localization of mRNAs at the germinal granules, which is achieved within 48 h after injection.

A transgenic approach is useful for analyzing mechanisms of the whole processes of mRNA localization during oogenesis and development in vertebrates as well as in invertebrates. Zebrafish *dazl* mRNA is localized at the Balbiani body in stage I oocytes and then moves to the vegetal cortex by stage III (Kosaka et al. 2007; Marlow and Mullins 2008). As mentioned earlier, the mRNA migrates to the blastoderm through cytoplasmic streaming after fertilization and is localized as four aggregates formed at the distal end of cleavage furrows at the four-cell stage (see Fig. 13.2). A reporter mRNA transcribed from the transgene containing the full-length sequence of *dazl* 3′ UTR recaptures the whole localizations of endogenous mRNA from the beginning of oogenesis to the cleavage stages (Kosaka et al. 2007). By generating many transgenic lines, Kosaka et al. (2007) showed that the localizations of *dazl* mRNA at the Balbiani body, vegetal cortex, and distal end of cleavage furrows are directed by distinct *cis*-acting elements. Furthermore, a *cis*-acting element in the middle region of the 3′ UTR, which is responsible for localization at the Balbiani body, is prerequisite for a *cis*-acting element in a proximal region to direct the vegetal cortex localization. Similarly, the *cis*-acting element responsible for localization at the vegetal cortex is prerequisite for an upstream *cis*-acting element to direct the localization at the distal end of cleavage furrows. These findings show that mRNA localization is regulated by well-ordered mechanisms, in which different *cis*-acting elements function in distinct and multistep processes.

The mechanisms of *cyclin B1* mRNA localization in oocytes have been analyzed by a transgenic approach in zebrafish (Kotani et al. 2006; Yasuda et al. 2010, 2013). Localization of *cyclin B1* mRNA at the animal polar cytoplasm is divided into, at least, two steps. The early step is translocation to the animal pole, and the late step is anchoring at their final destination, the cortical cytoplasm. By stage III, animal polar localization of *cyclin B1* mRNA is directed by the sequences of its 3′ UTR. A Pumilio-binding element (PBE) in the 3′ UTR of *cyclin B1* mRNA has been shown to be one of the *cis*-acting elements responsible for localization at the animal polar cytoplasm, because a mutation in this sequence prevents binding of Pumilio protein and causes mRNA distribution throughout the animal hemisphere of oocytes (Kotani et al. 2013). In stage IV fully grown oocytes, a *cis*-acting element in the coding region is required for the anchoring of *cyclin B1* mRNA at the cortical cytoplasm (Yasuda et al. 2010). This anchoring is dependent on the actin filaments (Kondo et al. 2001). The nine-nucleotide sequence CAGGAGACC in the coding region, which is almost completely conserved in vertebrates irrespective of the fact that the third nucleotide of each codon can substitute for others to encode the identical amino acids, is responsible for the anchoring (Yasuda et al. 2013). This process is mediated by a *trans*-acting factor(s) binding to the *cis*-acting element. The precious mechanisms underlying *cyclin B1* mRNA localization including identification of *trans*-acting factors remain to be resolved.

Zebrafish *syntabulin* mRNA is localized at the vegetal pole of oocytes through a Balbiani body-dependent pathway (Nojima et al. 2010) as in the case of *dazl* mRNA. However, in contrast to the *dazl* mRNA localization, which is dependent solely on the 3′ UTR, localization of *syntabulin* mRNA requires the intron-exon structure, because mRNAs transcribed from a transgene containing sequences of the coding region and the 3′ UTR do not localize, while mRNAs transcribed from a transgene containing exons and internal introns localize at the vegetal cortex. This finding suggests that factors assembled on the mRNA such as components of the exon-exon junction complex (EJC) during splicing are involved in the *syntabulin* mRNA localization after being exported to the oocyte cytoplasm, as in the case of *oskar* mRNA in *Drosophila* (Hachet and Ephrussi 2001, 2004). Synthesized Syntabulin protein from the localized mRNA is distributed at the vegetal pole of eggs shortly after fertilization and then translocates to one side of vegetal pole by the two-cell stage. Syntabulin functions in transporting dorsal determinants to the prospective dorsal side of the embryo, because maternal-effect mutants deficient in a *syntabulin* transcript, *tokkaebi*, show defects in the dorsal determination, resulting in a ventralized embryo (Nojima et al. 2010). This means that experiences in the process of splicing in the nucleus are important for proper mRNA localization in the oocyte cytoplasm and specification of the dorsal-ventral axis in embryos.

13.3.2 Mechanisms of the Translational Regulation of mRNAs Stored in Oocytes

13.3.2.1 The *cis*-Acting Elements and *trans*-Acting Factors

Mechanisms of the translational repression of mRNAs and their temporal activation are important for regulating the timing when protein products are provided. Independent of the analyses for mRNA localizations, mechanisms of translational regulation have been biochemically analyzed mainly during oocyte maturation using *Xenopus*. Most of the translationally repressed mRNAs, including *cyclin B1*, *mos*, *wee1*, and *emi2*, are stored in the oocyte cytoplasm with short poly (A) tails (~20–40 nucleotides). Cytoplasmic polyadenylation of these dormant mRNAs drives the translational activation of the mRNAs in *Xenopus* and mouse oocytes (Barkoff et al. 2000; McGrew et al. 1989; Sheets et al. 1994; Stutz et al. 1998; Tay et al. 2000; Vassalli et al. 1989). The cytoplasmic polyadenylation element (CPE) was identified as a *cis*-acting element required for the cytoplasmic polyadenylation of dormant mRNAs (Fox et al. 1989; McGrew et al. 1989; McGrew and Richter 1990). Its consensus sequence is UUUUAUU or UUUUAAU. The CPE-binding protein CPEB functions as a *trans*-acting factor directing both the translational repression and cytoplasmic polyadenylation of mRNAs (Hake and Richter 1994; Stebbins-Boaz et al. 1996). Putative CPEs have been found in thousands of mRNAs. In addition, we know that temporal regulation of mRNA translation requires complex mRNA-protein interactions. The temporal mRNA translation directed by *cis*-acting elements and *trans*-acting factors occurs, for example, in *Xenopus cyclin B1* mRNA.

Translational repression of *Xenopus cyclin B1* mRNA in immature oocytes is directed by sequences of its 3′ UTR including multiple CPEs. Three of these CPEs mainly mediate the repression of mRNA translation, because a reporter mRNA containing three CPEs is translationally repressed at a level similar to that of a reporter mRNA containing the full-length 3′ UTR (de Moor and Richter 1999). CPEBs bind to the CPEs and a protein called Maskin (Stebbins-Boaz et al. 1999). Maskin binds to the 5′ cap-binding initiation factor eIF4E, which prevents binding of eIF4G to the eIF4E-cap complex. Since binding of eIF4G to the eIF4E-cap complex is required for recruitment of the 40S ribosomal subunit, the Maskin-eIF4E complex prevents translation of the mRNA at the initiation step. After initiation of oocyte maturation, CPEB recruits cleavage and polyadenylation specificity factor (CPSF) to the polyadenylation hexanucleotide, AAUAAA, located at the end of the 3′ UTR (Mendez et al. 2000). CPSF brings the poly(A) polymerase (PAP) to the mRNA. Elongation of the poly(A) tail by PAP results in recruitment of poly(A)-binding protein (PABP), which promotes dissociation of Maskin from eIF4E, resulting in recruitment of eIF4G to eIF4E (Cao and Richter 2002). The eIF4E-eIF4G-PABP interactions recruit the 40S ribosomal subunit to the 5′ cap, leading to subsequent translational activation of the mRNA. In addition, the "closed loop" complex circularized by the eIF4E-eIF4G-PABP complex facilitates successive translation by polysomes (de Moor et al. 2005; Mendez and Richter 2001; Radford et al. 2008).

Since, in many vertebrate species other than *Xenopus*, Maskin homologs, transforming acidic coiled-coil (TACC) domain proteins, do not contain an eIF4E-binding motif and are not predicted to interact with eIF4E (de Moor et al. 2005), it is controversial whether TACCs in other species play the same roles as Maskin in *Xenopus*. Recent studies have shown that a highly conserved eIF4E-binding protein, eIF4E-Transporter (4E-T), interacts with CPEB and eIF4E, which prevents binding of eIF4G to eIF4E (Kamenska et al. 2014; Minshall et al. 2007; Villaescusa et al. 2006). Nevertheless, these models explain how the dormant mRNAs are stored in oocytes and translated during oocyte maturation. However, a large number of mRNAs including *mos*, *wee1*, and *emi2* in addition to *cyclin B1* contain CPEs in their 3′ UTR and are translated at different timings after resumption of the meiotic cell cycle to properly promote oocyte maturation. Considering this, there must be other mechanisms for achieving temporally ordered translational activation of the distinct mRNAs. One of these mechanisms relies on mRNA specificity factors such as Pumilio and Musashi.

Xenopus cyclin B1 mRNA contains Pumilio-binding element (PBE) in its 3′ UTR (Nakahata et al. 2001). The PBE consensus sequence is UGUANAUA (in which N is any nucleotide). Pumilio is a member of the PUF family of sequence-specific RNA-binding proteins that is highly conserved from yeast to mammals. *Xenopus* Pumilio1 binds to *cyclin B1* mRNA and CPEB but not to *mos* mRNA (Nakahata et al. 2001; Ota et al. 2011; Pique et al. 2008). Pumilio1 has been shown to play a role in determining the timing of *cyclin B1* translational activation but not that of *mos*, because inhibition of Pumilio1 by specific antibodies accelerates the translational activation of *cyclin B1* mRNA but not that of *mos* mRNA (Nakahata et al. 2003; Ota et al. 2011). Zebrafish and mouse *cyclin B1* mRNAs also contain PBEs in their 3′ UTR. In both species, Pumilio1 binds to *cyclin B1*, but not *mos*, mRNA in vivo (Kotani et al. 2013). This binding is required for the assembly of *cyclin B1* RNA granules, a structure functioning as a regulatory mechanism in achieving temporal regulation of mRNA translation. This mechanism will be described below.

Xenopus mos mRNA is translated at an early timing prior to the translation of *cyclin B1* mRNA. *Xenopus mos* mRNA contains Musashi-binding elements (MBEs) in addition to the CPE in its 3′ UTR (Charlesworth et al. 2002, 2006). The MBE consensus sequence is (G/A)U_{1-3}AGU. Musashi is a sequence-specific RNA-binding protein that is highly conserved from the fly to mammals. *Xenopus* Musashi binds to the MBE of *mos* mRNA and directs translational activation of the mRNAs in an early period. In contrast to *Xenopus mos*, mouse *Mos* and human *MOS* mRNAs do not contain an MBE. This is consistent with the fact that mouse *Mos* mRNA is translated at timing similar to that of *Cyclin B1* in this species. Indeed, a reporter mRNA containing the human *MOS* 3′ UTR is translated later than that containing the *Xenopus mos* 3′ UTR in *Xenopus* oocytes, suggesting species-specific differences dependent on the *cis*-acting elements of the mRNAs (Prasad et al. 2008). Translation of mouse *Mos* mRNA is mainly directed by CPEs (Gebauer and Richter 1996; Gebauer et al. 1994). Zebrafish *mos* mRNA contains a putative MBE in addition to CPEs and is polyadenylated during an early period of oocyte maturation (Horie and Kotani 2016). Whether the MBE in the zebrafish *mos* mRNA functions as a binding site of Musashi is unknown, and further studies are needed to determine the participation of Musashi in the translation regulation of mRNA in this species.

Xenopus emi2 and *wee1* mRNAs are translated at the end of meiosis I and after entry into meiosis II, respectively. The *emi1* and *wee1* mRNAs contain three CPEs in their 3′ UTR, all of which function as binding sites of CPEBs. These CPEs direct the translational repression and activation of mRNAs as in the case of *cyclin B1* mRNA (Charlesworth et al. 2000; Tung et al. 2007). However, *emi2* mRNA is translated in a period later than that of *cyclin B1* mRNA and earlier than that of *wee1* mRNA to ensure the transition from meiosis I to meiosis II by omitting an intervening period of DNA replication (Nakajo et al. 2000; Ohe et al. 2007). Pumilio1 binds to *cyclin B1* mRNA and mediates translational activation of *cyclin B1* mRNA in a period that is earliest in those of the three mRNAs. There may be additional factors or mechanisms regulating the translational activation of *emi2* and *wee1* mRNAs at distinct timings.

13.3.2.2 Comprehensive Changes in Translational Regulation in Mouse Oocytes

Studies of temporal regulation of dormant mRNAs have focused on individual mRNAs and RNA-binding proteins or some populations of mRNAs. Recently, comprehensive changes in translational regulation of whole mRNAs during oocyte maturation have been shown in mouse oocytes (Chen et al. 2011). In mouse oocytes, large numbers of maternal mRNAs including key components of the cell cycle are accumulated and regulated at the level of translation as in the case of fish and frog oocytes. These mRNAs are recruited to polysomes in a manner dependent on meiotic stages during the progression of oocyte maturation. Approximately 7600 mRNAs have been shown to be translated during oocyte maturation by microarray analysis of RNAs purified from the polysomes of oocytes at the GV, metaphase I, and metaphase II stages (Chen et al. 2011). These mRNAs are classified into three categories according to the patterns of polysome recruitment in GV- and metaphase II-stage oocytes. Approximately 4800 mRNAs are constitutively recruited on polysomes (class I). In contrast, the amount of other 1500 mRNAs recruited to polysomes decreases more than twofold during oocyte maturation (class II), and the amount of other 1300 mRNAs on polysomes increases by more than twofold during maturation (class III).

Class II mRNAs start to be translationally repressed during the progression of oocyte maturation by, at least, two mechanisms. One is mRNA degradation, and the other is dissociation from polysomes without degradation. The total amount of 800 mRNAs of class II decreases after initiation of oocyte maturation, limiting the translation of these mRNAs. The timings of mRNA degradation are divided into two types: an early period before metaphase I and a late period from metaphase I to metaphase II. On the other hand, the total amount of 400 mRNAs in class II remains constant during oocyte maturation, but the amount in the polysome-bound fraction decreases. These findings show that the release from polysomes and degradation are directed in mRNA-specific manner and are temporally regulated.

Class III mRNAs are stable during oocyte maturation and are translationally activated in distinct periods. Some of these mRNAs encode cell cycle regulators such as Cyclin B1 and Mos and components of the anaphase-promoting complex/cyclosome (APC/C) and the spindle checkpoint. mRNAs encoding components required for chromosome attachment and segregation are also included in this class. These class III mRNAs contain several motifs in their 3′ UTRs. The most abundant motif enriched in class III mRNAs is the cytoplasmic polyadenylation element (CPE). Consensus-binding elements for Pumilio, Musashi, and DAZ family of proteins are also enriched in the 3′ UTRs of class III mRNAs. The amounts of most of the mRNAs that do not contain a CPE or DAZL-binding element decrease in the polysome fraction during oocyte maturation (class II), whereas the amounts of mRNAs containing two or more CPEs or DAZL-binding elements increase in the polysome fraction during oocyte maturation (class III). Some of these mRNAs are regulated synergistically by CPEB1 and DAZL (Chen et al. 2011; Sousa Martins et al. 2016). *Dazl* mRNA itself is one of the mRNAs regulated by CPEB1 and DAZL. CPEB1 binds to *Dazl* mRNA in immature oocytes and activates translation of the mRNA in the early period after initiation of oocyte maturation. The synthesized DAZL protein facilitates the translation of *Dazl* mRNA itself, and subsequently the accumulated DAZL activates the translation of a subset of class III mRNAs, which is critical for the normal progression of oocyte maturation including bipolar spindle formation, chromosome segregation, and metaphase II arrest (Chen et al. 2011). *Tex19.1* mRNA is one of the CPEB1 and DAZL target mRNAs (Sousa Martins et al. 2016). Similar to *Dazl* mRNA, CPEB1 and DAZL promote the initial phase of translation of *Tex19.1* mRNA, whereas accumulated DAZL continues to play a role in facilitating translation during the late period of oocyte maturation. These findings indicate that interplay between CPEB1 and DAZL is an essential component for translational regulation during oocyte maturation.

13.3.2.3 Localization and Granule Formation

The comprehensive analysis of mRNA translation in mouse shows that translation of thousands of mRNAs is temporally regulated during oocyte maturation in a well-ordered manner. Furthermore, recent studies have suggested that very large complexes of mRNA-protein and protein-protein interactions are involved in the regulation of mRNAs in *Xenopus* oocyte (Minshall et al. 2007; Nakamura et al. 2010). However, it is still controversial whether mRNA translation is absolutely regulated by interactions between *cis*-acting elements and *trans*-acting factors and by some additional associating proteins. Very recent studies have indicated a new mode of translational regulation, that is, localization itself and formation of physiological structures such as RNA granules, which regulate accurate timing of translational activation of dormant mRNAs (Horie and Kotani 2016; Kotani et al. 2013; Yasuda et al. 2013).

As mentioned above, zebrafish *cyclin B1* mRNA localization at the animal polar cytoplasm during oocyte growth is divided into two steps, translocation and anchoring. The translocation is solely directed by the sequences of its 3′ UTR. In contrast, the anchoring is directed by the *cis*-acting element consisting of nine nucleotides in the coding region (Yasuda et al. 2010, 2013). The anchoring is most likely mediated by a *trans*-acting factor(s) binding to the *cis*-element. Translational repression in immature oocytes and activation during oocyte maturation are solely directed by the 3′ UTR. An intriguing finding is that anchoring of *cyclin B1* mRNA at the cortical cytoplasm is required for the regulation of translational activation at the accurate timing, because a reporter mRNA carrying full-length *cyclin B1* mRNA is anchored and translated at the accurate timing, while a reporter mRNA carrying a mutation in the *cis*-acting element consisting of nine nucleotides in the coding region is not anchored and is precociously translated in an early period after resumption of oocyte maturation (Yasuda et al. 2013). Although the possibility that the *cis*-acting element of the coding region and the *trans*-acting factor(s) directly participate in translational regulation cannot be ruled out, this finding suggests that anchoring of the mRNA at its final destination is important for precise regulation of its translation probably due to the requirement of properly assembled surroundings or local environment.

In addition to localization at a specific region, formation of subcellular structures such as RNA granules appears to function as a regulatory mechanism of temporal translation. The *cyclin B1* mRNA forms microscopically visible RNA granules at the animal polar cytoplasm beneath the micropyle in zebrafish oocytes (Fig. 13.3) (Horie and Kotani 2016; Kotani et al. 2013; Nukada et al. 2015). After initiation of oocyte maturation, *cyclin B1* RNA granules dissociate at the timing at which the mRNAs are polyadenylated (Fig. 13.3). At a similar timing, translation of *cyclin B1* mRNA is activated, and synthesized Cyclin B1 protein promotes MPF activation,

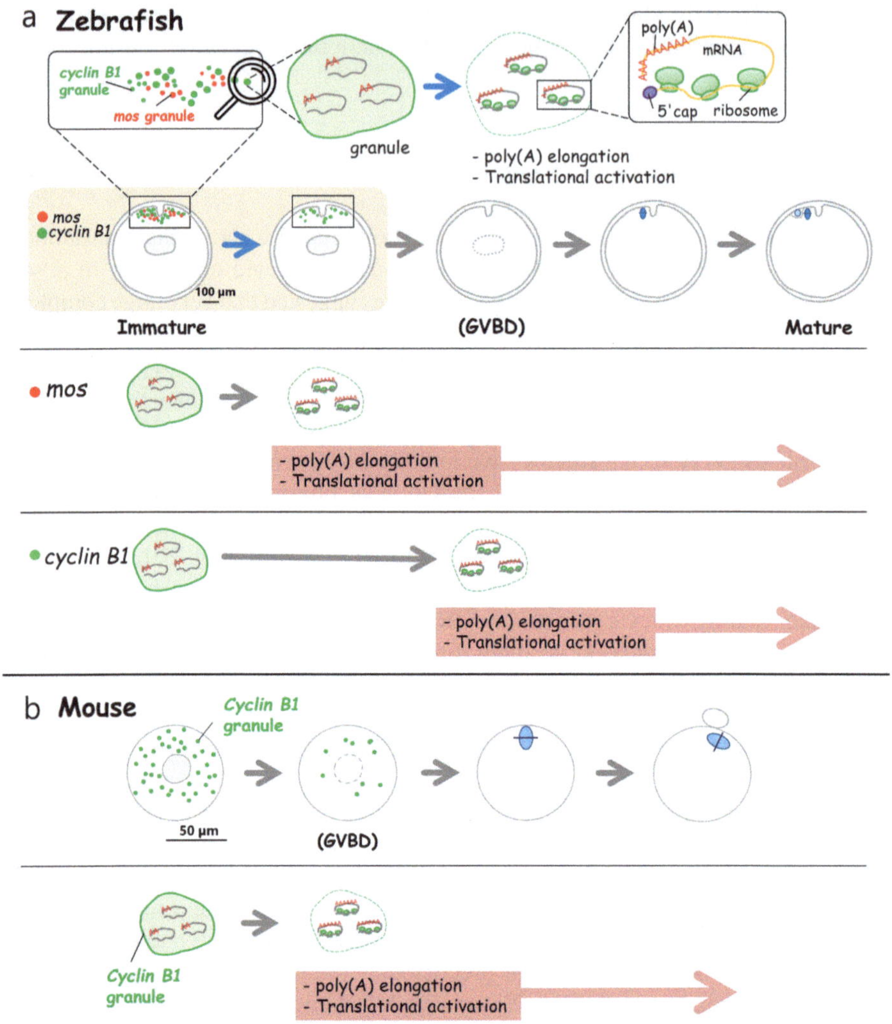

Fig. 13.3 Assembly of RNA granules in immature oocytes and translational regulation of the mRNAs through formation and disassembly of the granules. (**a**) Translationally repressed *cyclin B1* (*green*) and *mos* (*red*) mRNAs form distinct RNA granules at the animal polar cytoplasm of immature zebrafish oocytes. These granules disassemble after initiation of oocyte maturation at the timings at which the mRNAs are polyadenylated. These timings coincide with translational activation of the distinct mRNAs. Although both *cyclin B1* and *mos* RNA granules behave in similar manners, the timings and patterns of granule disassembly are different. *mos* RNA granules gradually disassemble in an early period of oocyte maturation, whereas *cyclin B1* RNA granules rapidly and simultaneously disassemble before GVBD. Cyclin B1 and Mos proteins synthesized at the accurate timings properly promote oocyte maturation processes including GVBD, spindle formation, transitions between meiosis I and meiosis II, and arrest at metaphase II. (**b**) *cyclin B1* mRNA forms RNA granules in the mouse cytoplasm (*green*). As in zebrafish oocytes, these granules disassemble after initiation of oocyte maturation at the timing at which the mRNA is polyadenylated. Synthesized Cyclin B1 protein promotes oocyte maturation

resulting in GVBD and spindle formation. Formation of these granules is dependent on the assembly of actin filaments on the surface of oocytes (Kotani et al. 2013; Nukada et al. 2015). In addition, Pumilio1 mediates the formation of *cyclin B1* RNA granules, probably through its self-aggregation into large complexes or conformational changes of the mRNAs by binding to the PBE, which would promote the assembly of some components required for the granule formation.

Assembly of zebrafish *cyclin B1* RNA granules does not appear to be required for the translational repression of mRNAs. Rather, it is required for regulating the accurate timing of translational activation of mRNAs (Kotani et al. 2013). The first evidence was provided by a transgenic approach showing that a reporter mRNA carrying a mutation in the PBE did not form granules and translated at a timing earlier than that of wild-type reporter mRNA. Evidence was also provided by pharmacological stimulation of oocytes showing that depolymerization of actin filaments causes dispersion of *cyclin B1* RNA granules and translational activation of the mRNAs immediately after initiation of oocyte maturation. Furthermore, inhibition of filament depolymerization causes stabilization of the RNA granules and prevents translational activation of the mRNAs. Further evidence was also provided by overexpression of the N-terminal region of Pumilio1, which resulted in stabilization of *cyclin B1* RNA granules and a delay in the timing of translational activation.

Regulation of temporal translation through formation and disassembly of RNA granules is not specific to zebrafish *cyclin B1* mRNA. In zebrafish oocytes, translationally repressed *mos* mRNA is also localized at the animal polar cytoplasm as the RNA granules (Fig. 13.3) (Horie and Kotani 2016). Although both *cyclin B1* and *mos* mRNAs contain CPEs in their 3′ UTRs and are distributed in similar regions, they are assembled into different RNA granules. In addition, distribution patterns are different, i.e., *cyclin B1* RNA granules form dense clusters, and *mos* RNA granules are dispersed within the cytoplasm. These differences are probably linked to the differences in regulation, i.e., *mos* RNA granules are gradually disassembled in the early period of oocyte maturation, while *cyclin B1* RNA granules are rapidly and simultaneously disassembled at a timing later than that of *mos* RNA granules (Fig. 13.3). Furthermore, translational regulation through formation and disassembly of RNA granules is not specific to zebrafish oocytes. Granules of *Cyclin B1* mRNA are also found in mouse oocytes. These granules disassemble at a timing coinciding with that of translational activation of the mRNA after initiation of oocyte maturation (Fig. 13.3) (Kotani et al. 2013). These findings suggest that temporal regulation of mRNA translation by the formation and disassembly of RNA granules is conserved in mouse.

Formation of RNA granules has been found in various cell types and many species. Stress granules and processing bodies (P-bodies) are distinct types of RNA granules that are assembled in the cytoplasm of cells from yeast to mammals (Kedersha et al. 2005; Sheth and Parker 2003). Neuronal granules consist of several types of RNA granules that are assembled and transported to dendrites in neuronal cells (Baez et al. 2011; Knowles et al. 1996; Vessey et al. 2006). Germ cell granules assembled in *C. elegance*, *Drosophila*, and *Xenopus* oocytes contain certain RNAs

(Boag et al. 2008; Kloc et al. 2002; Nakamura et al. 2001; Noble et al. 2008; Schisa et al. 2001). These subcellular structures are thought to function in mRNA storage, remodeling, and decay, although the strict functions are not clear. Some of the components of these granules are the same (Schisa 2012). For example, the DEAD-box RNA helicase DDX6 is a component of stress granules, P-bodies, neuronal granules, and germ cell granules. CPEB is also a component of stress granules and P-bodies (Wilczynska et al. 2005). A recent comprehensive study has shown that a neuronal granule containing Barentsz consists of 84 proteins and another type of neuronal granule containing Staufen2 consists of 65 proteins and that both neuronal granules share 29 proteins (Fritzsche et al. 2013). The formation of distinct RNA granules of *cyclin B1* and *mos* mRNAs suggests that the granules of *cyclin B1* and *mos* mRNAs consist of different and shared molecules, for example, Pumilio1 is specific for *cyclin B1* RNA granules, and CPEB is a component of both granules. Identification of all of the components will contribute to an understanding of the mechanisms of temporal regulation of mRNA translation through formation and disassembly of distinct granules.

13.4 Conclusion

Regulation of translationally repressed mRNAs in oocytes and embryos is extremely important for promoting oogenesis and development. Overall, the mechanisms of mRNA localization and temporal translation are complex and require chain of events beginning from transcription, followed by splicing, exporting from the nucleus, and multistep processes in the cytoplasm. Both localization and temporal translation are directed by complex interactions of mRNAs and proteins. In addition, mechanisms of localization are coupled with mechanisms of temporal translation. Furthermore, assembly and dissociation of subcellular structures such as RNA granules coordinate accurate timings of translational activation of dormant mRNAs. Relatively little is known about the mechanisms of temporal regulation of mRNA translation after fertilization. During embryonic cell cycles, *cyclin B1* mRNA is polyadenylated and deadenylated in a cell cycle-dependent manner, promoting mitotic cleavages in *Xenopus* (Groisman et al. 2002). In the dorsal-ventral axis formation of *Xenopus* embryos, translation of *wnt11* mRNA is regulated by cytoplasmic polyadenylation (Schroeder et al. 1999). These findings suggest that cytoplasmic polyadenylation is important for translational regulation after fertilization as well as during oocyte maturation. However, majority of these regulatory pathways remain unknown. Genetic analyses using zebrafish together with biochemical analyses will facilitate the understanding of mechanisms of mRNA localization and temporal translation during oogenesis and development. In addition, development of new technologies including live imaging of RNA molecules (Bertrand et al. 1998; Campbell et al. 2015) and translational activation of the mRNAs (Pichon et al. 2016; Rodriguez et al. 2006; Yasuda et al. 2010) is important for elucidating the roles and mechanisms of mRNA regulations.

Acknowledgments We would like to thank the past and present laboratory members. This work was supported by the Grants-in-Aid for Scientific Research (16K07242 to T.K.) from the Ministry of Education, Culture, Sports, Science and Technology, Japan.

References

Araki K, Naito K, Haraguchi S, Suzuki R, Yokoyama M, Inoue M, Aizawa S, Toyoda Y, Sato E (1996) Meiotic abnormalities of c-mos knockout mouse oocytes: activation after first meiosis or entrance into third meiotic metaphase. Biol Reprod 55:1315–1324

Bachvarova R, Burns JP, Spiegelman I, Choy J, Chaganti RSK (1982) Morphology and transcriptional activity of mouse oocyte chromosomes. Chromosoma 86:181–196

Baez MV, Luchelli L, Maschi D, Habif M, Pascual M, Thomas MG, Boccaccio GL (2011) Smaug1 mRNA-silencing foci respond to NMDA and modulate synapse formation. J Cell Biol 195:1141–1157

Barkoff AF, Dickson KS, Gray NK, Wickens M (2000) Translational control of cyclin B1 mRNA during meiotic maturation: coordinated repression and cytoplasmic polyadenylation. Dev Biol 220:97–109

Belting HG, Hauptmann G, Meyer D, Abdelilah-Seyfried S, Chitnis A, Eschbach C, Soll I, Thisse C, Thisse B, Artinger KB, Lunde K, Driever W (2001) spiel ohne grenzen/pou2 is required during establishment of the zebrafish midbrain-hindbrain boundary organizer. Development 128:4165–4176

Belting HG, Wendik B, Lunde K, Leichsenring M, Mossner R, Driever W, Onichtchouk D (2011) Pou5f1 contributes to dorsoventral patterning by positive regulation of vox and modulation of fgf8a expression. Dev Biol 356:323–336

Bertrand E, Chartrand P, Schaefer M, Shenoy SM, Singer RH, Long RM (1998) Localization of ASH1 mRNA particles in living yeast. Mol Cell 2:437–445

Besse F, Ephrussi A (2008) Translational control of localized mRNAs: restricting protein synthesis in space and time. Nat Rev Mol Cell Biol 9:971–980

Boag PR, Atalay A, Robida S, Reinke V, Blackwell TK (2008) Protection of specific maternal messenger RNAs by the P body protein CGH-1 (Dhh1/RCK) during Caenorhabditis elegans oogenesis. J Cell Biol 182:543–557

Braude P, Bolton V, Moore S (1988) Human gene expression first occurs between the four- and eight-cell stages of preimplantation development. Nature 332:459–461

Campbell PD, Chao JA, Singer RH, Marlow FL (2015) Dynamic visualization of transcription and RNA subcellular localization in zebrafish. Development 142:1368–1374

Cao Q, Richter JD (2002) Dissolution of the maskin-eIF4E complex by cytoplasmic polyadenylation and poly(A)-binding protein controls cyclin B1 mRNA translation and oocyte maturation. EMBO J 21:3852–3862

Cha BJ, Koppetsch BS, Theurkauf WE (2001) In vivo analysis of Drosophila bicoid mRNA localization reveals a novel microtubule-dependent axis specification pathway. Cell 106:35–46

Charlesworth A, Welk J, MacNicol AM (2000) The temporal control of Wee1 mRNA translation during Xenopus oocyte maturation is regulated by cytoplasmic polyadenylation elements within the 3′-untranslated region. Dev Biol 227:706–719

Charlesworth A, Ridge JA, King LA, MacNicol MC, MacNicol AM (2002) A novel regulatory element determines the timing of Mos mRNA translation during Xenopus oocyte maturation. EMBO J 21:2798–2806

Charlesworth A, Wilczynska A, Thampi P, Cox LL, MacNicol AM (2006) Musashi regulates the temporal order of mRNA translation during Xenopus oocyte maturation. EMBO J 25:2792–2801

Chen J, Melton C, Suh N, Oh JS, Horner K, Xie F, Sette C, Blelloch R, Conti M (2011) Genome-wide analysis of translation reveals a critical role for deleted in azoospermia-like (Dazl) at the oocyte-to-zygote transition. Genes Dev 25:755–766

Choi T, Fukasawa K, Zhou R, Tessarollo L, Borror K, Resau J, Vande Woude GF (1996) The Mos/mitogen-activated protein kinase (MAPK) pathway regulates the size and degradation of the first polar body in maturing mouse oocytes. Proc Natl Acad Sci USA 93:7032–7035

Davydenko O, Schultz RM, Lampson MA (2013) Increased CDK1 activity determines the timing of kinetochore-microtubule attachments in meiosis I. J Cell Biol 202:221–229

de Moor CH, Richter JD (1999) Cytoplasmic polyadenylation elements mediate masking and unmasking of cyclin B1 mRNA. EMBO J 18:2294–2303

de Moor CH, Meijer H, Lissenden S (2005) Mechanisms of translational control by the 3′ UTR in development and differentiation. Semin Cell Dev Biol 16:49–58

Dosch R, Wagner DS, Mintzer KA, Runke G, Wiemelt AP, Mullins MC (2004) Maternal control of vertebrate development before the midblastula transition: mutants from the zebrafish I. Dev Cell 6:771–780

Dupre A, Jessus C, Ozon R, Haccard O (2002) Mos is not required for the initiation of meiotic maturation in Xenopus oocytes. EMBO J 21:4026–4036

Fox CA, Sheets MD, Wickens MP (1989) Poly(A) addition during maturation of frog oocytes: distinct nuclear and cytoplasmic activities and regulation by the sequence UUUUUAU. Genes Dev 3:2151–2162

Fritzsche R, Karra D, Bennett KL, Ang FY, Heraud-Farlow JE, Tolino M, Doyle M, Bauer KE, Thomas S, Planyavsky M, Arn E, Bakosova A, Jungwirth K, Hormann A, Palfi Z, Sandholzer J, Schwarz M, Macchi P, Colinge J, Superti-Furga G, Kiebler MA (2013) Interactome of two diverse RNA granules links mRNA localization to translational repression in neurons. Cell Rep 5:1749–1762

Furuno N, Nishizawa M, Okazaki K, Tanaka H, Iwashita J, Nakajo N, Ogawa Y, Sagata N (1994) Suppression of DNA replication via Mos function during meiotic divisions in Xenopus oocytes. EMBO J 13:2399–2410

Gebauer F, Richter JD (1996) Mouse cytoplasmic polyadenylylation element binding protein: an evolutionarily conserved protein that interacts with the cytoplasmic polyadenylylation elements of c-mos mRNA. Proc Natl Acad Sci USA 93:14602–14607

Gebauer F, Xu W, Cooper GM, Richter JD (1994) Translational control by cytoplasmic polyadenylation of c-mos mRNA is necessary for oocyte maturation in the mouse. EMBO J 13:5712–5720

German SD, Lee JH, Campbell KH, Sweetman D, Alberio R (2015) Actin depolymerization is associated with meiotic acceleration in cycloheximide-treated ovine oocytes. Biol Reprod 92: 103

Groisman I, Jung MY, Sarkissian M, Cao Q, Richter JD (2002) Translational control of the embryonic cell cycle. Cell 109:473–483

Gross SD, Schwab MS, Taieb FE, Lewellyn AL, Qian YW, Maller JL (2000) The critical role of the MAP kinase pathway in meiosis II in Xenopus oocytes is mediated by p90(Rsk). Curr Biol 10:430–438

Gupta T, Marlow FL, Ferriola D, Mackiewicz K, Dapprich J, Monos D, Mullins MC (2010) Microtubule actin crosslinking factor 1 regulates the Balbiani body and animal-vegetal polarity of the zebrafish oocyte. PLoS Genet 6:e1001073

Hachet O, Ephrussi A (2001) Drosophila Y14 shuttles to the posterior of the oocyte and is required for oskar mRNA transport. Curr Biol 11:1666–1674

Hachet O, Ephrussi A (2004) Splicing of oskar RNA in the nucleus is coupled to its cytoplasmic localization. Nature 428:959–963

Hake LE, Richter JD (1994) CPEB is a specificity factor that mediates cytoplasmic polyadenylation during Xenopus oocyte maturation. Cell 79:617–627

Hamatani T, Carter MG, Sharov AA, Ko MS (2004) Dynamics of global gene expression changes during mouse preimplantation development. Dev Cell 6:117–131

Harvey SA, Sealy I, Kettleborough R, Fenyes F, White R, Stemple D, Smith JC (2013) Identification of the zebrafish maternal and paternal transcriptomes. Development 140:2703–2710
Hashimoto Y, Maegawa S, Nagai T, Yamaha E, Suzuki H, Yasuda K, Inoue K (2004) Localized maternal factors are required for zebrafish germ cell formation. Dev Biol 268:152–161
Heim AE, Hartung O, Rothhamel S, Ferreira E, Jenny A, Marlow FL (2014) Oocyte polarity requires a Bucky ball-dependent feedback amplification loop. Development 141:842–854
Hill RS, Macgregor HC (1980) The development of lampbrush chromosome-type transcription in the early Diplotene oocytes of Xenopus-Laevis - an electron-microscope analysis. J Cell Sci 44:87–101
Hochegger H, Klotzbucher A, Kirk J, Howell M, le Guellec K, Fletcher K, Duncan T, Sohail M, Hunt T (2001) New B-type cyclin synthesis is required between meiosis I and II during Xenopus oocyte maturation. Development 128:3795–3807
Horie M, Kotani T (2016) Formation of mos RNA granules in the zebrafish oocyte that differ from cyclin B1 RNA granules in distribution, density and regulation. Eur J Cell Biol 95:563–573
Houston DW, King ML (2000) A critical role for Xdazl, a germ plasm-localized RNA, in the differentiation of primordial germ cells in Xenopus. Development 127:447–456
Howley C, Ho RK (2000) mRNA localization patterns in zebrafish oocytes. Mech Dev 92:305–309
Ihara J, Yoshida N, Tanaka T, Mita K, Yamashita M (1998) Either cyclin B1 or B2 is necessary and sufficient for inducing germinal vesicle breakdown during frog (Rana japonica) oocyte maturation. Mol Reprod Dev 50:499–509
Iwabuchi M, Ohsumi K, Yamamoto TM, Sawada W, Kishimoto T (2000) Residual Cdc2 activity remaining at meiosis I exit is essential for meiotic M-M transition in Xenopus oocyte extracts. EMBO J 19:4513–4523
Jesuthasan S, Stahle U (1997) Dynamic microtubules and specification of the zebrafish embryonic axis. Curr Biol 7:31–42
Kajiura-Kobayashi H, Yoshida N, Sagata N, Yamashita M, Nagahama Y (2000) The Mos/MAPK pathway is involved in metaphase II arrest as a cytostatic factor but is neither necessary nor sufficient for initiating oocyte maturation in goldfish. Dev Genes Evol 210:416–425
Kamenska A, Lu WT, Kubacka D, Broomhead H, Minshall N, Bushell M, Standart N (2014) Human 4E-T represses translation of bound mRNAs and enhances microRNA-mediated silencing. Nucleic Acids Res 42:3298–3313
Kanayama N, Miyano T, Lee J (2002) Acquisition of meiotic competence in growing pig oocytes correlates with their ability to activate Cdc2 kinase and MAP kinase. Zygote 10:261–270
Kane DA, Kimmel CB (1993) The zebrafish midblastula transition. Development 119:447–456
Kedersha N, Stoecklin G, Ayodele M, Yacono P, Lykke-Andersen J, Fritzler MJ, Scheuner D, Kaufman RJ, Golan DE, Anderson P (2005) Stress granules and processing bodies are dynamically linked sites of mRNP remodeling. J Cell Biol 169:871–884
Kim-Ha J, Kerr K, Macdonald PM (1995) Translational regulation of oskar mRNA by bruno, an ovarian RNA-binding protein, is essential. Cell 81:403–412
Kishimoto Y, Koshida S, Furutani-Seiki M, Kondoh H (2004) Zebrafish maternal-effect mutations causing cytokinesis defect without affecting mitosis or equatorial vasa deposition. Mech Dev 121:79–89
Kloc M, Etkin LD (1995) Two distinct pathways for the localization of RNAs at the vegetal cortex in Xenopus oocytes. Development 121:287–297
Kloc M, Bilinski S, Pui-Yee Chan A, Etkin LD (2000) The targeting of Xcat2 mRNA to the germinal granules depends on a cis-acting germinal granule localization element within the 3′ UTR. Dev Biol 217:221–229
Kloc M, Dougherty MT, Bilinski S, Chan AP, Brey E, King ML, Patrick CW Jr, Etkin LD (2002) Three-dimensional ultrastructural analysis of RNA distribution within germinal granules of Xenopus. Dev Biol 241:79–93
Kloc M, Jedrzejowska I, Tworzydlo W, Bilinski SM (2014) Balbiani body, nuage and sponge bodies--term plasm pathway players. Arthropod Struct Dev 43:341–348

Knowles RB, Sabry JH, Martone ME, Deerinck TJ, Ellisman MH, Bassell GJ, Kosik KS (1996) Translocation of RNA granules in living neurons. J Neurosci 16:7812–7820

Kondo T, Yanagawa T, Yoshida N, Yamashita M (1997) Introduction of cyclin B induces activation of the maturation-promoting factor and breakdown of germinal vesicle in growing zebrafish oocytes unresponsive to the maturation-inducing hormone. Dev Biol 190:142–152

Kondo T, Kotani T, Yamashita M (2001) Dispersion of cyclin B mRNA aggregation is coupled with translational activation of the mRNA during zebrafish oocyte maturation. Dev Biol 229: 421–431

Kosaka K, Kawakami K, Sakamoto H, Inoue K (2007) Spatiotemporal localization of germ plasm RNAs during zebrafish oogenesis. Mech Dev 124:279–289

Kotani T, Kawakami K (2008) Misty somites, a maternal effect gene identified by transposon-mediated insertional mutagenesis in zebrafish that is essential for the somite boundary maintenance. Dev Biol 316:383–396

Kotani T, Yamashita M (2002) Discrimination of the roles of MPF and MAP kinase in morphological changes that occur during oocyte maturation. Dev Biol 252:271–286

Kotani T, Yoshida N, Mita K, Yamashita M (2001) Requirement of cyclin B2, but not cyclin B1, for bipolar spindle formation in frog (Rana japonica) oocytes. Mol Reprod Dev 59:199–208

Kotani T, Nagayoshi S, Urasaki A, Kawakami K (2006) Transposon-mediated gene trapping in zebrafish. Methods 39:199–206

Kotani T, Yasuda K, Ota R, Yamashita M (2013) Cyclin B1 mRNA translation is temporally controlled through formation and disassembly of RNA granules. J Cell Biol 202:1041–1055

Ku M, Melton DA (1993) Xwnt-11: a maternally expressed Xenopus wnt gene. Development 119:1161–1173

Lachnit M, Kur E, Driever W (2008) Alterations of the cytoskeleton in all three embryonic lineages contribute to the epiboly defect of Pou5f1/Oct4 deficient MZspg zebrafish embryos. Dev Biol 315:1–17

Ledan E, Polanski Z, Terret ME, Maro B (2001) Meiotic maturation of the mouse oocyte requires an equilibrium between cyclin B synthesis and degradation. Dev Biol 232:400–413

Lee MT, Bonneau AR, Takacs CM, Bazzini AA, DiVito KR, Fleming ES, Giraldez AJ (2013) Nanog, Pou5f1 and SoxB1 activate zygotic gene expression during the maternal-to-zygotic transition. Nature 503:360–364

Leichsenring M, Maes J, Mossner R, Driever W, Onichtchouk D (2013) Pou5f1 transcription factor controls zygotic gene activation in vertebrates. Science 341:1005–1009

Levesque JT, Sirard MA (1996) Resumption of meiosis is initiated by the accumulation of cyclin B in bovine oocytes. Biol Reprod 55:1427–1436

Lu FI, Thisse C, Thisse B (2011) Identification and mechanism of regulation of the zebrafish dorsal determinant. Proc Natl Acad Sci USA 108:15876–15880

Lunde K, Belting HG, Driever W (2004) Zebrafish pou5f1/pou2, homolog of mammalian Oct4, functions in the endoderm specification cascade. Curr Biol 14:48–55

MacNicol MC, Cragle CE, Arumugam K, Fosso B, Pesole G, MacNicol AM (2015) Functional Integration of mRNA Translational Control Programs. Biomolecules 5:1580–1599

Madgwick S, Hansen DV, Levasseur M, Jackson PK, Jones KT (2006) Mouse Emi2 is required to enter meiosis II by reestablishing cyclin B1 during interkinesis. J Cell Biol 174:791–801

Maegawa S, Yasuda K, Inoue K (1999) Maternal mRNA localization of zebrafish DAZ-like gene. Mech Dev 81:223–226

Marlow FL, Mullins MC (2008) Bucky ball functions in Balbiani body assembly and animal-vegetal polarity in the oocyte and follicle cell layer in zebrafish. Dev Biol 321:40–50

Martin KC, Ephrussi A (2009) mRNA localization: gene expression in the spatial dimension. Cell 136:719–730

Masui Y, Clarke HJ (1979) Oocyte maturation. Int Rev Cytol 57:185–282

McGrew LL, Richter JD (1990) Translational control by cytoplasmic polyadenylation during Xenopus oocyte maturation: characterization of cis and trans elements and regulation by cyclin/MPF. EMBO J 9:3743–3751

McGrew LL, Dworkin-Rastl E, Dworkin MB, Richter JD (1989) Poly(A) elongation during Xenopus oocyte maturation is required for translational recruitment and is mediated by a short sequence element. Genes Dev 3:803–815
Mendez R, Richter JD (2001) Translational control by CPEB: a means to the end. Nat Rev Mol Cell Biol 2:521–529
Mendez R, Murthy KG, Ryan K, Manley JL, Richter JD (2000) Phosphorylation of CPEB by Eg2 mediates the recruitment of CPSF into an active cytoplasmic polyadenylation complex. Mol Cell 6:1253–1259
Minshall N, Reiter MH, Weil D, Standart N (2007) CPEB interacts with an ovary-specific eIF4E and 4E-T in early Xenopus oocytes. J Biol Chem 282:37389–37401
Moore GPM, Linternm S, Peters H, Faber M (1974) Rna-synthesis in mouse oocyte. J Cell Biol 60:416–422
Murakami MS, Vande Woude GF (1998) Analysis of the early embryonic cell cycles of Xenopus; regulation of cell cycle length by Xe-wee1 and Mos. Development 125:237–248
Murakami MS, Copeland TD, Vande Woude GF (1999) Mos positively regulates Xe-Wee1 to lengthen the first mitotic cell cycle of Xenopus. Genes Dev 13:620–631
Naito K, Hawkins C, Yamashita M, Nagahama Y, Aoki F, Kohmoto K, Toyoda Y, Moor RM (1995) Association of p34cdc2 and cyclin B1 during meiotic maturation in porcine oocytes. Dev Biol 168:627–634
Nakahata S, Katsu Y, Mita K, Inoue K, Nagahama Y, Yamashita M (2001) Biochemical identification of Xenopus Pumilio as a sequence-specific cyclin B1 mRNA-binding protein that physically interacts with a Nanos homolog, Xcat-2, and a cytoplasmic polyadenylation element-binding protein. J Biol Chem 276:20945–20953
Nakahata S, Kotani T, Mita K, Kawasaki T, Katsu Y, Nagahama Y, Yamashita M (2003) Involvement of Xenopus Pumilio in the translational regulation that is specific to cyclin B1 mRNA during oocyte maturation. Mech Dev 120:865–880
Nakajo N, Yoshitome S, Iwashita J, Iida M, Uto K, Ueno S, Okamoto K, Sagata N (2000) Absence of Wee1 ensures the meiotic cell cycle in Xenopus oocytes. Genes Dev 14:328–338
Nakamura A, Amikura R, Hanyu K, Kobayashi S (2001) Me31B silences translation of oocyte-localizing RNAs through the formation of cytoplasmic RNP complex during Drosophila oogenesis. Development 128:3233–3242
Nakamura Y, Tanaka KJ, Miyauchi M, Huang L, Tsujimoto M, Matsumoto K (2010) Translational repression by the oocyte-specific protein P100 in Xenopus. Dev Biol 344:272–283
Newport J, Kirschner M (1982) A major developmental transition in early Xenopus embryos: I. characterization and timing of cellular changes at the midblastula stage. Cell 30:675–686
Noble SL, Allen BL, Goh LK, Nordick K, Evans TC (2008) Maternal mRNAs are regulated by diverse P body-related mRNP granules during early Caenorhabditis elegans development. J Cell Biol 182:559–572
Nojima H, Rothhamel S, Shimizu T, Kim CH, Yonemura S, Marlow FL, Hibi M (2010) Syntabulin, a motor protein linker, controls dorsal determination. Development 137:923–933
Nukada Y, Horie M, Fukui A, Kotani T, Yamashita M (2015) Real-time imaging of actin filaments in the zebrafish oocyte and embryo. Cytoskeleton 72:491–501
Nurse P (1990) Universal control mechanism regulating onset of M-phase. Nature 344:503–508
Oh D, Houston DW (2017) RNA localization in the vertebrate oocyte: establishment of oocyte polarity and localized mRNA assemblages. In: Kloc M (ed) Oocytes: maternal information and functions. Springer, Cham
Ohe M, Inoue D, Kanemori Y, Sagata N (2007) Erp1/Emi2 is essential for the meiosis I to meiosis II transition in Xenopus oocytes. Dev Biol 303:157–164
Ota R, Kotani T, Yamashita M (2011) Biochemical characterization of Pumilio1 and Pumilio2 in Xenopus oocytes. J Biol Chem 286:2853–2863
Pelegri F, Knaut H, Maischein HM, Schulte-Merker S, Nusslein-Volhard C (1999) A mutation in the zebrafish maternal-effect gene nebel affects furrow formation and vasa RNA localization. Curr Biol 9:1431–1440

Pelegri F, Dekens MP, Schulte-Merker S, Maischein HM, Weiler C, Nusslein-Volhard C (2004) Identification of recessive maternal-effect mutations in the zebrafish using a gynogenesis-based method. Dev Dyn 231:324–335
Pichon X, Bastide A, Safieddine A, Chouaib R, Samacoits A, Basyuk E, Peter M, Mueller F, Bertrand E (2016) Visualization of single endogenous polysomes reveals the dynamics of translation in live human cells. J Cell Biol 214:769–781
Pique M, Lopez JM, Foissac S, Guigo R, Mendez R (2008) A combinatorial code for CPE-mediated translational control. Cell 132:434–448
Prasad CK, Mahadevan M, MacNicol MC, MacNicol AM (2008) Mos 3′ UTR regulatory differences underlie species-specific temporal patterns of Mos mRNA cytoplasmic polyadenylation and translational recruitment during oocyte maturation. Mol Reprod Dev 75:1258–1268
Radford HE, Meijer HA, de Moor CH (2008) Translational control by cytoplasmic polyadenylation in Xenopus oocytes. Biochim Biophys Acta 1779:217–229
Reijo R, Lee TY, Salo P, Alagappan R, Brown LG, Rosenberg M, Rozen S, Jaffe T, Straus D, Hovatta O et al (1995) Diverse spermatogenic defects in humans caused by Y chromosome deletions encompassing a novel RNA-binding protein gene. Nat Genet 10:383–393
Reim G, Brand M (2006) Maternal control of vertebrate dorsoventral axis formation and epiboly by the POU domain protein Spg/Pou2/Oct4. Development 133:2757–2770
Reim G, Mizoguchi T, Stainier DY, Kikuchi Y, Brand M (2004) The POU domain protein spg (pou2/Oct4) is essential for endoderm formation in cooperation with the HMG domain protein casanova. Dev Cell 6:91–101
Rodriguez AJ, Shenoy SM, Singer RH, Condeelis J (2006) Visualization of mRNA translation in living cells. J Cell Biol 175:67–76
Rosbash M, Ford PJ (1974) Polyadenylic Acid-Containing Rna in Xenopus-Laevis Oocytes. J Mol Biol 85:87–101
Ruggiu M, Speed R, Taggart M, McKay SJ, Kilanowski F, Saunders P, Dorin J, Cooke HJ (1997) The mouse Dazla gene encodes a cytoplasmic protein essential for gametogenesis. Nature 389: 73–77
Sagata N, Daar I, Oskarsson M, Showalter SD, Vande Woude GF (1989) The product of the mos proto-oncogene as a candidate "initiator" for oocyte maturation. Science 245:643–646
Sakamoto I, Takahara K, Yamashita M, Iwao Y (1998) Changes in cyclin B during oocyte maturation and early embryonic cell cycle in the newt, Cynops pyrrhogaster: requirement of germinal vesicle for MPF activation. Dev Biol 195:60–69
Schisa JA (2012) New insights into the regulation of RNP granule assembly in oocytes. Int Rev Cell Mol Biol 295:233–289
Schisa JA, Pitt JN, Priess JR (2001) Analysis of RNA associated with P granules in germ cells of C. elegans adults. Development 128:1287–1298
Schroeder KE, Condic ML, Eisenberg LM, Yost HJ (1999) Spatially regulated translation in embryos: asymmetric expression of maternal Wnt-11 along the dorsal-ventral axis in Xenopus. Dev Biol 214:288–297
Selman K, Wallace RA, Sarka A, Qi XP (1993) Stages of Oocyte Development in the Zebrafish, Brachydanio-Rerio. *J Morphol* 218:203–224
Sheets MD, Fox CA, Hunt T, Vande Woude G, Wickens M (1994) The 3′-untranslated regions of c-mos and cyclin mRNAs stimulate translation by regulating cytoplasmic polyadenylation. Genes Dev 8:926–938
Sheth U, Parker R (2003) Decapping and decay of messenger RNA occur in cytoplasmic processing bodies. Science 300:805–808
Sousa Martins JP, Liu X, Oke A, Arora R, Franciosi F, Viville S, Laird DJ, Fung JC, Conti M (2016) DAZL and CPEB1 regulate mRNA translation synergistically during oocyte maturation. J Cell Sci 129:1271–1282
St Johnston D (2005) Moving messages: the intracellular localization of mRNAs. Nat Rev Mol Cell Biol 6:363–375

Stebbins-Boaz B, Hake LE, Richter JD (1996) CPEB controls the cytoplasmic polyadenylation of cyclin, Cdk2 and c-mos mRNAs and is necessary for oocyte maturation in Xenopus. EMBO J 15:2582–2592

Stebbins-Boaz B, Cao Q, de Moor CH, Mendez R, Richter JD (1999) Maskin is a CPEB-associated factor that transiently interacts with elF-4E. Mol Cell 4:1017–1027

Stutz A, Conne B, Huarte J, Gubler P, Volkel V, Flandin P, Vassalli JD (1998) Masking, unmasking, and regulated polyadenylation cooperate in the translational control of a dormant mRNA in mouse oocytes. Genes Dev 12:2535–2548

Susor A, Kubelka M (2017) Translational regulation in the mammalian oocyte. In: Kloc M (ed) Oocytes: maternal information and functions. Springer, Cham

Suzuki H, Maegawa S, Nishibu T, Sugiyama T, Yasuda K, Inoue K (2000) Vegetal localization of the maternal mRNA encoding an EDEN-BP/Bruno-like protein in zebrafish. Mech Dev 93: 205–209

Suzuki H, Tsukahara T, Inoue K (2009) Localization of c-mos mRNA around the animal pole in the zebrafish oocyte with Zor-1/Zorba. Biosci Trends 3:96–104

Tanaka T, Yamashita M (1995) Pre-MPF is absent in immature oocytes of fishes and amphibians except Xenopus. Develop Growth Differ 37:387–393

Tao Q, Yokota C, Puck H, Kofron M, Birsoy B, Yan D, Asashima M, Wylie CC, Lin X, Heasman J (2005) Maternal wnt11 activates the canonical wnt signaling pathway required for axis formation in Xenopus embryos. Cell 120:857–871

Tay J, Hodgman R, Richter JD (2000) The control of cyclin B1 mRNA translation during mouse oocyte maturation. Dev Biol 221:1–9

Tran LD, Hino H, Quach H, Lim S, Shindo A, Mimori-Kiyosue Y, Mione M, Ueno N, Winkler C, Hibi M, Sampath K (2012) Dynamic microtubules at the vegetal cortex predict the embryonic axis in zebrafish. Development 139:3644–3652

Tung JJ, Padmanabhan K, Hansen DV, Richter JD, Jackson PK (2007) Translational unmasking of Emi2 directs cytostatic factor arrest in meiosis II. Cell Cycle 6:725–731

Vassalli JD, Huarte J, Belin D, Gubler P, Vassalli A, O'Connell ML, Parton LA, Rickles RJ, Strickland S (1989) Regulated polyadenylation controls mRNA translation during meiotic maturation of mouse oocytes. Genes Dev 3:2163–2171

Verlhac MH, Kubiak JZ, Weber M, Geraud G, Colledge WH, Evans MJ, Maro B (1996) Mos is required for MAP kinase activation and is involved in microtubule organization during meiotic maturation in the mouse. Development 122:815–822

Vessey JP, Vaccani A, Xie Y, Dahm R, Karra D, Kiebler MA, Macchi P (2006) Dendritic localization of the translational repressor Pumilio 2 and its contribution to dendritic stress granules. J Neurosci 26:6496–6508

Villaescusa JC, Allard P, Carminati E, Kontogiannea M, Talarico D, Blasi F, Farookhi R, Verrotti AC (2006) Clast4, the murine homologue of human eIF4E-Transporter, is highly expressed in developing oocytes and post-translationally modified at meiotic maturation. Gene 367: 101–109

Vincent JP, Oster GF, Gerhart JC (1986) Kinematics of gray crescent formation in Xenopus eggs: the displacement of subcortical cytoplasm relative to the egg surface. Dev Biol 113:484–500

Wagner DS, Dosch R, Mintzer KA, Wiemelt AP, Mullins MC (2004) Maternal control of development at the midblastula transition and beyond: mutants from the zebrafish II. Dev Cell 6: 781–790

Wang Q, Latham KE (1997) Requirement for protein synthesis during embryonic genome activation in mice. Mol Reprod Dev 47:265–270

Wang Q, Latham KE (2000) Translation of maternal messenger ribonucleic acids encoding transcription factors during genome activation in early mouse embryos. Biol Reprod 62:969–978

Webster PJ, Liang L, Berg CA, Lasko P, Macdonald PM (1997) Translational repressor bruno plays multiple roles in development and is widely conserved. Genes Dev 11:2510–2521

Wilczynska A, Aigueperse C, Kress M, Dautry F, Weil D (2005) The translational regulator CPEB1 provides a link between dcp1 bodies and stress granules. J Cell Sci 118:981–992
Yamashita M, Kajiura H, Tanaka T, Onoe S, Nagahama Y (1995) Molecular mechanisms of the activation of maturation-promoting factor during goldfish oocyte maturation. Dev Biol 168: 62–75
Yasuda K, Kotani T, Ota R, Yamashita M (2010) Transgenic zebrafish reveals novel mechanisms of translational control of cyclin B1 mRNA in oocytes. Dev Biol 348:76–86
Yasuda K, Kotani T, Yamashita M (2013) A cis-acting element in the coding region of cyclin B1 mRNA couples subcellular localization to translational timing. Dev Biol 382:517–529
Zhou Y, King ML (1996) Localization of Xcat-2 RNA, a putative germ plasm component, to the mitochondrial cloud in Xenopus stage I oocytes. Development 122:2947–2953

Chapter 14
Switches in Dicer Activity During Oogenesis and Early Development

Mandy Yu Theng Lim and Katsutomo Okamura

Abstract Dicer is a versatile protein regulating diverse biological processes via the production of multiple classes of small regulatory RNAs, including microRNAs (miRNAs) and small interfering RNAs (siRNAs). In this chapter, we will discuss roles for Dicer in driving temporal changes in activity of individual small RNA classes to support oogenesis and early embryogenesis. Genetic strategies that perturb particular functions of Dicer family proteins, such as ablation of individual Dicer paralogs or their binding partners as well as introduction of point mutations to individual domains, allowed the dissection of Dicer functions in diverse small RNA pathways. Evolutionary conservation and divergence of the mechanisms highlight the importance of Dicer versatility in supporting rapid changes in gene expression during oogenesis and early development. Furthermore, we will discuss potential roles of Dicer in transgenerational inheritance of small RNA-mediated gene regulation.

14.1 Introduction

The central dogma of molecular biology dictates the general flow of genetic information from DNA to RNA to protein. However, there are exceptions of this classical definition, and we now know that the transfer of information is not unidirectional from gene sequence to protein product. Retroviruses encode reverse transcriptase that can produce DNA from its RNA genome. Epigenetic marks, such as DNA methylation and histone modifications, are found "above" the encoded genetic information and code for heritable changes without alteration of the underlying DNA sequence. Non-coding RNAs that do not encode proteins represent a large portion of the human genome (Lappalainen et al. 2013). With an expanding list of non-coding RNA classes, these functional RNAs are proving to be more than

M.Y.T. Lim • K. Okamura (✉)
Temasek Life Sciences Laboratory, National University of Singapore, 1 Research Link, Singapore 117604, Singapore

School of Biological Sciences, Nanyang Technological University, 60 Nanyang Drive, Singapore 639798, Singapore
e-mail: okamurak@tll.org.sg

M. Kloc (ed.), *Oocytes*, Results and Problems in Cell Differentiation 63,
DOI 10.1007/978-3-319-60855-6_14

just transcriptional noise and act as key regulators of diverse cellular processes (Cech and Steitz 2014). Eukaryotic small regulatory non-coding RNAs are defined by their size (~20–30 nt) and association with effector Argonaute (AGO) proteins to mediate downstream silencing processes (Kim et al. 2009). Major classes of small RNAs include microRNAs (miRNAs), small interfering RNAs (siRNAs) and Piwi-interacting RNAs (piRNAs).

The best-characterized function of Dicer is perhaps its role in production of regulatory small RNAs, processing hairpin-like structures and long double-stranded RNAs (dsRNAs) into miRNAs and siRNAs, respectively. The ribonuclease III enzyme Dicer was first characterized as a functional component of the RNA interference (RNAi) machinery in *Drosophila* (Bernstein et al. 2001). The phenotypic resemblance between *dcr-1* mutants and the classic heterochronic miRNA mutants in nematodes provided the first evidence into the involvement of Dicer in miRNA processing (Ketting et al. 2001; Grishok et al. 2001; Lee et al. 1993; Wightman et al. 1993; Reinhart et al. 2000). In contrast, piRNAs are generally produced by Dicer-independent mechanisms (Houwing et al. 2007; Vagin et al. 2006).

In animals, the majority of miRNAs are processed by the canonical miRNA processing pathway (Fig. 14.1a). MiRNAs are transcribed by RNA polymerase II as long primary transcripts (primary miRNA), which are cleaved by Drosha/DGCR8 (Drosha/Pasha in invertebrates) microprocessor complex in the nucleus to give hairpin-like structures (precursor miRNAs) (Ha and Kim 2014). Precursor miRNAs are exported to the cytoplasm by exportin-5. In the miRNA processing pathway, Dicer is involved in the second cleavage step to release mature miRNA duplexes. Dicer binds partner dsRNA binding proteins (TRBP in mammals and Loqs in *Drosophila*), which are required for efficient cleavage at precise sites for a subset of miRNA genes. After the production of the small RNA duplex, one of these strands (guide strand) is selectively loaded onto AGO proteins. miRNAs often bind target mRNA 3′-UTRs through imperfect complementarity, with only the 5′ seed region fully paired with its targets. AGO proteins then recruit other protein factors to mediate downstream translational repression and mRNA decay. In addition, a variety of non-canonical miRNA processing pathways, including the Drosha-independent mirtron pathway (miRNA precursors directly produced by intron splicing) and the Dicer-independent *mir-451* pathway (short miRNA processed by Drosha and AGO2), have been described (Fig. 14.1a). Therefore, loss-of-function mutations in core miRNA processing factors result in removal of specific subsets of miRNAs (Yang et al. 2014).

Long dsRNAs can derive from exogenous (for exo-siRNAs) or endogenous (for endo-siRNAs) sources (Fig. 14.1b). Processing of siRNAs does not require Drosha and is carried out by cytoplasmic Dicer, with one strand from the resulting duplex preferentially loaded onto AGO. These regulatory non-coding RNAs guide AGO to target mRNAs with perfect complementarity to mediate post-transcriptional gene silencing via target cleavage (Okamura and Lai 2008).

The structural architecture of Dicer is highly conserved across metazoans consisting of an N-terminal helicase domain, DUF283 (domain of unknown function 283), PAZ

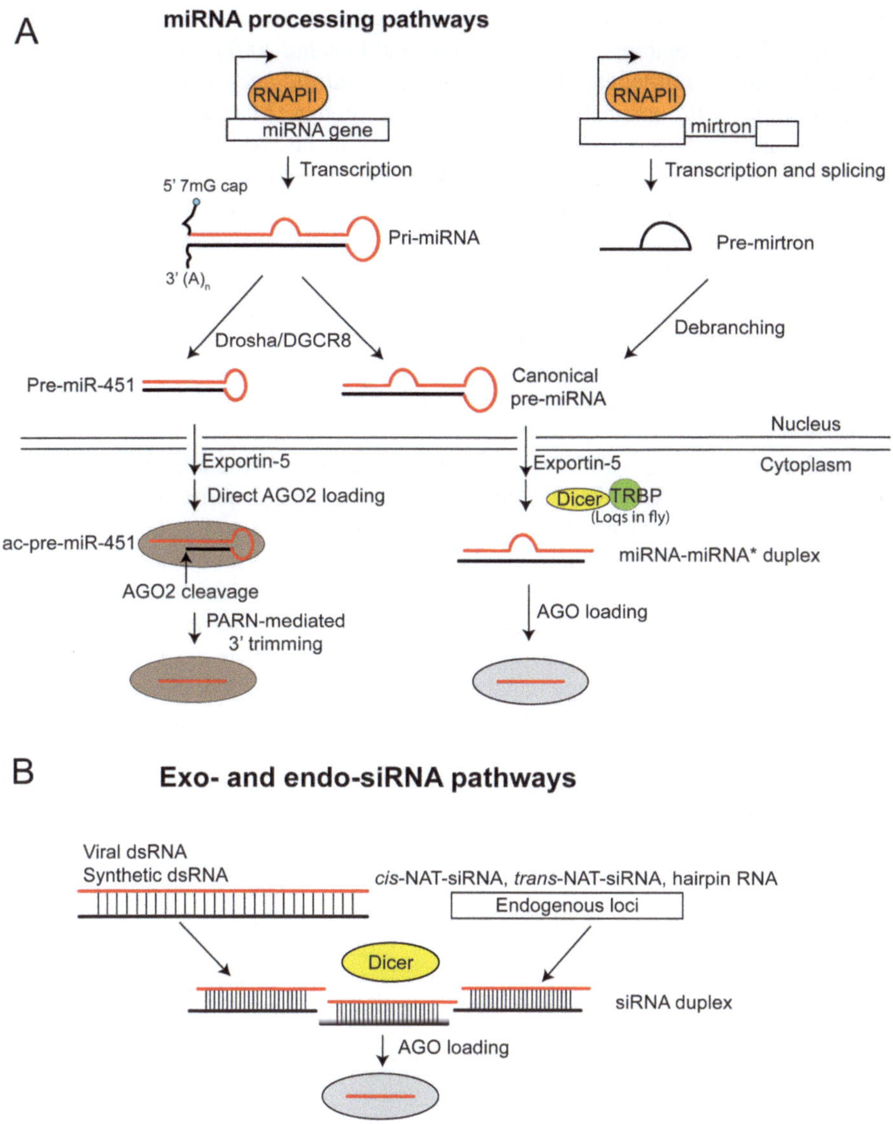

Fig. 14.1 Dicer-dependent small RNA pathways. (**a**) Canonical and non-canonical miRNA processing pathways. The Drosha-independent mirtron pathway and the Dicer-independent *mir-451* pathway are shown as representative examples of non-canonical pathways. (**b**) Exo- and endo-siRNA processing pathways. SiRNAs can derive from exogenous or endogenous sources. The exo-siRNA pathway is important for viral defence, and endo-siRNA pathway is important for endogenous gene regulation. Endo-siRNAs can derive from a variety of genomic sources, including transposons, overlapping transcripts, and from long inverted repeat transcripts known as hairpin RNAs. In both pathways, Dicer processively cleaves long siRNA precursors to produce siRNA duplexes. Note that Drosha is not involved in the production of siRNAs

(Piwi-Argonaute-Zwille) domain, two tandem catalytic RNase III domains and a C-terminal dsRNA-binding domain (MacRae and Doudna 2007) (Fig. 14.2). The first crystal structure of full-length Dicer was solved in *Giardia intestinalis*, where only the PAZ and RNase III domains constitute the active protein (Macrae et al. 2006, 2007). This structural analysis together with biochemical studies provided insights into the "molecular ruler" mechanism by which Dicer family enzymes produce small RNA duplexes of discrete size (Tsutsumi et al. 2011; Lau et al. 2012; Park et al. 2011; Zhang et al. 2004). Some organisms including *C. elegans* and mammals rely on a single Dicer gene to process both miRNAs and siRNAs (Filipowicz et al. 2008). Two Dicers, Dicer-1 and Dicer-2, have evolved to take on specialized roles in processing miRNAs and siRNAs, respectively, in *Drosophila* (Lee et al. 2004). Plants lack Drosha homologs and rely on four partially redundant Dicer-like proteins to process diverse small RNAs (Gasciolli et al. 2005; Kurihara and Watanabe 2004; Henderson et al. 2006; Jones-Rhoades et al. 2006).

AGO protein family can be broadly subdivided into two clades in eukaryotes: AGO clade and PIWI clade (Meister 2013). Eukaryotic genomes encode various numbers of AGO genes, ranging from 5 in *D. melanogaster*, to 8 in humans, and up to 27 in *C. elegans* (Hutvagner and Simard 2008; Yigit et al. 2006). Of the four mammalian AGO clade proteins, only AGO2 encodes catalytic "slicer" activity that can mediate mRNA cleavage (Liu et al. 2004). In *D. melanogaster*, both AGO1 and AGO2 possess slicer activity, but AGO1 is generally regarded as a poor slicer with weak endonucleolytic cleavage capability (Miyoshi et al. 2005; Forstemann et al. 2007). miRNAs often bind to AGO1 to mediate mRNA degradation and translational repression, while siRNAs typically bind the catalytic "slicer" AGO2 to mediate target cleavage (Valencia-Sanchez et al. 2006; Liu et al. 2004).

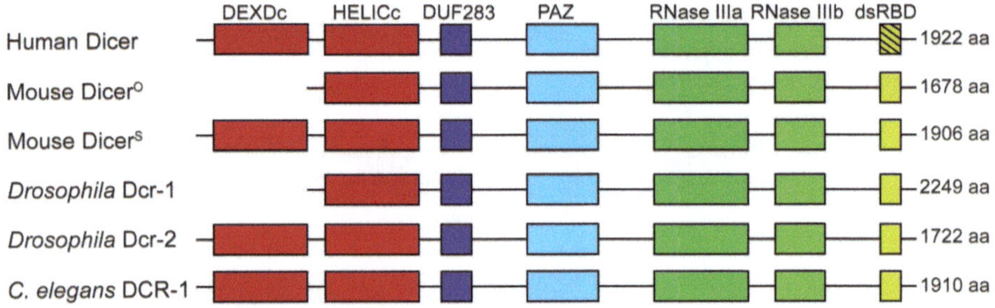

Fig. 14.2 Domain structures of Dicer homologs. Dicer homologs generally have N-terminal helicase, DUF283, PAZ, RNase III and C-terminal dsRNA-binding domains. DUF283 is an atypical dsRNA-binding domain, the PAZ domain binds 2-nt 3′ overhangs of dsRNA substrates, and the two RNase III domains make up the catalytic core of the enzyme. The tandemly arranged DEXDc and HELICc domains constitute the functional helicase domain. The dsRBD of human Dicer contains a non-canonical nuclear localization signal (*shaded*), allowing it to shuttle between the nucleus and cytoplasm. In mouse, two Dicer isoforms with different cleavage activities are found in the oocyte (Dicer^{O}) and somatic cells (Dicer^{S}). Dicer^{O} is an N-terminally truncated form that lacks the DExD helicase domain. In *Drosophila*, Dcr-1 cleavage is ATP-independent, whereas Dcr-2 cleavage is catalysed by ATP hydrolysis. Fly Dcr-1 lacks a functional DEXDc domain, which is important for ATP binding

Sexual reproduction in animals relies on gametes, specialized haploid cells that carry hereditary genetic and epigenetic information to the progeny. Fusion between the sperm and egg during fertilization creates a totipotent zygote, the beginning of a new organism. Animal development is a complex process that involves interplay of many factors, and the underlying developmental mechanisms encompass massive molecular changes required for growth and maturation. Before zygotic genome activation, early embryonic development relies heavily on maternally deposited factors in the oocyte, including RNA and protein molecules. Therefore, the process requires coordinated post-transcriptional regulation, and many RNA-mediated regulatory mechanisms were discovered through molecular studies of oogenesis and early developmental processes (Lasko 2011).

Since its initial characterization in 2001, novel and unexpected roles of Dicer have continued to emerge across disciplines. In this chapter, we will focus on the roles of Dicer and its small RNA products in promotion of oogenesis and early embryogenesis. Roles of Dicer in gametogenesis and transgenerational effects of Dicer products will be discussed in separate sections.

14.2 Roles for Dicer in Animal Gametogenesis

Diverse roles for Dicer have been proposed in gonadal development and oogenesis. In this section, we summarize studies regarding Dicer and its partner proteins in female gonads of three major model animals: *Drosophila*, mouse and *C. elegans*. We also describe individual small RNA genes that were implicated in the Dicer-dependent biological processes, when such information is available.

14.2.1 Roles for Dicer Complexes in Drosophila *Oogenesis*

Early studies provided several lines of evidence that the miRNA Dicer, Dcr-1, is involved in *Drosophila* female germline development. Since whole animal mutants of *dcr-1* are lethal, studies were performed using genetic mosaics removing the gene activity only in particular tissues (Dang and Perrimon 1992). Using this technique, Hatfield et al. generated germline stem cell (GSC) clones lacking Dcr-1 activity and found that the number of egg chambers per GSC was reduced in mutant clones (Hatfield et al. 2005). On the other hand, no defects in maintenance of GSCs were observed, suggesting that loss of stem cells was not the cause of the decrease in egg chamber production (Fig. 14.3a). Using molecular markers of cell cycle, the authors showed that the rate of GSC division was reduced in *dcr-1* mutant with an elevated frequency of cyclin E-positive cells, suggesting that cell cycle was delayed at the G1/S transition (Fig. 14.3a). Genetic interaction analysis indicated the involvement of the cyclin-dependent kinase inhibitor Dacapo (Dap) in the cell cycle phenotype of *dcr-1* mutant GSCs.

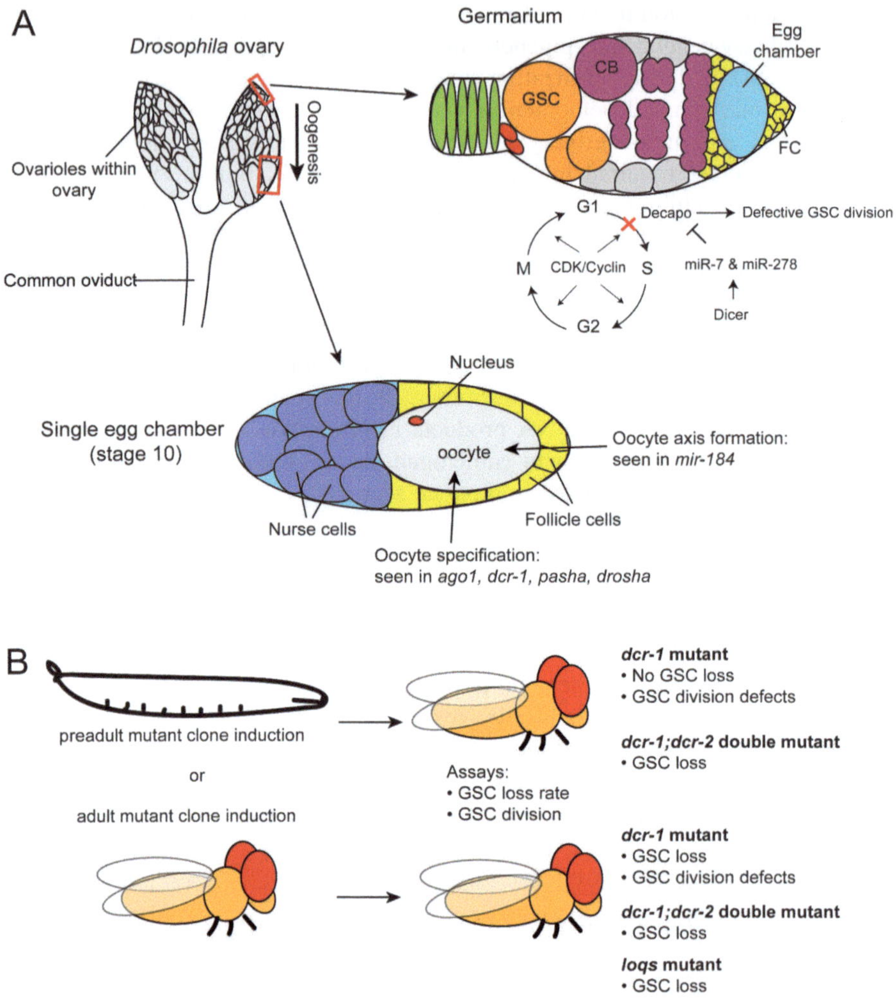

Fig. 14.3 Roles of miRNA/siRNA processing factors in *Drosophila* oogenesis. (**a**) Drosophila oogenesis. The adult female has a pair of ovaries, each containing about 18 ovarioles. The germarium is found at the apical tip of each ovariole, and egg chambers progressively mature and travel down the ovariole as oogenesis proceeds, with the mature egg found at the distal end. Germline stem cells (GSCs) are located at the anterior tip of the germarium as shown in the simplified representation (*upper right*). When a GSC divides, it produces another stem cell and a daughter cell known as the cystoblast (CB). The cystoblast undergoes four mitotic divisions with incomplete cytokinesis, resulting in 16 interconnected cells known as a cyst. Maturing cysts move posteriorly along the germarium. Only 1 of the 16 cells will eventually become the oocyte, while the other 15 become nurse cells that support egg maturation. Oocytes develop in individual egg chambers, which are surrounded by somatic follicle cells (FC). A combination of various cyclin-dependent kinases (CDKs) and cyclins drives progression though the cell cycle. miR-7 and miR-278 were found to underlie G1/S transition delay of *dcr-1* mutant GSCs through regulation of GSC division by repressing a CDK inhibitor known as Dacapo. Development of the egg chamber can be divided into 14 stages, and a single egg chamber at stage 10 is represented here. Mutants of core miRNA processing factors, including Ago1, Dcr-1, Pasha and Drosha, result in defective oocyte specification. miR-184 was found to play a role in oocyte patterning, effected by K10 and leading to defective GRK localization. (**b**) Schematic representation of experimental

Regulation of *dap* by Dcr-1 in the female germline was dependent on the 3′-UTR of *dap* mRNA, raising the possibility that *dap* is post-transcriptionally regulated by miRNAs. Through a computational search and experimental validation, miR-278 and miR-7 were identified as candidate miRNAs that mediate regulation of *dap* (Yu et al. 2009). Both miRNAs could downregulate a reporter carrying *dap* 3′-UTR in cultured cells, and deletion of the *mir-278* or *mir-7* locus caused mild defects in cell division or expression of cell cycle markers, respectively (Fig. 14.3a). While these results support the notion that the two miRNAs are involved in this process, the phenotypes do not sufficiently explain the reduced cell division observed with *dcr-1* mutant and suggest that additional miRNAs might be involved.

In addition to regulation of cell division, Dcr-1 plays a stage-specific role in maintenance of GSCs (Fig. 14.3b). After the initial study that revealed roles of Dcr-1 in GSC division (Hatfield et al. 2005), two other studies reported seemingly contradictory results (Jin and Xie 2007; Shcherbata et al. 2007). These studies reported progressive loss of *dcr-1* mutant GSCs, a phenotype that was not documented in the original study (Hatfield et al. 2005). This apparent discrepancy stemmed from differences in the timing of mutant clone induction (Jin and Xie 2007; Shcherbata et al. 2007). Mutant GSC loss was seen only when the clone was induced in the adult stage but not in the larval/pupal stage (Shcherbata et al. 2007). Loss of Dcr-1 activity did not cause detectable upregulation of the key differentiation factor Bam (bag of marbles) in GSCs, and GSCs that lacked both *dcr-1* and *bam* still showed the stem cell loss phenotype, suggesting that Dcr-1 is essential for GSC maintenance by inhibiting differentiation through a Bam-independent pathway.

What makes "young" GSCs resilient to *dcr-1* deletion? Genetic interaction analysis revealed the involvement of the siRNA Dicer, Dcr-2 (Shcherbata et al. 2007). Unlike *dcr-1* single mutant, *dcr-1;dcr-2* double mutant GSCs were lost more quickly even when the mutant clones were induced in the larval/pupal stage (Fig. 14.3b). siRNAs involved in this process were not identified, but the results that Dcr-2 is essential for age-dependent resilience of stem cells suggest that parallel small RNA pathways ensure proper maintenance of stem cells during gonadal development and ageing. It is an interesting coincidence that age-dependent control of GSC niche in the *Drosophila* male gonad also requires gene regulation by both siRNA and miRNA pathways (Toledano et al. 2012). Intertwined mechanisms involving miRNA and siRNA pathway components were also proposed in ageing of the fly brain (Abe et al. 2014). Interplay between the two small RNA pathways may be a common theme underlying gene regulation during ageing.

Consistent with the notion that Dcr-1 functions together with its partner protein Loqs to generate miRNAs, ovaries lacking *loqs* showed a GSC loss phenotype

Fig. 14.3 (continued) design in the Shcherbata study (Shcherbata et al. 2007). *dcr-1* mutant GSCs were formed in the larval or adult stage, and the number of *dcr-1* mutant GSCs was examined in the adult

(Park et al. 2007; Jiang et al. 2005; Forstemann et al. 2005) (Fig. 14.3b). The *loqs* locus produces at least three isoforms by alternative splicing/transcriptional termination (Miyoshi et al. 2010; Hartig et al. 2009; Zhou et al. 2009). Two of the isoforms, Loqs-PA and Loqs-PB, bind to Dcr-1, but only Loqs-PB is able to enhance the miRNA processing activity of Dcr-1 (Park et al. 2007; Fukunaga et al. 2012). Isoform-specific rescue experiments indicated that only Loqs-PB is capable of rescuing the GSC loss phenotype, implying that the phenotype was direct result of miRNA processing defects (Park et al. 2007; Jiang et al. 2005; Forstemann et al. 2005; Fukunaga et al. 2012). Interestingly, miRNA profiling analysis in *loqs* mutant animals and cultured knockout cells indicated that only a small number of miRNAs require Loqs for efficient processing (Liu et al. 2007; Fukunaga et al. 2012; Lim et al. 2016). Because only ~40% of fly miRNAs require Loqs-PB (Lim et al. 2016), the results narrow down the list of candidate miRNAs involved in the GSC loss phenotype.

Then, which miRNAs are responsible for the GSC loss phenotype? The Bantam miRNA appears to play a role in the adult specific GSC maintenance (Shcherbata et al. 2007). Adult-generated *bantam* mutant GSCs were frequently lost, suggesting that Bantam may be the miRNA underlying the GSC loss phenotype seen in *dcr-1* or *loqs* mutant GSCs. However, previous biochemical data indicated that the expression level of Bantam was not strongly reduced in *loqs* knockout animals and cells, suggesting that Bantam is a Loqs-independent miRNA (Liu et al. 2007; Fukunaga et al. 2012; Lim et al. 2016). Bantam was also shown to be important for cell fate specification of primordial germ cells in fly larvae, which is somewhat inconsistent with the Shcherbata study (Yang et al. 2009). The study also suggested that the effects of the *bantam* mutation was cell nonautonomous, in contrast to the previous study that showed cell autonomous effects (Shcherbata et al. 2007; Yang et al. 2009). In addition, the molecular mechanism behind Bantam-mediated GSC regulation remains unclear (Shcherbata et al. 2007). Further studies will be needed to identify key targets of Bantam and additional miRNAs involved in GSC maintenance.

Besides regulation of stem cells, miRNA processing factors including Dcr-1 are required for oocyte formation. Developing egg chambers lacking core canonical miRNA processing factors showed abnormality with frequent occurrence of the egg chambers containing 4 or 8 cells instead of normal 16 cells (Azzam et al. 2012). In wild-type ovaries, 1 cell in each egg chamber is specified as an oocyte, while the other 15 cells become nurse cells (Fig. 14.3a). However, *ago1* mutant egg chambers often lacked cells positive for the oocyte marker Orb, suggesting that Ago1 is essential for oocyte specification (Fig. 14.3a). Mutants of *dcr-1* or microprocessor components (*pasha* and *drosha*) showed similar phenotypes, strongly suggesting that inactivation of canonical miRNAs was the major cause of the phenotypes (Azzam et al. 2012). However, *loqs* mutant was not tested in these studies, and the contribution of Loqs to this process remains to be tested.

A reverse genetics study identified *mir-184* as an essential regulator of oogenesis (Iovino et al. 2009). Although zygotic *mir-184* null mutant flies displayed no overt morphological defects, mutant females produced fewer progeny. The number of eggs produced per female was severely decreased in *mir-184* mutant, and microscopic observations revealed abnormal egg size and morphological defects with fused dorsal appendages, hinting at its roles in vitellogenesis and dorsoventral patterning (Fig. 14.3a). The dorsoventral patterning defects manifested in mislocalization of Gurken (GRK), a receptor tyrosine kinase ligand whose mRNA localization is essential for establishment of dorsal-ventral axis in the oocyte (Kugler and Lasko 2009). Localization of GRK mRNA at the dorso-anterior cap is established by a complex mechanism involving multiple factors (Kugler and Lasko 2009). Based on computational prediction followed by experimental validation using sensor assays, the study identified miR-184 target sites in the 3′-UTR of K10, a factor required for normal GRK localization. Indeed, in *mir-184* mutant ovaries, K10 expression started precociously during oogenesis, but unexpectedly, its expression level was reduced by stage 9 oocyte. It is unclear how the precocious expression of K10 eventually leads to downregulation of K10 expression in late stages. *Dcr-1* mutant and mutants of other core miRNA processing factors exhibit oocyte specification defects but not patterning defects as seen in *mir-184* mutant oocytes (Azzam et al. 2012). This apparent discrepancy could be explained if removal of many miRNAs by inactivation of a general processing factor somehow neutralizes the phenotype by eliminating miRNAs that are counteracting each other. Stronger phenotypes with mutants of individual miRNAs than the *dcr-1* deletion have been seen in other tissues (Bejarano et al. 2010), and it is also demonstrated that miRNA-target pairs with counteracting activity could neutralize the phenotypes (Choi et al. 2007). Therefore, phenotypes seen in mutants of miRNA processing factors should be interpreted with caution.

In summary, studies of the *Drosophila* female germline lacking Dicer activity revealed that Dicer family proteins are involved in various steps during female germline development and oogenesis. Comparing phenotypes of different mutants deleting individual miRNA processing factors may narrow down the list of candidate miRNAs that are responsible for the phenotype, because individual miRNAs often have distinctive biochemical dependencies. However, care should be taken when interpreting results because Dicer mutant phenotypes might not represent the full spectrum of functions of Dicer-dependent miRNAs due to the presence of miRNA genes with counteracting potentials. The experimental system or the mutant allele used in the study also influences the results, which may explain why some results appear inconsistent.

14.2.2 Endo-siRNAs in Developing Mouse Oocytes

Dicer plays essential roles in mouse oogenesis, as revealed by early studies using conditional knockouts specifically removing Dicer activity in the developing oocyte (Murchison et al. 2007; Tang et al. 2007). Adult mutant females were infertile, even though ovaries were morphologically normal and contain fully grown oocytes. GV oocytes isolated from Dicer mutants showed developmental arrest at meiosis I with multiple spindles and misaligned chromosomes (Fig. 14.4). Microarray data showed upregulation of many genes and misregulation of maternal transcript turnover in Dicer null oocytes. These phenotypes led to the speculation that there are miRNAs functioning in gene regulation during oogenesis. However, there was no direct evidence implicating functional loss of miRNAs in associated developmental defects of Dicer mutants. Follow-up studies revealed an unexpected switch in Dicer function.

Small RNA libraries of mouse oocytes revealed the presence of abundant piRNAs and endo-siRNAs that have functional importance in mediating repression of largely nonoverlapping groups of transposons (Watanabe et al. 2008; Tam et al. 2008). Sequencing data also indicated that endo-siRNAs, instead of miRNAs, constitute a major population of small RNAs in the oocyte (Watanabe et al. 2008). Dicer loss resulted in a corresponding increase in expression of endo-siRNA targets, providing evidence for endo-siRNA-mediated gene regulation (Tam et al. 2008).

Hence, Dicer cleavage produces two distinct classes of small RNAs in the mouse oocyte. Which small RNA family is responsible for the phenotypes observed in Dicer mutant oocytes? Two independent groups used different approaches to address this question (Ma et al. 2010; Suh et al. 2010) (Fig. 14.4). Biogenesis of canonical miRNAs is dependent on sequential Drosha/DGCR8 and Dicer cleavages, whereas endo-siRNA processing is Drosha-independent. Consequently, DGCR8 deletion would abolish all canonical miRNAs but leave endo-siRNAs intact. Zygotic DGCR8 mutant embryos arrest at E6.5, proving that the zygotic microprocessor activity is essential to support embryonic development (Wang et al. 2007). However, pre-implantation development appears unaffected in DGCR8 maternal-zygotic mutants, suggesting that maternal microprocessor effects are

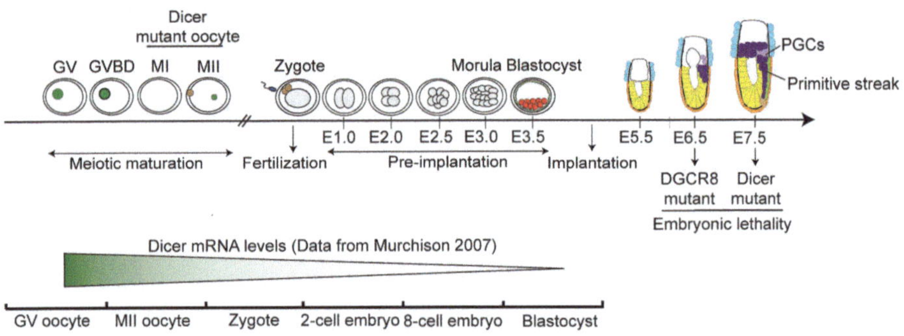

Fig. 14.4 Roles of miRNA/siRNA processing factors in mouse oogenesis. At birth, the germinal vesicle (GV) oocyte is arrested at prophase I. Under specific stimulation, germinal vesicle breakdown (GVBD) marks the resumption of meiosis. Wild-type oocytes progress through metaphase I (MI) and arrest at metaphase II (MII) until fertilization. Dicer mutant oocytes arrest at meiosis I with spindle defects (Murchison et al. 2007; Tang et al. 2007). DGCR8 and Dicer mutant mice exhibit embryonic lethality at E6.5 and E7.5, respectively (Wang et al. 2007; Bernstein et al. 2003). Dicer mRNA levels fluctuate during pre-implantation development, with peak levels measured in GV oocytes (data from Murchison et al. 2007)

dispensable for oogenesis and very early development (Suh et al. 2010). Despite global miRNA suppression in DGCR8-deficient oocytes, mice showed apparently normal maturation and produced healthy offspring when fertilized with wild-type sperm (Suh et al. 2010). This was in contrast to previous studies where Dicer mutant oocytes were arrested in meiotic maturation with spindle defects (Tang et al. 2007; Murchison et al. 2007). Furthermore, deletion of DGCR8 did not appear to perturb the transcriptome profile in oocytes. Hence, previously observed Dicer mutant phenotypes were likely resultant of endo-siRNA depletion instead of miRNA loss in oocytes. Lack of significant miRNA target site enrichment and presence of endo-siRNA target sites in upregulated transcripts of Dicer mutant oocytes supports the importance of endo-siRNAs in gene regulation during oogenesis (Suh et al. 2010; Ma et al. 2010). Intriguingly, reporter assays revealed that despite successful loading onto AGO proteins, miRNAs are functionally repressed and ineffective in mediating translational repression of reporter constructs during oocyte growth and meiotic maturation, whereas target mRNA cleavage activity was less strongly affected. More recently, Yang et al. showed that miRNA functions are only reactivated after the two-cell stage to regulate embryogenesis (Yang et al. 2016). The uncoupling of miRNA loading and its effector functions was proposed to be a mechanism used to achieve selective inhibition of the miRNA pathway in oocytes and a transitory switch in Dicer gene products from miRNA to endo-siRNA during oocyte-to-embryo progression (Ma et al. 2010; Svoboda 2010). It was proposed that global miRNA suppression in developing oocytes serves as a prelude for the extensive reprogramming of gene expression profiles during maternal-to-zygotic transition (MZT).

In mouse, endo-siRNAs were initially characterized in oocytes and embryonic stem cells and later also identified in male germ cells (Song et al. 2011; Tam et al. 2008; Babiarz et al. 2008; Watanabe et al. 2008). While *Drosophila* encodes for two Dicer proteins (Dicer-1 and Dicer-2) with distinct functions in miRNA and siRNA biogenesis, respectively, a single mammalian Dicer drives processing of small RNA classes including miRNAs and exo- and endo-siRNAs. How is endo-siRNA production specifically restricted to non-somatic cells? Flemr and colleagues recently characterized distinct Dicer isoforms present in mouse somatic cells and oocytes, which they termed Dicer^{S} and Dicer^{O}, respectively (Flemr et al. 2013) (Figs. 14.2 and 14.4). A retrotransposon insertion in the Dicer gene generates an alternative intronic promoter that produces a truncated isoform lacking the N-terminal helicase domain. Named Dicer^{O}, it is the predominant isoform expressed in oocytes that directs endo-siRNA production. As can be expected, Dicer^{O} deletion resulted in meiotic arrest of mutant oocytes and disruption of endo-siRNA production that is coupled to an upregulation of endo-siRNA targets. The identification of distinct Dicer isoforms with specialized roles in small RNA biogenesis presents a unique evolutionary event where a retrotransposon insertion drives oocyte-specific endo-siRNA production with functionally important roles both in oogenesis and transposon defence. Interestingly, the murid-specific insertion of the alternative promoter that drives Dicer^{O} expression appears to be accompanied by loss of the *piwi* family gene that is important in ovaries of other mammalian lineages (Hirano et al. 2014; Rosenkranz et al. 2015). This may imply that the emergence of Dicer^{O} isoform in the murid lineage drove the replacement of the piRNA-based genome defence system by the siRNA-based mechanism in the oocyte.

miRNAs and siRNAs exert their regulatory functions through the effector AGO protein (Meister 2013). Among the four mammalian AGO homologs, only AGO2 retains the slicer activity and is essential for functionality of siRNAs that regulate targets primarily via RNA cleavage. In contrast, other AGOs can compensate for the loss of AGO2 for miRNA functionality because miRNAs usually recruit downstream effectors to destabilize target mRNAs without directly cleaving the targets. AGO2 is essential for mouse development, even though the exact stage of developmental arrest remains controversial (Liu et al. 2004; Morita et al. 2007). The dispensability of other AGO proteins was attributed to the unique catalytic property of AGO2. By extension, this could suggest a role for the products of AGO2 catalytic cleavage in regulating developmental processes. In mouse, oocyte-specific deletion of Dicer or AGO2 resulted in similar phenotypes with developmental arrest at meiosis I (Murchison et al. 2007; Kaneda et al. 2009). In these studies, loss of Dicer or AGO2 was associated with failure to regulate maternal transcripts and dysregulation of miRNA expression patterns, respectively. However, the global suppression of miRNA expression in mouse oocytes prompted further investigation into the role of AGO2-enriched endo-siRNAs in meiosis. To address this, Stein and colleagues made use of an oocyte-specific catalytically inactive *Ago2* knock-in allele (Stein et al. 2015). The mutation of catalytic DDH to ADH motif specifically disrupts the endonucleolytic activity of AGO2 while retaining normal small RNA-binding capacity. *Ago2*^{ADH} oocytes also presented with severe defects in meiosis, providing definitive evidence that endo-siRNAs, but not miRNAs, are crucial in the female germline by controlling meiotic progression (Stein et al. 2015).

Small RNA libraries from pre-implantation mouse embryos revealed a class switch from siRNA/piRNA to miRNA (Ohnishi et al. 2010). Small RNA expression profiles are highly dynamic during oogenesis and early embryonic development in mouse, which seems pertinent considering the major reprogramming required in this process. Much of this reprogramming is dependent upon various classes of Dicer cleavage products, proving their roles as key regulatory molecules.

14.2.3 Roles of Dicer in Worm Gametogenesis

Dicer mutant worms are sterile with pleiotropic defects including malformation of oocytes (Knight and Bass 2001; Grishok et al. 2001; Ketting et al. 2001). Deep sequencing studies revealed a variety of germline-specific small RNA families, some of which are dependent on Dicer (Billi et al. 2014). In this section, we will discuss roles of Dicer-dependent small RNAs in the worm germline (Fig. 14.5).

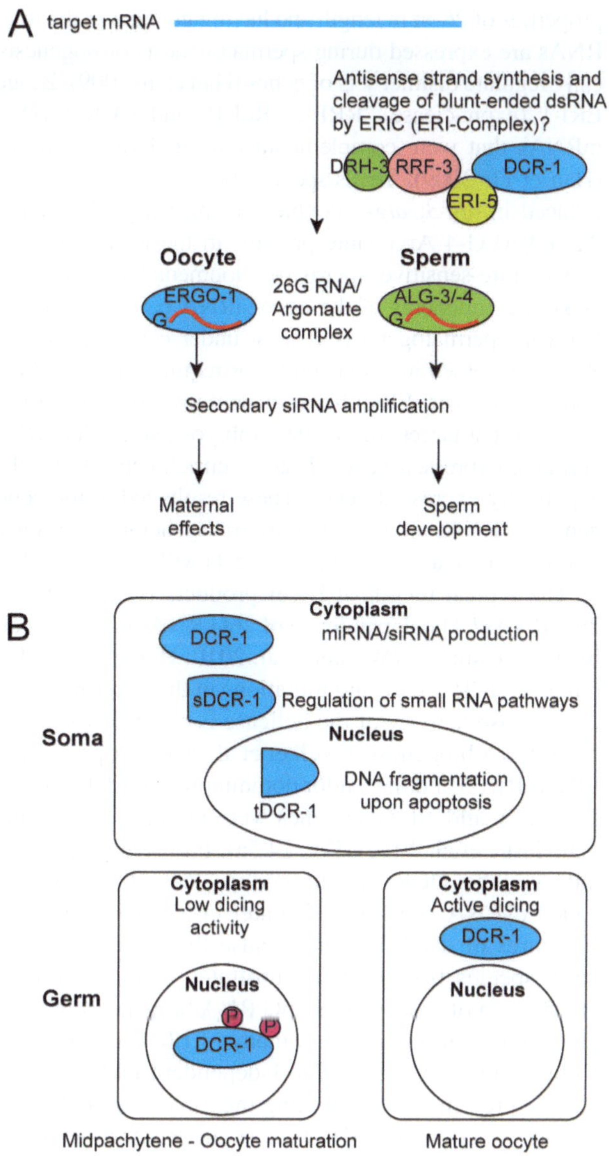

Fig. 14.5 Roles of Dicer in *C. elegans* soma and germline. (**a**) The primary endo-siRNA processing pathway. The ERIC (ERI complex) directs DCR-1 to the target mRNAs and triggers the production of 26G siRNAs. (**b**) Distinct roles of DCR-1 in the soma and germline. During the germline development, ERK-mediated phosphorylation regulates DCR-1 activity. In the soma, truncated DCR-1 (tDCR-1) and small DCR-1 (sDCR-1) are produced by proteolytic cleavage and act as a DNase that facilitates apoptosis and a regulator of DCR-1 activity, respectively

Endo-siRNAs were initially discovered in organisms possessing RNA-dependent RNA polymerase (RdRP) activity, such as fission yeast, plants and nematodes (Sijen et al. 2007; Ruby et al. 2006; Colmenares et al. 2007; Llave et al. 2002). In worms, siRNAs can be divided into two categories: primary siRNAs that are generally produced by Dicer-dependent mechanisms and secondary RNAs that are produced by RdRP-dependent amplification triggered by primary siRNAs (Billi et al. 2014). Endogenous primary siRNAs are commonly referred to as 26G RNAs, owing to their unique properties of 26-nt in length and having guanine as the first nucleotide (Fig. 14.5a). 26G RNAs are expressed during spermatogenesis or oogenesis/embryogenesis and specifically regulate distinct sets of genes (Han et al. 2009). Biogenesis of 26G RNAs requires ERI-1 (exonuclease), RRF-3 (RdRP) and DCR-1 (Dicer). In *eri-1* mutant, target mRNAs that were complementary to 26G RNAs showed upregulation in the testis (Han et al. 2009). Testis-specific but not oocyte-/embryo-specific 26G RNAs were reduced in *alg-3*; *alg-4* double mutant, suggesting that 26G RNAs are loaded onto ALG-3/ALG-4 Argonaute proteins in the testis. Mutations in ALG-3/ALG-4 led to temperature-sensitive sperm developmental defects as commonly seen in mutants lacking components of the endo-siRNA pathway, suggesting essential roles for 26G RNA in spermatogenesis at least under certain conditions. On the other hand, 26G RNAs expressed in oocyte/embryos required another AGO gene ERGO-1 (Yigit et al. 2006). However, ERGO-1 is dispensable for oogenesis, and computational analysis revealed that targets of oocyte-/embryo-specific 26G RNAs were underrepresented in germline expressed genes. Instead, enrichment of 26G RNA targets was observed in zygotically expressed genes. These results led to the conclusion that ERGO-1-dependent 26G RNAs are important for transgenerational transmission of endogenous RNAi information. Transgenerational effects will be discussed in Sect. 14.3.

The typical length of Dicer products is ~21–22 nt. How does DCR-1 produce 26G RNAs? The dependence of 26G RNA biogenesis on DCR-1 was first described by genetic studies (Welker et al. 2010; Han et al. 2009). Rescue experiments using mutated DCR-1 carrying mutations in the N-terminal helicase domain revealed that the ATPase activity of the helicase is required for production of 26G RNA but not for miRNA biogenesis (Welker et al. 2010). A proteomic study showed that DCR-1, ERI-1, RRF3 and the Tudor domain protein ERI-5 formed a complex (ERI complex or ERIC), and all these genes were required for normal production of 26G RNAs (Thivierge et al. 2011) (Fig. 14.5a). Furthermore, in vitro assays using worm lysate indicated that DCR-1 could produce 26-nt RNA fragments from the 5′-end of blunt-ended dsRNA substrates (Welker et al. 2011). This activity is dependent on the structures of dsRNA ends because the same assay failed to detect production of 26-nt species from dsRNA substrates with 3′ overhangs. The detailed molecular mechanism of endogenous 26G RNA biogenesis in vivo remains elusive. However, the results that mutations in many of ERIC components only affect the 26G RNA pathway but not other DCR-1-dependent pathways underscore the importance of partner proteins in determining the molecular activity of Dicer.

The function of DCR-1 in small RNA biogenesis and RNAi has been well studied since the first discoveries of small regulatory non-coding RNAs. More recently, a novel role of DCR-1 was unexpectedly uncovered in directing oocyte-to-embryo transition in *C. elegans*. MAP kinases mediate intracellular signalling and can be grouped into three main families: ERKs, JNKs and SAPKs (Morrison 2012). ERKs are perhaps the best-characterized MAP kinases. ERK (extracellular signal-regulated kinase)-mediated phosphorylation of DCR-1 on two residues served to inhibit Dcr-1 activities and trigger nuclear translocation. Subsequent dephosphorylation (reactivation of activities) of Dicer was found to promote embryogenesis (Drake et al. 2014) (Fig. 14.5b).

As expected from its broad roles in gene regulation via the miRNA pathway, DCR-1 is essential for oogenesis (*dcr-1* mutant worms can survive to adulthood due to maternal Dicer activity). *dcr-1* mutant adults are sterile, exhibiting defects in oogenesis with abnormal morphology of oocytes and DNA replication without cell division, a phenomenon known as endomitosis (Knight and Bass 2001; Grishok et al. 2001). These phenotypes were assumed to simply arise from loss of miRNA functions in *dcr-1* mutant until a recent study suggested that Dicer functions might be switched during worm oogenesis (Drake et al. 2014).

The study identified important phosphorylation sites of DCR-1 protein and functionally analysed DCR-1 in somatic and germline cells separately by using genetic mosaics. Mutant animals that retain DCR-1 activity only in the germline (referred to as "somatic *dcr-1* mutant") showed defects in oogenesis and ovulation similar to *dcr-1* null mutant, suggesting that somatic DCR-1 has cell nonautonomous effects regulating this process. On the other hand, mutant animals that lacked DCR-1 activity only in the germline cells (referred to as "germline *dcr-1* mutant") could support normal oogenesis but instead produced lethal embryos that died shortly after fertilization and the first cell division.

Rescue experiments using phosphomimetic- or phosphor-mutant at the identified phosphorylation sites of DCR-1 revealed distinct functions of DCR-1 in the somatic and germline tissues. The phosphomimetic mutant could restore ovulation defects but could not rescue the embryonic phenotypes, suggesting that DCR-1 needs to be dephosphorylated in embryos. In contrast, non-phosphorylatable DCR-1 mutants individually altering the two phosphorylation sites displayed more complex phenotypes. The alanine mutation at serine 1833 resulted in a small multiple oocyte phenotype, which is indicative of an elevated RAS/ERK pathway activity, suggesting that phosphorylated DCR-1 somehow represses the RAS/ERK activity in the germline. The alanine substitution at the other phosphorylation site, serine 1705, showed semidominant effects resulting in oogenesis defects distinct from the phenotypes in *dcr-1* null mutant. These suggested that this phospho-mutant was hyperactivated and the serine phosphorylation was essential for normal regulation of DCR-1 activity. Finally, the study showed that the double mutant replacing serine residues at these two positions with alanine resulted in additive phenotypes of these two single mutants. It remains unclear if the translocation of Dicer to the nucleus reflects distinct functions of Dicer in the two compartments or if it is a mechanism to spatially separate Dicer and its substrates including cytoplasmically localized miRNA/siRNA precursors to regulate dicing activity.

In summary, studies using *dcr-1* mutant worms revealed dynamic changes in the function of DCR-1 during gametogenesis. Post-translational modifications and partner proteins associated with DCR-1 are essential for proper spatio-temporal regulation of DCR-1 functions in the gonad.

14.3 Transgenerational Inheritance of Dicer Products

Oocytes carry hereditary information besides the information encoded in the nuclear genomic DNA sequence. Maternally deposited RNAs play important roles in patterning of embryos (Lasko 2011; Oh and Houston 2017). Furthermore, regulatory RNAs deposited in the eggs can also convey epigenetic information to the next generation. PiRNAs, a class of Dicer-independent small regulatory RNAs that are important for transposon silencing in germline cells, can act as transgenerational memory of transposon invasion, for example (Brennecke et al. 2008). In this section, we will discuss functional Dicer products that are deposited in the oocyte and their roles in gene regulation during embryogenesis.

14.3.1 Maternal Transmission of miRNAs

The fact that conditional knockout of DGCR8 or Drosha in mouse oocytes does not impair fertility and the resulting eggs support normal embryonic development when fertilized with wild-type sperms allows us to conclude that maternal miRNAs are dispensable for embryogenesis at least in some organisms (Suh et al. 2010; Yuan et al. 2014). Other organisms show oogenesis defects, which complicate functional analysis of maternally deposited miRNAs, as it is difficult to separate direct effects of maternal miRNAs and other indirect effects caused by oogenesis defects (Azzam et al. 2012). Several studies have profiled miRNA expression in oocytes from a variety of organisms, and a common conclusion of these studies is that the general abundance of miRNAs among all small RNA species is low in the mature oocyte, especially in zebrafish and frogs (Watanabe et al. 2005; Chen et al. 2005; Armisen et al. 2009; Ohnishi et al. 2010; Lee et al. 2014; Marco 2015). Furthermore, an evolutionarily conserved mechanism degrades maternal miRNAs by untemplated adenylation by the Wispy terminal nucleotide transferase, which presumably contributes to the clearance of miRNAs in late stage oocytes and early embryos (Armisen et al. 2009; Lee et al. 2014). These results may collectively suggest that maternal miRNAs play no or minor roles in the progeny.

A recent study using a population genomic approach also suggested that miRNAs present in oocytes play more important roles in oogenesis rather than in embryos. Egg activation and fertilization are two independent processes in *Drosophila* and allow for the dissection of maternal versus zygotic gene activities (Tadros and Lipshitz 2009). Sequencing of small RNAs from *Drosophila* unfertilized eggs identified a few highly abundant species including miR-92b, miR-184, Bantam and the *mir-310* cluster miRNAs (Marco 2015). Target sites of maternal miRNAs in maternal or zygotic transcripts differ, revealing a selective pressure against miRNA target sites in maternal transcripts, consistent with major roles of miRNAs in oogenesis rather than embryogenesis. However, there is an exception to this generalization in the maternally deposited miR-9c. miR-9c was found to have increased target sites in maternal transcripts, and thus it is speculated to function in maternal mRNA clearance during MZT (Marco 2015). miR-9c might play this role in concert with the zygotically expressed *mir-309* cluster miRNAs that is known as key factors in maternal mRNA clearance (Bushati et al. 2008). Interestingly, in mosquitos, expression of the *mir-309* cluster miRNAs is induced in ovaries upon stimulation by blood feeding (Zhang et al. 2016). Although the conservation of the roles for *mir-309* cluster miRNAs in MZT has not been tested in mosquitos, it is tempting to speculate that maternal miRNAs may somehow contribute to the clearance of maternal mRNAs in early embryos in insects at MZT.

A genetic study in *Drosophila* provided experimental evidence for the importance of maternal miRNAs including miR-9c (Kugler et al. 2013). Maternal loss of *mir-9c* or *mir-969* activity resulted in a reduced number of primordial germ cells (PGCs) in embryos (Kugler et al. 2013). Furthermore, those embryos from *mir-9c* or *mir-969* mutant mothers showed higher variations in the number of PGCs, suggesting that these miRNAs might contribute to the phenotypic robustness as proposed for some other miRNAs (Li et al. 2009). Targets of the two miRNAs and the molecular mechanism underlying these maternal mutant phenotypes remain unknown, and further studies will be needed to elucidate the importance of maternal miRNAs.

Soni et al. recently reported roles for maternal miR-34 in zebrafish. In zebrafish, miR-34 can be maternally loaded or zygotically expressed, and the authors made use of a knockdown method based on antisense oligonucleotides specifically targeting maternal or zygotic miR-34 to study their functions. Knockdown of maternal miR-34 did not alter the global mRNA expression profile of maternal or zygotic transcripts but resulted in derepression of specific targets Notch and Delta. Even though overall development did not appear to be affected with maternal miR-34 depletion, specific loss of miR-34 in the nervous system of developing embryos revealed brain morphological defects, likely effected via the Notch-Delta pathway (Soni et al. 2013). However, antisense technologies are prone to show off-target effects, and the results should be carefully interpreted (Kloosterman et al. 2007). Strict genetic tests will be required to support the findings.

In summary, studies reproducibly detected maternally deposited miRNAs, but their functionality currently remains obscure. The lack of molecular tools to study functions of individual maternally deposited miRNAs in embryos separately from their potential functions during oogenesis makes functional characterization of maternal miRNAs more difficult.

14.3.2 Transgenerational Inheritance of Small RNAs in Worms and Other Animals

Soon after the discovery of RNAi, its transgenerational effects were described (Grishok et al. 2000). This early study already revealed the complexity of inherited

RNAi that involves multiple distinct mechanisms for establishment of silencing and its transmission to the next generation. While some studies demonstrated the importance of siRNA amplification from targeted mRNAs and effects on the chromatin structures of target genes (Alcazar et al. 2008; Buckley et al. 2012; Houri-Ze'evi et al. 2016; Rechavi et al. 2014), other studies highlighted roles of inherited dsRNA or diced siRNA transported to oocytes (Marre et al. 2016; Devanapally et al. 2015). The silencing effects could be observed in the progeny from parents that either ingested or were injected with dsRNAs. Injection of fluorescently labelled dsRNAs and co-localization analysis with yolk protein showed that dsRNA molecules are incorporated into oocytes from the body cavity fluid by endocytosis (Marre et al. 2016). This study raised an interesting possibility that oocytes could receive information from other cells in the form of dsRNA through endocytosis (Kloc and Kubiak 2017).

Processed siRNAs can also carry transgenerational information. Dicer-dependent siRNAs, known as primary siRNAs, trigger the amplification of siRNAs using RdRPs and target RNAs as templates (Billi et al. 2014). Secondary siRNAs are known as 22G RNAs with characteristic 5′-guanosine enrichment and a strong 22-nt peak in the length. 22G RNAs are produced by a Dicer-independent mechanism and loaded to worm-specific AGO family proteins. Secondary siRNAs mediate transgenerational epigenetic inheritance of silencing that can last many generations (Billi et al. 2014). Although transgenerational silencing was first recognized in animals with artificial RNAi and antiviral defence, this phenomenon appears to reflect cellular machineries that transmit information about environmental and cellular conditions to the progeny.

Worms use the mechanisms of transgenerational RNAi as immunological memory (Rechavi et al. 2011). It was known that the worm exogenous siRNA pathway is involved in natural defence against viruses by producing siRNAs against the genomes of infecting viruses termed viRNAs (virus-derived siRNAs) (Felix et al. 2011). Under experimental conditions, the silencing effect against an artificial reporter that mimics viral replication could be inherited by the next generation in a non-Mendelian manner, indicating that the information was transmitted via extrachromosomal materials, most likely viRNAs (Rechavi et al. 2011). Production of primary viRNAs depends on the Dicer partner RDE-4 (Aoki et al. 2007) and the homozygous *rde-4* mutant progeny from heterozygous *rde-4* mutant parents contained primary viRNAs, providing genetic evidence that primary viRNAs can be transmitted to the next generation. The inheritance of the antiviral response for a longer duration (>5 generations) required the RRF-1 RdRP, suggesting that amplification of viRNAs is required for the long-term inheritance.

Heritable effects are not limited to antiviral responses. Endogenous siRNAs can also be transmitted to next generations via gametes in worms. Nutritional starvation influences gene expression, and in worms, the effects last for multiple generations (Rechavi et al. 2014). The heritable effects are mediated by inherited small RNAs as evidenced by the induction of endo-siRNAs targeting nutrient-related genes as well as up- or downregulation of target genes in the fed progeny.

It has not been tested if fitness benefits are conferred to animals by the heritable antivirus defence mechanism or the nutrient stress responses under more natural conditions in worms. Nevertheless, recent studies suggest that transgenerational inheritance of regulatory small RNAs via the gamete is not unique to worms but is a more common phenomenon.

In rodents, paternal epigenetic inheritance of diet-mediated metabolic changes has been proposed (Carone et al. 2010; Ng et al. 2010). Later, small RNAs derived from tRNA fragments were discovered as materials that are involved in this phenomenon (Sharma et al. 2016; Chen et al. 2016). Both studies showed that ~30–35nt fragments originating from the 5′ half of a subset of tRNAs were abundantly present in the mature sperm, and these tRNAs were capable of changing transcription profiles in embryos as demonstrated by injection of the tRNA fragments to fertilized eggs. The molecular mechanisms mediating gene regulation remain unclear, but transcriptional regulation of target mRNAs through the interaction with the endogenous retroelement MERVL is proposed. The expression changes of target genes were not accompanied by changes in the methylation status at the promoter regions, suggesting that the regulation does not involve DNA methylation. The involvement of Dicer for processing of this class of small RNAs is currently unknown, but it should be noted that some tRNA fragments are produced by a Dicer-dependent mechanism (Cole et al. 2009).

In the mouse sperm, miRNAs were also detected, and some of them are proposed to carry epigenetic information (Gapp et al. 2014). miRNA expression analysis in sperms from traumatically stressed mice detected differentially expressed miRNAs. Strikingly, injection of RNA isolated from sperms of traumatized mice could alter behaviours and metabolic profiles in the offspring. The RNA species responsible for these alterations are yet to be identified, but the differentially expressed miRNAs may be the materials carrying memory of traumatic stress.

Altogether, previous results support the notion that small RNAs are inherited by the next generation through gametes to alter gene expression profiles in the progeny. However, it is unclear how the long-term effects can be brought about by the small amount of inherited small RNAs especially in organisms that lack the small RNA amplification mechanisms by RdRP.

14.4 Conclusion

Oogenesis and early embryonic development are dynamic processes and involve complex post-transcriptional gene regulation. Organisms use diverse mechanisms to support these processes. This chapter summarized known roles for Dicer in a few organisms that represent vertebrates and invertebrates to highlight the fact that Dicer carries out various roles in small RNA pathways. A common conclusion that emerged from these studies is that the function of Dicer is temporally switched at certain stages of oogenesis or gonadal development. This can be achieved by use of paralogs (Dicer-1 and Dicer-2 in flies), mRNA isoforms from a single Dicer locus

($Dicer^S$ and $Dicer^O$ in rodents) and post-translational modifications (phosphorylation of Dicer in worms). Further complexity could be added by regulation of Dicer functions by their partner proteins (worm siRNA pathways). This is clearly the tip of the iceberg as the diverse organism-specific functions of Dicer were discovered by studies only in a small number of organisms. In addition, the roles of Dicer extend beyond production of small silencing RNAs, and many RNAi-independent functions have diversified. Dicer-2, central to RNAi responses in *Drosophila*, was reported to play small RNA-independent roles in regulation of immune signalling (Wang et al. 2015; Deddouche et al. 2008). Human Dicer can limit dsRNA accumulation to eliminate its cytotoxicity (White et al. 2014; Kaneko et al. 2011). In the worm soma, proteolytic cleavage can transform Dcr-1 to a regulator of dicing activity or even an apoptosis-promoting DNase (Nakagawa et al. 2010; Sawh and Duchaine 2013) (Fig. 14.5b). Furthermore, interaction with an assortment of protein cofactors and cellular RNA molecules seemingly expands the repertoire of Dicer substrates (Kurzynska-Kokorniak et al. 2015; Rybak-Wolf et al. 2014).

The dynamic changes in Dicer function and the diversity of the mechanisms in the three organisms discussed here suggest that there are more to be discovered in the future, especially in organisms where the mechanisms underlying oogenesis and early embryogenesis have not been molecularly studied.

Acknowledgements Research in K.O.'s group was supported by the National Research Foundation, Prime Minister's Office, Singapore, under its NRF Fellowship Programme (NRF2011NRF-NRFF001-042). The authors thank Dr. Yu Cai for the critical reading of the manuscript.

References

Abe M, Naqvi A, Hendriks GJ, Feltzin V, Zhu Y, Grigoriev A, Bonini NM (2014) Impact of age-associated increase in 2′-O-methylation of miRNAs on aging and neurodegeneration in Drosophila. Genes Dev 28(1):44–57. doi:10.1101/gad.226654.113

Alcazar RM, Lin R, Fire AZ (2008) Transmission dynamics of heritable silencing induced by double-stranded RNA in Caenorhabditis elegans. Genetics 180(3):1275–1288. doi:10.1534/genetics.108.089433

Aoki K, Moriguchi H, Yoshioka T, Okawa K, Tabara H (2007) In vitro analyses of the production and activity of secondary small interfering RNAs in C. elegans. Embo J 26(24):5007–5019

Armisen J, Gilchrist MJ, Wilczynska A, Standart N, Miska EA (2009) Abundant and dynamically expressed miRNAs, piRNAs, and other small RNAs in the vertebrate Xenopus tropicalis. Genome Res 19(10):1766–1775. doi:10.1101/gr.093054.109

Azzam G, Smibert P, Lai EC, Liu JL (2012) Drosophila Argonaute 1 and its miRNA biogenesis partners are required for oöcyte formation and germline cell division. Dev Biol 365(2):384–394. doi:10.1016/j.ydbio.2012.03.005

Babiarz JE, Ruby JG, Wang Y, Bartel DP, Blelloch R (2008) Mouse ES cells express endogenous shRNAs, siRNAs, and other microprocessor-independent, Dicer-dependent small RNAs. Genes Dev 22(20):2773–2785

Bejarano F, Smibert P, Lai EC (2010) miR-9a prevents apoptosis during wing development by repressing Drosophila LIM-only. Dev Biol 338(1):63–73. doi:10.1016/j.ydbio.2009.11.025

Bernstein E, Caudy A, Hammond S, Hannon G (2001) Role for a bidentate ribonuclease in the initiation step of RNA interference. Nature 409:363–366

Bernstein E, Kim SY, Carmell MA, Murchison EP, Alcorn H, Li MZ, Mills AA, Elledge SJ, Anderson KV, Hannon GJ (2003) Dicer is essential for mouse development. Nat Genet 35(3): 215–217

Billi AC, Fischer SE, Kim JK (2014) Endogenous RNAi pathways in C. elegans. WormBook:1–49. doi:10.1895/wormbook.1.170.1

Brennecke J, Malone CD, Aravin AA, Sachidanandam R, Stark A, Hannon GJ (2008) An epigenetic role for maternally inherited piRNAs in transposon silencing. Science 322(5906): 1387–1392

Buckley BA, Burkhart KB, SG G, Spracklin G, Kershner A, Fritz H, Kimble J, Fire A, Kennedy S (2012) A nuclear Argonaute promotes multigenerational epigenetic inheritance and germline immortality. Nature 489(7416):447–451. doi:10.1038/nature11352

Bushati N, Stark A, Brennecke J, Cohen SM (2008) Temporal reciprocity of miRNAs and their targets during the maternal-to-zygotic transition in Drosophila. Curr Biol 18:501–506. doi:10.1016/j.cub.2008.02.081

Carone BR, Fauquier L, Habib N, Shea JM, Hart CE, Li R, Bock C, Li C, Gu H, Zamore PD, Meissner A, Weng Z, Hofmann HA, Friedman N, Rando OJ (2010) Paternally induced transgenerational environmental reprogramming of metabolic gene expression in mammals. Cell 143(7):1084–1096. doi:10.1016/j.cell.2010.12.008

Cech TR, Steitz JA (2014) The noncoding RNA revolution-trashing old rules to forge new ones. Cell 157(1):77–94. doi:10.1016/j.cell.2014.03.008

Chen PY, Manninga H, Slanchev K, Chien M, Russo JJ, Ju J, Sheridan R, John B, Marks DS, Gaidatzis D, Sander C, Zavolan M, Tuschl T (2005) The developmental miRNA profiles of zebrafish as determined by small RNA cloning. Genes Dev 19(11):1288–1293

Chen Q, Yan M, Cao Z, Li X, Zhang Y, Shi J, Feng GH, Peng H, Zhang X, Zhang Y, Qian J, Duan E, Zhai Q, Zhou Q (2016) Sperm tsRNAs contribute to intergenerational inheritance of an acquired metabolic disorder. Science 351(6271):397–400. doi:10.1126/science.aad7977

Choi WY, Giraldez AJ, Schier AF (2007) Target protectors reveal dampening and balancing of Nodal agonist and antagonist by miR-430. Science 318(5848):271–274

Cole C, Sobala A, Lu C, Thatcher SR, Bowman A, Brown JW, Green PJ, Barton GJ, Hutvagner G (2009) Filtering of deep sequencing data reveals the existence of abundant Dicer-dependent small RNAs derived from tRNAs. RNA 15(12):2147–2160. doi:10.1261/rna.1738409

Colmenares SU, Buker SM, Buhler M, Dlakic M, Moazed D (2007) Coupling of double-stranded RNA synthesis and siRNA generation in fission yeast RNAi. Mol Cell 27(3):449–461

Dang DT, Perrimon N (1992) Use of a yeast site-specific recombinase to generate embryonic mosaics in Drosophila. Dev Genet 13(5):367–375

Deddouche S, Matt N, Budd A, Mueller S, Kemp C, Galiana-Arnoux D, Dostert C, Antoniewski C, Hoffmann JA, Imler JL (2008) The DExD/H-box helicase Dicer-2 mediates the induction of antiviral activity in drosophila. Nat Immunol 9(12):1425–1432

Devanapally S, Ravikumar S, Jose AM (2015) Double-stranded RNA made in C. elegans neurons can enter the germline and cause transgenerational gene silencing. Proc Natl Acad Sci USA 112(7):2133–2138. doi:10.1073/pnas.1423333112

Drake M, Furuta T, Suen KM, Gonzalez G, Liu B, Kalia A, Ladbury JE, Fire AZ, Skeath JB, Arur S (2014) A requirement for ERK-dependent Dicer phosphorylation in coordinating oocyte-to-embryo transition in C. elegans. Dev Cell 31(5):614–628. doi:10.1016/j.devcel.2014.11.004

Felix MA, Ashe A, Piffaretti J, Wu G, Nuez I, Belicard T, Jiang Y, Zhao G, Franz CJ, Goldstein LD, Sanroman M, Miska EA, Wang D (2011) Natural and experimental infection of Caenorhabditis nematodes by novel viruses related to nodaviruses. PLoS Biol 9(1): e1000586. doi:10.1371/journal.pbio.1000586

Filipowicz W, Bhattacharyya SN, Sonenberg N (2008) Mechanisms of post-transcriptional regulation by microRNAs: are the answers in sight? Nat Rev Genet 9(2):102–114

Flemr M, Malik R, Franke V, Nejepinska J, Sedlacek R, Vlahovicek K, Svoboda P (2013) A retrotransposon-driven dicer isoform directs endogenous small interfering RNA production in mouse oocytes. Cell 155(4):807–816. doi:10.1016/j.cell.2013.10.001

Forstemann K, Tomari Y, Du T, Vagin VV, Denli AM, Bratu DP, Klattenhoff C, Theurkauf WE, Zamore PD (2005) Normal microRNA maturation and germ-line stem cell maintenance requires Loquacious, a double-stranded RNA-binding domain protein. PLoS Biol 3(7):e236

Forstemann K, Horwich MD, Wee L, Tomari Y, Zamore PD (2007) Drosophila microRNAs are sorted into functionally distinct argonaute complexes after production by dicer-1. Cell 130(2): 287–297

Fukunaga R, Han BW, Hung JH, Xu J, Weng Z, Zamore PD (2012) Dicer partner proteins tune the length of mature miRNAs in flies and mammals. Cell 151(3):533–546. doi:10.1016/j.cell.2012.09.027

Gapp K, Jawaid A, Sarkies P, Bohacek J, Pelczar P, Prados J, Farinelli L, Miska E, Mansuy IM (2014) Implication of sperm RNAs in transgenerational inheritance of the effects of early trauma in mice. Nat Neurosci 17(5):667–669. doi:10.1038/nn.3695

Gasciolli V, Mallory AC, Bartel DP, Vaucheret H (2005) Partially redundant functions of Arabidopsis DICER-like enzymes and a role for DCL4 in producing trans-acting siRNAs. Curr Biol 15(16):1494–1500. doi:10.1016/j.cub.2005.07.024

Grishok A, Tabara H, Mello CC (2000) Genetic requirements for inheritance of RNAi in C. elegans. Science 287(5462):2494–2497

Grishok A, Pasquinelli A, Conte D, Li N, Parrish S, Ha I, Baillie DL, Fire A, Ruvkun G, Mello CC (2001) Genes and mechanisms related to RNA interference regulate expression of the small temporal RNAs that control *C. elegans* developmental timing. Cell 106:23–34

Ha M, Kim VN (2014) Regulation of microRNA biogenesis. Nat Rev Mol Cell Biol 15(8): 509–524. doi:10.1038/nrm3838

Han T, Manoharan AP, Harkins TT, Bouffard P, Fitzpatrick C, Chu DS, Thierry-Mieg D, Thierry-Mieg J, Kim JK (2009) 26G endo-siRNAs regulate spermatogenic and zygotic gene expression in Caenorhabditis elegans. Proc Natl Acad Sci USA 106(44):18674–18679. doi:10.1073/pnas.0906378106

Hartig JV, Esslinger S, Bottcher R, Saito K, Forstemann K (2009) Endo-siRNAs depend on a new isoform of loquacious and target artificially introduced, high-copy sequences. Embo J

Hatfield SD, Shcherbata HR, Fischer KA, Nakahara K, Carthew RW, Ruohola-Baker H (2005) Stem cell division is regulated by the microRNA pathway. Nature 435(7044):974–978

Henderson IR, Zhang X, Lu C, Johnson L, Meyers BC, Green PJ, Jacobsen SE (2006) Dissecting Arabidopsis thaliana DICER function in small RNA processing, gene silencing and DNA methylation patterning. Nat Genet 38(6):721–725

Hirano T, Iwasaki YW, Lin ZY, Imamura M, Seki NM, Sasaki E, Saito K, Okano H, Siomi MC, Siomi H (2014) Small RNA profiling and characterization of piRNA clusters in the adult testes of the common marmoset, a model primate. RNA 20(8):1223–1237. doi:10.1261/rna.045310.114

Houri-Ze'evi L, Korem Y, Sheftel H, Faigenbloom L, Toker IA, Dagan Y, Awad L, Degani L, Alon U, Rechavi O (2016) A tunable mechanism determines the duration of the transgenerational small RNA inheritance in C. elegans. Cell 165(1):88–99. doi:10.1016/j.cell.2016.02.057

Houwing S, Kamminga LM, Berezikov E, Cronembold D, Girard A, van den Elst H, Filippov DV, Blaser H, Raz E, Moens CB, Plasterk RH, Hannon GJ, Draper BW, Ketting RF (2007) A role for Piwi and piRNAs in germ cell maintenance and transposon silencing in Zebrafish. Cell 129 (1):69–82

Hutvagner G, Simard MJ (2008) Argonaute proteins: key players in RNA silencing. Nat Rev Mol Cell Biol 9(1):22–32. doi:10.1038/nrm2321

Iovino N, Pane A, Gaul U (2009) miR-184 has multiple roles in Drosophila female germline development. Dev Cell 17(1):123–133. doi:10.1016/j.devcel.2009.06.008

Jiang F, Ye X, Liu X, Fincher L, McKearin D, Liu Q (2005) Dicer-1 and R3D1-L catalyze microRNA maturation in Drosophila. Genes Dev 19:1674–1679

Jin Z, Xie T (2007) Dcr-1 maintains Drosophila ovarian stem cells. Curr Biol 17(6):539–544. doi:10.1016/j.cub.2007.01.050

Jones-Rhoades MW, Bartel DP, Bartel B (2006) MicroRNAS and their regulatory roles in plants. Annu Rev Plant Biol 57:19–53

Kaneda M, Tang F, O'Carroll D, Lao K, Surani MA (2009) Essential role for Argonaute2 protein in mouse oogenesis. Epigenetics Chromatin 2(1):9. doi:10.1186/1756-8935-2-9

Kaneko H, Dridi S, Tarallo V, Gelfand BD, Fowler BJ, Cho WG, Kleinman ME, Ponicsan SL, Hauswirth WW, Chiodo VA, Kariko K, Yoo JW, Lee DK, Hadziahmetovic M, Song Y, Misra S, Chaudhuri G, Buaas FW, Braun RE, Hinton DR, Zhang Q, Grossniklaus HE, Provis JM, Madigan MC, Milam AH, Justice NL, Albuquerque RJ, Blandford AD, Bogdanovich S, Hirano Y, Witta J, Fuchs E, Littman DR, Ambati BK, Rudin CM, Chong MM, Provost P, Kugel JF, Goodrich JA, Dunaief JL, Baffi JZ, Ambati J (2011) DICER1 deficit induces Alu RNA toxicity in age-related macular degeneration. Nature 471(7338):325–330. doi:10.1038/nature09830

Ketting R, Fischer S, Bernstein E, Sijen T, Hannon G, Plasterk RH (2001) Dicer functions in RNA interference and in synthesis of small RNAs involved in developmental timing in *C. elegans*. Genes Dev 15:2654–2659

Kim VN, Han J, Siomi MC (2009) Biogenesis of small RNAs in animals. Nat Rev Mol Cell Biol 10(2):126–139

Kloc M, Kubiak JZ (2017) Exogenous molecules and organelles delivery in oogenesis results and problems in cell differentiation. In: Kloc M (ed) Oocytes: maternal information and functions. Springer, Cham

Kloosterman WP, Lagendijk AK, Ketting RF, Moulton JD, Plasterk RH (2007) Targeted inhibition of miRNA maturation with morpholinos reveals a role for miR-375 in pancreatic islet development. PLoS Biol 5(8):e203. doi:10.1371/journal.pbio.0050203

Knight S, Bass BL (2001) A role for the RNAse III enzyme DCR-1 in RNA interference and germ line development in *Caenorhabditis elegans*. Science 293:2269–2271

Kugler JM, Lasko P (2009) Localization, anchoring and translational control of oskar, gurken, bicoid and nanos mRNA during Drosophila oogenesis. Fly (Austin) 3(1):15–28

Kugler JM, Chen YW, Weng R, Cohen SM (2013) Maternal loss of miRNAs leads to increased variance in primordial germ cell numbers in Drosophila melanogaster. G3 3(9): 1573–1576. doi:10.1534/g3.113.007591

Kurihara Y, Watanabe Y (2004) Arabidopsis micro-RNA biogenesis through Dicer-like 1 protein functions. Proc Natl Acad Sci USA 101(34):12753–12758

Kurzynska-Kokorniak A, Koralewska N, Pokornowska M, Urbanowicz A, Tworak A, Mickiewicz A, Figlerowicz M (2015) The many faces of Dicer: the complexity of the mechanisms regulating Dicer gene expression and enzyme activities. Nucleic Acids Res 43(9): 4365–4380. doi:10.1093/nar/gkv328

Lappalainen T, Sammeth M, Friedlander MR, t Hoen PA, Monlong J, Rivas MA, Gonzalez-Porta M, Kurbatova N, Griebel T, Ferreira PG, Barann M, Wieland T, Greger L, van Iterson M, Almlof J, Ribeca P, Pulyakhina I, Esser D, Giger T, Tikhonov A, Sultan M, Bertier G, MacArthur DG, Lek M, Lizano E, Buermans HP, Padioleau I, Schwarzmayr T, Karlberg O, Ongen H, Kilpinen H, Beltran S, Gut M, Kahlem K, Amstislavskiy V, Stegle O, Pirinen M, Montgomery SB, Donnelly P, McCarthy MI, Flicek P, Strom TM, Geuvadis C, Lehrach H, Schreiber S, Sudbrak R, Carracedo A, Antonarakis SE, Hasler R, Syvanen AC, van Ommen GJ, Brazma A, Meitinger T, Rosenstiel P, Guigo R, Gut IG, Estivill X, Dermitzakis ET (2013) Transcriptome and genome sequencing uncovers functional variation in humans. Nature 501(7468):506–511. doi:10.1038/nature12531

Lasko P (2011) Posttranscriptional regulation in Drosophila oocytes and early embryos. Wiley Interdiscip Rev RNA 2(3):408–416. doi:10.1002/wrna.70

Lau PW, Guiley KZ, De N, Potter CS, Carragher B, MacRae IJ (2012) The molecular architecture of human Dicer. Nat Struct Mol Biol 19(4):436–440. doi:10.1038/nsmb.2268

Lee RC, Feinbaum RL, Ambros V (1993) The *C. elegans* heterochronic gene *lin-4* encodes small RNAs with antisense complementarity to *lin-14*. Cell 75(5):843–854

Lee YS, Nakahara K, Pham JW, Kim K, He Z, Sontheimer EJ, Carthew RW (2004) Distinct roles for Drosophila Dicer-1 and Dicer-2 in the siRNA/miRNA silencing pathways. Cell 117(1): 69–81

Lee M, Choi Y, Kim K, Jin H, Lim J, Nguyen TA, Yang J, Jeong M, Giraldez AJ, Yang H, Patel DJ, Kim VN (2014) Adenylation of maternally inherited microRNAs by wispy. Mol Cell 56(5): 696–707. doi:10.1016/j.molcel.2014.10.011

Li X, Cassidy JJ, Reinke CA, Fischboeck S, Carthew RW (2009) A microRNA imparts robustness against environmental fluctuation during development. Cell 137(2):273–282

Lim MY, Ng AW, Chou Y, Lim TP, Simcox A, Tucker-Kellogg G, Okamura K (2016) The Drosophila Dicer-1 partner loquacious enhances miRNA processing from hairpins with unstable structures at the dicing site. Cell Rep. doi:10.1016/j.celrep.2016.04.059

Liu J, Carmell MA, Rivas FV, Marsden CG, Thomson JM, Song JJ, Hammond SM, Joshua-Tor L, Hannon GJ (2004) Argonaute2 is the catalytic engine of mammalian RNAi. Science 305 (5689):1437–1441

Liu X, Park JK, Jiang F, Liu Y, McKearin D, Liu Q (2007) Dicer-1, but not Loquacious, is critical for assembly of miRNA-induced silencing complexes. RNA 13(12):2324–2329

Llave C, Kasschau KD, Rector MA, Carrington JC (2002) Endogenous and silencing-associated small RNAs in plants. Plant Cell 14(7):1605–1619

Ma J, Flemr M, Stein P, Berninger P, Malik R, Zavolan M, Svoboda P, Schultz RM (2010) MicroRNA activity is suppressed in mouse oocytes. Curr Biol 20(3):265–270. doi:10.1016/j.cub.2009.12.042

MacRae IJ, Doudna JA (2007) Ribonuclease revisited: structural insights into ribonuclease III family enzymes. Curr Opin Struct Biol 17(1):138–145. doi:10.1016/j.sbi.2006.12.002

Macrae IJ, Zhou K, Li F, Repic A, Brooks AN, Cande WZ, Adams PD, Doudna JA (2006) Structural basis for double-stranded RNA processing by Dicer. Science 311(5758):195–198. doi:10.1126/science.1121638

MacRae IJ, Zhou K, Doudna JA (2007) Structural determinants of RNA recognition and cleavage by Dicer. Nat Struct Mol Biol 14(10):934–940. doi:10.1038/nsmb1293

Marco A (2015) Selection against maternal microRNA target sites in maternal transcripts. G3 5 (10):2199–2207. doi:10.1534/g3.115.019497

Marre J, Traver EC, Jose AM (2016) Extracellular RNA is transported from one generation to the next in Caenorhabditis elegans. Proc Natl Acad Sci USA 113(44):12496–12501. doi:10.1073/pnas.1608959113

Meister G (2013) Argonaute proteins: functional insights and emerging roles. Nat Rev Genet 14 (7):447–459. doi:10.1038/nrg3462

Miyoshi K, Tsukumo H, Nagami T, Siomi H, Siomi MC (2005) Slicer function of Drosophila Argonautes and its involvement in RISC formation. Genes Dev 19(23):2837–2848

Miyoshi K, Miyoshi T, Hartig JV, Siomi H, Siomi MC (2010) Molecular mechanisms that funnel RNA precursors into endogenous small-interfering RNA and microRNA biogenesis pathways in Drosophila. RNA. doi:10.1261/rna.1952110

Morita S, Horii T, Kimura M, Goto Y, Ochiya T, Hatada I (2007) One Argonaute family member, Eif2c2 (Ago2), is essential for development and appears not to be involved in DNA methylation. Genomics 89(6):687–696. doi:10.1016/j.ygeno.2007.01.004

Morrison DK (2012) MAP kinase pathways. Cold Spring Harb Perspect Biol 4(11). doi:10.1101/cshperspect.a011254

Murchison EP, Stein P, Xuan Z, Pan H, Zhang MQ, Schultz RM, Hannon GJ (2007) Critical roles for Dicer in the female germline. Genes Dev 21(6):682–693

Nakagawa A, Shi Y, Kage-Nakadai E, Mitani S, Xue D (2010) Caspase-dependent conversion of Dicer ribonuclease into a death-promoting deoxyribonuclease. Science 328(5976):327–334. doi:10.1126/science.1182374

Ng SF, Lin RC, Laybutt DR, Barres R, Owens JA, Morris MJ (2010) Chronic high-fat diet in fathers programs beta-cell dysfunction in female rat offspring. Nature 467(7318):963–966. doi:10.1038/nature09491

Oh D, Houston DW (2017) RNA localization in the vertebrate oocyte: establishment of oocyte polarity and localized mRNA assemblages. In: Kloc M (ed) Oocytes: maternal information and functions, Results and problems in cell differentiation. Springer, Cham

Ohnishi Y, Totoki Y, Toyoda A, Watanabe T, Yamamoto Y, Tokunaga K, Sakaki Y, Sasaki H, Hohjoh H (2010) Small RNA class transition from siRNA/piRNA to miRNA during pre-implantation mouse development. Nucleic Acids Res 38(15):5141–5151. doi:10.1093/nar/gkq229

Okamura K, Lai EC (2008) Endogenous small interfering RNAs in animals. Nat Rev Mol Cell Biol 9(9):673–678

Park JK, Liu X, Strauss TJ, McKearin DM, Liu Q (2007) The miRNA pathway intrinsically controls self-renewal of drosophila germline stem cells. Curr Biol 17(6):533–538

Park JE, Heo I, Tian Y, Simanshu DK, Chang H, Jee D, Patel DJ, Kim VN (2011) Dicer recognizes the 5′ end of RNA for efficient and accurate processing. Nature 475(7355):201–205. doi:10.1038/nature10198

Rechavi O, Minevich G, Hobert O (2011) Transgenerational inheritance of an acquired small RNA-based antiviral response in C. elegans. Cell 147(6):1248–1256. doi:10.1016/j.cell.2011.10.042

Rechavi O, Houri-Ze'evi L, Anava S, Goh WS, Kerk SY, Hannon GJ, Hobert O (2014) Starvation-induced transgenerational inheritance of small RNAs in C. elegans. Cell 158(2):277–287. doi:10.1016/j.cell.2014.06.020

Reinhart BJ, Slack F, Basson M, Pasquinelli A, Bettinger J, Rougvie A, Horvitz HR, Ruvkun G (2000) The 21-nucleotide *let-7* RNA regulates developmental timing in *Caenorhabditis elegans*. Nature 403:901–906

Rosenkranz D, Rudloff S, Bastuck K, Ketting RF, Zischler H (2015) Tupaia small RNAs provide insights into function and evolution of RNAi-based transposon defense in mammals. RNA 21(5):911–922. doi:10.1261/rna.048603.114

Ruby JG, Jan C, Player C, Axtell MJ, Lee W, Nusbaum C, Ge H, Bartel DP (2006) Large-scale sequencing reveals 21U-RNAs and additional microRNAs and endogenous siRNAs in C. elegans. Cell 127(6):1193–1207

Rybak-Wolf A, Jens M, Murakawa Y, Herzog M, Landthaler M, Rajewsky N (2014) A variety of dicer substrates in human and C. elegans. Cell 159(5):1153–1167. doi:10.1016/j.cell.2014.10.040

Sawh AN, Duchaine TF (2013) A truncated form of dicer tilts the balance of RNA interference pathways. Cell Rep 4(3):454–463. doi:10.1016/j.celrep.2013.07.013

Sharma U, Conine CC, Shea JM, Boskovic A, Derr AG, Bing XY, Belleannee C, Kucukural A, Serra RW, Sun F, Song L, Carone BR, Ricci EP, Li XZ, Fauquier L, Moore MJ, Sullivan R, Mello CC, Garber M, Rando OJ (2016) Biogenesis and function of tRNA fragments during sperm maturation and fertilization in mammals. Science 351(6271):391–396. doi:10.1126/science.aad6780

Shcherbata HR, Ward EJ, Fischer KA, JY Y, Reynolds SH, Chen CH, Xu P, Hay BA, Ruohola-Baker H (2007) Stage-specific differences in the requirements for germline stem cell maintenance in the Drosophila ovary. Cell Stem Cell 1(6):698–709. doi:10.1016/j.stem.2007.11.007

Sijen T, Steiner FA, Thijssen KL, Plasterk RH (2007) Secondary siRNAs result from unprimed RNA synthesis and form a distinct class. Science 315(5809):244–247

Song R, Hennig GW, Wu Q, Jose C, Zheng H, Yan W (2011) Male germ cells express abundant endogenous siRNAs. Proc Natl Acad Sci USA 108(32):13159–13164. doi:10.1073/pnas.1108567108

Soni K, Choudhary A, Patowary A, Singh AR, Bhatia S, Sivasubbu S, Chandrasekaran S, Pillai B (2013) miR-34 is maternally inherited in Drosophila melanogaster and Danio rerio. Nucleic Acids Res 41(8):4470–4480. doi:10.1093/nar/gkt139

Stein P, Rozhkov NV, Li F, Cardenas FL, Davydenko O, Vandivier LE, Gregory BD, Hannon GJ, Schultz RM (2015) Essential Role for endogenous siRNAs during meiosis in mouse oocytes. PLoS Genet 11(2):e1005013. doi:10.1371/journal.pgen.1005013

Suh N, Baehner L, Moltzahn F, Melton C, Shenoy A, Chen J, Blelloch R (2010) MicroRNA function is globally suppressed in mouse oocytes and early embryos. Curr Biol 20(3):271–277. doi:10.1016/j.cub.2009.12.044

Svoboda P (2010) Why mouse oocytes and early embryos ignore miRNAs? RNA Biol 7(5): 559–563

Tadros W, Lipshitz HD (2009) The maternal-to-zygotic transition: a play in two acts. Development 136(18):3033–3042. doi:10.1242/dev.033183

Tam OH, Aravin AA, Stein P, Girard A, Murchison EP, Cheloufi S, Hodges E, Anger M, Sachidanandam R, Schultz RM, Hannon GJ (2008) Pseudogene-derived small interfering RNAs regulate gene expression in mouse oocytes. Nature 453:534–538

Tang F, Kaneda M, O'Carroll D, Hajkova P, Barton SC, Sun YA, Lee C, Tarakhovsky A, Lao K, Surani MA (2007) Maternal microRNAs are essential for mouse zygotic development. Genes Dev 21(6):644–648. doi:10.1101/gad.418707

Thivierge C, Makil N, Flamand M, Vasale JJ, Mello CC, Wohlschlegel J, Conte D Jr, Duchaine TF (2011) Tudor domain ERI-5 tethers an RNA-dependent RNA polymerase to DCR-1 to potentiate endo-RNAi. Nat Struct Mol Biol 19(1):90–97. doi:10.1038/nsmb.2186

Toledano H, D'Alterio C, Czech B, Levine E, Jones DL (2012) The let-7-Imp axis regulates ageing of the Drosophila testis stem-cell niche. Nature 485(7400):605–610. doi:10.1038/nature11061

Tsutsumi A, Kawamata T, Izumi N, Seitz H, Tomari Y (2011) Recognition of the pre-miRNA structure by Drosophila Dicer-1. Nat Struct Mol Biol 18(10):1153–1158. doi:10.1038/nsmb.2125

Vagin VV, Sigova A, Li C, Seitz H, Gvozdev V, Zamore PD (2006) A distinct small RNA pathway silences selfish genetic elements in the germline. Science 313(5785):320–324

Valencia-Sanchez MA, Liu J, Hannon GJ, Parker R (2006) Control of translation and mRNA degradation by miRNAs and siRNAs. Genes Dev 20(5):515–524

Wang Y, Medvid R, Melton C, Jaenisch R, Blelloch R (2007) DGCR8 is essential for microRNA biogenesis and silencing of embryonic stem cell self-renewal. Nat Genet 39(3):380–385

Wang Z, Wu D, Liu Y, Xia X, Gong W, Qiu Y, Yang J, Zheng Y, Li J, Wang YF, Xiang Y, Hu Y, Zhou X (2015) Drosophila Dicer-2 has an RNA interference-independent function that modulates Toll immune signaling. Sci Adv 1(9):e1500228. doi:10.1126/sciadv.1500228

Watanabe T, Takeda A, Mise K, Okuno T, Suzuki T, Minami N, Imai H (2005) Stage-specific expression of microRNAs during Xenopus development. FEBS Lett 579(2):318–324

Watanabe T, Totoki Y, Toyoda A, Kaneda M, Kuramochi-Miyagawa S, Obata Y, Chiba H, Kohara Y, Kono T, Nakano T, Surani MA, Sakaki Y, Sasaki H (2008) Endogenous siRNAs from naturally formed dsRNAs regulate transcripts in mouse oocytes. Nature 453:539–543

Welker NC, Pavelec DM, Nix DA, Duchaine TF, Kennedy S, Bass BL (2010) Dicer's helicase domain is required for accumulation of some, but not all, C. elegans endogenous siRNAs. RNA 16(5):893–903. doi:10.1261/rna.2122010

Welker NC, Maity TS, Ye X, Aruscavage PJ, Krauchuk AA, Liu Q, Bass BL (2011) Dicer's helicase domain discriminates dsRNA termini to promote an altered reaction mode. Mol Cell 41(5):589–599. doi:10.1016/j.molcel.2011.02.005

White E, Schlackow M, Kamieniarz-Gdula K, Proudfoot NJ, Gullerova M (2014) Human nuclear Dicer restricts the deleterious accumulation of endogenous double-stranded RNA. Nat Struct Mol Biol 21(6):552–559. doi:10.1038/nsmb.2827

Wightman B, Ha I, Ruvkun G (1993) Posttranscriptional regulation of the heterochronic gene *lin-14* by *lin-4* mediates temporal pattern formation in *C. elegans*. Cell 75:855–862

Yang Y, Xu S, Xia L, Wang J, Wen S, Jin P, Chen D (2009) The bantam microRNA is associated with drosophila fragile X mental retardation protein and regulates the fate of germline stem cells. PLoS Genet 5(4):e1000444. doi:10.1371/journal.pgen.1000444

Yang JS, Smibert P, Westholm JO, Jee D, Maurin T, Lai EC (2014) Intertwined pathways for Argonaute-mediated microRNA biogenesis in Drosophila. Nucleic Acids Res 42(3): 1987–2002. doi:10.1093/nar/gkt1038

Yang Q, Lin J, Liu M, Li R, Tian B, Zhang X, Xu B, Liu M, Zhang X, Li Y, Shi H, Wu L (2016) Highly sensitive sequencing reveals dynamic modifications and activities of small RNAs in mouse oocytes and early embryos. Sci Adv 2(6):e1501482. doi:10.1126/sciadv.1501482

Yigit E, Batista PJ, Bei Y, Pang KM, Chen CC, Tolia NH, Joshua-Tor L, Mitani S, Simard MJ, Mello CC (2006) Analysis of the C. elegans Argonaute family reveals that distinct Argonautes act sequentially during RNAi. Cell 127(4):747–757

Yu JY, Reynolds SH, Hatfield SD, Shcherbata HR, Fischer KA, Ward EJ, Long D, Ding Y, Ruohola-Baker H (2009) Dicer-1-dependent Dacapo suppression acts downstream of Insulin receptor in regulating cell division of Drosophila germline stem cells. Development 136(9): 1497–1507. doi:10.1242/dev.025999

Yuan S, Ortogero N, Wu Q, Zheng H, Yan W (2014) Murine follicular development requires oocyte DICER, but not DROSHA. Biol Reprod 91(2):39. doi:10.1095/biolreprod.114.119370

Zhang H, Kolb FA, Jaskiewicz L, Westhof E, Filipowicz W (2004) Single processing center models for human Dicer and bacterial RNase III. Cell 118(1):57–68

Zhang Y, Zhao B, Roy S, Saha TT, Kokoza VA, Li M, Raikhel AS (2016) microRNA-309 targets the Homeobox gene SIX4 and controls ovarian development in the mosquito Aedes aegypti. Proc Natl Acad Sci USA 113(33):E4828–E4836. doi:10.1073/pnas.1609792113

Zhou R, Czech B, Brennecke J, Sachidanandam R, Wohlschlegel JA, Perrimon N, Hannon GJ (2009) Processing of Drosophila endo-siRNAs depends on a specific Loquacious isoform. RNA 15(10):1886–1895

Yang Y, Xu S, Xia L, Wang J, Wen S, Jin P, Chen D (2009) The bantam microRNA is associated with drosophila fragile X mental retardation protein and regulates the fate of germline stem cells. PLoS Genet 5(4):e1000444. doi:10.1371/journal.pgen.1000444

[illegible]

[illegible]

[illegible]

[illegible]

[illegible]

[illegible]

[illegible]

Part IV
Oocyte Specific Functions of Ubiquitous Molecules and Organelles

Chapter 15
The Regulation and Function of Cohesin and Condensin in Mammalian Oocytes and Spermatocytes

Jibak Lee

Abstract Germ cells, such as oocytes and spermatocytes, produce haploid gametes by a special type of cell division called meiosis. The reduction of chromosome number is achieved in meiosis I, in which homologous chromosomes (homologs) are paired and recombined with their counterparts and finally segregated from each other. How meiotic chromosomes behave in a different manner from mitotic chromosomes has been a fascinating problem for cellular and developmental biology. Cohesin and condensin are multi-subunit protein complexes that play central roles in sister chromatid cohesion and chromosome condensation (also segregation), respectively. Recent studies investigating the expression and function of cohesin and condensin in mammalian germ cells greatly advance our understanding of the molecular mechanism underlying the meiotic chromosomal events. Furthermore, accumulating evidence suggests that reduction of cohesin during prophase I arrest in mammalian oocytes is one of the major causes for age-related chromosome segregation error. This review focuses on the regulation and functions of cohesins and condensins during mammalian meiosis.

15.1 Introduction

Meiosis is a special type of cell division that produces gametes (sperms and eggs) from spermatocytes and oocytes in animals. Unlike mitosis in which a single round of DNA replication is followed by a single round of cell division, two successive rounds of cell division follow a single round of DNA replication in meiosis. The actual reduction of chromosome number is accomplished in the first meiotic division (meiosis I). In meiosis I, homologous chromosomes (homologs) recombine with their partners at prophase. As a result, sister chromatid cohesion distal to chiasmata make physical connection between homologs by metaphase I. At the

J. Lee (✉)
Laboratory of Developmental Biotechnology, Graduate School of Agricultural Science, Kobe University, 1-1 Rokkodai-cho, Nada-ku, Kobe 657-8501, Japan
e-mail: leej@tiger.kobe-u.ac.jp

M. Kloc (ed.), *Oocytes*, Results and Problems in Cell Differentiation 63,
DOI 10.1007/978-3-319-60855-6_15

onset of anaphase I, homologs separate from each other by resolution of inter-sister arm cohesion while sister chromatids (sisters) are kept attached at the centromere regions. Finally, in the second meiotic division (meiosis II), sisters separate from each other by resolution of the centromeric cohesion (Miyazaki and Orr-Weaver 1994).

15.2 Sister Chromatid Cohesion by Cohesin

Cohesin is a multi-subunit protein complex which establishes and maintains sister chromatid cohesion in mitosis and meiosis (Peters et al. 2008; Nasmyth and Haering 2009). The cohesin complex consists of four subunits: two structural maintenance of chromosome proteins, SMC1 and SMC3, a kleisin subunit RAD21 (also called SCC1), and either one of STAG1/SA1 or STAG2/SA2 (Table 15.1). Since cohesin complex forms a tripartite ringlike structure, it is believed that the complex might hold sisters together by embracing the two nucleosomes in its ring (Gruber et al. 2003; Nasmyth 2011) (Fig. 15.1). Although cohesin binds to chromatin throughout interphase, it can establish sister chromatid cohesion only during S phase (Uhlmann and Nasmyth 1998). Cohesin maintains the sister chromatid cohesion from S phase until the onset of anaphase. However, most of cohesin is released from chromosomes prior to sister chromatid separation in a separase-independent pathway, which involves cohesin-associated proteins, Wapl, Pds5, and Sororin as well as phosphorylation of cohesin subunits by mitotic kinases (Shintomi and Hirano 2009; Nishiyama et al. 2010). Therefore, just before anaphase, only a small population of cohesin remains on the chromosomes mainly at centromere regions. At anaphase onset, anaphase-promoting complex/cyclosome (APC/C) associating with CDC20 ubiquitinates the target proteins, securin and cyclin B, thereby inducing the destruction of them by proteasome (Yu 2007). This leads to the activation of separase that cleaves the RAD21 subunit of cohesin complex, resulting in sister chromatid separation (Uhlmann et al. 2000; Waizenegger et al. 2000).

15.3 Meiosis-Specific Cohesin Subunits

In addition to the canonical mitotic cohesin subunits, specific variants of cohesin subunit are expressed in meiosis. In 1999, yeast Rec8 was first identified as a meiosis-specific cohesin subunit that is required for the reductional cell division (Klein et al. 1999; Watanabe and Nurse 1999). Thereafter, several meiosis-specific cohesin subunits have been found in various species. So far, SMC1β (Revenkova et al. 2001), STAG3 (Prieto et al. 2001), REC8 (Eijpe et al. 2003; Lee et al. 2003), and RAD21L (Gutiérrez-Caballero et al. 2011; Ishiguro et al. 2011; Lee and Hirano 2011) have been found as meiosis-specific cohesin subunits in mammals

Table 15.1 Subunits of cohesin and condensin complexes in mammals

	Type	SMC subunits		Kleisin subunits	Other subunits	
Cohesin	Canonical	SMC1α	SMC3	RAD21	STAG1/SA1 or STAG2/SA2	
	Meiotic	SMC1β	SMC3	REC8 or RAD21L	STAG3/SA3	
Condensin	I	SMC2	SMC4	CAP-H	CAP-D2	CAP-G
	II	SMC2	SMC4	CAP-H2	CAP-D3	CAP-G2

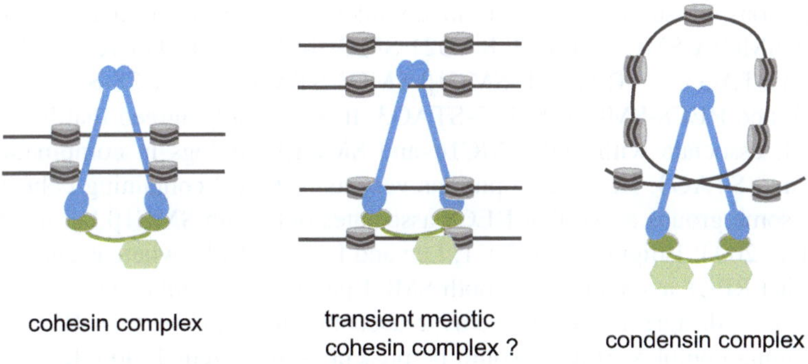

Fig. 15.1 Models of chromosome cohesion and condensation by cohesin and condensin complexes. A cohesin complex is thought to embrace two nucleosomes from sister chromatids, while a condensin complex is thought to connect the DNA segments from the same chromatid. A transiently expressed meiotic cohesin may embrace two nucleosomes from non-sister chromatids in this hypothetical model

(Table 15.1). Remarkably, canonical mitotic cohesin subunits are also expressed, at least in some stages of meiosis (Eijpe et al. 2000; Prieto et al. 2002; Xu et al. 2004; Parra et al. 2004) (Fig. 15.1). Although a great deal of effort has been made to examine the expression pattern of cohesin subunits during meiosis, there are some contradictions (see Suja and Barbero 2009). The contradictions might have arisen due to the differences in the fixation and labeling methods, antibody specificity, and/or sensitivity for detection of signals in the immunofluorescence analyses. Here, I will talk about kleisin subunits as one of the examples of contradictions. One paper reported that RAD21 is expressed almost throughout whole meiosis and localizes along axial element/lateral elements (AE/LEs) in prophase I and mainly at centromeres in metaphase I and metaphase II (Xu et al. 2004). Another paper shows RAD21 being present on the AE/LEs in prophase I and at inter-chromatid arm region and around centromeres in metaphase I (Parra et al. 2004). And another paper reports that RAD21 is expressed only during a short period of prophase I and localizes along LEs (Lee and Hirano 2011). As for RAD21L, some reports (Gutiérrez-Caballero et al. 2011; Ishiguro et al. 2011) show the localization on chromosomes from prophase I to metaphase I, while other reports do so only from leptotene to mid pachytene (Lee and Hirano 2011; Ishiguro et al. 2014). Although it is difficult to determine which of these observations is correct, functional analyses might help in making that judgment call. For example, RAD21, if it were present on the AEs at leptotene stage, would not contribute to AE formation, since AEs are not formed in *Rec8*/*Rad21L* double KO mice (Llano et al. 2012).

Meiotic Cohesin Complexes

Immunoprecipitation analyses using mouse testis extracts in combination with immunofluorescence analyses of localization suggest that there are several types of meiotic cohesin complexes in addition to the canonical cohesin complexes. The possible combinations of subunits in cohesin complexes so far reported are as follows: RAD21-SMC1α-SMC3-STAG3, RAD21-SMC1β-SMC3-STAG3, RAD21L-SMC1-α-SMC3-STAG3, RAD21L-SMC1β-SMC3-STAG3, REC8-SMC1α-SMC3-STAG3, and REC8-SMC1β-SMC3-STAG3. It is generally agreed that RAD21 and RAD21L associate with both SMC1α and SMC1β paralogs in combination with SMC3 and STAG3. However, opinions vary as to REC8-containing cohesin complexes: some groups argue that REC8 associates only with SMC1β but not SMC1α (Lee et al. 2003; Ishiguro et al. 2011; Lee and Hirano 2011), whereas another group insists that REC8 associates with both SMC1 paralogs (Revenkova et al. 2004). For settling this disagreement, it seems reasonable to suppose that REC8-SMC1-α-containing complex, if any, would be present at an extremely low level, because simultaneous depletion of SMC1β and RAD21L, which removes all the cohesin complexes with the exception of REC8-SMC1α-containing complex from chromosome axes at leptotene, almost completely abolishes the AE formation (Biswas et al. 2016).

15.4 Roles of Cohesins in Meiotic Prophase I

Prophase I proceeds in parallel with the assembly and disassembly of the synaptonemal complexes (SCs) (Page and Hawley 2004) (Fig. 15.2). After premeiotic S phase, axial elements (AEs) are formed along chromosomes at leptotene stage. The AEs on homologs start to be connected by transverse filaments by a mechanism called synapsis at zygotene stage. After synapsis the AEs are called the lateral elements (LEs). At pachytene stage, homologous AEs are connected along their entire length. Finally, the SC is disassembled at diplotene stage. Although the SC mediates the interaction between homologs transiently in prophase I, it is essential for the establishment of crossover recombination (de Vries et al. 2005), which holds association between homologs until anaphase I in conjunction with sister chromatid cohesion distal to chiasmata. The genetic studies using knockout (KO) mice showed that meiotic cohesins are involved in a variety of chromosomal events occurring in prophase I.

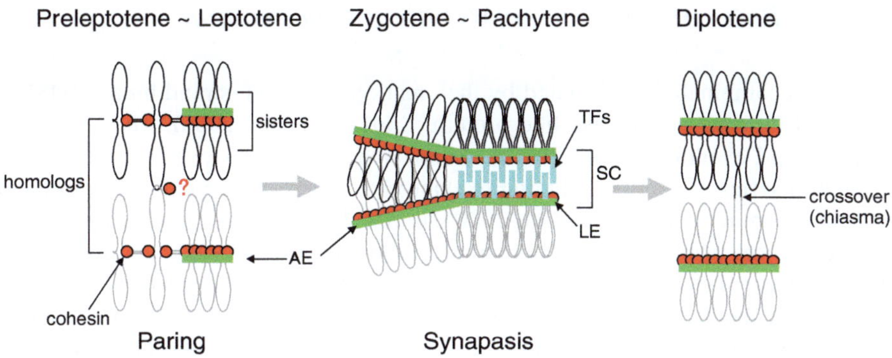

Fig. 15.2 Pairing and synapsis of homologs at prophase I. At the leptotene stage, cohesins are linearly localized along chromosomes concurrently with formation of axial elements (AEs). Homologs are juxtaposed through elusive phenomenon called pairing. At the zygotene stage, a pair of AEs on homologs start to be connected by transverse filaments (TFs). This phenomenon is so called synapsis. The AEs on synapsed region are called lateral elements (LEs). At the pachytene stage, synapsis is completed, and the synaptonemal complex (SC) is formed on each pair of homologs. At the diplotene stage, the SC is disassembled, but homologs are kept connected through crossover recombination

15.4.1 Roles of Cohesins in AE Formation

One of the common phenotypes observed in the mice depleted of any meiosis-specific cohesin subunit is the defect in AE formation although the severity is different among the mutants. In *Smc1β*-, *Rec8*-, or *Rad21L*-deficient mice, the defect is only partial: shorter AEs are formed (Bannister et al. 2004; Revenkova et al. 2004; Xu et al. 2005; Herrán et al. 2011). On the other hand, no AEs are formed in spermatocytes depleted of STAG3 (Winters et al. 2014). The difference would be explained by the presence or absence of cohesin subunit paralogs that play a nonredundant role in AE formation. Indeed, the paralogs of SMC1β, RAD21L, and REC8 are localized along AEs whereas STAG3 paralogs, i.e., STAG1 or STAG2, are absent on AEs. Further supporting of this explanation, *Rec8* and *Rad21L* double-knockout meiocytes completely abolished the AE formation (Llano et al. 2012). Thus, AE formation depends on several kinds of cohesins that function nonredundantly. Conversely, the linear assembly of cohesin on chromatin is not dependent on the AEs since thread-like signals of cohesin are detectable in *Sycp3* KO spermatocytes in which AEs are not formed (Pelttari et al. 2001). Therefore, it seems reasonable to suppose that cohesins create axial cores along chromosomes, onto which AEs are later assembled (Fig. 15.2).

15.4.2 Roles of Cohesins in Meiotic Recombination

Meiotic recombination is initiated by the creation of double-strand breaks (DSBs) mediated by a topoisomerase-like protein, Spo11 (Keeney 2001). The DSBs are repaired and eventually converted into crossover or noncrossover recombination through a series of events including exchange of repair- or recombination-related proteins (Santucci-Darmanin et al. 2000). In *Smc1β*-, *Rad21L*-, *Rec8*-, and *Stag3*-deficient mice, γ-H2AX signals (a marker for double-strand breaks) and RAD51 or RPA signals (markers for recombination intermediates) are detected on the chromosomes in both male and female meiocytes (Bannister et al. 2004; Revenkova et al. 2004; Xu et al. 2005; Herrán et al. 2011; Fukuda et al. 2014). Thus, cohesins are dispensable for creation and initiation of DSB processing. The subsequent processing of DSBs to crossover recombination, however, requires cohesins in male meiosis since MLH1 foci, a marker for crossover recombination, are not detected in the cohesin-deficient spermatocytes. In this regard, female meiosis is far more complex. In oocytes, formation of crossover requires REC8 and STAG3 (Xu et al. 2005; Fukuda et al. 2014) but neither SMC1β nor RAD21L (Revenkova et al. 2004; Herrán et al. 2011). This issue will be discussed in a later Sect. 15.6.

15.4.3 Roles of Cohesins in Homolog Pairing

Synapsis and recombination of homologs are preceded by an elusive chromosomal phenomenon called recognition or pairing of homologs. How do homologs find their partners for the initiation of physical interaction? It has been widely believed that 3′ single-stranded DNA overhangs resulting from the resection of 5′ ends of DSBs are utilized for homolog searching based on the nucleotide sequences (Keeney 2001) (Fig. 15.3). However, recent studies suggest that homolog searching and pairing occurs independently of DSBs (Boateng et al. 2013). Moreover, it is argued that the paring is mostly dependent on RAD21L but only partly on REC8, since initial association of homologs and the subsequent pairing were impaired in *Rad21L/Spo11* double KO but not in *Rec8/Spo11* double KO mice (Ishiguro et al. 2014). This view is supported by a recent super-resolution microscopic study showing that RAD21L and REC8 are present at the connection sites between lateral elements and transverse filaments with RAD21L being interior to REC8 (Rong et al. 2016). Furthermore, ectopically expressed RAD21L promote adjacency of homologs in the somatic cells (Rong et al. 2017). The accumulating data suggest that RAD21L is a special type of cohesin subunit dedicated for interaction between homologs. Interestingly, when either RAD21L or REC8 is depleted, synapsis occurs in an abnormal fashion, i.e., between sisters or between non-homologous chromosomes (Xu et al. 2005; Herrán et al. 2011), implying that several types of cohesin are required for the proper synapsis between homologs after pairing.

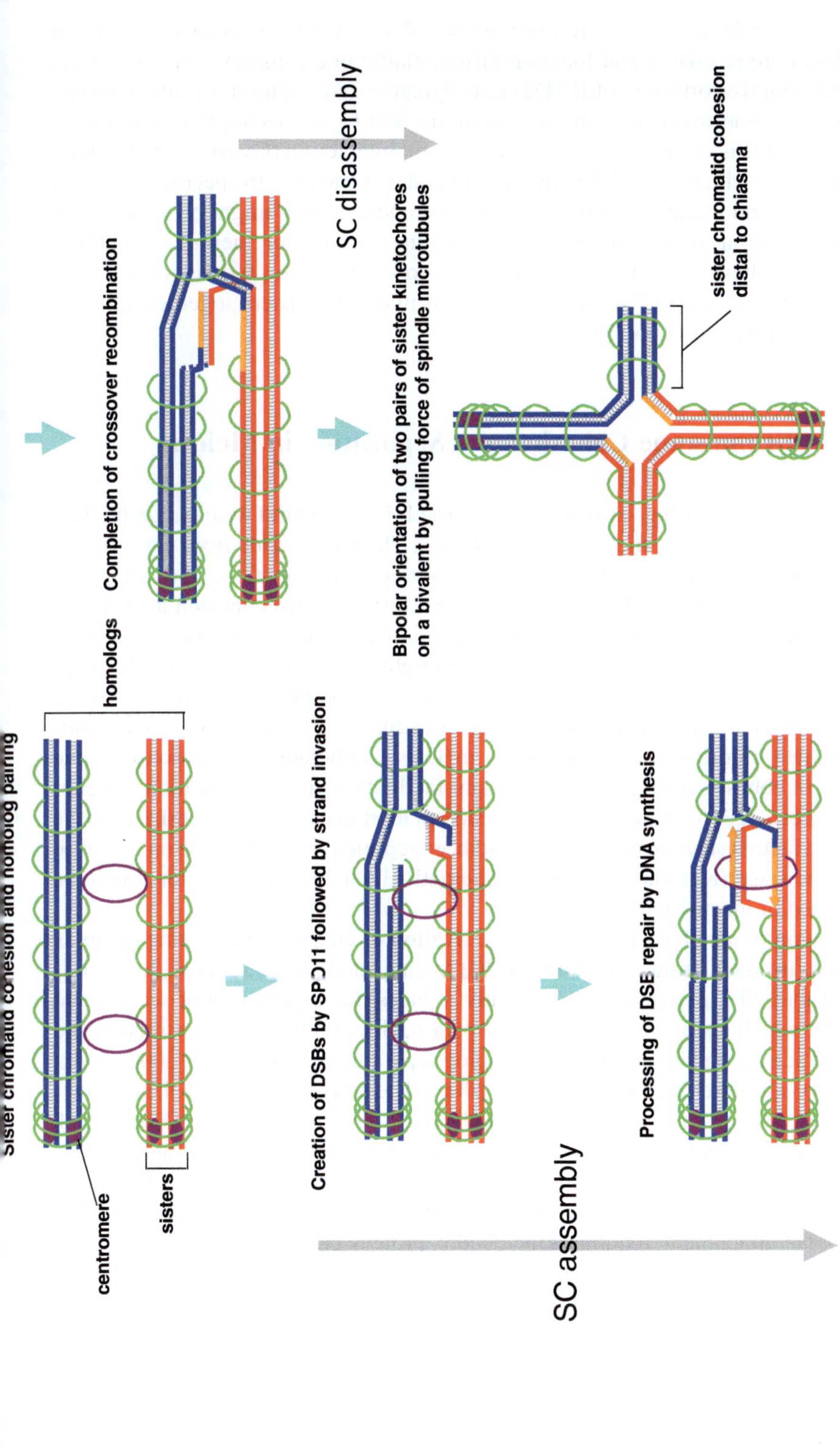

Fig. 15.3 Model of homolog association by two types of meiotic cohesin. In this model, there are two types of functionally different cohesins: one (*green ring*) contributes to sister chromatid cohesion, whereas another (*purple ring*) contributes to homolog association

In *Drosophila*, it has been recently proposed that two types of meiotic cohesin complexes are regulated and function differentially: one complex containing C(2) M, a presumptive ortholog of RAD21L, is dynamic and required for interhomolog interaction while another complex containing SOLO, a presumptive ortholog of REC8, is stable and required for sister chromatid interaction (Gyuricza et al. 2016). Therefore, contribution of different types of cohesin complex to specific process of establishing and maintaining meiotic chromosome cohesion might be a conserved mechanism in a wide variety of species. In mammalian meiosis, RAD21L-containing cohesin may transiently function to establish the interaction between homologs whereas REC8-containing cohesin may hold sisters together until the onset of anaphase II (Figs. 15.1 and 15.3).

15.5 Chromosome Cohesion and Separation in Meiosis

For the faithful homolog separation in meiosis I, it is essential to maintain the link between homologs until the onset of anaphase I. The link is made from the conjunction of sister chromatid cohesion and crossover recombination, both of which require meiotic cohesins (Fig. 15.3). The dissociation of arm cohesion leads to the homolog separation. As mentioned above, in mitotic prophase, most of cohesins are released from chromosome arms during sister chromatid resolution, which facilitates the subsequent sister chromatid separation. In contrast, total loss of cohesin from chromosome arms prior to metaphase I to anaphase I transition is deleterious for accurate chromosome separation in meiosis I although dissociation of some fraction of cohesin during prophase I might facilitate homolog separation (Brieño-Enríquez et al. 2016). Thus, meiotic cohesins, at least in part, must be resistant to the releasing factors such as Wapl, which had been reported to be localized on the SC at pachytene stage (Kuroda et al. 2005). Alternatively, the releasing factors might be suppressed or inactivated during meiosis.

Initiation of anaphase I requires the activation of anaphase-promoting complex (APC) by its associating activator CDC20 (Amanai et al. 2006; Jin et al. 2010). It is thought that APC activation subsequently induces the degradation of securin and cyclin B, which results in activation of separase, since both securin destruction and separase activation are required for homolog separation (Herbert et al. 2003; Terret et al. 2003; Kudo et al. 2006; Lee et al. 2006). The activated separase then cleaves REC8 on chromosome arms, thereby inducing homolog separation (Kudo et al. 2006, 2009) whereas REC8 on centromeres is protected by shugoshin (SGO2) (Lee et al. 2008). For the faithful chromosome separation in meiosis I, sister kinetochore should be oriented toward and connected to the same spindle poles. The so-called monopolar attachment requires kinetochore protein called MEIKIN, which has been recently identified as a functional homolog to budding yeast monopolin and fission yeast Moa1 (Kim et al. 2015). REC8-containing cohesin functions downstream to MEIKIN for mono-orientation of sister kinetochores; its cleavage at centromeres causes defect in mono-orientation (Tachibana-Konwalski et al. 2013).

15.6 Sexual Dimorphism in Cohesin KO Mice

Some of the meiosis-specific cohesin KO mice exhibit sexual dimorphism in the phenotype. Single knockout of either one of *Smc1β*, *Rad21L*, *Rec8*, or *Stag3* arrests the spermatocytes at prophase I due to severe defects in synapsis and/or recombination (DNA repair). In contrast, the phenotypes in oocytes differ among these knockout mice. SMC1β-depleted oocytes are highly error prone but proceed to metaphase II (Revenkova et al. 2004). RAD21L-deficient females are fertile but develop an age-dependent sterility (Herrán et al. 2011). On the other hand, REC8-null neonatal ovaries are devoid of oocytes and ovarian follicles, indicating that REC8-depleted oocytes never proceed beyond prophase I (Bannister et al. 2004; Xu et al. 2005). Likewise, ovarian follicles are not found in *Stag3* KO mice (Fukuda et al. 2014). Sexual dimorphism in meiotic cohesin mutants is partially attributed to the different stringency of control at pachytene checkpoint between spermatocytes and oocytes (Morelli and Cohen 2005; Burgoyne et al. 2009). Currently, however, the reason why variation in sexual dimorphism arises in a cohesin-subunit-dependent manner is unknown. Each cohesin subunit may have different degree of contribution to a particular event in male and female meiosis.

15.7 Age-Related Chromosome Error in Oocytes

It has been widely accepted that the rate of trisomy in human pregnancy increases with maternal age from ~2% among women under 25 years to ~35% among women over 40 years old (Hassold and Hunt 2001). Most of the trisomies originate from chromosome segregation errors during meiosis I in oocytes although there is variation in origin of trisomies depending on the type of chromosomes involved (Hassold and Hunt 2001). Recent studies using live cell imaging have investigated how chromosome segregation errors arise in aged oocytes and proposed two pathways. In one model, weakened centromere cohesion is a leading cause of premature sister chromatid separation in anaphase I in aged oocytes (Chiang et al. 2010). In another model, most of the errors are preceded by bivalent separation into univalent due to the intolerance of weakened arm cohesion against pulling force of spindle microtubules during metaphase I, which subsequently leads to premature separation of sisters in anaphase I (Sakakibara et al. 2015). In either case, aged deterioration of sister chromatid cohesion causes the premature separation of sisters in anaphase I. Thus, cohesin is one of the most probable molecules whose decay or destruction causes the age-related chromosome errors.

Several studies have approached the question whether cohesin is involved in the age-related meiotic chromosome errors. The first implication was provided by a study using conventional KO mice for *Smc1β*. The *Smc1β*-deficient female mice are sterile: the oocytes proceed to metaphase II but become aneuploidy due to premature loss of chromosome cohesion. In the *Smc1β*-deficient oocytes, the incidence of premature separation of both homologs and sisters is markedly increased as the mice ovulating grow up from 1 month old to 6 months old (Hodges et al. 2005). This suggests that depletion of SMC1β might accelerate or enhance the normal aging process. Thereafter, more direct evidence was provided by several studies using conditional KO strategies, which make meiotic cohesin genes being activated or inactivated at some critical stage(s) of oogenesis. Firstly, inactivation of *Smc1β* gene shortly after birth at dictyate arrest in oocytes didn't affect chiasma positions and sister chromatid cohesion (Revenkova et al. 2010). Secondly, activation of an ectopic *Rec8* transgene during the growing phase of the oocytes expressing artificially cleavable REC8 by TEV protease doesn't prevent TEV-mediated bivalent separation (Tachibana-Konwalski et al. 2010). Thirdly, a recent study using drug-inducible Cre system also shows that REC8 can establish cohesion when it is expressed only during meiotic S phase but not when expressed in later meiotic stages (Burkhardt et al. 2016). These studies suggest that only the meiotic cohesin expressed prior to meiotic S phase can establish and maintain sister chromatid cohesion and also imply that there is little or no cohesin turnover during the meiotic arrest at prophase I and thereafter. Indeed, immunofluorescence analyses show that chromosome-associated REC8 present on both chromosome arms and centromeres is reduced in aged oocytes compared with young oocytes, although total REC8 protein levels are similar in both types of oocytes (Chiang et al. 2010: Lister et al. 2010). Furthermore, chromosome-associated Sgo2, the protector of centromeric cohesin, is also reduced in the aged oocytes (Lister et al. 2010).

It seems indubitable that reduction of cohesin during prophase I arrest contributes to chromosome segregation errors in aged oocytes. However, factors other than cohesin degradation are also needed to explain the increase in maternal age-related aneuploidy. As has been pointed out by Hunt and Hassold (2010), the age-dependent increase in aneuploidy and decrease in cohesin are not completely synchronized: the increase in aneuploidy is only evident in reproductively senescent females although the studies using naturally aged mice report a linear age-related decline in chromosome-associated cohesin. Furthermore, provided that only the cohesin expressed during or prior to meiotic S phase plays a role, how can meiotic cohesins in human oocytes keep chromosome cohesion tens of times longer than those in mouse oocytes? We haven't got any clue to answer this question. In human oocytes, reduction of cohesin during prophase I arrest is still disputable (Garcia-Cruz et al. 2010; Tsutsumi et al. 2014).

15.8 Condensins in Mitosis

Condensin is a multi-subunit protein complex that is essential for chromosome condensation and segregation (Hirano 2016). Most of eukaryotes have two types of condensins, condensin I and condensin II. Both condensin complexes are composed of five subunits, that is, SMC2 and SMC4 common to condensin I and condensin II; CAP-D2, CAP-G, and CAP-H unique to condensin I; and CAP-D3,

CAP-G2, and CAP-H2 unique to condensin II (Table 15.1). Both condensin complexes are essential for construction of mitotic chromosomes and thus development of embryos in many species. In contrast to cohesin that connects the DNA segment from different chromatids, condensin is thought to connect the DNA segment from the same chromatid, thereby contributing to chromosome compaction (Fig. 15.4).

15.9 The Expression and Localization of Condensins During Mammalian Gametogenesis

Only several studies have been conducted to elucidate the expression and localization of condensins during mammalian gametogenesis. In the first study, it has been shown that condensin I is localized mainly around telomeres and to a lesser extent along chromosome axes (Viera et al. 2007). Thereafter, the expression of both subunits of condensin I and condensin II has been reported in pig oocytes, but their localization on chromosomes has not been examined (Lisková et al. 2010). The first comprehensive study on both condensin I and II in mammalian meiosis has been conducted in mouse oocytes (Lee et al. 2011). Although both condensins are localized on mitotic and meiotic chromosomes, their dynamics are slightly modified in the oocytes. Prior to germinal vesicle breakdown (GVBD) (nuclear envelop disassembly), condensin I is present in the cytoplasm whereas condensin II is localized in the germinal vesicle (GV) (oocyte nucleus). It has been reported in somatic HeLa cells that condensin I and condensin II are present in the cytoplasm and in the nucleus at interphase, respectively (Ono et al. 2003). Thus, the localizations of condensins prior to nuclear envelop disassembly are similar in somatic cells and oocytes. After GVBD, condensin I localizes mainly around centromere regions of bivalent chromosomes while condensin II localizes along chromatid axes. After anaphase I, both condensin I and II are localized on chromosome arms. Why is only condensin II, but not condensin I, stably localized on arm region of bivalent chromosomes? The meiotic cohesin along chromosome arms might affect the loading of condensin I. It has been also suggested that meiotic cohesin along arms might contribute to the solidity of bivalent chromosome in place of condensin I (Lee 2013). Indeed, it has been proposed that that meiotic cohesin contributes not only to the cohesion but also to the organization of chromosome axes and loop (Novak et al. 2008; Haering and Jessberger 2012).

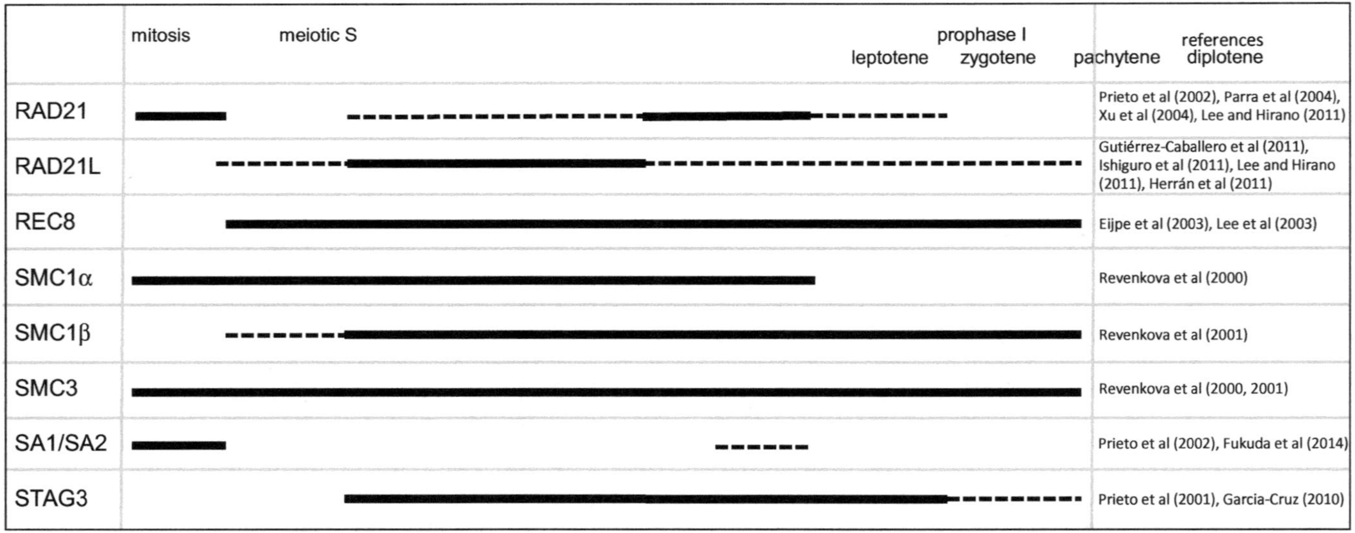

Fig. 15.4 Expression stages of cohesin subunits during mammalian meiosis. The bold lines indicate the expression stages of each cohesin subunit, whereas the dashed lines represent the expression stage for which conflicting results are reported

15.10 Role of Condensins in Construction of Bivalent Chromosomes in Oocytes

The role of condensins in mammalian oocytes has been investigated using two strategies. The first study using antibody injection suggests that both condensins I and II are essential for proper construction of bivalent chromosomes because injection of antibodies specific to condensin I or condensin II subunit affected the shape of chromosomes (Lee and Hirano 2011). However, in this strategy, the injected antibody might affect the other proteins localizing proximal to condensin I, thereby causing the defects in chromosome morphology indirectly. The second study using conditional knockout strategy has shown that knockout of condensin II subunit causes defects in chromosome compaction, chromatid disentanglement, and chromosome segregation during meiosis I while knockout of condensin I subunit hardly affects meiotic progression to metaphase II (Houlard et al. 2015). Therefore, it has been proposed that condensin I may function redundantly with condensin II and may be dispensable for meiotic maturation of oocytes. However, the conditional knockout strategy cannot exclude possibility that residual amount of condensin I due to insufficient depletion in the oocytes might influence the results. In spite of the discrepancy in the meiotic role of condensin I, both strategies argue that condensin II is essential for proper construction of bivalent chromosomes. The artificial inactivation of a condensin II subunit by TEV protease in oocytes further supports this view and suggests that condensin II is essential not only for formation but also for maintenance of bivalent chromosomes (Houlard et al. 2015).

15.11 Possible Roles of Condensins at Meiotic Prophase I

Although no studies have ever addressed the function of condensins at prophase I in mammals, the involvement of condensins in chromosome dynamics at prophase I has been reported in various species. In budding yeast, condensin I regulates various chromosome events including chromosome compaction, SC assembly, formation and processing of DSBs, repair at rDNA gene clusters, and resolution of recombination-dependent chromosome link (Yu and Koshland 2003; Li et al. 2014). In *C. elegans*, condensin I regulates the number and distribution of crossovers through construction of a higher-order chromosome structure (Mets and Meyer 2009). In a plant (*A. thaliana*), it has been suggested that condensin is involved in the regulation of crossover frequency (Smith et al. 2014). Thus, future studies are necessary to investigate possible roles of condensins at prophase I during mammalian gametogenesis.

15.12 Perspectives

In addition to the abovementioned roles, cohesins, condensins, and the related complexes are involved in multiple functions of chromosomes, such as genomic imprinting (Wendt et al. 2008), dosage compensation (Wood et al. 2010), and transvection (Hartl et al. 2008). Furthermore, the mutation of cohesin or condensin is one of the major causes for various congenital disorders in human (Peters et al. 2008; Martin et al. 2016). Notably, it has been reported that condensin I is recruited to chromosomes in a different manner in male and female pronuclei of mouse zygotes (Bomar et al. 2002). Thus, it is intriguing to search for unproved functions of cohesins and condensins in gametogenesis and early development of mammals.

References

Amanai M, Shoji S, Yoshida N, Brahmajosyula M, Perry AC (2006) Injection of mammalian metaphase II oocytes with short interfering RNAs to dissect meiotic and early mitotic events. Biol Reprod 75:891–898

Bannister LA, Reinholdt LG, Munroe RJ, Schimenti JC (2004) Positional cloning and characterization of mouse mei8, a disrupted allelle of the meiotic cohesin Rec8. Genesis 40:184–194

Biswas U, Hempel K, Llano E, Pendas A, Jessberger R (2016) Distinct roles of meiosis-specific cohesin complexes in mammalian spermatogenesis. PLoS Genet 12:e1006389

Boateng KA, Bellani MA, Gregoretti IV, Pratto F, Camerini-Otero RD (2013) Homologous pairing preceding SPO11-mediated double-strand breaks in mice. Dev Cell 24:196–205

Bomar J, Moreira P, Balise JJ, Collas P (2002) Differential regulation of maternal and paternal chromosome condensation in mitotic zygotes. J Cell Sci 115:2931–2940

Brieño-Enríquez MA, Moak SL, Toledo M, Filter JJ, Gray S, Barbero JL, Cohen PE, Holloway JK (2016) Cohesin removal along the chromosome arms during the first meiotic division depends on a NEK1-PP1g-WAPL axis in the mouse. Cell Rep 17:977–986

Burgoyne PS, Mahadevaiah SK, Turner JM (2009) The consequence of asynapsis form mammalian meiosis. Nat Rev Genet 10:207–216

Burkhardt S, Borsos M, Szydlowska A, Godwin J, Williams SA, Cohen PE, Hirota T, Saitoh M, Tachibana-Kowalski K (2016) Chromosome cohesion established by Rec8-cohesin in fetal oocytes is maintained without detectable turnover in oocytes arrested for months in mice. Curr Biol 26:678–685

Chiang T, Duncan FE, Schindler K, Schultz RM, Lampson MA (2010) Evidence that weakened centromere cohesion is a leading cause of age-related aneuploidy in oocytes. Curr Biol 20:1522–1528

de Vries FA, de Boer E, van den Bosch M, Baarends WM, Ooms M, Yuan L, Liu JG, van Zeeland AA, Heyting C, Pastin A (2005) Mouse Sycp1 functions in synaptonemal complex assembly, meiotic recombination, and XY body formation. Genes Dev 19:1376–1389

Eijpe M, Heyting C, Gross B, Jessberger R (2000) Association of mammalian SMC1 and SMC3 proteins with meiotic chromosomes and synaptonemal complexes. J Cell Sci 113:673–682

Eijpe M, Offenberg H, Jessberger R, Revenkova E, Heyting C (2003) Meiotic cohesin REC8 marks the axial elements of rat synaptonemal complexes before cohesins SMC1beta and SMC3. J Cell Biol 160:657–670

Fukuda T, Fukuda N, Agostinho A, Hernández-Hernández A, Kouznetsova A, Höög C (2014) STAG3-mediated stabilization of REC8 cohesin complexes promotes chromosome synapsis during meiosis. EMBO J 33:1243–1255

Garcia-Cruz R, Brieño MA, Roig I, Grossmann M, Velilla E, Pujol A, Cabero L, Pessarrodona A, Barbero JL, Garcia Caldés M (2010) Dynamics of cohesin proteins REC8, STAG3, SMC1β and SMC3 are consistent with a role in sister chromatid cohesion during meiosis in human oocytes. Hum Reprod 25:2316–2327
Gruber S, Haering CH, Nasmyth K (2003) Chromosomal cohesin forms a ring. Cell 112:765–777
Gutiérrez-Caballero C, Herrán Y, Sánchez-Martín MS, Suja JA, Barbero JL, Llano E, Pendáz AM (2011) Identification and molecular characterization of the mammalian kleisin RAD21L. Cell Cycle 10:1477–1487
Gyuricza MR, Manheimer KB, Apte V, Krishnan B, Joyce EF, Mckee BD, McKim KS (2016) Dynamic and stable cohesins regulate synaptonemal complex assembly and chromosome segregation. Curr Biol 26:1688–1698
Haering CH, Jessberger R (2012) Cohesin in determining chromosome architecture. Exp Cell Res 318:1386–1393
Hartl TA, Smith HF, Bosco G (2008) Chromosome alignment and transvection are antagonized by condensin II. Science 322:1384–1387
Hassold T, Hunt P (2001) To err (meiotically) is human: the genesis of human aneuploidy. Nat Rev Genet 2:280–291
Herbert M, Levasseur M, Homer H, Yallop K, Murdoch A, McDougall A (2003) Homologue disjunction in mouse oocytes requires proteolysis of securin and cyclin B1. Nat Cell Biol 5:1023–1025
Herrán Y, Gutiérrez-Caballero C, Sánchez-Martín M, Hernández T, Viera A, Barbero JL, de Álava E, de Rooij DG, Suja JA, Llano E, Pendás AM (2011) The cohesin subunit RAD21L functions in meiotic synapsis and exhibits sexual dimorphism in fertility. EMBO J 30:3091–3105
Hirano T (2016) Condensin-based chromosome organization from bacteria to vertebrates. Cell 164:847–857
Hodges CA, Revenkova E, Jessberger R, Hassold TJ, Hunt PA (2005) SMC1beta-deficient female mice provide evidence that cohesins are a missing link in age-related nondisjunction. Nat Genet 37:1351–1355
Houlard M, Godwin J, Metson J, Lee J, Hirano T, Nasmyth K (2015) Condensin confers the longitudinal rigidity of chromosomes. Nat Cell Biol 17:771–781
Hunt P, Hassold T (2010) Female meiosis: coming unglued with age. Curr Biol 20:R699–R702
Ishiguro K, Kim J, Fujiyama-Nakamura S, Kato S, Watanabe Y (2011) A new meiosis-specific cohesin complex implicated in the cohesin code for homologous pairing. EMBO Rep 12:267–275
Ishiguro K, Kim J, Shibuya H, Hernández-Hernández A, Suzuki A, Fukagawa T, Shioi G, Kiyonari H, Li XC, Schimenti J, Höög C, Watanabe Y (2014) Meiosis-specific cohesin mediates homolog recognition in mouse spermatocytes. Genes Dev 28:594–607
Jin F, Hamada M, Malureanu L, Jeganathan KB, Zhou W, Morbeck DE, van Deursen JM (2010) Cdc20 is critical for meiosis I and fertility of female mice. PLoS Genet 6:e1001147
Keeney S (2001) Mechanism and control of meiotic recombination initiation. Curr Top Dev Biol 52:1–53
Kim J, Ishiguro K-I, Nambu A, Akiyoshi B, Yokobayashi S, Kagami A, Ishiguro T, Pendas AM, Takeda N, Sakakibara Y, Kitajima T, Tanno Y, Sakuno T, Watanabe Y (2015) Meikin is a conserved regulator of meiosis-I-specific kinetochore function. Nature 512:466–471
Klein F, Mahr P, Galova M, Buonomo SB, Michelis C, Nairz K, Nasmyth K (1999) A central role for cohesins in sister chromatid cohesion, formation of axial elements, and recombination during yeast meiosis. Cell 98:91–103
Kudo NR, Wassmann K, Anger M, Schuh M, Wirth KG, Xu H, Helmhart W, Kudo H, McKay M, Maro B, Ellenberg J, de Boer P et al (2006) Resolution of chiasmata in oocytes requires separase-mediated proteolysis. Cell 126:135–146
Kudo NR, Anger M, Peters AHFM, Stemmann O, Theussi H-C, Helmhart W, Kudo H, Heyting C, Nasmyth K (2009) Role of cleavage by separase of the Rec8 kleisin subunit of cohesin during mammalian meiosis I. J Cell Sci 122:2686–2698

Kuroda M, Oikawa K, Ohbayashi T, Yoshida K, Yamada K, Mimura J, Matsuda Y, Fujii-Kuriyama Y, Mukai K (2005) A dioxin sensitive gene, mammalian WAPL, is implicated in spermatogenesis. FEBS Lett 579:167–172

Lee J (2013) Roles of cohesin and condensin in chromosome dynamics during mammalian meiosis. J Reprod Dev 59:431–436

Lee J, Hirano T (2011) RAD21L, a novel cohesin subunit implicated in linking homologous chromosomes in mammalian meiosis. J Cell Biol 192:263–276

Lee J, Iwai T, Yokota T, Yamashita M (2003) Temporally and spatially selective loss of Rec8 protein from meiotic chromosomes during mammalian meiosis. J Cell Sci 116:2781–2790

Lee J, Okada K, Ogushi S, Miyano T, Miyake M, Yamashita M (2006) Loss of Rec8 from chromosome arm and centromere region is required for homologous chromosome separation and sister chromatid separation, respectively, in mammalian meiosis. Cell Cycle 5:1448–1455

Lee J, Kitajima TS, Tanno Y, Yoshida K, Morita T, Miyano T, Miyake M, Watanabe Y (2008) Unified mode of centromeric protection by shugoshin in mammalian oocytes and somatic cells. Nat Cell Biol 10:42–52

Lee J, Ogushi S, Saitou M, Hirano T (2011) Condensins I and II are essential for construction of bivalent chromosomes in mouse oocytes. Mol Biol Cell 22:3465–3477

Li P, Jin H, H-G Y (2014) Condensin suppresses recombination and regulates double-strand break processing at the repetitive ribosomal DNA array to ensure proper chromosome segregation during meiosis in budding yeast. Mol Biol Cell 25:2934–2947

Lisková L, Susor A, Pivonkova K, Saskova A, Karabinova P, Kubelka M (2010) Detection of condensin I and II in maturing pig oocytes. Reprod Fertil Dev 22:644–652

Lister LM, Kouznetsova A, Hyslop LA, Kalleas D, Pace SL, Barel JC, Nathan A, Floros V, Adelfalk C, Watanabe Y, Jessberger R, Kirkwood TB, Höög C, Herbert M (2010) Age-related meiotic segregation errors in mammalian oocytes are preceded by depletion of cohesin and Sgo2. Curr Biol 20:1511–1521

Llano E, Herran Y, Garcia-Tunon I, Gutierrez-Caballero C, de Alava E, Barbero JL, Schimenti J, de Rooij DG, Sanchez-Martin M, Pendas AM (2012) Meiotic cohesin complexes are essential for the formation of the axial element in mice. J Cell Biol 197:877–885

Martin C-A, Murray JE, Carroll P, Leitch A, Mackenzie KJ, Halachev M, Fetit AE, Keith C, Bicknell LS, Fluteau A et al (2016) Mutations in genes encoding condensin complex proteins cause microcephaly through decatenation failure at mitosis. Genes Dev 30:2158–2172

Mets DG, Meyer BJ (2009) Condensins regulate meiotic DNA break distribution, thus crossover frequency, by controlling chromosome structure. Cell 139:73–86

Miyazaki WY, Orr-Weaver TL (1994) Sister-chromatid cohesion in mitosis and meiosis. Annu Rev Genet 28:167–187

Morelli MA, Cohen PE (2005) Not all germ cells are created equal: aspects of sexual dimorphism in mammalian meiosis. Reproduction 130:761–781

Nasmyth K (2011) Cohesin: a catenase with separate entry and exit gates? Nat Cell Biol 13:1170–1177

Nasmyth K, Haering CH (2009) Cohesin: its roles and mechanisms. Annu Rev Genet 43:525–558

Nishiyama T, Ladurner R, Schmitz J, Kreidl E, Schleiffer A, Bhaskara V, Bando M, Shirahige K, Hyman AA, Mechtler K, Peters JM (2010) Sororin mediates sister chromatid cohesion by antagonizing Wapl. Cell 143:737–749

Novak I, Wang H, Revenkova E, Jessberger R, Scherthan H, Hoog C (2008) Cohesin Smc1beta determines meiotic chromatin axis loop organization. J Cell Biol 180:83–90

Ono T, Losada A, Hirano M, Myers MP, Neuwald AF, Hirano T (2003) Differential contributions of condensin I and condensin II to mitotic chromosome architecture in vertebrate cells. Cell 115:109–121

Page SL, Hawley RS (2004) The genetics and molecular biology of the synaptonemal complex. Annu Rev Cell Dev Biol 20:525–558

Parra MT, Viera A, Gómez R, Page J, Benavente R, Santos JL, Rufas JS, Suja JA (2004) Involvement of the cohesin Rad21 and SCP3 in monopolar attachment of sister kinetochores during mouse meiosis I. J Cell Sci 117:1221–1234

Pelttari J, Hoja M-R, Yuan L, Liu J-G, Brundell E, Moens P, Santucci-Darmnin S, Jessberger R, Barbero JL, Heyting C, Höög C (2001) A meiotic chromosomal core consisting of cohesin complex proteins recruits DNA recombination proteins and promotes synapsis in the absence of an axial element in mammalian meiotic cells. Mol Cell Biol 21:5667–5677

Peters JM, Tedeschi A, Schmitz J (2008) The cohesin complex and its roles in chromosome biology. Genes Dev 22:3089–3114

Prieto I, Suja JA, Pezzi N, Kremer L, Martinez AC, Rufas JS, Barbero JL (2001) Mammalian STAG3 is a cohesin specific to sister chromatid arms in meiosis I. Nat Cell Biol 3:761–766

Prieto I, Pezzi N, Buesa J, Kremer L, Barthelemy I, Carreiro C, Roncal F, Martínez A, Gómez L, Fernández R, Martínez A, Barbero J (2002) STAG2 and Rad21 mammalian mitotic cohesins are implicated in meiosis. EMBO Rep 3:543–550

Revenkova E, Eijpe M, Heyting C, Gross B, Jessberger R (2001) Novel meiosis-specific isoform of mammalian SMC1. Mol Cell Biol 21:6984–6998

Revenkova E, Eijpe M, Heyting C, Hodges CA, Hunt PA, Liebe B, Scherthan H, Jessberger R (2004) Cohesin SMC1 beta is required for meiotic chromosome dynamics, sister chromatid cohesion and DNA recombination. Nat Cell Biol 6:555–562

Revenkova E, Herrmann K, Adelfalk C, Jessberger R (2010) Oocyte cohesin expression restricted to predictyate stages provides full fertility and prevents aneuploidy. Curr Biol 20:1529–1533

Rong M, Matsuda A, Hiraoka Y, Lee J (2016) Meiotic cohesin subunits RAD21L and REC8 are positioned at distinct regions between lateral elements and transverse filaments in the synaptonemal complex of mouse spermatocytes. J Reprod Dev 62:623–630

Rong M, Miyauchi S, Lee J (2017) Ectopic expression of meiotic cohesin RAD21L promotes adjacency of homologous chromosomes in somatic cells. J Reprod Dev. doi:10.1262/jrd.2016-171

Sakakibara Y, Hashimoto S, Nakaoka Y, Kouznetsova A, Höög C, Kitajima TS (2015) Bivalent separation into univalents precedes age-related meiosis I errors in oocytes. Nat Commun 6:7550

Santucci-Darmanin S, Walpita D, Lespinasse F, Desnuelle C, Ashley T, Paquis-Flucklinger V (2000) MSH4 acts in conjunction with MLH1 during mammalian meiosis. FASEB J 14:1539–1547

Shintomi K, Hirano T (2009) Releasing cohein from chromosome arms in early mitosis: opposing actions of Wapl-Pds5 and Sgo1. Genes Dev 23:2224–2236

Smith SJ, Osman K, Franklin CH (2014) The condensin complexes play distinct roles to ensure normal chromosome morphogenesis during meiotic division in Arabidopsis. Plant J 80:255–268

Suja JA, Barbero JL (2009) Cohesin complexes and sister chromatid cohesion in mammalian meiosis. Genome Dyn 5:94–116

Tachibana-Konwalski K, Godwin J, van der Weyden L, Champion L, Kudo NR, Adams DJ, Nasmyth K (2010) Rec8-containing cohesin maintains bivalents without turnover during the growing phase of mouse oocytes. Genes Dev 24:2505–2516

Tachibana-Konwalski K, Godwin J, Borsos M, Rattani A, Adams DJ, Nasmyth K (2013) Spindle assembly checkpoint of oocytes depends on a kinetochore structure determined by cohesin in meiosis I. Curr Biol 23:2534–2539

Terret ME, Wassmann K, Waizenegger I, Maro B, Peters JM, Verlhac MH (2003) The meiosis I-to-meiosis II transition in mouse oocytes requires separase activity. Curr Biol 13:1797–1802

Tsutsumi M, Fujiwara R, Nishizawa H, Ito M, Kogo H, Inagaki H, Ohye T, Kato T, Fujii T, Kurahashi H (2014) Age-related decrease of meiotic cohesins in human oocytes. PLoS One 9: e96710

Uhlmann F, Nasmyth K (1998) Cohesion between sister chromatids must be established during DNA replication. Curr Biol 8:1095–1101

Uhlmann F, Wernic D, Poupart MA, Koonin EV, Nasmyth K (2000) Cleavage of cohesin by the CD clan protease separin triggers anaphase in yeast. Cell 103:375–386

Viera A, Gomez R, Parra MT, Schmiesing JA, Yokomori K, Rufas JS, Suja JA (2007) Condensin I reveals new insights on mouse meiotic chromosome structure and dynamics. PLoS One 2:e783
Waizenegger IC, Hauf S, Meinke A, Peters JM (2000) Two distinct pathways remove mammalian cohesin from chromosome arms in prophase and from centromeres in anaphase. Cell 103:399–410
Watanabe Y, Nurse P (1999) Cohesin Rec8 is required for reductional chromosome segregation at meiosis. Nature 29:461–464
Wendt KS, Yoshida K, Itoh T, Bando M, Koch B, Schirghuber E, Tsutsumi S, Nagae G, Ishihara K, Mishiro T et al (2008) Cohesin mediates transcriptional insulation by CCCTC-binding factor. Nature 451:796–801
Winters T, McNicoll F, Jessberger R (2014) Meiotic cohesin STAG3 is required for chromosome axis formation and sister chromatid cohesion. EMBO J 33:1256–1270
Wood AJ, Severson AF, Meyer BJ (2010) Condensin and cohesin complexity: the expanding repertoire of functions. Nat Rev Genet 11:391–404
Xu H, Beasley M, Verschoor S, Inselman A, Handel MA, Mckay MJ (2004) A new role for the mitotic RAD21/SCC1 cohesin in meiotic chromosome cohesion and segregation in the mouse. EMBO Rep 5:378–384
Xu H, Beasley MD, Warren WD, van der Horst GT, McKay MJ (2005) Absence of mouse REC8 cohesin promotes synapsis of sister chromatids in meiosis. Dev Cell 8:949–961
Yu H (2007) Cdc20: a WD40 activator for a cell cycle degradation machine. Mol Cell 27:3–16
Yu HG, Koshland DE (2003) Meiotic condensin is required for proper chromosome compaction, SC assembly, and resolution of recombination-dependent chromosome linkages. J Cell Biol 163:937–947

Chapter 16
Supply and Demand of Energy in the Oocyte and the Role of Mitochondria

Wilding Martin

Abstract The sole purpose of any mammalian oocyte is to combine with a spermatozoon and form a viable embryo that implants into the uterus and forms a viable foetus. Most of the structures and mechanisms for this reside within the oocyte itself. The sperm limits itself to fertilisation of the oocyte; apart from this, its only contribution is the male genome and the centrosome, required for cell division. Both intrinsic and extrinsic factors determine the formation of a viable embryo. However, the fundamental necessity for successful reproduction resides within the capacity for the developing embryo to generate sufficient levels of energy for optimal development to occur. Energy is generated principally within mitochondria. In this chapter, we discuss some of the fundamental processes of preimplantation embryo development and the role of mitochondria in providing sufficient energy for the successful completion of these processes. We discuss mitochondrial genetics, replication and energy production. Ageing appears to affect the capacity of the mitochondrion to produce sufficient energy to balance the requirements of the embryo. We discuss some of the theories of the effect of maternal age on mitochondrial physiology and the role this plays in reproduction. We propose that maternal age has longer-term effects on individuals than simply on the efficiency of reproduction. We also discuss some of the procedures assisted reproduction has proposed to alleviate the effect of maternal age on reproduction.

16.1 Introduction

In order for mammalian reproduction to occur successfully, an egg must successfully combine with a spermatozoon and form an embryo, which implants into the uterus of a female recipient. Evidence suggests that it is principally the egg that determines the reproductive potential of the developing embryo; the sperm simply contributes an egg activation factor, the centrosome and the paternal genome (Sutovsky and Schatten 2000; Dale et al. 1985). For latest information on other

W. Martin (✉)
Centre for Reproduction and Advanced Technology (CREATE), 150 Cheapside, London EC2V 6ET, UK
e-mail: martinwilding@hotmail.com

M. Kloc (ed.), *Oocytes*, Results and Problems in Cell Differentiation 63,
DOI 10.1007/978-3-319-60855-6_16

molecules/organelles contributed by sperm, see Kloc and Kubiak (2017). So what is it in the egg that is so important for embryo development, when and where are these factors most critical and what can be done to assist either in increasing oocyte quality, or in selection of the highest potential embryos to implant?

16.2 How Does the Oocyte Become a Foetus?

Oocytes are the largest cells in the human body. These cells have a unique destiny and role: to be fertilised and start the developmental pathway to becoming a foetus. Developing mammalian embryos have a unique programme of development (Fig. 16.1). In the first 3 days after ovulation, a fertilised egg simply divides, forming a loosely attached ball of between 5 and 12 cells termed 'blastomeres'. The programme of development changes between days 3 and 6. Cell division continues, but the individual blastomeres start to attach together and communicate through gap junctions (Dale et al. 1991). Differentiation occurs, and the cell forms an internal vacuole, forming a structure consisting of an external layer of cells forming a balloon wall (the trophectoderm) and a small inner mass (termed the inner cell mass). The trophectoderm layer will form the placenta after implantation, whereas the inner cell mass forms the foetus. The new structure is termed the

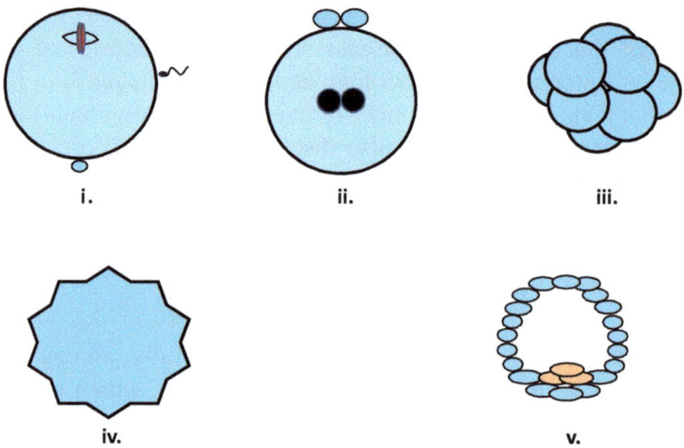

Fig. 16.1 Development of the human embryo. The figure illustrates the phases of human embryo development. (i). The mature egg is fertilised by the spermatozoon and development begins. (ii). On day 1, the egg has completed meiosis and two pronuclei are present, one derived from the maternal spindle and the other from the spermatozoon. (iii). Between days 1 and 3, the embryo simply divides by mitosis to form a ball of loosely attached blastomeres with an equivalent developmental fate. (iv). Between days 3 and 4, the developmental programme of the embryo changes. Cells compact together to form an amorphous mass. At this stage, the embryo starts to assimilate external nutrients. (v). By day 5, the embryo has differentiated into two tissues, the trophectoderm (*blue colour*) and inner cell mass (*peach colour*). The embryo has also formed an inner vacuole which enlarges, assisting the embryo in hatching and implantation

blastocyst. Without attachment to an external substrate, these embryos usually die after a maximum of 6 days, although it has been possible to maintain the embryos in culture up to 13 days post ovulation (Shahbazi et al. 2016). The usual substrate for embryo attachment and continued development is the uterine lining.

It is the period between the day 3 stage and the formation of the blastocyst that determines whether an oocyte will form a viable embryo. In order to develop beyond this stage, the embryo must achieve several goals. Firstly, the embryo must assimilate and use external nutrients to replace the depleted internal resources (Braude et al. 1988). The embryo must also activate genes, both to achieve assimilation of nutrients and to complete stages post cleavage (Braude et al. 1988). These processes are highly delicate, and many embryos stop development before successfully completing them. One of the most fundamental sources of embryo failure however may be the inability to produce sufficient energy for the completion of these steps, leading simply to exhaustion and arrest (Van Blerkom et al. 1995).

16.3 What Are Mitochondria?

Mitochondria are specialised organelles within the cytoplasm of all eukaryotes. The principle role of the mitochondrion in the cell is in the production of molecular energy in the form of adenosine trisphosphate (ATP) through oxidative phosphorylation (aerobic respiration). Oxidative phosphorylation is a highly efficient mechanism for energy production; $\Delta G = -686$ kcal are produced per mole of glucose (Krebs 1965). Cells also contain a second mechanism for energy production termed 'anaerobic respiration'. This mechanism is located within the cell cytoplasm and is far less efficient ($\Delta G = -47$ kcal/mole glucose). The two mechanisms are thought to be entirely responsible for the production of ATP within the cell, anaerobic respiration supplementing the more efficient aerobic mechanism. Additional roles for mitochondria such as steroid biosynthesis and calcium regulation have more recently been attributed to this organelle (Guantes et al. 2015; Muir et al. 2016). However, ATP production remains the principal role.

Mitochondria are distinguished by the presence of two sets of membranes. Particularly notable is the inner membrane which is highly folded with respect to the organelle itself. Experiments have shown that the outer membrane is highly permeable, whereas the inner membrane is impermeable. The permeability of these membranes is designed to enable efficient oxidative phosphorylation (Rich 2003). Within the core of the mitochondrion (termed the matrix) is a small circular molecule of DNA termed the mitochondrial DNA (mtDNA). This DNA molecule is often used to measure the copy number of mitochondria in cells, although it appears that the rule of one molecule per one mitochondrion may not be true (Cummins 2002).

Mitochondria follow cellular signals to maintain the correct copy number in the cell cytoplasm (Valero 2014). The copy number appears to be correlated with the energy requirement of the cell (Valero 2014). Therefore, each cell retains the optimum aerobic respiratory potential for its requirements. Mitochondria also follow their own cycle of replication and senescence of older organelles through mitophagy (Diot et al. 2016). This is necessary because the aerobic generation of ATP, although highly efficient, is also highly damaging to the organelles themselves, and these must be continually replaced in order to retain efficiency.

16.3.1 Mitochondria and the Oocyte

A typical human oocyte contains between 20,000 and 800,000 copies of mtDNA (Reynier et al. 2001; Barritt et al. 2002; Almeida-Santos et al. 2006). Although these figures neither indicate mitochondrial mass nor the efficiency of mitochondrial respiration, they could be related to the above, as mtDNA is required for mitochondrial function. The optimal copy number is still unknown. MtDNA copy number appears to be significantly lower in oocytes that remain unfertilised after assisted reproduction (Reynier et al. 2001). MtDNA and presumably mitochondrial numbers remain constant until the blastocyst stage in human embryos. During blastocyst formation, mtDNA initiates replication, and its levels within the embryo start to increase with the increase in embryo mass (St John et al. 2010; Eichenlaub-Ritter et al. 2011). Interestingly, the relative increase in mtDNA at the blastocyst stage may be an indicator of the health of the mtDNA in the whole embryo. It appears that there is inverse relationship between increase in mtDNA and embryo quality, because blastocyst mtDNA copy numbers were found to be significantly greater in embryos, which failed implantation (Fragouli et al. 2015).

Considering these contrasting data, a working hypothesis is that eggs with too few mitochondria are incompetent and will not fertilise due to the lack of ATP. The reasons for this could be multiple such as poor replication potential of mtDNA, poor-quality mitochondria, etc. Embryos with sufficient numbers of mitochondria fertilise and initiate development; however, those with low-efficiency mitochondria try to balance this after day 3 (when more energy is required) with an explosion of mtDNA replication and an increase in mitochondrial mass, leading to large mitochondria numbers in the blastocyst. Too many mtDNA in the cell cytoplasm at the blastocyst stage may therefore indicate the loss of control of mtDNA replication or again the presence of poor-quality mitochondria and the resulting need to excessively increase the mitochondria number in the cell cytoplasm.

Because the oocyte is formed within the follicular environment and oocyte helper cells such as the cumulus are vital to the development of oocyte competence, it is conceivable that mtDNA within the cumulus may also play a major role in oocyte competence. In fact, studies examining mtDNA within the cumulus have also shown a correlation between copy number and oocyte quality (Pawlak et al. 2016; Ogino et al. 2016).

16.3.2 Mitochondrial DNA

Mitochondria are unique organelles in that they contain units of nonnuclear DNA termed mtDNA. MtDNA is a small circular strand of DNA containing 16,569 base pairs (Anderson et al. 1981). MtDNA codes for many of the proteins needed in oxidative phosphorylation (Anderson et al. 1981). MtDNA is the only DNA species not centralised within the nucleus of eukaryotes. Interestingly, genes for many mitochondrial proteins have been relocated to the nucleus during evolution. It is therefore thought that the isolation of the remaining genes contained on the mtDNA confers a biological advantage (Björkholm et al. 2015).

Mitochondrial genetics differs quite significantly from nuclear genetics in terms of mode of inheritance and gene expression. Considering that a single nucleus exists per cell (two copies of each gene), most nuclear DNA throughout the body is homogeneous, although tissue-specific mutations can occur, leading to some degree of heterogeneity. Furthermore, nuclear DNA is inherited in a Mendelian fashion, i.e. one copy from the father and one from the mother, dominance ruling over recessive genes. MtDNA follows a completely different set of rules. MtDNA within a single oocyte is formed from a few species passed from mother to child. These few copies replicate to form the mitochondrial complement within the oocyte and consequently the entire body. Many thousands of copies occur within each cell, unlike that of nuclear DNA where two copies of each chromosome exist. Usually, the sequence of mtDNA throughout the cell is homogeneous. However, mutations within individual strands can occur leading to heterogeneity within cells, a phenomenon known as heteroplasmy. This means that each cell may contain variable quantities of normal mtDNA and therefore have variable numbers of fully functioning mitochondria. Individual cells or tissues therefore are thought to have unique levels of mitochondrial efficiency. Therefore, mitochondrial genetics operates in a non-Mendelian fashion, has variable outcomes depending on the proportion of mutated mtDNA in a cell and is inherited uniquely from the maternal lineage.

16.4 Mitochondria and Ageing: The 'Free Radical Theory'

The presence of numerous species of mtDNA in a cell, and in particular their location, is thought to explain the mechanism of ageing in mammals (Harman 1956, 2009). The effect of ageing on mtDNA is possible because one of the side effects of locating mtDNA within the mitochondrial matrix is that mtDNA is subject to free-radical-induced damage during the process of oxidative phosphorylation itself (Linnane et al. 1989; Wei et al. 1998; López-Otín et al. 2013). This occurs because active mitochondria produce oxygen free radicals during respiration. Although these species have a very short half-life, they are highly reactive and will oxidise anything they come into contact with (Fig. 16.2). Unfortunately for the

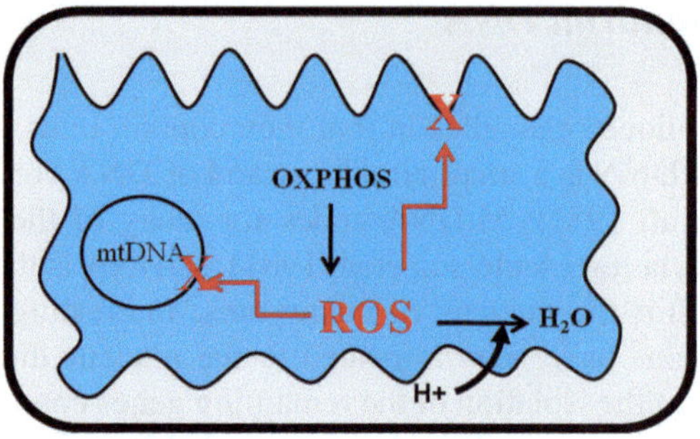

Fig. 16.2 Mode of damage of mitochondria by oxygen free radicals. The figure illustrates the location and mode of ROS (radical oxygen species) damage during oxidative phosphorylation (OXPHOS) in the mitochondria. The process of OXPHOS causes release of ROS. These species oxidise anything in their vicinity, which in the mitochondrial matrix includes both mtDNA and the inner mitochondrial membrane. ROS also combine with protons (H^+) to form water during the normal process of oxidative phosphorylation

mitochondria themselves, the mtDNA that encodes several components of the respiratory chain is present within these organelles. This would suggest that mitochondrial respiration damages the mtDNA. Interestingly, mtDNA repair mechanisms are absent or inefficient, further suggesting that damage is poorly repaired (Linnane et al. 1989, 1992). In rapidly dividing cells, the upkeep of mitochondrial quality is facilitated by the rapid turnover of cells and cytoplasmic components, and this eliminates poor-quality organelles by successively diluting them, therefore maintaining organelle (including mitochondrial) health. Problems however occur in slow or non-dividing cells because mitochondria are replaced slowly, and therefore levels of mutated mtDNA could build up to damaging levels. The best example of non-dividing tissues in mammals are neurons (brain tissue), and the free radical theory of ageing may therefore explain the onset of common age-related diseases including Alzheimer's (Adiele and Adiele 2016).

16.4.1 Is the 'Free Radical Theory of Ageing' Applicable to Reproduction?

Humans in particular are characterised by a decrease in the number of oocytes in the ovarian pool and decreased efficiency of these oocytes in producing a viable foetus, with respect to age (Navot et al. 1991). Within the oocyte, age is noted to correlate with an increase in the incidence of chromosomal abnormalities (Hassold and Chiu 1985; Dailey et al. 1996). Although these effects appear to explain the relationship between advanced maternal age and reproduction themselves, the data further suggest a more fundamental underlying mechanism considering that chromosome segregation relies on the availability of energy. The mechanism of ageing based on mitochondria may be central to the decrease in reproductive efficiency with respect to maternal age in human reproduction (Wilding et al. 2009). This is because mammalian follicles can be described as slow-dividing tissues. In mammals, the full complement of eggs is formed in the ovary prior to birth (Findlay et al. 2015). These eggs reside in the ovaries as dormant, primordial follicles; however, a low level of respiration occurs in these cells throughout their lifespan. In the short term, this low-level respiration may not cause significant damage; however, over the lifespan of the ovary (40–50 years in the human), the damage may build up to significant levels (Wilding et al. 2009). Considering that the egg contains most of the components, including all of the mitochondria, required to create a healthy embryo and implant into the uterus, damage to the egg will affect implantation. Furthermore, sperm cells are continually produced throughout the male lifespan (rapidly dividing cells), precluding the effect of free-radical-induced damage.

Interestingly, although the above theory appears sound, in reality there is a gross lack of evidence to support this hypothesis. In fact, although large mutations in the mitochondrial genome have been detected, for example, the delta4977 deletion (Barritt et al. 1999), no relationship with maternal age has been found, nor do numbers of mtDNA decrease with respect to maternal age in humans (Barritt et al. 2002). However, more recent research in other mammalian species suggests that mtDNA copy numbers do decrease with respect to female age (Rambags et al. 2014; Babayev et al. 2016; May-Panloup et al. 2016). Changes in the numbers of smaller mtDNA mutations such as point mutations have so far been more difficult to detect due to the large number of mtDNA species contained within the cell. More recently it has been possible to detect point mutations in mtDNA and quantification of mutation load (Hellebrekers et al. 2012). This has already been used in bovines and a relationship between ageing and an increase in mtDNA mutations found to be true (Hammond et al. 2016). With the advent of new mass sequencing techniques such as next-generation sequencing for mitochondrial mutations (Guo et al. 2012; Ancora et al. 2016), it is hoped that a resolution to the detection of small levels of mtDNA mutations is on the horizon.

16.5 Is Aerobic Respiration Relevant to Human Embryo Development?

Mitochondria generate ATP through aerobic respiration, and therefore a decrease in mitochondrial efficiency will correlate with a decrease in ATP production through this mechanism. In a developing embryo, ATP is required for the many cell processes that occur. A correlation between maternal age and mitochondrial activity, presumably affecting ATP production, has been shown (Wilding et al. 2001), and this was in turn correlated with the detection of chaotic mosaicism and poor oocyte spindle morphology in humans (Wilding et al. 2003). Therefore, a decrease in mitochondrial efficiency should lead to a decrease in embryo quality and lower implantation rates. However, this theory appears not to be true in some mammalian species. Treatment of mice embryos to completely eliminate aerobic respiration did not affect preimplantation development (Piko and Chase 1973). Mitochondria in preimplantation human embryos appear inactive, at least if morphology is considered (Bavister and Squirrell 2000). Furthermore, experiments looking at the level of the ATP content of preimplantation embryos showed a huge variance in levels and no significant difference with respect to age (Van Blerkom et al. 1995). However, metabolic experiments on human embryos developing in vitro suggested that the principal source of energy is derived from oxidative phosphorylation (Martin et al. 1993; Leese et al. 1993; Sturmey and Leese 2003; Leese 2012). Interestingly, but as to be expected, human preimplantation embryos produce large quantities of lactic acid during this stage, suggesting anaerobic respiration is occurring (Butcher et al. 1998). The ratio of oxidative phosphorylation to anaerobic respiration was estimated to be 56–44% (Butcher et al. 1998).

These data suggest that the principle source of energy in preimplantation mammalian embryos is anaerobic. However, these data are probably biased because of the maternal stores within the embryo (enabling development to continue in the absence of respiration), the inefficiency of anaerobic respiration that indicates that a small level of respiration will cause a large amount of lactic acid production, and perhaps the difference between in vitro and in vivo conditions, which may prefer anaerobic respiration as a consequence of stress or culture conditions. The most logical theory of the role of aerobic and anaerobic respiration in preimplantation mammalian embryo development is that both aerobic and anaerobic mechanisms are active in preimplantation embryo development (Wilding et al. 2005, 2009). Anaerobic respiration completes the deficit of ATP required and in the short term can substitute aerobic respiration although this is not sustainable over the long term due to the rundown of NAD^+ (Wilding et al. 2009). It is therefore quite possible that the deficit of aerobic respiration is expressed during blastocyst development or at a later stage and can arrest development and cause failure to implant or termination soon after implantation (Wilding et al. 2009).

16.6 Long-Term Consequences of the Maternal Age Effect?

The mitochondrial theory of the maternal age effect then suggests that mitochondria within the primordial oocytes tend to become damaged with respect to age, leaving progressively fewer primordial oocytes with the potential to develop and form a viable foetus. The data suggests that all primordial oocytes are damaged over time, and a threshold determines whether the oocyte is viable. If we accept the mitochondrial theory of the maternal age effect, we must also accept that oocytes from older females all have a degree of damage. Considering that the entire pool of mtDNA in adults is derived from the maternal pool, this could prove significant for the health of offspring born to older females since the mitochondrial pool of these individuals is based on a damaged population. In lower animals and mammals, a correlation between the lifespan of parents and siblings has been shown, and this is termed the 'Lansing effect' (Lansing 1947, 1948; Priest et al. 2002; Yonemura et al. 1991; Tarín et al. 1998, 2003, 2005; García-Palomares et al. 2009); however, the cause of this observation is currently unknown.

Although it is extremely difficult to determine the effect of this hypothesis on the individual, demographic studies have suggested a link between the two. In humans, several studies have suggested that a link exists between the lifespan of humans and their offspring (Wilding 2014, 2015; Wilding et al. 2014). Factors such as the lifespan of the parents and the age of conception appear to have an effect (Wilding 2014; Wilding et al. 2014). These data suggest that two confounding factors are expressed. The fact that the lifespan of the parents influences that of the offspring could indicate that the rate of mitochondrial damage (perhaps lower in individuals of a longer lifespan) varies, and this is inherited. The second factor is the age of conception, i.e. the amount of damage to the pre-existing oocytes that is transmitted to offspring (Wilding 2014; Wilding et al. 2014).

Although the above data may appear logical, it has to be stated here that the hypothesis remains a hypothesis. Many confounding factors were noted in the above studies, best examples being the definition of 'natural lifespan' (i.e. when can death from influenza be defined as 'natural death'?) and the many extrinsic forces that affect lifespan such as century of birth (Wilding et al. 2014). However, the fact that the age and lifespan of an individual can influence that of their offspring is interesting and confounds the hypothesis that each individual is equal at birth.

16.6.1 Attenuating the Maternal Age Effect

If the 'maternal age effect' is based on mitochondrial metabolism and is potentially inheritable, will it be possible to develop medical procedures to alter the level of mitochondrial metabolism in living eggs and therefore alleviate this effect? In theory, it is possible to alter the biological age of eggs during assisted reproduction by altering the mtDNA content of these cells. Changes in the mtDNA content of eggs have been attempted in assisted reproduction through two techniques—cytoplasmic transfer (Cohen et al. 1997) and maternal spindle/pronuclear transfer (Tachibana et al. 2009; Craven et al. 2010, Fig. 16.3).

Cytoplasmic transfer was developed by Dr. Cohen and colleagues in 1997 (Cohen et al. 1997). In this technique, a small amount of cytoplasm (5%) from a donor (younger) egg was injected into the recipient egg together with the spermatozoa. This appeared to increase the quality of the recipient egg to the point in which the resulting embryo produced a viable pregnancy. The component of the injected cytoplasm that was most likely to have changed the recipient egg was thought to be mitochondria. In fact, it was possible to detect mtDNA from the

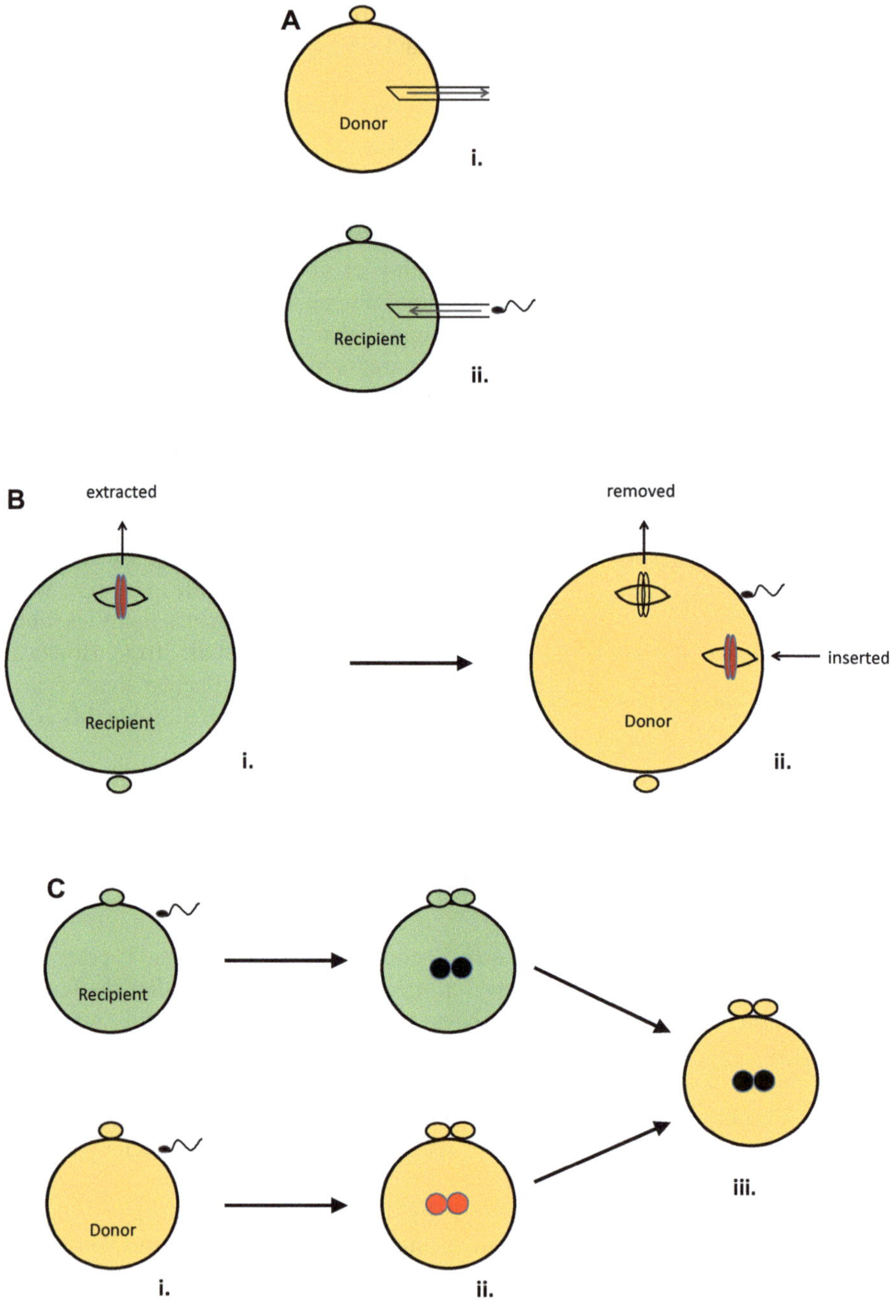

Fig. 16.3 Development of the mitochondrial replacement technique. The figure illustrates the three techniques for mitochondrial replacement. **(a)** Cytoplasmic transfer. This technique involves the aspiration of a portion (about 5%) of cytoplasm from a high-quality (donor) egg (i) and the injection of this material together with the sperm into the recipient (patient) egg (ii). The result is the original (recipient egg) supplemented with cytoplasm from the donor. **(b)** Maternal spindle transfer. This technique involves the removal of the maternal spindle from the recipients' egg

donated egg in children born with this technique, demonstrating survival of these organelles through to birth (Barritt et al. 2001a).

Although the technique was successful, fears for the safety of the technique, the idea of children with three parents and the fact that several children born with the technique had genetic conditions such as Turner's syndrome (Barritt et al. 2001b) caused the abrupt discontinuation of the procedure in the USA. The technique was not tested as a means to attenuate the maternal age effect.

More recently, new version of this technique, termed maternal spindle/pronuclear transfer, has been developed (Tachibana et al. 2009; Craven et al. 2010). With maternal spindle transfer, the chromosome content of the 'diseased' egg is transferred to a healthy egg. Pronuclear transfer operates on the same mechanism, but here the pronuclei are transferred. Whichever technique is used, the principle behind them is that the entire cytoplasmic content of the egg is replaced by healthy cytoplasm. These techniques are currently under examination, a baby has been reported to be born from the technique (New Scientist 2016), and the UK is preparing to licence the procedure. However, the technique is only being used to cure mitochondrial disease; its use for maternal age is not yet contemplated (Hyslop et al. 2016).

16.7 Conclusions

Human reproduction has until recently focused on the correlation between maternal age and chromosome abnormalities such as Down's syndrome and ways to prevent this. Mitochondria were always a poorly considered organelle in the egg cytoplasm. In recent years, the realisation that the quality of mitochondria can vary, the unique mode of inheritance of the organelle and the incidence of mitochondrial disease have caused an explosion of research into the role of this organelle in human reproduction. In this work, we have discussed the role of mitochondria, and in particular mtDNA, in the efficiency of human reproduction. We show that mitochondria are now and must continue to be in the forefront of research.

Manipulation of mitochondria in human eggs is possible, unlike that of individual chromosomes or genes on the human genome. This has generated a huge amount of interest in ways in which the mitochondrial content of eggs can be changed. Techniques are under development for the prevention of mitochondrial

Fig. 16.3 (continued) (i) and the insertion of this spindle into the donor egg, after removal of the donor spindle (ii). The egg is then fertilised. The result is an egg consisting of the recipient spindle and the entire cytoplasmic content of the donor egg. **(c)** Pronuclear transfer. Here, both recipient and donor egg are fertilised as normal (i). After the formation of pronuclei, the recipient pronuclei are removed from the recipient egg and replaced into the donor egg (ii). The pronuclei from the donor egg are removed, again leading to the recipient pronuclei and the entire cytoplasmic content of the donor egg (iii) disease. Whether these techniques will ever be tested for the cure of the 'maternal age effect' is debatable, due to the number of ethical concerns that have arisen from the manipulation of the egg cytoplasm.

References

Adiele RC, Adiele CA (2016) Mitochondrial regulatory pathways in the pathogenesis of alzheimer's disease. J Alzheimers Dis 53(4):1257–1270

Almeida-Santos T, El Shourbagy S, St. John JC (2006) Mitochondrial content reflects oocyte variability and fertilization outcome. Fertil Steril 85:584–591

Ancora M, Orsini M, Colosimo A, Marcacci M, Russo V, De Santo M, De Aurora M, Stuppia L, Barboni B, Camma C, Gstta V (2016) Complete sequence of human mitochondrial DNA obtained by combining multiple displacement amplification and next-generation sequencing on a single oocyte. Mitochondrial DNA A DNA Mapp. Seq Anal 24:1–2

Anderson S, Bankier AT, Barrell BG, de Bruijn MH, Coulson AR, Drouin J, Eperon IC, Nierlich DP, Roe BA, Sanger F et al (1981) Sequence and organization of the human mitochondrial genome. Nature 290:457–465

Babayev E, Wang T, Szigeti-Buck K, Lowther K, Taylor H, Horvath T, Seli E (2016) Reproductive ageing is associated with changes in oocyte mitochondrial dynamics, function and mtDNA quality. Maturitas 16:30145–30141

Barritt JA, Brenner CA, Cohen J, Matt DW (1999) Mitochondrial DNA rearrangements in human oocytes and embryos. Mol Hum Reprod 5(10):927–933

Barritt JA, Brenner CA, Malter HE, Cohen J (2001a) Mitochondria in human offspring derived from ooplasmic transplantation. Hum Reprod 16(3):513–516

Barritt J, Willadsen S, Brenner C, Cohen J (2001b) Cytoplasmic transfer in assisted reproduction. Hum Reprod Update 7(4):428–435

Barritt JA, Kokot M, Cohen J, Steuerwald N, Brenner CA (2002) Quantification of human ooplasmic mitochondria. Reprod Biomed Online 4(3):243–247

Bavister BD, Squirrell JM (2000) Mitochondrial distribution and function in oocytes and early embryos. Hum Reprod 15(Suppl 2):189–198

Björkholm P, Harish A, Hagström E, Ernst AM, Andersson SGE (2015) Mitochondrial genomes are retained by selective constraints on protein targeting. Proc Natl Acad Sci 112 (33):10154–10161

Braude P, Bolton V, Moore S (1988) Human gene expression first occurs between the four and eight-cell stages of preimplantation development. Nature 332:459–461

Butcher L, Coates A, Martin KL, Rutherford AJ, Leese HJ (1998) Metabolism of pyruvate by the early human embryo. Biol Reprod 58(4):1054–1056

Cohen J, Scott R, Schimmel T, Levron J, Willadsen S (1997) Birth of infant after transfer of anucleate donor oocyte cytoplasm into recipient eggs. Lancet 350(9072):186–187

Craven L, Tuppen HA, Greggains GD, Harbottle SJ, Murphy JL, Cree LM, Murdoch AP, Chinnery PF, Taylor RW, Lightowlers RN, Herbert M, Turnbull DM (2010) Pronuclear transfer in human embryos to prevent transmission of mitochondrial DNA disease. Nature 465 (7294):82–85

Cummins J (2002) The role of maternal mitochondria during oogenesis, fertilisation and embryogenesis. Reprod Biomed Online 4:176–182

Dailey T, Dale B, Cohen J, Munné S (1996) Association between nondisjunction and maternal age in meiosis-II human oocytes. Am J Hum Genet 59(1):176–184

Dale B, Defelice L, Ehrenstein G (1985) Injection of a soluble sperm extract into sea urchin eggs triggers the cortical reaction. Experentia 41:1068–1070

Dale B, Gualtieri R, Talevi R, Tosti E, Santella L, Elder K (1991) Intercellular communication in the early human embryo. Mol Reprod Dev 29:22–28

Diot A, Morten K, Poulton J (2016) Mitophagy plays a central role in mitochondrial ageing. Mamm Genome 27:381–395

Eichenlaub-Ritter U, Wieczorek M, Lüke S, Seidel T (2011) Age related changes in mitochondrial function and new approaches to study redox regulation in mammalian oocytes in response to age or maturation conditions. Mitochondrion 11:783–796

Findlay JK, Hutt KJ, Hickey M, Anderson RA (2015) How is the number of primordial follicles in the ovarian reserve established? Biol Reprod 93(5):111

Fragouli E, Spath K, Alfarawati S, Kaper F, Craig A, Michel CE, Kokocinski F, Cohen J, Munne S, Wells D (2015) Altered levels of mitochondrial DNA are associated with female age, aneuploidy, and provide an independent measure of embryonic implantation potential. PLoS Genet 11(6):e1005241

García-Palomares S, Pertusa JF, Miñarro J, García-Pérez MA, Hermenegildo C, Rausell F, Cano A, Tarín JJ (2009) Long-term effects of delayed fatherhood in mice on postnatal development and behavioral traits of offspring. Biol Reprod 80:337–342

Guantes R, Rastrojo A, Neves R, Lima A, Aguado B, Iborra FJ (2015) Global variability in gene expression and alternative splicing is modulated by mitochondrial content. Genome Res 25 (5):633–644

Guo Y, Cai Q, Samuels DC, Ye F, Long J, Li C, Winther J, Tawn E, Stovall M, Lahteenmaki P, Malila N, Levy S, Shaffer C, Shyr Y, Shu X, Boice J (2012) The use of next generation sequencing to study the effect of radiation therapy on mitochondrial DNA mutations. Mutat Res 744:154–160

Hammond ER, Green MP, Shelling AN, Berg MC, Peek JC, Cree LM (2016) Oocyte mitochondrial deletions and heteroplsmy in a bovine model of ageing and ovarian stimulation. Mol Hum Reprod 22:261–271

Harman D (1956) Aging: a theory based on free radical and radiation chemistry. J Gerontol 11 (3):298–300

Harman D (2009) Origin and evolution of the free radical theory of aging: a brief personal history, 1954–2009. Biogerontology 10(6):773–781

Hassold T, Chiu D (1985) Maternal age-specific rates of numerical chromosome abnormalities with special reference to trisomy. Hum Genet 70:11–17

Hellebrekers DM, Wolfe R, Hendrickx AT, de Coo IF, de Die CE, Geraedts JP, Chinnery PF, Smeets HJ (2012) PGD and heteroplasmic mitochondrial DNA point mutations: a systematic review estimating the chance of healthy offspring. Hum Reprod Update 18:341–349

Hyslop LA, Blakeley P, Craven L, Richardson J, Fogarty NM, Fragouli E, Lamb M, Wamaitha SE, Prathalingam N, Zhang Q, O'Keefe H, Takeda Y, Arizzi L, Alfarawati S, Tuppen HA, Irving L, Kalleas D, Choudhary M, Wells D, Murdoch AP, Turnbull DM, Niakan KK, Herbert M (2016) Towards clinical application of pronuclear transfer to prevent mitochondrial DNA disease. Nature 534(7607):383–386

Kloc M, Kubiak JZ (2017) Exogenous molecules and organelles delivery in oogenesis. In: Kloc M (ed) Oocytes: maternal information and functions. Springer, Cham

Krebs HA (1965) Gluconeogenesis. Expos Annu Biochim Med 26:13–30

Lansing AI (1947) A transmissible, cumulative and reversible factor in aging. J Gerontol 2:228–239

Lansing AI (1948) Evidence for aging as a consequence of growth cessation. Proc Natl Acad Sci 34:304–310

Leese HJ (2012) Metabolism of the preimplantation embryo: 40 years on. Reproduction 143 (4):417–427

Leese HJ, Conaghan J, Martin KL, Hardy K (1993) Early human embryo metabolism. Bioessays 15(4):259–264

Linnane AW, Marzuki S, Ozawa T et al (1989) Mitochondrial mutations as an important contributor to ageing and degenerative diseases. Lancet 1:642–645

Linnane AW, Zhang C, Baumer A, Nagley P (1992) Mitochondrial DNA mutations and the ageing process: bioenergy and pharmacological intervention. Mutat Res 275:195–208

López-Otín C, Blasco MA, Partridge L, Serrano M, Kroemer G (2013) The hallmarks of aging. Cell 153(6):1194–1217

Martin KH, Winston RML, Leese HJ (1993) Activity of enzymes of energy metabolism in single human preimplantation embryos. J Reprod Fertil 99:259–266

May-panloup P, Boucref L, Chao de la Barca J, Desquiret-Dumas V, Ferre-L'Hotelier V, Moriniere C, Descamps P, Procaccio V, Reynier P (2016) Ovarian ageing: the role of mitochondria in oocytes and follicles. Hum Reprod Update 22(6):725–743

Muir R, Diot A, Poulton J (2016) Mitochondrial content is central to nuclear gene expression: Profound implications for human health. BioEssays 38(2):150–156

Navot D, Bergh PA, Williams MA, Garrisi GJ, Guzman I, Sandler B, Grunfeld L (1991) Poor oocyte quality rather than implantation failure as a cause of age-related decline in female fertility. Lancet 337:1375–1377

New Scientist (2016) Exclusive: World's first baby born with new "3 parent" technique. https://www.newscientist.com/article/2107219-exclusive-worlds-first-baby-born-with-new-3-parent-technique/

Ogino M, Tsubamo H, Sakata K, Oohama N, Hyakawa H, Kojima T, Shigeta M, Shibahara H (2016) Mitochondrial DNA copy number in cumulus cells is a strong predictor of obtaining good quality embryos after IVF. J Assist Reprod Genet 33:367–371

Pawlak P, Chabowska A, Malyszka N, Lechniak D (2016) Mitochondria and mitochondrial DNA in porcine oocytes and cumulus cells—a search for developmental competence marker. Mitochondrion 27:48–55

Piko L, Chase DH (1973) Role of the mitochondrial genome during early development in mice. Effects of ethidium bromide and chloramphenicol. J Cell Biol 58:357–378

Priest NK, Mackowiak B, Promislow DEL (2002) The role of parental age effects on the evolution of aging. Evolution 56:927–935

Rambags BP, van Boxtel DC, Tharasanit Y, Lenstra JA, Colenbrander B, Stout TA (2014) Advancing maternal age predisposes to mitochondrial damage and loss during maturation of equine oocytes in vitro. Theriogenology 81:959–965

Reynier P, May-Panloup P, Chrétien MF, Morgan CJ, Jean M, Savagner F, Barrière P, Malthièry Y (2001) Mitochondrial DNA content affects the fertilizability of human oocytes. Mol Hum Reprod 7(5):425–429

Rich PR (2003) The molecular machinery of Keilin's respiratory chain. Biochem Soc Trans 31 (Pt 6):1095–1105

Shahbazi MN, Jedrusik A, Vuoristo S, Recher G, Hupalowska A, Bolton V, Fogarty NM, Campbell A, Devito LG, Ilic D, Khalaf Y, Niakan KK, Fishel S, Zernicka-Goetz M (2016) Self-organization of the human embryo in the absence of maternal tissues. Nat Cell Biol 18 (6):700–708

St John JC, Facucho-Oliveira J, Jiang Y, Kelly R, Salah R (2010) Mitochondrial DNA transmission, replication and inheritance: a journey from the gamete through the embryo and into offspring and embryonic stem cells. Hum Reprod Update 16:488–509

Sturmey RG, Leese HJ (2003) Energy metabolism in pig oocytes and early embryos. Reproduction 126(2):197–204

Sutovsky P, Schatten G (2000) Paternal contributions to the mammalian zygote: fertilization after sperm-egg fusion. Int Rev Cytol 195:1–65

Tachibana M, Sparman M, Sritanaudomchai H, Ma H, Clepper L, Woodward J, Li Y, Ramsey C, Kolotushkina O, Mitalipov S (2009) Mitochondrial gene replacement in primate offspring and embryonic stem cells. Nature 461(7262):367–372

Tarín JJ, Brines J, Cano A (1998) Long-term effects of delayed parenthood. Hum Reprod 13:2371–2376

Tarín JJ, Gómez-Piquer V, Manzanedo C, Miñarro J, Hermenegildo C, Cano A (2003) Long-term effects of delayed motherhood in mice on postnatal development and behavioural traits of offspring. Hum Reprod 18:1580–1587
Tarín JJ, Gómez-Piquer V, Rausell F, Navarro S, Hermenegildo C, Cano A (2005) Delayed motherhood decreases life expectancy of mouse offspring. Biol Reprod 72:1336–1343
Valero T (2014) Mitochondrial biogenesis: pharmacological approaches. Curr Pharm Des 20 (35):5507–5509
Van Blerkom J, Davis PW, Lee J (1995) ATP content of human oocytes and developmental potential and outcome after in-vitro fertilization and embryo transfer. Hum Reprod 10 (2):415–424
Wei YH, CY L, Lee HC, Pang CY, Ma YS (1998) Oxidative damage and mutation to mitochondrial DNA and age-dependent decline of mitochondrial respiratory function. Ann N Y Acad Sci 854:155–170
Wilding M (2014) Can we define maternal age as a genetic disease? Facts Views Vis Obgyn 6 (2):105–108
Wilding M (2015) Potential long-term risks associated with maternal aging (the role of the mitochondria). Fertil Steril 103(6):1397–1401
Wilding M, Dale B, Marino M, di Matteo L, Alviggi C, Pisaturo ML, Lombardi L, De Placido G (2001) Mitochondrial aggregation patterns and activity in human oocytes and preimplantation embryos. Hum Reprod 16(5):909–917
Wilding M, De Placido G, De Matteo L, Marino M, Alviggi C, Dale B (2003) Chaotic mosaicism in human preimplantation embryos is correlated with a low mitochondrial membrane potential. Fertil Steril 79(2):340–346
Wilding M, Di Matteo L, Dale B (2005) The maternal age effect: a hypothesis based on oxidative phosphorylation. Zygote 13(4):317–323
Wilding M, Coppola G, Dale B, Di Matteo L (2009) Mitochondria and human preimplantation embryo development. Reproduction 137(4):619–624
Wilding M, Coppola G, De Icco F, Arenare L, Di Matteo L, Dale B (2014) Maternal non-Mendelian inheritance of a reduced lifespan? A hypothesis. J Assist Reprod Genet 31 (6):637–643
Yonemura I, Motoyama T, Hasekura H, Boettcher B (1991) Cytoplasmic influence on the expression of nuclear genes affecting life span in Drosophila melanogaster. Heredity (Edinb) 66:259–264

[illegible]

Chapter 17
Functions of Vitellogenin in Eggs

Hongyan Li and Shicui Zhang

Abstract Our understanding of the functions of vitellogenin (Vtg) in reproduction has undergone an evolutionary transformation over the past decade. Primarily, Vtg was regarded as a female-specific reproductive protein, which is cleaved into yolk proteins including phosvitin (Pv) and lipovitellin (Lv), stored in eggs, providing the nutrients for early embryos. Recently, Vtg has been shown to be an immunocomponent factor capable of protecting the host against the attack by microbes including bacteria and viruses. Moreover, Pv and Lv that both are proteolytically cleaved products of maternal Vtg, as well as Pv-derived small peptides, all display an antibacterial role in developing embryos. In addition, both Vtg and yolk protein Pv possess antioxidant activity capable of protecting cells from damage by free radicals. Collectively, these data indicate that Vtg, in addition to being involved in yolk protein formation, also plays non-nutritional roles via functioning as immune-relevant molecules and antioxidant reagents.

17.1 Introduction

Nonplacental or non-trophotenic vertebrates and nearly all invertebrates are oviparous, with their eggs being fertilized externally (Jalabert 2005). Egg, or a haploid reproductive cell, which develops into a viable embryo after fertilization, is the final product of oocyte growth and differentiation (Lubzens et al. 2010). Generally, several steps are involved in egg development: formation of primordial germ cells (PGCs) and transformation of PGCs into oogonia and then into oocytes. Oocyte growth, particularly in oviparous species, is characterized by deposition of massive amount of maternal information and molecules needed for early embryo development, including RNAs, proteins, lipids, vitamins, and hormones (Lubzens et al. 2010; Patiño and Sullivan 2002). One of the most important proteins deposited in oocytes is vitellogenin (Vtg), a large precursor of the major yolk proteins (Yps).

H. Li • S. Zhang (✉)
Department of Marine Biology, Ocean University of China, 5 Yushan Road, Qingdao 266003, China
e-mail: sczhang@ouc.edu.cn

M. Kloc (ed.), *Oocytes*, Results and Problems in Cell Differentiation 63,
DOI 10.1007/978-3-319-60855-6_17

Vitellogenin (from latin *vitellus*, yolk, and *gener*, to produce) was first proposed over 48 years ago, by Pan et al. (1969), to describe the female-specific hemolymph protein precursor of egg yolk in insects. This term was later adopted for the yolk protein precursor of other egg-laying animals, regardless of its amino acid sequence and structure, and is currently widely used by the scientific community. Vtg is now known as a high molecular mass glycolipophosphoprotein circulating in the blood (vertebrates) or hemolymph (invertebrates) as a homodimer. In oviparous vertebrates, Vtgs are usually synthesized in the liver, mainly under estrogen (E2) control (Rosanova et al. 2002; Wallace 1985). Similarly, in invertebrates Vtgs are also synthesized in an extra-ovarian tissue such as hepatopancreas (in crustaceans) and the fat body (in insects; Girish et al. 2014; Mak et al. 2005; Meusy 1980; Tufail and Takeda 2008). Soon after synthesis, Vtgs are posttranslationally glycosylated and phosphorylated in the endoplasmic reticulum (ER) and Golgi complex before being tagged for export, secreted as homodimeric lipoprotein complexes, to the blood or circulating body fluid, and transported to the ovary, where they are internalized during vitellogenesis into growing oocytes via receptor (clathrin)-mediated endocytosis (Amano et al. 2008; Conner and Schmid 2003; Kolarevic et al. 2008; Sawaguchi et al. 2005; Wallace and Selman 1990; Williams et al. 2014).

Vtgs are multidomain proteins, belonging to the large lipid transfer protein (LLTP) superfamily. They typically contain three conserved domains, the LPD_N (also known as vitellogenin_N or LLT domain) at the N-terminus, the DUF1943 domain with unknown function, and the von Willebrand factor type D domain (vWD) at the C-terminus. Occasionally, a DUF1944 domain with unknown function is also present in between DUF1943 and vWD in some Vtg proteins from vertebrates such as chicken and fish (Hayward et al. 2010). Beginning at the N-terminus, a complete Vtg consists of a signal peptide, a lipovitellin (Lv) heavy chain (LvH), a phosphorylated serine-rich phosvitin (Pv), a lipovitellin light chain (LvL), and two Cys-rich C-terminal coding regions (β' and CT) that are homologous to the von Willebrand factor type D domain in mammals. The Pv region can be absent, as observed in zebrafish Vtg3 and most invertebrate Vtgs.

Vtg, as an egg yolk protein precursor, is present in the females of nearly all oviparous species including fish, amphibians, reptiles, birds, most invertebrates, and the platypus and thus was once regarded as a female-specific protein; however, its synthesis, albeit in smaller quantities, has been shown to occur in male and even sexually immature animals (Engelmann 1979; Piulachs et al. 2003). This suggests that Vtg presumably fulfills a more general role independent of a gender. Initially, Vtg was considered to be the energy source for the developing embryos. However, our understanding of the function of Vtg in reproduction has undergone a transformation over the past decade. A series of studies have demonstrated several non-nutritional roles of Vtg and its derivatives Yps such as Pv and Lv. For example, the multiplicity or loss of *vtg* family in vertebrates has been shown to have broad implications for the mode of reproduction (Babin et al. 2007; Brawand et al. 2008; Finn et al. 2009). The differential expression and deposition of Vtgs in the growing oocytes have been shown to be able to determine the cleavage pattern (Finn et al. 2009; Gilbert 2006). Likewise, the variable degradation of Vtg derivatives Yps during oocyte hydration, which makes the pelagic eggs buoyant, has been shown to be able to determine the egg type (Carnevali et al. 2001; Finn et al. 2002; Matsubara et al. 1999; Reith et al. 2001; Selman et al. 2001). Vtgs have also been shown to be a maternal immune-relevant factor in early embryos. In addition, Vtgs display antioxidant activity. In this chapter, we will mainly focus on the multifaceted functions of Vtgs in eggs and early embryos, especially of fishes.

17.2 Vtgs and Yps Are Determinants of Mode of Reproduction and Type of Eggs

The multiplicity or loss of *vtg* family in vertebrates has been argued to have broad implications for the type of the egg (pelagic or benthic), mode of reproduction (placental or nonplacental), and cleavage pattern (meroblastic or holoblastic). Vtgs are usually encoded by a multigene family in animals, and thus there are several isoforms of Vtg in a given species. For instance, six *vtg* genes have been identified in nematode (Blumenthal et al. 1984), four in *Xenopus* (Germond et al. 1984; Wahli et al. 1979), three in chicken (Schip et al. 1987; Silva et al. 1989), and eight in zebrafish (Wang et al. 2000, 2005). Vtg genes act in a dosage-dependent fashion, so that a correlation between gene copy number and speed of yolking can be observed. For example, fishes tend to possess multiple *vtg* gene copies and produce a larger amount of eggs in a shorter time than birds and reptiles (Buisine et al. 2002). It appears that the lineage-specific *vtg* gene duplications frequently determine the nature of fish eggs. There are primarily two types of eggs in fish: benthic and pelagic. It was shown that the differential expression of non-neofunctionalized and neofunctionalized *vtg* genes in *Acanthomorpha* teleosts is correlated with the benthic or pelagic type of spawned eggs (Finn et al. 2009).

The amount of Vtg and its derivatives Yps is also correlated with the cleavage symmetry and pattern. In general, yolk consisting of Yps inhibits cleavage. In eggs with relatively little Yps, i.e., isolecithal and mesolecithal eggs, cleavage is holoblastic, meaning that the cleavage furrow extends through the entire egg. Eggs containing large accumulations of Yps undergo meroblastic cleavage, wherein only a part of the cytoplasm is cleaved.

Yps are an evolutionary adaptation that enables an embryo to develop in the absence of an external food source. Eggs without large amount of Yps, such as sea urchin, usually develop into a feeding larva fairly rapidly. In mammals, around 30–70 million years ago (Mya), the ancestral Vtg-encoding genes were lost in all but the egg-laying monotremes, which have retained a single functional *vtg* gene. The loss of *vtg*-encoding genes in "placental" mammals is consistent with the gain of alternative nourishment resources, casein, for the mammalian offspring (Brawand et al. 2008). At the other extreme are the eggs of fishes, reptiles, and birds, which are full of Yps. These animals develop without a larval stage because Yps in their eggs are sufficient to nourish embryo and support early development.

17.3 Vtgs and Yps Are Sources of Free Amino Acids for Eggs and Embryos

Nutritional reserves that are stored in egg yolk are crucial for the development of nonmammalian oviparous vertebrates and invertebrates (Wahli 1988). An individual eukaryotic cell may express more than 2500 proteins at any one time. This huge biodiversity of cellular proteins has also been confirmed for fish oocytes, where more than 600 different proteins have been identified during oogenic stages of zebrafish and gilthead seabream (Ziv et al. 2008), and 1379 discrete proteins have been identified in the ovary of zebrafish (Groh et al. 2011). Despite such impressive number of proteins in the developing oocytes of fish, the most abundant and major proteins in all fishes are Vtgs (Groh et al. 2011; Ziv et al. 2008), of which three functional forms are known, VtgAa, vtgAb, and VtgC. In oviparous species, the nutrients of embryos largely derive from Vtgs and Yps stored in eggs. In mammals, the nutritive lactation with complex milk, coupled with a sophisticated placenta, which evolved in parallel with eutherian viviparity, effectively rendered embryonic nourishment through Vtg completely dispensable (Brawand et al. 2008). In most teleosts, a complete type Vtg has a pentapartite structure encoded by two major genes, *vtgAa* and *vtgAb*, while a minor *vtgC* gene encodes a smaller dipartite protein (Finn 2007a; Finn and Kristoffersen 2007). The majority of *vtg* genes are expressed according to season and reproductive cycle, and their products are secreted into the circulating plasma from the liver (Babin et al. 2007; Mommsen and Korsgaard 2008; Wallaert and Babin 1994), although some genes (*vtgC*) may be transcribed in the intestine (Wang et al. 2005). This latter observation is reminiscent of *vtg* gene transcription in nematodes where six Vtgs are secreted into the body fluid from the intestine, and one of which (Vtg6) is then cleaved before being incorporated into the oocytes (Spieth and Blumenthal 1985). In teleost, and particularly in cold-water marine fish, the VtgC synthetized by the intestine may be involved in the transport of highly unsaturated fatty acids (HUFA) and polyunsaturated fatty acids (PUFA) from the gut to the ovary, because cold-water fishes have limited abilities to synthesize phospholipids (Tocher et al. 2008). Such essential HUFAs and PUFAs must be supplied in the maternal diet (Harel et al. 1994) and then be transported by lipoproteins to the growing oocytes in order to later meet the physiological needs of the developing embryos and larvae.

The physiological significance of Vtgs is usually attributed to its role in the transport of amino acids, lipids, phosphorous, and calcium to the egg. Once secreted into the circulating plasma, Vtgs together with their phospholipids and other diverse cargo are internalized by the oocyte via clathrin-mediated endocytosis (Fig. 17.1). Vtg internalization occurs in coated pits of oocyte plasma membrane; they rapidly

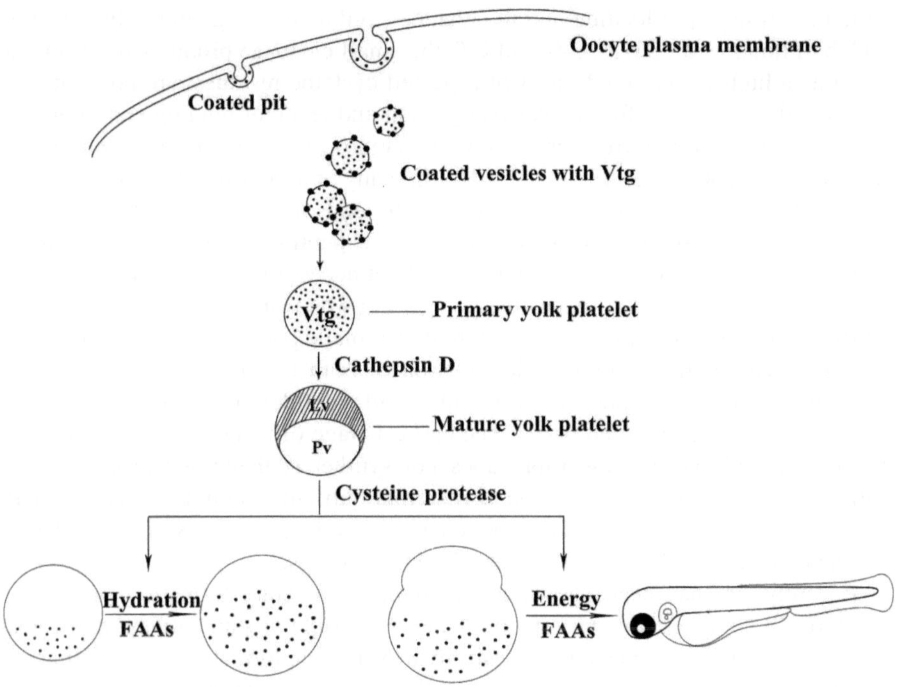

Fig. 17.1 Formation of yolk platelets from endocytosed Vtg and fate of Vtg-derived Yps Pv and Lv. Vtgs in the circulating plasma are taken in by the oocyte via clathrin-mediated endocytosis, forming coated vesicles in the cortical oocyte cytoplasm. The coated vesicles containing Vtg lose their clathrin coat and coalesce to form the primordial yolk globules, which are subsequently transformed into yolk platelets. Vtgs in the yolk platelets are primarily cleaved into Lv and Pv. In many marine fishes with pelagic eggs, an additional proteolysis of Yps occurs at the end of oocyte growth just before maturation, generating smaller peptides and free amino acids (FAAs), which form part of the osmotic gradient necessary for water uptake during the "hydration phase" that makes the pelagic eggs buoyant. During embryonic development, Lv and Pv are further cleaved into small peptides and FAAs, providing the nutrients for developing embryos

detach and give rise to coated vesicles in the cortical oocyte cytoplasm; the latter rapidly lose their clathrin coat and coalesce to form the primordial yolk globules subsequently transformed into yolk platelets (Ghiara et al. 1968; Limatola and Filosa 1989; Neaves 1972; Wallace 1985). Once packed in yolk platelets, Vtgs will undergo primary degradation. During degradation, Vtgs are cleaved by an aspartic proteinase called cathepsin D, into lipovitellins (Lv), phosvitin (Pv), β-component (β-C), and C-terminal coding region (CT) of different number and size (Komazaki and Hiruma 1999; Retzek et al. 1992). The Vtg-degraded products have different roles in oocytes and developing embryos. Lv, the largest yolk protein derived from the proteolytic processing of Vtgs, is an apoprotein delivering mainly phospholipids into developing oocytes (Romano et al. 2004; Yilmaz et al. 2015). Pv, the smallest yolk protein derived from the proteolytic process, largely consists of phosphorylated serine residues and is thought to be able to stabilize nascent Vtg structure during lipid loading and to enhance solubility of Vtgs in the blood (Finn 2007b; Yilmaz et al. 2015). β-C and CT, the small cleavage products of vWD that contains a highly conserved motif of repeated cysteine residues, are postulated to stabilize the Vtg dimer for cellular recognition and receptor binding and to protect Vtg or its product Yps from premature or inappropriate proteolysis (Finn 2007b; Reading et al. 2009; Williams et al. 2014). In many marine fishes with pelagic eggs, an additional proteolysis of Yps occurs at the end of oocyte growth just before maturation. This proteolysis generates smaller peptides and free amino acids (FAAs), which form part of the osmotic gradient necessary for water uptake during the "hydration phase" that makes the pelagic eggs buoyant (Finn and Fyhn 2010). During embryonic development, a different enzymatic pool, reported to be cysteine proteases, causes the secondary degradation of both Lv and Pv. This secondary cleavage of Yps generates a supply of small peptides and FAAs, which are important for developing embryos. The earliest stage of fish embryos, particularly before the mid-blastula transition, does not synthetize their own protein and is entirely dependent on the provision of maternal nutrients and transcripts or signaling factor that are produced during oogenesis. Therefore, the Vtgs-degraded FAAs are the essential energy supply for developing embryos.

Recently, Pv has been shown to be involved in bone formation in chicken embryos. In cultured calvarial osteoblasts, Pv is capable of stimulating the differentiation of osteoblasts, collagen synthesis, hydroxyproline formation, and biomineralization. Moreover, Li et al. (2014) showed that lysophophatidic acid, a signaling molecule derived from phosphatidylcholines of Pv (Ohlendorf et al. 1977), is able to regulate hemangioblast formation and primitive hematopoiesis in zebrafish. These data suggest that Pv and the smaller molecules derived from Pv also play important roles during embryonic development.

17.4 Vtgs and Yps Are Maternal Immune-Relevant Factors in Eggs and Embryos

Embryos of most mammalian species including humans develop in the uterus inside the mother's body and are thus well protected from external pathogenic attacks. In sharp contrast, eggs of most fish and aquatic invertebrates are released and fertilized externally, and therefore the resulting embryos are exposed to a hostile aquatic environment full of potential pathogens, which are capable of causing various types of diseases, even leading to death. For instance, it was shown that exposure of salmon fry and juveniles to the Gram-negative bacterium *Yerinia ruckeri* caused occurrence of enteric redmouth disease, resulting in 60% mortality (Haig et al. 2011). In addition, during the early stages of development, their embryos have little or limited ability to synthesize immune-relevant molecules endogenously, and their immune-relevant cells and tissues are not yet fully formed (Ellis 1988; Liang et al. 2009; Magnadóttir et al. 2004). How the embryos of fish and aquatic invertebrates survive the pathogenic attacks at this stage has received great attention in the past two decades. It is well known that fishes and aquatic invertebrates produce eggs endowed with all the nutrients and protective systems allowing the development of a fish embryo in an aquatic environment. It has been shown that embryo protection is partly ensured by Vtgs and their derived proteins Yps.

Accumulating data show that Vtgs are associated with antibacterial activity of the host against microbes including bacteria and virus (Garcia et al. 2010; Shi et al. 2006; Zhang et al. 2011). In zebrafish, Vtg has been shown to function as an acute phase protein in vivo, capable of leading to elimination of invading bacteria such as *Escherichia coli* and *Staphylococcus aureus* (Tong et al. 2010). Shi et al. (2006) showed that intraperitoneal injection of *E. coli* was able to enhance the level of serum Vtg in male rosy barb *Puntius conchonius*; and Lu et al. (2012, 2013) showed that the challenge with Gram-negative bacterium *Citrobacter freundii* induced upregulated expression of *vtg* in the skin of zebrafish. Bacterial challenge also caused a significant increase in *vtg* expression in scallop and insect (Nachappa et al. 2012; Wu et al. 2015). All these data suggest that Vtgs play an active role in protecting the host from infection. There is solid evidence that Vtgs are involved in the antimicrobial defense of the host against broad-spectrum bacteria and virus. Vtgs have been shown to act as a multivalent pattern recognition receptor capable of binding to lipopolysaccharide, lipoteichoic acid, peptidoglycan, glucan, and virons, a bactericidal molecule capable of damaging bacterial cell walls, and an opsonin capable of enhancing the phagocytosis of bacteria by macrophages (Garcia et al. 2010; Li et al. 2008; Sun and Zhang 2015; Zhang et al. 2011, 2015). Notably, Vtgs are shown to remain uncleaved in the oocytes of amphioxus and nematode (Sharrock 1983; Sun and Zhang 2001). Thus, it is possible that Vtgs in these animals may protect their oocytes and embryos against pathogenic attack.

Lv and Pv are the major Yps generated by the proteolytic cleavage of Vtg. As parental Vtg is an immune-competent molecule, it is thus reasonable to speculate that Lv and Pv also have similar immune activities. In fact, Pv has been proven to play a critical role in the immunity of zebrafish embryos via acting as a pattern recognition receptor and an antimicrobial effector molecule (Wang et al. 2011). Also smaller peptides derived from Pv have antibacterial activity (Wang et al. 2011; Zhang et al. 2015). In line with this, hen egg yolk Pv was also shown to be able to inhibit the growth of the Gram-negative *E. coli* and the Gram-positive bacterium *S. aureus* under thermal stress (Sattar Khan et al. 2000; Ma et al. 2013). Moreover, Pv from zebrafish was demonstrated to possess antiviral activity via inhibiting the formation of the cytopathic effect in lymphocystis disease virus-infected cells and reducing the virus quantities in the virus-infected cells and host. This indicates that Pv is a maternal immune-relevant factor capable of protecting developing embryos from virus attack (Sun et al. 2013). Like Pv, both LvH and LvL are also maternal factors capable of protecting early embryos and larvae (Liang et al. 2016; Zhang et al. 2011). These show that Pv and Lv are both maternally derived proteins involved in immune defense in embryos and larvae in fishes.

17.5 Vtgs and Yps Are Antioxidant Reagents in Eggs and Embryos

Oxidation is a chemical reaction that can produce free radicals, resulting in chain reactions that can cause extensive damage to DNA, proteins, and lipids. Therefore, antioxidant defense is thought to be important throughout an organism's life. This is also the case during embryo development and growth because intense embryonic metabolism entails massive production of oxidizing molecules. How rapidly growing embryos protect themselves from damage by free radicals is an interesting question in ecological evolutionary studies and in animal production disciplines (Ebrahimi et al. 2012; Müller et al. 2012; Romano et al. 2008; Saino et al. 2002, 2003; Selim et al. 2012; Surai 2002). It has been shown that the eggs of oviparous animals contain large amounts of antioxidants of maternal origin (Surai 2002). Mothers equip their eggs with various antioxidants such as vitamin A, vitamin E, and β-carotene (Barim-Oz and Sahin 2016; Dale et al. 2017). Among them, the specific antioxidants present in egg yolk are important for embryonic development (Ayala et al. 2006; Gao et al. 2013; Romano et al. 2008; Saino et al. 2003).

Another unique role of Vtgs is antioxidant/reducing activity. It was first shown that Vtg from eel (*Anguilla japonica*) was able to resist the copper-induced oxidation and seemed to chelate low concentrations of copper ion and could protect the very low-density lipoprotein (VLDL) against copper-induced oxidation (Ando and Yanagida 1999). This was the first report that Vtg is antioxidant reagent and serves to depress the free-radical reactions in fish oocytes. Later, it was demonstrated that Vtg has a role to protect other cellular components from oxidation because of its metal binding capacity in the nematode *C. elegans* (Nakamura et al. 1999). In the honeybee, Vtg was also shown to be able to reduce oxidative stress by scavenging free radicals, thus increasing the life span of the sterile honeybee workers and queen honeybee. The honeybee Vtg was also capable of recognizing cell damage through membrane binding and shielding living cells from reactive oxygen species (Havukainen et al. 2013). It is clear that Vtg can protect cells from free radical damage in both invertebrates and vertebrates.

Like Vtgs, the yolk protein Pv also shows antioxidant ability. It is well known that hen Pv serves as an antioxidant to inhibit metal-catalyzed phospholipid oxidations and hydroxyl radical formation (Guérin-Dubiard et al. 2002; Ishikawa et al. 2004; Lu and Baker 1986). Similarly, zebrafish recombinant Pv was also shown to be an antioxidant agent capable of inhibiting the oxidation of the linoleic acid and scavenging the DPPH radical (Hu et al. 2015). It is possible that antioxidant function of Pv can protect rapidly developing embryos from damage by free radicals. It is worthwhile to explore if other components derived from Vtgs such as Lv, β-C, and CT have antioxidant activity.

17.6 Conclusions

Our understanding of the functions of Vtgs in animal reproduction has undergone a revolutionary transformation over the past decade. This chapter summarized the state-of-the-art knowledge of Vtgs in animal reproduction. Primarily, Vtg was regarded as a female-specific reproductive protein, which is cleaved into Yps such as Pv and Lv, stored in eggs, providing the nutrients for developing embryos. However, recently, the Vtg was also shown to be an immune-relevant molecule involved in the defense of the host against the microbes including bacteria and viruses. Furthermore, Pv and Lv that both are proteolytically cleaved products of maternal Vtg, as well as Pv-derived small peptides, display an antibacterial activity in developing embryos. Finally, both Vtg and yolk protein Pv are an antioxidant reagents capable of protecting cells from damage by free radicals.

Acknowledgments We thank Mr. Xuwen Cao for the preparation of the references. During manuscript preparation, the authors were supported by the grants (U1401211; 31601862) of Natural Science Foundation of China.

References

Amano H, Fujita TN, Kagawa H, Matsubara T, Sullivan CV, Hara A (2008) Multiple vitellogenin-derived yolk proteins in gray mullet (*Mugil cephalus*): disparate proteolytic patterns associated with ovarian follicle maturation. Mol Reprod Dev 75:1307–1317

Ando S, Yanagida K (1999) Susceptibility to oxidation of copper-induced plasma lipoproteins from Japanese eel: protective effect of vitellogenin on the oxidation of very low density lipoprotein. Comp Biochem Physiol C Pharmacol Toxicol Endocrinol 123:1–7

Ayala RMD, Martinelli R, Saino N (2006) Vitamin E supplementation enhances growth and condition of nestling barn swallows (*Hirundo rustica*). Behav Ecol Sociobiol 60:619–630

Babin PJ, Carnevali O, Lubzens E, Schneider WJ (2007) Molecular aspects of oocyte vitellogenesis in fish. In: Babin PJ, Cerdà J, Lubzens E (eds) The fish Oocyte. Springer, Netherlands

Barim-Oz O, Sahin H (2016) The influence of dietary antioxidant on ovarian eggs and levels of vitamin E, C, A, astaxanthin, β-carotene and oxidative stress in tissues of *Astacus leptodactylus* (Eschscholtz) during reproduction. Cell Mol Biol 62:1–10

Blumenthal T, Squire M, Kirtland S, Cane J, Donegan M, Spieth J, Sharrock W (1984) Cloning of a yolk protein gene family from *Caenorhabditis elegans*. J Mol Biol 174:1–18

Brawand D, Wahli W, Kaessmann H (2008) Loss of egg yolk genes in mammals and the origin of lactation and placentation. PLoS Biol 6:e63

Buisine N, Trichet V, Wolff J (2002) Complex evolution of vitellogenin genes in salmonid fishes. Mol Gen Genomics 268:535–542

Carnevali O, Mosconi G, Cambi A, Ridolfi S, Zanuy S, Polzonetti-Magni AM (2001) Changes of lysosomal enzyme activities in sea bass (*Dicentrarchus labrax*) eggs and developing embryos. Aquaculture 202:249–256

Conner SD, Schmid SL (2003) Regulated portals of entry into the cell. Nature 422:37–44

Dale K, Rasinger J, Thorstensen K, Penglase S, Ellingsen S (2017) Vitamin E reduces endosulfan-induced toxic effects on morphology and behavior in early development of zebrafish (*Danio rerio*). Food Chemical Toxicol 101:84–93

Ebrahimi MR, Ahangari YJ, Zamiri MJ, Akhlaghi A, Atashi H (2012) Does preincubational in ovo injection of buffers or antioxidants improve the quality and hatchability in long-term stored eggs? Poult Sci 91:2970–2976

Ellis AE (1988) Fish vaccination. Academic Press, New York

Engelmann F (1979) Insect vitellogenin: identification, biosynthesis, and role in vitellogenesis. Adv Insect Physiol 14:49–108

Finn RN (2007a) Maturational disassembly and differential proteolysis of paralogous vitellogenins in a marine pelagophil teleost: a conserved mechanism of oocyte hydration. Biol Reprod 76:936

Finn RN (2007b) Vertebrate yolk complexes and the functional implications of phosvitins and other subdomains in vitellogenins. Biol Reprod 76:926–935

Finn RN, Fyhn HJ (2010) Requirement for amino acids in ontogeny of fish. Aquac Res 41:684–716

Finn RN, Kristoffersen BA (2007) Vertebrate vitellogenin gene duplication in relation to the "3R hypothesis": correlation to the pelagic egg and the oceanic radiation of teleosts. PloS One 2: e169

Finn RN, GC Ø, Norberg B, Fyhn HJ (2002) In vivo oocyte hydration in Atlantic halibut (*Hippoglossus hippoglossus*); proteolytic liberation of free amino acids, and ion transport, are driving forces for osmotic water influx. J Exp Biol 205:211–224

Finn RN, Kolarevic J, Kongshaug H, Nilsen F (2009) Evolution and differential expression of a vertebrate vitellogenin gene cluster. BMC Evol Biol 9:2

Gao YY, Xie QM, Ma JY, Zhang XB, Zhu JM, Shu DM, Sun BL, Jin L, Bi YZ (2013) Supplementation of xanthophylls increased antioxidant capacity and decreased lipid peroxidation in hens and chicks. Br J Nutr 109:1–7

Garcia J, Munro ES, Monte MM, Fourrier MC, Whitelaw J, Smail DA, Ellis AE (2010) Atlantic salmon (*Salmo salar* L.) serum vitellogenin neutralises infectivity of infectious pancreatic necrosis virus (IPNV). Fish Shellfish Immunol 29:293

Germond J-E, Walker P, ten Heggeler B, Brown-Luedi M, de Bony E, Wahli W (1984) Evolution of vitellogeain genes: comparative analysis of the nucleotide sequences downstream of the transcription initiation site of four *Xenopus laevis* and one chicken gene. Nucleic Acids Res 12:8595–8609

Ghiara G, Limatola E, Filosa S (1968) Ultrastructural aspects of nutritive process in growing oocytes of lizard. Electron Microscopy 2:331–332

Gilbert SF (2006) Developmental biology. Sinauer, Sunderland, MA

Girish BP, Ch S, Reddy PS (2014) Hepatopancreas but not ovary is the site of vitellogenin synthesis in female fresh water crab, *Oziothelphusa senex senex*. Biochem Biophys Res Commun 447:323–327

Groh KJ, Nesatyy VJ, Segner H, Eggen RIL, Suter JF (2011) Global proteomics analysis of testis and ovary in adult zebrafish (*Danio rerio*). Fish Physiol Biochem 37:619–647

Guérin-Dubiard C, Anton M, Dhene-Garcia A, Martinet V, Brulé G (2002) Hen egg and fish egg phosvitins: composition and iron binding properties. Eur Food Res Technol 214:460–464

Haig SJ, Davies RL, Welch TJ, Reese RA, Verner-Jeffreys DW (2011) Comparative susceptibility of Atlantic salmon and rainbow trout to *Yersinia ruckeri*:relationship to O antigen serotype and resistance to serum killing. Vet Microbiol 147:155–161

Harel M, Tandler A, Kissil GW, Applebaum SW (1994) The kinetics of nutrient incorporation into body tissues of gilthead seabream (*Sparus aurata*) females and the subsequent effects on egg composition and egg quality. Br J Nutr 72:45–58

Havukainen H, Münch D, Baumann A, Zhong S, Halskau Ø, Krogsgaard M, Amdam GV (2013) Vitellogenin recognizes cell damage through membrane binding and shields living cells from reactive oxygen species. J Biol Chem 288:28369–28381

Hayward A, Takahashi T, Bendena WG, Tobe SS, Hui JH (2010) Comparative genomic and phylogenetic analysis of vitellogenin and other large lipid transfer proteins in metazoans. FEBS Lett 584:1273–1278

Hu L, Sun C, Luan J, Lu L, Zhang S (2015) Zebrafish phosvitin is an antioxidant with non-cytotoxic activity. Acta Biochim Biophys Sin 47:349–354

Ishikawa S, Yano Y, Arihara K, Itoh M (2004) Egg yolk phosvitin inhibits hydroxyl radical formation from the fenton reaction. Biosci Biotechnol Biochem 68:1324–1331
Jalabert B (2005) Particularities of reproduction and oogenesis in teleost fish compared to mammals. Reprod Nutr Dev 45:261–279
Kolarevic J, Nerland A, Nilsen F, Finn RN (2008) Goldsinny wrasse (*Ctenolabrus rupestris*) is an extreme vtgAa-type pelagophil teleost. Mol Reprod Dev 75:1011–1020
Komazaki S, Hiruma T (1999) Degradation of yolk platelets in the early amphibian embryo is regulated by fusion with late endosomes. Develop Growth Differ 41:173–181
Li H, Yue R, Wei B, Gao G, Du J, Pei G (2014) Lysophosphatidic acid acts as a nutrient-derived developmental cue to regulate early hematopoiesis. EMBO J 33(12):1383–1396
Li Z, Zhang S, Liu Q (2008) Vitellogenin functions as a multivalent pattern recognition receptor with an opsonic activity. PLoS One 3:e1940
Liang J, Liu X, Wu FF, Li QW (2009) Progress of adaptive immunity system of agnathan vertebrates. Hereditas 31:969–976
Liang X, Hu Y, Feng S, Zhang S, Zhang Y, Sun C (2016) Heavy chain (LvH) and light chain (LvL) of lipovitellin (Lv) of zebrafish can both bind to bacteria and enhance phagocytosis. Dev Comp Immunol 63:47–55
Limatola E, Filosa S (1989) Exogenous vitellogenesis and micropinocytosis in the lizard, *Podarcis sicula*, treated with follicle-stimulating hormone. Gen Comp Endocrinol 75:165–176
Lu C-L, Baker RC (1986) Characteristics of egg yolk phosvitin as an antioxidant for inhibiting metal-catalyzed phospholipid oxidations. Poult Sci 65:2065–2070
Lu A, Hu X, Xue J, Zhu J, Wang Y, Zhou G (2012) Gene expression profiling in the skin of zebrafish infected with *Citrobacter freundii*. Fish Shellfish Immunol 32:273–283
Lu A, Hu X, Wang Y, Shen X, Zhu A, Shen L, Ming Q, Feng Z (2013) Comparative analysis of the acute response of zebrafish *Danio rerio* skin to two different bacterial infections. J Aquat Anim Health 25:243–251
Lubzens E, Young G, Bobe J, Cerdà J (2010) Oogenesis in teleosts: how eggs are formed. Gen Comp Endocrinol 165:367
Ma J, Wang H, Wang Y, Zhang S (2013) Endotoxin-neutralizing activity of hen egg phosvitin. Mol Immunol 53:355
Magnadóttir B, Lange S, Steinarsson A, Gudmundsdóttir S (2004) The ontogenic development of innate immune parameters of cod (*Gadus morhua* L). Comp Biochem Physiol B Biochem Mol Biol 139:217–224
Mak AS, Choi CL, Tiu SH, Hui JH, He JG, Tobe SS, Chan SM (2005) Vitellogenesis in the red crab *Charybdis feriatus*: hepatopancreas-specific expression and farnesoic acid stimulation of vitellogenin gene expression. Mol Reprod Dev 70:288–300
Matsubara T, Ohkubo N, Andoh T, Sullivan CV, Hara A (1999) Two forms of vitellogenin, yielding two distinct lipovitellins, play different roles during oocyte maturation and early development of barfin flounder, *Verasper moseri*, a marine teleost that spawns pelagic eggs. Dev Biol 213:18–32
Meusy JJ (1980) Vitellogenin, the extraovarian precursor of the protein yolk in Crustacea: a review. Reprod Nutr Dev *20*:1–21
Mommsen TP, Korsgaard B (2008) Vitellogenesis. In: Rocha MJ, Arukwe A, Kapoor BG (eds) Fish reproduction. Science Publishers, Enfield, NH
Müller W, Vergauwen J, Eens M, Blount JD (2012) Environmental effects shape the maternal transfer of carotenoids and vitamin E to the yolk. Front Zool 9:1–11
Nachappa P, Levy J, Tamborindeguy C (2012) Transcriptome analyses of *Bactericera cockerelli* adults in response to "*Candidatus Liberibacter solanacearum*" infection. Mol Gen Genomics 287:803–817
Nakamura A, Yasuda K, Adachi H, Sakurai Y, Ishii N, Goto S (1999) Vitellogenin-6 is a major carbonylated protein in aged nematode, *Caenorhabditis elegans*. Biochem Biophys Res Commun 264:580
Neaves WB (1972) The passage of extracellular tracers through the follicular epithelium of lizard ovaries. J Exp Zool 179:339–363

Ohlendorf DH, Barbarash GR, Trout A, Kent C, Banaszak LJ (1977) Lipid and polypeptide components of the crystalline yolk system from *Xenopus laevis*. J Biol Chem 252:7992
Pan ML, Bell WJ, Telfer WH (1969) Vitellogenic blood protein synthesis by insect fat body. Science 165(3891):393–394
Patiño R, Sullivan CV (2002) Ovarian follicle growth, maturation, and ovulation in teleost fish. Fish Physiol Biochem 26:57–70
Piulachs MD, Guidugli KR, Barchuk AR, Cruz J, Simoes ZLP, Belles X (2003) The vitellogenin of the honey bee, *Apis mellifera*: structural analysis of the cDNA and expression studies. Insect Biochem Mol Biol 33:459–465
Reading BJ, Hiramatsu N, Sawaguchi S, Matsubara T, Hara A, Lively MO, Sullivan CV (2009) Conserved and variant molecular and functional features of multiple egg yolk precursor proteins (vitellogenins) in white perch (*Morone americana*) and other teleosts. Mar Biotechnol 11:169–187
Reith M, Munholland J, Kelly J, Finn RN, Fyhn HJ (2001) Lipovitellins derived from two forms of vitellogenin are differentially processed during oocyte maturation in haddock (*Melanogrammus aeglefinus*). J Exp Zool 291:58–67
Retzek H, Steyrer E, Sanders EJ, Nimpf J, Schneider WJ (1992) Molecular cloning and functional characterization of chicken cathepsin D, a key enzyme for yolk formation. DNA Cell Biol 11:661–672
Romano M, Rosanova P, Anteo C, Limatola E (2004) Vertebrate yolk proteins: a review. Mol Reprod Dev 69:109–116
Romano M, Caprioli M, Ambrosini R, Rubolini D, Fasola M, Saino N (2008) Maternal allocation strategies and differential effects of yolk carotenoids on the phenotype and viability of yellow-legged gull (*Larus michahellis*) chicks in relation to sex and laying order. J Evol Biol 21:1626–1640
Rosanova P, Romano M, Marciano R, Anteo C, Limatola E (2002) Vitellogenin precursors in the liver of the oviparous lizard, *Podarcis sicula*. Mol Reprod Dev 63:349
Saino N, Bertacche V, Ferrari RP, Martinelli R, Møller AP, Stradi R (2002) Carotenoid concentration in barn swallow eggs is influenced by laying order, maternal infection and paternal ornamentation. Proc Biol Sci 269:1729–1733
Saino N, Ferrari R, Romano M, Martinelli R, Møller AP (2003) Experimental manipulation of egg carotenoids affects immunity of barn swallow nestlings. Proc Biol Sci 270:2485
Sattar Khan MA, Nakamura S, Ogawa M, Akita E, Azakami H, Kato A (2000) Bactericidal action of egg yolk phosvitin against *Escherichia coli* under thermal stress. J Agric Food Chem 48:1503–1506
Sawaguchi S, Ohkubo N, Koya Y, Matsubara T (2005) Incorporation and utilization of multiple forms of vitellogenin and their derivative yolk proteins during vitellogenesis and embryonic development in the mosquitofish, *Gambusia affinis*. Zool Sci *22*:701
Schip FDVH, Samallo J, Broos J, Ophuis J, Mojet M, Gruber M, Geert AB (1987) Nucleotide sequence of a chicken vitellogenin gene and derived amino acid sequence of the encoded yolk precursor protein. J Mol Biol 196:245–260
Selim SA, Gaafar KM, Elballal SS (2012) Influence of in-ovo administration with vitamin E and ascorbic acid on the performance of Muscovy ducks. Emirates J Food Agric 24:264–271
Selman K, Wallace RA, Cerdà J (2001) Bafilomycin A1 inhibits proteolytic cleavage and hydration but not yolk crystal disassembly or meiosis during maturation of sea bass oocytes. J Exp Zool 290:265–278
Sharrock WJ (1983) Yolk proteins of *Caenorhabditis elegans*. Dev Biol 96:182–188
Shi X, Zhang S, Pang Q (2006) Vitellogenin is a novel player in defense reactions. Fish Shellfish Immunol 20:769
Silva R, Fischer AH, Burch JB (1989) The major and minor chicken vitellogenin genes are each adjacent to partially deleted pseudogene copies of the other. Mol Cell Biol 9:3557–3562
Spieth J, Blumenthal T (1985) The *Caenorhabditis elegans* vitellogenin genefamily includes a geneencoding a distantly related protein. Mol Cell Biol 5:2495–2501

Sun X, Zhang S (2001) Purification and characterization of a putative vitellogenin from the ovary of amphioxus (*Branchiostoma belcheri tsingtaunese*). Comp Biochem Physiol B Biochem Mol Biol 129:121–127

Sun C, Zhang S (2015) Immune-relevant and antioxidant activities of vitellogenin and yolk proteins in fish. Nutrients 7:8818–8829

Sun C, Hu L, Liu S, Gao Z, Zhang S (2013) Functional analysis of domain of unknown function (DUF) 1943, DUF 1944 and von Willebrand factor type D domain (VWD) in vitellogenin2 in zebrafish. Dev Comp Immunol 41:469–476

Surai PF (2002) Natural antioxidants in avian nutrition and reproduction. Nottingham University Press, Nottingham

Tocher DR, Bendiksen EÅ, Campbell PJ, Bell JG (2008) The role of phospholipids in nutrition and metabolism of teleost fish. Aquaculture 280:21–34

Tong Z, Li LR, Zhang S (2010) Vitellogenin is an acute phase protein with bacterial-binding and inhibiting activities. Immunobiology 215:898–902

Tufail M, Takeda M (2008) Molecular characteristics of insect vitellogenins. J Insect Physiol 54:1447

Wahli W (1988) Evolution and expression of vitellogenin genes. Trends Genet 4:227–232

Wahli W, Dawid IB, Ryffel GU, Wyler T, Jaggi RB, Weber R (1979) Vitellogenin in *Xenopus laevis* is encoded in a small family of genes. Cell 16:535–549

Wallace RA (1985) Vitellogenesis and oocyte growth in nonmammalian vertebrates. Dev Biol 1:127–177

Wallace RA, Selman K (1990) Ultrastructural aspects of oogenesis and oocyte growth in fish and amphibians. J Electron Microsc Tech 16:175

Wallaert C, Babin PJ (1994) Age-related, sex-related, and seasonal changes of plasma lipoprotein concentrations in trout. J Lipid Res 35:1619–1633

Wang H, Yan T, Tan JT, Gong Z (2000) A zebrafish vitellogenin gene (*vg3*) encodes a novel vitellogenin without a phosvitin domain and may represent a primitive vertebrate vitellogenin gene. Gene 256:303–310

Wang H, Tan JT, Emelyanov A, Korzh V, Gong Z (2005) Hepatic and extrahepatic expression of vitellogenin genes in the zebrafish, *Danio rerio*. Gene 356:91–100

Wang S, Wang Y, Ma J, Ding Y, Zhang S (2011) Phosvitin plays a critical role in the immunity of zebrafish embryos via acting as a pattern recognition receptor and an antimicrobial effector. J Biol Chem 286:22653–22664

Williams VN, Reading BJ, Hiramatsu N, Amano H, Glassbrook N, Hara A, Sullivan CV (2014) Multiple vitellogenins and product yolk proteins in striped bass, *Morone saxatilis*: molecular characterization and processing during oocyte growth and maturation. Fish Physiol Biochem 40:395–415

Wu B, Liu Z, Zhou L, Ji G, Yang A (2015) Molecular cloning, expression, purification and characterization of vitellogenin in scallop *Patinopecten yessoensis* with special emphasis on its antibacterial activity. Dev Comp Immunol 49:249

Yilmaz O, Prat F, Ibañez AJ, Amano H, Koksoy S, Sullivan CV (2015) Estrogen-induced yolk precursors in European sea bass, *Dicentrarchus labrax*: Status and perspectives on multiplicity and functioning of vitellogenins. Gen Comp Endocrinol 221:16–22

Zhang S, Wang S, Li H, Li L (2011) Vitellogenin, a multivalent sensor and an antimicrobial effector. Int J Biochem Cell Biol 43:303–305

Zhang S, Dong Y, Cui P (2015) Vitellogenin is an immunocompetent molecule for mother and offspring in fish. Fish Shellfish Immunol 46:710

Ziv T, Gattegno T, Chapovetsky V, Wolf H, Barnea E, Lubzens E, Admon A (2008) Comparative proteomics of the developing fish (zebrafish and gilthead seabream) oocytes. Comp Biochem Physiol D Genomics Proteomics 3:12–35

Chapter 18
Lipids in Insect Oocytes: From the Storage Pathways to Their Multiple Functions

Leonardo L. Fruttero, Jimena Leyria, and Lilián E. Canavoso

Abstract In insect physiology, the mechanisms involved in the buildup and regulation of yolk proteins in developing oocytes have been thoroughly researched during the last three decades. Comparatively, the study of lipid metabolism in oocytes had received less attention. The importance of this issue lies in the fact that lipids make up to 40% of the dry weight of an insect egg, being the most important supply of energy for the developing embryo. Since the oocyte has a very limited capacity to synthesize lipids de novo, most of the lipids in the mature eggs arise from the circulation. The main lipid carriers in the insect circulatory system are the lipoproteins lipophorin and vitellogenin. In some species, the endocytosis of lipophorin and vitellogenin may account for about 10% of the lipids present in mature eggs. Thus, most of the lipids are transferred by a lipophorin-mediated pathway, in which the lipoprotein unloads its lipid cargo at the surface of oocytes without internalization. This chapter recapitulates the current status on lipid storage and its utilization in insect oocytes and discusses the participation of key factors including lipoproteins, transfer proteins, lipolytic enzymes, and dynamic organelles such as lipid droplets. The new findings in the field of lipophorin receptors are presented in the context of lipid accumulation during egg maturation, and the roles of lipids beyond energy source are summarized from the perspective of oogenesis and embryogenesis. Finally, prospective and fruitful areas of future research are suggested.

L.L. Fruttero
Instituto do Cerebro (InsCer). Pontifícia Universidade Católica do Rio Grande do Sul, Porto Alegre, RS, Brazil

J. Leyria • L.E. Canavoso (✉)
Departamento de Bioquímica Clínica, Centro de Investigaciones en Bioquímica Clínica e Inmunología (CIBICI-CONICET), Facultad de Ciencias Químicas, Universidad Nacional de Córdoba, Córdoba CP 5000, Argentina
e-mail: lcanavo@fcq.unc.edu.ar

M. Kloc (ed.), *Oocytes*, Results and Problems in Cell Differentiation 63,
DOI 10.1007/978-3-319-60855-6_18

Abbreviations

α-ATPase	α subunit of the F_1F_O adenosine triphosphate synthase complex
apoA-I	apolipoprotein A-I
apoLp-I	apolipophorin I
apoLp-II	apolipophorin II
ATPase	F_1F_O adenosine triphosphate synthase
β-ATPase	β subunit of the F_1F_O adenosine triphosphate synthase complex
bHLH-PAS	basic helix-loop-helix Per/Arnt/Sim
CaM	calmodulin
COX	cyclooxygenases
DAG	diacylglycerol
DGAT	acyl-CoA:diacylglycerol *O*-acyltransferase
EcR	ecdysone receptor
FAs	fatty acids
FFAs	free fatty acids
GPAT	glycerol-3-phosphate acyltransferase
HCs	hydrocarbons
HDL	high-density lipoproteins
HDLp	high-density lipophorin
IDL	intermediate-density lipoproteins
IP_2	phosphatidylinositol 4,5-biphosphate
IP_3	inositol 1,4,5-trisphosphate
JH	juvenile hormones
LA	LDLR class A ligand-binding domain
LDL	low-density lipoproteins
LDLp	low-density lipophorin
LDLR	low-density lipoprotein receptor
LOX	lipoxygenases
LPL	lipoprotein lipase
LpR	lipophorin receptor
LSD-1	lipid storage droplet 1
LSD-2	lipid storage droplet 2
LTP	lipid transfer particle
MAG	monoacylglycerol
Met	Methoprene-tolerant
mVg	microvitellogenin
PGs	prostaglandins
PLC	phospholipase C
PLs	phospholipids
SCRB15	scavenger receptor class B member 15
SREBP	sterol regulatory element-binding protein
TAG	triacylglycerol
TOR	target of rapamycin

USP	ultraspiracle
VgR	vitellogenin receptor
VHDLp	very high-density lipophorin
VLDL	very low-density lipoproteins
YPPs	yolk protein precursors
YPs	yolk polypeptides

18.1 Introduction

Insects belong to the largest class of the phylum Arthropoda in which the body, covered by an exoskeleton, is divided into head, thorax, and abdomen. The thorax carries three pairs of legs and usually one or two pairs of wings. Insects are the most diverse form of animal life on the planet with about 1 million of annotated species and 5 to 10 million of estimated living species. In many ways, insects became the most successful group of animals, being adapted to every environment where nutrients are available (Foottit and Adler 2009). Although they are usually associated with the transmission of diseases or damage of agricultural crops, many insects are highly beneficial playing important roles in nature such as plant pollination, honey and silk production, or decomposition of organic material, among others (Law and Wells 1989).

From a scientific point of view, some insects became valuable models in biological research since their full genome sequences are available, their rearing can be easily standardized in laboratory, and they are amenable for molecular manipulation, allowing the use of this group to address fundamental aspects of experimental biology (Schmidt-Ott and Lynch 2016). It is not surprising that many metabolic pathways and regulatory mechanisms involved in lipid metabolism are conserved in the fruit fly *Drosophila melanogaster* and humans. Thus, *D. melanogaster* has emerged as a useful model organism for lipid metabolism and energy homeostasis studies (Baker and Thummel 2007; Liu and Huang 2013; Ugur et al. 2016). However, insects also evolved specific mechanisms and strategies in response to their diverse behavior, source of nutrition, and mode of reproduction, making them unique individuals to understand key processes in the biology of reproduction such as egg maturation at a biochemical and molecular level.

In addition to their widespread adaptability, the success of living insect species in colonizing different environments may be attributed to the development of specialized, highly efficient modes of reproduction (Tufail et al. 2005). Like any oviparous animal, maturation of insect oocytes requires the organized recruitment of large amounts of proteins and lipids in specialized subcellular structures, the yolk bodies, and the lipid droplets (LDs). This complex material, termed yolk, also includes carbohydrates and other minor nutrients that will be used later to support embryonic development. Currently, significant progress has been made to understand the events involved in protein uptake by insect oocytes, from biosynthesis and processing of yolk protein precursors (YPP) to hormonal regulation of YPP gene expression (Raikhel 2005; Bellés and Piulachs 2015). Conversely, our knowledge on how lipids with remarkable diversity of chemical structures are stored in developing insect oocytes is still limited. In the context of insect reproductive physiology, lipids accumulated in the oocytes are critical to sustain energetic demands during embryogenesis, to provide precursors for the synthesis of new cell membranes in the developing embryo and to act as mediators in cell signaling (Fernandis and Wenk 2007). Research on few insect models provided increasing evidence about the convergence of mechanisms that ensure the buildup of lipid reserves in oocytes. However, this knowledge cannot be extrapolated to all the insect species. The relevance of endocytic lipoprotein receptors in insect oocytes, the role of lipoprotein lipases in the transfer of lipids, their intracellular transport, and the machinery implicated in lipolysis regulation during embryogenesis are some of the important issues that need to be addressed.

In this chapter, we will focus on the most recent contributions on the role of lipoprotein receptors on the uptake of lipids by insect oocytes. For readers not familiarized with insect reproduction, we will provide a short overview of the structure of ovarian follicles and compile information about lipid composition of ovarian tissue and oocytes. The chapter will also emphasize promising areas of research on lipid metabolism of insect oocytes beyond the classical function as energetic source.

18.2 Insect Ovaries and the Ovarian Follicle

The reproductive system of insect females is composed of a pair of ovaries organized into separate elongated tubes or ovarioles, which open into the oviduct. The ovariole is the fundamental unit of egg production, and its number in each ovary varies widely depending on the particular insect species.

The gross examination of an ovariole allows the distinction of the terminal filament in the apical region, the germarium, which is placed beneath the terminal filament and the vitellarium in the basal part (Büning 2006; Bellés and Piulachs 2015). There are two types of ovarioles, panoistic and meroistic (Fig. 18.1). The first one can be found in dragonflies (Odonata), stoneflies (Plecoptera), crickets (Orthoptera), and cockroaches (Blattodea) as well as in insects belonging to the basal orders. In the panoistic ovariole, the oocyte is surrounded by a monolayer of follicular cells conforming an ovarian follicle. Conversely, meroistic ovarioles present the oocytes surrounded by follicular cells and nurse cells, which synthesize and transport macromolecules and organelles to the oocytes. Meroistic ovarioles are classified into telotrophic and polytrophic, depending on their anatomical organization and the relationships between oocytes and nurse cells. In the meroistic telotrophic type of true bugs (Hemiptera), beetles (Coleoptera), snakeflies (Raphidioptera), and alderflies (Megaloptera), the nurse cells are located in the apical region of the ovariole, the trophic chamber, or tropharium and are connected

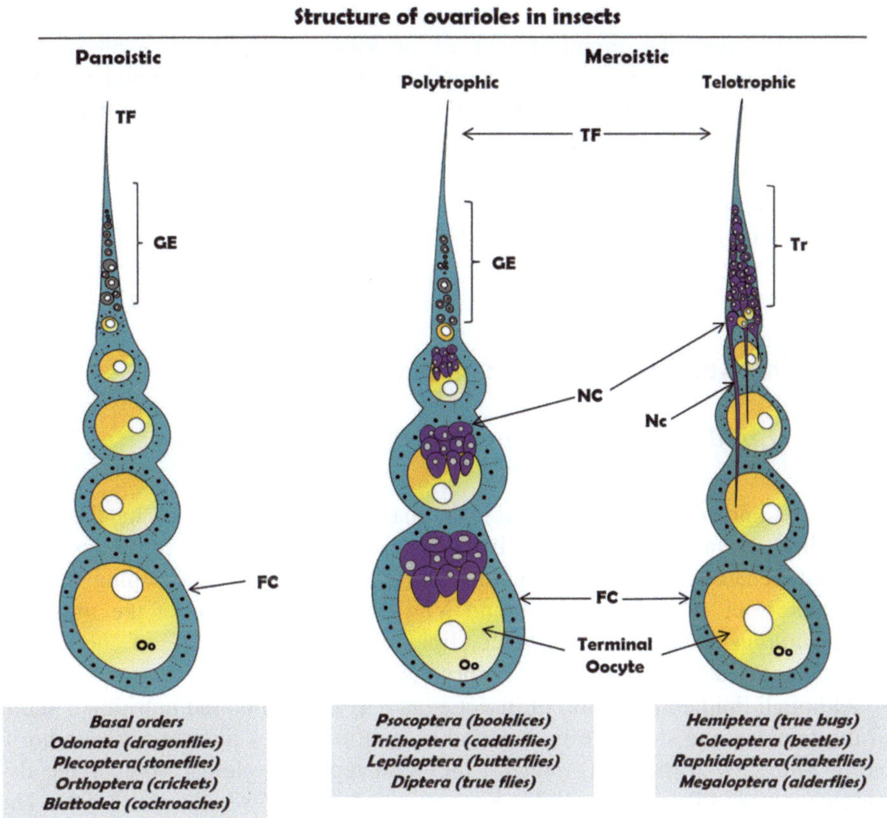

Fig. 18.1 Types of insect ovarioles. Panoistic and meroistic ovarioles. Meroistic ovarioles are subclassified into polytrophic and telotrophic. *Oo* oocyte, *FC* follicular cell, *TF* terminal filament, *GE* germarium, *NC* nurse cells, *Nc* nutritive chords, *Tr* tropharium. Insect orders associated with each type of ovariole are shown in *gray text boxes*. Original figure from the authors based on the references in the corresponding section

to early stage oocytes by trophic cords. In the meroistic polytrophic type, found in booklice (Psocoptera), bees (Hymenoptera), caddisflies (Trichoptera), butterflies (Lepidoptera), and true flies such as *D. melanogaster* (Diptera), the ovariole has a row of ovarian follicles (containing nurse cells and oocytes) (Bellés and Piulachs 2015).

Vitellogenesis is a central event in egg maturation and defines the period during which oocyte grows rapidly due to the deposition of yolk (Raikhel and Dhadialla 1992; Raikhel 2005). In insects, oogenesis and egg maturation are under endocrine regulation, and depending on the species analyzed, they are directed by juvenile hormones (JH) or ecdysteroids (Hagedorn 1985; Bellés and Piulachs 2015), although the involvement of neuropeptides in the control of oogenesis was also demonstrated (Swevers et al. 2005). The pioneering studies of V. W. Wigglesworth in the blood-sucking bug *Rhodnius prolixus* established the principles of the role of

JH in ovarian follicles, which was then extended to most insect species (Davey 2007). JH drives the rapid growth of oocytes by triggering expression of vitellogenic genes in extra-ovarian tissues and by facilitating the uptake of YPP by the oocyte, causing the formation of wide spaces in the follicular epithelium (Bellés 2005; Raikhel et al. 2005). In flies and mosquitoes, the ovaries synthesize 20-hydroxyecdysone, which in turn regulates YPP synthesis and egg maturation. However, in the dipteran *Aedes aegypti*, JH participates by preparing reproductive tissues for vitellogenesis (Wang et al. 2005).

18.3 The Buildup of Lipid Stores in Oocytes

18.3.1 Lipid Composition and Synthesis De Novo

Lipids are known to play a key role in maximizing the rates of hatched eggs (fecundity), larval growth, and survival (Angel-Dapa et al. 2010). In insects, lipids are vital because they are the main energy supply of the developing embryo (Beenakkers et al. 1981; Ziegler and Van Antwerpen 2006). In the mosquito *Culex quinquefasciatus*, for instance, lipids provide about 90% of the energy needed for the development of the embryo (Van Handel 1993).

Although lipid composition of insect eggs may depend on the nutritional status of the females, the stockpile is remarkable, and in species like the tobacco hornworm *Manduca sexta* or *A. aegypti*, they can represent almost 40% of the dry weight of eggs (Troy et al. 1975; Kawooya and Law 1988; Briegel 1990). The stores are conformed mainly by triacylglycerol (TAG), but other lipids such as phospholipids (PLs), cholesterol, or even hydrocarbons (HCs) are present (Fan et al. 2002; Ziegler and Van Antwerpen 2006). In *R. prolixus*, TAG represents about 60% of neutral lipids in the oocytes and accumulates as the cell grows (Santos et al. 2011). Currently, matrix-assisted laser desorption/ionization mass spectrometry (MALDI-MS) allows to obtaining lipid fingerprinting of a single insect oocyte for characterizing the profiles of TAG and PLs during oogenesis (Ferreira et al. 2010).

It has been demonstrated that enzymes involved in the biosynthetic pathways of TAG are active in insect oocytes (Van Antwerpen et al. 2005). However, their ability to synthesize fatty acids (FAs) de novo is very limited. In *A. aegypti* (Ziegler 1997) and *M. sexta* (Kawooya et al. 1988), two widely employed model species, only 1% of the egg lipids originate in the ovary. Therefore, the remaining lipids present in the eggs have to be imported from specialized circulating lipid carriers (Van Antwerpen et al. 2005; Ziegler and Van Antwerpen 2006). During the last years, significant progress has been achieved in *R. prolixus*, a model species broadly used to assess the characteristics of lipid metabolism. Several genes involved in TAG synthesis have been reported from the genome of this hematophagous insect (Alves-Bezerra and Gondim 2012; Alves-Bezerra et al. 2016). Thus, in oocytes of *R. prolixus*, transcriptional expression of genes that code for a glycerol-3-phosphate acyltransferase (GPAT, which catalyzes the acylation of glycerol-3-phosphate) and for diacylglycerol acyltransferase (responsible for the final esterification step of TAG synthesis) increases at early vitellogenesis and diminishes with oocyte development (Alves-Bezerra and Gondim 2012). More recently, a second GPAT gene was identified but its function remains unclear (Majerowicz et al. 2017). Transcript level of a long-chain acyl-CoA synthetase 2, the enzyme that activates long-chain FAs to enter metabolic pathways, is increased in ovaries, and gene knockdown significantly reduced female oviposition and egg hatching (Alves-Bezerra et al. 2016). Altogether, these results provide molecular evidence that formation of TAG pools in *R. prolixus* oocytes proceeds from FAs and diacylglycerol (DAG) transferred from circulating lipoproteins. Now, proteomic and metabolomic approaches are needed to validate the genomic information.

18.3.2 Lipoproteins in Lipid Delivery to Oocytes: The Role of Lipophorin and Vitellogenin

As stated previously, the bulk of lipids recruited in the oocytes must be imported from circulating lipoproteins (Canavoso et al. 2001; Van Antwerpen et al. 2005). These macromolecular complexes are conformed by a hydrophobic core of neutral lipids and an outer shell of amphipathic PLs, free cholesterol, and proteins. In mammals, the proteins present in the lipoprotein structure, apolipoproteins or apoproteins, help particle solubilization, serve as ligands for receptors, and are activators or inhibitors of enzymes involved in lipoprotein metabolism (Brown and Goldstein 1986; Van der Horst and Rodenburg 2010a; Remaley et al. 2015).

Mammals have a complex array of lipoprotein classes. The transport of bulk lipids, which in this group mainly consists of TAG, is achieved by intestine-derived chylomicrons and liver-derived very low-density lipoproteins (VLDL). Circulatory remodeling of lipoproteins yields the formation of remnant particles as well as low- and intermediate-density lipoproteins (LDL and IDL), which are ultimately taken up by receptor-mediated endocytosis and further degraded (Havel 1987; Tulenko and Sumner 2002; Goldstein and Brown 2009). Insects, on the contrary, rely on a single multifunctional lipoprotein termed lipophorin, the main lipid carrier in the hemolymph, the analog of mammalian blood. Lipophorin has two structural apolipoproteins, apolipophorin I (apoLp-I, M_r ~230–250 kDa) and apolipophorin II (apoLp-II, ~70–85 kDa). Even though the lipid composition is dynamic and varies with the species, the developmental stage, and the nutritional status, the main lipid class transported by lipophorin is DAG, although lesser amounts of sterols, FAs, carotenoids, HCs, and monoacylglycerols (MAG) can be present (Van der Horst et al. 2009). The exceptions to this rule are members of the dipteran infraorder Culicomorpha, such as mosquitoes and black flies, in which TAG is the major lipid carried by lipophorin (Pennington and Wells 2002). One distinguishing feature of lipophorin functioning is its ability to deliver different lipid classes to specific target tissues. For example, lipophorin transfers DAG to the fat body and oocytes, HCs to the cuticle, or carotenoids to the silk gland in the case of the silkworm *Bombyx mori* (Canavoso et al. 2001).

The fat body, an insect organ with functions that combine those of vertebrate liver and adipose tissue, is the site of lipophorin synthesis (Arrese and Soulages 2010). Once synthesized, lipophorin is secreted to the circulation where it performs several cycles of loading and unloading of lipids, without internalization and further degradation of its apolipoprotein matrix. Thus, Downer and Chino (1985) coined the expression "reusable shuttle" to point out the role of lipophorin in the transport of lipids between the sites of absorption, storage, and utilization. Although many studies were devoted to understand the molecular process of lipophorin-mediated lipid transfer, several key aspects of its delivery mechanism are still not completely understood. Part of the reusable shuttle mechanism implies that lipophorin has to interact with factor(s) present in the plasma membrane in order to transfer and/or take up lipids to and from target cells. Lipophorin binding sites have been described in different tissues such as midgut, fat body, and ovaries (Tsuchida and Wells 1990; Dantuma et al. 1996; Machado et al. 1996; Fruttero et al. 2009, 2011; Entringer et al. 2013). Several laboratories, including our own, have characterized the binding behavior of those sites toward lipophorin. In general, lipophorin binding sites are of protein nature, show high affinity by lipophorin and other lipoproteins, and display a pH dependence, and some of them require divalent cations to bind their ligands (Tsuchida and Wells 1990; Dantuma et al. 1996; Gondim and Wells 2000).

In most physiological conditions, lipophorin is a high-density lipoprotein (also known as HDLp), with a density of ~1.15 g/ml and a lipid content that varies between 30 and 50%. Nevertheless, under intense energy demands such as flight or oogenesis, lipophorin increases its lipid content by taking up additional DAG, generating low-density lipophorin particles (LDLp, density ~1.03 g/ml). Once fully loaded, LDLp delivers its lipid cargo to the target organs (Van der Horst and Rodenburg 2010b). Another class of lipophorin was first reported in *M. sexta* by Kawooya et al. (1988). The authors concluded that very high-density lipophorin (VHDLp) is formed when lipophorin is taken up by oocytes, deposited in yolk bodies, and stripped from its lipids by the action of a lipase, rendering a particle of higher density (~1.238 g/ml).

Vitellogenin is a phospholipoglycoprotein with a density of ~1.28 g/ml, thus falling in the range of very high-density lipoproteins. Vitellogenin is synthesized in the fat body and secreted to the hemolymph after complex steps of posttranslational modifications. Unlike lipophorin, which is present in both sexes and throughout the life span of the individual, vitellogenin is highly specific of females and is the most important YPP (Snigirevskaya and Raikhel 2005; Tufail and Takeda 2008; Tufail et al. 2014). During vitellogenesis, large amounts of vitellogenin are taken up by receptor-mediated endocytosis, stored in yolk bodies as vitellin to be digested, and used to fuel the embryo's development (Swevers et al. 2005). The estimated content of lipids present in vitellogenin ranges between 8 and 15%. It has been proposed that even though its high rate of internalization by the oocytes, the small lipid cargo of vitellogenin would represent a minor contributor to the final lipid deposits in the eggs (Ziegler and Van Antwerpen 2006). In general terms, about 5% of lipids in the oocytes can be provided by vitellogenin, whereas lipophorin takes the leading role in the process, accounting to approximately 95% of lipid reserves in mature eggs (Canavoso et al. 2001).

18.3.3 The Involvement of Lipophorin and Vitellogenin Receptors in the Accumulation of Lipids in Oocytes

Some lipophorin-binding proteins were identified and their functions described. Tsuchida and Sakudoh (2015) reported that Cameo2 and scavenger receptor class B member 15 (SCRB15), two transmembrane proteins belonging to the CD36 family, are responsible for the lipophorin binding and lipophorin-mediated transfer of specific carotenoid molecules to the silk gland. Nevertheless, neither of these putative lipophorin receptors seems to be involved in the bulk transfer of neutral lipids, and its expression in the ovarian tissue was not investigated.

On the other hand, an endocytic lipophorin receptor (LpR) was described for the first time in the fat body of *Locusta migratoria* (Dantuma et al. 1999). This receptor belongs to the family of low-density lipoprotein receptors (LDLR) (Van Hoof et al. 2005), and since its discovery, several LpRs from various insect species were reported. All members of the LDLR family are composed of five conserved structural domains: clusters of cysteine-rich repeats that constitute the ligand-binding domain, epidermal growth factor precursor homology domain that modulates the acid-dependent dissociation of ligands, an *O*-linked sugar domain with unknown function, a transmembrane domain that anchors the receptor in the plasma membrane, and a cytoplasmic domain that is responsible by the clustering of the receptor into the coated pits (Go and Mani 2012).

In agreement with the widespread function of lipophorin, LpRs are expressed in many tissues, including the ovary, testis, fat body, midgut, muscles, brain, and Malpighian tubules (Tufail and Takeda 2009). In the Madeira cockroach *Leucophaea maderae* (Tufail and Takeda 2009) and in the mosquito *A. aegypti* (Cheon et al. 2001), the ovarian isoforms of the LpR transcripts are expressed throughout the gonadotrophic cycle, increasing its levels during vitellogenesis, in coincidence with their roles in the uptake of lipids. The cockroach *Blatella germanica*, on the other hand, presents two ovarian LpR isoforms, L and S, the first one being highly expressed during the pre-vitellogenic period and diminishing during vitellogenesis. The S isoform shows a different profile: very low levels during pre-vitellogenesis, increasing slightly in mid-vitellogenesis, and diminishing again afterward (Ciudad et al. 2007). In *L. maderae*, the LpR protein was expressed in the germarium and in the oocytes, presenting cytoplasmic and cortical distributions, depending on the development stage of the oocyte (Tufail and Takeda 2009). In all the species analyzed so far, the transcripts and proteins of these LpRs were expressed in the ovarian tissue before they were required functionally (Cheon et al. 2001;

Gopalapillai et al. 2006; Ciudad et al. 2007; Tufail and Takeda 2009). Similarly, vitellogenin receptor (VgR), also a member of the LDLR family, was reported in several species or predicted from some genomes. In contrast to LpR, VgR seems to be expressed mainly in the ovary (Tufail and Takeda 2009; Roy-Zokan et al. 2015; Zhang et al. 2015; Upadhyay et al. 2016).

Up to date, evidences from several insect species demonstrated that lipophorin can be incorporated into the oocytes by endocytosis (Sun et al. 2000; Cheon et al. 2001; Atella and Shahabuddin 2002; Atella et al. 2005; Ximenes et al. 2008; Fruttero et al. 2011), thus confronting with its paradigmatic function as a reusable shuttle and leading researchers to seek for the physiological relevance of LpRs. The hematophagous triatomine *Panstrongylus megistus* served as a model for showing that lipid transfer to developing oocytes is accomplished by endocytosis of lipophorin and by the classic extracellular lipophorin shuttle mechanism (Fruttero et al. 2011). Therefore, the role of LpR in recruiting lipophorin converges with the non-endocytic mechanism of lipophorin lipid delivery in oocytes during the highly demanding period of vitellogenesis, ensuring enough lipids in the oocytes to support embryogenesis. It was also observed that, to some extent, lipids stored in the oocyte LDs originated from endocytosed vitellogenin (Fig. 18.2). Knocking down LpR in females of *B. germanica* led to a lesser lipophorin uptake by the growing oocytes, although no significant changes were observed in the fertility when compared to controls (Ciudad et al. 2007). Conversely, suppression of LpR in the tsetse fly *Glossina morsitans* led to delayed larval development (Benoit et al. 2011). In *D. melanogaster*, two lipophorin receptors termed LpR1 and LpR2 are needed for the accumulation of neutral lipids in oocytes and imaginal disc cells. Although the absence of those genes did not affect the viability and body lipid content of the insects, the eggs of females lacking the *lpr2* gene presented lower lipid content and reduced fertility, thus suggesting a relevant role for LpR2 during vitellogenesis (Parra-Peralbo and Culi 2011). Most LDLR members, including LpR, mediate endocytosis, but in *D. melanogaster*, LpR2 seems to stabilize lipophorin extracellularly in the proximity of the plasma membrane of the oocyte, facilitating lipid transfer from the lipoprotein to the cell (Parra-Peralbo and Culi 2011). LpR may also play a significant functional role for minor lipid components carried by lipophorin that could not reach the oocyte by aqueous diffusion through the cell membrane, so they must necessarily be taken up by endocytosis, as suggested for the fat body of *L. migratoria* (Dantuma et al. 1997).

The F_1F_O adenosine triphosphate synthase (ATPase) is a multiprotein complex whose most documented function is to catalyze the synthesis of ATP from ADP and Pi employing the chemiosmotic proton force to drive this conversion. The ATPase complex is composed by a F_O region embedded in the membrane and a soluble F_1 region that is formed by three copies of the β subunit (β-ATPase) and three of the α subunit (α-ATPase), among others. The catalytic sites are located in the F_1 portion. If the gradient is inverted, the ATP synthase complex will also perform ATP hydrolysis to fuel the transport of protons to the other side of the membrane (Walker 1998; Meier et al. 2011). ATPase is located in the inner membrane of mitochondria, in the thylakoid membrane of chloroplasts, and in the plasma membrane of bacteria

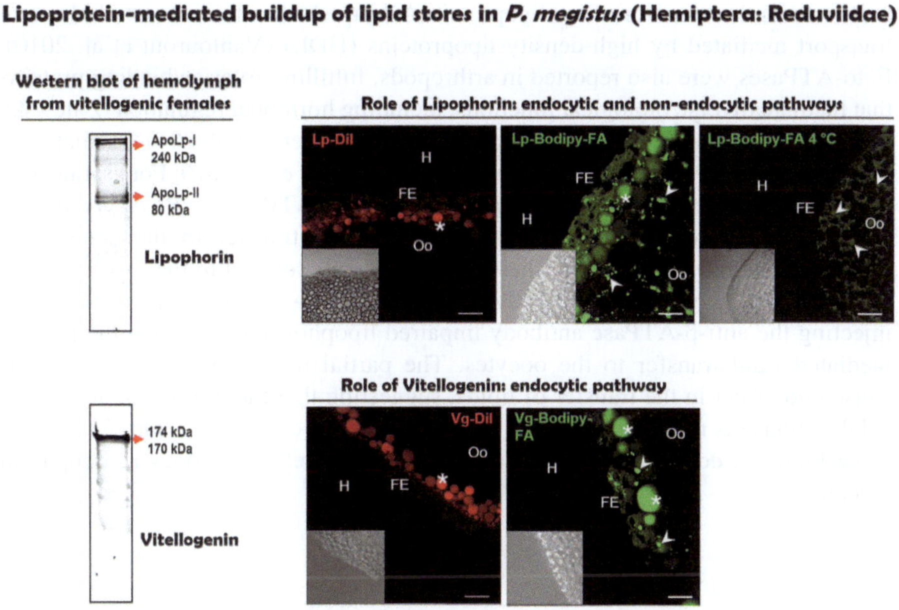

Fig. 18.2 Role of lipophorin (Lp) and vitellogenin (Vg) in lipid transfer to developing oocytes in *Panstrongylus megistus* (Hemiptera, Reduviidae). *Left panel*: western blots of hemolymph from vitellogenic females at third day after blood feeding. Polyclonal anti-Lp and anti-Vg antibodies detected apolipophorin I (apoLp-I, 240 kDa) and apolipophorin II (apoLp-II, 80 kDa) present in Lp and the two major subunits of Vg (174 and 170 kDa) (*red arrows*). *Right panel*: the pathways of Lp and Vg in the storage of lipids in developing oocytes. Females were injected with Lp and Vg labeled with different fluorophores: DiI (1,1-dioctadecyl-3,3,3′,3′-tetramethylindocarbocyanine) which follows the pathway of the entire particles (Lp-DiI or Vg-DiI) and with Bodipy-FA (4,4-difluoro-5,7-dimethyl-4-bora-3a,4a-diaza-s-indacene-3-hexadecanoic acid), employed to follow the fate of exchangeable lipids in the lipoproteins (Lp-Bodipy-FA or Vg-Bodipy-FA). Ovaries dissected out after injections were processed for confocal laser microscopy. *Upper panel*: fluorescence patterns for Lp-DiI (*red*) and Lp-Bodipy-FA (*green*). Whereas Lp-DiI was found in yolk bodies (*asterisk*), the signal for Bodipy-FA was distributed in yolk bodies (as Lp-Bodipy-FA) and lipid droplets (*asterisk* and *arrowheads*, respectively). In insects maintained at 4 °C (control for impaired endocytosis), Bodipy-FA was found exclusively in lipid droplets (*arrowheads*). *Lower panel*: fluorescence patterns of Vg-DiI and Vg-Bodipy-FA. The injection experiments showed that Vg-DiI entered the oocyte to be stored in yolk bodies. In addition, the label of Vg-Bodipy-FA was found in yolk bodies as vitellin-Bodipy-FA (Vt-Bodipy-FA, *asterisks*) as well as in lipid droplets (as Bodipy-FA, *arrowheads*). The insert shows the corresponding DIC images. *FE* follicular epithelium, *Oo* oocyte, *H* hemolymph. Bars, 25 μm. Reproduced in part from Fruttero et al. (2011), with permission

(Arechaga and Jones 2001; Andries et al. 2005). Reports from the last two decades have placed the ATPase complex, or at least some of its subunits, in the plasma membrane. In particular, the subunit β of the complex located on the surface, also known as ecto-β-ATPase, displays biological roles related to cholesterol homeostasis, endothelial cell proliferation, and tumor recognition (Chi and Pizzo 2006). The most thoroughly described biological function of ecto-β-ATPase in mammals is its role as the receptor of apolipoprotein A-I (apoA-I) in the reverse cholesterol transport mediated by high-density lipoproteins (HDL) (Vantourout et al. 2010). Ecto-ATPases were also reported in arthropods, fulfilling roles as binding proteins that mediate diverse biological functions, including hormonal regulation (Zalewska et al. 2009), phytoplasma transmission specificity (Galetto et al. 2011), hematopoiesis (Lin et al. 2009), and virus acquisition (Paingankar et al. 2010; Fongsaran et al. 2014). More recently, it was reported that ecto-β-ATPase in *P. megistus* was involved in bidirectional lipophorin-mediated lipid transfer to the enterocytes (Fruttero et al. 2014). Later on, these findings were extended to the ovaries of the same species (Fruttero et al. 2017). Therefore, in vivo blocking experiments injecting the anti-β-ATPase antibody impaired lipophorin binding and lipophorin-mediated lipid transfer to the oocytes. The partial inhibition of ATP synthesis caused no effect in the transfer of lipids, suggesting that the enzymatic activity of ATP synthase is not related to lipid transfer (Fruttero et al. 2017). On the basis of these findings, ecto-β-ATPase has emerged as a novel non-endocytic lipophorin receptor.

18.3.4 Other Factors Involved in Lipoprotein-Mediated Lipid Delivery to Insect Oocytes

Circulating lipophorin and their plasma membrane receptors are not the only factors involved in lipid transfer to insect oocytes. Several lines of evidence indicated the participation of transfer proteins and lipolytic enzymes in both the lipoprotein-mediated lipid transfer to the oocyte and the intracellular interconversion of lipoproteins. In this context, a very high-density lipoprotein (density ~1.23 g/ml) called lipid transfer particle (LTP) was reported for the first time in the hemolymph of larval *M. sexta* (Ryan et al. 1986). The LTP protein composition is unknown in most insect species, but in *B. mori*, the cDNA and deduced amino acid sequences showed that three LTP apoproteins are encoded by two genes, *apoLTP-II/I* and *apoLTP-III*. The *apoLTP-II/I* gene is a novel member of the apoB/large lipid transfer protein gene family (Yokoyama et al. 2013). Although the physiological function of LTP is not completely understood, gathered data point out a catalytic activity for LTP, mediating the transfer of lipids from tissues to lipophorin or vice versa (Van Heusden and Law 1989; Ryan and Van der Horst 2000; Canavoso and Wells 2001; Canavoso et al. 2004). There are also reports indicating a relevant role for LTP in the accumulation of lipids in the oocyte. A factor present in homogenates of ovarian tissue from *M. sexta* catalyzed the conversion of HDLp to VHDLp, and this reaction was abolished by an anti-LTP antibody (Liu and Ryan 1991). Moreover, the injection of such an antibody in *B. mori* pupae led to a 75% diminution in the content of lipids accumulated by the oocytes (Jouni et al. 2003). Recent reports in *D. melanogaster* established LTP as a ligand of a subset of LpR isoforms that contain a specific LDLR class A (LA) ligand-binding domain (Rodríguez-Vázquez et al. 2015). Surprisingly, LTP presented a higher affinity for LpR than lipophorin, and the authors showed that the LpR-LTP binding process was lipophorin-independent. From a physiological perspective, LpR promotes the recruitment of LTP to the plasma membrane of target cells to bind and stabilize lipophorin, thus facilitating the transfer of lipids. A combination of molecular and biochemical approaches demonstrated that LTP is required for the accumulation of lipids by vitellogenic follicles of *D. melanogaster* because low circulating LTP levels led to diminished lipid content in ovaries. Even though more experiments are needed to unravel the molecular mechanism involved in LTP-mediated lipid transfer, the authors favor a possibility in which a ternary complex is transiently formed. Once the complex LTP-LpR-lipophorin is made, lipids are transferred from lipophorin to LTP with the ensuing detachment of lipophorin. Later on, lipids are transferred from LTP to the oocyte plasma membrane and from there to the cytoplasm (Rodríguez-Vázquez et al. 2015).

As stated previously, neutral lipids transported by mammalian lipoproteins are hydrolyzed, while they are circulating through the blood vessels, by a lipoprotein lipase (LPL), which is attached to the luminal side of the endothelial cells by heparan sulfated proteoglycans (Van der Horst and Rodenburg 2010a). LPLs are also present in insect tissues such as flight muscles, where they play a pivotal role during TAG-fueled long flights. In ovaries, a membrane-bound and a soluble LTP were reported, each performing different roles (Ziegler and Van Antwerpen 2006). Similar to the vertebrate LPL, the ovarian membrane-bound LPL is anchored to the glycosaminoglycans of the extracellular space of follicles, as demonstrated by Telfer (1979) and Telfer et al. (1982). This LPL is responsible for the hydrolysis of the DAG cargo of lipophorin, generating FAs that are subsequently taken up by the oocyte. LPL presents a higher affinity to LDLp than to HDLp, but since HDLp concentration in the hemolymph is higher than LDLp, its contribution to the buildup of oocyte lipid reserves could be still important (Van Antwerpen et al. 1998). The soluble LPL was purified from the oocytes and eggs of *M. sexta*, and it is unclear whether this LPL is the same as the membrane-bound LPL, since they displayed some differences in their enzymatic properties (Van Antwerpen and Law 1992). The function of this soluble LPL may be related to the conversion of endocytosed HDLp to VHDLp and to the delipidation of endocytosed vitellogenin, producing lipid-depleted vitellin (Chino et al. 1977; Chinzei et al. 1981; Kawooya et al. 1988; Van Antwerpen and Law 1992).

18.4 Lipid Droplets in Insect Oocytes

The intracellular storage of lipids occurs in specialized structures called LDs, recently acknowledged as organelles, which are present in most organisms and cell types. The LDs contain a hydrophobic core of neutral lipids, mainly TAG, and a hydrophilic monolayer surface of PLs, free cholesterol, and proteins that segregates its core from the aqueous cytosol (Thiam et al. 2013; Arrese et al. 2014).

Besides being actively involved in lipid storage and lipid homeostasis (Ducharme and Bickel 2008), LDs are highly dynamic organelles (Suzuki et al. 2011). Biosynthesis and expansion of LDs are driven by an integrated mechanism, which involves interactions with other organelles and the recruitment of enzymes necessary for the expansion of the lipid core and for the modulation of the phospholipid monolayer composition (Yang et al. 2012; Barneda and Christian 2017). The fruit fly *D. melanogaster* represents a biological system that has revealed a wide spectrum of LDs functions (Kühnlein 2011). However, aside from research performed in this and other model species, the dynamic of LDs in insect oocytes has been scarcely studied.

LDs purified from whole ovarian preparations of *M. sexta* contained high quantities of apoLp-I and apoLp-II. Taking into account that lipophorin transports mainly DAG and that TAG is the major lipid stored in the LDs in insect oocytes, the association of apolipophorins with LDs may facilitate the conversion of DAG into TAG and/or the subsequent transference of TAG to the core of LDs (Soulages et al. 2012). Vitellogenin and microvitellogenin (mVg), both having lipid-binding properties, were also found associated with the LDs in oocytes of *M. sexta* and have been linked to structural functions (Soulages et al. 2012).

In the moth, *Hyalophora cecropia*, the cortical cytoplasm of the oocyte is the site of LD assembly (Wiemerslage 1976), whereas in *D. melanogaster*, it was reported that during oogenesis, LDs originate in nurse cells and are then transferred to oocytes *via* actin-mediated cytoplasmic streaming (Welte 2015). The biochemical pathway for LDs formation in nurse cells is not well characterized, but it has been shown that flies express two acyl-CoA:diacylglycerol *O*-acyltransferase enzymes, DGAT1 and DGAT2, which catalyze the final step of TAG biosynthesis and mediate LDs size control (Welte 2015).

In *D. melanogaster*, LDs sequester histones allowing embryos to build up extranuclear histone stores for chromatin assembly during the high demanding period of embryogenesis (Li et al. 2012). Interestingly, proteomics of LDs from *M. sexta* ovaries revealed the presence of histones but not lipogenic enzymes (Soulages et al. 2012).

The surface of LDs contains proteins of the PAT family, which in mammals is composed for five members collectively named perilipins. These proteins interact with lipases and regulate lipid homeostasis in response to metabolic stimuli (Kory et al. 2016). In insects, only two perilipins were reported, PLIN1/LSD-1 and PLIN2/LSD-2 (Bickel et al. 2009). These proteins do not have a known enzymatic activity; however, it was reported that LSD-1 phosphorylation is linked to lipolysis control, whereas phosphorylation of LSD-2 seems to promote neutral lipid storage (Bi et al. 2012). *M. sexta* ovaries showed low levels of LSD-1, consistent with the notion that ovaries and developing oocytes are rather accumulating than mobilizing lipids (Soulages et al. 2012).

Currently, the reports aimed to understand the dynamics of LDs during the reproductive cycle of insects are scarce. The first studies were carried out in pre-vitellogenic ovaries of *R. prolixus*, where relatively few LDs were found in the pre-follicular tissue (Huebner and Anderson 1972a). In a closely related species,

Dipetalogaster maxima, a scarce amount of lipids was detected in the ovaries at the pre-vitellogenic stage. They were mainly observed as small LDs at the periphery of the oocyte (Leyria et al. 2014). Clifton and Noriega (2012) reported that pre-vitellogenic ovaries of *A. aegypti* fed with high sucrose concentrations contained higher amounts of neutral lipids than those fed with lower ones, probably to support the energetic demands of endocytosis and protein synthesis in growing oocytes at the beginning of vitellogenesis. In the mosquito *C. quinquefasciatus*, the accumulation of LDs in vitellogenic oocytes was observed concomitantly with the formation of yolk bodies (Cardoso et al. 2010). In vitellogenic oocytes of the triatomines *R. prolixus*, *P. megistus*, and *D. maxima*, LDs were abundant, large, and homogeneously distributed, occupying the spaces left by yolk bodies, in a pattern expected for a period of high rate of lipid accumulation (Huebner and Anderson 1972b; Fruttero et al. 2011; Leyria et al. 2014). In follicular atresia induced by depriving *D. maxima* females of blood meal during post-vitellogenesis, oocyte LDs became remarkably smaller, and high amounts of free FAs (FFAs) were detected, suggesting the lipolysis of lipid stores. At early atresia, the small LDs were located at the periphery of most of the developed oocytes, although large LDs were observed in the tropharium of late atretic ovarioles (Leyria et al. 2014). Morphology and distribution of LDs in terminal oocytes of *D. maxima* are shown in Fig. 18.3. In the endoparasitoid *Pteromalus puparum*, the transition from vitellogenesis to oocyte resorption proceeded with a decrease in size of LDs, which were also found at the periphery of the oocytes. Such an organization of the lipid store was attributed to a transfer of the ovarian nutrients to the hemolymph, aimed to provide resources for enhancing the survival of the insect (Guo et al. 2011).

18.5 Hormonal Signaling During Accumulation of Lipids in Oocytes

The ecdysone receptor (EcR) is a nuclear receptor found in arthropods, which controls development and participates in reproduction. The receptor is a non-covalent heterodimer of two proteins, EcR and ultraspiracle (USP). EcR is activated upon binding ecdysteroids (Riddiford et al. 2000), mediating oocyte lipid accumulation mainly by promoting the expression and activation of SREBP (Sieber and Spradling 2015). In mammals, SREBP is a helix-loop-helix transcription factor that is retained in the endoplasmic reticulum in the presence of high levels of intracellular lipids. Once lipid levels drop, SREBP is transported to the Golgi apparatus, cleaved by proteases, and its DNA binding domain moves to the nucleus to induce the expression of many genes involved in lipid synthesis and uptake, including the LDL receptor (Shao and Espenshade 2012). It was reported that SREBP controls lipid uptake and accumulation in the *D. melanogaster* germline, in part, by regulating the LpR expression (Sieber and Spradling 2015).

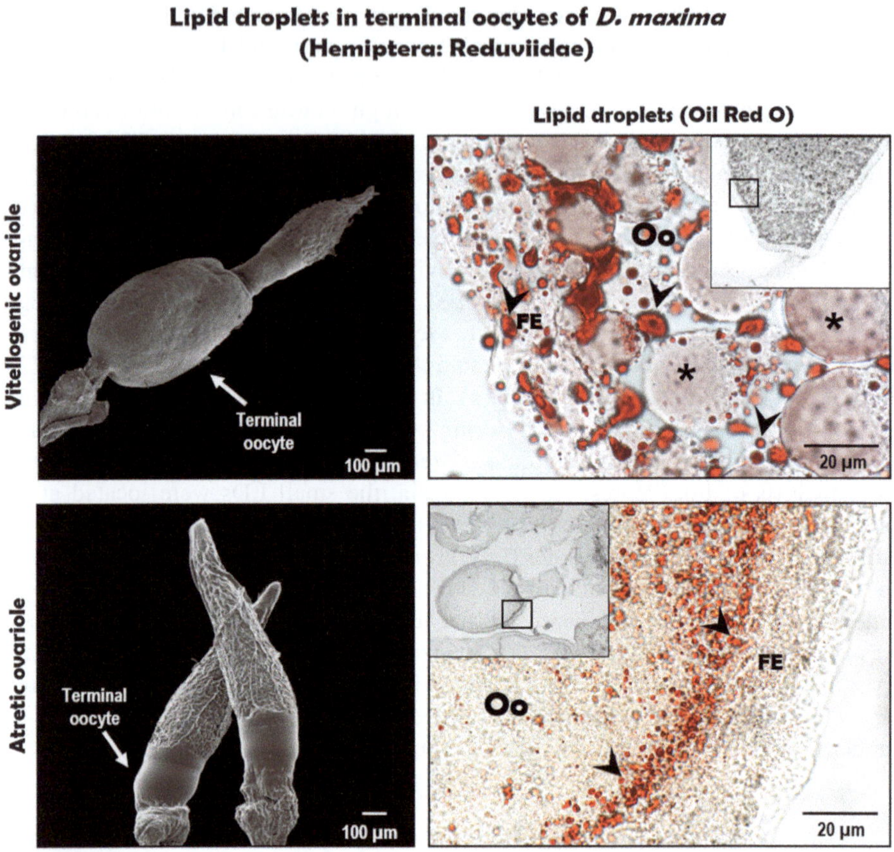

Fig. 18.3 Morphology and distribution of lipid droplets (LDs) in the ovarioles of *Dipetalogaster maxima* (Hemiptera, Reduviidae). Scanning electron micrographs and tissue sections of ovarioles at two stages of the reproductive cycle stained with Oil Red O. *Upper panel*: vitellogenic oocyte at fifth day after blood feeding. Tissue section of a terminal oocyte exhibiting numerous, large, and irregularly shaped LDs (*arrowheads*). LDs are also observed in the follicular epithelium (*asterisks*). *Lower panel*: atretic ovariole at day 10–12 post-vitellogenesis. Tissue section of a terminal oocyte showing small LDs, distributed mainly at the cortical area (*arrowheads*). *Oo* oocytes, *FC* follicular epithelium. Reproduced in part from Leyria et al. (2014), with permission

JH, a sesquiterpenoid hormone synthesized and secreted by the corpora allata, governs important aspects of the insect's life history by preventing metamorphosis and directing reproductive maturation in females, as described above (Raikhel 2005; Riddiford 2008). The molecular mechanisms of JH action began to be elucidated when Methoprene-tolerant (Met), a transcription factor belonging to a basic helix-loop-helix Per/Arnt/Sim (bHLH-PAS) family, was discovered as the JH receptor. Thereafter, several factors orchestrating JH signaling were identified (Wilson and Fabian 1986; Miura et al. 2005; Jindra et al. 2013, 2015). It is known that JH promotes the accumulation of lipids in the oocytes by stimulating the synthesis and release of lipid carrier proteins, mainly vitellogenin, from the fat body (Wyatt et al. 1996; Engelmann and Mala 2000; Lu et al. 2016) and by inducing large gaps or patency in the follicular epithelium, allowing insect lipoproteins to have a better access to the oocyte surface (Raikhel and Dhadialla 1992). In addition, JH seems to stimulate the expression of VgR at the oocyte plasma membrane although its effect on LpR has been documented only in few species (Chen et al. 2004; Lu et al. 2015). In *B. germanica*, adult topical treatment of females with JH III showed that the expression of LpR is posttranscriptionally regulated by this hormone (Ciudad et al. 2007). Currently, the limited status of knowledge on endocrine regulation of LpR makes it impossible to draw general conclusions, and future studies are needed to look deeper into this issue.

18.6 The Functions of Lipids During Oogenesis and Embryogenesis

Lipids imported from circulation or synthesized de novo in mature oocytes are committed to many functions to sustain the development of the embryo including promotion of metabolic processes, plasticity of cell boundaries, intracellular signaling, and cuticle formation (Atella et al. 2005).

During embryogenesis, the main lipid stored in mature insect eggs, TAG, must be broken down by lipases, to originate FFAs and glycerol. FFAs will be driven to β-oxidation to fuel the energetic demands of the embryo and to serve as metabolic intermediates for the synthesis of new embryonic membranes. The biochemical and molecular machinery devoted to TAG cleavage in insect mature eggs during embryogenesis has received little attention. Lipases, generally defined as triacylglycerol hydrolases (EC 3.1.1.3), act at the surface of emulsified lipid substrates by cleaving the carboxylester linkages in TAG, DAG, galactolipids, and PLs (Nomura and Casida 2016). A comparative study of lipase families based on the genome of four holometabolous insect orders revealed that, in higher Diptera, a large lineage of neutral lipases without catalytic activity is recruited as yolk polypeptides (YPs) in oocytes (in higher Diptera, yolk protein precursors are YPs instead of vitellogenin) (Horne et al. 2009). On the other hand, in the moth *Plodia interpunctella*, the predominant yolk proteins are acid lipase-like proteins, which are synthesized in ovarian follicle cells and may be involved in mobilization of lipids from follicle cells to the oocyte during oogenesis (Shirk and Perera 1998). Pistillo et al. (1998) characterized three genes from *D. melanogaster* whose translational products show a high level of similarity with mammalian lipases but lack significant homology with neutral lipases. It was proposed that *lip 1* gene, which codes for a protein similar to digestive lipases, is expressed in ovaries and early embryos and may be involved in lipid mobilization (Pistillo et al. 1998).

During embryogenesis, cells undergo proliferation, cell death, and differentiation allowing the transformation of the embryo into an adult organism (Singh and

Cuervo 2011). In particular, autophagy has been recently proposed as a crucial mechanism for the cellular remodeling that occurs during the development of multicellular organisms (Di Bartolomeo et al. 2010). During insect embryogenesis, a constant supply of nutrients is required to provide the energy necessary for growth, metabolism, and survival, and autophagy plays an important role in this process. Autophagy, also termed macroautophagy, represents a complex process involving the formation of structures surrounded by double membranes or autophagosomes that engulf cytoplasmatic contents and deliver them to lysosomes for degradation (Singh and Cuervo 2011). Many genes that regulate autophagy, known as *Atg*, are evolutionarily conserved from yeast to invertebrates and mammals. Specific *Atg* genes, together with their regulators, control different stages of autophagosome formation and maturation. The protein kinase target of rapamycin (TOR) is a crucial upstream regulator of autophagy. The contribution of autophagy to lipid metabolism has been recently elucidated through the initial discovery of a process known as macrolipophagy (Singh et al. 2009). This process occurs by de novo formation of a limiting membrane that sequesters LDs, forming an autophagosome, which then fuses with a lysosome to degrade the lipids from LDs. In addition to mobilization of TAG from LDs to generate FFAs for mitochondrial oxidation and energy production by cytosolic or ER-associated lipases, intracellular lipids can also reach the lysosomal lumen through sequestration inside autophagosomes (Singh et al. 2009). During *D. melanogaster* oogenesis, the autophagic process participates in programmed cell death in the ovary at two stages during oogenesis: the germarium and mid-oogenesis (McCall 2004). These stages represent two nutrient status checkpoints where egg chambers and/or the environment are monitored before investing energy into egg production (Drummond-Barbosa and Spradling 2001). In the hematophagous *D. maxima*, deprivation of blood meal during post-vitellogenesis triggers degeneration of some follicles into an atretic stage and leads to oocyte resorption (Aguirre et al. 2011). In this species, deprivation of nutrients enhances autophagy and promotes a decrease in the amount of ovarian lipid stores at late atretic stages (Aguirre et al. 2013; Leyria et al. 2014). Taking into account that *D. maxima*, like all triatomines, is an obligatory blood feeder throughout the whole life cycle, autophagy at late atresia may render the ovaries less competent to accomplish a new oviposition cycle, prioritizing the utilization of resources to increase female life span.

DAG functions not only as a building block for the synthesis of more complex lipid but also as a source of FFAs. DAG is a neutral glycerolipid with distinctive functions as a basic component of membranes, as intermediate in lipid metabolism and as a key element in lipid-mediated signaling (Carrasco and Mérida 2007). The relationship between DAG-dependent metabolism and signaling functions is essential for the maintenance of homeostasis during cell growth and development. This interconnection impacts on the appearance of the several mechanisms that control DAG levels in eukaryotic cells. Among them, members of the protein kinase C (PKC) family were described as the main effectors of cellular DAG-mediated signaling. In the context of insect vitellogenesis, studies in the milkweed bug *Oncopeltus fasciatus* showed that calcium-binding protein calmodulin (CaM, ~17 kDa) passes through gap junctions from the follicle cells, where it is synthesized, to the oocytes promoting active endocytosis of yolk proteins (Brooks and Woodruff 2004). In this species as well as in *D. melanogaster*, the signaling cascade is initiated when CaM activates membrane-bound phospholipase C (PLC) to transform phosphatidylinositol 4,5-biphosphate (IP_2) into inositol 1,4,5-trisphosphate (IP_3) and DAG. IP_3 triggers the release of Ca^{+2} from the internal pool, which along with DAG stimulates PKC. PKC is necessary for endocytic uptake of yolk proteins into new yolk bodies as well as for its clustering and fusion into mature yolk bodies (Brown et al. 2010).

FFAs, derived from lipid stores in oocytes, supply the embryo with most of the energy required for development. On the other hand, the PLs, imparting flexibility on cell membranes and membrane-surrounding organelles, are precursors of cell signaling molecules and act as biological lubricants. In neonate larvae of *A. aegypti* and *Anopheles gambiae* mosquitoes, FFAs and PLs from the mother, which were accumulated in the oocyte during vitellogenesis, are differentially segregated to supply critical functions. In this sense, it was proposed that FFAs distributed along the side of the body of larval neonates provide energy for their initial movements when seeking for food, whereas most maternal PLs may emulsify the ingested food facilitating its assimilation (Atella and Shahabuddin 2002).

Cholesterol is the major sterol found in insects that functions as a structural component of cell membranes and as a precursor for steroid hormones (Canavoso et al. 2001). Insects do not synthesize cholesterol, and most species are adapted to transform dietary sterols into cholesterol (Canavoso et al. 2001). Ecdysteroids, which are synthesized from cholesterol in the prothoracic gland, play a major role in regulating molting and metamorphosis (Niwa and Niwa 2014). However, in the ovaries and eggs, ecdysteroids are also important for early embryonic development. In the ovary of many insect species, ecdysteroids synthesized de novo in follicle cells are stored as inactive phosphoric esters (Lagueux and Hoffmann 1984). Studies performed in *B. mori* helped to unveil the dynamics of the process. During ovarian development of this species, most of ecdysteroids synthesized by ecdysteroid hydroxylases in follicle cells are transferred to oocytes and phosphorylated by an ecdysteroid 22-kinase in the ooplasm, generating inactive ecdysteroid phosphates (Ito et al. 2008). Ecdysteroid phosphates bind to vitellin to form a complex that will be stored in yolk bodies. The action of ecdysteroid phosphate phosphatase is required for dephosphorylation when free ecdysteroids are needed in early embryo development (Yamada and Sonobe 2003).

HCs, mainly normal and methyl-branched alkanes and alkenes, are the major cuticular lipids in insects. Their hydrophobic nature contributes to water retention, thus preventing insects from lethal desiccation. Cuticular HCs also alter the penetration of chemicals and bring protection against microorganism invasion or abrasive injuries (Howard and Blomquist 2005). In the insect species analyzed so far, large amounts of HCs are stored in oocytes during vitellogenesis, since embryos and newborn individuals need extensive waterproofing. HCs stored in the developing basal oocytes are not metabolized by the embryos but are retained in the hemocoel and cuticle of the first two nymphal instars of *B. germanica* (Fan et al. 2003). Moreover, all ovarian HCs are transferred to the egg case, demonstrating that HCs were localized within the oocytes and not in follicle cells. Embryos of the viviparous *Diploptera punctata* display a prominent waxy coating from the day of their oviposition into the brood sac, which consists of HCs (>80%), long-chain alcohols, wax esters, and aldehydes (Nelson et al. 2004). In this species, accumulated HCs are mainly of maternal origin since the embryos synthesized HCs at very low rates. Studies in the cockroaches *Periplaneta americana* and *B. germanica* showed that HCs synthesis is restricted to oenocytes, which are large cells of the integument associated with abdominal epidermis (Fan et al. 2003). Oenocytes contain abundant smooth endoplasmic reticulum suggestive of a role in the synthesis, processing, and/or secretion of lipids and express a battery of lipid-synthesizing and catabolizing enzymes and other proteins such as lipophorin receptors (Makki et al. 2014). In the blood-sucking *Triatoma infestans*, unlaid eggs or oviposited eggs metabolize radiolabeled propionate, a specific precursor of methyl-branched chains, into methyl-branched HCs implying that biosynthesis of HCs takes place in the eggs (Juárez 1994). Transport of HCs through the hemolymph is accomplished by lipophorin, and in many insects, this lipid class may constitute up to 50% of the lipophorin lipid cargo (Sevala et al. 1999). LTP also carries HCs (Blacklock and Ryan 1994), and in *P. americana*, they comprised about 40% of its lipid content (Takeuchi and Chino 1993). The mechanism underlying HCs exchange either between the integument and lipophorin or between lipophorin and the ovary remains largely unknown. Whereas LTP does not seem to play a relevant role in such a process (Fan et al. 2002), contributions from LpR and endocytic pathway of lipophorin need to be examined further.

Eicosanoids are a family of local lipid mediators synthesized from polyunsaturated FAs, mainly arachidonic acid, which is released from membrane phospholipids via activation of a phospholipase A_2 (Harizi et al. 2008). Upon releasing, arachidonic acid follows different enzymatic oxygenation pathways involving cyclooxygenases (COX) to generate prostaglandins (PGs) and thromboxanes and lipoxygenases (LOX) yielding lipoxins and leukotrienes (Lone and Tasken 2013). Eicosanoid and PGs were intensively studied in the context of human health and physiology because they exert important biological actions in almost all tissues and body fluids (Stanley et al. 2002; Lone and Tasken 2013). In insects, eicosanoids have been implicated in the regulation of immunity (Stanley et al. 2012), host-parasite relationships (Stanley et al. 2002; Stanley 2006), ion transport physiology (Stanley and Kim 2011), and reproduction (Atella et al. 2005). In the context of reproductive physiology, several reports indicated that PGs affect the oviposition behavior of some but not all species (Atella et al. 2005). In *R. prolixus*, it was demonstrated that products of the COX pathway led to a decrease in the yolk protein uptake by developing oocytes (Medeiros et al. 2002), while products of the LOX pathway acted as positive effectors of yolk uptake (Medeiros et al. 2004), supporting a role for eicosanoids in the regulation of oogenesis. It was also reported by Medeiros et al. (2004) that the inhibition of yolk uptake induced by COX products was accompanied by an increase in the ovarian levels of cAMP. In *H. cecropia*, the addition of membrane-permeable analogs of cAMP or cGMP to follicles inhibited vitellogenin uptake, an effect that was interpreted as a signal for triggering the termination of vitellogenesis in the follicles (Wang and Telfer 1996). In vertebrates, eicosanoids bind to G-protein-coupled receptors, thus regulating intracellular concentrations of cAMP (Negishi et al. 1995). The relatively few studies in invertebrates seem to confirm the involvement of cyclic nucleotides in eicosanoid-mediated modulation (Stanley 2006; Machado et al. 2007). More recently, it was demonstrated in *D. melanogaster* that PGs are implicated in eggshell formation (Tootle et al. 2011) and also in the regulation of cytoskeleton remodeling during oogenesis (Spracklen et al. 2014).

18.7 Concluding Remarks

In this chapter, we have presented an overview on lipid storage and functions in insect oocytes. A large body of evidence assigned a leading role to lipophorin and vitellogenin as the main lipid donors to insect oocytes. Whereas the lipophorin shuttle mechanism for transferring lipids accounts for most of lipid recruited in the oocytes, modern approaches showed that endocytosis of lipophorin plays an important, yet not restrictive, role in such a process. Research on the subject has to cope with fast growing advances, which give insight into the role of old actors such as LTP, in lipophorin-mediated lipid transfer to oocytes.

Our intention in placing together the major pieces involved in the recruitment of lipid in insect oocytes is presented as an integrated model depicted in Fig. 18.4. Most molecular participants in the model arise from findings in a few insect species, usually those of easy rear and handle in the laboratory. Therefore, caution should be exercised in drawing general conclusions. Since insects display a complex array of mechanisms and strategies in response to their diverse behavioral patterns, this model surely needs to be adjusted to fit the lipid metabolism in oocytes of a particular species. The availability of new experimental tools offers the way to match genomic, proteomic, and metabolomic information to succeed in this challenge.

The model points out many issues currently unknown on lipid storage in insect oocytes that need to be addressed. Up to date, several LpRs and VgRs have been characterized at a molecular level. However, how nutritional and/or endocrine regulation may be interconnected to govern the physiological functioning of LpRs remains largely unanswered. More efforts should be devoted to identify the roles of proteins that might take part in the transmembrane lipid transfer and in the lipid trafficking in oocyte cytoplasm. The dynamics between biosynthesis and remodeling of LDs in response to metabolic requirements of insect oocytes is undoubtedly an area of fruitful research.

Lipids in insect oocytes play key roles that span from energy supply for developing embryos to their role as cell signaling mediators. Such remarkable physiological functions, summarized in Fig. 18.5, rely in part on their diverse

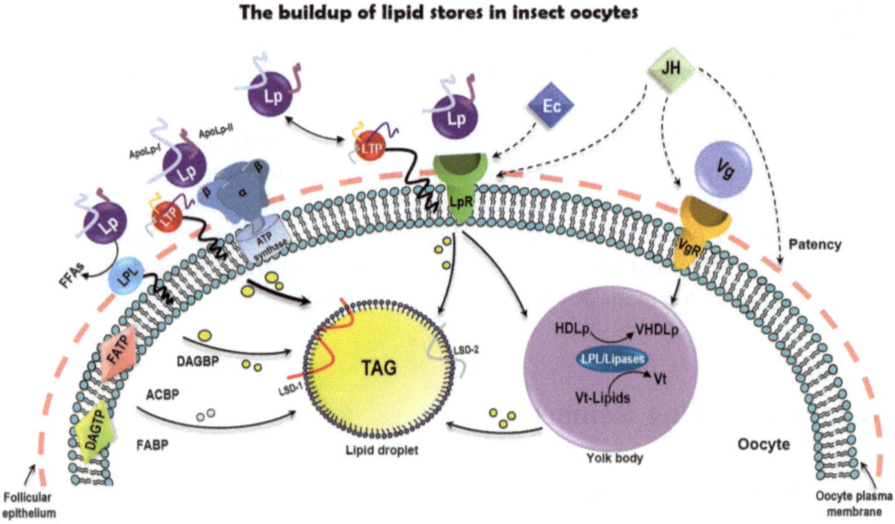

Fig. 18.4 Integrated model for lipid uptake and storage in developing oocytes. Lipids are delivered to the oocytes by two main circulating lipoproteins, lipophorin (Lp) and vitellogenin (Vg). Juvenile hormone (JH) promotes lipid uptake by inducing the shrinkage of follicle cells (patency) and the formation of intercellular spaces, which facilitate the access of Lp and Vg to the oocyte surface. JH also upregulates the expression of endocytic Vg and Lp receptors (VgR and LpR, respectively). Ecdysteroids (Ec) seem to stimulate the expression of LpR. Internalized Lp (as high-density lipophorin, HDLp) and Vg traffic to yolk bodies where they are stripped from their lipids by a lipoprotein lipase (LPL), generating very high-density Lp (VHDLp) and vitellin (Vt). The lipid cargo of endocytosed lipoproteins is relocated from yolk bodies to lipid droplets by an unknown mechanism. The bulk of neutral lipids in oocytes is transferred by Lp via a non-endocytic pathway. This mechanism is not yet fully understood, but partial evidence involves the β subunit of the ATP synthase complex (β-ATPase) in Lp binding and Lp-mediated lipid transfer to the oocytes. LpR contributes to the process promoting the recruitment of lipid transfer particle (LTP) to the plasma membrane, thus stabilizing Lp to facilitate lipid transfer. A membrane-bound LPL that hydrolyzes the diacylglycerol (DAG) carried by Lp also accounts for the process. The transfer of DAG and/or fatty acids (FAs) and its cytoplasmic trafficking mediated by putative specific lipid carriers were also considered in this integral model. The lipids from the lipoproteins and from the yolk bodies are esterified in lipid droplets, where triacylglycerol (TAG) is the major hydrophobic component. Proteins of the PAT family, LSD-1 and LSD-2, regulate at least in part the dynamic metabolism of lipid droplets. FATP and DAGTP, fatty acid and diacylglycerol transport proteins, respectively; ACBP, acyl-CoA-binding protein; FABP and DAGBP, fatty acid and diacylglycerol binding proteins, respectively. Dashed lines indicate hormonal regulation. Original drawing from the authors based on the references in the text

chemical structures. Neutral lipids are broken down to fuel energetic demands during embryonic development. However, little is known about lipases and regulation of lipolysis during this developmental period. The contribution of emerging concepts such as macrolipophagy to the energetic balance during oogenesis and embryogenesis needs to be examined in more insect species. Currently, it is difficult to establish to what extent the dynamics of ecdysteroids between follicle cells and the oocyte reported in *B. mori* can be extended to other insect species. Large

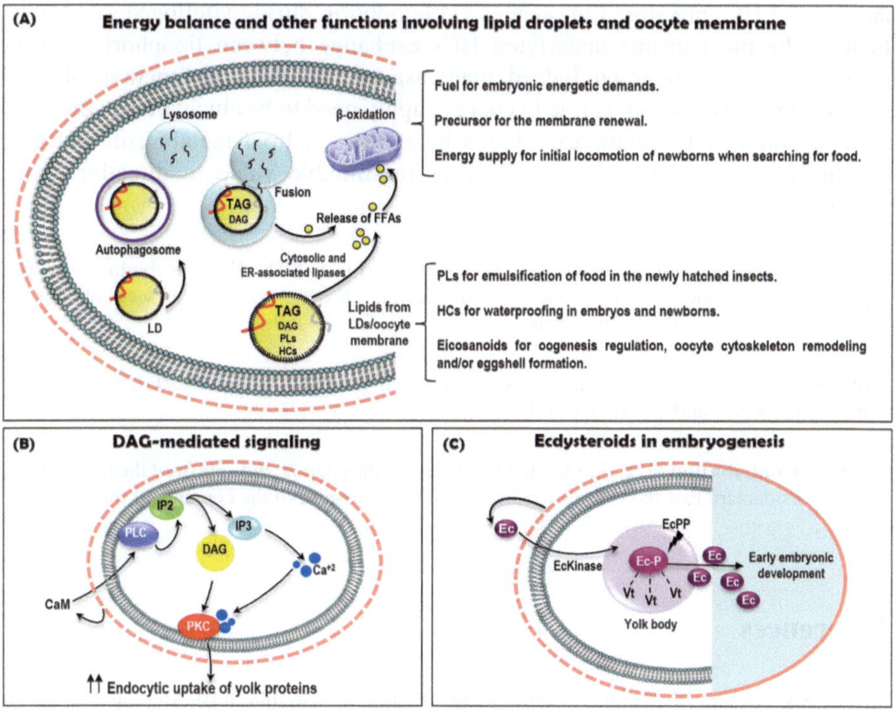

Fig. 18.5 The functions of lipids in oogenesis and embryogenesis. **(a)** Energy balance and other functions involving lipid droplets (LDs) and oocyte cell membrane. During embryogenesis, constant supply of energy is accomplished by breakdown of triacylglycerol (TAG) stores in lipid droplets of mature eggs by cytosolic or ER-associated lipases. The process yields free fatty acids (FFAs), which are driven to β-oxidation. FFAs are also generated by macroautophagy, involving sequestration of LDs in autophagosomes, which fuse with lysosomes for degradation of their content. Phospholipids (PLs) contained in the LDs seem to serve as emulsifier of ingested food by newborn individuals. Large amounts of hydrocarbons (HCs) stored in oocytes during vitellogenesis might be compartmentalized in LDs. These highly hydrophobic molecules are used for extensive waterproofing of embryos and newborns. Eicosanoids, a family of local lipid mediators synthesized mainly from arachidonic acid released from membrane PLs, regulate oogenesis and participate in oocyte cytoskeleton remodeling and in eggshell formation among other functions. **(b)** DAG-mediated signaling. The signaling cascade is initiated when the calcium-binding protein calmodulin (CaM), synthesized in follicle cells, activates the membrane-bound phospholipase C (PLC) to transform phosphatidylinositol 4,5-biphosphate (IP_2) into inositol 1,4,5-trisphosphate (IP_3) and DAG. IP_3 triggers the release of Ca^{+2} from the internal pool, which along with DAG stimulates protein kinase C (PKC). PKC is required for endocytic uptake of yolk proteins into new yolk bodies and for its clustering and fusion into mature yolk bodies. **(c)** Ecdysteroids in embryogenesis. Ecdysteroids (Ec) synthesized de novo in follicle cells are stored as inactive phosphoric esters via ecdysteroid kinase (EcKinase). These molecules bind to vitellin (Vt) to form complexes that are stored in yolk bodies. The action of an ecdysteroid phosphate phosphatase (EcPP) is required for dephosphorylation when free Ec are needed in early embryo development. Original drawing from the authors based on the references in the corresponding section

amounts of HCs are stored in oocytes via lipophorin during vitellogenesis. Nevertheless, the mechanisms underlying HCs exchange between lipophorin and the ovarian tissue are unknown. Indeed, many aspects regarding the functions of lipids in insect oocytes beyond its role in energy supply need to be elucidated. A thorough understanding of the complex relationships between lipid metabolism and their functions in insect oocytes would be of inestimable value in directing future research in insect reproduction.

Acknowledgments Authors thank Dr. B. P. Settembrini and Dr. E. R. Rubiolo for helpful suggestions and critical reading. Unpublished scanning electron micrographs were obtained in collaboration with Dr. B. P. Settembrini.

Funding Work in the L.E.C. laboratory is supported by grants from SECyT-UNC, FONCyT (PICT 2013-0626), and CONICET (PIP 0159).

Authors' Contribution L.L.F., J.L., and L.E.C. wrote the chapter and approved the final version. The authors declare that there is no conflict of interest in regard to the contents of this chapter.

References

Aguirre SA, Fruttero LL, Leyria J, Defferrari MS, Pinto PM, Settembrini BP, Rubiolo ER, Carlini CR, Canavoso LE (2011) Biochemical changes in the transition from vitellogenesis to follicular atresia in the hematophagous *Dipetalogaster maxima* (Hemiptera: Reduviidae). Insect Biochem Mol Biol 41:832–841

Aguirre SA, Pons P, Settembrini BP, Arroyo D, Canavoso LE (2013) Cell death mechanisms during follicular atresia in *Dipetalogaster maxima*, a vector of Chagas' disease (Hemiptera: Reduviidae). J Insect Physiol 59:532–541

Alves-Bezerra M, Gondim KC (2012) Triacylglycerol biosynthesis occurs via the glycerol-3-phosphate pathway in the insect *Rhodnius prolixus*. Biochim Biophys Acta 1821:1462–1471

Alves-Bezerra M, De Paula IF, Medina JM, Silva-Oliveira G, Medeiros JS, Gäde G, Gondim KC (2016) Adipokinetic hormone receptor gene identification and its role in triacylglycerol metabolism in the blood-sucking insect *Rhodnius prolixus*. Insect Biochem Mol Biol 69:51–60

Andries K, Verhasselt P, Guillemont J, Göhlmann HW, Neefs JM, Winkler H, Van Gestel J, Timmerman P, Zhu M, Lee E (2005) A diarylquinoline drug active on the ATP synthase of *Mycobacterium tuberculosis*. Science 307:223–227

Angel-Dapa MA, Rodríguez-Jaramillo C, Cáceres-Martínez CJ, Saucedo PE (2010) Changes in lipid content of oocytes of the penshell *Atrina mauraas* a criterion of gamete development and quality: a study of histochemistry and digital image analysis. J Shellfish Res 29:407–413

Arechaga I, Jones PC (2001) Quick guide: ATP synthase. Curr Biol 11:R117

Arrese EL, Soulages JL (2010) Insect fat body: energy, metabolism, and regulation. Annu Rev Entomol 55:207–225

Arrese EL, Saudale FZ, Soulages JL (2014) Lipid droplets as signaling platforms linking metabolic and cellular functions. Lipid Insights 7:7–16

Atella GC, Shahabuddin M (2002) Differential partitioning of maternal fatty acid and phospholipid in neonate mosquito larvae. J Exp Biol 205:3623–3630

Atella GC, Gondim KC, Machado EA, Medeiros MN, Silva-Neto MA, Masuda H (2005) Oogenesis and egg development in triatomines: a biochemical approach. An Acad Bras Cienc 77:405–430

Baker KD, Thummel CS (2007) Diabetic larvae and obese flies-emerging studies of metabolism in *Drosophila*. Cell Metab 6:257–266

Barneda D, Christian M (2017) Lipid droplet growth: regulation of a dynamic organelle. Curr Opin Cell Biol 47:9–15

Beenakkers AMT, Van der Horst DJ, Van Marrewijk WJA (1981) Role of lipids in energy metabolism. In: Downer RGH (ed) Energy metabolism in insects. Plenum, New York, pp 53–100

Bellés X (2005) Vitellogenesis directed by juvenile hormone. In: Raikhel AS (ed) Progress in vitellogenesis. Reproductive biology of invertebrates, vol 12, part B. Science Publishers, Enfield, pp 157–197

Bellés X, Piulachs MD (2015) Ecdysone signalling and ovarian development in insects: from stem cells to ovarian follicle formation. Biochim Biophys Acta 1849:181–186

Benoit JB, Yang G, Krause TB, Patrick KR, Aksoy S, Attardo GM (2011) Lipophorin acts as a shuttle of lipids to the milk gland during tsetse fly pregnancy. J Insect Physiol 57:1553–1561

Bi J, Xiang Y, Chen H, Liu Z, Grönke S, Kühnlein RP, Huang X (2012) Opposite and redundant roles of the two *Drosophila* perilipins in lipid mobilization. J Cell Sci 125:3568–3577

Bickel PE, Tansey JT, Welte MA (2009) PAT proteins, an ancient family of lipid droplet proteins that regulate cellular lipid stores. Biochim Biophys Acta 1791:419–440

Blacklock BJ, Ryan RO (1994) Hemolymph lipid transport. Insect Biochem Mol Biol 24:855–873

Briegel H (1990) Metabolic relationship between female body size, reserves, and fecundity of *Aedes aegypti*. J Insect Physiol 36:165–172

Brooks RA, Woodruff RI (2004) Calmodulin transmitted through gap junctions stimulates endocytic incorporation of yolk precursors in insect oocytes. Dev Biol 271:339–349

Brown MS, Goldstein JL (1986) A receptor-mediated pathway for cholesterol homeostasis. Science 232:34–47

Brown PT, Herbert P, Woodruff RI (2010) Vitellogenesis in *Oncopeltus fasciatus*: PLC/IP$_3$, DAG/PK-C pathway triggered by CaM. J Insect Physiol 56:1300–1305

Büning J (2006) Ovariole structure supports sistergroup relationship of Neuropterida and Coleoptera. Arthropod Syst Phylogeny 64:115–126

Canavoso LE, Wells MA (2001) Role of lipid transfer particle in delivery of diacylglycerol from midgut to lipophorin in larval *Manduca sexta*. Insect Biochem Mol Biol 31:783–790

Canavoso LE, Jouni ZE, Karnas KJ, Pennington JE, Wells MA (2001) Fat metabolism in insects. Annu Rev Nutr 21:23–46

Canavoso LE, Yun HK, Jouni ZE, Wells MA (2004) Lipid transfer particle mediates the delivery of diacylglycerol from lipophorin to fat body in larval *Manduca sexta*. J Lipid Res 45:456–465

Cardoso AF, Cres RL, Moura AS, Almeida F, Bijovsky AT (2010) *Culex quinquefasciatus* vitellogenesis: morphological and biochemical aspects. Mem Inst Oswaldo Cruz 105:254–262

Carrasco S, Mérida I (2007) Diacylglycerol, when simplicity becomes complex. Trends Biochem Sci 32:27–36

Chen ME, Lewis DK, Keeley LL, Pietrantonio PV (2004) cDNA cloning and transcriptional regulation of the vitellogenin receptor from the imported fire ant, *Solenopsis invicta* Buren (Hymenoptera: Formicidae). Insect Mol Biol 13:195–204

Cheon HM, Seo SJ, Sun J, Sappington TW, Raikhel AS (2001) Molecular characterization of the VLDL receptor homolog mediating binding of lipophorin in oocyte of the mosquito *Aedes aegypti*. Insect Biochem Mol Biol 31:753–760

Chi SL, Pizzo SV (2006) Cell surface F_1F_O ATP synthase: a new paradigm? Ann Med 38:429–438

Chino H, Downer RGH, Takahashi K (1977) The role of diacylglycerol-carrying lipoprotein I in lipid transport during insect vitellogenesis. Biochim Biophys Acta 487:508–516

Chinzei Y, Chino H, Wyatt GR (1981) Purification and properties of vitellogenin and vitellin from *Locusta migratoria*. Insect Biochem 11:1–7

Ciudad L, Bellés X, Piulachs MD (2007) Structural and RNAi characterization of the German cockroach lipophorin receptor and the evolutionary relationships of lipoprotein receptors. BMC Mol Biol 8:53

Clifton ME, Noriega FG (2012) The fate of follicles after a blood meal is dependent on previtellogenic nutrition and juvenile hormone in *Aedes aegypti*. J Insect Physiol 58:1007–1019

Dantuma NP, Van Marrewijk WJA, Wynne HJ, Van der Horst DJ (1996) Interaction of an insect lipoprotein with its binding site at the fat body. J Lipid Res 37:1345–1355

Dantuma NP, Pijnenburg MA, Diederen JH, Van der Horst DJ (1997) Developmental down-regulation of receptor-mediated endocytosis of an insect lipoprotein. J Lipid Res 38:254–265

Dantuma NP, Potters M, De Winther MPJ, Tensen CP, Kooiman FP, Bogerd J, Van der Horst DJ (1999) An insect homolog of the vertebrate very low density lipoprotein receptor mediates endocytosis of lipophorins. J Lipid Res 40:973–978

Davey K (2007) From insect ovaries to sheep red blood cells: a tale of two hormones. J Insect Physiol 53:1–10

Di Bartolomeo S, Nazio F, Cecconi F (2010) The role of autophagy during development in higher eukaryotes. Traffic 11:1280–1289

Downer RGH, Chino H (1985) Turnover of protein and diacylglycerol components of lipophorin in insect haemolymph. Insect Biochem 15:627–630

Drummond-Barbosa D, Spradling AC (2001) Stem cells and their progeny respond to nutritional changes during *Drosophila* oogenesis. Dev Biol 231:265–278

Ducharme NA, Bickel PE (2008) Lipid droplets in lipogenesis and lipolysis. Endocrinology 149:942–949

Engelmann F, Mala J (2000) The interactions between juvenile hormone (JH), lipophorin, vitellogenin, and JH esterases in two cockroach species. Insect Biochem Mol Biol 30:793–803

Entringer PF, Grillo LA, Pontes EG, Machado EA, Gondim KC (2013) Interaction of lipophorin with *Rhodnius prolixus* oocytes: biochemical properties and the importance of blood feeding. Mem Inst Oswaldo Cruz 108:836–844

Fan Y, Chase J, Sevala VL, Schal C (2002) Lipophorin-facilitated hydrocarbon uptake by oocytes in the German cockroach *Blatella germanica*. J Exp Biol 205:781–790

Fan Y, Zurek L, Dykstra M, Schal C (2003) Hydrocarbon synthesis by enzymatically dissociated oenocytes of the abdominal integument of the German cockroach, *Blattella germanica*. Naturwissenschaften 90:121–126

Fernandis AZ, Wenk MR (2007) Membrane lipids as signaling molecules. Curr Opin Lipidol 18:121–128

Ferreira CR, Saraiva SA, Catharino RR, Garcia JS, Gozzo FC, Sanvido GB, Santos LF, Lo Turco EG, Pontes JH, Basso AC, Bertolla RP, Sartori R, Guardieiro MM, Perecin F, Meirelles FV, Sangalli JR, Eberlin MN (2010) Single embryo and oocyte lipid fingerprinting by mass spectrometry. J Lipid Res 51:1218–1227

Fongsaran C, Jirakanwisal K, Kuadkitkan A, Wikan N, Wintachai P, Thepparit C, Ubol S, Phaonakrop N, Roytrakul S, Smith DR (2014) Involvement of ATP synthase β subunit in chikungunya virus entry into insect cells. Arch Virol 159:3353–3364

Foottit RG, Adler PH (2009) Introduction. In: Foottit RG, Adler HP (eds) Insect biodiversity: science and society, 1st edn. Wiley-Blackwell, Oxford, pp 1–6

Fruttero LL, Rubiolo ER, Canavoso LE (2009) Biochemical and cellular characterization of lipophorin-midgut interaction in the hematophagous *Panstrongylus megistus* (Hemiptera: Reduviidae). Insect Biochem Mol Biol 39:322–331

Fruttero LL, Frede S, Rubiolo ER, Canavoso LE (2011) The storage of nutritional resources during vitellogenesis of *Panstrongylus megistus* (Hemiptera: Reduviidae): the pathways of lipophorin in lipid delivery to developing oocytes. J Insect Physiol 57:475–486

Fruttero LL, Demartini DR, Rubiolo ER, Carlini CR, Canavoso LE (2014) β-chain of ATP synthase as a lipophorin binding protein and its role in lipid transfer in the midgut of *Panstrongylus megistus* (Hemiptera: Reduviidae). Insect Biochem Mol Biol 52:1–12

Fruttero LL, Leyria J, Ramos FO, Stariolo R, Settembrini BP, Canavoso LE (2017) The process of lipid storage in insect oocytes: the involvement of β-chain of ATP synthase in lipophorin-mediated lipid transfer in the chagas' disease vector *Panstrongylus megistus* (Hemiptera: Reduviidae). J Insect Physiol 96:82–92

Galetto L, Bosco D, Balestrini R, Genre A, Fletcher J, Marzachì C (2011) The major antigenic membrane protein of "Candidatus Phytoplasma asteris" selectively interacts with ATP synthase and actin of leafhopper vectors. PLoS One 6:e22571

Go GW, Mani A (2012) Low-density lipoprotein receptor (LDLR) family orchestrates cholesterol homeostasis. Yale J Biol Med 85:19–28

Goldstein JL, Brown MS (2009) The LDL receptor. Arterioscler Thromb Vasc Biol 29:431–438

Gondim KC, Wells MA (2000) Characterization of lipophorin binding to the midgut of larval *Manduca sexta*. Insect Biochem Mol Biol 30:405–413

Gopalapillai R, Kadono-Okuda K, Tsuchida K, Yamamoto K, Nohata J, Ajimura M, Mita K (2006) Lipophorin receptor of *Bombyx mori*: cDNA cloning, genomic structure, alternative splicing, and isolation of a new isoform. J Lipid Res 47:1005–1013

Guo JY, Dong SZ, Ye GY, Li K, Zhu JY, Fang Q, Hu C (2011) Oosorption in the endoparasitoid, *Pteromalus puparum*. J Insect Sci 11:90

Hagedorn HH (1985) The role of ecdysteroids in reproduction. In: Kerkut GA, Gilbert LI (eds) Comprehensive insect physiology, Biochemistry and pharmacology, vol 8. Pergamon Press, Oxford, pp 205–261

Harizi H, Corcuff JB, Gualde N (2008) Arachidonic-acid-derived eicosanoids: roles in biology and immunopathology. Trends Mol Med 14:461–469

Havel RJ (1987) Lipid transport function of lipoproteins in blood plasma. Am J Physiol 253:E1–E5

Horne I, Haritos VS, Oakeshott JG (2009) Comparative and functional genomics of lipases in holometabolous insects. Insect Biochem Mol Biol 39:547–567

Howard RW, Blomquist GJ (2005) Ecological, behavioral, and biochemical aspects of insect hydrocarbons. Annu Rev Entomol 50:371–393

Huebner E, Anderson E (1972a) A cytological study of the ovary of *Rhodnius prolixus*. I. The ontogeny of the follicular epithelium. J Morphol 136:459–493

Huebner E, Anderson E (1972b) A cytological study of the ovary of *Rhodnius prolixus*. II. Oocyte differentiation. J Morphol 137:385–415

Ito Y, Yasuda A, Sonobe H (2008) Synthesis and phosphorylation of ecdysteroids during ovarian development in the silkworm, *Bombyx mori*. Zoolog Sci 25:721–727

Jindra M, Palli SR, Riddiford LM (2013) The juvenile hormone signaling pathway in insect development. Annu Rev Entomol 58:181–204

Jindra M, Bellés X, Shinoda T (2015) Molecular basis of juvenile hormone signaling. Curr Opin Insect Sci 11:39–46

Jouni ZE, Takada N, Gazard J, Maekawa H, Wells MA, Tsuchida K (2003) Transfer of cholesterol and diacylglycerol from lipophorin to *Bombyx mori* ovarioles in vitro: role of the lipid transfer particle. Insect Biochem Mol Biol 33:145–153

Juárez MP (1994) Hydrocarbon biosynthesis in *Triatoma infestans* eggs. Arch Insect Biochem Physiol 25:193–206

Kawooya JK, Law JH (1988) Role of lipophorin in lipid transport to the insect egg. J Biol Chem 263:8748–8753

Kawooya JK, Osir EO, Law JH (1988) Uptake of the major hemolymph lipoprotein and its transformation in the insect egg. J Biol Chem 263:8740–8747

Kory N, Farese RV, Walther TC (2016) Targeting fat: mechanisms of protein localization to lipid droplets. Trends Cell Biol 26:535–546

Kühnlein RP (2011) The contribution of the *Drosophila* model to lipid droplet research. Prog Lipid Res 50:348–356

Lagueux M, Hoffmann JA (1984) Ecdysteroids in ovaries and embryos of *Locusta migratoria*. In: Hoffman JA, Porchet M (eds) Biosynthesis, metabolism and mode of actions of invertebrates hormones. Springer, Berlin, pp 168–180

Law JH, Wells MA (1989) Insects as biochemical models. J Biol Chem 264:16335–16338

Leyria J, Fruttero LL, Aguirre SA, Canavoso LE (2014) Ovarian nutritional resources during the reproductive cycle of the hematophagous *Dipetalogaster maxima* (Hemiptera: Reduviidae): focus on lipid metabolism. Arch Insect Biochem Physiol 87:148–163
Li Z, Thiel K, Thul PJ, Beller M, Kühnlein RP, Welte MA (2012) Lipid droplets control the maternal histone supply of *Drosophila* embryos. Curr Biol 22:2104–2113
Lin X, Kim YA, Lee BL, Söderhäll K, Söderhäll I (2009) Identification and properties of a receptor for the invertebrate cytokine astakine, involved in hematopoiesis. Exp Cell Res 315:1171–1180
Liu Z, Huang X (2013) Lipid metabolism in *Drosophila*: development and disease. Acta Biochim Biophys Sin (Shanghai) 45:44–50
Liu H, Ryan RO (1991) Role of lipid transfer particle in transformation of lipophorin in insect oocytes. Biochim Biophys Acta 1085:112–118
Lone AM, Tasken K (2013) Proinflammatory and immunoregulatory roles of eicosanoids in T cells. Front Immunol 4:130
Lu K, Shu Y, Zhou J, Zhang X, Zhang X, Chen M, Yao Q, Zhou Q, Zhang W (2015) Molecular characterization and RNA interference analysis of vitellogenin receptor from *Nilaparvata lugens* (Stål). J Insect Physiol 73:20–29
Lu K, Chen X, Liu WT, Zhang XY, Chen MX, Zhou Q (2016) Nutritional signaling regulates vitellogenin synthesis and egg development through juvenile hormone in *Nilaparvata lugens* (Stål). Int J Mol Sci 17:269
Machado EA, Atella GC, Gondim KC, de Souza W, Masuda H (1996) Characterization and immunocytochemical localization of lipophorin binding sites in the oocytes of *Rhodnius prolixus*. Arch Insect Biochem Physiol 31:185–196
Machado E, Swevers L, Sdralia N, Medeiros MN, Mello FG, Iatrou K (2007) Prostaglandin signaling and ovarian follicle development in the silkmoth, *Bombyx mori*. Insect Biochem Mol Biol 37:876–885
Majerowicz D, Calderón-Fernández GM, Alves-Bezerra M, De Paula IF, Cardoso LS, Juárez MP, Atella GC, Gondim KC (2017) Lipid metabolism in *Rhodnius prolixus*: lessons from the genome. Gene 596:27–44
Makki R, Cinnamon E, Gould AP (2014) The development and functions of oenocytes. Annu Rev Entomol 59:405–425
McCall K (2004) Eggs over easy: cell death in the *Drosophila* ovary. Dev Biol 274:3–14
Medeiros MN, Oliveira DMP, Paiva-Silva GO, Silva-Neto MAC, Romero A, Bozza M, Masuda H, Machado EA (2002) The role of eicosanoids on *Rhodnius* heme-binding protein (RHBP) endocytosis by *Rhodnius prolixus* ovaries. Insect Biochem Mol Biol 32:537–545
Medeiros MN, Mendonça LH, Hunter AL, Paiva-Silva GO, Mello FG, Henze IP, Masuda H, Maya-Monteiro CM, Machado EA (2004) The role of lipoxygenase products on the endocytosis of yolk proteins in insects: participation of cAMP. Arch Insect Biochem Physiol 55:178–187
Meier T, Faraldo-Gomez J, Börsch B (2011) ATP synthase-a paradigmatic molecular machine. In: Frank J (ed) Molecular machines in biology. Cambridge University Press, Cambridge, pp 208–238
Miura K, Oda M, Makita S, Chinzei Y (2005) Characterization of the *Drosophila* Methoprene-tolerant gene product. FEBS J 272:1169–1178
Negishi M, Sugimoto Y, Ichikawa A (1995) Molecular mechanisms of diverse actions of prostanoid receptors. Biochim Biophys Acta 1259:109–120
Nelson DR, Hines H, Stay B (2004) Methyl-branched hydrocarbons, major components of the waxy material coating the embryos of the viviparous cockroach *Diploptera punctata*. Comp Biochem Physiol B Biochem Mol Biol 138:265–276
Niwa R, Niwa YS (2014) Enzymes for ecdysteroid biosynthesis: their biological functions in insects and beyond. Biosci Biotechnol Biochem 78:1283–1292
Nomura DK, Casida JE (2016) Lipases and their inhibitors in health and disease. Chem Biol Interact 259:211–222

Paingankar MS, Gokhale MD, Deobagkar DN (2010) Dengue-2-virus-interacting polypeptides involved in mosquito cell infection. Arch Virol 155:1453–1461

Parra-Peralbo E, Culi J (2011) *Drosophila* lipophorin receptors mediate the uptake of neutral lipids in oocytes and imaginal disc cells by an endocytosis independent mechanism. PLoS Genet 7: e1001297

Pennington JE, Wells MA (2002) Triacylglycerol-rich lipophorins are found in the dipteran infraorder Culicomorpha, not just in mosquitoes. J Insect Sci 2:15

Pistillo D, Manzi A, Tino A, PiloBoyl P, Graziani F, Malva C (1998) The *Drosophila melanogaster* lipase homologs: a gene family with tissue and developmental specific expression. J Mol Biol 276:877–855

Raikhel AS (2005) Vitellogenesis of disease vectors, from physiology to genes. In: Marquardt WC (ed) Biology of disease vectors. Elsevier, Academic Press, London, pp 329–346

Raikhel AS, Dhadialla TS (1992) Accumulation of yolk proteins in insect oocytes. Annu Rev Entomol 37:217–225

Raikhel AS, Brown M, Belles X (2005) Endocrine control of reproductive processes. In: Gilbert L, Gill S, Iatrou K (eds) Comprehensive molecular insect science. Elsevier Press, Oxford, pp 433–491

Remaley AT, Rifai N, Russell Warnick G (2015) Lipids, Lipoproteins, apolipoproteins and other cardiac risk factors. In: Burtis CA, Bruns DE (eds) Fundamentals of clinical chemistry and molecular diagnostics, 7th edn. Saunders, St. Louis, pp 338–411

Riddiford LM (2008) Juvenile hormone action: a 2007 perspective. J Insect Physiol 54:895–901

Riddiford LM, Cherbas A, Truman W (2000) Ecdysone receptors and their biological actions. Vitam Horm 60:1–73

Rodríguez-Vázquez M, Vaquero D, Parra-Peralbo E, Mejía-Morales JE, Culi J (2015) *Drosophila* lipophorin receptors recruit the lipoprotein ltp to the plasma membrane to mediate lipid uptake. PLoS Genet 11:e1005356

Roy-Zokan EM, Cunningham CB, Hebb LE, McKinney EC, Moore AJ (2015) Vitellogenin and vitellogenin receptor gene expression is associated with male and female parenting in a subsocial insect. Proc Biol Sci 282:20150787

Ryan RO, Van der Horst DJ (2000) Lipid transport biochemistry and its role in energy production. Annu Rev Entomol 45:233–260

Ryan RO, Wells MA, Law JH (1986) Lipoprotein interconversions in an insect, *Manduca sexta*. Evidence for a lipid transfer factor in the hemolymph. J Biol Chem 261:563–568

Santos R, Rosas-Oliveira R, Saraiva FB, Majerowicz D, Gondim KC (2011) Lipid accumulation and utilization by oocytes and eggs of *Rhodnius prolixus*. Arch Insect Biochem Physiol 77:1–16

Schmidt-Ott U, Lynch JA (2016) Emerging developmental genetic model systems in holometabolous insects. Curr Opin Genet Dev 39:116–128

Sevala V, Shu S, Ramaswamy SB, Schal C (1999) Lipophorin of female *Blattella germanica* (L.): characterization and relation to hemolymph titers of juvenile hormone and hydrocarbons. J Insect Physiol 45:431–441

Shao W, Espenshade PJ (2012) Expanding roles for SREBP in metabolism. Cell Metab 16:414–419

Shirk PD, Perera OP (1998) 5′ coding region of the follicular epithelium yolk polypeptide 2 cDNA in the moth, *Plodia interpunctella*, contains an extended coding region. Arch Insect Biochem Physiol 39:98–108

Sieber MH, Spradling AC (2015) Steroid signaling establishes a female metabolic state and regulates SREBP to control oocyte lipid accumulation. Curr Biol 25:993–1004

Singh R, Cuervo AM (2011) Autophagy in the cellular balance. Cell Metabolism 13:495–504

Singh R, Kaushik S, Wang Y, Xiang Y, Novak I, Komatsu M, Tanaka K, Cuervo AM, Czaja MJ (2009) Autophagy regulates lipid metabolism. Nature 458:1131–1135

Snigirevskaya ES, Raikhel AS (2005) Receptor-mediated endocytosis of yolk proteins in insect oocytes. In: Raikhel AS (ed) Progress in vitellogenesis. Reproductive biology of invertebrates, vol 12, part B. Adiyodi AG, Adiyodi RG (eds) Science Publishers, Enfield, pp 198–227

Soulages JL, Firdaus SJ, Hartson S, Chen X, Howard AD, Arrese EL (2012) Developmental changes in the protein composition of *Manduca sexta* lipid droplets. Insect Biochem Mol Biol 42:305–320

Spracklen AJ, Kelpsch DJ, Chen X, Spracklen CN, Tootle TL (2014) Prostaglandins temporally regulate cytoplasmic actin bundle formation during *Drosophila* oogenesis. Mol Biol Cell 25:397–411

Stanley D (2006) Prostaglandins and other eicosanoids in insects: biological significance. Annu Rev Entomol 51:25–44

Stanley D, Kim Y (2011) Prostaglandins and their receptors in insect biology. Front Endocrinol (Lausanne) 2:105

Stanley DW, Nor Aliza AR, Tunaz H, Putnam SM, Park Y, Bedick JC (2002) FORUM Eicosanoids in insect biology. Neotrop Entomol 31:341–350

Stanley D, Haas E, Miller J (2012) Eicosanoids: exploiting insect immunity to improve biological control programs. Insects 3:492–510

Sun J, Hiraoka T, Dittmer NT, Cho K-H, Raikhel AS (2000) Lipophorin as a yolk protein precursor in the mosquito, *Aedes aegypti*. Insect Biochem Mol Biol 30:1161–1171

Suzuki M, Shinohara Y, Ohsaki Y, Fujimoto T (2011) Lipid droplets: size matters. J Electron Microsc (Tokyo) 1:S101–S116

Swevers L, Raikhel AS, Sappington TW, Shirk P, Iatrou K (2005) Vitellogenesis and post-vitellogenic maturation of the insect ovarian follicle. In: Gilbert LI, Iatrou K, Gill SS (eds) Comprehensive molecular insect science, Reproduction and development, vol 1. Elsevier, Pergamon, pp 87–155

Takeuchi N, Chino H (1993) Lipid transfer particle in the hemolymph of the American cockroach: evidence for its capacity of transfer hydrocarbons between lipophorin particles. J Lipid Res 34:543–551

Telfer WH (1979) Sulfate and glucosamine labelling of the intercellular matrix in vitellogenic follicles of a moth. Wilehm Roux Arch Dev Biol 185:347–362

Telfer WH, Huebner E, Smith DS (1982) The cell biology of vitellogenic follicles in *Hyalophora* and *Rhodnius*. In: King RC, Akai H (eds) Insect ultrastructure, vol 1. Plenum, New York, pp 118–149

Thiam AR, Farese Jr RV, Walther TC (2013) The biophysics and cell biology of lipid droplets. Nat Rev Mol Cell Biol 14:775–786

Tootle TL, Williams D, Hubb A, Frederick R, Spradling A (2011) *Drosophila* eggshell production: identification of new genes and coordination by Pxt. PLoS One 6:e19943

Troy S, Anderson WA, Spielman A (1975) Lipid content of maturing ovaries of *Aedes aegypti* mosquitoes. Comp Biochem Physiol B 50:457–461

Tsuchida K, Sakudoh T (2015) Recent progress in molecular genetic studies on the carotenoid transport system using cocoon-color mutants of the silkworm. Arch Biochem Biophys 572:151–157

Tsuchida K, Wells MA (1990) Isolation and characterization of a lipoprotein receptor from the fat body of an insect, *Manduca sexta*. J Biol Chem 265:5761–5767

Tufail M, Takeda M (2008) Molecular characteristics of insect vitellogenins. J Insect Physiol 54:1447–1458

Tufail M, Takeda M (2009) Insect vitellogenin/lipophorin receptors: molecular structures, role in oogenesis, and regulatory mechanisms. J Insect Physiol 55:87–103

Tufail M, Raikhel AS, Takeda M (2005) Biosynthesis and processing of insect vitellogenins. In: Raikhel AS (ed) Progress in vitellogenesis. Reproductive biology of invertebrates, vol 12, part B. Adiyodi AG, Adiyodi RG (eds) Science Publishers, Enfield, pp 1–32

Tufail M, Nagaba Y, Elgendy AM, Takeda M (2014) Regulation of vitellogenin genes in insects. Entomol Science 17:269–282

Tulenko TN, Sumner AE (2002) The physiology of lipoproteins. J Nucl Cardiol 9:638–649

Ugur B, Chen K, Bellen HJ (2016) *Drosophila* tools and assays for the study of human diseases. Dis Model Mech 9:235–244

Upadhyay SK, Singh H, Dixit S, Mendu V, Verma PC (2016) Molecular characterization of vitellogenin and vitellogenin receptor of *Bemisia tabaci*. PLoS One 11:e0155306

Van Antwerpen R, Law JH (1992) Lipophorin lipase from the yolk of *Manduca sexta* eggs: identification and partial characterization. Arch Insect Biochem Physiol 20:1–12

Van Antwerpen R, Salvador K, Tolman K, Gentry C (1998) Uptake of lipids by developing oocytes of the hawkmoth, *Manduca sexta*. The possible role of lipoprotein lipase. Insect Biochem Mol Biol 28:399–408

Van Antwerpen R, Pham DQ-D, Ziegler R (2005) Accumulation of lipids in insect oocytes. In: Raikhel AS (ed) Progress in vitellogenesis. Reproductive biology of invertebrates, vol 12, part B. Adiyodi AG, Adiyodi RG (eds) Science Publishers, Enfield pp 265–287

Van der Horst DJ, Rodenburg KW (2010a) Lipoprotein assembly and function in an evolutionary perspective. BioMol Concepts 1:165–183

Van der Horst DJ, Rodenburg KW (2010b) Locust flight activity as a model for hormonal regulation of lipid mobilization and transport. J Insect Physiol 56:844–853

Van der Horst DJ, Roosendaal SD, Rodenburg KW (2009) Circulatory lipid transport: lipoprotein assembly and function from an evolutionary perspective. Mol Cell Biochem 326:105–119

Van Handel E (1993) Fuel metabolism of the mosquito (*Culex quinquefasciatus*) embryo. J Insect Physiol 39:831–833

Van Heusden MC, Law JH (1989) An insect lipid transfer particle promotes lipid loading from fat body to lipoprotein. J Biol Chem 264(29):17287–17292

Van Hoof D, Rodenburg KW, Van der Horst DJ (2005) Receptor-mediated endocytosis and intracellular trafficking of lipoproteins and transferrin in insect cells. Insect Biochem Mol Biol 35:117–128

Vantourout P, Radojkovic C, Lichtenstein L, Pons V, Champagne E, Martinez LO (2010) Ecto-F_1-ATPase: a moonlighting protein complex and an unexpected apoA-I receptor. World J Gastroenterol 16:5925–5935

Walker JE (1998) ATP synthesis by rotary catalysis. Angew Chem Int Ed 37:2309–2319

Wang Y, Telfer WH (1996) Cyclic nucleotide-induced termination of vitellogenin uptake by *Hyalophora cecropia* follicles. Insect Biochem Mol Biol 26:85–94

Wang SF, Zhu J, Martin D, Raikhel AS (2005) Regulation of vitellogenin gene expression by ecdysteroids. In: Raikhel AS (ed) Progress in vitellogenesis. Reproductive biology of invertebrates, vol 12, part B. Adiyodi AG, Adiyodi RG (eds) Science Publishers, Enfield, pp 70–93

Welte MA (2015) As the fat flies: the dynamic lipid droplets of *Drosophila* embryos. Biochim Biophys Acta 1851:1156–1185

Wiemerslage LJ (1976) Lipid droplets formation during vitellogenesis in the cecropia moth. J Insect Physiol 22:41–50

Wilson TG, Fabian JA (1986) *Drosophila melanogaster* mutant resistant to a chemical analog of juvenile hormone. Dev Biol 118:190–201

Wyatt GR, Braun RP, Zhang J (1996) Priming effect in gene activation by juvenile hormone in locust fat body. Arch Insect Biochem Physiol 32:633–640

Ximenes AA, Oliveira GA, Bittencourt-Cunha P, Tomokyo M, Leite DB, Folly E, Golodne DM, Atella GC (2008) Purification, partial characterization and role in lipid transport to developing oocytes of a novel lipophorin from the cowpea weevil, *Callosobruchus maculatus*. Braz J Med Biol Res 41:18–25

Yamada R, Sonobe H (2003) Purification, kinetic characterization and molecular cloning of a novel enzyme ecdysteroid phosphate phosphatase. J Biol Chem 278:26365–26373

Yang H, Galea A, Sytnyk V, Crossley M (2012) Controlling the size of lipid droplets: lipid and protein factors. Curr Opin Cell Biol 24:509–516

Yokoyama H, Yokoyama T, Yuasa M, Fujimoto H, Sakudoh T, Honda N, Fugo H, Tsuchida K (2013) Lipid transfer particle from the silkworm, *Bombyx mori*, is a novel member of the apoB/large lipid transfer protein family. J Lipid Res 54:2379–2390

Zalewska M, Kochman A, Estève JP, Lopez F, Chaoui K, Susini C, Ozyhar A, Kochman M (2009) Juvenile hormone binding protein traffic. Interaction with ATP synthase and lipid transfer proteins. Biochim Biophys Acta 1788:1695–1705

Zhang W, Ma L, Xiao H, Xie B, Smagghe G, Guo Y, Liang G (2015) Molecular characterization and function analysis of the vitellogenin receptor from the cotton bollworm, *Helicoverpa armigera* (Hübner) (Lepidoptera, Noctuidae). PLoS One 11:e0155785

Ziegler R (1997) Lipid synthesis by ovaries and fat body of *Aedes aegypti* (Diptera: Culicidae). Eur J Entomol 94:385–391

Ziegler R, Van Antwerpen R (2006) Lipid uptake by insect oocytes. Insect Biochem Mol Biol 36:264–272

Chapter 19
Parthenogenesis in Insects: The Centriole Renaissance

Maria Giovanna Riparbelli, Marco Gottardo, and Giuliano Callaini

Abstract Building a new organism usually requires the contribution of two differently shaped haploid cells, the male and female gametes, each providing its genetic material to restore diploidy of the new born zygote. The successful execution of this process requires defined sequential steps that must be completed in space and time. Otherwise, development fails. Relevant among the earlier steps are pronuclear migration and formation of the first mitotic spindle that promote the mixing of parental chromosomes and the formation of the zygotic nucleus. A complex microtubule network ensures the proper execution of these processes. Instrumental to microtubule organization and bipolar spindle assembly is a distinct non-membranous organelle, the centrosome. Centrosome inheritance during fertilization is biparental, since both gametes provide essential components to build a functional centrosome. This model does not explain, however, centrosome formation during parthenogenetic development, a special mode of sexual reproduction in which the unfertilized egg develops without the contribution of the male gamete. Moreover, whereas fertilization is a relevant example in which the cells actively check the presence of only one centrosome, to avoid multipolar spindle formation, the development of parthenogenetic eggs is ensured, at least in insects, by the de novo assembly of multiple centrosomes.

Here, we will focus our attention on the assembly of functional centrosomes following fertilization and during parthenogenetic development in insects. Parthenogenetic development in which unfertilized eggs are naturally depleted of centrosomes would provide a useful experimental system to investigate centriole assembly and duplication together with centrosome formation and maturation.

19.1 Parthenogenesis: Sexual Reproduction Without Males

Parthenogenesis is a mode of sexual reproduction in which the embryonic development takes place in the egg cell without fertilization (Suomalainen 1950). It is clearly different from asexual modes of reproduction like budding or fragmentation that

M.G. Riparbelli • M. Gottardo • G. Callaini (✉)
Department of Life Sciences, University of Siena, Via A. Moro 2, 53100 Siena, Italy
e-mail: callaini@unisi.it

M. Kloc (ed.), *Oocytes*, Results and Problems in Cell Differentiation 63,
DOI 10.1007/978-3-319-60855-6_19

typically involve somatic cells. Parthenogenesis is often reported for various flowering plant species (Asker and Jerling 1992) where it is usually associated with sterile or absent pollen. In the animal kingdom, it occurs naturally in several invertebrate species. In vertebrates the parthenogenesis is rare and typically requires the fusion of male and female gametes without fusion of their nuclei (Bell 1982). Artificial parthenogenesis has been experimentally induced in animals through chemical agents (e.g., hypertonic or hypotonic solutions) or by physical stimuli (e.g., heating, and manipulation of the egg cell) (Tyler 1941).

The cytological mechanisms by which the parthenogenetic embryo develops can be divided into two main types. The apomictic (or ameiotic) parthenogenesis takes place without meiosis. It is the simplest form of parthenogenesis, characterized by the absence of recombination of alleles (Suomalainen et al. 1987). The resulting offspring that develops from unfertilized eggs are considered "full clones" and are genetically identical to their mother. Examples of apomixis can be commonly found in invertebrates such as nematodes (Trudgill and Blok 2001), arthropods (decapod crustaceans; Seitz et al. 2005; Buřič et al. 2011), and in various insects (Blackman and Spence 1996; Schwander and Crespi 2009).

In the automictic (or meiotic) parthenogenesis, the oocyte undergoes meiosis, and the egg nucleus has a halved number of chromosomes. In some rare cases, these unfertilized eggs can generate haploid individuals like in the males of ants and bees (Heimpel and de Boer 2008). Nevertheless, most commonly haploid individuals are not viable so that diploidy must be restored during the parthenogenetic development. This is accomplished by various means, for example, during meiosis by the fusion of first division nuclei (central fusion) or the second division nuclei (terminal fusion) to form a diploid zygote nucleus (Stenberg and Saura 2009). Other mechanisms of diploidization include the doubling of chromosomes during oogenesis, which occurs without nucleus and cell division, or the fusion of the first two blastomeres. The automictic parthenogenesis usually involves crossing over by homologous recombination of chromosomes during prophase I. This has important genetic implications, as the parthenogenetic offspring are considered "half clones" and they differ from one another and from their mother. The automictic reproduction is a widespread cytological mechanism of parthenogenesis (Cook 1993) with several examples found among the insects (Pijnacker 1969; Nur 1972; Marescalchi and Scali 2003; Sekine and Tojo 2010).

Depending on the sex of the offspring, three main types of parthenogenesis can be distinguished: thelytoky, arrhenotoky, and deuterotoky. In thelytoky, only females are produced from unfertilized eggs. It is the most common type of parthenogenesis, found, for example, in arthropods and in some vertebrates including fish, amphibians, and reptiles. Strictly thelytokous species are found, for example, in the insect orders Thysanoptera (thrips) and Phasmatodea (stick insects) (Normark and Kirkendall 2009; Li et al. 2014; Vershinina and Kuznetsova 2016). In arrhenotoky, the unfertilized eggs develop exclusively into males. This reproductive mode, also known as haplodiploidy, is found, for instance, in the drones of the honeybee and more generally is recognized as the typical sex-determination system of hymenopteran insects (Heimpel and de Boer 2008). Finally, deuterotoky occurs in the case that either males or females develop from unfertilized eggs, and it is found in some species of insects and mites that usually reproduced mainly by thelytoky (Suomalainen and Saura 1993).

Cyclical parthenogenesis refers to an alternation of different reproductive modes during the succession of generations of a given species. It occurs in over 15,000 animal species representing different taxonomic groups. A well-known example is offered by the complex reproductive cycle of aphids, a type of homopteran insects on which the first observations of parthenogenesis date back to the 1740s by the Swiss naturalist Charles Bonnet (Normark and Kirkendall 2009). These insects use thelytoky associated with viviparous reproduction throughout the summer. This enables them to quickly exploit resources and to produce multiple generations building up large populations. In autumn aphids use amphigony along with oviparous reproduction to produce overwintering eggs (Davis 2012). Branchiopod crustaceans of the genus *Daphnia* (Cladocera) represent another example of cyclical parthenogenesis. Similarly to aphids, daphnids exhibit parthenogenesis in order to grow the population during the favorable season. In the case of critical environmental conditions, e.g., food shortage or shortening photoperiod, they switch to amphigonic reproduction by producing fertilized and diapausing eggs (Hebert 1981; Thielsch et al. 2009).

Accidental and rudimentary parthenogenesis occur when the unfertilized eggs undergo autoactivation, but the embryos fail to develop completely and eventually die before hatching. This is a rather common phenomenon in the animal kingdom. It has been mainly studied in some domesticated birds including chickens and turkeys (Olsen 1967; Sarvella 1970).

Facultative parthenogenesis is characteristic of species that usually reproduce by amphigony, but if the males are absent, the unmated females can produce a viable parthenogenetic offspring. This type of parthenogenesis is widespread throughout the animal kingdom, and it has an important adaptive significance. It is common in at least ten insect orders, including mayflies (Odonata), crickets and grasshoppers (Orthoptera), cockroaches and termites (Blattodea), and stick insects (Phasmatodea) (Went 1982). In vertebrates, facultative parthenogenesis has been reported in captive bred sharks (Feldheim et al. 2010) and in some squamate reptiles including snakes and lizards (Groot et al. 2003; Watts et al. 2006; Booth and Schuett 2011). Facultative parthenogenesis is usually characterized by the presence of developmental abnormalities and low offspring viability (Schuett et al. 1997; Reynolds et al. 2012).

Parthenogenesis is obligate when it constitutes the only mode of reproduction of a species. Obligately parthenogenetic lineages often arise from species that show cyclical parthenogenesis, as exemplified in aphids and daphnids (Innes and Hebert 1988; Simon et al. 1996; Rispe and Pierre 1998; Sullender and Crease 2001).

19.2 The Centrosome: The Master of Spindle Assembly

More than a century ago, Flemming (1875) and Van Beneden (1876) first observed small bodies at the poles of the cleaving eggs of the nematode *Ascaris megalocephala*. Boveri (1887), believing the central role of these bodies in the cell life, coined the name "centrosome." Although enormous progress has been made from the early studies by iron hematoxylin as a major cytological staining technique, the centrosome still represents an attractive research field that offers new aspects and often delivers surprises.

The centrosome is composed of a pair of orthogonal barrel-shaped organelles, the centrioles, surrounded by a matrix of pericentriolar material (PCM) that dictates the functional properties of the centrosome itself (Kloc et al. 2014). Microtubule nucleation depends on γ-tubulin-containing nucleation complexes, which amount amplifies during interphase/prophase transition to provide support for the increased need for microtubules during the formation of the bipolar spindle. It has been demonstrated that the accumulation of γ-tubulin at the centrosome follows a general pathway requiring the sequential recruitment of various PCM proteins that surround the proximal region of the centrioles (Conduit et al. 2014). Since the centrosome is the major MT-organizing center (MTOC) in animal cells (Bornens 2002), it has been implicated in various aspects of the cell life, namely, cell migration, cell division, adhesion, vesicle trafficking, and signaling (Bettencourt-Dias et al. 2011; Arquint et al. 2014). The proper segregation of chromosomes to the daughter cells during the cell division relies on the correct assembly and orientation of the bipolar mitotic spindle (Bettencourt-Dias and Glover 2007). Therefore, the proper centrosome duplication represents a crucial step in the cell life, and deregulation of centrosome structure and number may lead to multipolar spindles, with increasing merotelic attachments of chromatids and ensuing abnormal chromosome segregation. This abnormal condition may contribute to aneuploidy and chromosome fragmentation, forms of genomic instability that represent an established hallmark of cancer cells (Nigg 2002, 2006; Pellman 2007; Basto et al. 2008; Ganem et al. 2009; Zyss and Gergely 2009; Bettencourt-Dias et al. 2011; Crasta et al. 2012; Vitre and Cleveland 2012; Pihan 2013; Godinho and Pellman 2014). Therefore, the number of centrosomes must be tightly monitored during cell division to ensure that the centrosome duplication occurs once and only once per cell cycle (Nigg and Stearns 2011).

The centrioles that play crucial roles to spatially organize the proteins involved in the recruitment of γ-TuRC complexes (Conduit et al. 2015) are instrumental to the assembly of a functional centrosome and to determine its replication status. Moreover, the centriole number determines the number of centrosomes within the cell (Sluder and Rieder 1985). Thus, basic to the knowledge of centrosome biogenesis is to understand the early steps of centriole assembly and duplication.

19.2.1 At the Heart of the Centrosome: The Centriole

The centriole is a small cylindrical organelle with a diameter of 230–250 nm and a length ranging from 130 to 500 nm. Centrioles are widely distributed within the animal kingdom, and their wall is usually built by triplet microtubules, a complete A-tubule and incomplete B- and C-tubules, arranged in a beautiful and evolutionarily conserved ninefold symmetry. Although it is largely accepted that nine triplet microtubules represent the widely evolutionarily conserved architecture of the centriole, some remarkable exceptions have been reported. Centrioles with 14 or up to 70–80 microtubule doublets are found during male meiosis of the basal hexapod *Acerentomonon microrhinus* (Riparbelli et al. 2009) and of the fungus gnat *Sciara coprophila* (Phillips 1967). These unusual centrioles assemble cilium-like structures during the male spermatogenesis and motile sperm axonemes. Interestingly, the parents of the giant centrioles observed in early primary spermatocytes in both *Sciara* and proturans have a conventional ninefold symmetry. A special feature of *Drosophila* development is that certain tissues have centrioles with doublets whereas others with triplets (Gonzalez et al. 1998; Carvalho-Santos et al. 2011). These structural differences seem to be correlated with different centriolar lineages in the *Drosophila* tissues. Somatic centrioles are, indeed, composed by microtubule doublets, whereas triplets characterize the centrioles of the germ cell line (Gottardo et al. 2015a).

An internal skeleton, the cartwheel, ensures the proper correct symmetrical disposition of the nine triplets (Nigg and Stearns 2011; Bornens 2012; Brito et al. 2012; Gönczy 2012). The cartwheel is formed by a central tube, the hub, and nine radial spokes linked to the peripheral A-tubules. The hub is composed of the Sas-6 protein that in vitro assembles in cartwheel-like structures, suggesting its involvement in centriole symmetry (Kitagawa et al. 2011; van Breugel et al. 2011). Other proteins such as Cep135/Bld10 and Sas-5/Ana2/STIL contribute to the final organization of the cartwheel (Brito et al. 2012; Avidor-Reiss and Gopalakrishnan 2013; Hirono 2014). This geometric scaffold is highly conserved among eukaryotes and is usually restricted to the basal end of the centriole. However, in vertebrate cells, the cartwheel is transiently present during the earlier stages of the daughter centriole formation and is lost before duplication in the S phase (Alvey 1986; Strnad and Gönczy 2008; Azimzadeh and Marshall 2010; Hatch and Stearns 2011). By contrast, the cartwheel is always present in early *Drosophila* embryos (Callaini and Riparbelli 1990), somatic tissues (Callaini et al. 1997; Vidwans et al. 2003), and cultured cells (Bettencourt-Dias et al. 2005; Dzhindzhev et al. 2010) where it crosses the whole length of the short centrioles. Interestingly, the cartwheel is restricted to the proximal region of the extraordinarily long centrioles found during male meiosis in *Drosophila* (Tates 1971; Fritzi-Niggli and Suda 1972; Riparbelli et al. 2012). However, the length of the cartwheel of the spermatocyte centrioles is the same of that observed in the shorter centrioles of germline stem cells and spermatogonia (Gottardo et al. 2015a). Therefore, during male gametogenesis in *Drosophila*, the centrioles increase ten times their length, but the cartwheel dimensions do not change, suggesting that this geometric framework may represent a structural scaffold needed to maintain the integrity of the centriole during its elongation. Although, the cartwheel seems to be essential only for centriole assembly, recent observations suggest that cartwheel removal during centriole duplication may have severe impact on centriole/centrosome conversion (Izquierdo et al. 2014).

The cartwheel appears as a common intermediate during procentriole formation although the unusual *Caenorhabditis elegans* centriole that is composed by nine single microtubules may lack a true cartwheel (Pelletier et al. 2006). However, ultrastructural analysis raises the possibility that a cartwheel-like structure may also exist in this species (Malone et al. 2003). Although a core module of evolutionary conserved proteins dictates the formation of daughter centrioles in the presence of preexisting mothers or during de novo assembly (Gönczy 2012), conflicting evidences have been reported concerning the time and the morphological steps preceding the appearance of the cartwheel in different systems. A distinct cartwheel marks the early stage of procentriole formation in *Tetrahymena* and vertebrate cells (Allen 1969; Culver et al. 2009; Guichard et al. 2010). An amorphous disklike structure, the germinative disk, mainly dependent on Sas-4 (Gogendeau et al. 2011), appears in *Paramecium* before the formation of a distinct cartwheel (Dippel 1968). In *Chlamydomonas* the onset of centriole formation is characterized by the presence of an amorphous ring, but the addition of peripheral microtubules occurs after the formation of the cartwheel (Cavalier-Smith 1974; Dutcher 2007). Disk- or ringlike structures have never been observed at the onset of procentriole formation in higher animal cells, but the cartwheel appears as the first visible sign of procentriole formation. However, an "annulus" of amorphous material might be present in vertebrate-multiciliated cells before the cartwheel begins to form (Anderson and Brenner 1971).

Centriole duplication in somatic cells requires the presence of preexisting centrioles that establish the site of the daughter assembly. However, the classical model of centriole duplication in which a preexisting centriole acts as genuine template for the daughter has been overcome by the idea that the mother centriole does not give structural information to its daughter but represents a platform for the recruitment of centriole assembly factors (Rodrigues-Martins et al. 2008; Sluder and Khodjakov 2010). The proximal end of the mother centriole represents, indeed, a suitable environment in which centriolar proteins accumulate. The giant basal body provided by the *Drosophila* sperm is formed by nine triplet microtubules, whereas the centrioles that support the embryonic divisions and the forthcoming somatic mitoses are small and built by nine doublets (Callaini and Riparbelli 1990). This observation strengthens the current idea that the mother centriole does not represent a template for daughter formation, but rather organizes a restricted domain in which specific centriole assembling factors are concentrated (Sluder and Khodjakov 2010; Brito et al. 2012). A general model proposes that Asl/Cep152, Sak/Plk4, Sas-6, Ana2/Still, and Sas-4/CPAP are sequentially recruited at the basal region of the mother centriole to originate the daughter one (Azimzadeh and Marshall 2010; Carvalho-Santos et al. 2010; Gönczy 2012; Jana et al. 2014; Conduit et al. 2015; Fu et al. 2016; Novak et al. 2014, 2016). An individual microtubule triplet among the nine triplets of the centriole wall might exhibit special characteristics that dictate the site of daughter formation (O'Toole and Dutcher 2014). The pericentrin/Dplp, a scaffold protein involved in centrosome organization, forms around the mother centriole a cylindrical network that is interrupted in correspondence of a specific triplet (Lawo et al. 2012; Mennella et al. 2012). It is possible that this gap may concentrate the molecules needed for centriole biogenesis thus limiting the number of forming procentrioles (Pearson 2014). A distinct stalk has been reported to connect the nascent daughter centriole to the proximal wall of the mother in both human-cultured cells (Guichard et al. 2010) and butterfly spermatocytes (Gottardo et al. 2014). This stalk process connecting the forming daughter with the wall of the mother centriole ensures that only one centriole may be assembled at each time for each mother. Another possibility is that the cartwheel of the mother centriole moves from its proximal region to template the assembly of the daughter (Fong et al. 2014). Centriole amplification in multiciliated cells depends on the simultaneous assembly of procentrioles at the proximal end of the mother and around the deuterostomes, clusters of dense material formed at the basis of the daughter centrioles (Al Jord et al. 2014). Thus, the deuterostome pathway of centriole assembly also needs a centriole to start. Therefore, although different, all these models require a preexisting centriole to assemble a daughter.

Besides the canonical duplication cycle, in which only one daughter forms at the proximal end of the mother, the centrioles can also form de novo in the absence of preexisting mothers, under experimental conditions (Kallenbach and Mazia 1982; Khodjakov et al. 2002; La Terra et al. 2005; Peel et al. 2007; Rodrigues-Martins et al. 2007; Uetake et al. 2007) or during natural parthenogenesis (Callaini et al. 1999; Riparbelli et al. 2005, 2010).

19.2.2 Parent Centrioles: A Generational Diversity

Vertebrate cells usually begin interphase with a centrosome that contains one pair of centrioles that have distinctive morphologies and different abilities to nucleate microtubules (Piel et al. 2000; Bornens 2012). During G1/S, the centrioles start to duplicate, and each forms a short daughter, the procentriole, that elongates during G2/M. Although of similar size, mother and daughter are not functionally equivalent. The mother centriole recruits, indeed, more centrosomal material and displays characteristics distal appendages involved in microtubule binding and membrane docking (Nigg and Stearns 2011). Only the mother centriole can assemble the primary cilium in nondividing cells. The daughter centriole will acquire the distal appendages at the end of the interphase and the ability to organize a primary cilium a cell cycle later (Fırat-Karalar and Stearns 2014). Therefore, only the cell receiving the older mother centriole is able to form a primary cilium earlier in the cell cycle (Anderson and Stearns 2009). The different age between the parent centrioles has been involved in multiple cellular and developmental processes (Pelletier and Yamashita 2012).

By contrast, parent centrioles in the *Drosophila* centrosome lack a distinct structural dimorphism (Callaini et al. 1997; Gonzalez et al. 1998) and may be only recognized by markers such as centrobin (Januschke et al. 2013) and asterless (Novak et al. 2014) or by the specific accumulation to the mothers of some centrosomal proteins (Conduit and Raff 2010; Fu and Glover 2012; Mennella et al. 2012). In addition to an intrinsic asymmetry, the parent centrioles often display a functional asymmetry and an asymmetric inheritance that is relevant in stem cell division of both *Drosophila* and vertebrate cells (Pelletier and Yamashita 2012; Reina and Gonzalez 2014). However, the lack of centrioles in *Sas-4 Drosophila* mutants does not affect the asymmetric division of larval neuroblasts (Basto et al. 2006) and male germline stem cells (Riparbelli and Callaini 2011) suggesting that the role of the centrioles may be redundant in fly stem cells.

Centrioles have also the additional role as the platforms for the assembly of cilia and flagella, specialized structures mainly involved in promoting several key cellular processes, namely, signal transduction, cell division, and cell motility (Chavali et al. 2014; Nachury 2014). During interphase or in quiescent cells, the centriole pair migrates to the periphery, and the mother docks to the cell membrane and converts into a basal body that nucleates the ciliary axoneme. Since the structure of the axoneme depends on the centriole architecture (Gottardo et al. 2013), eventual abnormalities of the centriole would ultimately give rise to specific cell defects. Various human diseases and pathologies, including ciliopathies, developmental defects, male sterility, and microcephalies, are likely to reflect structural defects in the underlying centrioles from which ciliary and flagellar axonemes are nucleated (Nigg and Raff 2009).

19.2.3 Control of Centriole Duplication: A Key Process in Centrosome Biology

Having only one centriole may result in monopolar spindles and mitotic arrest (Kirkham et al. 2003), whereas the absence of centrioles may have consequences on the proper orientation of the mitotic spindle (Basto et al. 2008). By contrast, too many centrioles are dangerous to animal development and may result in multipolar spindles and defects in the formation of the primary cilia (Holland et al. 2010; Mahjoub and Stearns 2012; Godinho et al. 2014). Therefore, in dividing somatic cells, the centriole duplication is subjected to strict temporal and numerical controls, so that the centrioles duplicate exactly once during the cell cycle concurrently with DNA synthesis and only one daughter is assembled for each mother (Loncarek and Khodjakov 2009). Thus, the assembly of additional centrioles has to be prevented once the first one is formed.

Basic to avoid harmful centriole duplication during mitosis is the control of their disengagement at metaphase-to-anaphase transition by separase and Plk1 (Tsou and Stearns 2006; Tsou et al. 2009; Sluder and Khodjakov 2010; Wang et al. 2011). The loss of orthogonal configuration of the parent centrioles is thought, indeed, to represent the licensing factor that allows the assembly of a new centriole, the procentriole, during the G1/S phase (Agircan et al. 2014). The procentriole elongates during G2/M enabling the duplication of the centrosome and the forthcoming formation of the bipolar spindle. However, it has been recently reported in human-cultured cells that the loss of orthogonal orientation between the centrioles during anaphase does not prevent centriole reduplication (Shukla et al. 2015). Rather, the centriole replication block relies on the close association between parent centrioles, which is established at the time of duplication. Therefore, the centriole duplication arrest might be already lost before metaphase/anaphase transition (Shukla et al. 2015). In agreement, with this hypothesis, orthogonally oriented centrioles may duplicate in *Drosophila* wing imaginal disk cells (Vidwans et al. 2003), and up to four procentrioles form at the proximal end of the still-engaged centrioles in butterfly primary spermatocytes (Gottardo et al. 2014).

Centriole duplication may be uncoupled from the DNA replication cycle, and the rule of "only one centriole forming per mother" may be overcome. This is the case of mammalian multiciliated epithelial cells such as those in the respiratory and reproductive tracts in which hundred basal bodies arise from a single centriole or from deuterostomes (Dawe et al. 2007). Another emblematic case occurs during male gametogenesis of several organisms, human included, in which the centriole duplicates from meiosis I to meiosis II in the absence of DNA replication. By contrast, in insects centrioles duplicate only one time during the first prophase concurrently to DNA replication. Consequently, the sperm receives only one centriole, but the egg needs one pair of centrosomes each containing a centriole pair to assemble the poles of the first mitotic spindle. Thus, the centriole provided by the sperm must duplicate twice in the absence of DNA synthesis. However, it has been shown in echinoderms and *Drosophila* that centrosome continues to duplicate when DNA replication is inhibited by aphidicolin (Nagano et al. 1981; Raff and Glover 1988).

19.2.4 Life Without Centrioles: Are They Dispensable for Development?

The critical functions of the centrioles may explain why these organelles are widely distributed within the eukaryotic tree (Debec et al. 2010). However, there are various examples of cell systems lacking centrioles, including higher plant cells and the meiotic spindle of most animal oocytes (Szollosi et al. 1972; Theurkauf and Hawley 1992; Lambert 1993; McKim and Hawley 1995; Balczon 1996; Matthies et al. 1996). *Drosophila Sas-4* (Basto et al. 2006) and *Sak/Plk4* (Bettencourt-Dias et al. 2005) mutants develop up to the adult stage in almost total or partial absence of centrioles. Moreover, the *Drosophila* cell line (1182–1184) that constitutively lacks centrioles does not recruit PCM proteins but assemble functional spindles and divides quite well in culture (Debec and Abbadie 1989). All these observations point to alternative processes of spindle organization. However, the centrioles also play an important role in the assembly of the ciliated structures, including the axonemes of sperm cells and sensory neurons. Thus, the flies lacking centrioles are sterile and uncoordinated. Moreover, the centrioles are essential to the early embryonic division cycles and to the proper execution of male meiosis in *Drosophila* (Stevens et al. 2007; Riparbelli and Callaini 2011). The spindles can form in both acentriolar embryos and spermatocytes, but they are abnormal. Therefore, centrioles are dispensable for cell division in some tissues of the fly, but they are essential in others (Rodrigues-Martins et al. 2008).

When centrioles are absent, the centrosomes cannot be organized. Thus, a chromatin-mediated pathway of microtubule organization serves as a default mode that ensures bipolar spindle formation. It has been demonstrated in *Xenopus* egg cytoplasmic extracts that bipolar spindles form in the complete absence of centrosomes upon DNA beads (Heald et al. 1996). This process depends on a Ran gradient that surrounds the chromatin and activates regulatory proteins involved in microtubule nucleation (Carazo-Salas et al. 1999; Wilde and Zheng 1999; Gruss et al. 2002; Trieselmann et al. 2003; Tsai et al. 2003; Forbes et al. 2015). The acentrosomal pathway of spindle formation is also present in mitotic cells, but this mechanism is masked by the centrosomes that represent the main site for spindle organization (Meraldi and Nigg 2002; Wadsworth and Khodjakov 2004; Hinchcliffe 2014; Heald and Khodjakov 2015).

The constant presence of the centriole in ciliated/flagellated cells and its redundant role to organize the bipolar mitotic spindles in various systems suggests that the centriole presumably evolved from an ancestral basal body primarily involved in axoneme nucleation (Bornens and Azimzadeh 2007; Carvalho-Santos et al. 2011; Azimzadeh 2014). Only later the centriole will be the center for recruitment of the components involved in microtubule organization. Thus, the centriole becomes the master for centrosome organization and dynamic, presumably giving some selective advantages to the cell. However, the acentrosomal spindle assembly pathway remains. Remarkably, the planarian *Schmidtea mediterranea* regenerates damaged or lost tissues without intervening centrioles that, albeit present, are only used to assemble ciliary structures (Azimzadeh et al. 2012).

19.3 Fertilization: Only One Centrosome at the Beginning of Development

The centrosome is instrumental to organize the astral array of microtubules that drives pronuclear migration and apposition during egg fertilization (Briedis and Elinson 1982; Schatten 1982; Schatten and Schatten 1986; Callaini and Riparbelli

1996; Riparbelli et al. 2000). Moreover, the duplicated centrosome organizes the first bipolar mitotic spindle that supports the proper distribution of the parental chromosomes to the daughter cells. Because centrosomes are present during male and female gametogenesis, their number must be controlled at fertilization to avoid multipolar spindle formation and abnormal chromosome segregation that may lead to developmental failures (Sathananthan et al. 2006; Sun and Schatten 2007; Scheer 2014).

Since at fertilization the sperm provides its basal body from which the zygotic centrosome will be organized, most animals have developed different strategies to inactivate the centrioles of the female germline cells and avoid the formation of supernumerary centrosomes. Usually the centrioles degenerate with the consequential dispersal of the pericentriolar material and the loss of their reproductive potential. The oocytes of a large group of organisms including mollusks, annelids, and echinoderms eliminate their centrioles during meiosis (Krioutchkova and Onishchenko 1999; Manandhar et al. 2005; Crowder et al. 2015). The poles of the meiotic spindles are astral, and the centrioles are extruded during polar body formation. In echinoderms, the mature egg inherits one centriole from meiosis II, but it is inactivated before the first mitosis losing its capacity to duplicate (Sluder et al. 1989; Uetake et al. 2002; Shirato et al. 2006). It has been demonstrated in starfish oocytes that the remnant centrioles are the daughters and that the mother centrioles are selectively discarded into the polar bodies by a dynein-driven transport (Borrego-Pinto et al. 2016). Although present in the egg, the daughter centriole loses its ability to recruit centrosomal material from the cytoplasm being unable to organize a functional centrosome (Borrego-Pinto et al. 2016). By contrast, the oocytes of insects, worms, and mammalians lose their centrioles before the onset of meiosis and shift toward an acentrosomal pathway of meiotic spindle organization (Huettner 1933; Mahowald and Strassheim 1970; Mikeladze-Dvali et al. 2012). Interestingly, centriole elimination in the *Drosophila* female germline has been correlated with the loss of pericentriolar material and the downregulation of Polo kinase (Pimenta-Marques et al. 2016).

19.3.1 The Sperm Basal Body: An Attracting Center for Maternal Centrosomal Components

Boveri first (1901) had the brilliant intuition that the zygotic centrosome was a chimeric organelle assembled by the contribution of both the gametes. This idea was later confirmed in various organisms by a growing number of evidences (Schatten 1994; Schatten and Sun 2011). The sperm carries, indeed, a naked basal body that had lost its pericentriolar material involved in microtubule nucleation through a process called in mammals "centrosome reduction" (Manandhar et al. 2000). Therefore, the sperm does not provide a functional centrosome at fertilization, but its basal body has to recruit the pericentriolar material from the egg cytoplasm to assemble an active microtubule-organizing center (Delattre and Gonczy 2004). How the sperm basal body attracts the maternal proteins to build a functional centrosome has remained for a long time a mystery. Only recently, the molecular mechanisms driving centrosome assembly in interphase and centrosome maturation during mitosis of somatic cells have been clarified. A small module of proteins conserved from flies to humans has been implicated in the process that leads to γ-tubulin accumulation and to the increase in microtubule nucleation. Among them, there are proteins tightly associated to the centriole wall such as Asl/Cep152 (Dzhindzhev et al. 2010) and Sas-4/CPAP (Gopalakrishnan et al. 2011), proteins that form a scaffold around the centriole such as the *Drosophila* orthologue pericentrin-like protein (Dplp) (Martinez-Campos et al. 2004; Mennella et al. 2012), and proteins that help to maintain γ-tubulin at the centrosome such as cnn and Spd-2 (Dix and Raff 2007; Giansanti et al. 2008; Conduit et al. 2014). It has been demonstrated in *Drosophila* somatic cells that Asl initiates the recruitment of Spd-2 and cnn at the pericentriolar material (PCM) to help the recruitment of γ-tubulin to the developing centrosome (Fu and Glover 2012; Conduit et al. 2014, 2015; Fu et al. 2016). This mechanism seems rather conserved from worms to humans even if slight modifications have been described. For example, in vertebrate cells, pericentrin plays a main role in assembling interphase centrosomes and performs critical functions during centrosome maturation at mitosis (Doxsey et al. 1994; Dictenberg et al. 1998; Zimmerman et al. 2004). By contrast, in flies, Dplp seems to play minor roles in centrosome biogenesis, and the function of Dplp is not essential for centrosome maturation at mitosis (Martinez-Campos et al. 2004). However, depletion of Dplp affects the recruitment of interphase PCM components in fly-cultured cells (Mennella et al. 2012), and its association with cnn is required to drive centrosome separation during the early embryonic divisions (Lerit et al. 2015). We do not know if the sequential recruitment of centrosomal proteins as described in *Drosophila* somatic cells may be working during the reconstitution of the first functional zygotic centrosome. However, in agreement with this possibility, it has been shown that antibodies against Asl label the zygotic centrioles in *Drosophila* just before the centrosomes are properly assembled (Blachon et al. 2014).

19.3.2 The Assembly of the First Zygotic Spindle: How Many Centrioles the Egg Inherits at Fertilization?

The centrioles of male germ cell in most animals duplicate during both meiosis I and II, albeit DNA replication does not. Thus, the spermatids inherit two centrioles that may be transferred by the sperm to the oocyte at fertilization (Krioutchkova and Onishchenko 1999).

However, like somatic cells in which the assembly of primary cilia reflects the functional asymmetry between parent centrioles, in differentiating spermatids only one centriole is able to organize the axoneme that will support sperm cell motility.

If the centrioles are coaxial as in Hydrozoans (Afzelius 1971), Xiphosurans, and some Arachnids (Jamieson 1987; Michalik and Ramírez 2014), only the distal one nucleates the axoneme, whereas the proximal centriole is very short and reduced. This disposition is similar to that found in type 1 sensory neurons of the insects that have ciliary structures in which the distal centriole is retained to be the mother and the proximal one the daughter (Gottardo et al. 2015c). It is actually unclear if both centrioles are functionally equivalent and are able to duplicate properly in the egg cytoplasm. In particular, in non-rodent mammals, including humans, the proximal centriole remains intact, whereas the distal one becomes modified losing the nine-triplet structure of microtubules (Sathananthan et al. 1991, 1996; Manahndar et al. 2000). Parent centrioles may be, indeed, highly modified during the process of spermiogenesis (Manandhar et al. 2005; Avidor-Reiss et al. 2015).

Otherwise, the secondary spermatocytes of nematodes, earthworms, and insects have one centriole at each spindle pole, and the spermatids inherit only one basal body from which the flagellar axoneme elongates (Favard 1961; Anderson et al. 1967; Friedländer and Wahrman 1971). In particular, both the parent centrioles in insects organize cilium-like structures during spermatogenesis and are able to nucleate a distinct sperm axoneme. By contrast, the daughter centrioles in most animal male germ cells are unable to nucleate an axoneme when the spermatid elongates.

When only one functional centriole is supplied to the oocyte, it must duplicate twice to give origin to the centrosome pair that will support the formation of the poles of the first zygotic spindle. Recent reports based on conventional immunofluorescence observations have revealed the presence of a dot-like structure close to the proximal region of the centriole in *Drosophila* spermatids (Blachon et al. 2009). Since this structure contains the typical centriolar proteins Ana1 and Bld10/Cep135 and its assembly depends on Sak/Plk4 and Sas-6, it has been suggested that it may represent an early intermediate structure in centriole assembly (PCL). Moreover, it has been proposed that the PCL persists during spermatid differentiation and that PCL supplied by the sperm to the oocyte drives the origin of the second centriole at fertilization (Blachon et al. 2014; Avidor-Reiss et al. 2015). These observations represent an unexpected feature in centriole biogenesis and may open an interesting debate on centriole duplication. However, conflicting evidences have been reported on the presence of the PLC in mature sperm. Ultrastructural observations revealed, indeed, the presence of a distinct procentriole that assembles next to the basal body at the beginning of spermatid elongation, but this incomplete centriole is no longer visible in mature sperm (Gottardo et al. 2015b). This finding agrees with similar observations in the lepidopteran *Pieris* where some procentrioles assembled during prophase of the first meiosis but disappear from mature sperm (Gottardo et al. 2014). However, it is possible that the sperm provides a morphologically undetectable precursor that could become active at fertilization (Avidor-Reiss et al. 2015). The centriole structure varies within the different hexapod groups (Dallai 2014). Exceptions to the single basal body condition, due to a nonconventional spermatogenesis, have been reported in the spermatids of some insect species in which two functional centrioles, both able to nucleate their own axoneme, are present. The sperm flagellum of Anoplura and Mallophaga (Baccetti et al. 1969), as well as related Psocoptera (King and Ahmed 1989) and gall-midge *Semudobia* (Dallai and Mazzini 1989), contains, indeed, twin axonemes. Interestingly, the early spermatids of the thrips *Haplothrips* (Thysanoptera) show three parallel centrioles, and the sperm flagellum at the beginning of its elongation is composed of three axonemes (Paccagnini et al. 2006). The extreme condition is represented by the multiflagellated sperm cell of the termite *Mastotermes darwiniensis* that contains hundred centrioles each able to nucleate an axoneme during spermatid differentiation (Baccetti and Dallai 1978; Riparbelli et al. 2010). The fate and contribution of the supernumerary basal bodies at fertilization is unknown. Conversely, some insect species have aflagellate sperm that maintained distinct centrioles (Dallai et al. 2016). The spermatids of Rhinotermitidae and Termitidae (Baccetti et al. 1981) and the coccid *Matsucoccus* (Paoli et al. 2015) have, indeed, distinct centrioles but lack axonemal microtubules. Even if unable to nucleate axonemal microtubules, the centrioles might be involved in the assembly of the first zygotic spindle. All these findings suggest that insect male gametogenesis may represent a useful model for the study of the assembly and function of the basal body during axoneme formation and fertilization.

A common feature from insects to mammals is that the entire sperm enters the egg during fertilization (Karr 1991; Simerly et al. 1993; Schatten 1994). The sperm centriole remains linked to the apical end of the sperm tail through early development, and, at least in *Drosophila*, it is able to duplicate and recruit maternal components to assemble a functional centrosome (Riparbelli and Callaini 2010).

Rodent sperm are motile, but their basal bodies are unable to duplicate and to organize functional centrosomes. In mouse the centrioles are inherited at fertilization with the sperm, but they become soon unidentifiable by electron microscopy (Manandhar et al. 1998). Thus, the early development of mouse embryos occurs in the absence of centrioles (Maro et al. 1985; Manandhar et al. 1998; Courtois et al. 2012). Several acentriolar MTOCs appear in the oocyte cytoplasm at the end of meiosis. These structures have an important role in the assembly of the anastral mitotic spindles that drive the nuclear divisions of the early mouse embryo (Szollosi et al. 1972; Maro et al. 1985; Manandhar et al. 1998). The origin of the acentriolar MTOCs is unclear. They have been proposed to form at the onset of the first meiosis though a microtubule self-organization process (Schuh and Ellenberg 2007) or from preexisting inactive centrosome precursors (Calarco 2000). It has been shown that the large microtubule asters close to the nuclear envelope at the beginning of cleavage are derived by the gradual fusion of small PCM foci previously dispersed in the cytoplasm of the growing oocytes (Łuksza et al. 2013). Centrioles appear later at the blastocyst stage (Magnuson and Epstein 1984; Szollosi and Ozil 1991; Gueth-Hallonet et al. 1993) by an unknown mechanism. It is possible that the assembly of centrioles in the mouse early embryos follows unconventional routes. Plk4, a protein kinase known as a master regulator of centriole formation, is essential for spindle pole organization in mice embryos, but not for centriole assembly (Coelho et al. 2013).

19.3.3 Too Many Sperms in the Egg: The Case of Polyspermy

Many sperms reach the oocyte during the fertilization process. This can be dangerous. If more than one sperm enters the egg, the supernumerary asters can be assembled leading to embryo death. To solve this problem, most animals have evolved different mechanisms to restrict the fertilization process to only one sperm and to protect the egg from polyspermy (Wong and Wessel 2006). However, several sperms may enter the large yolky eggs of some fishes, amphibians, reptiles, and birds without detrimental effects on embryonic development (Snook et al. 2011; Iwao 2012). Polyspermy seems to be a characteristic of yolky eggs that due to their large surface area are unable to organize a functional cortical reaction to block the entry of extra sperm. However, physiological polyspermic fertilization has been also described in the small eggs of the Ctenophoran *Beroe ovata* (Gould and Stephano 2003) and of the basal insect *Tetrodontophoora bielanensis* (Jura and Krzysztofowicz 1992). These findings raise the further question of how the ability to assemble the first zygotic spindle is restricted to only one basal body and how the supernumerary centrioles are inactivated. Presumably, the sperm nucleus with the larger aster and closer to female pronucleus might have advantage in the formation of the zygote nucleus. Observations on amphibian eggs indicate that DNA synthesis is faster in the male pronucleus nearest to the female one and the bipolar spindle is assembled first in the zygotic nucleus than in the supernumerary sperm nuclei (Wakimoto 1979; Iwao et al. 1993). Moreover, the centrosomes do not separate, and the sperm nuclei degenerate before the first cleavage division. It has been suggested that the arrest in the cell cycle and the further degeneration of the supernumerary sperm nuclei in amphibian eggs might be correlated to a reduced level of the maturation-promoting factor (MPF) activity at the egg periphery (Iwao and Elinson 1990; Iwao et al. 2002). In birds, the degradation of the supernumerary sperm may be the result of the DNase activity of the oocyte (Stepinska and Bakst 2007).

In *Drosophila melanogaster*, where polyspermy represents a rare event (Hildreth and Lucchesi 1963; Snook and Karr 1998), the supernumerary centrioles organize bipolar spindles that allow haploid chromosome congression to the metaphase plate but do not support metaphase-anaphase transition (Callaini and Riparbelli 1996; Riparbelli et al. 2010). Interestingly, centrosomes duplicate at the spindle poles despite the halted chromosome movement and the arrest at the beginning of anaphase. In some *Drosophila* species of the *obscura* group, a large proportion of the eggs may be fertilized by many sperms (Snook and Markow 2001).

19.4 Parthenogenesis: No Need for the Male Gamete

The male gamete plays a crucial role during the development of most organisms since it supplies half of the chromosomes needed to restore diploidy and provides the active centriole involved in the assembly of the first zygotic centrosome. However, the paternal chromosomes are not essential for proper development since the oocyte of several species may restore diploidy by unconventional meiosis, fusion of two polar bodies or gamete duplication, and in some cases the oocytes develop as haploids (Wright 1973; Lamb and Willey 1987; Suomalainen et al. 1987; Mirzaghaderi and Hörandl 2016). Since the oocytes lose or inactivate their centrosomes, the embryonic development without the inheritance of the sperm centriole at fertilization would be unexpected. However, there are several examples of animal species that naturally reproduce in the absence of the male centriole contribution. This kind of development raises the question if the sperm centriole plays essential roles in the egg development. Female meiosis proceeds in most organisms such as humans, mouse, and *Drosophila* by a centrosome-independent pathway of spindle organization (Dumont and Desai 2012; Bennabi et al. 2016; Severson et al. 2016) suggesting that the egg may accomplish chromosome separation also in the absence of centrioles. Some animal cell types also may assemble mitotic spindles and undergo several rounds of mitosis in the absence of functional centrosomes (Debec et al. 2010) by chromatin-based Ran and Augmin pathways (Clarke and Zhang 2008; Goshima and Kimura 2010; Heald and Khodjakov 2015). However, *Drosophila Sas-4* mutants that lack centrioles can develop to adulthood (Basto et al. 2006), but the early embryonic divisions fail (Stevens et al. 2007). This suggests that at least in *Drosophila* the acentrosomal pathway of spindle organization may support proper mitotic divisions only after the syncytial blastoderm stage. This absolute need of centrioles during earlier stages of development could explain why unfertilized eggs that assemble anastral bipolar spindles do not develop further. For example, unfertilized eggs from the dipteran *Sciara* resume meiosis after artificial activation and organize anastral bipolar spindles that partially support chromosome movement and sister chromatid separation (De Saint Phalle and Sullivan 1998). However, these spindles soon collide giving rise to developmental failures. In wild type fertilized *Drosophila* eggs, meiosis resumes and four haploid complements are formed, but only the most internal complement will give rise to the female pronucleus. The other complements remain at the egg surface in association with acentrosomal monastral arrays of microtubules and undergo several round of DNA duplication without division (Callaini and Riparbelli 1990). By contrast, unfertilized *polo* mutant eggs resume meiosis and form four distinct haploid complements that have the same potential and divide by means of bipolar anastral spindles (Riparbelli et al. 2000). However, the bipolar spindles are unable to support proper chromosome segregation and irregular mitotic figures with abnormal chromatin clusters soon appear. Anastral mini spindles have been also observed in activated *nod* (Theurkauf and Hawley 1992) and γ*TUB37C* (Tavosanis et al. 1997) mutant *Drosophila* oocytes. Attempts to induce parthenogenetic development have been reported in several organisms, including mammals. However, in the absence of centrioles, the development arrests at early stages of embryogenesis (Schatten 1994; Manandhar et al. 2005; Paffoni et al. 2007; de Fried et al. 2008).

All these examples may be attractive to investigate the process of spindle assembly in the absence of centrioles, but they are not relevant to explain naturally occurring parthenogenetic development in which functional mitotic spindles are assembled. Therefore, the centrioles represent a hallmark of the early embryonic development, and their absence leads to developmental failures. Indeed, the cell cycle may be resumed in *Xenopus* eggs by pricking with glass needles, but cleavage does not occur without functional centrosomes (Hara et al. 1980; Bornens et al. 1987). Accordingly, the injection of purified centrosomes mimics fertilization and leads to tadpole formation (Tournier et al. 1989). Interestingly, foreign centrosomes from various cell types can induce successful parthenogenetic development in *Xenopus* (Klotz et al. 1990).

How can eggs lacking centrosomes assemble functional mitotic spindles and properly develop in the absence of inherited centrosomes at fertilization is a central question in the mechanism of parthenogenesis. Moreover, how can a cell devoted to prevent the formation of supernumerary centrosomes enable de novo centrosome assembly? How does this cell avoid formation of multipolar spindles? Why cells known to contain large stockpiles of centrosomal subunits do not usually form multiple centrosomes? What are the conditions to override the control mechanisms involved in checking the correct centrosome number? How common is the potential for maternal centrosome formation in other organisms? However, despite the large number of organisms that undergo parthenogenetic development, only a few papers have until now analyzed the process of the first mitotic spindle formation in the absence of centrioles.

19.4.1 The Case of Hymenopterans: A Tale of Cytoplasmic Asters

Insect species belonging to the order Hymenoptera, i.e., ants, wasps, and bees, are characterized by haplodiploidy, a specialized mode of facultative parthenogenesis in which males arise from haploid unfertilized eggs, whereas females emerge from fertilized diploid eggs. There are, however, examples of hymenopteran eggs infected by bacteria of the genus *Wolbachia* in which the diploidy is restored and the eggs develop as females without fertilization (Stouthamer and Werren 1993; Stouthamer et al. 1999). When bacteria were removed with tetracycline or high temperature, the unfertilized eggs develop as haploid and give origin to males. Centrosome inheritance has been examined in unfertilized haploid eggs of *Nasonia vitripennis* that give origin to males (Tram and Sullivan 2000) and in *Wolbachia*-infected *Muscidifurax uniraptor* eggs from which females emerge (Riparbelli et al.

1998). Both these studies revealed the same mechanism of zygotic spindle assembly in the absence of inherited centrosomes whether or not these embryos are haploid or diploid and despite the different sex of the offspring. Female meiosis resumes after the oocyte passes throughout the oviduct and leads to the formation of four haploid complements the innermost of which represent the female pronucleus, whereas the others remain at the surface and form the polar bodies. Chromosome separation is supported by two tandem-aligned anastral meiotic spindles that are orthogonally oriented to the egg membrane. The twin spindles are separated by a prominent astral array of microtubules that does not contain centrioles and soon disappears after completion of meiosis before the zygotic spindle is assembled. The transition from metaphase to anaphase of the first meiosis is underlined in *Muscidifurax uniraptor* unfertilized eggs by the reorganization of the cortical microtubule network in heterogeneous small clusters in which the microtubules appear to radiate from central foci (Riparbelli et al. 1998). As meiosis progresses, the cortical asters increase in size, and their number amplifies to about three hundred at the beginning of prophase of the first mitosis when two of them interact with the female pronucleus (Riparbelli et al. 1998). The asters have a transient nature, and their number decreases dramatically during progression through the first zygotic division and disappear after the third nuclear division. Real-time observations of *Nasonia vitripennis* unfertilized eggs demonstrated that the cortical asters gradually move to the inner cytoplasm, and two of them reach in rapid succession the female pronucleus (Tram and Sullivan 2000). As the first aster contacted the nucleus, the second aster is captured within a few minutes to organize the first mitotic spindle (Tram and Sullivan 2000). Although the interaction of the first two asters with the female pronucleus appears a stochastic event, all the eggs develop successfully.

19.4.2 **Drosophila mercatorum***: An Attempt for Parthenogenetic Development*

Drosophila mercatorum reproduces sexually, although some female-only strains have been described (Kramer and Templeton 2001). In this case, the parthenogenetic development has low efficiency, and only 8–10% of the eggs develop properly pointing to a stochastic phenomenon rather than to the presence of operative mechanisms. The eggs of *Drosophila mercatorum*, like those of most insects, are arrested at metaphase of the first meiosis and do not retain maternal centrosomes that are lost early during oogenesis. Only one giant microtubule aster, likely involved in maintaining the orientation of the meiotic apparatus, is present between the twin meiotic spindles during the second meiosis (Fig. 19.1). It contains centrosomal components and gamma-tubulin but not centrioles. None of the centrosomal proteins are found at the spindle poles. The central aster does not take part in the assembly of the zygotic spindle and disappears at the end of meiosis

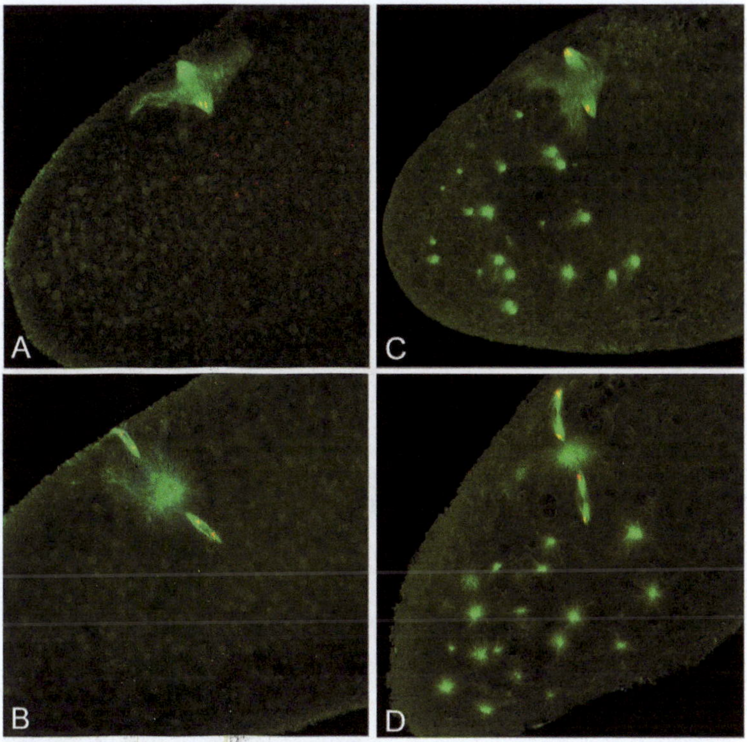

Fig. 19.1 Meiosis in unfertilized dioic (**a**, **b**) and parthenogenetic (**c**, **d**) *Drosophila mercatorum* oocytes. A large microtubule cluster is present in the middle of the anaphase I spindle in both dioic (**a**) and parthenogenetic (**c**) oocytes. This cluster transforms during meiosis II in a distinct microtule aster placed between the twin spindles (**b**, **d**). The anterior half of the parthenogenetic oocyte displays many scattered asters of different size. Microtubules, *green*; DNA, *red*

(Riparbelli and Callaini 2003). A variable number of distinct comet-like microtubule-based structures (from 1 to 20) appear in the anterior region of the parthenogenetic eggs during anaphase of the first meiosis. Soon the comet-like structures transform in discrete asters with a distinct central focus, and their number notably increases during the following mitoses. Interestingly, the average number of the asters roughly doubles from telophase of the second meiosis to the first and second mitoses, suggesting a possible duplication cycle synchronized with mitotic progression (Riparbelli and Callaini 2003; Eisman and Kaufman 2007). This is consistent with the observation of several aster pairs during prophase. The cytoplasmic asters are visible in the anterior region of the embryo until the twelfth nuclear division cycle (Fig. 19.2). The formation of the first mitotic spindle requires the close proximity of one or two asters with the female pronucleus and implies the restoration of diploidy that usually occurs by gamete duplication, i.e., replication of the pronucleus and fusion of the two daughter nuclei (Carson 1973; Eisman and Kaufman 2007). The low probability to accomplish both these processes explains

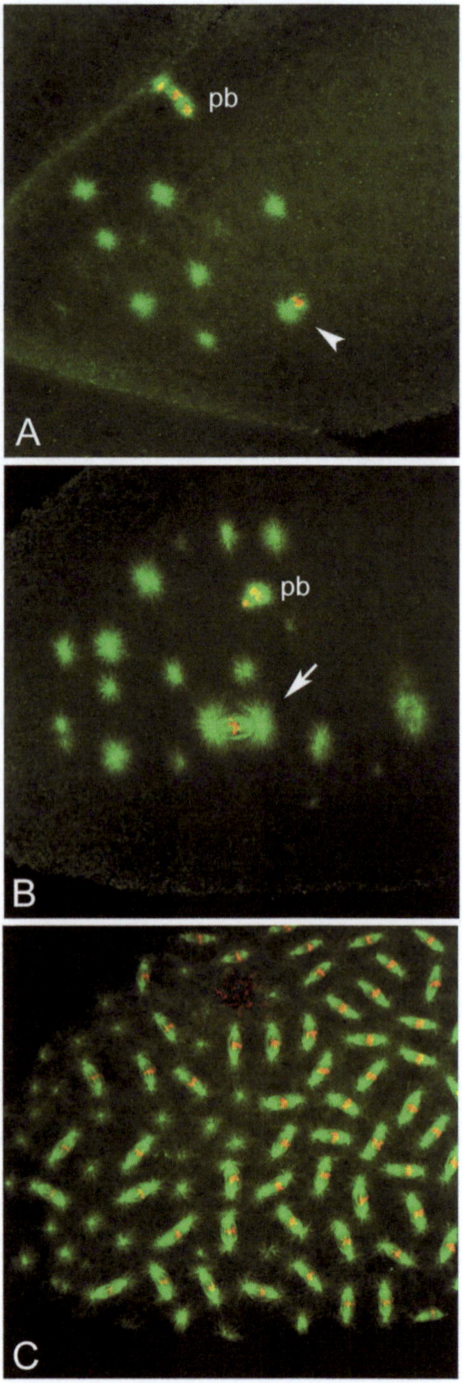

Fig. 19.2 Early development of the parthenogenetic *Drosophila mercatorum* egg. Contact between a cytoplasmic aster and the female pronucleus (**a**, *arrowhead*). First mitotic spindle during metaphase (**b**, *arrow*). Syncytial blastoderm embryo with many remnant asters in the anterior region (**c**). Microtubules, *green*; DNA, *red*. *pb* polar bodies

the small developmental rate observed in the parthenogenetic strain of *Drosophila mercatorum*. An additional problem that restricts the successful embryonic development comes from the observation that the formation of the asters is restricted to about half of the unfertilized eggs, whereas the others are unable to organize astral arrays of microtubules (Riparbelli and Callaini 2003; Eisman and Kaufman 2007). Moreover, multiple asters may interact with the chromatin leading to abnormal spindles and developmental failures (Eisman and Kaufman 2007). This indicates that in *Drosophila mercatorum*, unlike Aphids or Hymenopterans in which the contact with the female pronucleus is restricted to only one or two asters, a mechanism limiting aster interaction with the female pronucleus does not work. Rather, spindles with multiple poles, monopolar spindles and anastral bipolar spindles are often found (Riparbelli and Callaini 2003; Eisman and Kaufman 2007) suggesting that acentrosomal and centrosomal routes for spindle assembly are working whether cytoplasmic asters are present or not (Riparbelli and Callaini 2008). Therefore, the *Drosophila mercatorum* egg assembles cytoplasmic asters de novo, but it is unable to accurately control their behavior to ensure that only one or two asters may contact the female pronucleus. This suggests that some *Drosophila mercatorum* strains incidentally tried to develop without the male gametes and developed mechanisms leading to parthenogenesis. However, the various abnormal phenotypes observed and the reduced offspring indicate that the parthenogenetic process in *Drosophila mercatorum* is largely inefficient. The large eggs compared with those of other insects that fully exploit parthenogenesis may represent an important limiting factor to properly monitor the process required for successful development (Riparbelli and Callaini 2003). Moreover, only the anterior cytoplasm of about half of the unfertilized eggs have the potential to assemble microtubule asters suggesting that parthenogenesis is a recent acquisition in *Drosophila mercatorum*.

19.4.3 Does Egg Activation Induce the Formation of Cytoplasmic Asters? The Aphid Story

A rise of intracellular calcium that occurs at fertilization is sufficient to trigger oocyte activation and resumption of meiosis in most animal oocytes (Sartain and Wolfner 2013). However, fertilization is not necessary to activate the insect oocyte that is activated during its transit through the female reproductive tract (Horner and Wolfner 2008). Studies in *Drosophila* have suggested, indeed, that this travel triggers the influx of calcium from the external environment, thereby initiating egg activation and resumption of meiosis (Kaneuchi et al. 2015). The cytoplasmic asters appear soon after resumption of meiosis suggesting that oocyte activation may be also involved in triggering biochemical processes leading to aster assembly. By contrast, the activation of the oocyte in parthenogenetic viviparous aphids does not require neither the travel through the female reproductive tract nor fertilization

(Blackman 1987). The progeny in viviparous aphids is not produced from laid eggs but develops inside small ovarian follicles and is born alive as nymphs. Remarkably, the developing embryos have already their daughters inside them like Russian dolls. Diploidy is restored by a modified meiosis in which the oocyte undergoes a single maturation division. The analysis of the embryonic development in the parthenogenetic viviparous pea aphid *Acyrthosiphon pisum* reveals that from one to three small asters assemble in the cortical cytoplasm of the oocyte as soon as meiosis resumes (Fig. 19.3; Riparbelli et al. 2005). Two of them are seen in contact with the female pronucleus to organize the first mitotic spindle. However, due to the lack of in vivo analysis, we cannot exclude the possibility that the aster pair was the result of a duplication process. The number of the asters roughly doubles at the beginning of the first mitosis and decreased afterwards. However, in contrast to hymenopterans, the asters are also detected in later stages of embryonic development.

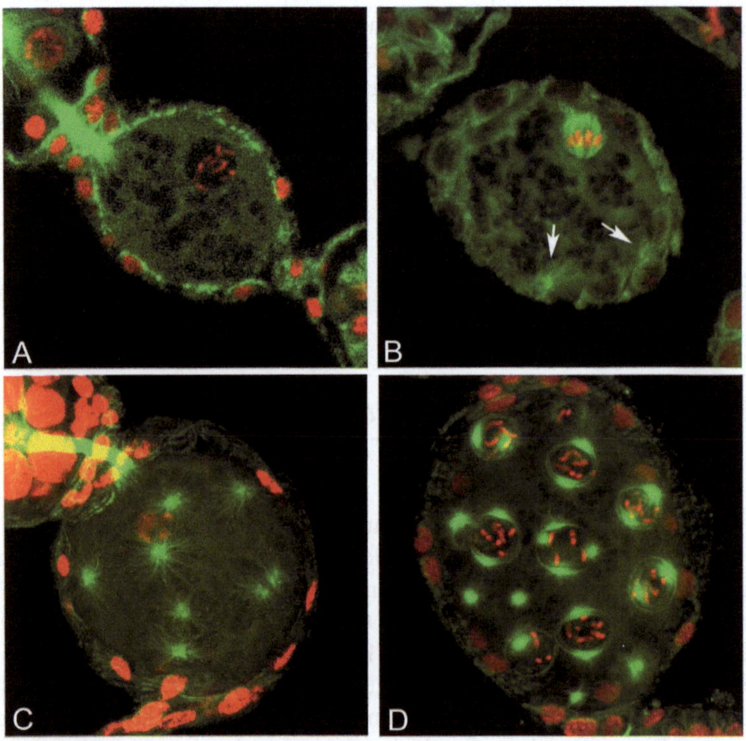

Fig. 19.3 Parthenogenetic development of the viviparous pea aphid *Acyrthosiphon pisum*. Prophase (**a**) and metaphase (**b**) of the first meiosis: the cytoplasmic asters (*arrows*) appear during metaphase. Some asters contact the female pronucleus to organize the first mitotic spindle (**c**). Prometaphase of the third mitosis with some remnant asters (**d**). Microtubules, *green*; DNA, *red*

19.4.4 Development of the Unfertilized Folsomia *Egg: The Conservation of a Common Mechanism?*

Parthenogenesis is also present in the insect basal order Collembola. Among these insects, *Folsomia candida* reproduces in the absence of male contribution and seems to restore diploidy by an aberrant meiosis in which the oocyte undergoes maturation division with the elimination of only one polar body (Palévody 1973). In this species, also the assembly of the meiotic spindle follows a centrosome-independent pathway. However, as meiosis resumes, some astral arrays of microtubules appear in the egg cytoplasm (Fig. 19.4). One of them contacts the female pronucleus, whereas the others disappear (Riparbelli et al. 2006). During prophase of the first mitosis, this aster enlarges, and two close microtubule foci become evident at its center. This suggests that the focus of the aster undergoes duplication. The presence of centrioles within the asters supports this duplication process. These foci gradually increase dimension, and two large asters are visible at the opposite poles of the first mitotic spindle (Riparbelli et al. 2006). The findings that the basal insect *Folsomia candida* organizes the first zygotic spindle by means of self-assembled cytoplasmic asters suggest that the formation of these structures is a conserved phenomenon that in insects may support the embryonic development in the absence of the sperm basal body.

19.4.5 The Case of Stick Insects: An Exception to the General Rule?

Although the aster self-assembly appears to be a general mechanism driving parthenogenetic development in insects, at least two remarkable exceptions have

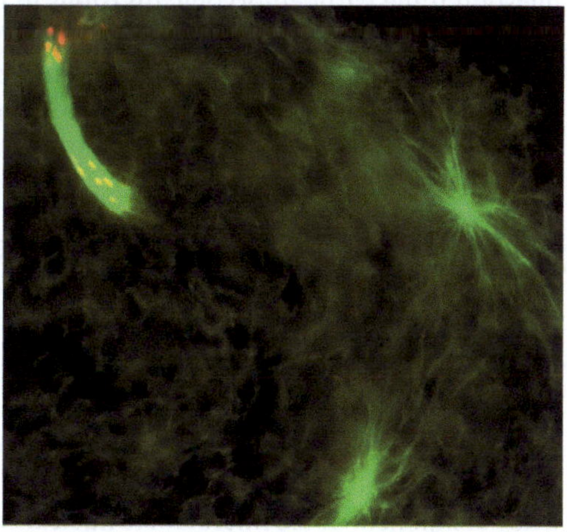

Fig. 19.4 Parthenogenesis in *Folsomia candida*. Detail of an unfertilized oocyte during meiotic anaphase showing two large microtubule asters. Microtubules, *green*; DNA, *red*

been reported: the fungus gnat *Sciara* and the stick insect *Bacillus*. Unfertilized *Sciara* eggs assemble bipolar spindles after artificial activation and initiate development in the absence of male contribution and self-assembled cortical asters (de Saint Phalle and Sullivan 1998). However, the development of the *Sciara* unfertilized eggs could not be considered true parthenogenesis since never the embryos develop and the spindles that lack functional centrosomes do not support chromosome segregation. Moreover, the syncytial nuclei are not evenly distributed as during normal development and fail to migrate to the egg cortex and undergo fusion. Bipolar anastral spindles that do not support embryonic development have been also described in some *Drosophila* mutants that display problems in centrosome assembly or have DNA replication defects. Moreover, *Drosophila Sas-4* mutants can assemble mitotic spindles by an acentrosomal route and can develop to adult stage (Basto et al. 2006), but the early embryos fail to develop in the absence of functional centrosomes (Stevens et al. 2007). This suggests that the early embryonic development strictly requires the presence of centrosomes to assemble functional spindles. Stick insects represent one of the best examples of organisms that reproduce by parthenogenesis. Some of them, including the genus *Bacillus*, display either facultative or obligatory parthenogenesis (Mantovani and Scali 1992). Despite the interest that this model inspires, only one study tries to understand the assembly of the first zygotic spindle and the progression through the following syncytial mitotic divisions in parthenogenetic eggs (Marescalchi et al. 2002). This study does not reveal cytoplasmic asters, and the mitotic spindles seem to be assembled without functional centrosomes. However, the eggs of stick insects are large, yolky with a thick chorion, and very difficult to fix. In this condition, the immunostaining is not easy, and it is possible, therefore, that thin asters may escape detection or that the fixation protocol was not suitable to preserve the astral microtubules. Moreover, due to the difficulty in analysis of eggs at earlier stages of parthenogenetic development, most of the results describe later embryonic stages. Barrel-shaped anastral mitotic spindles have been reported in embryos at early germ anlage, but the assembly of the zygotic spindle is unclear. This raises the question if the early embryonic divisions in *Bacillus* need some maternal contribution to assemble centrosomes and then shift to an acentrosomal route of spindle assembly as it occurs in *Sas-4 Drosophila* embryos or if the first zygotic spindle assembles without centrosomes like the mouse embryo.

19.4.6 Asters or Centrioles? An Egg-Chicken Enigma

When the parthenogenetic eggs resume meiosis, they undergo a dramatic rearrangement of the cortical microtubules and form the aster-like microtubule-based structures in the cytoplasm. In the parthenogenetic *Drosophila mercatorum*, the main centrosomal proteins are not detected within the cytoplasm and along the meiotic spindles before oocyte activation. In contrast, at the end of meiosis, γ-tubulin, a key component of the microtubule nucleation machinery (Oakley 2000); cnn, a core component needed to recruit γ-tubulin at the centrosome (Conduit et al. 2015); and abnormal spindle (asp), a protein playing a main role in centrosome organization and microtubule bundling at the spindle poles (Avides and Glover 1999; Riparbelli et al. 2002), are found at the foci of the astral microtubules. Moreover, centrin-associated material at the center of the aster-like structures is stained with antibodies against the human centrin HsCen1p (Paoletti et al. 1996). These findings suggest that the foci of the astral arrays of microtubules contain true centrosomes that might be actively involved in microtubule nucleation and point to the presence of centrioles that are, indeed, found by transmission electron microscope analysis (Riparbelli and Callaini 2003). Therefore, the rapid appearance of the cytoplasmic asters might be associated with the sudden self-assembly of centrioles that soon accumulate γ-tubulin and centrosomal components. Thus, the functional centrosomes can nucleate the astral arrays of microtubules. However, not all the asters examined by transmission electron microscopy in the *Drosophila mercatorum* parthenogenetic eggs contain centrioles. Moreover, the asters differ in size, and the smaller ones do not cross-react with any of the anti-centrosomal antibodies and do not contain γ-tubulin. Additionally, the smaller asters do not duplicate at the time the larger do it, pointing to the lack of centrioles that are the master of centrosome replication. Centrioles were also found by transmission electron microscopy in self-assembled asters of artificially activated sea urchin (Dirksen 1961). However, in agreement with the observations in parthenogenetic *Drosophila* eggs, the asters that are just assembled after activation do not contain centrioles that appear several hours later (Dirksen and Crocker 1966). Small cortical asters have also been described in artificially activated *Drosophila* oocytes from *γTub37CDRS42* mutants (Wilson and Borisy 1998), suggesting that γ-tubulin is not required for their assembly. These findings, therefore, raise important questions on the origin of the cytoplasmic asters and strongly support the possibility that their increase in dimension is mediated by centrosomal material, but their assembly does not require functional centrosomes. It has been demonstrated in some cell systems that under certain conditions randomly arranged microtubules might self-organize in astral arrays via minus-end-directed motors (Nedelec et al. 1997, 2003; Karsenti 2008). We can speculate, therefore, that motor proteins can crosslink cortical microtubules and focus their minus ends to form the asters of smaller dimension. The microtubule-stabilizing drug taxol induces the formation of multiple asters in *Xenopus* extracts that do not assemble from preformed centers but rather by the cross-linking of microtubules (Buendia et al. 1990). This rearrangement requires the cytoplasmic dynein (Verde et al. 1991). The report that microtubule asters spontaneously assemble in newly laid *Drosophila* unfertilized *Laborc*^{D} eggs, a dominant-negative dynein mutation, supports the possibility that motor proteins can be indeed involved in aster formation (Belecz et al. 2001).

The peripheral cytoplasm of the parthenogenetic eggs might represent a unique environment in which motor proteins operate to focus cortical microtubules in distinct aster-like structures. Accordingly, the first asters form in the cortical cytoplasm in both *Nasonia* (Tram and Sullivan 2000) and *Drosophila* (Riparbelli and Callaini 2003). The centrosomal proteins may be gradually recruited from the surrounding cytoplasm along the astral microtubules and actively transported to their focus by motor proteins to enable aster growth. This possibility is supported by the finding of centrosomal proteins along the aster-like structures observed during anaphase of the first meiosis in *Drosophila mercatorum* parthenogenetic eggs (Riparbelli and Callaini 2003). In this way, the focus of the asters becomes a suitable environment in which specific proteins may reach a concentration proper to warrant centriole self-assembly. Then the centrioles can recruit pericentriolar material to assemble functional centrosomes. However, to unravel this point, it is critical to analyze the temporal localization patterns of some key centriolar proteins such as Asl, Sak/Plk4, Sas6, and Sas-4.

An alternative hypothesis has been proposed to explain the formation of the hundred microtubule-organizing centers found in the parthenogenetic eggs of the hymenopteran *Muscidifurax* and *Nasonia* (Ferree et al. 2006). The oocytes of these species are characterized by specific organelles, the accessory nuclei, that form by budding off from the oocyte nucleus and contain high concentration of γ-tubulin. During late oogenesis, the accessory nuclei disintegrate leaving free small γ-tubulin foci. It has been suggested, therefore, that cytoplasmic asters assemble from maternal centrosomes derived from the accessory nuclei (Ferree et al. 2006). Although there is a close correlation between the numbers of accessory nuclei and the cytoplasmic asters that form in the parthenogenetic *Muscidifurax* egg, a mechanistic relationship between these structures has not been demonstrated. Nor a correlation between the γ-tubulin foci released by the fragmentation of the accessory nuclei and aster assembly has been showed. Moreover, accessory nuclei are not detected in *Drosophila mercatorum* and aphid oocytes that also assemble cytoplasmic asters. We cannot exclude, however, that the hymenopteran eggs could assemble the centrosomes by a mechanism distinct from that working in other parthenogenetic species.

19.4.7 Egg Activation: The Aster Big Bang

The process of canonical centriole assembly from a preexisting mother has been considered as the main mechanism of centrioles propagation. However, this mechanism does not explain the rapid appearance of multiple centrioles that occurs in a wide variety of organisms under particular conditions. It is well known since the work of Morgan (1899) and Loeb (1913) that the artificial activation of unfertilized sea urchin eggs leads to the formation of cytoplasmic asters. In sea urchin eggs, the cytoplasmic asters also form following treatment with various parthenogenetic agents, such as hypertonic solutions, NH_3, urea, alcohol, heavy water, microtubule-stabilizing drug taxol, hexyleneglycol, and procaine (Kallenbach 1983; Kuryiama and Borisy 1983; Mazia 1987; Harris and Clason 1992; Schatten 1982; Schatten et al. 1992). Ultrastructural analysis showed that the cytoplasmic asters contain distinct centrioles (Dirksen 1961; Sachs and Anderson, 1970; Kato and Sugiyama 1971; Mazia 1987; Miki-Noumura 1977; Kallenbach 1982, 1983; Kallenbach and Mazia 1982). Although the asters may contain up to eight centrioles at their focus (Kuriyama and Borisy 1983), no centriole pairs with orthogonally arranged parents were found. Rather, the centrioles are scattered within the center of the asters. This supports a reiterate process of de novo assembly rather than the duplication of the firstly formed centrioles. Remarkably, the centrioles are absent in eggs activated with seawater containing procaine (Moy et al. 1977) or ammonia (Paweletz and Mazia 1979) where astral microtubules are focused on clusters of osmiophilic bodies. This suggests that different activation processes may influence centriole assembly, leaving the formation of the cytoplasmic asters unaffected. Ammonia-activated unfertilized sea urchin eggs also contained bipolar anastral mini spindles organized by the chromatin (Henson et al. 2016). The de novo formation of centrioles has been also observed in other models, including mammalian culture cells (Zorn et al. 1979; Khodjakov et al. 2002; La Terra et al. 2005; Uetake et al. 2007), *Chlamydomonas* (Marshall et al. 2001), and *Drosophila* (Peel et al. 2007; Rodrigues-Martins et al. 2007; Dzhindzhev et al. 2010). Therefore, de novo centriole assembly may represent a common feature in a wide variety of organisms. However, in all the systems examined, this process occurs under specific experimental conditions that can modify the normal cell metabolism. Conversely, parthenogenesis represents a natural process in which centrioles form de novo without external interferences. Centriole duplication by the canonical pathway of procentriole assembly and de novo formation during both artificial activation and natural parthenogenesis represent distinct processes. However, it is possible that these processes utilize a common protein set that triggers the assembly and defines the structure of the centrioles (Brito et al. 2012). Accordingly, the overexpression of Sak/Plk4, which plays a main role in the assembly of the procentrioles and in the sequential recruitment of some centriole-associated proteins, results in the formation of many centrosomal foci in the unfertilized *Drosophila* eggs (Rodrigues-Martins et al. 2007). Small centrosomal foci have been also observed in unfertilized *Drosophila* eggs following the overexpression of the scaffold protein Asl that is needed to recruit Sak/Plk4 (Dzhindzhev et al. 2010) and of the centriole duplication proteins Sas-4 and Sas-6 (Peel et al. 2007). However, although all the cytoplasmic foci contain typical centrosomal proteins, distinct centrioles have been detected only in eggs overexpressing Sak/Plk4 and Asl. These findings suggest that raising the concentration of the proteins involved in the canonical pathway of centriole duplication leads to the de novo formation of many centrioles in unfertilized *Drosophila* eggs independently from the presence of preexisting centrioles. Although appealing, this model cannot explain the centriole biogenesis in parthenogenetic eggs. Firstly, the parthenogenetic eggs contain a large stockpile of the proteins needed for the assembly/duplication of the centrioles, but the concentration of these proteins does not exceed that found in eggs that have to be fertilized and that can organize up to 15,000 centrioles in the *Drosophila* early embryo. Second, the asters first appeared at the beginning of the parthenogenetic development and then the centrioles appear at their centers, whereas in eggs overexpressing the centriolar proteins, the asters appear after the centrioles are formed. Therefore, raising the concentration of the centriolar proteins leads to the de novo assembly of the centrioles that in turn recruit centrosomal material and finally nucleate the microtubule asters. By contrast, in the parthenogenetic eggs, the cytoplasmic asters form first, and then they recruit pericentriolar material to allow their increase in size. The active recruitment of the centriolar proteins by the astral microtubules may constitute a restrict domain at the center of the aster in which these components reach the suitable concentration to enable the self-organization of the centrioles.

19.4.8 Too Many Asters: Why Only One Spindle Forms?

All the asters in the cortical region have the same potential to interact with the female pronucleus, but only one or two of them win excluding the other competitors. Are specific characteristics required to give advantage in this competition? The assembly of the first mitotic spindle during parthenogenetic development is presumably due to a stochastic event that involves the interaction of some peripheral cytoplasmic asters with the female pronucleus. It is possible, however, that different species have independently evolved diverse mechanisms to ensure that only one or two asters can interact with the female pronucleus. The peripheral asters in the hymenopteran eggs have the same size but different position in respect to the inner female pronucleus. Thus, the asters in the more anterior and posterior halves of the oocyte have fewer chances to interact with the female pronucleus (Tram and Sullivan 2000). In the parthenogenetic aphid oocyte, two asters interact with the female pronucleus and then move to its opposite poles to assemble the first mitotic spindle. The small size of the oocyte may facilitate the interaction of those asters that have the prominent arrays of microtubules (Riparbelli et al. 2005). The length of the microtubules might also represent a preferential way to help the interaction with the female pronucleus of two independently assembled asters in the *Drosophila mercatorum* oocyte, where different-sized asters are, indeed, found.

One main problem at fertilization is the control of the centriole number, since the first mitotic spindle only needs two pairs of centrioles to be assembled. Thus, different species have evolved various mechanisms to eliminate or inactivate during oogenesis or slightly later their own centrioles and ensure that the egg inherits only one centriole or a centriole pair from the sperm. Therefore, the simultaneous presence of several asters within the egg cytoplasm raises the question of how the egg may restrict their number to properly organize the first bipolar spindle. Such control prevents the formation of multipolar spindles and the ensuing developmental defects that are seen during polyspermic fertilization in which multiple sperm asters result in aberrant spindles and developmental failure (Schatten 1994; Sluder and Hinchcliffe 1998; Snook et al. 2011). Too many centrosomes have been also associated with defective mitosis leading to aneuploidy and chromosome instability in somatic cells, both hallmarks of cancer cells (Brinkley 2001; Pihan 2013). The mechanism preventing simultaneous interaction of supplementary asters with the female pronucleus is actually unknown. It has been proposed that in the parthenogenetic *Nasonia* eggs, the first two centrosomes that reached the female pronucleus are stabilized by the interaction with the nuclear envelope and then with the chromatin. Following this contact, the centrosomes may nucleate many astral microtubules that prevent and exclude the interaction with additional centrosomes (Tram and Sullivan 2000). A mechanism that maintains the proper spacing among the cytoplasmic asters may be also present in the *Drosophila mercatorum* oocyte, since the fusion between neighboring asters rarely occurs. The spatial exclusion of the asters may occur in the aphid oocyte that has few asters and therefore low probability that they may come in contact. However, this hypothesis does not fully explain why the microtubules of the supernumerary asters do not interact with kinetochores to give rise to merotelic attachments and therefore to abnormal spindles. The mitotic spindles in the syncytial *Drosophila* blastoderm embryos may collide in the absence of the cortical microfilaments that spatially isolate spindles (Riparbelli et al. 2007). It is actually unknown if such microfilament scaffold also exists in the parthenogenetic eggs. Alternatively, the length of the astral microtubules may be so reduced that they can only interact with the chromatin at very short distance. The number of the supernumerary-free asters observed in the parthenogenetic hymenopteran eggs rapidly decrease and soon disappear confirming that the interaction with the chromatin is crucial to maintain only the maternally inherited centrosomes that organize the spindle poles (Riparbelli et al. 1998). Presumably, the same mechanism that excludes additional asters operates during fertilization of parthenogenetic *Drosophila mercatorum* eggs when the male centrosomes prevent maternal ones from approaching and interacting with the forming zygotic nucleus.

The asters disappear at different times in different parthenogenetic species. *Muscidifurax* eggs lose their asters during the third mitosis. By contrast, *Drosophila mercatorum* and *Acyrthosiphon pisum* eggs retain some asters during later cleavage divisions. However, the remnant asters do not interfere with development in aphids, whereas several multipolar spindles have been found in *Drosophila*. These observations strengthen the possibility that parthenogenetic development in *Drosophila* is a recent acquisition and that in this species the establishment of control mechanisms ensuring proper spindle assembly in the presence of multiple asters is still in progress. Remarkably, the free cytoplasmic asters in the *Drosophila* eggs seem to duplicate in a cell cycle-dependent manner: single asters have been observed during interphase, whereas distinct pairs are found in prometaphase. Close aster pairs are also visible during late anaphase/telophase, the time the centrioles disengage at the spindle poles. These finding suggest that the duplication of the asters is not related to DNA replication, but depends on cytoplasmic factors. Similarly, during normal development of the early *Drosophila* syncytial embryo, the centrosomes of the yolk region duplicate independently of chromosome replication (Riparbelli and Callaini 2003). Surprisingly, these factors allow duplication of the cytoplasmic asters but do not influence the behavior of the cortical asters associated with the polar bodies.

19.4.9 *Sperm Basal Body Versus Microtubule Clustering: Does Fertilization of Parthenogenetic Eggs Inhibit the Aster Self-Assembly?*

The oocytes from the dioic strains of *Drosophila mercatorum* and *Acyrthosiphon pisum* do not contain cortical asters, and the only aster present is that organized at fertilization by the sperm basal body. Therefore, the sperm entry might exert an additional control on the mechanism of spontaneous aster assembly. However, it has been shown that in the parthenogenetic strain of *Drosophila mercatorum*, the formation of the asters was not influenced by the sperm. When parthenogenetic females were crossed with males, the fertilized eggs assembled cytoplasmic asters (Riparbelli and Callaini 2003). Therefore, the mechanism involved in microtubule polymerization and/or aster self-assembly was not silenced by the basal body introduced by the sperm. It has been suggested that in cultured vertebrate cells, there is a limited amount of centriolar proteins that is sequestered by the duplicated centrioles, thus preventing assembly of the supernumerary centrioles (Song et al. 2008). By contrast, the insect eggs have a large store of material needed to assemble thousands of centrioles, and the sperm centriole has a limited consequence on the recruitment of the centriolar components. However, the formation of the zygotic spindle is always driven by the sperm centriole. It is unclear how the newly assembled sperm aster wins the competition with the cytoplasmic asters.

19.5 The Mechanism of Parthenogenesis: Hypothesis from Insects

The analysis of the parthenogenetic development in some insect orders revealed that the sudden assembly of cytoplasmic asters represents a shared mechanism to drive the organization of the first mitotic spindle in the absence of male contribution. Therefore, the formation of the cortical asters may represent a preferred mechanism to ensure the reproduction without fertilization. Since it is conceivable that all the components needed for the assembly of the cytoplasmic asters are present in both dioic and parthenogenetic eggs, an intriguing question is why dioic eggs do not organize microtubule asters. Because the asters organize by preassembled microtubule bundles, one possibility could be that the dioic eggs lack cortical microtubules. A common feature, indeed, of the parthenogenetic oocytes examined so far, is a dense network of randomly arranged cortical microtubules. Moreover, despite the different numbers of asters observed in the different species, the triggering event seems to be always correlated with the metaphase/anaphase transition of the first meiosis. Hymenopteran oocytes in which the microtubule bundles fill the whole peripheral cytoplasm assemble a rather uniform cortical layer of asters. By contrast, the microtubule bundles are only present in the anterior half of the *Drosophila mercatorum* oocyte, and the asters are restricted to this region. Therefore, the cortical microtubule network may represent the main source for cytoplasmic asters assembly and might be the hallmark of parthenogenetic eggs. However, such peripheral microtubules have been also described in fertilized and unfertilized *Drosophila melanogaster* eggs that do not reproduce by parthenogenesis, although cytoplasmic asters do not form (Fahmi et al. 2014). Remarkably, distinct asters were observed when unfertilized oocytes of *Drosophila melanogaster* were artificially activated (Wilson and Borisy 1998). These findings suggest that the *Drosophila* egg has all the components to organize the peripheral asters, but microtubule clustering does not occur during normal development and only the artificial activation may trigger microtubule clustering. How do these controversial results may be explained? The simple explanation might be that artificial and natural activation follow different routes. It is generally assumed that the insect oocyte is not activated by its interaction with the male gamete but by physical compression during the transit through the female reproductive tract (Heifetz et al. 2001). Nevertheless, a calcium wave that occurs in most organisms in which the activation is mediated by interactions with the male gamete has been also found in *Drosophila* oocyte, and it is likely induced by the pressure (Kaneuchi et al. 2015). The increased cytosolic calcium levels have been also found in artificially activated oocytes. However, microtubule polymerization does not seem to be correlated with the Ca++ rise (Fahmy et al. 2014; Kaneuchi et al. 2015). It is possible that the chemical treatment during artificial activation may lead to side effects including excess of tubulin polymerization and microtubule clustering. Increasing the concentration of motor proteins in vitro leads to aster assembly with motors at their focus (Nédèlec et al. 2003). It has been, indeed, demonstrated that under normal conditions, the minus-end motor *ncd* is critically important for cortical microtubule polymerization in the anterior half of the early *Drosophila melanogaster* embryo (Fahmy et al. 2014). Therefore, the limiting factor to assemble cortical asters may be the activity of the motor proteins. Thus, aster self-assembly may be regulated by cytoplasmic factors monitoring the local dynamics of the motor proteins.

The formation of cortical asters in parthenogenetic insect eggs may result by the up or downregulation of processes that are usually strictly controlled during fertilization. Since both asters and associated centrioles/centrosomes originate from the cytoplasm pool of stored material, all the insect eggs are potentially parthenogenetic. Parthenogenetic development may arise when the controls preventing the self-assembly of the asters from maternal stores do not work properly. Aphids, hymenopterans, collembolans, and dipterans have independently overcome these controls. Hymenopterans assemble a large number of asters, whereas aphids and collembolans only a few. However, in both these insect groups, the eggs develop properly in the absence of males. By contrast, the asters assemble only in the anterior region of about half of the eggs in the parthenogenetic strain of *Drosophila mercatorum*. Moreover, only a few eggs develop. This suggests that the mechanisms monitoring aster self-assembly may be active in the posterior region of the egg, but does not work properly in the anterior region.

The ability to assembly cytoplasmic asters is not limited to insects. The unfertilized eggs of many organisms can form microtubule asters when they are artificially activated. It has been proposed that the highly conserved property to organize scattered cortical microtubules in distinct aster-like structures might be inherited by a polyspermic ancestor in the lineage leading to Bilateria (Salinas-Saavedra and Vargas 2011). However, the sudden assembly of cytoplasmic asters is not sufficient to support the parthenogenetic development, and artificially activated unfertilized eggs arrest very early. Despite the potential interest in centriole/centrosome biology, the early parthenogenetic development has been, until now, carefully examined only in a few insect species. The mechanism operating in formation of the first mitotic spindle in the absence of the male gamete in noninsect eggs is unclear, and it is still unknown if this process requires the formation of cytoplasmic asters and the assembly of centrioles de novo.

References

Afzelius BA (1971) Fine structure of the spermatozoon of *Tubularia larynx* (Hydrozoa, Coelenterata). J Ultrastruct Res 37:679–689

Agircan FG, Schiebel E, Mardin BR (2014) Separate to operate: control of centrosome positioning and separation. Philos Trans R Soc Lond B Biol Sci 369:20130461

Al Jord A, Lemaître A-I, Delgehyr N, Faucourt M, Spassky N, Meunier A (2014) Centriole amplification by mother and daughter centrioles differs in multiciliated cells. Nature 516:104–107

Allen RD (1969) The morphogenesis of basal bodies and accessory structures of the cortex of the ciliated protozoan *Tetrahymena pyriformis*. J Cell Biol 40:716–733

Alvey PL (1986) Do adult centrioles contain cartwheels and lie at right angles to each other? Cell Biol Int Rep 10:589–598

Anderson RG, Brenner RM (1971) The formation of basal bodies (centrioles) in the *Rhesus* monkey oviduct. J Cell Biol 50:10–34

Anderson CT, Stearns T (2009) Centriole age underlies asynchronous primary cilium growth in mammalian cells. Curr Biol 19:1498–1502

Anderson WA, Weissman A, Ellis RA (1967) Cytodifferentiation during spermiogenesis in *Lumbricus terrestris*. J Cell Biol 32:11–26

Arquint C, Gabryjonczyk A-M, Nigg EA (2014) Centrosomes as signalling centres. Philos Trans R Soc Lond B Biol Sci 369:20130464

Asker SE, Jerling L (1992) Apomixis in plants. CRC Press, Boca Raton, FL

Avidor-Reiss T, Gopalakrishnan J (2013) Building a centriole. Curr Opin Cell Biol 25:72–77

Avidor-Reiss T, Khire A, Fishman EL, Jo KH (2015) Atypical centrioles during sexual reproduction. Front Cell Dev Biol 3:21

Azimzadeh J (2014) Exploring the evolutionary history of centrosome. Philos Trans R Soc Lond B 369:20130453

Azimzadeh J, Marshall WF (2010) Building the centriole. Curr Biol 20:R816–R825

Azimzadeh J, Wong ML, Downhour DM, Sánchez Alvarado A, Marshall WF (2012) Centrosome loss in the evolution of planarians. Science 335:461–463

Baccetti B, Dallai R (1978) The spermatozoon of Arthropoda. XXX. The multiflagellate spermatozoon in the termite *Mastotermes darwiniensis*. J Cell Biol 76:569–576

Baccetti B, Dallai R, Rosati F (1969) The spermatozoon of Arthropoda. IV. Corrodentia, Mallophaga and Thysanoptera. J Microsc 8:249–262

Baccetti B, Dallai R, Callaini G (1981) The spermatozoon of arthropoda: *Zootermopsis nevadensis* and isopteran sperm phylogeny. Int J Invert Reprod 3:87–99

Balczon R (1996) The centrosome in animal cells and its functional homologs in plant and yeast cells. Int Rev Cytol 169:25–82

Basto R, Lau J, Vinogradova T, Gardiol A, Woods CG, Khodjakov A, Raff JW (2006) Flies without centrioles. Cell 125:1375–1386

Basto R, Brunk K, Vinadogrova T, Peel N, Franz A, Khodjakov A, Raff JW (2008) Centrosome amplification can initiate tumorigenesis in flies. Cell 133:1032–1042

Belecz I, Gonzalez C, Puro J, Szabad J (2001) Dominant-negative mutant dynein allows spontaneous centrosome assembly, uncouples chromosome and centrosome cycles. Curr Biol 11:136–140

Bell G (1982) The masterpiece of nature: the evolution and genetics of sexuality. University of California Press, Berkeley, CA

Bennabi I, Terret M-E, Verlhac M-H (2016) Meiotic spindle assembly and chromosome segregation in oocytes. J Cell Biol 215:611–619

Bettencourt-Dias M, Glover DM (2007) Centrosome biogenesis and function: centrosomics brings new understanding. Nat Rev Mol Cell Biol 8:451–463

Bettencourt-Dias M, Rodrigues-Martins A, Carpenter L, Riparbelli M, Lehmann L, Gatt MK, Carmo N, Balloux F, Callaini G, Glover DM (2005) SAK/PLK4 is required for centriole duplication and flagella development. Curr Biol 15:2199–2207

Bettencourt-Dias M, Hildebrandt F, Pellman D, Woods G, Godinho SA (2011) Centrosomes and cilia in human disease. Trends Genet 27:307–315

Blachon S, Cai X, Roberts KA, Yang K, Polyanovsky A, Church A et al (2009) A proximal centriole-like structure is present in *Drosophila* spermatids and can serve as a model to study centriole duplication. Genetics 182:133–144

Blachon S, Khire A, Avidor-Reiss T (2014) The origin of the second centriole in the zygote of *Drosophila melanogaster*. Genetics 197:199–205

Blackman RL (1987) Reproduction, cytogenetics and development. In: Minks AK, Harrewijn P (eds) Aphids: their biology, natural enemies and control. Elsevier, Amsterdam, pp 163–195

Blackman RL, Spence JM (1996) Ribosomal DNA is frequently concentrated on only one X chromosome in permanently apomictic aphids, but this does not inhibit male determination. Chromosome Res 4:314–320

Booth W, Schuett GW (2011) Molecular genetic evidence for alternate reproductive strategies in North American pitvipers (Serpentes, Viperidae): long-term sperm storage and facultative parthenogenesis. Biol J Linn Soc 104:934–942

Bornens M (2002) Centrosome composition and microtubule anchoring mechanisms. Curr Opin Cell Biol 14:25–34

Bornens M (2012) The centrosome in cells and organisms. Science 335:422–426

Bornens M, Azimzadeh J (2007) Origin and evolution of the centrosome. In: Jékely G (ed) Eukaryotic endomembrances and cytoskeleton: origins and evolution. Springer, New York

Bornens M, Paintrand M, Berges J, Marty MC, Karsenti E (1987) Structural and chemical characterization of isolated centrosomes. Cell Motil Cytoskeleton 8:238–249

Borrego-Pinto J, Somogyi K, Karreman MA, König J, Müller-Reichert T, Bettencourt-Dias M, Gönczy P, Schwab Y, Lénárt P (2016) Distinct mechanisms eliminate mother and daughter centrioles in meiosis of starfish oocytes. J Cell Biol 212:815–827

Boveri T (1887) Ueber den Antheil des Spermatozoon an der Teilung des Eies. Sitzungsber Ges Morph Physiol München 3:151–164

Boveri T (1901) Zellen-Studien: Über die Natur der Centrosomen. Jena. Germany: Fisher Z Med Naturw 28:1–220

Briedis A, Elinson RP (1982) Suppression of male pronuclear movement in frog eggs by hydrostatic pressure and deuterium oxide yields androgenetic haploids. J Exp Zool 222:45–57

Brinkley BR (2001) Managing the centrosome numbers game: from chaos to stability in cancer cell division. Trends Cell Biol 11:18–21

Brito DA, Gouveia SM, Bettencourt-Dias M (2012) Deconstructing the centriole: structure and number control. Curr Opin Cell Biol 24:4–13

Buendia B, Bré MH, Griffiths G, Karsenti E (1990) Cytoskeletal control of centrioles movement during the establishment of polarity in Madin-Darby canine kidney cells. J Cell Biol 110:1123–1135

Buřič M, Hulák M, Kouba A, Petrusek A, Kozák P (2011) A successful crayfish invader uses facultative parthenogenesis: a novel reproductive mode in decapod crustaceans. PLoS ONE 6: e20281

Calarco PG (2000) Centrosome precursors in the acentriolar mouse oocyte. Microsc Res Tech 49:428–434

Callaini G, Riparbelli MG (1990) Centriole and centrosome cycle in the early *Drosophila* embryo. J Cell Sci 97:539–543

Callaini G, Riparbelli MG (1996) Fertilization in *Drosophila melanogaster*: centrosome inheritance and organization of the first mitotic spindle. Dev Biol 176:199–208

Callaini G, Whitfield WG, Riparbelli MG (1997) Centriole and centrosome dynamics during the embryonic cell cycles that follow the formation of the cellular blastoderm in *Drosophila*. Exp Cell Res 234:183–190

Callaini G, Riparbelli MG, Dallai R (1999) Centrosome inheritance in insects: fertilization and parthenogenesis. Biol Cell 91:355–366

Carazo-Salas RE, Guarguaglini G, Gruss OJ, Segref A, Karsenti E, Mattaj IW (1999) Generation of GTP-bound Ran by RCC1 is required for chromatin-induced mitotic spindle formation. Nature 400:178–181

Carson HL (1973) The genetic system in parthenogenetic strains of *Drosophila mercatorum*. Proc Natl Acad Sci USA 70:1772–1774

Carvalho-Santos Z, Machado P, Branco P, Tavares-Cadete F, Rodrigues-Martins A, Pereira-Leal JB, Bettencourt-Dias M (2010) Stepwise evolution of the centriole-assembly pathway. J Cell Sci 123:1414–1426

Carvalho-Santos Z, Azimzadeh J, Pereira-Leal JB, Bettencourt-Dias M (2011) Evolution: tracing the origins of centrioles, cilia, and flagella. J Cell Biol 194:165–175

Cavalier-Smith T (1974) Basal body and flagellar development during the vegetative cell cycle and the sexual cycle of *Chlamydomonas reinhardtii*. J Cell Sci 16:529–556

Chavali PL, Pütz M, Gergely F (2014) Small organelle, big responsibility: the role of centrosomes in development and disease. Philos Trans R Soc Lond B 369:20130468

Clarke PR, Zhang C (2008) Spatial and temporal coordination of mitosis by Ran GTPase. Nat Rev Mol Cell Biol 9:464–477

Coelho PA, Bury L, Sharif B, Riparbelli MG, Fu J, Callaini C, Glover DM, Zernicka-Goetz M (2013) Spindle formation in the mouse embryo requires Plk4 in the absence of centrioles. Dev Cell 27:586–597

Conduit PT, Raff JW (2010) Cnn dynamics drive centrosome size asymmetry to ensure daughter centriole retention in *Drosophila* neuroblasts. Curr Biol 20:2187–2192

Conduit PT, Feng Z, Richens JH, Baumbach J, Wainman A, Bakshi SD, Dobbelaere J, Johnson S, Lea SM, Raff JW (2014) The centrosome-specific phosphorylation of Cnn by Polo/Plk1 drives Cnn scaffold assembly and centrosome maturation. Dev Cell 28:659–669

Conduit PT, Wainman A, Raff JW (2015) Centrosome function and assembly in animal cells. Nat Rev Mol Cell Biol 16:611–624

Cook JM (1993) Sex determination in Hymenoptera: a review of models and evidence. Heredity 71:421–435

Courtois A, Schuh M, Ellenberg J, Hiiragi T (2012) The transition from meiotic to mitotic spindle assembly is gradual during early mammalian development. J Cell Biol 198:357–370

Crasta K, Ganem NJ, Dagher R, Lantermann AB, Ivanova EV, Pan Y, Nezi L, Protopopov A, Chowdhury D, Pellman D (2012) DNA breaks and chromosome pulverization from errors in mitosis. Nature 482:53–58

Crowder ME, Flynn JR, McNally KP, Cortes DB, Price KL, Kuehnert PA, Panzica MT, Andaya A, Leary JA, McNally FJ (2015) Dynactin-dependent cortical dynein and spherical spindle shape correlate temporally with meiotic spindle rotation in *Caenorhabditis elegans*. Mol Biol Cell 26:3030–3046

Culver BP, Meehl JB, Giddings TH, Winey M (2009) The two SAS-6 homologs in *Tetrahymena thermophila* have distinct functions in basal body assembly. Mol Biol Cell 20:1865–1877

Dallai R (2014) Overview on spermatogenesis and sperm structure of Hexapoda. Arthropod Struct Dev 43:257–290

Dallai R, Mazzini M (1989) The spermatozoon of the gall-midge Oligotrophidi (Diptera, Cecidomyiidae). Boll Zool 56:13–27

Dallai R, Gottardo M, Beutel RG (2016) Structure and evolution of insect sperm: new interpretations in the age of phylogenomics. Ann Rev Entomol 61(1):1–1.23

Davis GK (2012) Cyclical parthenogenesis and viviparity in aphids as evolutionary novelties. J Exp Zool (Mol Dev Evol) 318:448–459

Dawe HR, Farr H, Gull K (2007) Centriole/basal body morphogenesis and migration during ciliogenesis in animal cells. J Cell Sci 120:7–15

de Fried EP, Ross P, Zang G, Divita A, Cunniff K, Denaday F, Salamone D, Kiessling A, Cibelli J (2008) Human parthenogenetic blastocysts derived from noninseminated cryopreserved human oocytes. Fertil Steril 89:943–947

de Saint Phalle B, Sullivan W (1998) Spindle assembly and mitosis without centrosomes in parthenogenic *Sciara* embryos. J Cell Biol 141:1383–1391

Debec A, Abbadie C (1989) The acentriolar state of the *Drosophila* cell lines 1182. Biol Cell 67:307–311

Debec A, Sullivan W, Bettencourt-Dias M (2010) Centrioles: active players or passengers during mitosis? Cell Mol Life Sci 67:2173–2194

Delattre M, Gonczy P (2004) The arithmetic of centrosome biogenesis. J Cell Sci 117:1619–1630

Dictenberg JB, Zimmerman W, Sparks CA, Young A, Vidair C, Zheng Y, Carrington W, Fay FS, Doxsey SJ (1998) Pericentrin and gamma-tubulin form a protein complex and are organized into a novel lattice at the centrosome. J Cell Biol 141:163–174

Dippel RV (1968) The development of basal bodies in *Paramecium*. Proc Natl Acad Sci USA 61:461–468

Dirksen ER (1961) The presence of centrioles in artificially activated sea urchin eggs. J Cell Biol 11:244–247

Dirksen ER, Crocker TT (1966) Centriole replication in differentiating ciliated cells of mammalian respiratory epithelium. An electron microscope study. J Micr 5:639–644

Dix CI, Raff JW (2007) *Drosophila* Spd-2 recruits PCM to the sperm centriole, but is dispensable for centriole duplication. Curr Biol 17:1759–1764

do Carmo Avides M, Glover DM (1999) Abnormal spindle protein, Asp, and the integrity of mitotic centrosomal microtubule organizing centers. Science 283:1733–1735

Doxsey SJ, Stein P, Evans L, Calarco PD, Kirschner M (1994) Pericentrin, a highly conserved centrosome protein involved in microtubule organization. Cell 76:639–650

Dumont J, Desai A (2012) Acentrosomal spindle assembly and chromosome segregation during oocyte meiosis. Trends Cell Biol 22:241–249

Dutcher SK (2007) Finding treasures in frozen cells: new centriole intermediates. Bioessays 29:630–634

Dzhindzhev NS, QD Y, Weiskopf K, Tzolovsky G, Cunha-Ferreira I, Riparbelli M, Rodrigues-Martins A, Bettencourt-Dias M, Callaini G, Glover DM (2010) Asterless is a scaffold for the onset of centriole assembly. Nature 467:714–718

Eisman R, Kaufman TC (2007) Cytological investigation of the mechanism of parthenogenesis in *Drosophila mercatorum*. Fly (Austin) 1:317–329

Fahmy K, Akber M, Cai X, Koul A, Hayder A, Baumgartner S (2014) αTubulin 67C and Ncd are essential for establishing a cortical microtubular network and formation of the Bicoid mRNA gradient in *Drosophila*. PLoS One 9:e112053

Favard P (1961) Évolution des ultrastructures cllulaires au cours de la spermatogenese de l'Ascaris. Ann Sci Nat Zool Ser 12:53–152

Feldheim KA, Chapman DD, Sweet D, Fitzpatrick S, Prodöhl PA, Shivji MS, Snowden B (2010) Shark virgin birth produces multiple, viable offspring. J Hered 101:374–377

Ferree PM, McDonald K, Fasulo B, Sullivan W (2006) The origin of centrosomes in parthenogenetic hymenopteran insects. Curr Biol 16:801–817

Fırat-Karalar EN, Stearns T (2014) The centriole duplication cycle. Philos Trans R Soc Lond B Biol Sci 369(1650):pii: 2013046

Flemming W (1875) Studien über die Entwicklungsgeschichte der Najaden. Sitzungsber Akad Wissensch Wien 71:81–147

Fong CS, Kim M, Yang TT, Liao JC, Tsou MF (2014) SAS-6 assembly templated by the lumen of cartwheel-less centrioles precedes centriole duplication. Dev Cell 30:238–245

Forbes DJ, Travesa A, Nord MS, Bernis C (2015) Nuclear transport factors: global regulation of mitosis. Curr Opin Cell Biol 35:78–90

Friedländer M, Wahrman J (1971) The number of centrioles in insect sperm. A study two kinds of differentiating silkworm spermatids. J Morphol 134:383–398

Fritz-Niggli H, Suda T (1972) Bildung und bedeutung der zentriolen: eine studie und neuinterpretation der meiose von *Drosophila*. Cytobiologie 5:12–41

Fu J, Glover DM (2012) Structured illumination of the interface between centriole and pericentriolar material. Open Biol 2:120104

Fu J, Lipinszki Z, Rangone H, Min M, Mykura C, Chao-Chu J, Schneider S, Dzhindzhev NS, Gottardo M, Riparbelli MG et al (2016) Conserved molecular interactions in centriole-to-centrosome conversion. Nat Cell Biol 18:87–99

Ganem NJ, Godinho SA, Pellman D (2009) A mechanism linking extra centrosomes to chromosomal instability. Nature 460:278–282

Giansanti MG, Bucciarelli E, Bonaccorsi S, Gatti M (2008) *Drosophila* SPD-2 is an essential centriole component required for PCM recruitment and astral-microtubule nucleation. Curr Biol 18:303–309

Godinho SA, Pellman D (2014) Causes and consequences of centrosome abnormalities in cancer. Philos Trans R Soc B 369:20130467

Godinho SA, Picone R, Burute M et al (2014) Oncogene-like induction of cellular invasion from centrosome amplification. Nature 510:167–171

Gogendeau D, Hurbain I, Raposo G, Cohen J, Koll F, Basto R (2011) Sas-4 proteins are required during basal body duplication in Paramecium. Mol Biol Cell 22:1035–1044

Gönczy P (2012) Towards a molecular architecture of centriole assembly. Nat Rev Mol Cell Biol 13:425–435

Gonzalez C, Tavosanis G, Mollinari C (1998) Centrosomes and microtubule organisation during *Drosophila* development. J Cell Sci 111:2697–2706

Gopalakrishnan J, Mennella V, Blachon S, Zhai B, Smith AH, Megraw TL, Nicastro D, Gygi SP, Agard DA, Avidor-Reiss T (2011) Sas-4 provides a scaffold for cytoplasmic complexes and tethers them in a centrosome. Nat Commun 2:359

Goshima G, Kimura A (2010) New look inside the spindle: microtubule-dependent microtubule generation within the spindle. Curr Opin Cell Biol 22:44–49

Gottardo M, Callaini G, Riparbelli MG (2013) The cilium-like region of the *Drosophila* spermatocyte: an emerging flagellum? J Cell Sci 126:5441–5452

Gottardo M, Callaini G, Riparbelli MG (2014) Procentriole assembly without centriole disengagement—a paradox of male gametogenesis. J Cell Sci 127:3434–3439

Gottardo M, Callaini G, Riparbelli MG (2015a) The *Drosophila* centriole—conversion of doublets into triplets within the stem cell niche. J Cell Sci 128:2437–2442

Gottardo M, Callaini G, Riparbelli MG (2015b) Structural characterization of procentrioles in *Drosophila* spermatids. Cytoskeleton 72:576–584

Gottardo M, Pollarolo G, Llamazares S, Reina J, Riparbelli MG, Callaini G, Gonzalez C (2015c) Loss of centrobin enables daughter centrioles to form sensory cilia in *Drosophila*. Curr Biol 25:2319–2324

Gould MC, Stephano JL (2003) Polyspermy prevention in marine invertebrates. Microsc Res Tech 61:379–388

Groot TVM, Bruins E, Breeuwer JAJ (2003) Molecular genetic evidence for parthenogenesis in the Burmese python, *Python molurus bivittatus*. Heredity 90:130–135

Gruss OJ, Wittmann M, Yokoyama H, Pepperkok R, Kufer T, Sillje H, Karsenti E, Mattaj IW, Vernos I (2002) Chromosome-induced microtubule assembly mediated by TPX2 is required for spindle formation in HeLa cells. Nat Cell Biol 4:871–879

Gueth-Hallonet C, Antony C, Aghion J, Santa-Maria A, Lajoie-Mazenc I, Wright M, Maro B (1993) y-Tubulin is present in acentriolar MTOCs during mouse development. J Cell Sci 105:157–166

Guichard P, Chrétien D, Marco S, Tassin AM (2010) Procentriole assembly revealed by cryo-electron tomography. EMBO J 29:1565–1572

Hara K, Tydeman P, Kirschner M (1980) A cytoplasmic clock with the same period as the division cycle in *Xenopus* eggs. Proc Natl Acad Sci USA 77:462–466

Harris P, Clason EL (1992) Conditions for assembly of tubulin-based structures in unfertilized sea urchin eggs. Spirals, monoasters and cyasters. J Cell Sci 102:557–556

Hatch E, Stearns T (2011) The life cycle of centrioles. Cold Spring Harb Symp Quant Biol LXXV:425–431

Heald R, Khodjakov A (2015) Thirty years of search and capture: the complex simplicity of mitotic spindle assembly. J Cell Biol 211:1103–1111

Heald R, Tournebize R, Blank T, Sandaltzopoulos R, Becker P, Hyman A, Karsenti E (1996) Self-organization of microtubules into bipolar spindles around artificial chromosomes in *Xenopus* egg extracts. Nature 382:420–425

Hebert PDN (1981) Obligate asexuality in *Daphnia*. Am Nat 117:784–789

Heifetz Y, Yu J, Wolfner MF (2001) Ovulation triggers activation of *Drosophila* oocytes. Dev Biol 234:416–424

Heimpel GE, de Boer JG (2008) Sex determination in the Hymenoptera. Annu Rev Entomol 53:209–230

Henson JH, Buckley MW, Yeterian M, Weeks RM, Simerly CR, Shuster CB (2016) Central spindle self-organization and cytokinesis in artificially activated sea urchin eggs. Biol Bull 230:85–95

Hildreth EP, Lucchesi CJ (1963) Fertilization in *Drosophila*. I. Evidence for the regular occurrence of monospermy. Dev Biol 6:262–278

Hinchcliffe EH (2014) Centrosomes and the art of mitotic spindle maintenance. Int Rev Cell Mol Biol 313:179–217

Hirono M (2014) Cartwheel assembly. Phil Trans R Soc B 369:20130458

Holland AJ, Lan W, Niessen S, Hoover H, Cleveland DW (2010) Polo-like kinase 4 kinase activity limits centrosome overduplication by autoregulating its own stability. J Cell Biol 188:191–198

Horner VL, Wolfner MF (2008) Mechanical stimulation by osmotic and hydrostatic pressure activates *Drosophila* oocytes in vitro in a calcium-dependent manner. Dev Biol 316:100–109

Huettner AF (1933) Continuity of the centrioles in *Drosophila melanogaster* Z. Zellforsch Mikrosk Anat 19:119–134

Innes DJ, Hebert PND (1988) The origin and genetic basis of obligate parthenogenisis in *Daphnia pulex*. Evolution 42:1024–1035

Iwao Y (2012) Egg activation in physiological polyspermy. Reproduction 144:11–12

Iwao Y, Elinson RP (1990) Control of sperm nuclear behavior in physiologically polyspermic newt eggs: possible involvement of MPF. Dev Biol 142:301–312

Iwao Y, Sakamoto N, Takahara K, Yamashita M, Nagahama Y (1993) The egg nucleus regulates the behavior of sperm nuclei as well as cycling of MPF in physiologically polyspermic newt eggs. Dev Biol 160:15–27

Iwao Y, Murakawa T, Yamaguchi J, Yamashita M (2002) Localization of γ-tubulin and cyclin B during early cleavge in physiologically polyspermic newt eggs. Dev Growth Differ 44:489–499
Izquierdo D, Wang W-J, Uryu K, Tsou M-FB (2014) Stabilization of cartwheel-less centrioles for duplication requires CEP295-mediated centriole-to-centrosome conversion. Cell Rep 8:957–965
Jamieson BGM (1987) The ultrastructure and phylogeny of insect Spermatozoa. Cambridge University Press, Cambridge, p 320
Jana SC, Marteil G, Bettencourt-Dias M (2014) Mapping molecules to structure: unveiling secrets of centriole and cilia assembly with near-atomic resolution. Curr Opin Cell Biol 26:96–106
Januschke J, Reina J, Llamazares S, Bertran T, Rossi F, Roig J, Gonzalez C (2013) Centrobin controls mother-daughter centriole asymmetry in *Drosophila* neuroblasts. Nat Cell Biol 15:241–248
Jura C, Krzysztofowicz A (1992) Initiation of embryonic development in *Tetrodontophora bielanensis* (Waga) (Collembola) eggs: meiosis, polyspermy, union of gametes and the first cleavage. Int J Insect Morphol Embryol 21:87–94
Kallenbach RJ (1982) De novo centrioles originate at sites associated with annulate lamellae in sea urchin eggs. Biosci Rep 2:959–966
Kallenbach RJ (1983) The induction of de novo centrioles in sea urchin eggs: a possible common mechanism for centriolar activation among parthenogenetic procedures. Eur J Cell Biol 30:159–166
Kallenbach RJ, Mazia D (1982) Origin and maturation of centrioles in association with the nuclear envelope in hypertonic-stressed sea urchin eggs. Eur J Cell Biol 28:68–76
Kaneuchi T, Sartain CV, Takeo S, Horner VL, Buehner NA, Aigaki T, Wolfner MF (2015) Calcium waves occur as *Drosophila* oocytes activate. Proc Natl Acad Sci USA 112:791–796
Karr TL (1991) Intracellular sperm/egg interactions in *Drosophila*: a threedimensional structural analysis of a paternal product in the developing egg. Mech Dev 34:101–111
Karsenti E (2008) Self-organization in cell biology: a brief history. Nat Rev Mol Cell Biol 9:255–262
Kato KH, Sugiyama M (1971) On the de novo formation of the centriole in the activated sea urchin egg. Dev Growth Differ 13:359–366
Khodjakov A, Rieder CL, Sluder G, Cassels G, Sibon O, Wang CL (2002) De novo formation of centrosomes in vertebrate cells arrested during S phase. J Cell Biol 158:1171–1181
King PE, Ahmed KS (1989) Sperm structure in Psocoptera. Acta Zool 70:57–61
Kirkham M, Müller-Reichert T, Oegema K, Grill S, Hyman AA (2003) SAS-4 is a *C. elegans* centriolar protein that controls centrosome size. Cell 112:575–587
Kitagawa D, Vakonakis I, Olieric N, Hilbert M, Keller D, Olieric V, Bortfeld M, Erat MC, Fluckiger I, Gonczy P et al (2011) Structural basis of the 9-fold symmetry of centrioles. Cell 144:364–375
Kloc M, Kubiak JZ, Li XC, Ghobrial RM (2014) The newly found functions of MTOC in immunological response. J Leukoc Biol 95:417–430
Klotz C, Dabauvalle MC, Paintrand M, Weber T, Bornens M, Karsenti E (1990) Parthenogenesis in *Xenopus* eggs requires centrosomal integrity. J Cell Biol 110:405–415
Kramer MG, Templeton AR (2001) Life-history changes that accompany the transition from sexual to parthenogenetic reproduction in *Drosophila mercatorum*. Evolution 55:748–761
Krioutchkova MM, Onishchenko GE (1999) Structural and functional characteristics of the centrosome in gametogenesis and early embryogenesis of animals. Int Rev Cytol 185:107–186
Kuriyama R, Borisy GG (1983) Cytasters induced within unfertilized sea-urchin eggs. J Cell Sci 61:175–189
La Terra S, English CN, Hergert P, McEwen BF, Sluder G, Khodjakov A (2005) The de novo centriole assembly pathway in HeLa cells: cell cycle progression and centriole assembly/maturation. J Cell Biol 168:713–722

Lamb RY, Willey RB (1987) Cytological mechanisms of thelytokous parthenogenesis in insects. Genome 29:367–369

Lambert AM (1993) Microtubule-organizing centers in higher plants. Curr Opin Cell Biol 5:116–122

Lawo S, Hasegan M, Gupta GD, Pelletier L (2012) Subdiffraction imaging of centrosomes reveals higher-order organizational features of pericentriolar material. Nat Cell Biol 14:1148–1158

Lerit DA, Jordan HA, Poulton JS, Fagerstrom CJ, Galletta BJ, Peifer M, Rusan NM (2015) Interphase centrosome organization by the PLP-Cnn scaffold is required for centrosome function. J Cell Biol 210:79–97

Li X-W, Fail J, Wang P, Feng J-N, Shelton AM (2014) Performance of arrhenotokous and thelytokous *Thrips tabaci* (Thysanoptera: Thripidae) on onion and cabbage and its implications on evolution and pest management. J Econ Entomol 107:1526–1534

Loeb J (1913) Artificial parthenogenesis and fertilization. University of Chicago Press, Chicago, IL

Loncarek J, Khodjakov A (2009) Ab ovo or de novo? Mechanisms of centriole duplication. Mol Cells 27:135–142

Łuksza M, Queguigner I, Verlhac M-H, Brunet S (2013) Rebuilding MTOCs upon centriole loss during mouse oogenesis. Dev Biol 382:48–56

Magnuson T, Epstein C (1984) Oligosyndactyly: a lethal mutation in the mouse that results in mitotic arrest very early in development. Cell 38:823–833

Mahjoub MR, Stearns T (2012) Supernumerary centrosomes nucleate extra cilia and compromise primary cilium signaling. Curr Biol 22:1628–1634

Mahowald AP, Strassheim JM (1970) Intercellular migration of centrioles in the germarium of *Drosophila melanogaster*: an electron microscopic study. J Cell Biol 45:306–320

Malone CJ, Misner L, Le Bot N, Tsai MC, Campbell JM, Ahringer J, White JG (2003) The *C. elegans* hook protein, ZYG-12, mediates the essential attachment between the centrosome and nucleus. Cell 115:825–836

Manandhar G, Sutovsky P, Joshi HC, Stearns T, Schatten G (1998) Centrosome reduction during mouse spermiogenesis. Dev Biol 203:424–434

Manandhar G, Simerly C, Schatten G (2000) Centrosome reduction during mammalian spermiogenesis. Curr Top Dev Biol 49:343–363

Manandhar G, Schatten H, Sutovsky P (2005) Centrosome reduction during gametogenesis and its significance. Biol Reprod 72:2–13

Mantovani B, Scali V (1992) Hybridogenesis and androgenesis in the stick-insect *Bacillus rossius-grandii benazzii* (Insecta, Phasmatodea). Evolution 46:783–796

Marescalchi O, Scali V (2003) Automictic parthenogenesis in the diploid-triploid stick insect *Bacillus atticus* and its flexibility leading to heterospecific diploid hybrids. Invertebr Reprod Dev 43:163–172

Marescalchi O, Zauli C, Scali V (2002) Centrosome dynamics and inheritance in related sexual and parthenogenetic *Bacillus* (Insecta Phasmatodea). Mol Reprod Dev 63:89–95

Maro B, Howlett SK, Webb M (1985) Non-spindle microtubule organizing centers in metaphase II arrested mouse oocytes. J Cell Biol 101:1665–1672

Marshall WF, Vucica Y, Rosenbaum JL (2001) Kinetics and regulation of de novo centriole assembly: implications for the mechanism of centriole duplication. Curr Biol 11:308–317

Martinez-Campos M, Basto R, Baker J, Kernan M, Raff JW (2004) The *Drosophila* pericentrin-like protein is essential for cilia/flagella function, but appears to be dispensable for mitosis. J Cell Biol 165:673–683

Matthies HJG, McDonald HB, Goldstein LSB, Theurkauf WE (1996) Anastral meiotic spindle morphogenesis: role of the non-claret disjunctional kinesin-like protein. J Cell Biol 134:455–464

Mazia D (1987) The chromosome cycle and the centrosome cycle in the mitotic cycle. Int Rev Cytol 100:49–92

McKim KS, Hawley RS (1995) Chromosomal control of meiotic cell division. Science 270:1595–1601

Mennella V, Keszthelyi B, McDonald KL, Chhun B, Kan F, Rogers GC, Huang B, Agard DA (2012) Subdiffraction-resolution fluorescence microscopy reveals a domain of the centrosome critical for pericentriolar material organization. Nat Cell Biol 14:1159–1168

Meraldi P, Nigg EA (2002) The centrosome cycle. FEBS Lett 521:9–13

Michalik P, Ramírez MJ (2014) Evolutionary morphology of the male reproductive system, spermatozoa and seminal fluid of spiders (Araneae, Arachnida)—current knowledge and future directions. Arthropod Struct Dev 43:291–322

Mikeladze-Dvali T, von Tobel L, Strnad P, Knott G, Leonhardt H, Schermelleh L et al (2012) Analysis of centriole elimination during *C. elegans* oogenesis. Development 139:1670–1679

Miki-Noumura T (1977) Studies on the de novo formation of centrioles: aster formation in the activated eggs of sea urchin. J Cell Sci 24:203–216

Mirzaghaderi G, Hörandl E (2016) The evolution of meiotic sex and its alternatives. Proc R Soc Lond B Biol Sci 283:20161221

Morgan TH (1899) The action of salt-solutions on the unfertilized and fertilized eggs of *Arbacia*, and of other animals. Arch Entwicklungsmech Org 8:448–539

Moy AW, Brandriff B, Vacquier VD (1977) Cytasters from sea urchin eggs parthenogenetically activated by procaine. J Cell Biol 73:788–793

Nachury MV (2014) How do cilia organize signalling cascades? Philos Trans R Soc Lond B 369:20130465

Nagano H, Hirai S, Okano K, Ikegami S (1981) Achromosomal cleavage of fertilized starfish eggs in the presence of aphidicolin. Dev Biol 85:409–415

Nedelec FJ, Surrey T, Maggs AC, Leibler S (1997) Self-organization of microtubules and motors. Nature 389:305–308

Nedelec F, Surrey T, Karsenti E (2003) Self-organisation and forces in the microtubule cytoskeleton. Curr Opin Cell Biol 15:118–124

Nigg EA (2002) Centrosome aberrations: cause or consequence of cancer progression? Nat Rev Cancer 2:815–825

Nigg EA (2006) Origins and consequences of centrosome aberrations in human cancers. Int J Cancer 119:2717–2723

Nigg EA, Raff JW (2009) Centrioles, centrosomes, and cilia in health and disease. Cell 139:663–678

Nigg EA, Stearns T (2011) The centrosome cycle: centriole biogenesis, duplication and inherent asymmetries. Nat Cell Biol 13:1154–1160

Normark BB, Kirkendall LR (2009) Parthenogenesis. In: Resh VH, Cardé RT (eds) Encyclopedia of insects, 2nd edn. Academic Press, Amsterdam, pp 753–757

Novak ZA, Conduit PT, Wainman A, Raff JW (2014) Asterless licenses daughter centrioles to duplicate for the first time in *Drosophila* embryos. Curr Biol 24:1276–1282

Novak ZA, Wainmann A, Gartenmann A, Raff JW (2016) Cdk1 phosphorylates *Drosophila* Sas-4 to recruit Polo to daughter centrioles and convert them to centrosomes. Dev Cell 37:545–557

Nur U (1972) Diploid arrhenotoky and automictic thelytoky in soft scale insects (Lecaniidae: Coccoidea: Homoptera). Chromosoma 39:381–401

O'Toole ET, Dutcher SK (2014) Site-specific basal body duplication in *Chlamydomonas*. Cytoskeleton 71:108–118

Oakley BR (2000) g-tubulin. In: Palazzo RE, Schatten GP (eds) The centrosome in cell replication and early development, Current topics in development biology, vol 49. Academic Press, San Diego, pp 27–54

Olsen MW (1967) Age as a factor influencing the level of parthenogenesis in eggs of turkeys. Proc Soc Exp Biol Med 124:617–619

Paccagnini E, De Marzo L, Giusti F, Dallai R (2006) The aberrant spermatogenesis of the *Haplothrips simplex* (Buffa) (Thysanoptera): ultrastructure study. Tissue Cell 38:177–186

Paffoni A, Brevini TA, Somigliana E, Restelli L, Gandolfi F, Ragni G (2007) In vitro development of human oocytes after parthenogenetic activation or intracytoplasmic sperm injection. Fertil Steril 87:77–82

Palévody C (1973) Etude cytologique de la parthémogenése chez *Folsomia candida* (Collembole, Isotomide). C R Acad Sci III 277:2501–2504

Paoletti A, Moudjou M, Paintrand M, Salisbury JL, Bornens M (1996) Most of centrin in animal cells is not centrosome-associated and centrosomal centrin is confined to the distal lumen of centrioles. J Cell Sci 109:3089–3102

Paoli F, Roversi PF, Gottardo M, Callaini G, Mercati D, Dallai R (2015) A microtubule organizing centre (MTOC) is responsible for the production of the sperm flagellum in Matsucoccus feytaudi (Hemiptera: Coccoidea). Arthropod Struct Dev 44:237–442

Paweletz N, Mazia D (1979) Fine structure of the mitotic cycle of unfertilized sea urchin eggs activated by ammoniacal sea water. Eur J Cell Biol 20:37–44

Pearson CG (2014) Choosing sides – asymmetric centriole and basal body assembly. J Cell Sci 127:2803–2810

Peel N, Stevens NR, Basto R, Raff JW (2007) Overexpressing centriole-replication proteins in vivo induces centriole overduplication and de novo formation. Curr Biol 17:834–843

Pelletier L, Yamashita YM (2012) Centrosome asymmetry and inheritance during animal development. Curr Opin Cell Biol 24:541–546

Pelletier L, O'Toole E, Schwager A, Hyman AA, Müller-Reichert T (2006) Centriole assembly in *Caenorhabditis elegans*. Nature 444:619–623

Pellman D (2007) Cell biology: aneuploidy and cancer. Nature 446:38–39

Phillips DM (1967) Giant centriole formation in *Sciara*. J Cell Biol 33:73–92

Piel M, Meyer P, Khodjakov A, Rieder CL, Bornens M (2000) The respective contributions of the mother and daughter centrioles to centrosome activity and behavior in vertebrate cells. J Cell Biol 149:317–329

Pihan GA (2013) Centrosome dysfunction contributes to chromosome instability, chromoanagenesis, and genome reprograming in cancer. Front Oncol 3:277

Pijnacker LP (1969) Automictic parthenogenesis in the stick insect *Bacillus rossius* Rossi (Cheleutoptera, Phasmidae). Genetica 40:393–399

Pimenta-Marques A, Bento I, Lopes CA, Duarte P, Jana SC, Bettencourt-Dias M (2016) A mechanism for the elimination of the female gamete centrosome in *Drosophila* melanogaster. Science 353:aaf4866

Raff JW, Glover DM (1988) Nuclear and cytoplasmic mitotic cycles continue in *Drosophila* embryos in which DNA synthesis is inhibited with aphidicolin. J Cell Biol 107:2009–2019

Reina J, Gonzalez C (2014) When fate follows age: unequal centrosomes in asymmetric cell division. Phil Trans R Soc B 369:20130466

Reynolds RG, Booth W, Schuett GW, Fitzpatrick BM, Burghardt GM (2012) Successive virgin births of viable male progeny in the checkered gartersnake, *Thamnophis marcianus*. Biol J Linn Soc 107:566–572

Riparbelli MG, Callaini G (2003) *Drosophila* parthenogenesis: a model for de novo centrosome assembly. Dev Biol 260:298–313

Riparbelli MG, Callaini G (2008) *Drosophila* parthenogenesis: a tool to decipher centrosomal vs acentrosomal spindle assembly pathways. Exp Cell Res 314:1617–1625

Riparbelli MG, Callaini G (2010) Detachment of the basal body from the sperm tail is not required to organize functional centrosomes during *Drosophila* embryogenesis. Cytoskeleton 67:251–258

Riparbelli MG, Callaini G (2011) Male gametogenesis without centrioles. Dev Biol 349:427–439

Riparbelli MG, Stouthamer R, Dallai R, Callaini G (1998) Microtubule organization during the early development of the parthenogenetic egg of the hymenopteran *Muscidifurax uniraptor*. Dev Biol 195:88–99

Riparbelli MG, Callaini G, Glover DM (2000) Failure of pronuclear migration and repeated divisions of polar body nuclei associated with MTOC defects in polo eggs of *Drosophila*. J Cell Sci 113:3341–3350

Riparbelli MG, Callaini G, Glover DM, Avides MDC (2002) A requirement for the abnormal spindle protein to organise microtubules of the central spindle for cytokinesis in *Drosophila*. J Cell Sci 115:913–922
Riparbelli MG, Tagu D, Bonhomme J, Callaini G (2005) Aster self-organization at meiosis: a conserved mechanism in insect parthenogenesis? Dev Biol 278:220–230
Riparbelli MG, Giordano R, Callaini G (2006) Centrosome inheritance in the parthenogenetic egg of the collembolan *Folsomia candida*. Cell Tissue Res 326:861–872
Riparbelli MG, Callaini G, Schejter E (2007) Microtubule-dependent organization of subcortical microfilaments in the early *Drosophila* embryo. Dev Dyn 236:662–670
Riparbelli MG, Dallai R, Mercati D, Bu Y, Callaini G (2009) Centriole symmetry: a big tale from small organisms. Cell Motil Cytoskeleton 66:1100–1105
Riparbelli MG, Dallai R, Callaini G (2010) The insect centriole: a land of discovery. Tissue Cell 42:69–80
Riparbelli MG, Callaini G, Megraw TL (2012) Assembly and persistence of primary cilia in dividing *Drosophila* spermatocytes. Dev Cell 23:425–432
Rispe C, Pierre JS (1998) Coexistence between cyclical parthenogens, obligate parthenogens, and intermediates in a fluctuating environment. J Theor Biol 195:97–110
Rodrigues-Martins A, Riparbelli MG, Callaini G, Glover DM, Bettencourt-Dias M (2007) Revisiting the role of the mother centriole in centriole biogenesis. Science 316:1046–1050
Rodrigues-Martins A, Riparbelli M, Callaini G, Glover DM, Bettencourt-Dias M (2008) From centriole biogenesis to cellular function: centrioles are essential for cell division at critical developmental stages. Cell Cycle 7:11–16
Sachs MI, Anderson E (1970) A cytological study of artificial parthenogenesis in the sea urchin *Arbacia punctulata*. J Cell Biol 47:140–158
Salinas-Saavedra M, Vargas AO (2011) Cortical cytasters: a highly conserved developmental trait of Bilateria with similarities to Ctenophora. Evodevo 2:23
Sartain CV, Wolfner MF (2013) Calcium and egg activation in *Drosophila*. Cell Calcium 53:10–15
Sarvella P (1970) Sporadic occurrences of parthenogenesis in poultry. J Hered 61:215–219
Sathananthan AH, Kola I, Osborne J, Trounson A, Ng SC, Bongso A, Ratnam SS (1991) Centrioles in the beginning of human development. Proc Natl Acad Sci USA 88:4806–4810
Sathananthan AH, Ratnam SS, Ng SC, Tarin JJ, Gianaroli L, Trounson A (1996) The sperm centriole: its inheritance, replication and perpetuation in early human embryos. Hum Reprod 11:345–356
Sathananthan AH, Selvaraj K, Girijashankar ML, Ganesh V, Selvaraj P, Trounson AO (2006) From oogonia to mature oocytes: inactivation of the maternal centrosome in humans. Microsc Res Tech 69:396–407
Schatten G (1982) Motility during fertilization. Int Rev Cytol 79:59–163
Schatten G (1994) The centrosome and its mode of inheritance: the reduction of the centrosome during gametogenesis and its restoration during fertilization. Dev Biol 165:299–335
Schatten H, Schatten G (1986) Centrosomes and microtubule organization during mouse and sea urchin fertilization. Prog Clin Biol Res 217B:91–94
Schatten H, Sun QY (2011) Centrosome dynamics during mammalian oocyte maturation with a focus on meiotic spindle formation. Mol Reprod Dev 78:757–768
Schatten H, Walter M, Biessmann H, Schatten G (1992) Activation of maternal centrosomes in unfertilized sea urchin eggs. Cell Motil Cytoskeleton 23:61–70
Scheer U (2014) Historical roots of centrosome research: discovery of Boveri's microscope slides in Würzburg. Phil Trans R Soc B 369:20130469
Schuett GW, Fernandez PJ, Gergits WF, Casna NJ, Chiszar D, Smith HM, Mitton JB, Mackessy SP, Odum RA, Demlong MJ (1997) Production of offspring in the absence of males: evidence for facultative parthenogenesis in bisexual snakes. Herpetological. Nat History 5:1–10
Schuh M, Ellenberg J (2007) Self-organization of MTOCs replaces centrosome function during acentrosomal spindle assembly in live mouse oocytes. Cell 130:484–498

Schwander T, Crespi BJ (2009) Multiple direct transitions from sexual reproduction to apomictic parthenogenesis in *Timema* stick insects. Evolution 63:84–103

Seitz R, Vilpoux K, Hopp U, Harzsch S, Maier G (2005) Ontogeny of the Marmorkrebs (marbled crayfish): a parthenogenetic crayfish with unknown origin and phylogenetic position. J Exp Zool 303A:393–405

Sekiné K, Tojo K (2010) Automictic parthenogenesis of a geographically parthenogenetic mayfly, *Ephoron shigae* (Insecta: Ephemeroptera, Polymitarcyidae). Biol J Linn Soc 99:335–343

Severson AF, von Dassow G, Bowerman B (2016) Oocyte meiotic spindle assembly and function. Curr Top Dev Biol 116:65–98

Shirato Y, Tamura M, Yoneda M, Nemoto S (2006) Centrosome destined to decay in starfish oocytes. Development 133:343–350

Shukla A, Kong D, Sharma M, Magidson V, Loncarek J (2015) Plk1 relieves centriole block to reduplication by promoting daughter centriole maturation. Nat Commun 6:8077

Simerly CR, Hecht NB, Goldberg E, Schatten G (1993) Tracing the incorporation of the sperm tail in the mouse zygote and early embryo using an anti-testicular alpha-tubulin antibody. Dev Biol 158:536–548

Simon JC, Martinez-Torres D, Latorre A, Moya A, Hebert PD (1996) Molecular characterization of cyclic and obligate parthenogens in the aphid *Rhopalosiphum padi* (L.) Proc R Soc Lond B Biol Sci 263:481–486

Sluder G, Hinchcliffe EH (1998) The apparent linkage between centriole replication and the S phase of the cell cycle. Cell Biol Int 22:3–5

Sluder G, Khodjakov A (2010) Centriole duplication: analogue control in a digital age. Cell Biol Int 34:1239–1245

Sluder G, Rieder CL (1985) Experimental separation of pronuclei in fertilized sea urchin eggs: chromosomes do not organize a spindle in the absence of centrosomes. J Cell Biol 100:897–903

Sluder G, Miller FJ, Lewis K, Davison ED, Rieder CL (1989) Centrosome inheritance in starfish zygotes: selective loss of the maternal centrosome after fertilization. Dev Biol 131:567–579

Snook RR, Karr TL (1998) Only long sperm are fertilization-competent in six sperm-heteromorphic *Drosophila* species. Curr Biol 8:291–294

Snook RR, Markow TA (2001) Mating systemevolution in sperm heteromorphic *Drosophila*. J Insect Physiol 47:957–964

Snook RR, Hosken DJ, Karr TL (2011) The biology and evolution of polyspermy: insights from cellular and functional studies of sperm and centrosomal behavior in the fertilized egg. Reproduction 142:779–792

Song MH, Miliaras NB, Peel N, O'Connell KF (2008) Centrioles: some self-assembly required. Curr Opin Cell Biol 20:688–693

Stenberg P, Saura A (2009) Cytology of asexual animals. In: Schoen I, Martens K, van Dijk P (eds) Lost sex—the evolutionary biology of parthenogenesis. Springer, Heidelberg, pp 63–74

Stepinska U, Bakst MR (2007) Fertilization. In: Jamieson BGM (ed) Reproductive biology and phylogeny of birds: phylogeny, morphology, hormones, fertilization, vol 6A. Science Publishers, Enfield, NH, pp 553–587

Stevens NR, Raposo AA, Basto R, St Johnston D, Raff JW (2007) From stem cell to embryo without centrioles. Curr Biol 17:1498–1503

Stouthamer R, Werren JH (1993) Microbes associated with parthenogenesis in wasps of the genus *Trichogramma*. J Inv Pathol 61:6–9

Stouthamer R, Breeuwer JAJ, Hurst GDD (1999) *Wolbachia pipientis*: microbial manipulator of arthropod reproduction. Annu Rev Microbiol 53:71–102

Strnad P, Gönczy P (2008) Mechanisms of procentriole formation. Trends Cell Biol 18:389–396

Sullender B, Crease T (2001) The behavior of a *Daphnia pulex* transposable element in cyclically and obligately parthenogenetic populations. J Mol Evol 53:63–69

Sun Q-Y, Schatten H (2007) Centrosome inheritance after fertilization and nuclear transfer in mammals. Adv Exp Med Biol 591:58–71

Suomalainen E (1950) Parthenogenesis in animals. Adv Genet 3:193–253
Suomalainen E, Saura A (1993) Patterns of reproduction in insects. In: Wrensch DL, Ebbert MA (eds) Evolution and diversity of sex ratio in insects and mites. Chapman & Hall, New York, pp 100–117
Suomalainen E, Saura A, Lokki J (1987) Cytology and evolution in parthenogenesis. CRC Press, Boca Raton, FL
Szollosi D, Ozil J-P (1991) De novo formation of centrioles in parthenogenetically activated, diploidized rabbit embryos. Biol Cell 72:61–66
Szollosi D, Calarco P, Donahue RP (1972) Absence of centrioles in the first and second meiotic spindles of mouse oocytes. J Cell Sci 11:521–541
Tates AD (1971) Cytodifferentiation during spermatogenesis in *Drosophila melanogaster*: an electron microscopy study. PhD thesis. Rijksuniversiteit de Leiden, Netherlands
Tavosanis G, Llamazares S, Goulielmos G, Gonzalez C (1997) Essential role for γ-tubulin in the acentriolar female meiotic spindle of *Drosophila*. EMBO J 16:1809–1819
Theurkauf WE, Hawley RS (1992) Meiotic spindle assembly in *Drosophila* females: behavior of nonexchange chromosomes and the effects of mutations in the nod kinesin-like protein. J Cell Biol 116:1167–1180
Thielsch A, Brede N, Petrusek A, De ML, Schwenk K (2009) Contribution of cyclic parthenogenesis and colonization history to population structure in *Daphnia*. Mol Ecol 18:1616–1628
Tournier F, Karsenti E, Bornens M (1989) Parthenogenesis in *Xenopus* eggs injected with centrosomes from synchronized human lymphoid cells. Dev Biol 136:321–329
Tram U, Sullivan W (2000) Reciprocal inheritance of centrosomes in the parthenogenetic Hymenopteran *Nasonia vitripennis*. Curr Biol 10:1413–1419
Trieselmann N, Armstrong S, Rauw J, Wilde A (2003) Ran modulates spindle assembly by regulating a subset of TPX2 and kid activities including Aurora A activation. J Cell Sci 116:4791–4798
Trudgill DL, Blok VC (2001) Apomictic polyphagous root-knot nematodes: exceptionally successful and damaging biotrophic root pathogens. Annu Rev Phytopathol 39:53–77
Tsai MY, Wiese C, Cao K, Martin O, Donovan P, Ruderman J, Prigent C, Zheng YX (2003) A Ran signalling pathway mediated by the mitotic kinase Aurora A in spindle assembly. Nat Cell Biol 5:242–248
Tsou MF, Stearns T (2006) Mechanism limiting centrosome duplication to once per cell cycle. Nature 442:947–951
Tsou MF, Wang WJ, George KA, Uryu K, Stearns T, Jallepalli PV (2009) Polo kinase and separase regulate the mitotic licensing of centriole duplication in human cells. Dev Cell 17:344–354
Tyler A (1941) Artificial parthenogenesis. Biol Rev Camb Philos Soc 16:291–336
Uetake Y, Kato KH, Washitani-Nemoto S, Nemoto Si S (2002) Nonequivalence of maternal centrosomes/centrioles in starfish oocytes: selective casting-off of reproductive centrioles into polar bodies. Dev Biol 247:149–164
Uetake Y, Loncarek J, Nordberg JJ, English CN, La Terra S, Khodjakov A, Sluder G (2007) Cell cycle progression and de novo centriole assembly after centrosomal removal in untransformed human cells. J Cell Biol 176:173–182
van Beneden E (1876) Contribution al'histoire de la vesiculaire germinative et du premier embryonnaire. Bull Acad R Belg 42:35–97
van Breugel M, Hirono M, Andreeva A, Yanagisawa HA, Yamaguchi S, Nakazawa Y, Morgner N, Petrovich M, Ebong IO, Robinson CV et al (2011) Structures of SAS-6 suggest its organization in centrioles. Science 331:1196–1199
Verde F, Berrez JM, Antony C, Karsenti E (1991) Taxol-induced microtubule asters in mitotic extracts of *Xenopus* eggs: requirement for phosphorylated factors and cytoplasmic dynein. J Cell Biol 112:1177–1187
Vershinina AO, Kuznetsova VG (2016) Parthenogenesis in hexapoda: entognatha and non-holometabolous insects. J Zool Syst Evol Res 54:257–268

Vidwans SJ, Wong ML, O'Farrell PH (2003) Anomalous centriole configurations are detected in *Drosophila* wing disc cells upon Cdk1 inactivation. J Cell Sci 116:137–143
Vitre BD, Cleveland DW (2012) Centrosomes, chromosome instability (CIN) and aneuploidy. Curr Opin Cell Biol 24:809–815
Wadsworth P, Khodjakov A (2004) E pluribus unum: towards a universal mechanism for spindle assembly. Trends Cell Biol 14:413–419
Wakimoto BT (1979) DNA synthesis after polyspermic fertilization in the axolotl. J Embryol Exp Morphol 52:39–48
Wang WJ, Soni RK, Uryu K, Bryan Tsou MF (2011) The conversion of centrioles to centrosomes: essential coupling of duplication with segregation. J Cell Biol 193:727–739
Watts PC, Buley KR, Sanderson S, Boardman W, Ciofi C, Gibson R (2006) Parthenogenesis in komodo dragons. Nature 444(7122):1021–1022
Went DF (1982) Egg activation and parthenogenetic reproduction in insects. Biol Rev 57:319–344
Wilde A, Zheng Y (1999) Stimulation of microtubule aster formation and spindle assembly by the small GTPase Ran. Science 284:1359–1362
Wilson PG, Borisy GG (1998) Maternally expressed γTub37CD in *Drosophila* is differentially required for female meiosis and embryonic mitosis. Dev Biol 199:273–290
Wong J, Wessel GM (2006) Defending the zygote: search for the ancestral animal block to polyspermy. Curr Top Dev Biol 72:1–151
Wright JW (1973) Evolution of the X_1X_2Y sex chromosome mechanism in the scincid lizard *Soincella laterale* (Say). Chromosoma 43:101–108
Zimmerman WC, Sillibourne J, Rosa J, Doxsey SJ (2004) Mitosis-specific anchoring of gamma tubulin complexes by pericentrin controls spindle organization and mitotic entry. Mol Biol Cell 15:3642–3657
Zorn GA, Lukdas JJ, Kates JR (1979) Purification and characterization of regenerating mouse L929 karyoplasts. Cell 18:659–672
Zyss D, Gergely F (2009) Centrosome function in cancer: guilty or innocent? Trends Cell Biol 19:334–346

Part V
Maternal Factors: Origin, Evolution and Application in Genetic Engineering

Chapter 20
The Origin and Evolution of Maternal Genes

Antonio Marco

Abstract Proteins and RNA molecules are deposited into the developing egg by the mother. These gene products will drive the first stages of development and are coded by maternal genes. Maternal genes are essential, yet, despite their importance, their evolutionary dynamics is largely unknown. Here I review the current knowledge of maternal gene evolution. The evolutionary origin of maternal genes tends to be more recent than that of zygotic genes. Some studies support the theoretical prediction that maternal genes evolve faster than zygotic genes. However, most studies were done on a limited set of species and genes. I also discuss the way forward to understand the evolution of maternal genes by combining high-throughput genomics and theoretical evolutionary approaches.

Abbreviations

mRNA messenger RNA
Myr Million years
UTR Untranslated Region

20.1 Introduction

Maternal effect genes are genes expressed in females whose phenotype is manifested in their offspring. These genes are essential, particularly during development. Early embryologists already noticed that the cytoplasm of the oocyte contained products that were involved in the first stages of development. For instance, Ziegler (1898) described that enucleated sea urchin embryos initiate cleavage (cell division of the zygote), suggesting that early development is driven by non-nuclear (cytoplasmatic) factors. Likewise, segmentation is observed in chromosome-free frog eggs (Briggs et al. 1951) [reviewed in Marco (2017)]. The first cases of maternal inheritance were reported in 1912 in the colour and shape of

A. Marco (✉)
School of Biological Sciences, University of Essex, Wivenhoe Park, Colchester CO4 3SQ, UK
e-mail: amarco.bio@gmail.com

M. Kloc (ed.), *Oocytes*, Results and Problems in Cell Differentiation 63,
DOI 10.1007/978-3-319-60855-6_20

silkworm eggs (Toyama 1914) [cited in Dobzhansky (1937)]. Other well-known example is the direction of coiling in the sea snail *Lymnaea peregra* (Boycott and Diver 1923), which only depends on the genotype of the mother (Sturtevant 1923). Nowadays we know that there are many maternal-effect genes that are essential for the development and function of the organism.

An important type of maternal-effect genes are those whose products (RNA and/or proteins) are deposited by the mother into the egg, determining the first stages of development, as embryonic genes are expressed after fertilization. These are generally called *maternal* genes, in contrast to *zygotic* genes that are those expressed from embryonic genes (Wolpert et al. 2006). The evolutionary dynamics of maternal genes is peculiar because their phenotype is expressed in the following generation. This generational uncoupling between genotype and phenotype results in faster-evolving genes, when compared to zygotic genes (as I discuss here). Other types of maternal-effect genes have been described, for instance, genes that affect the womb environment (which in turn will influence embryonic development), mitochondrial genes or genes affecting the behaviour of the mother in terms of parental care. In this review, I will focus exclusively on the evolution of maternal genes in animals (as opposed to zygotic genes) and their importance in the early stages of development.

20.2 Evolutionary Origin of Maternal Genes

20.2.1 Maternal Genes in Drosophila

If a geneticist had to use a single word to describe how important is a mother, this word would probably be ‘bicoid’. Bicoid is the product of the gene *bicoid* (*bcd*) in the fruit fly *Drosophila melanogaster*. The position of *bcd* products in the egg determines where the head of the fly will develop (Lawrence 1992). Transcripts from the maternal *bcd* are deposited in the anterior pole of the developing egg. This transcript is transported towards the posterior pole and subsequently translated (Lipshitz 2009), producing a gradient of Bicoid (Bcd) protein (Fig. 20.1a). This gradient of Bcd protein will affect the expression of other maternal and zygotic genes, determining the anteroposterior axis of the animal. Indeed, embryos from females with mutations in *bicoid* do not develop head or thorax (Frohnhöfer and Nüsslein-Volhard 1986).

bcd has become the paradigm of maternal genes, and its evolution has been thoroughly studied. Since *bcd* is essential for early patterning of the embryo, one may expect it to be broadly conserved. However, the identification of *bcd* in other model insects (such as flour beetle and honeybee) was not successful [reviewed in McGregor (2005)]. It turned out that *bcd* and its zygotic paralog *zerknüllt* (*zen*) originated from a duplication of the gene *Hox3* (often referred to as *Hox3/zen*) within flies (Stauber et al. 1999), specifically at the lineage leading to higher flies (Cyclorrhapha; Fig. 20.1b). That means that *bcd* is only about 115 years old (Ding et al. 2015). The question is are other maternal genes also relatively young?

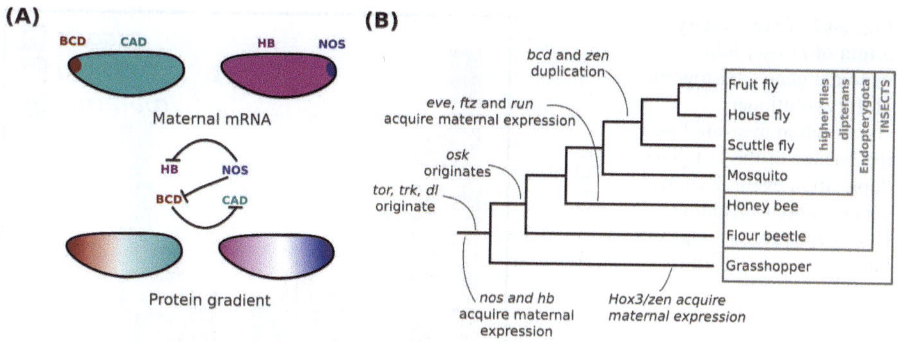

Fig. 20.1 Maternal genes in *Drosophila melanogaster*. (**a**) Location of maternal transcripts in unfertilized eggs (*top*). After translation starts in embryos, the regulation of translation by maternal gene products (*mid*) gives rise to a protein gradient (*bottom*). The anterior part of the embryos is on the *left*. Genes in the figure are *bicoid* (*bcd*), *hunchback* (*hb*), *nanos* (*nos*) and *caudal* (*cad*). Information compiled from (Lawrence 1992; Wolpert et al. 2006). (**b**) Recent origin of maternal genes or maternal functions within the insect lineage. See main text for details and references

Another iconic maternal gene in *Drosophila* is *oskar* (*osk*), whose RNA is localized in the posterior of the developing oocyte, and determines where the germline will be formed (Ephrussi and Lehmann 1992). The gene *osk* probably originated in holometabolous insects (Fig. 20.1b), and it is associated with the emergence of the maternal inheritance of germline determination (Lynch et al. 2011). In *Drosophila* and other holometabolous insects, maternal *oskar* triggers the establishment of the germline cells during early development. In other insects (and probably in the common ancestor of all insects), germ cells are determined by epigenetic mechanisms (Extavour and Akam 2003). Other maternal genes such as *torso* (*tor*), *trunk* (*trk*) and *dorsal* (*dl*) appear to be also restricted to insects, and they have further functionally diverged in the *Drosophila* lineage (Abzhanov and Kaufman 2000; Lynch and Roth 2011; Schoppmeier and Schröder 2005).

Some maternal genes are broadly conserved in animals. *Hunchback* (*hb*) is a maternal gene encoding a zinc-finger protein involved in the establishment of the anteroposterior axis of the embryo (Fig. 20.1a). But zygotic Hunchback protein is also produced during early development. In that sense, *hb* is both zygotic and maternal. *hb* is a highly conserved gene which was probably present in the common ancestor of most animals (John et al. 2009). Although a maternal function of hunchback has only been studied in insects, its presence in unfertilized embryos has also been detected in other invertebrates such as annelids (Pinnell et al. 2006), which may also indicate a conserved maternal role of this gene. The gene Nanos (*nos*) is also highly conserved among animals. Nanos mRNA is deposited by the mother in the posterior part of the egg (Fig. 20.1a). Maternal Nanos has a role in specifying the posterior identity, and its protein product will repress the translation of Hunchback and Bicoid transcripts by binding to their 3′ UTRs (Murata and Wharton 1995). Zygotic Nanos have additional functions in neural development and germ cells (Deshpande et al. 1999). Work in cnidarians revealed that *nos*

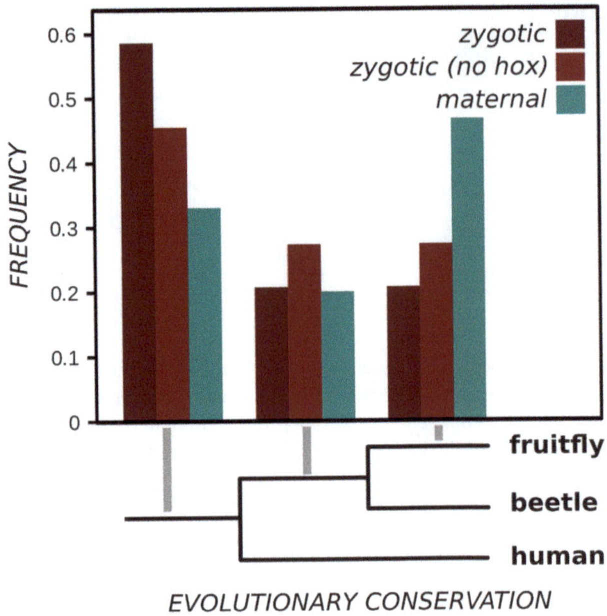

Fig. 20.2 Evolutionary origin of *Drosophila* maternal genes. Histogram of the evolutionary conservation measured as the lineage of the last gene duplication event. Zygotic genes without including HOX genes are in a separate bin

zygotic functions are conserved in animals (Kanska and Frank 2013; Mochizuki et al. 2000), but the maternal role in posterior patterning is probably an evolutionary novelty in insects (Kanska and Frank 2013).

If we compare the evolutionary age, measured as the time since the last gene duplication event according to TreeFam (Ruan et al. 2007), between selected *Drosophila* maternal and zygotic genes [as listed in page 85 in Wolpert et al. (2006)], we observe that maternal genes tend to be evolutionarily younger than zygotic genes (Fig. 20.2). This observation may be biased by the fact that a significant fraction of zygotic genes are homeotic genes, which are highly conserved in all animals. After removing homeotic genes from the analysis, the pattern is still observed: maternal genes are younger than zygotic genes (Fig. 20.2). However, this search does not take into account that many of those genes are both maternal and zygotic. Also, the evolutionary analysis is not exhaustive, and many gene duplication events may have not been appropriately identified. Thus, we have to be cautious with the observation that maternal genes are younger than zygotic until we have more evidence.

20.2.2 Maternal Genes in Other Species

Maternal genes in mammals have been only recently described [reviewed by (Li et al. 2010)]. One of the first maternal genes reported in mouse was *Mater*, a maternal antigen required by the embryo (Tong et al. 2000). This gene was found to be also maternal in humans (Tong et al. 2002). *Mater*, or its human ortholog *NLRP5*, originated in mammals (Tian et al. 2009). The other gene originally described as maternal in mammals was *HSF1*, a transcription factor essential for development (Christians et al. 2000). Although *HSF1* is generally seen as broadly conserved in eukaryotes, human and mouse *Hsf1* are the product of gene duplication in vertebrates (treefam.org; accession number TF330401). Thus, mammalian *Hsf1* genes may have functionally diverged from other eukaryotic *Hsf* genes. Whilst knock-down mutations in the homologs *Hsf* in *Drosophila* and yeast are lethal in physiological conditions (Jedlicka et al. 1997; Zarzov et al. 1997), mouse *Hsf1* zygotic mutants can survive (Christians 2011; McMillan et al. 1998), although its maternal expression is required (Christians et al. 2000).

Thanks to the development of high-throughput sequencing technologies, it has been possible to compare the oocyte transcriptome of humans (Kocabas et al. 2006) and mice (Hamatani et al. 2004; Zeng et al. 2004). About 30% of human maternal genes are also maternal in mice (Kocabas et al. 2006). However, most of these transcripts are of unknown function. When limiting the set to those genes present in oocytes and in embryonic stem cells, a core group of 66 transcripts are identified (Kocabas et al. 2006). At this stage it is not clear whether the high number of common maternal genes is due to a high conservation of maternal products and functions or due to genes highly expressed during oocyte formation (likely to be conserved), whose products remain in the egg but have no maternal function (probably a combination of both). A more recent research suggests that maternal genes are broadly conserved in animals (Liu et al. 2014), but the data provided in that report is insufficient to perform appropriate statistical tests, and further investigations are required.

20.2.3 Maternal microRNAs

Among maternal products we also find non-coding RNAs. MicroRNAs are of particular relevance, as they are major gene regulators in animals and plants (Axtell et al. 2011; Bartel 2009; Lim and Okamura 2017). Early high-throughput experiments in *Drosophila* already suggested that microRNAs may be maternal since they are abundant in very early developmental stages (Ruby et al. 2007). Multiple experiments show that, indeed, microRNAs are abundant in the unfertilised egg of insects (Lee et al. 2004, 2014; Marco 2015; Ninova et al. 2015; Votruba 2009). Maternal microRNAs are often encoded within maternal protein-coding genes (Marco 2014, 2015), so they should be evolutionarily younger than the host gene (except in the rare cases in which a microRNA translocates from another genomic locations to an intron). Other maternal microRNA genes like the *mir-310/311/312/313* cluster and *mir-14* are specific to insects (Kozomara and Griffiths-Jones 2014; Marco et al. 2013; Ninova et al. 2014). Some maternal microRNAs derive from the cluster *mir-92a/mir-92b*, which is believed to be conserved in animals. However, it may have actually been originated in insects (Marco et al. 2013), further supporting the idea that maternal genes, including microRNAs, are relatively young in evolutionary terms. On the other hand, other maternal microRNAs appear to be broadly conserved such as *bantam* or *mir-184*. Whether the maternal function of the evolutionary conserved microRNAs is lineage specific has not been explored.

In summary, maternal genes seem to be, generally, lineage specific, and many conserved maternal genes may have acquired their maternal function later on during evolution. That is, current evidence suggests a relatively recent origin of maternal genes.

20.3 Evolutionary Dynamics of Maternal Genes

20.3.1 Population Genetics and Molecular Evolution

The impact of maternal effects on selection has been of particular interest to animal breeders, for instance, to quantify the impact of the maternal genotype on the size of the litter [Falconer (1964), cited in Wright(1984)]. These effects, mostly explored by quantitative geneticists (Falconer 1981; Lynch and Walsh 1998), have been largely ignored in classic population genetics treatises [as an exception Wright (1968, pp. 57–59) has a brief section on maternal inheritance]. That may explain why general evolution textbooks rarely cover this topic (Wade 1998). Maternal effects appeared to have been an 'issue' rather than an important biological phenomenon to take into account for early population geneticists, as their mathematical treatment is more complex.

The population genetics of maternal effects was mostly developed during the 1980s and after [reviewed in Wade (1998)]. Theoretical work has shown that the equilibrium frequency of deleterious alleles is double for genes expressed in one of the sexes than for genes expressed in both sexes (Whitlock and Wade 1995). In other words, selection is half as efficient in maternal genes (expressed only in the mother) than in genes expressed in both sexes (Wade 1998). This case was explored for *bcd*, the best-studied maternal gene. Indeed, it was shown that *bcd* was about twice as diverse (as in nucleotide diversity) in the coding regions than its paralog *zen* in *Drosophila* populations (Barker et al. 2005). Also, *bcd* evolves faster than *zen* (Demuth and Wade 2007), an observation which also agrees with theoretical predictions (Demuth and Wade 2007; Wade 1998). Figure 20.3 shows the accelerated evolution of maternal *bcd* with respect to other homologous sequences. These studies also provide an explanation to the 'developmental hourglass', that is, the observation that genes expressed during early and late development evolve faster than genes expressed during mid-development (Davis et al. 2005; Raff 1996). According to this model, the base of the hourglass is enriched in maternal genes and, therefore, of genes evolving faster than zygotic genes (Cruickshank and Wade 2008).

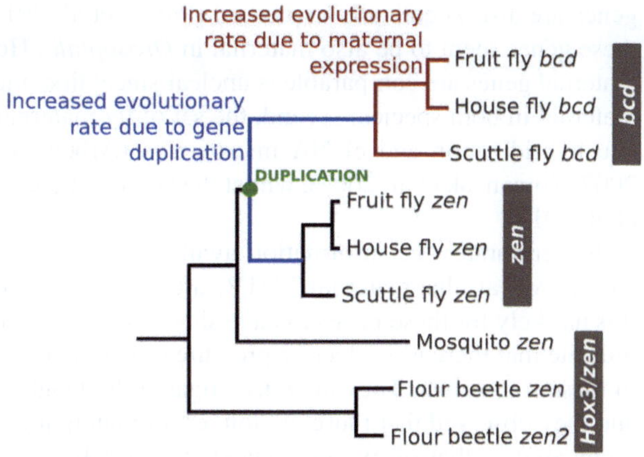

Fig. 20.3 Fast evolution of maternal *bcd*. Phylogenetic tree showing the accelerated evolutionary rate of *bcd* after gene duplication and the acquisition of a maternal function. The tree is a schematic simplified version of the tree in Demuth and Wade (2007)

20.3.2 Functional Diversification of Maternal Genes

The fast evolution of *bcd* is reflected in its functional diversity. Since *bcd* and *zen* emerged from a gene duplication of *Hox3*, *bcd* has taken over the anterior determination in higher flies and even acquired a novel molecular function: binding to RNA. It has been even called 'a hopeful monster of a protein' due to the dramatic functional changes in a short evolutionary time span (Dearden and Akam 1999). The function of *bcd* has diverged in different lineages. RNAi experiments in the housefly suggest that *bcd* may not conserve in this species the regulatory role in thoracic development that it has in *Drosophila* (Shaw et al. 2001). Indeed, it is not only the divergence of *bcd* between these two species but also the fast evolutionary changes at Bicoid-dependent promoters what may explain the functional differences (McGregor et al. 2001). In the scuttle fly (*Megaselia abdita*), the expression of Bicoid is broader than in *Drosophila*, having an impact in the expression of other gap genes during early development (Wotton et al. 2015a, b). In this species, the gene *caudal* (*cad*), which is maternal in *Drosophila*, is not maternally expressed, and the maternal function of *hb* has diverged compared with *Drosophila* (Wotton et al. 2015b). In general, maternal genes in *Drosophila melanogaster* are likely to be also maternal in other *Drosophila* species, although they get degraded at different rates (Lott et al. 2014), probably due to divergent post-transcriptional regulatory sequences.

The evolution of maternal genes in insects other than flies is less understood. In the grasshopper the gene *Hox3/zen* has acquired maternal expression (Dearden et al. 2000; Fig. 20.1b). In honeybee the pair-rule genes *even-skipped* (*eve*), *fushi tarazu* (*ftz*) and *runt* (*run*) (zygotic genes regulated by gap genes) have also acquired a maternal role (Wilson and Dearden 2012). These genes are widely conserved, indicating that evolutionarily old genes can evolve maternal functions (as we discussed in the previous section for *Drosophila* genes *hb* and *nos*). In the flour beetle, a maternal transcriptome has been described, and many of the maternal genes are also zygotically expressed (Preuss et al. 2012). A significant fraction of these genes seem to be also maternal in *Drosophila*. However, whether both list of maternal genes are comparable is unclear since: first, more than half of all genes are maternal in both species; second, the set of fly maternal genes was derived from in situ hybridization and cDNA microarrays (Arbeitman et al. 2002; Renzis et al. 2007; Tomancak et al. 2007), whilst the beetle set comes from tilling arrays (Preuss et al. 2012).

In vertebrates the information available is very limited. A comparison of six species reveals that maternal 3′UTRs are underrepresented for short lengths, that is, it is unlikely for these genes to have short 3′UTRs (Shen-Orr et al. 2010). That may indicate that there is a selective pressure to maintain post-transcriptional regulation in these transcripts. They also investigated the length of the 5′ intergenic regions, and they observed that these are longer for maternal genes in mammals but shorter (compared to that of zygotic genes) in invertebrates, fish and chicken (Shen-Orr et al. 2010). They proposed that maternal genes in viviparous species have a complex transcriptional regulation, whilst maternal genes in egg-laying species are less regulated in terms of transcription. This indicates, from a comparative genomics point of view, that the evolution of maternal genes depends on life-history strategies. The relationship between evolution of maternal genes and life history is well known in population biology and ecology (Mousseau and Fox 1998), but it has not being accounted for in most of genomics studies.

20.4 Open Questions and Perspectives

Maternal genes are essential during development. However, their evolution is less understood than that of zygotic genes. On the one hand, they have been largely neglected in early evolutionary studies. On the other hand, their function and evolution have not been systematically explored, and only a few species, mostly invertebrates, have been investigated. Thus, a number of questions remain open. Are maternal genes younger than zygotic genes? Here I show some evidence that this may be the case. However, the existing analyses are limited and mostly focused on individual genes of certain relevance. High-throughput studies are far from convincing. Also, some maternal genes are also zygotic (pleiotropic), adding further complexity to this issue. Another open question is do maternal genes evolve fast? Again, evidence coming from the pair *bcd*/*zen* indicates that this may be the case. In addition, the functional diversity of *bcd* within flies reflects a fast evolution pattern. But whether this is the norm is not clear. Indeed, early work on evolutionary rates in *Drosophila* suggested that maternal genes actually evolve slowly (Davis et al. 2005). Despite recent advances, the evolutionary origin and dynamics of maternal genes are still not well understood.

The number of genome sequences from different species and populations, as well as transcriptome data for hundreds of model and non-model species, is increasing at an unprecedented level. This will undoubtedly contribute to a broader view of maternal gene evolution. However, an additional aspect is needed: the integration of comparative genomic analysis with theoretical evolutionary models. Some important work has been made in that direction (Barker et al. 2005; Demuth and Wade 2007), although it has been mostly centred on *Drosophila bcd*. Only by combining genomics and theoretical biology we will be able to understand the origin and evolution of maternal genes, and research in this area should move in that direction.

Acknowledgements I am very grateful to Jordi Paps and an anonymous reviewer for critical reading of the manuscript and to Jordi Paps and Cristina Guijarro for discussion on the evolution of *HSF1*. Work in my lab is funded by the Wellcome Trust [200585/Z/16/Z].

References

Abzhanov A, Kaufman TC (2000) Crustacean (malacostracan) Hox genes and the evolution of the arthropod trunk. Dev Camb Engl 127:2239–2249

Arbeitman MN, Furlong EEM, Imam F, Johnson E, Null BH, Baker BS, Krasnow MA, Scott MP, Davis RW, White KP (2002) Gene expression during the life cycle of Drosophila melanogaster. Science 297:2270–2275. doi:10.1126/science.1072152

Axtell MJ, Westholm JO, Lai EC (2011) Vive la différence: biogenesis and evolution of microRNAs in plants and animals. Genome Biol 12:221. doi:10.1186/gb-2011-12-4-221

Barker MS, Demuth JP, Wade MJ (2005) Maternal expression relaxes constraint on innovation of the anterior determinant, bicoid. PLOS Genet 1:e57. doi:10.1371/journal.pgen.0010057

Bartel DP (2009) MicroRNAs: target recognition and regulatory functions. Cell 136:215–233. doi:10.1016/j.cell.2009.01.002

Boycott AE, Diver C (1923) On the inheritance of sinistrality in limnaea peregra. Proc R Soc Lond B Biol Sci 95:207–213. doi:10.1098/rspb.1923.0033

Briggs R, Green EU, King TJ (1951) An investigation of the capacity for cleavage and differentiation in Rana pipiens eggs lacking "functional" chromosomes. J Exp Zool 116:455–499. doi:10.1002/jez.1401160307

Christians E (2011) Hsf1 knock-out. J Biol Chem 286:le26–le26. doi:10.1074/jbc.L111.221796

Christians E, Davis AA, Thomas SD, Benjamin IJ (2000) Maternal effect of Hsf1 on reproductive success. Nature 407:693–694. doi:10.1038/35037669

Cruickshank T, Wade MJ (2008) Microevolutionary support for a developmental hourglass: gene expression patterns shape sequence variation and divergence in Drosophila. Evol Dev 10:583–590. doi:10.1111/j.1525-142X.2008.00273.x

Davis JC, Brandman O, Petrov DA (2005) Protein evolution in the context of Drosophila development. J Mol Evol 60:774–785. doi:10.1007/s00239-004-0241-2

Dearden P, Akam M (1999) Developmental evolution: axial patterning in insects. Curr Biol 9: R591–R594

Dearden P, Grbic M, Falciani F, Akam M (2000) Maternal expression and early zygotic regulation of the Hox3/zen gene in the grasshopper Schistocerca gregaria. Evol Dev 2:261–270. doi:10.1046/j.1525-142x.2000.00065.x

Demuth JP, Wade MJ (2007) Maternal expression increases the rate of bicoid evolution by relaxing selective constraint. Genetica 129:37–43. doi:10.1007/s10709-006-0031-4

Deshpande G, Calhoun G, Yanowitz JL, Schedl PD (1999) Novel functions of nanos in downregulating mitosis and transcription during the development of the Drosophila germline. Cell 99:271–281

Ding S, Li X, Wang N, Cameron SL, Mao M, Wang Y, Xi Y, Yang D (2015) The phylogeny and evolutionary timescale of muscoidea (Diptera: Brachycera: Calyptratae) inferred from mitochondrial genomes. PloS One 10:e0134170. doi:10.1371/journal.pone.0134170

Dobzhansky T (1937) Genetics and the origin of species. Columbia University Press, New York

Ephrussi A, Lehmann R (1992) Induction of germ cell formation by oskar. Nature 358:387–392. doi:10.1038/358387a0

Extavour CG, Akam M (2003) Mechanisms of germ cell specification across the metazoans: epigenesis and preformation. Development 130:5869–5884. doi:10.1242/dev.00804

Falconer D (1964) Maternal effects and selective responses. Proceedings of the XI International Congress on Genetics

Falconer DS (1981) Introduction to quantitative genetics. Longman, New York

Frohnhöfer HG, Nüsslein-Volhard C (1986) Organization of anterior pattern in the Drosophila embryo by the maternal gene bicoid. Nature 324:120–125. doi:10.1038/324120a0

Hamatani T, Carter MG, Sharov AA, Ko MSH (2004) Dynamics of global gene expression changes during mouse preimplantation development. Dev Cell 6:117–131

Jedlicka P, Mortin MA, Wu C (1997) Multiple functions of Drosophila heat shock transcription factor in vivo. EMBO J 16:2452–2462. doi:10.1093/emboj/16.9.2452

John LB, Yoong S, Ward AC (2009) Evolution of the Ikaros gene family: implications for the origins of adaptive immunity. J Immunol 182:4792–4799. doi:10.4049/jimmunol.0802372

Kanska J, Frank U (2013) New roles for Nanos in neural cell fate determination revealed by studies in a cnidarian. J Cell Sci 126:3192–3203. doi:10.1242/jcs.127233

Kocabas AM, Crosby J, Ross PJ, Otu HH, Beyhan Z, Can H, Tam W-L, Rosa GJM, Halgren RG, Lim B, Fernandez E, Cibelli JB (2006) The transcriptome of human oocytes. Proc Natl Acad Sci 103:14027–14032. doi:10.1073/pnas.0603227103

Kozomara A, Griffiths-Jones S (2014) miRBase: annotating high confidence microRNAs using deep sequencing data. Nucleic Acids Res 42:D68–D73. doi:10.1093/nar/gkt1181

Lawrence PA (1992) The making of a fly: the genetics of animal design. Blackwell Scientific, Oxford

Lee YS, Nakahara K, Pham JW, Kim K, He Z, Sontheimer EJ, Carthew RW (2004) Distinct roles for Drosophila Dicer-1 and Dicer-2 in the siRNA/miRNA silencing pathways. Cell 117:69–81. doi:10.1016/S0092-8674(04)00261-2

Lee M, Choi Y, Kim K, Jin H, Lim J, Nguyen TA, Yang J, Jeong M, Giraldez AJ, Yang H, Patel DJ, Kim VN (2014) Adenylation of maternally inherited microRNAs by wispy. Mol Cell 56:696–707. doi:10.1016/j.molcel.2014.10.011

Li L, Zheng P, Dean J (2010) Maternal control of early mouse development. Dev Camb Engl 137:859–870. doi:10.1242/dev.039487

Lim MYT, Okamura K (2017) Switches in Dicer activity during oogenesis and early development. In: Kloc M (ed) Oocytes: maternal information and functions. Springer, Cham

Lipshitz HD (2009) Follow the mRNA: a new model for Bicoid gradient formation. Nat Rev Mol Cell Biol 10:509–512. doi:10.1038/nrm2730

Liu MM, Davey JW, Jackson DJ, Blaxter ML, Davison A (2014) A conserved set of maternal genes? Insights from a molluscan transcriptome. Int J Dev Biol 58:501–511. doi:10.1387/ijdb.140121ad

Lott SE, Villalta JE, Zhou Q, Bachtrog D, Eisen MB (2014) Sex-specific embryonic gene expression in species with newly evolved sex chromosomes. PLoS Genet 10:e1004159. doi:10.1371/journal.pgen.1004159

Lynch JA, Roth S (2011) The evolution of dorsal–ventral patterning mechanisms in insects. Genes Dev 25:107–118. doi:10.1101/gad.2010711

Lynch M, Walsh B (1998) Genetics and analysis of quantitative traits. Sinauer, Sunderland, MA

Lynch JA, Özüak O, Khila A, Abouheif E, Desplan C, Roth S (2011) The phylogenetic origin of oskar coincided with the origin of maternally provisioned germ plasm and pole cells at the base of the holometabola. PLOS Genet 7:e1002029. doi:10.1371/journal.pgen.1002029

Marco A (2014) Sex-biased expression of microRNAs in Drosophila melanogaster. Open Biol 4:140024. doi:10.1098/rsob.140024
Marco A (2015) Selection against maternal microRNA target sites in maternal transcripts. G3 5:2199–2207. doi:10.1534/g3.115.019497
Marco A (2017) Clearance of maternal RNAs: not a mummy's embryo anymore. In: Lee K (ed) Zygotic genome activation: methods and protocols, Methods in molecular biology. Springer, New York
Marco A, Ninova M, Ronshaugen M, Griffiths-Jones S (2013) Clusters of microRNAs emerge by new hairpins in existing transcripts. Nucleic Acids Res 41:7745–7752. doi:10.1093/nar/gkt534
McGregor AP (2005) How to get ahead: the origin, evolution and function of bicoid. BioEssays News Rev Mol Cell Dev Biol 27:904–913. doi:10.1002/bies.20285
McGregor A, Shaw P, Hancock J, Bopp D, Hediger M, Wratten N, Dover G (2001) Rapid restructuring of bicoid-dependent hunchback promoters within and between Dipteran species: implications for molecular coevolution. Evol Dev 3(407):397
McMillan DR, Xiao X, Shao L, Graves K, Benjamin IJ (1998) Targeted disruption of heat shock transcription factor 1 abolishes thermotolerance and protection against heat-inducible apoptosis. J Biol Chem 273:7523–7528
Mochizuki K, Sano H, Kobayashi S, Nishimiya-Fujisawa C, Fujisawa T (2000) Expression and evolutionary conservation of nanos-related genes in Hydra. Dev Genes Evol 210:591–602
Mousseau TA, Fox CW (1998) Maternal effects as adaptations. Oxford University Press, Oxford
Murata Y, Wharton RP (1995) Binding of pumilio to maternal hunchback mRNA is required for posterior patterning in Drosophila embryos. Cell 80:747–756
Ninova M, Ronshaugen M, Griffiths-Jones S (2014) Fast-evolving microRNAs are highly expressed in the early embryo of Drosophila virilis. RNA 20:360–372. doi:10.1261/rna.041657.113
Ninova M, Ronshaugen M, Griffiths-Jones S (2015) MicroRNA evolution, expression and function during short germband development in Tribolium castaneum. Genome Res. doi:10.1101/gr.193367.115
Pinnell J, Lindeman PS, Colavito S, Lowe C, Savage RM (2006) The divergent roles of the segmentation gene hunchback. Integr Comp Biol 46:519–532. doi:10.1093/icb/icj054
Preuss KM, Lopez JA, Colbourne JK, Wade MJ (2012) Identification of maternally-loaded RNA transcripts in unfertilized eggs of Tribolium castaneum. BMC Genomics 13:671. doi:10.1186/1471-2164-13-671
Raff RA (1996) The shape of life: genes, development, and the evolution of animal form. University of Chicago Press, Chicago
Renzis SD, Elemento O, Tavazoie S, Wieschaus EF (2007) Unmasking activation of the zygotic genome using chromosomal deletions in the Drosophila embryo. PLOS Biol 5:e117. doi:10.1371/journal.pbio.0050117
Ruan J, Li H, Chen Z, Coghlan A, Coin LJM, Guo Y, Heriche J-K, Hu Y, Kristiansen K, Li R, Liu T, Moses A, Qin J, Vang S, Vilella AJ, Ureta-Vidal A, Bolund L, Wang J, Durbin R (2007) TreeFam: 2008 update. Nucleic Acids Res 36:D735–D740. doi:10.1093/nar/gkm1005
Ruby JG, Stark A, Johnston WK, Kellis M, Bartel DP, Lai EC (2007) Evolution, biogenesis, expression, and target predictions of a substantially expanded set of Drosophila microRNAs. Genome Res 17:1850–1864. doi:10.1101/gr.6597907
Schoppmeier M, Schröder R (2005) Maternal torso signaling controls body axis elongation in a short germ insect. Curr Biol 15:2131–2136. doi:10.1016/j.cub.2005.10.036
Shaw PJ, Salameh A, McGregor AP, Bala S, Dover GA (2001) Divergent structure and function of the bicoid gene in Muscoidea fly species. Evol Dev 3:251–262
Shen-Orr SS, Pilpel Y, Hunter CP (2010) Composition and regulation of maternal and zygotic transcriptomes reflects species-specific reproductive mode. Genome Biol 11:R58. doi:10.1186/gb-2010-11-6-r58
Stauber M, Jäckle H, Schmidt-Ott U (1999) The anterior determinant bicoid of Drosophila is a derived Hox class 3 gene. Proc Natl Acad Sci USA 96:3786–3789

Sturtevant AH (1923) Inheritance of direction of coilling in Limnaea. Science 58:269–270. doi:10.1126/science.58.1501.269
Tian X, Pascal G, Monget P (2009) Evolution and functional divergence of NLRPgenes in mammalian reproductive systems. BMC Evol Biol 9:202. doi:10.1186/1471-2148-9-202
Tomancak P, Berman BP, Beaton A, Weiszmann R, Kwan E, Hartenstein V, Celniker SE, Rubin GM (2007) Global analysis of patterns of gene expression during Drosophila embryogenesis. Genome Biol 8:R145
Tong ZB, Gold L, Pfeifer KE, Dorward H, Lee E, Bondy CA, Dean J, Nelson LM (2000) Mater, a maternal effect gene required for early embryonic development in mice. Nat Genet 26:267–268. doi:10.1038/81547
Tong Z-B, Bondy CA, Zhou J, Nelson LM (2002) A human homologue of mouse Mater, a maternal effect gene essential for early embryonic development. Hum Reprod Oxf Engl 17:903–911
Toyama K (1914) On certain characteristics of the Silk-worm which are apparently non-Mendelian. Z Für Indukt Abstamm- Vererbungslehre 12:247–247. doi:10.1007/BF01837303
Votruba SM (2009) MicroRNAS in the Drosophila egg and early embryo. University of Toronto, Canada
Wade M (1998) The evolutionary genetics of maternal effects. In: Mousseau TA, Fox CW (eds) Maternal effects as adaptations. Oxford University Press, New York
Whitlock MC, Wade MJ (1995) Speciation: founder events and their effects on X-linked and autosomal genes. Am Nat 145:676–685. doi:10.1086/285762
Wilson MJ, Dearden PK (2012) Pair-rule gene orthologues have unexpected maternal roles in the honeybee (Apis mellifera). PLoS One 7:e46490. doi:10.1371/journal.pone.0046490
Wolpert L, Smith J, Lawrence P, Robertson E, Meyerowitz E (2006) Principles of development. Oxford University Press, Oxford
Wotton KR, Jiménez-Guri E, Crombach A, Janssens H, Alcaine-Colet A, Lemke S, Schmidt-Ott U, Jaeger J (2015a) Quantitative system drift compensates for altered maternal inputs to the gap gene network of the scuttle fly Megaselia abdita. eLife 4:e04785. doi:10.7554/eLife.04785
Wotton KR, Jiménez-Guri E, Jaeger J (2015b) Maternal co-ordinate gene regulation and axis polarity in the scuttle fly Megaselia abdita. PLoS Genet 11:e1005042. doi:10.1371/journal.pgen.1005042
Wright S (1968) Evolution and the genetics of populations, Theory of gene frequencies, vol 2. University of Chicago Press, Chicago
Wright S (1984) Evolution and the genetics of populations, Experimental results and evolutionary deductions, vol 3. University of Chicago Press, Chicago
Zarzov P, Boucherie H, Mann C (1997) A yeast heat shock transcription factor (Hsf1) mutant is defective in both Hsc82/Hsp82 synthesis and spindle pole body duplication. J Cell Sci 110 (Pt 16):1879–1891
Zeng F, Baldwin DA, Schultz RM (2004) Transcript profiling during preimplantation mouse development. Dev Biol 272:483–496. doi:10.1016/j.ydbio.2004.05.018
Ziegler HE (1898) Experimentelle Studien über die Zelltheilung. Arch Für Entwicklungsmechanik Org 6:249–293. doi:10.1007/BF02152958

Chapter 21
Noninheritable Maternal Factors Useful for Genetic Manipulation in Mammals

Takayuki Sakurai, Takayuki Shindo, and Masahiro Sato

Abstract Mammalian early embryogenesis is supported by maternal factors, such as messenger RNA (mRNA) and proteins, produced and accumulated during oogenesis at least up to the stage when zygotic activation commences. These maternal factors are involved in biologically important events such as epigenetic activation, reprogramming, and mitochondrial growth. Most of these maternal mRNAs are degraded by the 2-cell to 4 ~ 8-cell stages. Maternal proteins, which are produced during oogenesis or by the maternal mRNAs, are degraded by the 4 ~ 8-cell stage. In other words, the maternal factors exist during specific stages of early embryogenesis. In this chapter, we will briefly summarize the property of these maternal factors and mention possible applications of these factors for developing new reproduction engineering-related technologies and producing genetically modified animals. More specifically, we will show the usefulness of maternally accumulated Cas9 protein as a promising tool for CRISPR-/Cas9-based simultaneous genetic modification of multiple loci in mammals.

21.1 Maternal Factors Important for Early Mammalian Embryogenesis

Maternal factors supporting early embryogenesis are usually accumulated as maternal mRNAs and proteins during oogenesis (Fig. 21.1; Rosenthal and Ruderman 1987; Schultz 1993; Telford et al. 1990; Wilt 1973). These oocyte-specific maternal effect genes, such as *Zar1* (zygote arrest 1-like) and *Mater* (maternal effect), are mostly

T. Sakurai (✉) • T. Shindo
Department of Cardiovascular Research, Graduate School of Medicine, Shinshu University, 3-1-1 Asahi, Matsumoto, Nagano 390-8621, Japan

Basic Research Division for Next-Generation Disease Models and Fundamental Technology, Research Center for Next Generation Medicine, Shinshu University, 3-1-1 Asahi, Matsumoto, Nagano 390-8621, Japan
e-mail: tsakurai@shinshu-u.ac.jp

M. Sato
Section of Gene Expression Regulation, Frontier Science Research Center, Kagoshima University, 8-35-1 Sakuragaoka, Kagoshima, Kagoshima 890-8544, Japan
e-mail: masasato@m.kufm.kagoshima-u.ac.jp

M. Kloc (ed.), *Oocytes*, Results and Problems in Cell Differentiation 63,
DOI 10.1007/978-3-319-60855-6_21

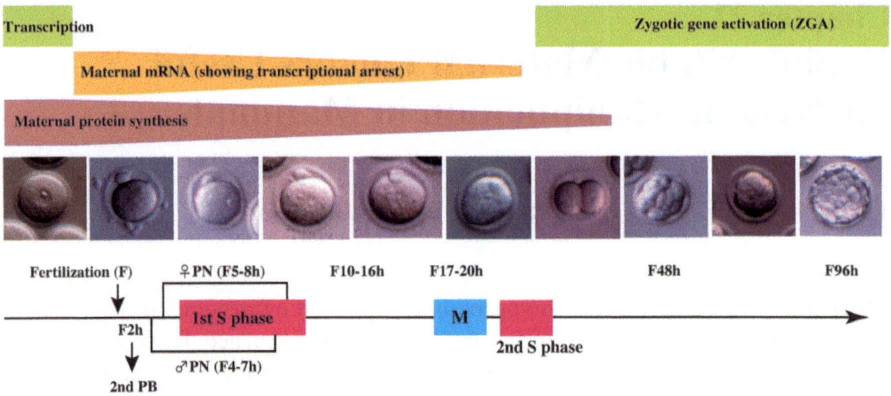

Fig. 21.1 Schematic representation for synthesis, accumulation, and degradation of maternal mRNA/protein in association with mouse preimplantation development. Two hours (F2h) after fertilization (F), the second polar body (second PB) is released. Then, male pronucleus (♂PN) and female pronucleus (♀PN) appear 4–7 hours (F4–7 h) or 5–8 hours (F5–8 h), respectively, after fertilization. During this stage, the first S phase occurs. At 17–20 hours after fertilization (17–20 h) and onward, mitotic phase (M) and the second S phase occur. By this period, almost all maternal mRNAs are degraded and zygotic gene activation (ZGA) commences at two-cell stage and onward

housekeeping genes (Bachvarova et al. 1985; Hamatani et al. 2004; Piko and Clegg 1982; Rothstein et al. 1992; Tong et al. 2000; Wu et al. 2003). They are generally degraded when zygotic activation commences, around the 2-cell to 4 ~ 8-cell stage in mice (Alizadeh et al. 2005; Bachvarova et al. 1985; Clegg and Piko 1983; Flach et al. 1982; Oh et al. 2000; Paynton et al. 1988), although the timing of degradation differs for each gene. Maternal mRNAs and proteins may function in the certain specific stages owing to the inability of de novo synthesis of transcripts during fertilization and subsequent early development and inability within a very short period to produce proteins required for cell cleavage. It remains unknown how maternal mRNAs are translated and then degraded at preimplantation stage. Accumulated data in various animals, including *Drosophila*, amphibians, and mammals, show that elongation of poly (A) tails, known as "cytoplasmic polyadenylation," can regulate translation of maternal mRNAs (Charlesworth et al. 2013; Cui et al. 2013; Radford et al. 2008). In mice, maternal mRNAs with long poly(A) tails (approx. 7–140 nucleotides) are accumulated in oocytes (Sakurai et al. 2005). Extension of poly(A) tails proceeds along with embryonic development, probably in response to oocyte maturation, syngamy, and egg cleavage (Huarte et al. 1987; Oh et al. 2000; Zhou and Dean 2015). Elongation of poly(A) tails leads to increased protein translation activity. This elongation event is associated with the presence of a conserved sequence in the 3′ untranslated region (UTR) of the maternal mRNAs, as exemplified by cytoplasmic polyadenylation elements (CPEs), and some proteins capable of binding to the CPE sequences, such as polyadenylation element-binding protein (CPEBP), poly(A)-binding protein cytoplasmic (PABPC), and poly(A)-binding protein interacting protein 1 (PAIP1) (Charlesworth et al. 2013). The translational potential largely depends on the poly(A) tail extension, whereas stage-specific degradation of maternal RNAs is related to the shortening of the poly(A) tail (Oh et al. 2000). Interestingly, this poly(A) tail extension does not occur in an entire set of maternal mRNAs; rather it occurs in each individual mRNA, and its timing and extent differ among mRNAs. As for the extension of maternal mRNA poly (A) tails, we also reported that individual maternal mRNAs each exhibited a different pattern of poly(A) tail extension/degradation during preimplantation development in mice (Sakurai et al. 2005). We also observed a relationship between stage-specific poly (A) tail extension of these maternal mRNAs and their translational activity (Sakurai et al. unpublished results). However, the precise mechanism underlying stage-specific extension of poly(A) tails of maternal mRNAs and their degradation is still a challenge and a subject for future research. Interestingly, the machinery regulating translational activity through the length of the poly(A) tail is limited to early embryo development such as that seen in zebrafish and frog embryos (Subtelny et al. 2014).

It is known that maternal proteins are involved in the phosphorylation of maternal-derived cyclin-dependent kinase 1 (CDK1). The formation of a complex between CDK1 and cyclin B1 is needed for initiation of M phase during oocyte maturation (Tay et al. 2000). In mice, 1 ~ 2-cell embryos are developmentally totipotent. When spermatozoa fertilize oocytes, maternal proteins accumulated in oocytes activate the embryonic genome by triggering the transition from maternal to embryonic gene expression patterns, thus leading to the formation of zygotes with totipotency (Zhou and Dean 2015). Reprogramming of the gene expression pattern involves the reconstitution of the genome, including demethylation of DNA and histone modification. In regard to DNA demethylation, immediately after fertilization, the male genome undergoes "active" demethylation of DNA, whereas the female genome undergoes "passive" demethylation of DNA, with demethylation completed by the blastocyst stage (Iqbal et al. 2011). Experiments using genetically modified mice demonstrated that these reprogramming events are preceded by translation of the maternal proteins ten-eleven translocation 3 (Tet3) and PGC7 (also known as Dppa3 or Stella), which have been stored in ovarian oocytes (Gu et al. 2011; Nakamura et al. 2007; Sato et al. 2002). In the case of histone modification, protamine, associated with the sperm genome, is exchanged by the maternal-derived histone after fertilization (Adenot et al. 1991; Burton and Torres-Padilla 2010). Furthermore, maternal-derived histone variants, particularly H2A.X and H3.3/H3.1 (predominantly expressed at the 1 ~ 2-cell embryonic stage), are known to be involved in reprogramming activity of both gametes (Zhou and Dean 2015). These maternal proteins are degraded through autophagy and the ubiquitin-proteasome system (Shin et al. 2012; Tsukamoto et al. 2008).

21.2 Examples for Application of Maternal Factors

21.2.1 Improvement of mRNA Synthesis and Translation Efficiency After Injection of mRNA that Mimics the Maternal mRNA's Structure

As mentioned above, the maternal mRNAs/proteins present during oogenesis to early embryogenesis (from fertilization to 2 ~ 4-cell stages) function in a stage-specific manner (Fig. 21.1). To examine the function of each maternal mRNA/protein, micromanipulator-based microinjection of synthesized mRNA or protein into zygotes or later-stage embryos has been carried out. The molecule of synthesized eukaryotic mRNA is composed of m^7G cap structure at the 5′ terminal, a sequence containing an open-reading frame, and a polyadenylation sequence at the 3′ terminal. In some cases, to guarantee the stability of mRNA itself, and the degradative and improved translational abilities, the 5′ UTR and 3′ UTR are included in the synthesized mRNA molecule. As mentioned previously, maternal mRNAs have 7–140 nucleotide long poly(A) tails at their 3′ end. Research shows that mRNA with polyadenylated sequences at the 3′ end exhibited better translation activity than those with no poly(A) tails (Aida et al. 2001; Vassalli et al. 1989). Generally, addition of poly(A) sequences to the synthesized mRNA was performed by in vitro synthesis using poly(A) polymerase. Briefly, plasmid DNA carrying a sequence coding for mRNA was produced. Next, addition of m^7G cap was carried out using this plasmid as a template in a four-ribonucleotide solution containing m^7G cap analog and T3 RNA polymerase. Finally, polyadenylation was performed using poly(A) polymerase. The resulting mRNA can be used for microinjection and then it is translated into proteins. However, this process has two demerits: the procedure itself is laborious, and the length of the resulting poly(A) tails is not uniform. To improve this, we constructed a poly(A) plasmid for in vitro synthesis of mRNA with uniform length of poly(A) tails [Fig. 21.2, presented at the 24th Annual Meeting of the Molecular Biology Society of Japan (Yokohama 2001) by Sakurai et al.]. In this plasmid, the stretch of 85 dA/dT residues and a unique enzymatic site *Bsm* BI (creating an asymmetric cleaved structure) were placed at the 3′ end of the sequence coding for the target mRNA (Fig. 21.2a). Using this plasmid, it was possible to synthesize mRNA with uniform length of poly(A) tails and relatively high efficiency in the absence of poly (A) polymerase.

We describe here experiments on zygote microinjection of synthesized mRNA produced by the method mentioned above as a typical example. *Enhanced green fluorescent protein* (*EGFP*) cDNA was added upstream of the 3′ UTR of *stage-specific embryonic clone-D* (*SSECD*) gene, which is predominantly expressed in early embryos (Rothstein et al. 1992), and downstream of this core structure, we added the stretch of 4, 85, and 350 dA/dT residues. The resulting mRNA was called ESA4, ESA85, and ESA350, respectively (Fig. 21.2b). Cytoplasmic injection of the

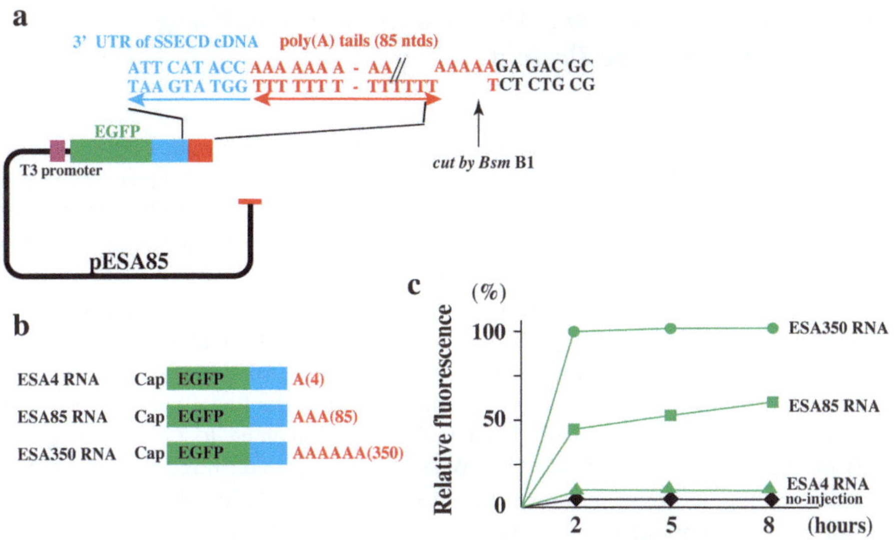

Fig. 21.2 Strength of EGFP-derived fluorescence after zygote microinjection with mRNA with various lengths of poly(A) tails. (**a**) Structure of a poly(A) vector (called pESA85) with 85 dA/dT residues (shown by *red*) downstream of the 3′ UTR of SSECD cDNA (shown by *blue*). (**b**) Structure of mRNAs that are synthesized in vitro from three types of poly(A) vectors with different lengths of poly(A) tails. The synthesized products called ESA4 RNA, ESA85 RNA, and ESA350 RNA have poly(A) tails comprised of 4, 85, and 350 nucleotides, respectively. The 5′ end of each mRNA has a capped structure (Cap). (**c**) Translational activity of each in vitro-synthesized mRNA [shown in (**b**)] 2, 5, and 8 h after zygote microinjection. Zygotes were injected with 2–5 pL of each mRNA (65 pM). EGFP-derived fluorescence in the mRNA-injected zygotes ($N = 20$–30 for each group) was measured using an Olympus IX70 inverted fluorescence microscope (Olympus, Tokyo, Japan). Data are expressed relative to the fluorescence intensity (100%) of zygotes 2 h after injection with ESA350 mRNA

synthesized mRNA into zygotes was performed and the strength of EGFP-derived fluorescence was measured at 2 h after injection and onward. In this case, the level of fluorescence in the sample injected with ESA350 was considered as 100%. We found that the time for translation and its efficiency increased depending on the length of poly(A) tails (Fig. 21.2c). Therefore, we inserted sequences containing the *Cas9* gene, a stretch of 95 dA/dT residues, and *Sap* I enzyme recognition site into the poly(A) plasmid to synthesize mRNA. The resulting mRNA was called 5′m7G-Cas9-polyA95 tail (Sakurai et al. 2014, 2016). Zygote microinjection of this synthesized mRNA resulted in successful generation of genome-edited mice (Sakurai et al. 2014, 2016), probably because the mRNAs possess both 5′m7G capping structure and poly(A) tails with a uniform length of 95 adenylic acids, which will be efficiently converted to Cas9 proteins in zygotes after microinjection.

21.2.2 Conditional Gene Modification in Mice, Using Maternal Cre Protein

We have observed that oocytes isolated from the female green mice (Tg/+, heterozygous for the EGFP transgene) (Okabe et al. 1997) expressed EGFP protein/mRNA even in the absence of the transgene (Sakurai et al., unpublished results). This result suggests that gene products from the transgene can be used as "maternal factors" that are maintained throughout very early embryogenesis.

Sakai and Miyazaki (1997) and Lallemand et al. (1998) used this phenomenon to apply the Cre-/*loxP*-based conditional gene modification in mice. Sakai and Miyazaki (1997) first constructed a plasmid (pCAG-Cre) carrying a CAG promoter (comprised of cytomegalovirus enhancer and chicken β-actin promoter), the *Cre* gene, and poly(A) sites from rabbit β-globin gene to create transgenic (Tg) mouse lines carrying CAG-Cre. When CAG-Cre male mice (Tg/+, heterozygous for the Cre transgene) were mated to CAG-CAT-Z female mice [carrying a transgene, comprised by CAG promoter, *loxP*-flanked *chloramphenicol acetyltransferase* (*CAT*) gene and *lacZ* gene, as a heterozygous state], only the pups carrying CAG-Cre transgene exhibited lacZ expression in the tissues tested. This was a result of Cre-mediated removal of the *loxP*-flanked *CAT* gene in the CAG-CAT-Z transgene. On the other hand, mating between CAG-Cre female mice (Tg/+, heterozygous for the *Cre* transgene) and CAG-CAT-Z male mice (Tg/+, heterozygous for the CAG-CAT-Z transgene) led to production of pups showing systemic lacZ expression, irrespective of the presence of *Cre* gene. *LacZ* activity in these embryos at stages before the two-cell stage suggests that Cre-mediated excision might have occurred in the *Cre* gene-free oocytes during oogenesis. Lallemand et al. (1998) constructed the pPGK-Cre plasmid, in which expression of the *Cre* gene is controlled by the mouse phosphoglycerate kinase (PGK) promoter, and the pGLD plasmid that carries a transgene comprised of the PGK promoter, *loxP*-flanked *lacZ* gene and *distal-less homeobox 2* (*Dlx2*) gene, and produced Tg lines carrying one transgene only in their genome. They mated GLD female Tg mice (Tg/+, heterozygous for the GLD transgene) to PGK-Cre^{m} male Tg mice (Tg/+, heterozygous for the PGK-Cre transgene). As a result, all of the pups (including those having no *Cre* gene) exhibited Cre-mediated excision of *loxP*-flanked *lacZ* gene in the GLD transgene. These two reports clearly suggest that the maternal *Cre* (mRNA and/or protein) existing in ovarian oocytes and embryos at very early stage of embryogenesis might function to remove the *loxP*-flanked sequence in the transgene.

Besides the Cre-*loxP* system, there are several systems for genome modification, which include Flp-FRT system (Raymond and Soriano 2007), PhiC31 integrase-based recombination (Groth et al. 2000; Raymond and Soriano 2007; Watanabe et al. 2011), Dre-rox system (Anastassiadis et al. 2009), VCre-VloxP system (Suzuki and

Nakayama 2011), SCre-SloxP system (Suzuki and Nakayama 2011), Nigri-nox system (Karimova et al. 2016), and Panto-pox system (Karimova et al. 2016). These systems appear to work as well as the abovementioned Cre-*loxP* system if they are used as sources for maternal factors like maternal *Cre*. Similarly, the recently developed genome editing systems, such as zinc finger nuclease (ZFN), transcriptional activator-like effector nucleases (TALEN), and clustered regularly interspaced short palindromic repeat (CRISPR)/CRISPR-associated protein 9 (CRISPR/Cas9), are also promising candidates for genome modification through maternal factors. Among those systems, we thought that the CRISPR/Cas9 system could be an appropriate system for this purpose, since it requires only two components, namely, Cas9 endonuclease and single guide (sg) RNA. If Tg mice expressing Cas9 systemically are produced, it is possible to perform genome editing using Cas9 gene-free zygotes (which are thought to accumulate Cas9 mRNA and/or protein as maternal factors), when they are isolated through mating between Cas9 Tg female mice (Tg/+, heterozygous for the Cas9 transgene) and normal males. In the next chapter, we will show that maternal Cas9 mRNA can be useful for efficient in vivo genome editing.

21.3 Production of Genome-Edited Animals, Using Maternal Cas9 mRNA

21.3.1 CRISPR/Cas9-Based Genome Editing

CRISPR/Cas9 is one of the adaptive immunological systems in bacteria and archaea that developed to exclude pathogens such as viruses and plasmids (Jinek et al. 2012). As mentioned previously, this system requires sgRNA and Cas9, which creates double-strand breaks (DSBs) in the chromosome. sgRNA is composed of two RNA units, CRISPR-coded RNA (crRNA) and trans-activating crRNA (tracrRNA). The crRNA has a sequence of 20 nucleotides corresponding to a target DNA sequence and 3 nucleotides (NGG) at the 5′ end, called the protospacer-adjacent motif (PAM). The PAM causes the Cas9 protein to induce DSBs at three base pairs (bp) upstream of the PAM (Cong et al. 2013; Mali et al. 2013). The resulting DSBs are repaired by the two important pathways employed by eukaryotic cells, the so-called nonhomologous DNA end joining (NHEJ) and the homology-directed recombination (HDR). The highly efficient genome editing ability of CRISPR/Cas9 system has dramatically simplified the production of genetically modified animals. For example, it does not require the laborious construction of a targeting vector, its gene delivery into mouse embryonic stem (ES) cells, the isolation and characterization of gene-targeted ES clones and the production of chimeric mice.

21.3.2 Production of Tg Mice Overexpressing Systemic Cas9

The Jaenisch group first reported a one-step generation of mice carrying mutations in single or multiple genes through zygote injection of Cas9 mRNA and sgRNAs (Wang et al. 2013; Yang et al. 2013). Using similar approaches, genetically modified animals of various species have been generated (Hai et al. 2014; Hwang et al. 2013; Li et al. 2013a, b; Niu et al. 2014). However, these approaches often resulted in generation of animals with mosaic mutations, probably due to the delay of translation after mRNA injection (Oliver et al. 2015; Yen et al. 2014). In fact, it has been reported that translation of injected mRNA in zygotes usually occurs several hours after injection (Aida et al. 2001; Fig. 21.2c). Notably, the use of Cas9 protein is known to decrease the occurrence of mosaic genome editing, and in fact zygote injection of those materials together with sgRNAs resulted in increased efficiency for production of genome-edited animals with homozygous KO phenotype (Aida et al. 2015).

We have seen that the CRISPR/Cas9 system can modify multiple genes simultaneously. Injection of Cas9 mRNA/proteins might interfere with the introduction of multiple sgRNAs, probably due to the limited capacity of the pronucleus and cytoplasm in view of both the physical and functional aspects. Therefore, we hypothesized that another approach was to use Tg mice expressing systemic Cas9 as a source of "maternal Cas9." Since zygotes derived from Cas9 Tg mice should contain abundant Cas9 mRNA and/or protein (Fig. 21.3), multiple sgRNAs (at least nine, Sakurai et al. 2016) can be introduced into a zygote at once, which will lead to efficient production of individuals with mutated multiple target loci. Cas9 Tg mice have other merits. Researchers can obtain genome-edited primary cultured cells after transfection with sgRNA alone (see below). Also, in vivo genome editing of organs/tissues is possible through in vivo gene delivery of sgRNA alone (see below). Therefore, these Cas9 Tg mice can be a valuable resource for CRISPR/Cas9-based genome engineering in mice.

To perform "maternal Cas9"-based genome editing at a zygote level, we produced nine Tg lines by introducing a transgene containing CAG promoter, Cas9 gene [with FLAGtag at its 5′ end and nuclear location signal (NLS) at both ends], and poly(A) sites into the mouse genome (Sakurai et al. 2016). Of these, one line exhibiting the highest expression of Cas9 was selected for further assays and named "sCAT" (systemic Cas9 expressed Tg mice) (Sakurai et al. 2016). We found no abnormality in sCAT in genetical, morphological, or physiological characteristics (Sakurai et al. 2016). Chromosomal mapping and transgene copy number estimation revealed that sCAT had nine copies with transgenes arrayed in tandem in exon 9 of *GTP-binding protein 10* (*Gtpbp10*) gene located on chromosome 5 (Sakurai et al. 2016). The transmission of these transgenes to the offspring followed Mendelian law. Notably, offspring homozygous for the transgenes were healthy and viable (Sakurai et al. 2016). Real-time (RT)-PCR analysis of sCAT demonstrated ubiquitous expression of Cas9 mRNA among major organs examined with relatively high levels of Cas9 mRNA in the heart,

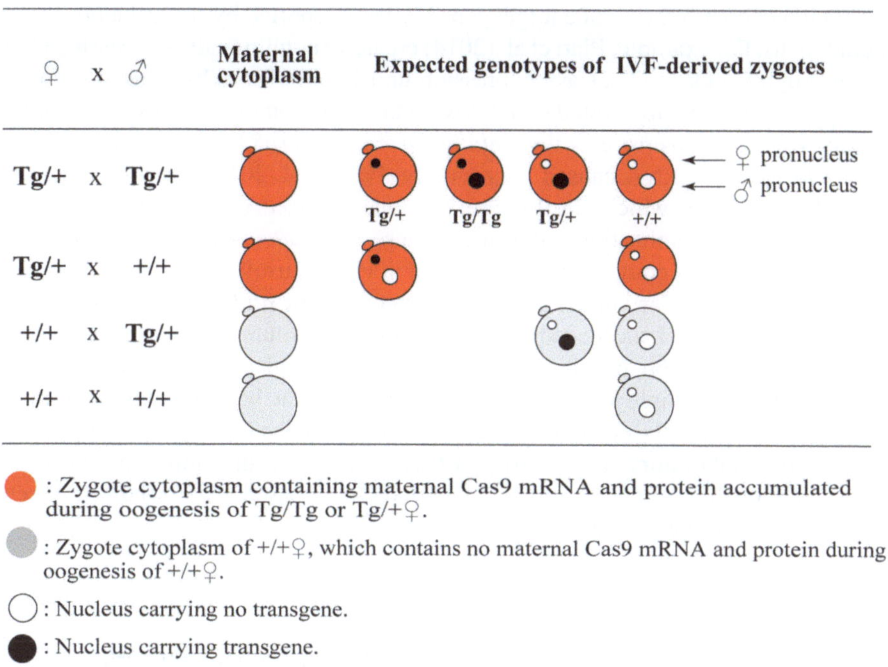

Fig. 21.3 Generation of noninheritable maternal Cas9-containing zygotes after IVF with different combinations of oocytes and sperm of sCAT mice. *Black* and *white circle* in zygote shows Tg and wild-type pronuclei, respectively

skeletal muscle, and testis (Sakurai et al. 2016). Since the establishment of sCAT in 2013, the pattern of Cas9 expression, as revealed by RT-PCR analysis, remained unchanged (Sakurai et al., unpublished results). We next examined whether sCAT has the ability to edit the target gene after transfection with sgRNAs under in vitro and in vivo conditions. For example, primary fibroblasts isolated from sCAT tails were transfected in vitro with sgRNA that targeted *α-1,3-galactosyltransferase* (*Ggtal*) gene, a gene responsible for synthesis of α-Gal epitope, a cell surface carbohydrate. As a result, the transfected cells exhibited loss of α-Gal epitope expression, as revealed by staining with a lectin capable of binding to α-Gal epitope specifically (Sakurai et al. 2016). Regarding in vivo genome editing, we attempted to destroy the endogenous albumin gene in sCAT by a hydrodynamics-based tail vein-mediated gene transfer of sgRNA. As a result, we succeeded in creating mutations in the target gene in hepatocytes after delivery of sgRNA alone (Sakurai et al. 2016).

Based on these findings, it was concluded that Cas9 overexpressed in the sCAT has the ability to edit the target genes efficiently in vitro and in vivo. At present, besides sCAT, several mouse lines expressing Cas9 have been produced worldwide (Carroll et al. 2016; Chu et al. 2016; Dow et al. 2015; Nakao et al. 2016; Platt et al. 2014; Sakurai et al. 2016). Notably, no evident abnormality has been described in any of the reported Cas9 mouse lines. Therefore, the usefulness of these Cas9 mice for in vivo genome editing of a target gene has been reported by several laboratories including us. For example, Platt et al. (2014) created a conditionally expressing Cas9 line through knock-in of a Cas9 expression unit into the Rosa26 locus at chromosome 6. This line demonstrated that in vivo administration of viral DNA containing *Kirsten-ras (Kras)*, *transformation-relating protein53 (Trp53)*, and *serine/threonine kinase 11 (Stk11)*-gRNA caused lung tumors (Platt et al. 2014). Dow et al. (2015) created Tg lines in which Cas9 expression was induced upon administration of doxycycline and demonstrated that using these lines *adenomatous polyposis coli (Apc)* and *p53* genes could be genome-edited in vivo. Carroll et al. (2016) produced Tg lines expressing Cas9 under a cardiomyocyte-specific *alpha myosin heavy chain (Myh6)* promoter and succeeded in inducing cardiac failure and hypertrophy after in vivo administration of viral DNA carrying *Myh6*-gRNA. Chu et al. (2016) demonstrated that highly efficient genome editing can be achieved in primary cultured B cells, T cells, and macrophages isolated from the Cas9 Tg mice, produced by their own laboratory, by inserting a Cas9 expression unit into Rosa26 locus. However, there has been no report on "maternal Cas9"-based genome editing except our report (Sakurai et al. 2016).

21.3.3 Maternal Cas9 Exhibits a High Degree of Genome Editing Activity

It was expected that zygotes derived from sCAT mice would have abundant Cas9 mRNA and protein as maternal factors, which may allow highly efficient genome editing in those embryos (Fig. 21.3). Therefore, for inducing CRISPR-/Cas9-based mutation in a target gene of a zygote, introduction of sgRNA alone (not Cas9-related nucleic acids) is prerequisite. To test this point, we microinjected R2-gRNA targeted to *receptor activity modifying protein 2* (*Ramp2*) gene into pronuclei of zygotes obtained through various mating combinations. The injected embryos were cultured to blastocysts. These blastocysts were next subjected to a single embryo assay for genotyping and mutation detection (Sakurai et al. 2016). As a result, microinjection of sgRNA into zygotes obtained through in vitro fertilization (IVF) between Tg/+ females and Tg/+ males or between Tg/+ females and +/+ males caused high efficiency of insertion and deletion (indel) mutations ranging from 92 to 94%. In addition, genotyping of a single blastocyst derived from those sgRNA-injected zygotes revealed that embryos having no Cas9 transgene exhibited indel mutations with high efficiency. Thus, microinjection of sgRNA into +/+ zygotes obtained through IVF between Tg/+ females and Tg/+ males or between Tg/+ females and +/+ males resulted in generation of all +/+ blastocysts with mutations in the *Ramp2* locus. On the other hand, IVF between +/+ females and Tg/+ males resulted in generation of embryos with low efficiency of indel mutations (23%). This low efficiency of genome editing activity is thought to be derived from Cas9 expressed

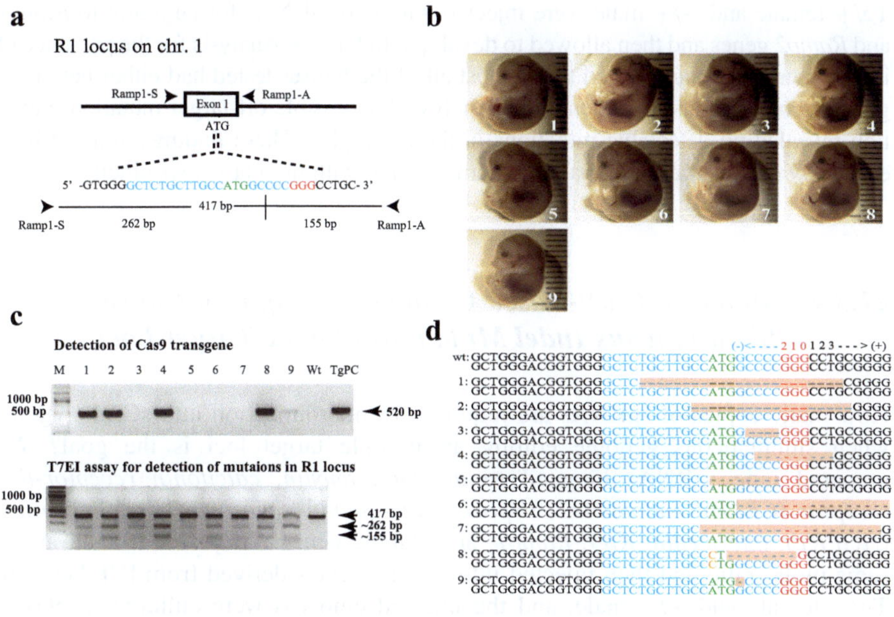

Fig. 21.4 Occurrence of indel mutations in the offspring of sCAT mice after microinjection with sgRNA at zygote stage. (**a**) Schema of R1-gRNA-mediated genome editing at exon 1 in the murine *Rampl* gene. The sequence encoded by R1-gRNA is shown in *blue*. Start codon ATG is shown in *green*. PAM sequence is shown in *red*. (**b**) Normal morphology of nine fetuses at day 15.5 derived from fertilized eggs (obtained through a combination of Tg/+♀ × +/+♂) injected with R1-gRNA alone. (**c**) Gel images of Cas9 genotyping and indel mutations in the fetuses shown in (**b**). Lanes Wt (wild type) and Tg PC indicate PCR products obtained using wild-type sample and a Cas9 transgene, respectively. Fetuses numbered 1, 2, 4, 8, and 9 were identified as Tg/+ for Cas9 transgene. Those numbered 3, 5, 6, and 7 were identified as +/+. Indel mutations were detected in all nine fetuses tested. Notably, indel mutations occur in the +/+ fetuses (nos. 1, 2, 4, 8, and 9). (**d**) Sequencing of PCR products. Indel mutations in a region recognized by R1-gRNA were detected in all nine fetuses tested. "wt" indicates the wild-type sequence. Base deletion is indicated by "–". Pink line marked as "–" indicates base deleted or substituted. "C" shown in *yellow* in No. 8 sequence indicates base substitution from "A" to "C"

after zygotic activation, since zygotic gene expression does not occur at the zygotic stage but becomes active at two-cell stage and onward (Oh et al. 2000).

Next, we examined whether genome-edited fetuses could be obtained when sCAT-derived zygotes were used for microinjection of sgRNAs. Zygotes derived from IVF between Tg/+ female and +/+ male were injected with sgRNA (called R1-gRNA, Fig. 21.4a) targeted to *receptor activity modifying protein 1* (*Rampl*) gene. As a result, nine fetuses at day 15.5 were successfully recovered (Fig. 21.4b). Genotyping of these fetal samples found that four were Tg/+ and five were +/+. Analysis for the presence of indel mutations revealed that all samples tested had mutations at the target region (Fig. 21.4c, d). These findings indicate that it is possible to obtain genome-edited "transgene-free" mice using maternal Cas9 mRNA.

Finally, we attempted simultaneous genome editing in two target genes, *Rampl* and *Ramp2*, using sCAT zygotes (Sakurai et al. 2016). Zygotes derived from IVF between

Tg/+ female and +/+ male were injected with two sgRNAs for targeting to *Rampl* and *Ramp2* genes and then allowed to develop into fetuses. Analysis for the presence of indel mutations demonstrated that almost all of the fetuses tested had either heterozygous or homozygous KO phenotype. Analysis for possible off-target mutation induction revealed no detectable mutations in these samples. These findings indicate high efficiency of maternal Cas9-based genome editing with high target specificity.

21.3.4 Maternal Cas9-Based Genome Editing Can Cause Simultaneous Indel Mutations at Nine Target Loci

How many copies of sgRNAs are required for microinjection into sCAT zygotes when simultaneous genome editing at multiple target loci is the goal? We prepared nine sgRNAs targeted to *albumin*, *amylin*, *calcitonin receptor-like receptor* (*CRLR*), *endothelin1*, *hypoxanthine-guanine phosphoribosyltransferase* (*HGPRT*), *intermedin*, *Klf5*, *Rampl*, and *Ramp2* genes. They were mixed at the same molar ratio and then injected into sCAT zygotes derived from IVF between Tg/+ female and +/+ male, and the injected embryos were cultured to blastocysts (Sakurai et al. 2016). The rate for development to blastocyst was about 46%. A single blastocyst assay for PCR-based genotyping and analysis for the presence of indel mutations demonstrated that ~20% of blastocysts that were derived from Tg/+ zygotes or those carrying maternal Cas9 (+/+) exhibited simultaneous genome editing at the nine loci. An average of 7.2 loci per embryo showed indel mutations. On the other hand, injection of Cas9 mRNA (100 ~ 200 ng/μL) and the nine sgRNAs into wild-type zygotes as control resulted in low developmental rate to blastocysts (11% on average), but the number of loci showing indel mutations per embryo was almost the same with the experimental group mentioned above. These data indicate that maternal Cas9-based genome editing is useful for one-step production of genetically modified animals with multiple mutated loci.

21.4 Summary

Early embryogenesis in mammals is supported by the maternal mRNA/proteins accumulated during oogenesis at least up to the period of initiation of zygotic gene expression. In vitro-synthesized mRNA that mimics the structure of maternal mRNA exhibits efficient translational activity at least during the periods of ovulated oocytes to late zygotes. The transgene-derived products (mRNA/protein) accumulated in the oocytes of Tg animals are considered as "exogenous" maternal factors and their fate follows that of internal maternal factors. This transient nature of maternal factors can be used for genome modification in animals, as exemplified by "maternal Cre"-based recombination of a *loxP*-flanked transgene and "maternal Cas9"-based genome editing of endogenous target genes. The latter system is particularly useful for creating mice with multiple targeted loci through a one-step procedure. Notably, several groups, including ours, succeeded in producing mice overexpressing Cas9 systemically and found no evident abnormality in them based on various criteria. This point is very important based on two reasons: one is that continuous expression of *Cas9* and accumulated Cas9 mRNA/protein are not harmful for cell survival and its function, and the other is that Cas9 Tg animals can offer an opportunity to manipulate the genome of individuals in a simpler manner than is ordinarily done, since only gRNA alone is introduced into Cas9-derived zygotes/tissues/cells for that purpose.

Acknowledgments T.Sa. was supported by JSPS KAKENHI Grants (Nos. 26293085 and 16K15233) from the Japanese Ministry of Education, Culture, Sports, Science, and Technology. This study is also supported by the Core Research for Evolutionary Science and Technology (CREST), Japan Science and Technology Agency (JST), and AMED Core Research for Evolutionary Medical Science and Technology (AMED-CREST), Japan Agency for Medical Research and Development (AMED).

References

Adenot PG, Szollosi MS, Geze M, Renard JP, Debey P (1991) Dynamics of paternal chromatin changes in live one-cell mouse embryo after natural fertilization. Mol Reprod Dev 28(1):23–34

Aida T, Oda S, Awaji T, Yoshida K, Miyazaki S (2001) Expression of a green fluorescent protein variant in mouse oocytes by injection of RNA with an added long poly(A) tail. Mol Human Reprod 7(11):1039–1046

Aida T, Chiyo K, Usami T, Ishikubo H, Imahashi R, Wada Y, Tanaka KF, Sakuma T, Yamamoto T, Tanaka K (2015) Cloning-free CRISPR/Cas system facilitates functional cassette knock-in in mice. Genome Biol 16:87

Alizadeh Z, Kageyama S, Aoki F (2005) Degradation of maternal mRNA in mouse embryos: selective degradation of specific mRNAs after fertilization. Mol Reprod Dev 72(3):281–290

Anastassiadis K, Fu J, Patsch C, Hu S, Weidlich S, Duerschke K, Buchholz F, Edenhofer F, Stewart AF (2009) Dre recombinase, like Cre, is a highly efficient site-specific recombinase in E. coli, mammalian cells and mice. Dis Model Mech 2(9–10):508–515

Bachvarova R, De Leon V, Johnson A, Kaplan G, Paynton BV (1985) Changes in total RNA, polyadenylated RNA, and actin mRNA during meiotic maturation of mouse oocytes. Dev Biol 108(2):325–331

Burton A, Torres-Padilla ME (2010) Epigenetic reprogramming and development: a unique heterochromatin organization in the preimplantation mouse embryo. Brief Funct Genomics 9(5–6): 444–454

Carroll KJ, Makarewich CA, McAnally J, Anderson DM, Zentilin L, Liu N, Giacca M, Bassel-Duby R, Olson EN (2016) A mouse model for adult cardiac-specific gene deletion with CRISPR/Cas9. Proc Natl Acad Sci USA 113(2):338–343

Charlesworth A, Meijer HA, de Moor CH (2013) Specificity factors in cytoplasmic polyadenylation. Wiley Interdiscip Rev 4(4):437–461

Chu VT, Graf R, Wirtz T, Weber T, Favret J, Li X, Petsch K, Tran NT, Sieweke MH, Berek C, Kuhn R, Rajewsky K (2016) Efficient CRISPR-mediated mutagenesis in primary immune cells using CrispRGold and a C57BL/6 Cas9 transgenic mouse line. Proc Natl Acad Sci USA 113(44):12514–12519

Clegg KB, Piko L (1983) Poly(A) length, cytoplasmic adenylation and synthesis of poly(A)+ RNA in early mouse embryos. Dev Biol 95(2):331–341

Cong L, Ran FA, Cox D, Lin S, Barretto R, Habib N, Hsu PD, Wu X, Jiang W, Marraffini LA, Zhang F (2013) Multiplex genome engineering using CRISPR/Cas systems. Science 339(6121):819–823

Cui J, Sartain CV, Pleiss JA, Wolfner MF (2013) Cytoplasmic polyadenylation is a major mRNA regulator during oogenesis and egg activation in Drosophila. Dev Biol 383(1):121–131

Dow LE, Fisher J, O'Rourke KP, Muley A, Kastenhuber ER, Livshits G, Tschaharganeh DF, Socci ND, Lowe SW (2015) Inducible in vivo genome editing with CRISPR-Cas9. Nat Biotechnol 33(4): 390–394

Flach G, Johnson MH, Braude PR, Taylor RA, Bolton VN (1982) The transition from maternal to embryonic control in the 2-cell mouse embryo. EMBO J 1(6):681–686

Groth AC, Olivares EC, Thyagarajan B, Calos MP (2000) A phage integrase directs efficient site-specific integration in human cells. Proc Natl Acad Sci USA 97(11):5995–6000

Gu TP, Guo F, Yang H, Wu HP, Xu GF, Liu W, Xie ZG, Shi L, He X, Jin SG, Iqbal K, Shi YG, Deng Z, Szabo PE, Pfeifer GP, Li J, GL X (2011) The role of Tet3 DNA dioxygenase in epigenetic reprogramming by oocytes. Nature 477(7366):606–610

Hai T, Teng F, Guo R, Li W, Zhou Q (2014) One-step generation of knockout pigs by zygote injection of CRISPR/Cas system. Cell Res 24(3):372–375

Hamatani T, Carter MG, Sharov AA, Ko MS (2004) Dynamics of global gene expression changes during mouse preimplantation development. Dev Cell 6(1):117–131

Huarte J, Belin D, Vassalli A, Strickland S, Vassalli JD (1987) Meiotic maturation of mouse oocytes triggers the translation and polyadenylation of dormant tissue-type plasminogen activator mRNA. Genes Dev 1(10):1201–1211

Hwang WY, Fu Y, Reyon D, Maeder ML, Tsai SQ, Sander JD, Peterson RT, Yeh JR, Joung JK (2013) Efficient genome editing in zebrafish using a CRISPR-Cas system. Nat Biotechnol 31(3):227–229

Iqbal K, Jin SG, Pfeifer GP, Szabo PE (2011) Reprogramming of the paternal genome upon fertilization involves genome-wide oxidation of 5-methylcytosine. Proc Natl Acad Sci USA 108(9):3642–3647

Jinek M, Chylinski K, Fonfara I, Hauer M, Doudna JA, Charpentier E (2012) A programmable dual-RNA-guided DNA endonuclease in adaptive bacterial immunity. Science 337(6096): 816–821

Karimova M, Splith V, Karpinski J, Pisabarro MT, Buchholz F (2016) Discovery of Nigri/nox and Panto/pox site-specific recombinase systems facilitates advanced genome engineering. Sci Rep 6:30130

Lallemand Y, Luria V, Haffner-Krausz R, Lonai P (1998) Maternally expressed PGK-Cre transgene as a tool for early and uniform activation of the Cre site-specific recombinase. Transgenic Res 7(2):105–112

Li D, Qiu Z, Shao Y, Chen Y, Guan Y, Liu M, Li Y, Gao N, Wang L, Lu X, Zhao Y, Liu M (2013a) Heritable gene targeting in the mouse and rat using a CRISPR-Cas system. Nat Biotechnol 31(8):681–683

Li W, Teng F, Li T, Zhou Q (2013b) Simultaneous generation and germline transmission of multiple gene mutations in rat using CRISPR-Cas systems. Nat Biotechnol 31(8):684–686

Mali P, Yang L, Esvelt KM, Aach J, Guell M, DiCarlo JE, Norville JE, Church GM (2013) RNA-guided human genome engineering via Cas9. Science 339(6121):823–826

Nakamura T, Arai Y, Umehara H, Masuhara M, Kimura T, Taniguchi H, Sekimoto T, Ikawa M, Yoneda Y, Okabe M, Tanaka S, Shiota K, Nakano T (2007) PGC7/Stella protects against DNA demethylation in early embryogenesis. Nature Cell Biol 9(1):64–71

Nakao H, Harada T, Nakao K, Kiyonari H, Inoue K, Furuta Y, Aiba A (2016) A possible aid in targeted insertion of large DNA elements by CRISPR/Cas in mouse zygotes. Genesis 54(2): 65–77

Niu Y, Shen B, Cui Y, Chen Y, Wang J, Wang L, Kang Y, Zhao X, Si W, Li W, Xiang AP, Zhou J, Guo X, Bi Y, Si C, Hu B, Dong G, Wang H, Zhou Z, Li T, Tan T, Pu X, Wang F, Ji S, Zhou Q, Huang X, Ji W, Sha J (2014) Generation of gene-modified cynomolgus monkey via Cas9/RNA-mediated gene targeting in one-cell embryos. Cell 156(4):836–843

Oh B, Hwang S, McLaughlin J, Solter D, Knowles BB (2000) Timely translation during the mouse oocyte-to-embryo transition. Development 127(17):3795–3803

Okabe M, Ikawa M, Kominami K, Nakanishi T, Nishimune Y (1997) 'Green mice' as a source of ubiquitous green cells. FEBS Lett 407(3):313–319

Oliver D, Yuan S, McSwiggin H, Yan W (2015) Pervasive Genotypic mosaicism in founder mice derived from genome editing through pronuclear injection. PLoS One 10(6):e0129457

Paynton BV, Rempel R, Bachvarova R (1988) Changes in state of adenylation and time course of degradation of maternal mRNAs during oocyte maturation and early embryonic development in the mouse. Dev Biol 129(2):304–314

Piko L, Clegg KB (1982) Quantitative changes in total RNA, total poly(A), and ribosomes in early mouse embryos. Dev Biol 89(2):362–378

Platt RJ, Chen S, Zhou Y, Yim MJ, Swiech L, Kempton HR, Dahlman JE, Parnas O, Eisenhaure TM, Jovanovic M, Graham DB, Jhunjhunwala S, Heidenreich M, Xavier RJ, Langer R, Anderson DG, Hacohen N, Regev A, Feng G, Sharp PA, Zhang F (2014) CRISPR-Cas9 knockin mice for genome editing and cancer modeling. Cell 159(2):440–455

Radford HE, Meijer HA, de Moor CH (2008) Translational control by cytoplasmic polyadenylation in Xenopus oocytes. Biochim Biophys Acta 1779(4):217–229

Raymond CS, Soriano P (2007) High-efficiency FLP and PhiC31 site-specific recombination in mammalian cells. PLoS One 2(1):e162

Rosenthal ET, Ruderman JV (1987) Widespread changes in the translation and adenylation of maternal messenger RNAs following fertilization of Spisula oocytes. Dev Biol 121(1): 237–246

Rothstein JL, Johnson D, DeLoia JA, Skowronski J, Solter D, Knowles B (1992) Gene expression during preimplantation mouse development. Genes Dev 6(7):1190–1201

Sakai K, Miyazaki J (1997) A transgenic mouse line that retains Cre recombinase activity in mature oocytes irrespective of the cre transgene transmission. Biochem Biophys Res Commun 237(2):318–324

Sakurai T, Sato M, Kimura M (2005) Diverse patterns of poly(A) tail elongation and shortening of murine maternal mRNAs from fully grown oocyte to 2-cell embryo stages. Biochem Biophys Res Commun 336(4):1181–1189

Sakurai T, Watanabe S, Kamiyoshi A, Sato M, Shindo T (2014) A single blastocyst assay optimized for detecting CRISPR/Cas9 system-induced indel mutations in mice. BMC Biotechnol 14:69

Sakurai T, Kamiyoshi A, Kawate H, Mori C, Watanabe S, Tanaka M, Uetake R, Sato M, Shindo T (2016) A non-inheritable maternal Cas9-based multiple-gene editing system in mice. Sci Rep 6:20011

Sato M, Kimura T, Kurokawa K, Fujita Y, Abe K, Masuhara M, Yasunaga T, Ryo A, Yamamoto M, Nakano T (2002) Identification of PGC7, a new gene expressed specifically in preimplantation embryos and germ cells. Mech Dev 113(1):91–94

Schultz RM (1993) Regulation of zygotic gene activation in the mouse. Bioessays 15(8):531–538

Shin SW, Shimizu N, Tokoro M, Nishikawa S, Hatanaka Y, Anzai M, Hamazaki J, Kishigami S, Saeki K, Hosoi Y, Iritani A, Murata S, Matsumoto K (2012) Mouse zygote-specific proteasome assembly chaperone important for maternal-to-zygotic transition. Biol Open 2(2):170–182

Subtelny AO, Eichhorn SW, Chen GR, Sive H, Bartel DP (2014) Poly(A)-tail profiling reveals an embryonic switch in translational control. Nature 508(7494):66–71

Suzuki E, Nakayama M (2011) VCre/VloxP and SCre/SloxP: new site-specific recombination systems for genome engineering. Nucleic Acids Res 39(8):e49

Tay J, Hodgman R, Richter JD (2000) The control of cyclin B1 mRNA translation during mouse oocyte maturation. Dev Biol 221(1):1–9

Telford NA, Watson AJ, Schultz GA (1990) Transition from maternal to embryonic control in early mammalian development: a comparison of several species. Mol Reprod Dev 26(1): 90–100

Tong ZB, Gold L, Pfeifer KE, Dorward H, Lee E, Bondy CA, Dean J, Nelson LM (2000) Mater, a maternal effect gene required for early embryonic development in mice. Nat Genet 26(3): 267–268

Tsukamoto S, Kuma A, Murakami M, Kishi C, Yamamoto A, Mizushima N (2008) Autophagy is essential for preimplantation development of mouse embryos. Science 321(5885):117–120

Vassalli JD, Huarte J, Belin D, Gubler P, Vassalli A, O'Connell ML, Parton LA, Rickles RJ, Strickland S (1989) Regulated polyadenylation controls mRNA translation during meiotic maturation of mouse oocytes. Genes Dev 3(12B):2163–2171

Wang H, Yang H, Shivalila CS, Dawlaty MM, Cheng AW, Zhang F, Jaenisch R (2013) One-step generation of mice carrying mutations in multiple genes by CRISPR/Cas-mediated genome engineering. Cell 153(4):910–918

Watanabe S, Nakamura S, Sakurai T, Akasaka K, Sato M (2011) Improvement of a phiC31 integrase-based gene delivery system that confers high and continuous transgene expression. New Biotechnol 28(4):312–319

Wilt FH (1973) Polyadenylation of maternal RNA of sea urchin eggs after fertilization. Proc Natl Acad Sci USA 70(8):2345–2349

Wu X, Viveiros MM, Eppig JJ, Bai Y, Fitzpatrick SL, Matzuk MM (2003) Zygote arrest 1 (Zar1) is a novel maternal-effect gene critical for the oocyte-to-embryo transition. Nat Genet 33(2): 187–191

Yang H, Wang H, Shivalila CS, Cheng AW, Shi L, Jaenisch R (2013) One-step generation of mice carrying reporter and conditional alleles by CRISPR/Cas-mediated genome engineering. Cell 154(6):1370–1379

Yen ST, Zhang M, Deng JM, Usman SJ, Smith CN, Parker-Thornburg J, Swinton PG, Martin JF, Behringer RR (2014) Somatic mosaicism and allele complexity induced by CRISPR/Cas9 RNA injections in mouse zygotes. Dev Biol 393(1):3–9

Zhou LQ, Dean J (2015) Reprogramming the genome to totipotency in mouse embryos. Trends Cell Biol 25(2):82–91

MIX
Papier aus verantwortungsvollen Quellen
Paper from responsible sources
FSC® C105338

If you have any concerns about our products,
you can contact us on
ProductSafety@springernature.com

In case Publisher is established outside the EU,
the EU authorized representative is:
Springer Nature Customer Service Center GmbH
Europaplatz 3, 69115 Heidelberg, Germany

Printed by Libri Plureos GmbH
in Hamburg, Germany